# Special Operations Success

# Special Operations Success

## Balancing Capabilities and Control

James D. Kiras

OXFORD
UNIVERSITY PRESS

OXFORD
UNIVERSITY PRESS

Great Clarendon Street, Oxford, OX2 6DP,
United Kingdom

Oxford University Press is a department of the University of Oxford.
It furthers the University's objective of excellence in research, scholarship,
and education by publishing worldwide. Oxford is a registered trade mark of
Oxford University Press in the UK and in certain other countries



Published in the United States of America by Oxford University Press
198 Madison Avenue, New York, NY 10016, United States of America

British Library Cataloguing in Publication Data
Data available

Library of Congress Control Number: 2024941649

ISBN 9780198902065

DOI: 10.1093/9780198902096.001.0001

Printed and bound by
CPI Group (UK) Ltd, Croydon, CR0 4YY

*To Colin S. Gray—professor, mentor, and friend.*

# Foreword

This excellent book, written by one of the Western world's most perceptive scholars of special operations, brings context and clarity to a set of issues that has irritated and perplexed practitioners and theorists for decades. In developing his theory, illustrating it through historical case studies, and testing its application to the contemporary environment, James Kiras has made a valuable contribution to an important debate. It is worth briefly noting how this debate arose.

By the last years of the twentieth century, special operations had evolved away from their roots in the unconventional reconnaissance, sabotage, and support-to-resistance work of the Second World War, which had seen the birth both of modern military special forces and of the covert-action arms of civilian intelligence services. Likewise, special operations had moved far beyond the guerrilla and counter-guerrilla campaigns of the Cold War, fought across the developing world by the surrogates of superpowers reluctant to risk thermonuclear war through direct confrontation.

By the 1990s, the Cold War had given way to a string of peacekeeping, peacemaking, and stabilization tasks in which weak states, failing states, and non-state actors—rather than powerful state adversaries—posed the primary threat. Political decision-makers, like the broader public (encouraged by media depictions of special forces as super-soldiers), came to see special operations as any missions that were low profile, politically sensitive, or conducted by specially selected, trained, and equipped forces.

In the United States, special operations were seen through the lens of post-Vietnam small-scale advisory missions (notably in Latin America) and raiding operations (in Africa and the Middle East). In Britain and Australia, special operations thinking focused on long-range covert reconnaissance behind enemy lines during conventional conflict (as in the Scud-hunting operations by UK special forces during the Gulf War) and on counterterrorism, epitomized by the near-flawless takedown of a terrorist cell by the British Special Air Service in London during the Iranian Embassy Siege of 1980, which spurred the growth of specialized domestic counterterrorism units across many special forces organizations.

The stealthy, masked-up, black-clad operators of the 1990s were sometimes known as 'ninjas'—a reductionist perception, defining the whole through one of its parts, that missed an enormous range of activities in the broader category of special operations.

There were two additional problems with this perception, one theoretical and the other practical. The theoretical problem was that the very concept of special operations was becoming self-referential: special operations were 'operations conducted by special forces', while special operations forces (SOF) were simply 'forces that conduct special operations'. The core characteristics of special operations—unconventionality, complexity, technical difficulty, and political sensitivity—and the forces that conduct them (specially selected, trained, and equipped, as a risk-reduction measure, to handle those core characteristics) became blurred, while a plethora of new mission types in emerging domains such as cyberspace were added to an already extensive task list.

The practical issue, exacerbated by the can-do attitudes of special operators themselves, was a growing preference by political leaders for the use of special rather than general-purpose forces in any and every important task (especially in the land domain), whether it involved the core characteristics of special operations or not. This resulted, over time, in stressed and over-stretched special forces, demoralized and underused general-purpose forces, and resentment between them.

The concerns of conventional commanders, who saw SOF encroaching on their turf, caused special forces to feel a threat to their existence, giving them an institutional imperative to prove their value. This in turn exacerbated their tendency, under the relatively peaceful conditions of the 1990s, to elbow aside conventional forces whenever challenging, interesting, or sensitive work came up. In effect, conventional-force concerns about SOF 'mission creep' were themselves a key driver of that very mission creep.

Less than nine months into the new century, this picture was radically disrupted by the September 2001 al-Qaeda attacks on the United States, which launched the Global War on Terrorism (GWOT) in response. Within weeks, special forces from the US and its key military allies (including the UK and Australia), alongside paramilitary special activities teams of the CIA, were in Afghanistan performing tasks—support to the irregular forces of the Northern Alliance, forward control of air strikes, reconnaissance and intelligence collection, and support-to-resistance networks—that looked a lot more like their Office of Strategic Services or Special Operations Executive predecessors from the 1940s than the black-clad ninjas of the 1990s.

In retrospect, the early war on terrorism—the era of horse soldiers, working intimately on the ground with local commanders, embedded in teams of indigenous guerrillas and enabled by B-52s, satellites, and drones—can be clearly seen as a missed opportunity. Instead of following the unconventional path pioneered in Afghanistan, the GWOT was quickly derailed by the ill-judged invasion and morally conflicted occupation of Iraq, while special operations became increasingly centred on kinetic direct action in support of large-scale conventional campaigns.

Elite-conventional rather than unconventional, such direct action—close-quarter battle, night raids, door-kicking, explosive breaching, 'warheads on foreheads'—became a defining task for special forces during much of the war on terror, receiving enormous amounts of funding, attention, and energy. Unfortunately, such activity often lacked a larger strategic purpose, with exquisitely trained, expensively equipped troops being thrown repeatedly into a kill cycle that achieved astounding tactical success, but thereby allowed political leaders to avoid the need to rethink strategic direction. With notable exceptions—the Village Stability Operations programme in Afghanistan, the Sons of Anbar in Iraq, or the development of Jedburgh teams in other theatres—strategically focused unconventional warfare played second fiddle to direct-action raiding.

As a consequence of its prominence, direct action exercised a magnetic effect on special forces, with organizations recasting their roles as kinetic and direct, rather than enabling and indirect. This was true for general-purpose forces too: even as special forces focused on raiding, conventional troops became more 'SOF-like', with capabilities and missions previously associated with special forces now appearing in, for example, conventional infantry battalions. Likewise, civilian intelligence services—notably the CIA with its Counterterrorism Pursuit Teams and other paramilitary forces—developed complementary (or competing) direct-action capabilities.

As the war on terrorism came to its ignominious conclusion (which we might roughly date to the fall of Kabul in mid-August 2021), special operators found themselves at something of an existential crossroads. The capabilities developed since 2001 represented a deep investment in one particular subset of special operations, but other organizations—conventional forces at one end of the spectrum, and intelligence services at the other—were now operating, and competing for resources, in that same capability space. At the same time, in a new era of great-power competition, unconventional (rather than elite-conventional) tasks such as support to resistance, cognitive warfare, and the space–cyber–SOF nexus were becoming more prominent in hybrid conflict, triggering concerns about control of far-flung teams in ambiguous, complex conflicts which could have nuclear consequences. Should special operations organizations double down on the capabilities built since 9/11, or should they reorient to grow new capabilities for the emerging mission space of great-power competition?

In order to answer this question, we must first ask ourselves exactly what special operations are, how their enduring nature and changing character relate to the forces that carry them out, and how capabilities and control intersect to form the vertical and horizontal dimensions of a complex special operations space.

These three elements—capabilities, control, and complexity—lie at the heart of James Kiras' argument in this book. In tying capability and control together as an integrated theory, linking them to historical case studies drawn from British, American, and South African experience, and exploring how his theory might apply in the complexity of the emerging operating environment, James has performed a valuable service for anyone who thinks deeply about special operations, does them for a living, or relies on their success. His book is essential reading for all of us who seek to grapple with these issues in the search for special operations success.

David Kilcullen

*Denver, Colorado*
*May 2024*

# Acknowledgements

This work represents the culmination of a twenty-year journey, one starting in the Pentagon in 2003. I was working in the Office of the Assistant Secretary of Defense for Special Operations and Low Intensity Conflict (OASD SO/LIC) as part of a special study group looking at the global composition, disposition, and employment of American special operations forces (SOF) while at the same time completing my doctoral dissertation on special operations. The group was asked to present at an upcoming conference on the subject held at the Royal Military College (RMC) in Kingston, Ontario, Canada. The specific subject was the transformation of American special operations on the heels of recent successes in Afghanistan and Iraq. Once selected for the task, I developed a PowerPoint presentation and submitted a paper, which subsequently appeared in edited form as a chapter in a conference proceedings volume entitled *Choice of Force*, the second in a series published by McGill-Queen's University Press.

That chapter comprised a kaleidoscope of ideas, some influenced by my dissertation research and others by my experiences 'in the building' working on special operations policy and strategy. The central insight of the chapter was that a new special operations triad was forming between conventional military forces, SOF, and paramilitary forces identified with intelligence agencies. I argued for greater synergy between them. I lacked the lexicon, depth, and experience, however, to take it to its logical conclusions. Subsequent study, presentations, publications, interactions and conversations, and experience over the past two decades have filled in some of the gaps. Panel chairs and audiences at various conferences over the years have listened patiently and provided excellent feedback. Portions of this argument, particularly in Chapter 3, have appeared in print elsewhere, including chapters in the following works: Gitte Hjostrup Christensen, ed., *Interdisciplinary Perspectives on Special Operations Forces*, Conference Proceedings No. 4 (a) (Copenhagen: Royal Danish Defence College, 2017) and Jessica Turnley, Kobi Michael, and Eyal Ben-Ari, eds., *Special Operations Forces in the 21st Century: Perspectives from the Social Sciences* (London: Routledge, 2018). I thank the publishers for providing permission to reproduce the copyrighted material.

The journey has not been a solitary one. Along the way, I have had opportunities provided by access, some privileged, to deepen my knowledge and understanding of the field of special operations. This includes entrée into and perspectives from sharp and dedicated minds on special operations primarily in the United States, but also in Canada, Denmark, Sweden, Singapore, and the United Kingdom. In the United States, I would like to thank Tom Duckworth, Tom Miller, Greg Jannerone, Dick Basehart, Steve Grenier, Tom Searle, Ken Poole, Kalev Sepp, Grey Welborn, Glenn Harned, Gordon Ettenson, and others whose names would fill a dozen pages. Ashley

Luckie, who bends bureaucracies to her will, knows why she is mentioned here. Christopher Marsh, a force of nature, has been an intellectual partner and friend who swept me up in his vision for a professional peer-reviewed journal on special operations. In addition, this work owes much to the special operations students I have had the good fortune of mentoring, advising, and getting to know over the years, including Rob Masaitis, Pat Wnetrzak, Erick Turasz, Mike Curry, Pat Dierig, Rob Powell, Nick Dipoma, Mark Hamilton, Brett Cassidy, Josh Petry, Marc Church, Kaveri Crum, Erik Towal, Tyler Martin, Joe Tomczak, and many others over the past two decades. Our conversations on a range of topics have deepened my understanding of much that would otherwise remain obscure in the special operations world.

Outside of the United States, other figures have influenced the course of this work. In the UK, the late Colin Gray, to whom this work is dedicated, hectored me over the years to write a special operations theory. We shared this goal; his sadly went unrealized. Also in the UK, Chris Luck, Charles Antelme, and Andrew Massey made spirited but ultimately unsuccessful efforts to reconnect with the British special operations community. The staff of the National Archives at Kew and the Imperial War Museum assisted in accessing what few records are available on special operations in Northern Ireland and the Falklands. North of the American border, Bernd Horn and Emily Spencer provided their own inspiration and support, the former in his inimical, 'special' way. Martijn Kitzen in the Netherlands has been a pillar, supporter, and friend for almost as long as the Dutch had an empire, and Funs Titulaer and Martijn van der Vorm will recognize their influences in the text. In Poland, Dariusz Dachowitz has always been there with a friendly smile, advice, and support. In the Nordic countries, Ulrica Petterson in Sweden, Tone Danielson in Norway, and Anja Dalgaard-Nielsen in Denmark each contributed in important ways of which they were perhaps unaware. Interviews in South Africa would not have been possible without the efforts of Abel Esterhuyse, a scholar I look up to figuratively as well as literally. Thanks to him, Chris Serfontein and Christo Roelfse of the South African Special Forces Association opened doors that otherwise would have remained closed.

Closer to home, this book would not exist without the support and assistance of my friends and colleagues at Maxwell Air Force Base, specifically at what Secretary of Defense James Mattis termed 'the crown jewel of professional military education', the School of Advanced Air and Space Studies (SAASS). A succession of School commandants provided the time and resources to pursue my professional research over the years. My colleagues at SAASS not only supported me but also made sacrifices to allow this work to develop, such as assuming my thesis load and shielding me from administration at a crucial time. To you all, I give my deepest thanks. Derrick Frazier and Tom Hughes did so particularly enthusiastically, and Rich Muller, Mark Conversino, John Maurer, Lena Svedin, and Rob Hutchinson have provided friendship and intellectual and moral support at crucial junctures of the writing process. To Rob, I am especially grateful. He reviewed each chapter in draft form, some very rough indeed, and not only provided questions and observations but also pored over the text with a keen editorial eye to pare away the undergrowth to ensure the trees of the argument stand more visible. S. Mike Pavelec's friendship is sterling; few friends are as thoughtful and supportive. In good times and bad, he has always been there regardless of where his own journeys have taken him. Sheila McKitt, who has since moved on from SAASS, is a force of nature. I know, even if she never mentioned it specifically, what she did to shield me from administrativia, allowing me to focus my energy and effort. The staff of the Air University Library and the Historical Research Agency pursued my requests for archival and interlibrary loan materials with alacrity.

At home, my wife Michelle bore the burdens of a project of this scope and scale. Her unwavering support in every category imaginable and seemingly endless patience for my flights of fancy and absences, real, intellectual, and emotional, provided the foundation on which this work could be built. She does more than a spouse, partner, and friend could or should each and every day, in so many ways. Doris Kiras started before and picked up where Colin left off in terms of prodding, in a manner only a mother can, to write my second book and then became a tireless cheerleader for me to finish once under contract. My in-laws from the Spencer clan, especially Jim and Dorothy, have been equally supportive while this book was underway. You are unaware just how much I have appreciated your kindness and encouragement. In addition, mornings and afternoons on your '15' and 'the back 40' provided much-needed time and space to think.

To Dominic Byatt, Phoebe Aldridge-Turner, Purushothaman Govindhasamy, and Sally Evans-Darby team at Oxford University Press, my thanks can never be enough. Dominic has been everything an acquisitions editor should be and more. Both he and Phoebe have been patient and understanding in the face of the inevitable, unexpected delays. Purushoth managed the production process smoothly and Sally's sharp editorial eye saved me from numerous embarrassments. An unnamed reviewer provided outstanding suggestions to sharpen the arguments and improve their clarity. Any mistakes contained within are mine, and mine alone.

Finally, a necessary disclaimer. The views and opinions presented herein are those of the author and do not necessarily represent the views of Department of Defense (DoD) or its Components. Appearance of, or reference to, any commercial products or services does not constitute DoD endorsement of those products or services. The appearance of external hyperlinks does not constitute DoD endorsement of the linked websites, or the information, products or services therein. The appearance of US DoD visual information does not imply or constitute DoD endorsement. The information researched and presented in this manuscript is not critical or sensitive in that it could be harmful to the United States, its government, or its militaries. All DoD and US government policies and operating guidelines are accurately represented throughout this publication.

# Contents

# List of Figures

# List of Table

# List of Maps

# List of Abbreviations

| | |
|---|---|
| AAA | anti-aircraft artillery |
| AFP | Armed Forces of the Philippines |
| AFSOC | Air Force Special Operations Command (US) |
| AFSOD | Armed Forces Special Operations Division (India) |
| ALP | Afghan Local Police |
| ANA | Afghan National Army |
| ANAP | Afghan National Auxiliary Police |
| ANC | African National Congress |
| AOB | advanced operations base |
| AQI | al-Qaeda in Iraq |
| ASDS | Advanced SEAL Delivery System |
| ASG | Abu Sayyaf Group |
| ASU | active service unit |
| AUMF | Authorization to Use Military Force (US) |
| BAOR | British Army of the Rhine |
| BATT | British Army Training Team |
| BCE | before common era |
| BSAP | British South Africa Police |
| C2 | command and control |
| CA | civil affairs |
| CAD | Civil Aid Department |
| CAOC | Combined Air Operations Center |
| CAS | close air support |
| CASEVAC | casualty evacuation |
| CCB | Civil Cooperation Bureau |
| CCO | Clandestine Communist Organisation |
| CDS | Chief of the Defence Staff (UK and Commonwealth) |
| CE | common era |
| CENTCOM | Central Command |
| CERP | Commander's Emergency Response Program |
| CFACC | Combined Forces Air Component Commander |
| CIA | Central Intelligence Agency |
| CIDG | Civilian Irregular Defense Group |
| CIO | Central Intelligence Organisation (Rhodesia) |
| CJCS | Chairman of the Joint Chiefs of Staff (US) |
| CJOC | Combined Joint Operations Center |
| CJSOTF | Combined Joint Special Operations Task Force |
| CN | counternarcotics |
| COBR | Cabinet Office Briefing Room |
| COCOM | Combatant Command |
| COIN | counterinsurgency |

| | |
|---|---|
| COMREL | command relationships |
| CONOP | concept of operations |
| COP | close observation post |
| CP | counter-proliferation |
| CPA | Coalition Provisional Authority |
| CRW | counter-revolutionary wing |
| CSAR | combat search and rescue |
| CT | counterterrorism |
| CTS | Counter Terrorism Service (Iraq) |
| DA | direct action |
| DAT | District Augmentation Team |
| DIRLAUTH | direct liaison authority |
| DoD | Department of Defense (US) |
| DRFLA | Democratic Revolutionary Front for the Liberation of Arabistan |
| DSF | Directorate of Special Forces (UK) |
| DST | Directorate of Special Tasks (South Africa) |
| EO | executive order |
| ESG | Electronic Support Group (India) |
| EXORD | execute orders |
| F3EAD | find, fix, finish, exploit, analyse, disseminate |
| FAES | *Forces Armada de El Salvador* |
| FAPLA | *Forças Armadas Populares de Libertação de Angola* |
| FATA | Federally Administered Tribal Areas |
| FID | foreign internal defence |
| FMLN | *Frente Farabundo Marti de Liberacîon Nacional* |
| FMV | full motion video |
| FNLA | *Frente Nacional de Libertação de Angola* |
| FOO | forward observation officer |
| FRELIMO | *Frente de Libertação de Moçambique* |
| FRU | Force Research Unit |
| GCC | Geographic Combatant Command |
| GCSF | Global Security Contingency Fund |
| GIRoA | Government of the Islamic Republic of Afghanistan |
| GNA | Greek National Army |
| GOC | General Officer Commanding |
| GPS | global positioning system |
| HADR | Humanitarian and Disaster Relief |
| HVT | high-value target |
| IBT | Indonesian Border Terrorist |
| IDAD | internal defence and development |
| IFP | Inkatha Freedom Party (South Africa) |
| IPKF | India Peace Keeping Force |
| IR | infrared |
| IRA | Irish Republican Army |
| ISAF | International Stabilization Force |
| ISIS | Islamic State in Iraq and Syria |
| ISIS-GS | Islamic State in Iraq and Syria—Greater Sahara |

| | |
|---|---|
| ISR | intelligence, surveillance, and reconnaissance |
| IT | investigative team |
| JACE | joint air coordination element |
| JAGIC | Joint Air Ground Integration Center |
| JCET | joint combined exchange training |
| JFE | joint fires element |
| JIATF | Joint Interagency Task Force |
| JIB | Joint Intelligence Bureau (UK) |
| JMC | Joint Management Centre (South Africa) |
| JOC | Joint Operations Center |
| JSOTF | Joint Special Operations Task Force |
| JSOTF-P | Joint Special Operations Task Force—Philippines |
| JSOU | Joint Special Operations University (US) |
| JTAC | joint terminal air controller |
| KDP | Kurdish Democratic Party |
| LRA | Lord's Resistance Army |
| LRC | Light Reaction Companies (Philippines) |
| LRDG | Long Range Desert Group (UK) |
| LTTE | Liberation Tigers of Tamil Eelam |
| MAAG | Military Advisory Assistance Group |
| MACV | Military Assistance Command, Vietnam |
| MACVSOG | Military Assistance Command, Vietnam, Studies and Observation Group |
| MARCOS | Marine Commandos (India) |
| MCP | Malayan Communist Party |
| MCTF | Maritime Counter-Terrorist Force (UK) |
| MEW | Ministry of Economic Warfare |
| MID | Military Intelligence Directorate |
| MILF | Moro Islamic Liberation Front |
| MILGRP | military group |
| MISO | military information support operations |
| MIST | military information support team |
| MK | *uMkhonto we Sizwe* |
| MNLA | Malayan National Liberation Army |
| MNLF | Moro National Liberation Front |
| MPAJA | Malayan People's Anti-Japanese Army |
| MPLA | *Movimento Popular de Libertação de Angola* |
| MRF | military reaction force |
| MTT | military or mobile training team |
| NATO | North Atlantic Treaty Organization |
| NCM | National Coordinating Mechanism (South Africa) |
| NCO | non-commissioned officer |
| NGO | non-governmental organization |
| NILE | Northern Iraq Liaison Element (US) |
| NIS | National Intelligence Service (South Africa) |
| NJMC | National Joint Management Centre (South Africa) |
| NSCC | NATO SOF Coordination Centre |
| NSG | National Security Guard (India) |

| | |
|---|---|
| NSMS | National Security Management System |
| NVA | North Vietnamese Army |
| OASD SO/LIC | Office of the Assistant Secretary of Defense for Special Operations and Low Intensity Conflict |
| OAU | Organization of African Unity |
| OC | officer commanding |
| OCO | overseas contingency operations |
| ODA | Operational Detachment Alpha (US) |
| OEF | Operation ENDURING FREEDOM |
| OGP | Operation GALLANT PHOENIX |
| OIF | Operation IRAQI FREEDOM |
| OIR | Operation INHERENT RESOLVE |
| OOD | Operation ODYSSEY DAWN |
| OODA | observe-orient-decide-act |
| OP | observation post |
| OPCON | operational control |
| OPE | operation preparation of the environment |
| ORHA | Office of Reconstruction and Humanitarian Affairs |
| OSS | Office of Strategic Services (US) |
| PAC | Pan Africanist Congress |
| PFLOAG | Population Front for the Liberation of the Arabian Gulf |
| PFLP | Popular Front for the Liberation of Palestine |
| PIFWC | persons indicted for war crimes |
| PIRA | Provisional Irish Republican Army |
| PLAN | People's Liberation Army of Namibia |
| POG | Psychological Operations Group |
| PRC | People's Republic of China |
| PTSD | post-traumatic stress disorder |
| PUK | Patriotic Union of Kurdistan |
| PWE | Political Warfare Executive |
| RAF | Royal Air Force |
| RAW | Research and Analysis Wing (India) |
| REDCOM | Readiness Command |
| RENAMO | *Resistência Nacional Moçambicana* |
| RMBPD | Royal Marines Boom Patrol Detachment (UK) |
| RMC | Royal Military College (Canada) |
| RPA | remotely piloted aircraft |
| RSM | regimental sergeant major |
| RTU | returned to unit |
| RUC | Royal Ulster Constabulary |
| SAAF | South African Air Force |
| SAASS | School of Advanced Air and Space Studies (US) |
| SACP | South African Communist Party |
| SADF | South African Defence Forces |
| SADM | Special Atomic Demolition Munition |
| SAF | Sultan's Armed Forces |
| SAG | Special Action Group (India) |

| | |
|---|---|
| SAN | South African Navy |
| SAP | South African Police |
| SAS | Special Air Service |
| SASR | Special Air Service Regiment (Australia) |
| SATCOM | satellite communication |
| SBS | Special Boat Squadron or Section (UK) |
| SCG | Special Composite Group (India) |
| SDF | Syrian Democratic Forces |
| SF | Special Forces (US) |
| SFG | Special Forces Group (US) |
| SGT | sergeant |
| SIGINT | signals intelligence |
| SIS | Secret Intelligence Service (UK) |
| SN | strategic national |
| SOAR | Special Operations Aviation Regiment (US) |
| SOE | Special Operations Executive (UK) |
| SOF | special operations forces |
| SOG | Studies and Observation Group |
| SOLE | Special Operations Liaison Elements |
| SOP | standard operation procedure |
| SOS | Special Operations Squadron (US) |
| SOST | Special Operations Support Team |
| SOTF | Special Operation Task Force |
| SPT | Special Projects Team |
| SR | special reconnaissance |
| SRU | Special Reconnaissance Unit |
| SSC | State Security Council (South Africa) |
| SSE | sensitive site exploitation |
| SSG | staff sergeant |
| ST | strategic theatre |
| STEP | Syria Train and Equip Program |
| STS | Special Tactics Squadron (US) |
| SWAPO | South West Africa People's Organization |
| SWAT | special weapons and tactics |
| SWATF | South West Africa Territorial Force |
| TA | tactical |
| TCG | Tasking and Co-ordination Group |
| TEA | target engagement authority |
| TF | task force |
| TIP | target intelligence package |
| TSOC | Theater Special Operations Command |
| TTP | Tehrik-e Taliban Pakistan |
| UDA | Ulster Defence Association |
| UDF | United Democratic Front (South Africa) |
| UK | United Kingdom |
| UKSF | United Kingdom Special Forces |
| UMSA | Unity Movement of South Africa |

| | |
|---|---|
| UNITA | *União Nacional para a Independência Total de Angola* |
| US | United States |
| USAID | US Agency for International Development |
| USD(I) | Undersecretary of Defense for Intelligence |
| USSOCOM | United States Special Operations Command |
| USSR | Union of Soviet Socialist Republics (Soviet Union) |
| UVF | Ulster Volunteer Force |
| VBIED | vehicle-borne improvised explosive device |
| VEO | violent extremist organization |
| VSO | Village Stability Operations |
| VTC | video teleconference |
| WMD | weapons of mass destruction |
| YPG | *Yekîneyên Parastina Gel* |
| ZANLA | Zimbabwe African National Liberation Army |
| ZANU | Zimbabwe African National Union |
| ZAPU | Zimbabwe African People's Union |
| ZIPRA | Zimbabwe People's Revolutionary Army |

# PART I

# 1
# Introduction

Since the terrorist attacks on the World Trade Center in New York, and the Pentagon in Arlington, Virginia, on 11 September 2001, American special operations can claim several accomplishments. They include taking actions against terrorist groups overseas, killing or capturing more than 90 per cent of the senior leadership of the al-Qaeda, and toppling or playing a major role in overthrowing the governments responsible for harbouring al-Qaeda or suspected of supporting them, including the Taliban in Afghanistan and the regime of Saddam Hussein in Iraq. Battlefield successes include the killing or capturing of thousands of mid- and high-level terrorist group leaders and facilitators in more than a dozen countries, as well as the capture and exploitation of millions of pages of terrorist documents, providing crucial insights into how global terrorist networks operate. How special operations are conducted has changed, largely through a range of organizational and technical innovations that redefined the methodology for surveilling, targeting, and striking individual terrorists, often from thousands of miles away. The pace of special operations has been seemingly relentless, with the units charged to prosecute them spending as much overseas as they have at home recovering and preparing to go back.

The success of special operations reflects the confidence that national security leaders have in the leaders and forces conducting them. This confidence has been the result of the ability of American special operations to respond to unexpected crises confronting these leaders. Such crises include the resurgence or reformation of terrorist groups that threaten countries partnered or allied to the United States, including Iraq, Turkey, and the Philippines, among others, and rescuing American hostages and seizing materials transported illegally for building weapons of mass destruction. American, British, Canadian, and other national leaders know that when crises erupt, special operators have plans and capabilities in reaction to them ready to go almost immediately, while other government elements are slow to respond. Leaders translate this confidence in special operations responsiveness into trust, access, influence, and resources. In consequence, the American special operations community has grown organizationally by more than 33 per cent over a decade and expanded several units, adding new special operations capabilities.

Success, however, cannot be sustained forever and comes with its own costs. Less frequent but highly publicized failures in the field of operations, such as the ambush of an Army Special Forces team in Niger in 2017 (discussed in Chapter 4) or the killing of a hostage during a rescue operation, as was the case with Linda Norgrove in

*Special Operations Success*. James D. Kiras, Oxford University Press. © James D. Kiras (2024).
DOI: 10.1093/9780198902096.003.0001

Afghanistan in 2010, are one thing.[1] Most failures can be ascribed to bad luck, insufficient information or intelligence, or a tactical error while in contact with the enemy. They pale in number and consequence to incidents that call into question the core values or ethics of a national special operations community. Such incidents include: the murder by SEALs of a fellow operator and its cover-up in Africa; the abuse of alcohol by, rape allegations surrounding, command loss of confidence in, and rotation home of an entire platoon from SEAL Team 7; alleged drug use by operators in SEAL Team 10 that some termed 'staggering'; and the desecration of corpses by SEAL senior enlisted officer Eddie Gallagher, among others.[2] These incidents and investigations surrounding them, which have also occurred or are underway in Australia and the United Kingdom, are one reflection of a relentless, sustained pace of operations that has claimed a toll beyond moral, ethical, and integrity lapses. Within the American special operations forces (SOF) community, operators are committing suicides at high rates and families are broken apart through increasing numbers of divorces, in what may be a 'crisis' for the force.[3]

The once unshakeable confidence in special operations due to their mission focus and track record of successes has led to efforts to reassert presumed lost control over SOF. In specific, members of Congress took the extraordinary step in 2017 of writing into law the requirement that American SOF conduct a 'comprehensive review' of their culture and ethics based on the incidents described above. The resulting review, undertaken by the United States Special Operations Command (USSOCOM), identified sixteen action items across five 'key areas': force employment, force accountability, leader development, force structure, and assessment and selection.[4] The review participants determined that these and other incidents did not reflect systemic problems within SOF. Ultimately, their overarching conclusion was SOF's action and results orientation, or mission focus, occurs at the expense of 'leadership, discipline, and accountability', sacrifices opportunities for individual growth, and loses touch with the community's collective values.[5] These answers did not satisfy legislators' concerns. Their continued questions, combined with a reorientation in national security priorities that forces SOF to identify how to continue operations against irregular threats, compete against 'great powers' daily, and prepare for future conflict, has led to introspection within the American special operations community bordering on existential crisis.

[1] Julian Borger, 'Linda Norgrove: US Navy Seal Faces Disciplinary Action over Grenade Death', *The Guardian* (13 October 2010), www.theguardian.com/world/2010/oct/13/linda-norgrove-us-commando-disciplinary.

[2] Joanne Kimberline, 'Murder Trials. War Crimes. Drug Abuse. The Navy SEALs Want to Redeem Themselves', *The Virginian-Pilot* (27 December 2019), www.pilotonline.com/military/vp-nw-seal-troubles-20,191,227-fwlykcrsfnbcfexzzxmyzwy65m-story.html; David Martin, 'Navy SEAL Drug Use "Staggering", Investigation Finds', CBS News (11 April 2017), www.cbsnews.com/news/navy-seal-drug-use-staggering-investigation-finds.

[3] 'Are America's Special Operations Forces in Crisis?' *The Week*, 9 January 2015, https://theweek.com/articles/443,052/are-americas-special-operations-forces-crisis.

[4] USSOCOM Review Team, *United States Special Operations Command Comprehensive Review* (MacDill Air Force Base, FL: USSOCOM, January 2020), pp. 24–43.

[5] USSOCOM Review Team, *United States Special Operations Command Comprehensive Review*, pp. 4–7.

Scholastically, interest in special operations is growing. It includes an increasing number of academic dissertations on the subject, a peer-reviewed journal on special operations and irregular warfare (*Inter Populum*), and conferences and panels on the subject. The Joint Special Operations University, (possessive) USSOCOM's education arm, has a press that has published 213 reports and monographs since 2005. Its Canadian equivalent has been equally energetic in its output. There is scholastic interest in and desire to connect the subject of special operations to specific academic subdisciplines, including the social sciences (sociology, anthropology, political science, international relations, etc.) and the liberal arts (including history). Evidence of broader academic acceptance is available in a few conferences on the subject, including an annual event held at Yale University, as well as in a number of panels at mainstream academic conferences such as the International Studies Association, American Political Science Association, Inter-University Seminar on Armed Forces and Society, and the Society for Military History.

The body of scholarship, however, is uneven. Part of this is a function of information and access. Information is scarce and access often comes at a price. Scarce information leads to inferential leaps and an overreliance on personal accounts or experiences for familiar scholastic reasons, including identifying causation based on correlation. The price of access to special operations communities can compromise scholastic objectivity or integrity by associating too closely with your object of study and not wanting to 'burn sources' through critical inquiry who may not agree with or endorse findings. Those unfamiliar with institutional values and norms can lead scholars to knowingly or unwittingly advance their agendas through what one scholar has identified succinctly as 'cheerleading' for them.[6] Privileged experience and information results in axe-grinding and desire to right observed wrongs.[7] Advancing agendas leads to the perpetuation of institutional myths outside of the special operations community, specifically related to the misuse of SOF, as well as tribal feuds within it that reflect jockeying for influence, primacy, and resources internally.[8] Scholarly work is being produced but is often narrowly focused, unlikely to influence the broader SOF community because it lacks perceived utility and applicability, compromised due to access and agendas, or designed to influence change and redistribute resources or power in favour of a specific special operations community.

## Generational waves of special operations theory

Within this body of scholarship is a narrow subset: works attempting to codify a theory that explains special operations. This subset of works manifested in two waves.

[6] Anton Asklund Johnsen and Gitte Højstrup Christensen, 'Clarifying the Antisystemic Elements of Special Operations: A Conceptual Inquiry', *Special Operations Journal*, 2:2 (Fall 2016): 110.

[7] For example, see Hy Rothstein, *Afghanistan and the Troubled Future of Unconventional Warfare* (Annapolis, MD: Naval Institute Press, 2006); Charles Cleveland and Daniel Egel, *The American Way of Irregular War: An Analytic Memoir*, Report PE-A301-1 (Santa Monica, CA: RAND, 2020).

[8] See David Tucker and Christopher Lamb, *United States Special Operations Forces* (New York: Columbia University Press, 2007), chapter 7, 'Restructuring Special Operations Forces'. In a revised and updated edition, the authors removed this chapter in favour of more recent case studies.

The first occurred between 1994 and 2008. The second wave began in 2012 with an abortive attempt at drafting a theory of SOF power theory that led to Harry Yarger's monograph the following year and peaked in 2017 with the publication of Joint Special Operations University's (JSOU's) three-volume series on special operations theory.

At the risk of repetition, but in the interest of surveying the literature quickly, the following is a summary of first-wave efforts. Prior to 1994, several unpublished studies explored different types of special operations analytically based on the interest of the clients who contracted them. Edward Luttwak and two of his associates produced a study for the Pentagon in 1982 that reviewed 'commando (special) operations' and defined mission types, identified organizational characteristics, and derived planning principles based on an analysis of nine individual operations.[9] The same year, a team led by David Charters completed a study of 'surgical operations' for the Centre of Conflict Studies at the University of New Brunswick based on an analysis of six rescue operations or military interventions.[10] While not theoretical works per se, such studies contained intriguing insights useful for building theory. Their unpublished nature, however, led to limited reproduction and distribution. In 1994, Steven Lambakis published one of the first works on special operations theory, an article entitled 'Forty Selected Men Can Shake the World'.[11] This article, as well as four chapters in a book (1996) and article (1999) written by Colin Gray, resulted from the first of three studies conducted by them for the Pentagon submitted in 1992.[12] These articles and chapters explored special operations contribution to victory, as well as their strategic utility and effectiveness. Among their insights for theory are master and supporting claims about the value of special operations strategically. Although one might take issue with the scope of the definition used in the studies, as well as evidence limited largely to the Second World War, these claims remain unchallenged and are often cited in more recent works.

The other major theoretical work from this period that has cast an equally long shadow, based on his professional military education thesis, is William McRaven's *Spec Ops.* McRaven develops the concept of 'relative superiority' for a specific type of special operation, the raid or 'direct-action' mission. Based on six principles that form a model for such missions, McRaven's theory postulates that success or failure hinges on maintaining relative superiority for a critical period.[13] The popularity and acceptance of his work rose in the American special operations community with his rise in rank, ultimately as commander, USSOCOM.

[9] Edward Luttwak, Steven Canby, and David Thomas, 'A Systematic Review of "Commando" (Special) Operations, 1939–1980', unpublished report, C & L Associates, 24 May 1982.

[10] David Charters et al., 'Surgical Operations: Military Lessons from Selected Case Studies', unpublished report, Centre for Conflict Studies, Fall 1982.

[11] Steven Lambakis, '"Forty Selected Men Can Shake the World": The Contributions of Special Operations to Victory', *Comparative Strategy*, 13:2 (Summer 1994): 211–221.

[12] Colin Gray, *Explorations in Strategy* (Westport, CT: Praeger, 1996); Colin Gray, 'Handful of Heroes on Desperate Ventures: When Do Special Operations Succeed?', *Parameters*, 29:1 (Spring 1999): 2–24; Colin Gray, 'Special Operations: What Succeeds and Why? Lessons of Experience, Phase I', unpublished report, National Institute for Public Policy, 1992.

[13] William McRaven, *Spec Ops, Case Studies in Special Operations Warfare: Theory and Practice* (Novato, CA: Presidio Press, 1995), pp. 8–23.

Other first-wave theoretical works point out some of the limitations of Lambakis, Gray, and McRaven but largely accept their conclusions. These include Alastair Finlan's and my books (2009 and 2006, respectively) and Robert Spulak's 2008 monograph. Of these three, only Spulak claims to present a theory of special operations. Finlan harnesses the ideas of Thomas More, and more specifically his work *Utopia*, to develop a SOF-centric approach to state-based conflict. He does so to understand the appeal of special operations to policy makers while exploring the various boundaries that led to their ultimate misapplication and misuse.[14] I limited my inquiry to the relationship between military SOF and conventional forces during major conflicts, using classical military theory to answer the question of strategic effectiveness of special operations. I concluded SOF are most profitably used to enhance the strategic performance of conventional forces in campaigns of attrition, and not in the more appealing attempts at decisive strategic paralysis.[15] Spulak's theory of SOF is based on the attributes distinguishing them from their conventional military counterparts, ones that allow them to overcome the friction of war. A work taking a different tack is by Michelle Malvesti. Her unpublished 2002 doctoral dissertation on special operations and decision making explores counterterrorism operations through the lens of rational choice theory. This dissertation abstracts away much in doing so and focuses only on decision making in crises in which SOF were primary response mechanism for a specific mission set, rather than relationship of special operations to decision making.[16] The works of the first wave explore facets of theory, and provide useful insights, but fall short of the mark of developing a general theory of special operations.

The second wave of theoretical works were prompted by the growing influence and visibility of special operations. One of the first efforts to develop a general theory, specifically on SOF power, was ambitious. It sought to elevate thinking about special operations to the same level as military theories on land, sea, air, and space power. Spurred by USSOCOM commander McRaven, this theory was written by retired special operators who had a track record of publishing on SOF and related topics. In addition, each was chosen to represent within the theory the perspectives of Army, Navy, and Air Force special operations components. The resulting work by committee, intended for publication in 2012, lived up to T. Barnett Cocks' depiction of them as 'a cul-de-sac down which ideas are lured and quietly strangled'.[17] The work was disjointed, aphoristic, and descriptive, based on five 'immutable truths', and never passed internal review for publication before being shelved.[18]

[14] Alastair Finlan, *Special Forces, Strategy, and the War on Terror: Warfare by Other Means* (London: Routledge, 2009).

[15] James Kiras, *Special Operations and Strategy: From World War II to the War on Terrorism* (London: Routledge, 2006).

[16] Michelle Malvesti, 'Risk-Taking in Countering Terrorism: A Study of United States Presidential Decisions to Use Special Operations and Covert Action', unpublished PhD dissertation, June 2002.

[17] Quoted in Kevin McGrew, 'Is the Intellectual Functioning Component of AAIDD's 12th Manual Satisficing?', *Intellectual and Developmental Disabilities*, 59:5 (2021): 369.

[18] Joseph Celeski, Timothy Slemp, and John Jogerst, 'The Evolution of Special Operations Power: Purpose, Application, and Theory', unpublished draft report, Joint Special Operations University, December 2012, p. 139.

Much of the subsequent scholarship in this wave on special operations theory explored its boundaries, or argued for its necessity and feasibility, but did not develop theory per se. In 2013 Harry Yarger, a professor at the Army War College, identified principles and attributes of special operations that provided the precursor to theory but stopped there.[19] A debate regarding the prospects and desirability of theory, first in *Special Operations Journal* in 2015 that continued two years later at a conference sponsored by JSOU, resulted in proceedings and the release of two works discussed below.[20] More recently, in 2019 Finlan provided the basis for a theory centred around the analogue of anaphylaxis, or allergic shock to an enemy system provided by SOF.[21] More of an operating concept than a theory, Finlan's work falls into the same causal logic trap of his predecessors, advancing comparable ideas of strategic paralysis.[22]

Two further works by Richard Rubright and Tom Searle advance claims of developing a general unified theory of special operations, both published in 2017. Rubright's theory focuses on defining special operations and differentiating them from SOF. What his theory offers by way of the principle of Occam's Razor, or simplicity—it comprises just ten words—it lacks in explanatory power. His theory stated is 'Special Operations are extraordinary operations to achieve a specific effect'.[23] Searle, in his examination, points out that 'the second half of [Rubright's] definition is even less clear than the first'.[24] Searle's eponymous 'box' outside of which special operations operate consists of the range of activities, or roles and missions, of conventional operations that are defined by a number of Department of Defense statutes and outlined in a variety of authorities. He concludes that 'the operation remains special only so long as the conventional box does not expand to include that type of operation'.[25]

The primary value of Searle's brief exploration is outlining, in a clear and accessible manner, the legally mandated yet potentially changing relationship of special operations to conventional ones. One of the useful insights of his theory, which directly contradicts Rubright's assertion, is that the specialness of individual missions is irrelevant. What matters is 'the decision on whether or not to institutionalize a particular kind of SOF ... is based on the perceived enduring need for that capability'—in other words, if a specific special operations role and mission should become SOF legally

[19] Harry Yarger, *21st Century SOF: Toward an American Theory of Special Operations*, JSOU Report 13-1 (MacDill Air Force Base, FL: JSOU Press, April 2013).

[20] James Kiras, 'A Theory of Special Operations: "These Ideas Are Dangerous"', *Special Operations Journal*, 1:2 (2015): 75–88; Christopher Marsh, Mike Kenny, and Nathanael Joslyn, 'SO What? The Value of Scientific Inquiry and Theory Building in Special Operations Research', *Special Operations Journal*, 1:2 (2015): 89–104; Peter McCabe and Paul Lieber, eds., *Special Operations Theory*, JSOU Report 17-6 (MacDill Air Force Base, FL: JSOU Press, 2017).

[21] Alastair Finlan, 'A Dangerous Pathway? Toward a Theory of Special Forces', *Comparative Strategy*, 38:4 (Summer 2019): 255–275.

[22] Kiras, *Special Operations and Strategy*, pp. 16–34.

[23] Richard Rubright, *A Unified Theory for Special Operations*, JSOU Report 17-1 (MacDill Air Force Base, FL: JSOU Press, 2017), p. 15.

[24] Tom Searle, *Outside the Box: A New General Theory of Special Operations*, JSOU Report 17-4 (MacDill Air Force Base, FL: JSOU Press, 2017), p. 57.

[25] Searle, *Outside the Box*, p. 29.

outlined responsibility.[26] Searle's theory has explanatory power in the realm of distinguishing special from not. As a general theory, however, Searle's idea focuses on distinguishing between phenomena but does little to explain cause-and-effect relationships between or outside of them. In short, he answers the questions 'what?' and 'so what [does this mean for SOCOM]?' but not the most important question theory should address: 'why?'.

Efforts at developing special operations theory across the two waves of scholarship suffer from the same problems as others, including airpower: overpromising yet underdelivering.[27] The barriers to developing a theory of special operations remain and have been mapped out by those seeking to create it. Reflecting on his previous effort, Harry Yarger summarizes two of the primary problems admirably: 'providing sufficient depth for understanding the nature of special operations and SOF phenomena without plummeting into prescriptive dogma'.[28] The former can gravitate towards the ethereal and philosophical in the absence of sufficient evidence or experience, all but ensuring disinterest from pragmatic, problem-solving special operators. The latter, seeking practical application, can lose sight of the central phenomenon in its effort to codify useful advice.[29] Rubright reminds us in his exercise in lexical semantics to avoid the trap of using the terms 'special operations' and 'special operations forces' interchangeably and indiscriminately.

Given this reminder, and to meet the first function of theory outlined below, special operations are defined as extra-normative activities conducted by military or paramilitary forces that are costly, difficult, or impossible for others to perform well—to a sufficiently high standard in demanding conditions. Special operations are best performed by standing forces, with sufficient oversight, control, and support, by those readily available, equipped, and trained to do so: SOF. Special operations capabilities refer to the ability to perform historically identified functions established in Chapter 2 repeatably over time in a manner sustainable for SOF, in a range of different environments where others find it difficult to do so, given considerations of risk and political sensitivity.

## Theory: purpose, function, type

In its most basic sense, theory consists of a supposition or hypothesis intended to explain actions or behaviour based on systematic exploration of its nature through a series of codified, related propositions, backed by sufficient analysis and evidence. Theory should reduce complex phenomena down to an elegantly simply, abstract

[26] Ibid., p. 36.

[27] Colin Gray, *Airpower for Strategic Effect* (Maxwell Air Force Base, AL: Air University Press, 2012), pp. 183, 198.

[28] Harry Yarger, 'Two Special Operations/SOF Theory Challenges: Building Depth and Avoiding Prescription', in Peter McCabe and Paul Lieber, eds., *Special Operations Theory*, Report 17-6 (MacDill Air Force Base, FL: JSOU Press, 2017), p. 159.

[29] James Kiras, 'Do We Even Need a Theory?', in Peter McCabe and Paul Lieber, eds., *Special Operations Theory*, Report 17-6 (MacDill Air Force Base, FL: JSOU Press, 2017), pp. 136–139.

explanation. There are several ways to parse theory. One category, military theory, explores as its central phenomenon war or armed conflict. Armed conflict involves the use or the threat of use of violence to impose one's will against a dynamic, adaptive opponent seeking to deny you this goal and impose their will upon you. The debate currently within military theory is whether this central assumption about opponents and imposition of will is still valid given the range of actors and increasingly blurring lines between war and peace.[30] Most military theories, however, start from the assumption that conflict, in its most abstract form, is a rational undertaking for a political purpose and involves a dynamic between competing wills. In other words, the nature of armed struggle remains a reflection of and is heavily influenced by human nature.

Theory has both purpose and function. Its purpose has been identified in the previous section and is encapsulated in Yarger's summary of the problems of creating theory. Theory can provide insight about the nature of phenomena descriptively and it can prescribe actions to be taken. A common error in categorizing its purpose this way is assuming one must choose between the two. More often, the choice is between what its purpose will be primarily. Carl von Clausewitz started with prescriptions on how best to fight battles in different circumstances and conditions, but in subsequent revisions incorporated more thoughts on the nature of war still studied and discussed in staff and war colleges. His contemporary, Antoine de Baron Jomini, focused primarily on the prescriptive—elements which if understood sufficiently well laid out how and what to do—although there are descriptive aspects of his theory as well, including aspects Clausewitz overlooked such as the value of sea power for mobility and the implications of technology on battle. Descriptive theory attempts to answer the question of why a phenomenon occurs—to determine cause from effect and allow those with limited combat experience to put themselves in the mind of a decision maker and weigh the options. Prescriptive theories place a premium on simplifying the process of battle, identifying causal mechanisms for success, and suggesting how specific means used in particular ways will lead to desired outcomes.

Beyond what it does, a theory itself should fulfil specific functions. According to Harold Winton, there are five such functions upon which a theory can be evaluated for its comprehensiveness. Theory should define, categorize, explain, connect, and anticipate. Defining clarifies the subject under investigation by outlining what is comprised within it and what is not in a general sense—Searle's 'box'—and where special operations reside in relation to it. Definitions establish the boundaries for enquiry.[31] Theory's next function, categorizing, 'break[s] the field down into its constituent parts'.[32] An important and often neglected aspect of categorizing lies not just in identifying and separating out the constituent parts but relating them together.

[30] Emile Simpson, *War from the Ground Up: Twenty-First Century Combat as Politics* (Oxford: Oxford University Press, 2012), pp. 54–75.

[31] Harold Winton, 'An Imperfect Jewel: Military Theory and the Military Profession', *The Journal of Strategic Studies*, 34:6 (December 2011): 855–857.

[32] Harold Winton, 'An Imperfect Jewel: Military Theory and the Military Profession', paper presented at the Society of Military History conference, 2004, p. 2.

Next, theory should explain the phenomenon under investigation through 'repetitive observation and imaginative analysis' or Albert Einstein's 'intuition, supported by being sympathetically in touch with experience'.[33] For a theory to have explanatory power, it must explore both the scope and limitations of such operations relative to other methods or means. Winton concludes pithily, 'theory without explanation is like salt without savor—it is worthy only of the dung heap'.[34] Military theory can and should connect to other fields of study as well. As Winton argues, Clausewitz's genius lay not in divining the character of battle or identifying timeless principles of war but rather in connecting war to politics to determine its nature.[35] Lastly, theory can anticipate; this function is especially valuable as it should provide a clear direction for future military operations and guide the organizing, training, and equipping in times of great change and uncertainty.

Another way to categorize military theory beyond its approach is the scope of inquiry. Military theory, as Colin Gray observes and parses, can be general or specific.[36] General theories explore a phenomenon at its highest level, such as armed conflict, and border on the philosophical or meta level of analysis. Examples of general military theories are theories on war by Clausewitz, the art of war by Sun Tzu, Nicolo Machiavelli, and Jomini, or strategy by Edward Luttwak, Colin Gray, and Everett Dolman. Specific military theory, in contrast, explores subordinate phenomena within armed conflict, at a much narrower level, focusing on activities within operating environments, or what the military terms as physical 'domains' in which wars are fought, such as land, sea, air, or space. More recent specific theory has attempted to probe less tangible domains, such as the virtual one of cyberspace or the short-lived social 'human domain'. Other specific theory can be even more circumscribed in the scope of its inquiry. Specific theory can assess particular special operations organizations or units, as well as roles and missions, such as counterinsurgency and counterterrorism. McRaven's *Spec Ops* fits into this category, as does the 2015 article by Christopher Marsh, Mike Kenney, and Nathanael Joslyn centred around two roles and missions: special warfare and surgical strike.[37]

I assume here general theory for special operations can be developed. Underpinning general theory is the notion that it can have 'validity independent of all historical contextual detail', according to Colin Gray.[38] These ideas are in line with the concept that absolute truth about a phenomenon exists and can be discovered. Truth in this case means the philosophical, as opposed to scientific sense of the term, doubtless disappointing to followers of the method prescribed by Karl Popper. In addition, the theory developed in Chapter 5 pushes back against the postmodern notion that epistemic certainty, or understanding derived from certain truths reflecting the nature of a subject, is impossible. While special operations on a certain level are socially

[33] Winton, 'An Imperfect Jewel' (2011), p. 855.
[34] Ibid.
[35] Ibid., p. 856.
[36] Colin Gray, *The Strategy Bridge: Theory for Practice* (Oxford: Oxford University Press, 2010), pp. 22–27.
[37] Marsh et al., 'SO What?', pp. 89–104.
[38] Gray, *The Strategy Bridge*, p. 20.

determined, useful in explaining contextual differences in their understanding and use across time, they also have a common, rational, instrumental value.

This work subscribes to the notion that strategy is instrumental in nature. Stated differently, the nature of conflict and competition reflects both the strategies used in a specific context to accomplish unique objectives and the broader universal logic underpinning strategy—the root phenomenon. Despite its many flaws and contradictions, Clausewitz's *On War* identified this nature of war and strategy as well as its theoretical implications—his trinity mentioned previously—that stands the test of time and numerous assaults on its validity.[39] Gray accepts Clausewitz's continuing validity and relevance in creating a 'theoretical architecture' beginning with strategy and works from general theory to specific, or partial general theories of strategy, down to contextually unique grand, military, and instrumental military strategies.[40] This work acknowledges the universality of strategy but, to avoid confusion, omits 'partial' from the label of general theory of special operations.

## Methodology

Exploring a subject as complicated and elusive as special operations theoretically requires structured inquiry, or methodology. Methods serve several purposes for research projects, including structuring, bounding, and guiding the inquiry. The preceding section identified the limitations of most forays into special operations undertaken to date. The most common issue plaguing these forays has been a lack of bounded, structured inquiry. Most do not step beyond descriptive theory or historical narratives. Such theory limits itself to answering questions of *what* the phenomenon of special operations is and proceeds to examine it from a single context—most frequently, the state of contemporary American special operations. These theories are further limited by a lack of deep empirical exploration to test whether their assumptions hold true in a range of different time periods and conditions. Examples, when employed, tend to be illustrations employed in support of a broader point and selected based on the author's preference, before leaping into implications and recommendations. More problematically, they use such examples without explaining how they fit into the broader current of events and consideration, or as contextomy, in addition to employing them based on their outcome or in hindsight or outcome bias. Rubright's use of the Greek example of the Trojan Horse, as well as Frederick Funston's mission to capture Emilio Aguinaldo during the Philippine–American War (1899–1902), fall into this trap, as Searle suggests.[41]

[39] Two articles that summarize these assaults and argue for Clausewitz's continuing relevance are Hugh Smith, 'Clausewitz's Definition of War and Its Limits', *Military Strategy Magazine*, Special Edition (2021): 9–15 and Lukas Milevski, 'The Mirage of Post-Clausewitzianism: Understanding War and Politics on the Frontier of Clausewitzian Thought', *Military Strategy Magazine*, Special Edition (2021): 16–21.

[40] Gray, *The Strategy Bridge*, pp. 20–22.

[41] Christopher Marsh, James Kiras, and Patricia Blocksome, eds., *Special Operations: Out of the Shadows* (Boulder, CO: Lynne Rienner, 2020); Searle, *Outside the Box*, pp. 33, 35–36.

This theoretical examination avoids these traps by employing a different research methodology. The selection of a method should be driven first and foremost by the nature of the question asked. In the case of this work, the primary question is: to what extent have modern special operations accomplished desired political goals over a period of time, and why have they been successful in some cases in doing so but not in others? Other related but subordinate questions are: what factors contribute to special operations success or failure? Are the factors contributing to these outcomes inherent to special operations or outside of their span of control? Given these questions, I structure this inquiry along the lines of explanatory theory.

The primary question here is one of cause and effect, in this case success or failure, and the extent to which theory can provide insights or answers to this. In doing so, this theory avoids the pitfall of ascribing a single, overarching characteristic to success or failure, such as strategic culture or national style. According to this overly deterministic line of reasoning, American special operations cannot succeed in contemporary conflicts because American thinking about limited wars is fundamentally flawed.[42] An explanatory theory for special operations escapes over-determinism by first identifying several related dimensions of special operations inductively, connecting them together into a broader theory of special operations, and then testing the utility of the theory deductively according to various case studies that differ in time and geographic location to guard against several cognitive traps and biases.

Another factor influencing the nature of this theory is the evidence, or data, available on or about special operations. Experimental research is difficult, if not impossible, for this subject. Special operations may indeed be coming 'out of the shadows' as a subject of interest but its study academically, based on available information, remains problematic. Given its sensitive political and military nature, and to protect sources and operational methods, information about specific special operations is classified and often remains so for several decades, depending on a country's declassification procedures and guidelines. Individuals who are granted access to classified information on a 'need-to-know' basis, as I have, are bound to protect release of that information or face censure or severe penalties, including imprisonment.

Another potentially useful source of information about special operations, those who have conducted them, is equally problematic. The vast majority of special operators remain true to their unofficial motto as 'quiet professionals', and in consequence rarely talk about their actions. Those who appear willing to talk must be approached with caution by the researcher. Some operators seek to exploit their experiences for personal gain, in the form of money, exposure, and fame. Perhaps the best-known recent examples are the books published by former Navy SEALs Matt Bissonnette and Robert O'Neill, who published autobiographies capitalizing on their roles as

[42] The central argument of Donald Stoker, *Why America Loses Wars: Limited War and US Strategy from the Korean War to the Present* (Cambridge: Cambridge University Press, 2019).

team members responsible for the killing of Osama bin Laden.[43] In the best case, others amenable to discussing their experiences likely have a personal or institutional agenda or motive for doing so. In the worst, individuals invent or fictionalize accounts for any one of a number of reasons. Respectable persons such as Sir Ranulph Fiennes and Larry Cable fabricated narratives or their role as special operators in conflicts.[44] They share with other first-hand accounts the problems of accuracy and lack of verification, not to mention distortion of events based on their recollection, often enhancing or overstating their role in the process. Enough interview data of sufficient quality is necessary for sound qualitative analysis.

Given these issues, this book utilizes a qualitative, comparative case study analysis. It avoids the problems of prior theoretical efforts in several ways. First, the analysis assesses a range of different types of special operations over a period of time rather than extrapolating from a single event too broadly or focusing too narrowly on a specific set of special operations roles and missions. The specific method of inquiry here involves process tracing, to explore events deeply and connect them together to determine causal relationships. Second, the analysis looks at the rationale behind the use of special operations, as well as the intended outcome, in different geographical areas or regions. Doing so assists the analysis in breaking free of several cognitive biases inherent in works to date, which include anchoring, in-group bias, belief bias, and the availability heuristic.

In lieu of such a test, but to determine the sufficiency of special operations coherence as a general theory applicable across different national experiences, the remaining chapters explore three in particular: the United Kingdom (UK), the United States (US), and South Africa. To maintain a level of consistency between them, a two-decade timespan across each was chosen. Two of the time periods overlap: the 1970s through the 1990s for the UK and South Africa. Both featured special activities at different ends of the spectrum of conflict, from discreet covert actions and asymmetric responses to regional conventional war. The time periods allow for a historical evaluation of their coherence. British security and information release policies prevent more comprehensive analysis of this period in Chapter 6, leading to uneven source depth and quality. The South African case study in Chapter 7, given the historical nature of the apartheid regime for this period, allows for deeper analysis of the effectiveness of special operations across a full spectrum of their use from the vantage point of a non-Western perspective. The US case study is closer in time to the other two, allowing more familiar episodes to be evaluated differently given the theory's scope and elements. While more information is available on the US case study, it shares restrictions with the British one on specific information necessary for

[43] Mark Owen, pseudo. Matt Bissonnette, *No Easy Day: The Autobiography of a Navy SEAL* (New York: Dutton, 2012); Robert O'Neill, *The Operator: Firing the Shots That Killed Osama bin Laden and My Years as a SEAL Team Warrior* (New York: Scribner, 2018).

[44] Tony Rennell, 'How Ranulph Fiennes Cashed In on the "Murder" of SAS Heroes', *The Daily Mail* (16 July 2010), www.dailymail.co.uk/news/article-1,295,371/How-Ranulph-Fiennes-cashed-murder-SAS-heroes.html; Ben Steelman, 'Whatever Happened to Larry Cable?' *Wilmington Star-News* (9 July 2009), archived online at https://web.archive.org/web/20,191,108,003,326/http://www.myreporter.com/2009/07/whatever-happened-to-larry-cable.

comprehensive testing of the general theory. The theory has sufficient explanatory power, however, to derive unique insights from all three cases.

Too much of the writing on special operations and particularly of 'unified' or 'general' theory is held hostage to the dominant examples in the literature: the American, British, and Israeli. Some of this reliance is a function of access to available information in open, democratic societies and source material available in English. To be more critical, much of it reflects the tendency of authors in this literature to tap their experiences within specific organizations, to right perceived wrongs or serve institutional agendas, at the expense of more objective inquiry. Few have the interest or the inclination to look outside of what they already know well. For a broader, specific theory of special operations to have value and utility, it should be able to explain causal relationships regardless of organizational size or location, which is the goal here. In other words, this theory, like strategy, is context neutral within the boundaries of enquiry established. Third, and taking into consideration the issues with the availability of sufficient, accurate information, this work is based on a range of primary sources wherever possible, including official histories, government documents, first-hand accounts, and interviews. These primary sources, cross referenced wherever possible, are bolstered by a wide range of secondary source works to guarantee sufficient depth of information and analysis. Fourth, and last, this work draws upon existing academic works wherever possible, including theoretical models. A tremendous body of scholarly literature on areas involving special operations, particularly in the social sciences and more specifically related to organization and group behaviour, has become available thanks, in large part, to the involvement of so many national contingents in Iraq, Syria, and Afghanistan. Such works and models, where introduced, are caveated and explained to ensure their ideas are used responsibly and not merely employed for the sake of convenience. In doing so, I fulfil an often-overlooked function of military theory, which is connection to other fields of study from which to draw its insights. The next section provides a roadmap of the arguments in subsequent chapters and explains how they connect together.

## Roadmap of the argument

Although the case studies in this book focus on the modern period, or special operations since the Second World War, the examination in Chapter 2 fulfils an initial requirement of general theory: explanatory power across space and time. For special operations to have a general theory, there must be evidence of their use throughout history. In doing so, it expands the scope of understanding of special operations prior to the creation of modern SOF. It establishes the enduring roots functionally of special operations identifiable in the canon of classical literature on ancient, medieval, and pre-modern warfare and statecraft. The resulting survey identifies five persistent special operations functions historically: intelligence gathering and reconnaissance, treachery and subversion, direct and indirect shock, military assistance and advising, and asymmetric response. Identifying them occurred thanks to decades teaching theory to military officers and noticing identifiable patterns of activity historically. Exploring these functions and using them in subsequent chapters serves a useful purpose for theory. They help break study and analysis free from the dominant paradigm of special operations: the dozen core activities and terms used by USSOCOM that dominate discussion and stymie critical inquiry. By looking at the western European, South Asian, and East Asian experiences, this chapter argues there are functional equivalents to special operations in pre-modern history. In doing so, it establishes the validity of a general military theory on the subject, following Colin Gray's definition and taxonomy discussed previously.

Chapter 3 presents the first element of special operations theory: capability. This chapter begins at the most fundamental level of special operations, the capability to conduct them. The appeal of special operations for many resides primarily not in their audacity or seeming ubiquity but rather in the qualities of the operators who conduct them and their representative units or organizations. Like the sirens of Greek mythology, though, most authors cannot escape the call of special operators and often become beholden to the individuals and communities about which they write. This unfortunately leads to discussions about special operations capabilities in an unduly narrowly focused manner, and most pedantically in lists of units and their associated or defined roles and missions.

This analysis begins by defining special operations and moves on to recast special operations capability in a different manner, to account for the widest spectrum of activities possible. More specifically, Chapter 3 considers capability as integration between different organizations and forces from across government and the military. Inspired by terms developed by others but drawing upon model on insurgent cohesion, this chapter discusses capability as the sum product of horizontal integration of special operations units connected together with related government departments and agencies. Integration serves different purposes depending on the stakes of competition and conflict and the goals sought: to improve tactical performance and outcomes, to hedge against uncertainty and organizational friction, and to make up for individual or collective unit shortfalls by creating a coherent team. The level of horizontal integration is crucial in defining the range of special operations activities that are possible, in both senses: the spectrum of actions as well as their reach.

The chapter breaks free from these limitations by redefining special operations capability as a function of defined requirement or need, size, scope, reach, and sustainability. It also explores different lifecycle models of SOF capability. Capabilities identify what is possible for special operations to accomplish under a set of operational or functional conditions. The conceptual key to understanding what constitutes capability, however, is the level or degree of horizontal integration, within or between special operations units, conventional forces, and other government agencies. These ideas are subsequently explored in the case study of India and the development, use, and expansion of its special operations capability. India is a useful example for several reasons. Its SOF developed in a non-Western democracy, have expanded considerably over the past dozen years, and have been involved in a range of different activities including subversion and sabotage, indirect shock missions, and asymmetric response (countering insurgents and terrorists). Assessing the level of horizontal integration tests the validity of this concept in evaluating India's special operations capabilities as well as its operational successes and failures.

The second element of special operations theory, control, is introduced in Chapter 4. The military and civilian worlds frame control in different ways, the former as a function of leadership and command and the latter as a function of subservience to political authority. Military control is clearest at the tactical level and civilian at the national. Control is a central concept common to military theory, theories of insurgent behaviour and counterinsurgency, and the use of violence in violence in civil war, among others. These aspects are explored but connected to special operations in a specific way. Control of special operations is a product of the type of government or political system directing them. In addition, this chapter captures the relationship that exists between the collective agent, or special operations capability, as thinking, reactive entity, and the directive principal of that government or political system whose goals and objectives they are working towards. Within the academic disciplines of politics, political science, and sociology there is a healthy literature that explores civil–military relations, as well as the direction and overseeing of clandestine or covert activities within political systems.

Ultimately, the principal and agent relationship brings into focus integration in its vertical dimension. More specifically, this chapter examines the relationships that exist between different principals and agents, from the decision-making executive authority through levels of command and control to special operations capabilities. Vertical integration, or control, is a concept central to special operations theory for two reasons. First, control assists in breaking the analysis 'out of the box' of discussion focused unnecessarily on direct and indirect mission sets. Second, it identifies relationships and behaviours under a set of different objectives—political conditions.

Between these levels, at the strategic and operational, is where miscommunications and these different frames of reference, particularly regarding risk, can create friction and jeopardize success. By separating out levels, identifying different types of control, and assessing their degree through the prism of risk, we arrive at a new understanding of control. The new conceptual key resulting from this understanding is the extent or degree of vertical integration, where the locus or degree of control rests, and how effectively it is exercised. These ideas are subsequently explored in the example of the ambush and killing of four American Special Forces soldiers in Niger in 2017. This example suggests that even in cases where sufficient control and oversight mechanisms appear to be in place, the reality was different. SOF took advantage of a level of delegative control assigned to them and exercised a degree of agency, suggesting the trend apparent in Chapter 8 reached its logical conclusion.

Chapter 5, a short synthesis chapter, combines the elements identified in the two preceding ones into a simple explanatory theory. Capability and control, or the horizontal and vertical integrative dimension of special operations, are examined in the two preceding chapters statically. Doing so aids in a more objective inquiry but robs the analysis of its explanatory power. This chapter introduces a third, dynamic dimension of special operations: how the relationship between capability and control changes over time. The concept introduced that captures change is coherence. It has two different states. The first is where capabilities remain broadly aligned with the desired outcome and serve the political purposes of the principal directing them. This state is known as congruence. The second is where capabilities or control drift apart for any number of reasons examined here, including different frames of reference of risk and reward, desirable outcomes, and cost and benefit calculations, among others, leading to a possible second state: divergence.

The theory developed here, termed 'special operations coherence', allows special operations capabilities and control to be evaluated in two ways: forensic, or *post facto*, and predictive. It introduces new elements into horizontal and vertical integration that allows the model underpinning the theory to become dynamic and account for a range of different special operations. These elements include time, the stakes involved in competition or conflict, the influence of opponents and allies, and the bearing and calibration of special operations. From this theory, the degree and stability of vertical and horizontal integration can be evaluated dynamically when put to the test: using, maintaining, adapting, and controlling special operations over an extended duration, measured in years or decades. In addition, the model provides a way to evaluate the effectiveness and health of special operations capabilities and control regimes in a range of different conditions: domestic crisis response, regional conflict, and global power projection, sometimes simultaneously. The model is subsequently put to the test in three different national case study experiences outlined below.

The theory of special operations coherence first tills familiar ground, the British special operations experience from 1971 until 1991, in Chapter 6. It begins with a different form of control identified in Chapter 5 but explored in more detail here: collaborative control between civil authorities and the military. The differences reflect two factors: Britain's long history of modern special operations development and use, its processes for civilian control and the rather opaque workings of the machinery of government, and the institutional commitment to official secrecy. This chapter breaks new ground by evaluating the challenges to and successes of British special operations across a range of different contingencies in a variety of conditions: the last of the postcolonial wars in a protectorate, domestic counterterrorism, and two conventional wars (the Falklands in 1982 and the First Gulf War, 1991). These two decades witnessed several changes that stressed British special operations capabilities and their control: shifting from overseas to domestic operations, the development of new capabilities, the creation of controlling headquarters, and difficulties in integrating with conventional forces in major conflicts. The British case reveals moderate horizontal integration and shifting levels of vertical integration, reflecting the informal nature of British civilian control as well as the influence of key personalities.

Chapter 7 examines special operations used in South Africa during the apartheid era, from 1972 until 1992. The case study is important for the general theory for several reasons. First, South Africa is a non-Western and African case study, and therefore an 'outlier' relative to the more commonly discussed US and British ones.

If the general theory proves sufficient for this case, it should also be able to explain others ranging in size and scope. The second concerns source material and this case study has a range of accessible, English-language source material. The dissolution of the apartheid system in 1994 prompted a volume of personal memoirs, political and military, available in English. In addition, the new government created a commission to study and report on the prior administration's use of various instruments of state security for a wide range of activities. The Truth and Reconciliation Commission published its findings in five volumes, which offer insights into the use of special operations across the range of overt, covert, and clandestine activities. Third, South Africa provides a case of a different form of government: an exclusive one that exerted a high degree of directive civilian control and possessed a wide range of special operations capabilities in a highly integrated manner—yet one that struggled to translate operational success into strategic effectiveness. The theory attempts to explain why.

In addition to these methodological reasons, the South African case study broadens the appeal of the work by providing insights into a puzzle currently facing special operators and decision makers. While many nations and their partners accept the need for strategic competition, they are grappling with how to integrate and balance different capabilities. One reason for this struggle is the seemingly vastly different requirements for conventional and special operations, as well as the blurring lines between competition and conflict. The South African case study saw a wide range of activities and the integrated use of military and paramilitary activities, including conventional forces and special operations, into what one former operator has termed 'composite warfare'. This case study also saw another source of confusion in competition: the use of proxy forces, including armed militias and insurgent groups as well as third-party actors, such as East Germany and Cuba. Thanks to the publication of the KGB Mitrokhin Archive and new volumes of the American Foreign Relations of the United States, this case study sheds important light on how special operations contribute to local, regional, and global competition and conflict.

The subject of Chapter 8 is the best known but perhaps least understood: the contemporary American experience, or the two decades from 1994 until 2014. This seeming contradiction reflects successes and the narratives constructed institutionally to explain the outcome: the ascendence of American special operations as a premier and influential instrument of national policy. Success, however, masks a more interesting, instructive, and even cautionary tale: the politicization of the military and the ambivalence of civilian control. Because of a continuing string of successes, American special operations were able increasingly to shift the locus of control and direct both the expansion of capabilities and the degree of control. Despite this, a few events and some unique attributes of American governance meant this unhealthy shift was short-lived. The theory of special operations coherence allows the exploration in this period of several changes that occurred: the drastic expansion of special operations capabilities and their impact, growing civilian trust and military political influence, limited but important challenges in integrating with the conventional military in major conflict and with other government agencies in countering terrorism, and an expansion of authorities. The American case reveals a high degree of horizontal integration, and moderate vertical integration, which has shifted over time. Chapter 9 concludes with an overall assessment of identifiable themes and patterns from the case studies and considers the value of the theory of special operations coherence.

# 2
# Special Operations Precedents

Special operations appear to be a relatively recent phenomenon historically. Some contend the era of modern special operations began in the Second World War. Their reasons are sound enough. Operations by German forces of the 'Training and Construction Company 800 for Special Duties' of Admiral Wilhelm Canaris *Abwehr* seized critical infrastructure including bridges and tunnels in the early phases of the invasions of the Netherlands, Belgium, and France in May 1940, marking the beginning of 'first age' of modern, Western special operations forces (SOF), according to some.[1] By the war's end, the Allies and Axis had raised and fielded more three dozen distinct 'private armies' of uniformed special operators, ranging from the German Brandenburger Regiment and Japanese *Teishin Shudan* to the innocuous-sounding American 5307th Composite Unit, or 'Merrill's Marauders'. These units operated within every domain of activity during the conflict, including on land, on and below the sea, in the air, in the electromagnetic spectrum, and in what is currently labelled the 'cognitive' or psychological realm. Finally, several current special operations units trace their lineage directly back to the Second World War, including the British Special Air Service, Special Boat Squadron, and the Air Commandos of the American Air Force Special Operations Command.

Others cast the net farther back or more recently. Bernd Horn suggests special operations, specifically the Canadian heritage, commences with the War of the Austrian Succession (1740–1748). During various North American portions during and after that conflict, including King George's War (1744–1748) and Father Le Loutre's War (1749–1755), Captain John Goreham employed a small force comprising a mix of settlers and local indigenous tribesmen of rangers in Nova Scotia and New Brunswick in brutal frontier fighting against the Acadians and their local allies.[2] John Teirney Jr, in contrast, suggests American 'special warfare' began not with a specialized unit but rather with the first codified best practices in 1759.[3] These practices, the twenty-eight 'Rules of Ranging' of Major Robert Rogers for his unit of 'rangers' during operations against the French in the Seven Years' War (1756–1763; known as the 'French and Indian War' in North America), are identifiable as one of

[1] Yair Ansbacher and Ron Schleifer, 'The Three Ages of Modern Western Special Operations Forces', *Comparative Strategy*, 41:1 (2022): 32–33.

[2] Bernd Horn, '*We Will Find a Way*': *Understanding the Legacy of Canadian Special Operations Forces*, JSOU Report 12-2 (MacDill Air Force Base, FL: JSOU Press, February 2012), pp. 4–5; John Grenier, *The Far Reaches of Empire: War in Nova Scotia, 1710–1760* (Norman, OK: University of Oklahoma Press, 2008), pp. 124–125; 132–133; 140–141.

[3] John Tierney Jr, *Chasing Ghosts: Unconventional Warfare in American History* (Washington, DC: Potomac Books, Inc., 2006), p. 24.

*Special Operations Success*. James D. Kiras, Oxford University Press. © James D. Kiras (2024).
DOI: 10.1093/9780198902096.003.0002

the earliest foundational forms of contemporary army operations: doctrine. Other reasons argue in favour of Rogers' Rangers. The current American 75th Ranger Regiment traces its lineage back to them. In addition, much like contemporary SOF, Rogers' forces as well as other Ranger units raised during the American Revolutionary War (1777–1783) captured the imagination of the contemporary American public through their exploits.[4] Adrian Weale, while acknowledging Rogers' orders, sets the date differently: 1810. This year saw the first of a range of British activities that formed what came to be known as 'the Great Game', or 'strategic competition' in contemporary verbiage, between Great Britain and Tsarist Russia for influence over different regions in Central and South Asia.[5]

More recent claims fix the date for the birth of modern special operations from 1945 onwards. Peter Harclerode, in pursuing an argument about covert action and support for proxy forces that blows back on sponsoring states, begins with American special operations support for Vietnamese forces led by Ho Chi Minh in 1945 and ends with Osama bin Laden.[6] American 'Green Berets', in contrast, cite specifically either 19 May or 19 June 1952, the activation date of their first continuous Special Forces unit, 10th Group, at Fort Bragg, North Carolina.[7] Others suggest the modern era begins as recently as 1987, with the formation of the first unified command headquarters unit, the United States Special Operations Command, that brought together different Armed Service components to operate jointly and which several other countries have since emulated.[8] Unsurprisingly, contextually specific frames of reference, such as unit or level of command, doctrine, and orders, have also informed the study of special operations beginnings.

Regardless of the origin date affixed, the idea of 'modern' special operations poses a particular challenge for general theory development. This challenge exists for two reasons, one epistemological and the other evidentiary. For a general theory to be valid, it must account for the nature of a phenomenon across time and space. For special operations, this translates into linking them back through ancient and early modern conflict. The evidentiary problem occurs not as much from its lack as it is succumbing to three related cognitive and analytic biases: availability, selection, and confirmation. In seeking to validate general theory, there is a very real danger in establishing linkages that might not exist, conforming historical examples to modern interpretations, or, to channel a Greek myth, cutting off the feet to fit Procrustes' bed.

[4] David Hogan Jr., *Raiders or Elite Infantry? The Changing Role of the U.S. Army Rangers from Dieppe to Grenada* Contributions in Military Studies, Number 129 (Westport, CT: Greenwood Press, 1992), pp. 2–4.

[5] Adrian Weale, *Secret Warfare: Special Operations Forces from the Great Game to the SAS* (London: Coronet Books, 1998), p. 9.

[6] Peter Harclerode, *Fighting Dirty: The Inside Story of Covert Operations from Ho Chi Minh to Osama bin Laden* (London: Cassell, 2002).

[7] The difference in the dates reflects the formal activation of headquarters units, as opposed to when Colonel Aaron Bank formally assumed command of the Group. Alfred Paddock Jr, *U.S. Army Special Warfare: Its Origins*, rev. ed. (Lawrence, KS: University Press of Kansas, 2002), p. 145.

[8] United States Special Operations Command History and Research Office, *United States Special Operations Command History*, 6th ed. (MacDill Air Force Base, FL: USSOCOM, 2008), p. 7.

A prevalent example illustrates this point. The Trojan Horse is represented widely in SOF heraldry and literature. It was or is a symbol for a few special operations units and organizations, including the 10th Special Forces Group.[9] The episode involving its use has become a metaphor for cunning and guile and has been labelled in more popular works as 'a classic commando operation'[10] or 'truly a "special operation"', given its 'coordination ... timing ... [and] subtlety'.[11] Given their audience, such conclusions are understandable. Less so is that reached by analysts from the RAND Corporation, who conclude 'The occupants of the Trojan horse [sic] were no doubt the period SOF equivalent'.[12] An early anthology of readings on historical special operations, after asking a series of very practical questions of interest to both the operator and the scholar about its conduct, qualifies the event as 'probably apocryphal' and then connects it to works of literary fiction within the volume.[13] The episode of the Trojan Horse does not appear in Homer's primary work on the Trojan War, the *Iliad*, but rather as a passing incident in the subsequent work about Odysseus' epic voyage returning from the war, the *Odyssey*. As stirring as it is, this so-called historical event is derived from epic poetry written almost three millennia ago whose details cannot be substantiated. Classicists continue to debate both the nature and even existence of this 'gift horse' as a siege machine, a physical ruse, or perhaps even a 'magic device'.[14]

Scholars of various disciplines often pragmatically seek to make their research subjects and findings as relevant and potentially useful as possible. Special operations have been one of these subjects, particularly in the wake of successful operations in Afghanistan in 2001 and Iraq in 2003. In consequence, several score of books and articles have been published on subjects ranging from biblical to seventeenth- through nineteenth-century specialized light infantry or skirmishers, with claims, direct or ironic, that they are Special Forces, in pursuit of the claim of being 'first' or 'most decisive'.[15] Studies along these lines tend to reflect issues with more modern examinations of special operations. They focus specifically on individual units or episodes, including 'great raids', or survey broad swathes of history in such a manner

[9] Troy Sacquety, 'The Trojan Horse Badge: Distinctive Identity for the 10th Special Forces Group', *Veritas*, 5:4 (2005): 56–59.

[10] James Dunnigan, *The Perfect Soldier: Special Operations, Commandos, and the Future of War* (New York: Citadel Press, 2003), p. 16.

[11] Derek Leebaert, *To Dare and to Conquer: Special Operations and the Destiny of Nation, from Achilles to Al Qaeda* (New York: Little, Brown & Company, 2006), p. 44.

[12] Christopher Paul, Isaac R. Porche, and Elliot Axelband, *The Other Quiet Professionals: Lessons for Future Cyber Forces from the Evolution of Special Forces*, Report RR-780-A (Santa Monica, CA: RAND Corporation, 2014), p. 5.

[13] John Arquilla, ed., *From Troy to Entebbe: Special Operations in Ancient and Modern Times* (Lanham, MD: University Press of America, 1996), p. xviii.

[14] Matt Pickles, 'Did the Trojan Horse Exist? Classicist Tests Greek "Myths"', Oxford News Blog (25 July 2014), www.ox.ac.uk/news/arts-blog/did-trojan-horse-exist-classicist-tests-greek-myths; Everett Wheeler, *Strategem and the Vocabulary of Military Trickery* (Leiden: E.J. Brill, 1988), p. 30 and fn. 27.

[15] Benjamin D. Giffone, '"Special Forces": A Stereotype of Benjaminite Soldiers in the Deuteronomistic History and Chronicles', *Scandinavian Journal of the Old Testament*, 30:1 (2016): 16–29, DOI: 10.1080/09018328.2016.1122287; James Capua and Carl Kruger, *Jåger, Europe's First Special Operations Forces: History, Organization, Arms and Equipment of the Austro-Hungarian Empire's Elite Light Infantry to 1866* (New York: Signal Horn Publishers, 2020).

as to incorporate every elite military formation and their activities superficially or speciously. In addition to the focus of such works, their scholastic quality runs the gamut from hobbyist and popular to niche academic. Rarely are such works connected across fields of study or to the emerging, yet broader literature on special operations and relatively few link this literature to systematic exploration of their subjects. Happily, there are exceptions to this observation, including a broad historical survey by Derek Leebaert and a narrower case for medieval special operations made by Yuval Harari.[16] Uneven scholastic quality, combined with academic disciplinary preferences and prejudices, calls into question the value or necessity of inquiry into subjects as narrow as special operations, in contrast to conventional and nuclear war.

Probing the distant past raises questions of relevance from quarters outside of the academic. They include both the policy or decision-making level, as well as practitioner, or within the special operations community. Both see ancient history as quaint and perhaps mildly interesting, if they are historically inclined, but of questionable value.[17] The reasons for these views are not complicated. Decision makers and practitioners are confronted by a range of complex challenges, for which they must propose practical and working solutions. Often these challenges are occurring simultaneously, limiting the time and attention they have. While history should inform their judgement and decisions, unfortunately the presentation by historians of an overwhelming bounty of rich contextual detail chronologically, along with the use of specialist methods and language, can deter even the most eager of consumers, a malady also shared by social scientists.[18] For history to have value to such audiences, it must be presented in an analytically defensible yet systematic and accessible manner.

Before proceeding, a few caveats are in order. This chapter comprises a short analytic historical survey as opposed to an exhaustive and detailed chronological one. The reasons are simple enough: the purpose of this chapter first is to demonstrate that the rationale for and conduct of special operations has existed historically even when the highly specialized, dedicated forces of the present did not. The survey encompasses the period between the first written sources and the transition to the early modern period in the sixteenth century. Warfare after this period can be considered 'modern', as the discussion above suggests, but there are reasons for limiting the survey thusly.[19]

Early precedents based on function or activities prevalent in the past not only establish a framework for analysis but also in their aggregate establish a picture that

[16] Leebaert, *To Dare and To Conquer*; Yuval Noah Harari, *Special Operations in the Age of Chivalry, 1100–1550* (Woodbridge: The Boydell Press, 2007).

[17] James D. Kiras, 'A Theory of Special Operations: "These Ideas Are Dangerous"', *Special Operations Journal*, 1 (2015): 79.

[18] These challenges are outlined admirably by Michael Desch, *Cult of the Irrelevant: The Waning Influence of Social Science on National Security* (Princeton, NJ: Princeton University Press, 2019).

[19] Jeremy Black notes the issues related to the concept of 'modernity' and divides his study of war into ancient, early modern, and modern. He characterizes 'early modern' as the advent of professionalized and standing armies from 1490 until 1700. While open to debate, they serve the purpose of the analysis here by marking a division between ancient and modern. Jeremy Black, *War: Past, Present and Future* (New York: St Martin's Press, 2000), pp. 30; 46.

roughly overlays the general functions of special operations today, thereby establishing a degree of continuity historically. In other words, the survey establishes on the furthest temporal shore a bridge between the ancient past and the more recent experiences, all described by the general theory. I accept few of the polities discussed would recognize, distinguish, or even categorize the functions or activities discussed subsequently as special operations. Indeed, at different times and in specific historical contexts, these functions and activities may have been the norm in competition and conflict as opposed to the exception. I acknowledge that categorizing activities and functions as Harari does risks imposing contemporary ideas onto past experiences and individual contexts. In his review of Harari's work, Peter Konieczny rightly points out that while 'covert operations were undertaken by medieval armies and were always part of military strategy in the Middle Ages ... that does not make this the same thing as the modern view of special operations'.[20] Bearing this observation in mind and fully aware that academic specialists will take issue with the categorization contained herein, I argue the utility of certain activities is functionally universal even if they were not considered special operations during their time.

I also concede my limitations when it comes to the very discrete knowledge from fields such as classical history, anthropology, sociology, or regional/national specialization. In consequence, I rely on translations of source primary source accounts and material and the use of secondary sources, although every effort has been made to draw primarily from works regarded as classics in the field of military and strategic studies. Given the limitations outlined above, I hope other specialists and scholars will relate the ideas explored here in their own subsequent works and either support, modify, or refute them. The second purpose is to provide a historical analysis of precursors to modern special operations accessible to an audience beyond academic specialists. Doing so addresses the issue of relevance for policy and decision makers, as well as the special operations community, to convince them that the past contains parallels and meaningful insights to what are continuing operational and strategic challenges.

This work provides broad functional categories of recurring activities from the ancient and medieval world to modern special operations historically that are relatable to what we consider special operations in the most general sense today. My review of history suggests elements or functions relevant and relatable to roles and missions, but not directly equivalent to modern special operations that have occurred across time. The categories, which are much broader in scope than the twelve 'core activities' identified by United States Special Operations Command (USSOCOM), I identified and developed for two reasons.[21] The first is based on twenty years of teaching theory and history to military officers and noticing identifiable patterns

[20] Peter Konieczny, review of *Special Operations in the Age of Chivalry, 1100–1550* by Yuval Noah Harari, *Sixteenth Century Journal* (2009): 1201.

[21] USSOCOM's twelve core activities are: direct action, special reconnaissance, unconventional warfare, foreign internal defense, civil affairs operations, counterterrorism, military information support operations, counter-proliferation of weapons of mass destruction, security force assistance, counterinsurgency, hostage rescue and recovery, and foreign humanitarian assistance. 'Core Activities', USSOCOM, www.socom.mil/about/core-activities.

of activity, relatable but not directly equivalent to contemporary special operation within this literature. The second is to help break study and analysis free from the dominant paradigm: USSOCOM's core activities. Although these were developed and instituted for understandable reasons, their specific association either doctrinally or their fierce guardianship by specific special operations communities and units unfortunately leads to several problems. One is stymieing broader exploration by limiting it to these activities. Another, and more problematic, is directly equating what occurred in the ancient world with these activities, a problem I noticed in classroom discussion. For these reasons, and in line with the explanation contained in Chapter 1, I developed five broader functions in the ancient world that roughly equate to special operations today: intelligence gathering and reconnaissance, treachery and subversion, direct and indirect shock, military assistance and advising, and asymmetric response. The types of activities we do today, under the label of special operations, have always been done even if not considered as such. One important clarification is necessary. For the period covered here, the ruler and commander are often one and the same, obviating the need to distinguish between 'strategic'-level activities and 'tactical' ones. For analytic clarity, however, they are separated as appropriate given the different considerations inherent within them.

One last critical observation is required before proceeding. Other works focus exclusively on Western and indeed Greek, Roman, and European experiences. While some of this is understandable and can be put down to sourcing and the limitations outlined above, a broader analytic survey should include a diverse range of insights. Although what follows draws heavily in parts upon these experiences due largely to the volume of written and translated materials that exist, it also incorporates perspectives and examples from Asia, Africa, and Latin America, where available, for reasons scholars have identified.[22] Doing so has an additional benefit for general theory: guarding against overly deterministic arguments that privilege culture or nationality and suggest it influences strategic behaviour, including following 'ways of war' or the use of special operations throughout time.[23] The general theory of this work, as well as this chapter, supports the argument of 'culture as context', related to and influencing but not determining strategic behaviour.[24] Cultures privilege different attributes at different times and determine what is normal versus what is unconventional, but the utilitarian logic underpinning certain functions remains constant. In addition, it suggests that strategic behaviour, in this case the rationale and use of forces functionally that align with special operations, arose from pragmatic considerations that quite often eclipse cultural ones. For example, there is a tendency for some Western readers

[22] Black, *War*, p. 38.

[23] See, for example, works on Chinese strategic behaviour and military culture. For the former, see a realist interpretation of Ming Dynasty behaviour in Alastair Iain Johnston, *Cultural Realism: Strategic Culture and Grand Strategy in Chinese History* (Princeton, NJ: Princeton University Press, 1998). For military culture, James Pangelinan stretches a cultural argument in 'From Red Cliffs to Chosin: The Chinese Way of War', unpublished monograph, School of Advanced Military Studies, May 2010, https://apps.dtic.mil/sti/pdfs/ADA523450.pdf.

[24] Colin S. Gray, 'Strategic Culture as Context: The First Generation of Theory Strikes Back', *Review of International Studies*, 25:1 (January 1999): 54.

to seek out specific insights from Eastern sources based on outmoded, sweeping, and prejudiced ideas of 'oriental' attributes and approaches.[25]

## Intelligence gathering and reconnaissance

Unlike battle or combat, whose role and importance are stressed in numerous ancient texts, that of intelligence gathering remains murky and understudied. This state is even more curious given the level of importance some ancient writers place on it in their texts. The earliest and perhaps best known is by Chinese author Sun Tzu (~400 BCE), the *Bing Fa* or *Art of War*. While his existence some 2,500 years ago is the subject of some debate,[26] the value he places on intelligence is indisputable. In writing about warfare and statecraft during the earliest portion of the Warring States period of Ancient China (476–221 BCE), Sun Tzu links together two concepts related to intelligence in his first section, entitled '(Initial) Estimates', and the last one, 'Employment of Secret Agents (Spies)'.[27] In the first, he outlines the various elements about which a sovereign should be reasonably informed, such as the moral qualities of his opponent, including the degree to which his leader and his population are synchronized, the operating terrain, the weather, the general and his qualities, and the organization of the army. Samuel Griffith interprets these as 'doctrine' but Ralph Sawyer in more literary fashion describes them as 'laws' including 'the Tao of command'.[28]

Compiling information on these elements, along the lines of what is now known as 'net assessment', does more for the sovereign than just inform decision making. According to Sun Tzu, the net effect of this compiled information is a reasonable determination of the outcome, 'the strategist's keys to victory'.[29] Modern interpreters of his work often stop there or pass over the next step: divining the auguries to ensure that the gods favour the upcoming enterprise.[30] Within the other six works that comprise 'the seven military classics of ancient China', which range in time from 'T'ai Kung's Six Secret Teachings' (finalized ~447 BCE but perhaps as early as 1100 BCE) to 'Questions and Replies between T'ang T'ai-tsung and Li Wei-kung' (~627 CE), intelligence is implicit in the subjects under discussion or not mentioned due to more pressing issues of concern.[31] These issues include the qualities and characteristics of

[25] Porter makes a compelling case for the emptiness of such views. Patrick Porter, *Military Orientalism: Eastern War through Western Eyes* (New York: Columbia University Press, 2009), pp. 15–18.

[26] Sun Tzu, *The Illustrated Art of War*, trans. Samuel B. Griffith (Oxford: Oxford University Press, 2005), pp. 17–21.

[27] The titles in quotation marks are from the Griffith translation, while the parenthetical inserts or differences are used by Ralph Sawyer in *The Seven Military Classics of Ancient China, Including the Art of War* (Boulder, CO: Westview Press, 1993), pp. 157; 184.

[28] Griffith, *The Art of War*, p. 91; Sawyer, *Seven Classics*, p. 157.

[29] Griffith, *The Art of War*, p. 103.

[30] Sawyer, *The Seven Military Classics of Ancient China, Including the Art of War*, p. 159. Nor was this an exclusively Eastern phenomenon. By discounting omens and throwing the sacred chickens used to divine them into the sea, Roman consul Publius Cornelius Pulcher allegedly lost the naval battle of Drepana (249 BCE) during the First Punic War with Carthage. J.F. Lazenby, *The First Punic War: A Military History* (London: University College of London Press, 1996), pp. 132–134; 136.

[31] An exception is 'Wu-tzu', which contains a brief section on 'Evaluating the Enemy' and offers eight subjective conditions that can be assessed. Sawyer, *The Seven Military Classics of Ancient China, Including the Art of War*, pp. 201–214.

the sovereign or military leader, military organization, training, and formations, different strategies for force employment, and the nature and strength of the connection between the sovereign and the people or the general and the troops.[32]

A century and a half later, Indian author Chanakya, under the name Kautilya, wrote a guide to statecraft entitled the *Arthashastra* (the science of material gain), dating from between 300 and 250 BCE. Although this work shares remarkable similarities with Sun Tzu's, there are key contextual differences. Sun Tzu's work reflected a period of remarkable competition between states with no clear, single ruler. Each state was vying for expansion at the expense of the others and Sun Tzu was seeking to provide the best advice to the ruler of Wu, one of seven 'Warring States'. Kautilya, in contrast, was writing for the sovereign of the united Mauriyan Empire (317–185 BCE), whose authority extended over an estimated 50 million inhabitants. There are few subjects related to economic and political statecraft uncovered in a systematic manner in the *Arthashastra*, starting with the sources and branches of knowledge and ending with the means and ends of conquest, including the composition of an ideal army and the 'peaceful rule of acquired territory'. The attention to and level of detail Kautilya provides is overwhelming. For example, he discusses the specifics of the various sources of revenue for the states, the rules governing them, and the amount of taxes they should pay, including transaction taxes based on a series of standardized weights and measures.[33] Upholding and enforcing these rules and collecting the taxes is considerable machinery of government, which Kautilya envisions as reporting through to a collection of councillors, ministers, and *amatya* (officials) to the king.

With few but notable exceptions, the types of intelligence sovereigns require roughly mirror that of Sun Tzu. Kautilya is clear on structure and targets of collection but much less so on how the analytic process works. Two scholars, in arguing for Kautilya's contemporary relevance in terms of intelligence collection and processing, conclude he 'provides a detailed account of intelligence collection, processing, consumption, and covert operations' even if his preferred method of analysis is 'mostly not explicitly elaborated'.[34] One reason for this opacity is that he is talking about the state and its mechanisms and processes in largely idealized, conceptual terms. When put into practice, the king and his councillors and ministers must adjust these conceptual ideals depending on the target, such as enemy or ally, as well as which of the six methods of foreign policy, ranging along a spectrum from *samdhi* (peace) to *vigraha* (hostilities), they wish to pursue.[35]

Both Kautilya and Sun Tzu outline the different categories or classes of agents the sovereign requires, so they can be appropriately recruited, compensated, and employed. Reflecting cultural aspects of Chinese numerology, and in particular the

[32] Ibid., pp. 60–69; 96–105; 137–142; 214–224; 260–276; 292–306.

[33] Kautilya, *The Arthasashtra*, ed. and trans. L.N. Rangarajan (New Delhi: Penguin Books, 1992), pp. 263–264.

[34] Dany Shoham and Michael Liebig, 'The Intelligence Dimension of Kautilyan Statecraft and Its Implications for the Present', *Journal of Intelligence History*, 15:2 (2016): 119, DOI: 10.1080/16, 161,262.2015.1116330.

[35] Kautilya, *The Arthasashtra*, pp. 545; 548–549.

dimensions of luck and chance associated with the number 'five', Sun Tzu lists the types of agents as 'native, inside, doubled, expendable, and living'.[36] They can roughly be divided along three lines: individuals within an enemy state whose loyalty has been purchased from both the peasant and ruling classes (native and inside), enemy spies who have been turned (doubled), and one's own people who have been sent to the enemy to deceive (expendable) or collect information (living). He cannot stress their collective importance enough to the sovereign, to whom they directly report. According to Griffith's translation, Sun Tzu labels them '"The Divine Skein" and ... the treasure of a sovereign'.[37] Sun Tzu prized such information, and the spies that collected it. Given the expansion in the size and overall capabilities of contemporary armies, and the cutthroat nature of state competition in this period of Chinese history, the consequences of failure when states went to war were exceptionally high. Therefore, war occurred infrequently and, if waged, should be done expeditiously and economically by setting the conditions for and winning a decisive battle. To wage war for a prolonged period could ruin a state, and even if successful could make it vulnerable to conquest by one of its neighbours.

Kautilya, in contrast, separates his fourteen different categories of spies, divided into two classes, based on their function. The first class, fixed spies, were useful in maintaining order within the boundaries of the empire and are discussed more in the section that follows on treachery and subversion. The second class of spies, wandering ones, performed a wide variety of tasks in neighbouring states such as Yavanas (Greece) and Yukas (Seleucia), among others, during times of competition and conflict. Wandering spies could monitor conditions within these states and report back, through unspecified communications means, during times of *samdhi* and *asana* (staying quiet) and *yana* (preparing for war). Kautilya is highly prescriptive about both the types of agents and their operating environments. One of the keys to their success is operating in as unobtrusive a manner as possible, adopting nondescript forms of concealment as members of the lower classes of society, including farmers, herdsmen, and traders, a form of intelligence tradecraft known today as 'cover'. Appearing physically handicapped was another form of cover he specified. Their target, in both war and peace, included the sovereign and his highest officials in both adversary and allied kingdoms.[38] When war breaks out, spies in the wandering class take on a different importance. They are discussed in treachery and subversion as well as direct and indirect shock.

Shifting the focus of analysis westward, it is surprising how few of the Western 'classics' of statecraft and military art from Ancient Greece discuss matters related to information agents and collection in support of the sovereign. This level of absence of evidence, however, should not be interpreted as absence of either interest or importance. Aristotle alludes to two different functions spies can serve in *Politics*

[36] Griffith, *The Art of War*, p. 232.

[37] Ibid., p. 233. Sawyer translates them as 'their Tao', or 'spiritual methodology'. Sawyer, *The Seven Military Classics of Ancient China*, p. 185.

[38] Kautilya, *The Arthasashtra*, p. 506.

and Plato, more indirectly, as a simile in *Republic*.[39] Based on the elusive 'shadowy picture' of sources surveyed, J.A. Richmond confidently concludes nevertheless that 'there were many spies in Ancient Greece'.[40] They ranged from upper-class *proxeni* ('friend-guests'), *kerukes* (heralds), and emissaries to non-Greek states to low-level agents in markets and ports. These agents employed various communication methods, judging from the observations of Herodotus, Thucydides, and Strabo, as well as security measures outlined later by Aeneas Tacitus.[41] Culture and gender roles may have contributed to limiting discussion of spying and spies in Ancient Greece, given prevailing norms about the freedom of women to travel unaccompanied, but these social dimensions are difficult to prove definitely and exceptions existed in certain city-states.[42] A more plausible explanation is that the small size, intimate connections, and constant interactions that existed between Greek city-states provided a constant flow of information about conditions within their neighbours and later allies and colonies, only occasionally supplemented by the work of intelligence agents. Further muddying the waters is the staggering array of information sources, some of which were personal or private in nature and included itinerants, oracles, envoys and ambassadors, and those sympathetic to friends or relatives in other city-states.[43] Later Roman sources about the Greeks, including the histories of Polybius and Livy, repeat anecdotes about spying from Greek sources but often only add unsubstantiated embellishments.

In addition to a paucity of sources, the Greek language and its use of indefinite pronouns and passive verbs further impedes our understanding of specific intelligence actions and the actors that undertook them.[44] Biographers of Alexander the Great, including Arrian, focus much of their attention on the daring exploits and character of a conqueror seemingly out of the literary tradition of Homer and mythology. Unfortunately, they follow the fashion of opaque references and elliptical allusions and add little to our understanding of Macedonian intelligence activities beyond rather questionable anecdotes, such as the precocious young Alexander pumping Achaemenid emissaries from Persia (modern Iran) for specific information about roads and military strength he later used during his conquest of their empire.[45]

[39] Aristotle, *Aristotle in 23 Volumes*, vol. 21, trans. H. Rackham (Cambridge, MA: Harvard University Press, 1944), *Politics*, Book 5, Section 1313b, www.perseus.tufts.edu/hopper/text? doc=Perseus%3Atext%3A1999.01.0058%3Abook%3D5%3Asection%3D1313b; Plato, *Plato in Twelve Volumes*, vols. 5 and 6, trans. Paul Shorey (Cambridge, MA: Harvard University Press1969), *Republic*, Book 7, Section 539b, www.perseus.tufts.edu/hopper/text?doc=Perseus%3Atext%3A1999.01.0168%3Abook%3D7%3Asection%3D539b.

[40] J.A. Richmond, 'Spies in Ancient Greece', *Greece & Rome*, 45:1 (April 1998): 13.

[41] *Proxeni* served as trade facilitators between city-states and were drawn from the elite classes. They often had hereditary ties to families in other city-states. Richmond, 'Spies in Ancient Greece', pp. 6–13.

[42] Richmond suggests gender might have played a role in limiting espionage while Frank Russell argues for the opposite based on the case of Syracuse. Ibid., pp. 13–14; Frank Russell, *Information Gathering in Classical Greece* (Ann Arbor, MI: University of Michigan Press, 2000), p. 104.

[43] Russell, *Information Gathering in Classical Greece*, pp. 88–91; 98–99. Thucydides mentions in 1.72, 'There happened to be Athenian envoys in Sparta present on other business'. Robert Strassler, trans. and ed., *The Landmark Thucydides: A Comprehensive Guide to the Peloponnesian War* (New York: Free Press, 1996), 1.72, p. 41.

[44] Russell, *Information Gathering in Classical Greece*, pp. 101–102.

[45] Plutarch, *The Age of Alexander: Nine Greek Lives by Plutarch*, trans. Ian Scott-Kilvert (London: Penguin Books, 1973), p. 256.

The maintenance of empire, rather than its conquest, provides us with the first concrete example of specialized forces dedicated to collecting information in the Western world. Roman activities during the Period of Kings (625–510 BCE) and Republican Rome (510–31 BCE) offer little in terms of valuable insights and on how or indeed whether information was collected in a systematic manner, including what is tempting to call the strategic level, or political intelligence, about foreign empires, compared to the period of Imperial Rome (31 BCE–476 CE). Modern attempts to discern different types or levels of intelligence in classical history, however appealing, are problematic. Rose Mary Sheldon reminds us that for the Romans, '[p]olitical intelligence, useful to making decisions about foreign policy in which war was often an instrument, became military intelligence once the war had begun', and in consequence they did not distinguish between the two.[46] In addition, scholars remain divided on how much Imperial Romans valued intelligence. Sheldon, Hugh Elton, and others offer they did. In his assessment of seventy-five years of Roman warfare in Europe, Elton concludes that in sum, Roman intelligence about 'barbarian' tribes across the frontiers 'seems good'.[47] Other scholars take issue with such assessments, suggesting that Roman collection of meaningful advance information as part of a 'grand strategy' is a fiction, given the nature of Roman political institutions and cultural norms.[48] For example, Benjamin Issac, in looking specifically at the Roman legions in the eastern regions of the empire, caustically concludes, 'Roman intelligence was notoriously bad in the East. There is no lack of evidence to show the Romans would embark on major military campaigns without an elementary understanding of where they were going'.[49]

Although 'strategic intelligence' of the Romans remains a controversial subject open to debate, based on an institutional capacity to systemically process information, their gathering and use of military intelligence is not. Various sources of military information can be teased out of different Roman sources. For example, as Amiram Ezov points out, Julius Caesar's *De Bello Gallico* or *The Gallic War* mentions four sources of intelligence, mirroring, for the most part, the Greek ones mentioned above. They include prisoners, the dispatches of officials including legates, and that provided by neighbouring tribes and allies.[50] The fourth source is particularly relevant to the enquiry here: the establishment of an identifiable cadre of information gatherers, the *exploratores*.

*Exploratores* were more than just scouts operating ahead of advancing Roman legions. The Romans had a separate term for those scouts: *procursatores* and *speculatores*, although the leading experts on the subject remind us that sources are not

[46] Rose Mary Sheldon, *Intelligence Activities in Ancient Rome: Trust the Gods, but Verify* (London: Routledge, 2005), p. 5.

[47] Hugh Elton, *Warfare in Roman Europe, AD 350–425* (Oxford: Clarendon Press, 1997), p. 75.

[48] The work that began the lengthy debate about the existence of a grand strategy is Edward Luttwak, *The Grand Strategy of the Roman Empire: From the First Century A.D. to the Third* (Baltimore, MD: The Johns Hopkins University Press, 1976).

[49] Benjamin Issac, *The Limits of Empire: The Roman Army in the East* (Oxford: Clarendon Press, 1990), p. 402.

[50] Amiram Ezov, 'The "Missing Dimension" of C. Julius Caesar', *Historia: Zeitschrift für Alte Geschichte*, 45:1 (1996): 66–67; Caesar, *De Bello Gallico*, Libri I.50; II.1; II.4–5; V.45 in *Caesar Delphini* (Philadelphia, PA: Classics Press, 1804), pp. 36; 40; 41–43; 107.

always clear and distinctive, reflecting literary imperatives rather than technical ones, in their use of these terms.[51] *Exploratores*, mounted on horseback, rode far in advance of the legions, indeed much farther than the normal 12-kilometre scouting range of cavalry, in order to fulfil several different functions.[52] The first was to seek out and establish the position and learn whatever they could of the size and composition of enemy forces.[53] A second, given that legions were marching in unfamiliar territory and, in some cases, beyond the realms of known civilization, was rough topographical surveying. Without the benefit of maps, *exploratores* would seek out the best locations for camps or for future battles, as they did for Caesar prior to the battle of Gergovia, the chief town of Vercingetorix's Arveni tribe, in Gaul (52 BCE).[54] Although the information they provided was not always acted upon, *exploratores* could be crucial in changing commanders' understanding of the enemy. During the Roman Civil War (49–48 BCE), information provided by Pompey's *exploratores* was crucial in changing his perception of Caesar's sudden withdrawal from a potential field of battle at Asparagium.[55]

Not all *exploratores*' reports were accurate and could reflect very real human weaknesses such as anxiety and panic. Nonetheless, in weighing the evidence for and against them as a standing specialized force, Ezov concludes that 'the *exploratores* were a special operations intelligence body which by its very nature had to have specific professional skills, such as the abilities to observe and move with stealth. These required special training'.[56] As an example of how little institutionalized reconnaissance techniques had evolved over almost 2,000 years, Arthur Wellesley, Lord Wellington, commissioned and employed 'exploring officers' in Spain and Portugal to conduct the same missions, with a few noteworthy additions discussed below, during the Peninsular War (1808–1814 CE) against the forces of Joseph Bonaparte. These officers, selected for their horsemanship and language skills, operated in denied and contested areas in their often elaborate and colourful regimental uniforms collecting information on troop movements, providing terrain sketches, and liaising with Spanish guerrilla bands.[57]

Although we have little idea of the exact numbers of *exploratores* Roman legions employed, their activities exhibit several characteristics roughly equivalent to those of modern special reconnaissance missions conducted by SOF. These activities were conducted independently from but supportive of tactical or strategic objectives. In consequence, and as the examples above suggest, they could generate both tactical and strategic effects. The potential payoff, however, was accompanied

[51] N.J.E. Austin and N.B. Rankov, *Exploratio: Military and Political Intelligence in the Roman World from the Second Punic War to the Battle of Adrianople* (London: Routledge, 1995), p. 42.

[52] Austin and Rankov, *Exploratio*, pp. 44–45.

[53] Ezov, 'The "Missing Dimension" of C. Julius Caesar', p. 67.

[54] Austin and Rankov, *Exploratio*, p. 43; Caesar, *De Bello Gallico*, L. VII.44, pp. 160–161.

[55] Austin and Rankov, *Exploratio*, p. 48; Caesar, *De Bello Civili*, Liber III.41 in *Caesar Delphini* (Philadelphia, PA: Classics Press, 1804), p. 309.

[56] Ezov, 'The "Missing Dimension" of C. Julius Caesar', p. 75.

[57] Huw Davies, 'The Influence of Intelligence on Wellington's Art of Command', *Intelligence and National Security* 22:5 (October 2007): 633.

by considerable personal risk. *Exploratores* often operated in unknown territory occupied by hostile tribes or forces and some never returned.[58] Other sources suggest *exploratores* had specialized training and could operate in different modes, from individuals to large parties, in overt or covert manner, depending on the target.[59] While some assessments are based on educated inferences from a limited number of sources, they nevertheless point to an identifiable organization other than the *frumentarii* or domestic secret police, conducting a wide range of specialized reconnaissance activities during the Imperial Roman period (31 BCE–476 CE).[60]

Ensuring the security of the frontiers of empire could lead to use of other specialized reconnaissance and reaction forces. The Rhine River served as a natural but not impenetrable barrier between the Roman provinces which, by 70 CE, consisted of *Germania Superior* and *Inferior* and *Belgica* and was inhabited by three main barbarian tribes, which the Romans identified as the Franci, Burgundi, and Alamanni. The Romans built fortified towns on the river in their provinces, including *Argentoratum* (Strasbourg), *Mogontiacum* (Mainz), *Colonia* (Cologne), and *Noviomagus* (Nijmegen). To supplement the fortifications of the *Limes Germanicus* and to respond to and annihilate barbarian incursion on the frontiers of empire, the Romans modified their marines into a form of specialized riverine infantry. While Roman marines existed since the First Punic War (264–241 BCE) and were little more than an extension of the legions on boats, Derek Leebaert argues they modified their activities to operate more effectively on the Rhine. He further suggests that these *lembarii*, as they were known, bore attributes similar to but should not be considered SOF. He contends that although they used modified equipment for camouflage and tactical effectiveness, they were an ad hoc response to specific environmental conditions, not a force specifically trained in a wide range of skirmishing and survival skills.[61] As an interesting aside, one need not be a Roman citizen to serve in this capacity; subjects already possessing specialized skills in small-boat handling and seamanship were also recruited.[62] Although equating these forces with modern SOF is fraught with peril, the *lembarii* and other units on the Rhine call to mind more recent forces, based on their size and use. The parallels include the type of reconnaissance carried out by naval commandos employed by different flotillas of the Soviet Union during the Second World War, as well as the rapid response and strike operations of the US Navy's Mobile Riverine Force used on the Mekong River and its tributaries during the Vietnam War.[63]

[58] Austin and Rankov, *Exploratio*, p. 52.
[59] Austin and Rankov, *Exploratio*, p. 49.
[60] The *frumentarii* performed several domestic economic, security, and stability functions from the last first until the late third century. For details, see Stuart McCann, 'What's in a Name? The Evolving Role of the *Frvmentarii*', *The Classical Quarterly*, 69:1 (May 2019): 340–354.
[61] Leebaert, *To Dare and To Conquer*, pp. 94–95.
[62] Ibid.
[63] Iurii Strenkhnen, *Commandos from the Sea: Soviet Naval Spetsnaz in World War II*, trans. James Gebhardt (Annapolis, MD: Naval Institute Press, 1996); Thomas Cutler, *Brown Water, Black Berets: Coastal and Riverine Warfare in Vietnam* (Annapolis, MD: Naval Institute Press, 1988), pp. 209–266.

After the fall of the Western (476 CE) Roman Empire and the subsequent Dark Ages that swept the European continent, intelligence and reconnaissance activities were likely conducted but few surviving written sources discuss them. Most are concerned with reconnaissance in support of military campaigns. In the Eastern Roman Empire, or Byzantium, scouts played an important role in directing what Edward Luttwak has termed 'relational methods and tactics' to catch advancing foreign armies unaware, prevent surprise, and conduct raiding if necessary.[64] One Western Christian author, Christine de Pizan, writing in France in 1411 and outlining the attributes of an ideal commander in a manner not unlike Sun Tzu and Kautilya, lists the need 'to send surreptitiously spies and scouts here and there' to collect information on enemy dispositions and composition and local guides familiar with the area to prevent ambush or surprise.[65] During the Ayyubid Period (1171–1260 CE) of the medieval Sultanate of Egypt, the scouting units of Muslim armies were expanded and 'not only consisted of special reconnaissance troops and scouts to be sent ahead of the main army before hostilities broke out, but also included spies and informers' who relayed information to the commander.[66]

More than a century later in Renaissance Italy, Niccolo Machiavelli, writing in his celebrated advice to a Florentine prince, does not broach the subject of spies or intelligence collection and only mentions 'information' once, in an oblique way in a single example.[67] Both are implicit throughout the text, however, reflecting the nature of competition in mercantile city-states in Italy as well as the widespread employment of mercenary armies, which obviated developing specialized reconnaissance forces.[68] In his *Art of War*, for example, Machiavelli exhibits the prevailing trend in classics of military art and statecraft of this period: he cribs heavily from rediscovered ancient classics by authors such as Aristotle and Vegetius.[69] More specifically, Vegetius, writing in the late Roman period, alludes in two particular passages of *De Re Militari* to networks of spies and agents conducting reconnaissance and gathering information for field commanders once armies had crossed the frontier into enemy territory, if at a minimum to prevent surprise and ambush.[70]

[64] According to his translation and interpretation of Maurikios' (Maurice) and Nikephoros' works. Edward Luttwak, *The Grand Strategy of the Byzantine Empire* (Cambridge, MA: Belknap Press, 2009), pp. 299; 342.

[65] Christine de Pizan, *The Book of Deeds of Arms and of Chivalry*, trans. Sumner Willard and ed. Charity Cannon Willard (University Park, PA: Penn State Press, 1999), pp. 46; 45–50.

[66] David Nicolle, *Crusader Warfare, Volume II: Muslims, Mongols, and the Struggle against the Crusades* (London: Hambleton Continuum, 2007), p. 137.

[67] Nicolo Machiavelli, *The Prince*, trans. W.K. Marriott (Project Gutenberg eBook, 1998), chapter XIX, www.gutenberg.org/files/1232/1232-h/1232-h.htm#pref03.

[68] On the qualities and abilities of Swiss and German mercenaries as the premier infantry of their day, thanks to drill and skill employing the pike, see F.L. Taylor, *The Art of War in Italy, 1494–1529* (Cambridge: Cambridge University Press, 1921; reprinted Greenhill Books, 1993), pp. 29–39.

[69] As part of a Socratic dialogue in the Fifth Book of Machiavelli's *The Art of War*, one of his discussants, Fabrizio, regurgitates a portion of Book III of Vegetius' work, under the section 'Marches in the Neighborhood of the Enemy'. Machiavelli, *The Art of War*, trans. Neville, https://oll.libertyfund.org/title/machiavelli-the-art-of-war-neville-trans; Vegetius, *The Military Institutions of the Romans (De Re Militari)*, trans. John Clarke, www.digitalattic.org/home/war/vegetius.

[70] Vegetius, 'Military Institutions of the Romans', in Thomas Phillips, ed., *Roots of Strategy: A Collection of Military Classics* (London: John Lane the Bodley Head, 1943), p. 70.

Intelligence and reconnaissance activities were inherent to statecraft and warfare in subsequent centuries up to the modern period but they were neither widely written about nor were specialized units, beyond some frontier scouting units, formed to conduct them. Much of this can be attributed to warfare focused on the technical and logistical aspects of conducting sieges against fortifications.[71] More was written about subversion and treachery, for reasons the next section makes clear.

## Subversion and treachery

The subjects of subversion and treachery, while discussed extensively in ancient works, are rarely associated with special operations in the modern era. There are several possible explanations for this absence. One relates to the connotation of the words. Within Western democracies, subversion and treachery call to mind underhanded activities conducted by shadowy intelligence agencies to influence Western population by competitor states such as Russia and China.[72] Over the past century, Western democracies were targeted by subversion and treachery orchestrated by Communist ideologues in the Soviet Union seeking to spread socialist revolution, as outlined in the various works by Karl Marx and Vladimir Lenin and encompassed in the terms 'active measures' or 'political warfare'.[73] This experience within living memory made a forceful impression on several generations of Western citizens given shocking revelations of those within willing to betray or actively erode the foundations of democracy.[74] More recent generations now associate subversion and treachery with social media, misinformation, and election interference. In addition, as our vocabulary has expanded in parallel with the growth in the scale, complexity, and destructiveness of warfare, we have created specific terms with narrowly defined meanings within specialized military communities. In the military vocabulary, 'unconventional warfare' comprises a range of activities that include sabotage and subversion, while in the intelligence community, these activities and more are encapsulated with the term 'covert action'. Although a case can be made in an analytic survey of history to include a discussion on stratagems, ruses, trickery, and other forms of deception, there are two arguments against it here. The first keeps the survey focused. The second relates to the qualities and personality of individual

[71] De Pizan, *The Book of Deeds of Arms and of Chivalry*, pp. 115–139.

[72] See the initial examples and expansive definition of irregular warfare put forward by Seth Jones in *Three Dangerous Men: Russia, China, Iran, and the Rise of Irregular Warfare* (New York: W.W. Norton & Co., 2021), pp. 1–3; 11.

[73] For a useful history starting from 1917 onward, see Thomas Rid, *Active Measures: The Secret History of Disinformation and Political Warfare* (London: Farrar, Straus, & Giroux, 2020).

[74] Noteworthy agents working for the Soviet cause include Julius and Ethel Rosenberg and the 'Cambridge Five', including Kim Philby. Although most were uncovered during the early portion of the Cold War, some as early as 1945 thanks to Soviet defector Igor Gouzenko, later generations received a reminder when others were revealed as part of the release of the Mitrokhin archive in 1999. See, for example, the 'outing' of Melita Norwood in Christopher Andrew and Vasili Mitrokhin, *The Sword and the Shield: The Mitrokhin Archive and the Secret History of the KGB* (New York: Basic Books, 1999), pp. 397–398.

ancient commanders to use inspired or unorthodox approaches on the battlefield, or innovation, rather than a broader and more systematic use culturally or contextually.

A second connotation relates to suspicion of power within democracies and more specifically its abuse. Within Western nations and the United States specifically, enquiries into the manner and extent of activities conducted by intelligence agencies and secret or security services in destabilizing and overthrowing regimes, starting in the wake of the Vietnam War and most recently with Iraq in 2003, confirms for many the suspicion that power ultimately corrupts in liberal democracies.[75] Such activities, driven by a hubristic self-image of virtue, are contrary to Western democratic values and norms and should rarely if ever be undertaken. Both connotations relate to fears and anxieties about the institutional threat from within pursuing its own agenda of its own ends, or individuals being manipulated by the threat from without. They also point out the dual nature of subversion and treachery, which can be used offensively or offset defensively. For our purposes here, subversion is 'the undermining or detachment of the loyalties of significant political and social groups within the victimized [polity], and their transference, under ideal conditions, to the symbols and institutions of the aggressor'.[76] Subversion requires active engagement in the targeted polity to cause citizens or groups to switch their loyalties. Treachery, in contrast, suggests individuals or groups are willing to take action to betray their fellow citizens.

This dual nature of subversion and treachery is very much evident in ancient and classical writing on statecraft and military art. Indeed, the first written military classic we have available is in essence a manual of subversion and treachery. T'ai Kung's *Six Secret Teachings*, the first of the seven military classics of Ancient China, dates from as early as 1000 BCE.[77] T'ai Kung's work is secret for two reasons. The first relates to its ostensible context and purpose. The author provides a range of military and political advice to sovereigns of the Zhou Dynasty on the preparations they should make to overthrow the dominant Shang, or Yin, Dynasty (~1600–1045 BCE). Four of the six 'T'ao', or teachings, are familiar to students of the military art, given their focus on military preparations, organization, tactics, and optimal employment of combat arms. The first T'ao, the Civil T'ao, provides political advice to a sovereign in the form of a dialogue between T'ai Kung and King Wen and outlines the 'six thieves' and 'seven harms' of his subjects any ruler should know that subvert his authority.[78] The remaining teachings contained in the Martial T'ao offer twelve 'civil offenses' and answers to 'three doubts'.[79] The sum of these two teachings provides a guide to human weaknesses and foibles that can be exploited in order

[75] A popular history that captures this sentiment is Stephen Kinzer, *Overthrow: America's Century of Regime Change from Hawaii to Iraq* (New York: Times Books, 2006).

[76] Paul Blackstock, *The Strategy of Subversion: Manipulating the Politics of Other Nations* (Chicago, IL: Quadrangle Books, 1964), p. 56.

[77] Like many ancient works, its specific date, provenance, and author are the subject of healthy, if at times contentious, debate. For a useful if now dated overview of these debates, see Ralph Sawyer, *The Seven Military Classics of Ancient China*, pp. 35–37.

[78] Ibid., pp. 48–50.

[79] Ibid., pp. 56–57.

to subvert the ruler's virtue, as well as the bonds of trust and loyalty between officials and subjects, as covertly as possible. These weaknesses, codified later in the Christian tradition as capital vices or 'seven deadly sins', include vanity, lust, and greed.

The second reason for its secrecy, after Zhou was ultimately successful in internally corroding the Shang Dynasty and ultimately defeating them in a decisive battle at Muye (1046 BCE), has a practical as well as cultural dimension.[80] From a purely pragmatic standing, it did not benefit Zhou leaders to provide their enemies from outside their territory, much less within it, with a playbook for their own destruction once they were successful. Culturally, the amoral advice contained within the *Six Secret Teachings* amounts to the ends justifying any and every means, no matter how base or distasteful. The sovereigns of the Zhou and subsequent dynasties took great pains to reinforce their virtuous qualities and demonstrate they were deserving of 'heaven's mandate', a concept the Zhou implemented, and were therefore worthy to rule.[81]

Reflecting the unique vulnerabilities of authoritarian regimes, including tyrannies, feudal monarchies, and empires, considerably more energy and effort was devoted to detecting, quelling, or fomenting uprisings through treachery and subversion. In their discussions, the dual nature of subversion and treachery is evident in Sun Tzu, Kautilya, and other texts. The virtuous ruler should exploit the weaknesses of his mendacious opponents, while making every effort to identify, neutralize, or benefit from threats that exist internally. One of Sun Tzu's category of spies is 'double agents'. Their relationship to 'inside agents' reflects logically a version of the 'golden rule': do unto others as they do to you. More to the point, if a ruler can turn officials within a neighbouring state, as well as neutralizing their efforts to do the same while providing them with false information through their own source, then so much the better. Yet to do so, and to do so effectively without succumbing to moral corrosion, the ruler employing spies must be 'sage and wise, humane and just' in a manner that is 'delicate and subtle'.[82]

In the *Arthasastra*, Kautilya comprehensively distinguishes between preserving internal security and conducting offensive covert action against the sovereign. To conduct the former, Kautilya proposes to test high officials to assess their authorities and develop a hierarchy of governance that employs a surveillance network to detect threats. These include established local officials and agents, whose loyalty he presumes can be bought, as well as more mobile, clandestine ones. In essence, Kautilya outlines conceptually the model for the modern, authoritarian surveillance state in general terms. One can infer from his brief discussion of counter-espionage, comprising two points, that enemy agents will have a difficult time penetrating the theoretical

[80] A contemporary account of the Battle of Muye, written by Zhou authors, is the 'Shifu' of the *Yizhoushu*. Although its provenance was doubted by none other than Mencius, Edward Shaughnessy makes a compellingly detailed case for its authenticity. Edward Shaughnessy, '"New" Evidence on the Zhou Conquest', *Early China*, 6 (1980–1981): 57–79.

[81] 'Heaven's mandate' is outlined succinctly in Li Feng, *Early China: A Social and Cultural History* (Cambridge: Cambridge University Press, 2013), pp. 142–143.

[82] Griffith, *The Art of War*, p. 236.

web of reporting eyes and ears the king employs.[83] At least one modern author suggests Kautilya's guidance on this point is relevant to the range of domestic threats states currently face, including terrorism and rebellion.[84]

The rationale driving Kautilya's advice is the potential behind and power of subversion. In words that could have been uttered two millennia later by Allen Dulles, the first director of central intelligence, Kautilya begins his fourth section of part IX, 'Covert Operations', with the following statement: 'Miraculous results can be achieved by practicing the methods of subversion.'[85] The reasons for his confidence are simple and based on the same insight as T'ai Kung: people have a range of exploitable grievances and weaknesses, which include the angry, greedy, frightened, and haughty, regardless of their station in life.[86] Those who cannot be bribed, encouraged, or undermined should be killed. While acknowledging the weaknesses of the others, the king must ensure the cultivation of three qualities: *dharma*, *artha*, and *kama*, or spiritual, material, and aesthetic well-being—collectively virtue, by another name.[87] *Dharma* and *kama*, however, depend on *artha*, forcing the king to institute measures to oversee his subjects and coerce them through inducements or punishments, but to do so in a just manner in the eyes of his subjects. For both Kautilya and Sun Tzu, however, we have precious few details on how effective their ideas at fomenting subversion and treachery were.

Ancient Greece, however, offers several examples of the outcome of treachery and subversion. As in Ancient China and medieval India, Greek polities were in nearly constant competition with their rivals. Many of the details of Greek competition and conflict rely on available written sources, both cultural and more scholarly in nature, as well as insights and inferences derived from archaeology. The specific details are open to interpretation, mythicizing, and debate. The generalities of conflict during this period are clear enough, however. The main political unit was the city-state, and city-states often formed alliances with one another against their rivals and a few grew into empires. Navies, if available, had limited mobility away from shore, carried troops or supplies, and fought one another by ramming, hurling missiles, or boarding each other's ships and conducting combat at sea as on land. Few city-states could afford the cost of navies. Armies on the land, which were more common, consisting of professionals, citizens, and/or mercenaries, conducted military campaigns of short duration due to limitations in logistics or finances, the need to harvest crops, or domestic politics.

Conflicts were decided conventionally in one of a few manners: rather quickly by battle with one another if an immediate decision was sought, or if an opponent chose to do so, retiring behind defensive walls protecting the city and forcing a long, costly, and often inconclusive siege. In Homer's *Iliad*, the Trojans forced the Greeks into

[83] Kautilya, *The Arthasastra*, 1.12.24–25, pp. 506–507.

[84] Ni Made Sumaryani, 'Kautilya's Views on Secret Activities of Intelligence in *Arthasastra* and Its Current Relevance', *International Journal of Hindu Science and Religious Studies*, 5:1 (June 2021): 104–110.

[85] Kautilya, *The Arthasastra*, 13.1.21, p. 519.

[86] Ibid., 1.14.2–5, pp. 519–520.

[87] Ibid., 15.1.71–73, p. 100.

a siege, despite some customary individual combat between champions from both sides. According to him, one of the Greeks, the clever and cunning Odysseus, came up with the idea of presenting an offering to the Trojans in the form of a horse. Hidden within the horse were the best Greek warriors. The puzzled Trojans pulled the offering through the main gate and into their defensive walls, and later that evening the Greek warriors debarked, opened the gates to the comrades outside, surprised the Trojans, and sacked and plundered the city. An important question remains: assuming the Trojan Horse existed, and given its decisive success, why were versions of it not repeated historically to overcome the sieges and the power of the defensive?

A literal reading of Homer suggests the behaviour of the Greek warriors within the walls of Troy after the special operation, involving widespread sacking, looting, raping, and taking slaves, so angered the gods that they needed to be taught a moral lesson. Scholars of Greek warfare offer a different and considerably less dramatic answer, one relying less on the whims of the Greek deities and more on human frailties. The activity conducted most successfully and repeatedly during the age of Greek warfare, and while cities and towns relied on walls and battlements for defence, was treachery. In its simplest and notional form, one or more disgruntled individuals within the besieged city offer to do almost exactly what the Greek warriors were alleged to have done at Troy: open the gates to the besieging forces. Doing so would allow the besiegers to pass through the defences. This objective could just as easily be done by coopting disgruntled besieged inhabitants intimately familiar with the city's defences, including its weaknesses. As Hans van Wees describes, 'Horse-style stratagems might work from time to time, but more, perhaps most, cities and forts were taken as a result of treason. We have evidence from treachery from the moment we have detailed evidence for sieges.'[88]

The historical evidence of treachery, however, can be challenging to interpret for several reasons. In strictly numerical terms from a single source, *The Peloponnesian War*, Thucydides mentions eighteen different sieges in his history, as opposed to seven instances of cities taken by treachery.[89] The seeming simplicity of the approach outlined above, inducing or cajoling those sympathetic to you to open the city gates, belies the often-complex coordination necessary to pull off such a *coup de main*. The fall of the port of Nisaea (424 BCE), in the city of Megara, provides an exceptional example, involving elements of siege and treachery. Paul Kern labels it '[o]ne of the more colorful examples of victory through ruse and treachery', enabled by Megarian traitors engaged in a 'nightly charade' of passing through the gates with a boat on a cart, whose routine pattern lured the gate guards into complacency.[90] At a pre-coordinated time, Athenian hoplites concealed in a ditch moved with speed when the Megarian traitors blocked the gate from closing with the cart. In Derek Leebaert's words, Nisaea's capture provides 'a stunning example of simultaneous nighttime infiltration by land and sea, coordination between strike teams big and

[88] Hans van Wees, *Greek Warfare: Myths and Realities* (London: Duckworth, 2004), p. 138.
[89] Strassler, *The Landmark Thucydides*, pp. 692; 792.
[90] Paul Kern, *Ancient Siege Warfare* (Bloomington, IN: Indiana University Press, 1999), p. 118.

small, synchronization with a main force, and acute intelligence work in knowing how to leverage the cunning friends on the inside'.[91]

Seizing the city was one thing and securing it quite another. According to Thucydides, a political misunderstanding between the Athenians and Megarian traitors gave time for the ruling oligarchs to recover and consolidate their grip on the city. Upon hearing of the fall of the port, Spartan commander Brasidas rushed to the city with his army and won a political victory on two levels, by taking advantage of the discord between the small numbers of Athenians and Megarians, without having to fight—the very pinnacle or acme of skill in war, according to Sun Tzu.[92]

Betraying a city through treason had both a direct physical and indirect psychological effect characteristic of many special operations, as an assessment of the foremost ancient author on sieges, Aeneas Tacitus, suggests. Van Wees offers, 'The spectre of the enemy within haunted [him], who devoted more space to the prevention of treason to fighting off an assault.'[93] Tacitus provides cautionary examples of successful treachery in the Greek city-states of Chios, Argos, Heracleia Pontica, and Corcyra. He further outlines a range of methods the leaders of the besieged should adopt to counter treachery in a section entitled 'On Maintaining Morale and Discipline and General Measures for Thwarting Treachery and Revolution'.[94] In short, treachery could force democratic city-states into suppressing the freedom of their citizenry and subverting their political values, which echoes the modern liberal distrust of intelligence operations in democracies.

Treachery in the ancient world could also occur outside of sieges, affecting the outcome of battles. The most famous classical act of betrayal in this context was done by Ephialtes, at the Battle of Thermopylae in 480 BCE. A Spartan contingent numbering 300, under King Leonidas, used advantages of terrain and tactics to hold up the advance of the vastly superior invading Persian Army under the command of Xerxes. Apocryphal incidents and statements that occurred during the battle and recorded by Herodotus, later embellished by Plutarch and others, have become the stuff of legend. In addition, they have inspired political and military rallying cries (*Molon labe*, or 'Come and get them', referring to the Spartan taunt for the Persians to take their weapons), as well as a graphic novel and a film.[95] Having failed over the course of three days to overwhelm the Spartans, Xerxes welcomed the information Ephialtes provided on the Anopaia Path, which led to a mountain summit flanking the pass the Spartans were holding.[96] After pushing aside a detachment of Phocians guarding

[91] Leebaert, *To Dare and To Conquer*, p. 57.

[92] Strassler, trans. and ed., *The Landmark Thucydides*, p. 262; Griffith, *The Art of War*, p. 115.

[93] Van Wees, *Greek Warfare*, p. 138.

[94] Aeneas Tacitus, *Siege Warfare*, trans. Illinois Greek Club, Loeb Classical Library (Cambridge, MA: Harvard University Press, 1928), pp. 53–73, https://penelope.uchicago.edu/Thayer/E/Roman/Texts/Aeneas_Tacticus/home.html.

[95] The Greek neo-Nazi party 'Golden Dawn' recreates the battle annually, following the example first set by the Nazis in June 1941. Eleftheria Ioannidou, 'Mythologies of Genesis and Neo-Nazi Palingenesis: Commemorating the Battle of Thermopylae in the Political Rites of the Golden Dawn'. *Humanities* 11:88 (2022): 1–21, https://doi.org/10.3390/h11040088.

[96] Robert Strassler, ed., *The Landmark Herodotus: The Histories*, trans. Andrea Purvis (New York: Pantheon, 2007), 7.213–215, p. 588.

the path, the Persians seized the summit with archers and, after Leonidas rejected an offer to surrender, cut the remaining Spartans down with arrows.

The details Herodotus recounts about Ephialtes illuminate aspects of Ancient Greek views towards treachery and subversion. These views have their roots in Homer's *Iliad* and *Odyssey* and, more specifically, the contrast between Achilles and Odysseus. Ancient Greek martial virtue prized strength, courage, and bravery most often associated with the word 'standing', as in facing danger squarely. In the *Iliad*, Achilles is the paragon of these virtues, armed with his magical armour, spear, and sword and seemingly oblivious to threats from the enemy. He positively relishes the idea of confronting the enemy on the field of battle; if he dies, it is with honour. In contrast, Odysseus is characterized as crouching. His weapon is the bow, which can kill the enemy without coming in range of their weapons. In other words, the Greeks had strong views on what constituted proper, honourable military behaviour and actions as well as the improper and dishonourable. Ephialtes is described as physically deformed; a hunchback, who cannot help but crouch. The Persians cannot best the Spartans in 'standing' combat. They have to resort to treachery and masses of archers, who could fire so many arrows that 'the sun would be blocked out by their number'.[97]

Echoes of these Greek paradigms of honourable and dishonourable, Achilles and Odysseus, resound today in the self-image those within conventional armed forces have of themselves and how they view SOF. Nowhere was this more in evidence than in disagreements over the legality and propriety of American SOF wearing non-standard uniforms, including civilian clothing, in Afghanistan, a declared theatre of active armed conflict.[98] Within various special operations communities, the Trojan Horse may be more alluring because it an act of cunning conducted by warriors with whom special operators can more readily self-identify. This allure stands in contrast to cultivating traitors which runs counter to modern military values and ethics and is more likely the purview of, and more frequently associated with activities of, intelligence agencies. The broader point of this diversion is that what is 'special' should be determined by both the prevailing norm of the day and the most common and successfully repeatable unorthodox method of overcoming it, as opposed to more attractive outlying examples by exception. The allure of the Trojan Horse as a masterstroke continues to appeal to contemporary special operators, however, despite treachery far better, and more frequently, fulfilling one of the modern 'master claims' put forward by Colin Gray for the strategic utility of special operations discussed in Chapter 1: an exercise in economy of force.[99]

Despite numerous examples of revolts and rebellions, it is challenging to connect any of them specifically to long-term efforts at subversion by external actors. Much

[97] Strassler, ed., *The Landmark Herodotus*, 7.226, p. 593.

[98] 'Special Ops Ordered to Shave', ABC News (13 September 2002), https://abcnews.go.com/WNT/story? id=130,114&page=1. For a legal and historical examination of the subject, see W. Hays Parks, 'Special Forces' Wear of Non-Standard Uniforms', *Chicago Journal of International Law*, 4:2 (2003): 493–560.

[99] Colin S. Gray, *Explorations in Strategy* (Boulder, CO: Westview Press, 1996), pp. 168–174.

like clandestine operations today, if conducted in a professional manner the sponsor's role remains hidden. Several Greek revolts were inspired by enterprising tyrants seeking to gain power for themselves. In one case, Athens sent support to the Ionian Revolt (499–498 BCE) once it was underway after local tyrant Aristagoras' appeal, much to their later regret. Most Roman revolts were inspired by excesses, either by the Romans themselves (Boudica [60/61 CE], the Batavi Revolt [69–70 CE], etc.) or their local representatives (the Great Jewish Revolt [66–73 CE]).[100] Later political philosophers would be inclined to agree. Machiavelli, for example, suggests in *Discourses on Livy* the seeds of revolt are planted due to unjust rule, particularly calumny.[101] In other words, the idea linking together Chinese, Indian, Greek, and Roman authors on statecraft and the military arts is that a prosperous society, ruled justly and virtuously and guarded vigilantly, need not fear domestic upheaval, much less foreign interference. In the medieval period, subversion took on other overtones when heterodoxy and heresy were added into the mix, culminating in the Spanish Inquisition.[102]

From the late Middle Ages onward, thanks to increased literary and record keeping, we have numerous examples of subversion to draw upon, including the three plots and rebellions in late sixteenth-century England to assassinate or depose Elizabeth in favour of her cousin, Mary, Queen of Scots. These efforts included the Northern Rebellion (1569) and the Ridolfi (1571), Throckmorton (1583), and finally Babington Plot (1586). They involved a mixture of individuals, *agent provocateurs*, financiers, and Catholic noblemen within England and Scotland as well as Spain. All were ultimately discovered, the plots more specifically thanks to the brutal efficiency and cunning of Sir Francis Walsingham, who employed the sorts of agents and methods outlined by Kautilya.[103] Time and space, unfortunately, do not permit further exploration of subversion. In any event, the theoretical foundations of treachery and subversion were well established in Ancient China, Medieval India, and Ancient Egypt, Greece, and Rome.

## Direct and indirect shock

The history of warfare is replete with examples of treachery and subversion. These served as a means of minimizing costs and avoiding direct confrontation on a battlefield. Their immediate effect was to surprise opponents, create a window of opportunity to seize the initiative, and overcome superior defences or numbers. William McRaven developed a conceptual explanation for success and failure for

[100] Stephen L. Dyson, 'Native Revolts in the Roman Empire', *Historia: Zeitschrift für Alte Geschichte*, Bd. 20, H. 2/3 (2nd Qtr., 1971): 239–274.

[101] Niccolo Machiavelli, *Discourses on Livy*, trans. Harvey Mansfield and Nathan Tarcov (Chicago, IL: University of Chicago Press, 1996), pp. 26–27.

[102] Henry Kamen, *The Spanish Inquisition: A Historical Revision* (London: Weidenfeld & Nicolson, 1997).

[103] Robert Hutchinson, *Elizabeth's Spy Master: Francis Walsingham and the Secret War That Saved England* (London: Weidenfeld & Nicholson, 2006), pp. 83–145.

these special operations in his theory of relative superiority.[104] McRaven's theory, however, looks exclusively at 'direct action' special operations, known historically as 'raids'. Shock is a function of the combination of surprise, speed, purpose, security, repetition, and simplicity that allows relative superiority to be sustained or lost.[105] Harlan Ullman and James Wade Jr, who created the concept of 'shock and awe', explicitly use as an illustration the tactics of a specific type of special operation, hostage rescue, as an analogue to achieve wider battlefield shock.[106] This section explores a different type of shock related to special operations: the battlefield use of elite forces in a more direct manner. Under investigation here is whether these historical elites form the first links in the chain leading to contemporary SOF. In addition to these forms of direct battlefield shock, this section also examines the genesis and examples of more indirect forms by raiding an opponent's supplies, supporting infrastructure, or leadership.

Chinese writer Sun Tzu seemingly was the first to make a distinction between regular and more elite fielded forces. He identifies *ch'i*, or 'extraordinary', forces in the fifth section of *The Art of War* entitled, 'Shih', which Griffith translates as 'Energy' and Sawyer 'Strategic Military Power'.[107] More specifically, he says that, 'in battle, use the normal force to engage; use the extraordinary to win'.[108] Ordinary forces fix an opponent, in position, attention, and focus. Extraordinary forces manoeuvre into a position and strike on the opponent's flanks or rear.[109] They have their possible roots in the various teachings outlined in the Leopard book of T'ai Kung's *Six Secret Teachings*, in the form of various tactics such as 'explosive warfare', 'shaking invaders', the 'crow and cloud formation', and 'the few and the many'.[110] Some modern readers interpret 'extraordinary' or 'unorthodox' forces literally as standing special operations ones.[111] They do so in error, however, for philosophical as well as contextual reasons.

Philosophically, Sun Tzu's distinction between *cheng* and *ch'i* reflects the foundational dualism of Chinese Taoist thought. The two forces are inextricably connected; they are opposites but inseparable. Derek Yuen makes the point that the relationship of *cheng* to *ch'i* is never fixed. Quoting Sawyer, who translates *cheng* and *ch'i* as 'orthodox' and 'unorthodox', Yuen points out:

> The essence of the unorthodox and orthodox does not lie in seeking the unorthodox out of the orthodox. Rather, it is in reaching a realm where 'there are none

[104] William McRaven, *Spec Ops, Case Studies in Special Operations Warfare: Theory and Practice* (Novato, CA: Presidio Press, 1995), pp. 4–8.

[105] McRaven, *Spec Ops*, p. 11.

[106] Harlan Ullman and James Wade Jr, *Shock and Awe: Achieving Rapid Dominance* (Washington, DC: National Defense University Press, November 1996), p. 57.

[107] Griffith, *The Illustrated Art of War*, p. 135; Sawyer, *The Seven Military Classics of Ancient China*, p. 164.

[108] Griffith, *The Illustrated Art of War*, p. 137.

[109] Ibid., p. 136, fn. 5.

[110] Sawyer, *The Seven Military Classics of Ancient China*, pp. 89–95.

[111] An error most recently committed by Dr Nasir Mehmood and Usman Haider in 'Understanding India's "Surgical Strike" Special Operations and Pakistan's Response', *Journal of Security and Strategic Analyses*, 8:1 (Summer 2022): 52.

> that are not orthodox, none that are not unorthodox, so they cause the enemy never to be able to fathom them. Thus, with the orthodox they are victorious, with the unorthodox they are also victorious'. This is what is meant by 'being formless', and such logic can best be demonstrated by yin and yang. Hence the level of Heaven and Earth is about the attainment of formlessness or 'oneness' through fully grasping the logic of yin and yang.[112]

*Cheng* and *ch'i* are intertwined and inseparable; they are integrated in command and function, in the ruler and commander's mind, to the highest degree possible. In addition, their qualities and combinations are constantly changing to keep an opponent uncertain as to which force is which. In other words, *cheng* and *ch'i* should be constantly changing relative to one another in form and employment. In this way, capabilities can achieve levels of surprise and inflict battlefield shock against an opponent. Once an enemy intuits or comprehends *ch'i* forces, they lose their unorthodox nature and power; they become *cheng*.[113] Sun Pin, who also wrote a treatise on war and its art, is alleged to have used *ch'i* forces, in the form of several thousand men armed with crossbow, to ambush a Wei army at the battle of Ma Ling (341 BCE).[114] Any element or portion of the army could become *ch'i*, depending on the circumstances. Liu Po-wen provides the example from the Chien-an reign period (207 CE) of Ts'ao Ts'ao at Mount Pai-lang. Ts'ao Ts'ao, vastly outnumbered against invading barbarian tribes and with his own forces strung out over a long distance, used his small vanguard to charge the unorganized barbarians, routing them and seizing thousands of prisoners, turning *cheng* into *ch'i*.[115] The theme of combinations of forces inflicting shock is enduring, appearing in modern Chinese military writings including *Unrestricted Warfare* (1999) and *The Science of Strategy* (2015).[116]

Moving from the philosophical to the real, and to place *cheng* and *ch'i* in their appropriate context, the distinction between the forces reflected the recent revolutionary changes in warfare that occurred during China's Warring States period. The introduction of iron weapons and the crossbow, the expansion in numbers of chariots, and the fielding of mass armies created considerable command and control challenges but also opened up new possibilities in war fighting. In short, mass, manoeuvre, and missiles allowed a range of new options for engaging and destroying enemy forces, compared to how battle was conducted during the late Zhou Dynasty and early Warring States period, which was much more limited and highly stylized and ritualistic. These options and possibilities are what led Sun Tzu to characterize the relationship of *cheng* and *ch'i* in a number of elliptical and opaque but

112 Derek Yuen, *Deciphering Sun Tzu: How to Read 'The Art of War'* (Oxford: Oxford University Press, 2014), p. 35.

113 A point made by Griffith in a footnote on the subject. *The Illustrated Art of War*, p. 136, fn. 5.

114 Sun Pin, *The Art of Warfare*, trans. D.C. Lay and Roger Ames (New York: Ballantine, 1996), p. 26.

115 Liu Po-wen, *Unorthodox Strategies: 100 Lessons in the Art of War*, trans. Ralph Sawyer (Boulder, CO: Westview Press, 1996), pp. 26–27.

116 Qiao Liang and Wang Xiangsui, *Unrestricted Warfare*, trans. Foreign Broadcast Information Service (Beijing: People's Liberation Army Literature and Arts Publishing House, February 1999), pp. 101–104; Xiao Tianliang, ed., 战略学 (*The Science of Strategy*), trans. SAIC, U//FOUO (Beijing: National Defense University Press, 2015), p. 313.

nevertheless accurate ways, including 'their combinations are limitless' and 'mutually reproductive; their interaction as endless as that of interlocking rings. Who can determine where one ends and the other begins?'.[117]

Chinese dynasties, some before or during Sun Tzu's time and others much later, developed and fielded standing elite forces to inflict shock, to break and defeat opposing armies. These forces included specially selected, trained, and equipped infantry units such as the Wei Wuzou developed by Wu Qi during the spring and autumn eras of the Warring States period (770–476 BCE).[118] Others employed highly skilled cavalry formations, including the Tiger and Leopard Cavalry of Cao Cao of the Eastern Han Dynasty (25–220 CE) mentioned in the popular work *The Romance of the Three Kingdoms* and the Beiwei Army of the Song Dynasty (1127–1279 CE). These units reflected another trend beginning from Sun Tzu onward, the professionalization of military matters and specialization of combat means in a bid to replace the superstition and divination that guided decision making with sound military advice.[119] Superior morale and discipline, combined with unique equipment and employed by effective leaders, could defeat forces ten times their number. Another Ancient Chinese military classic, the writings of Wu-Tzu, outlines the methods by which an elite force can be raised and sustained, primarily by screening talented individuals and rewarding them, and their families, according to their accomplishments. Sawyer concludes the wise commander assigns men based on their aptitude to different 'combat specialities', and rewards accord to relative merit, which would result in a 'disciplined, spirited, strongly motivated force'.[120] To further improve morale, distinctive banners and uniforms also accompanied their unique monikers.

Other empires throughout the world also developed elite units of soldiers or cavalry with shared or similar attributes. The details and exploits of many, particularly those of ancient Persia, Greece, and Macedon, are already well known and frequently appear in histories of special operations. These ancient elites include 'the Immortals' of the Achaemenid Empire (550 BCE–330 BCE), the 'Sacred Band' of Thebes (fourth century BCE), responsible, in part, for defeating Sparta and known for the unique recruiting and military bonding process of couples of male soldiers, and Philip II and later Alexander of Macedon's 'Companion' horse cavalry.[121] Outside of Greece and the Near East, other elites include the Aztec or Triple Alliance 'jaguar knights' and 'eagle warriors', drawn from the ranks of commoners and carefully screened and selected for qualities and attributes, including discipline and stamina, at the *telpochcalli* or Aztec military academy.[122] Leebaart sees elites less in these Aztec formations than with the Spanish conquistadors, led by Hernán Cortés, whose few

[117] Griffith, *The Illustrated Art of War*, p. 138.

[118] https://daydaynews.cc/en/history/38,369.html.

[119] Robin D.S. Yates, 'New Light on Ancient Chinese Military Texts: Notes on Their Nature and Evolution, and the Development of Military Specialization in Warring States China', *T'oung Pao*, 74:4/5 (1988): 223.

[120] Sawyer, *The Seven Military Classics of Ancient China*, p. 204.

[121] Leebaert, *To Dare and to Conquer*, chapters 2 and 3; Lloyd, *Special Forces*, pp. 20–22.

[122] Claude-François Baudez and Nicolas Latsanopoulos, 'Political Structure, Military Training and Ideology at Chichen Itza', *Ancient Mesoamerica*, 21:1 (Spring 2010): 9.

numbers managed to offset Aztec superiority thanks to a number of attributes of modern special operators including bluff and 'intimidation' that 'becomes part of the leverage exerted by the commando.'[123] Aztec elite forces were not overcome so much by superior technology or Spanish elites tactically as they were strategically by 'the protracted nature of conquest, the role of indigenous allies, and the disparate ideologies of war and society between the Spanish and indigenous.'[124]

Elite shock units also performed a range of other duties. The eagle and jaguar warriors, for example, served in important social and political capacities in Aztec society, including capturing prisoners for ritual sacrifice to appease the sun god, as well as providing domestic security and stability. Other military elites protected emperors at home, during travel, as well as in wartime, and include the bodyguard of Inca emperor Tupac Yupanqui (1471–1493 CE), the Roman Praetorian Guard (27 BCE–312 CE), the Hittite *mesedi*, the Mongol *kheshig*, and perhaps one of the longest-serving ones, the Korean *naegeumwi* or Joseon royal guards (1407–1910 CE). Their numbers could range anywhere from a dozen in the case of the *mesedi* to several thousand.

Rather than direct use of elites in a pitched battle, whose outcomes were often uncertain and consequences dire, there was another, less direct option available to ancients: raiding. Azar Gat concludes that before the introduction of horses and firearms, among various tribes the 'generally preferred destructive raid' was the predominant if not normative form of war.[125] The reasons are simple enough. Raids occur quickly and often involve limited if any casualties on the side of the raider, due to their surprising nature. Uncivilized battle could be ritual in nature, reflecting 'the fear of being killed rather than killing.'[126] Within uncivilized groups, raiding could be conducted by necessity, as a means of survival. In pastoral or early agricultural societies, livestock seized in a raid could provide a continuous supply of resources, including offspring, meat, dairy, wool, hides, and others, not to mention enhancing the prestige and social standing of their new owner. Lawrence Keeley, for example, details persistent livestock raiding and its impact on culture and conflict among different tribes in southern Africa, including the Khoikhoi, San, Nguni, and Bantu.[127] Cattle raiding, along with the abduction of women, constituted the two primary reasons for going to war in the Greek heroic literary tradition, according to Peter Walcot.[128]

This tradition carried over into the conduct of Greek warfare and beyond. Ancient Greeks used raids used as a means to draw enemies out from behind fortifications, conduct 'tit-for-tat' reprisals, and punish.[129] Herodotus, for example, identifies

[123] Leebaert, *To Dare and to Conquer*, p. 159.

[124] Thomas Brinkerhoff, 'Reexamining the Lore of the "Archetypal Conquistador": Hernán Cortés and the Spanish Conquest of the Aztec Empire, 1519–1521', *The History Teacher*, 49:2 (February 2016): 169.

[125] Azar Gat, *War in Human Civilization* (Oxford: Oxford University Press, 2006), p. 121.

[126] Gat, *War in Human Civilization*, p. 123.

[127] Lawrence Keeley, *War before Civilization: The Myth of the Peaceful Savage* (Oxford: Oxford University Press, 1996), pp. 132–133.

[128] Peter Walcot, 'Cattle Raiding, Heroic Tradition, and Ritual: The Greek Evidence', *History of Religions*, 18:4 (May 1979): 350.

[129] Archer Jones, *The Art of War in the Western World* (New York: Barnes and Noble Press, 1997), pp. 55–56.

two different types of raids in Athens' war with Aegina (~505 BCE). Using their naval power, the Aeginetans raided Attica while the Athenians were engaged in their war with the Boeotians, doing 'great damage'.[130] In another example, this time along 'tit-for-tat' lines and to preserve a tribute agreement, the Athenians landed a raiding party in Aegina to seize olive wood statues that Aeginetans had seized from the Epidaurians, following the former's revolt from the latter.[131] In the Eastern Roman or Byzantine Empire (330 CE–1453 CE), writers such as Maurice thought the rationale behind raids or other surprise attacks was self-evident enough to constitute a maxim: 'assault the enemy without ourselves suffering any injury', whose consideration commanders should always give 'high priority'.[132] As late as the sixteenth century, communities of refugees and adventurers, collectively known as 'uskoks', pursued raiding in the target-rich waters of the Adriatic as their sole means of subsistence and created a political economy around it in the process.[133]

Raiding is often interpreted within Western military circles as a distraction or tactical diversion. For these reasons, raids are often overlooked in military and more specifically ancient history. Fortunately, some modern historians have recognized this shortcoming and explored the consequences extended raiding could have environmentally and militarily on various combatants, as in the American southwest in the nineteenth century. In this example, successive raids and reprisals by Apache bands and Mexicans denuded the environment, depopulated the area, weakened both combatants, and led to the development of long-standing grievances, retribution for which was pursued long past the point of rational gain.[134] One explanation downplaying raiding is cultural, reflecting 'the allure of battle' prevalent within the popular and military imagination, so aptly examined by Cathal Nolan.[135] Another is possibly linked to the frustrations that raids generate among powerful militaries, where soldiers must guard against, remain on the defensive, and deny them the opportunity to be defeated in a fair fight on their terms. There is an interesting corollary that appears within the special operations literature, the oversold notion of 'the great raid' that has independent, decisive consequences.[136] Equating special operations exclusively with raiding occurred in some of the earliest literature on the subject.[137]

[130] Strassler, ed., *The Landmark Herodotos*, 5.80, p. 401.

[131] Thomas Figueira, 'Herodotus on the Early Hostilities between Aegina and Athens', *The American Journal of Philology*, 106:1 (Spring 1985): 72–73.

[132] Maurice, *Maurice's Strategikon: Handbook of Byzantine Military Strategy*, trans. George Dennis (Philadelphia, PA: University of Pennsylvania Press, 1984), p. 93.

[133] Catherine Wendy Bracewell, *The Uskoks of Senj: Piracy, Banditry, and Holy War in the Sixteenth-Century Adriatic* (Ithaca, NY: Cornell University Press, 1992), pp. 89–117.

[134] Brian DeLay, *War of the Thousand Deserts: Indian Raids and the U.S.–Mexican War* (New Haven, CT: Yale University Press, 2008), pp. xv–xvii.

[135] Cathal Nolan, *The Allure of Battle: A History of How Wars Have Been Won and Lost* (Oxford: Oxford University Press, 2017).

[136] This notion is the departure point for James D. Kiras, *Special Operations and Strategy: From World War II to the War on Terrorism* (London: Routledge, 2006), pp. 1–15.

[137] 'Special operations ... have a long and rich tradition ... This volume provides a representative sampling of instances of raiding the *coup de main* from antiquity to the Cold War'. Arquilla, *From Troy to Entebbe*, p. ix.

To look upon episodic raiding as something other than war is a mistake, according to Steven LeBlanc. Some scholars, he argues, suggest raiding 'is not "real" warfare ... but is something more akin to game playing', which is rather ironic given the 'tournament-like form' that battles, as opposed to raids, assume in pre-civilized societies such as the Yanomamo living in the Amazon rainforest.[138] Le Blanc concludes that '[s]uch an approach or attitude, however, confuses the method of war with the results of war'.[139] Although the focus of his definition of strategy is unduly narrow, Archer Jones nevertheless intuits two types of military strategy, the first of which is raiding. To conquer or occupy, one must employ a persisting strategy. At times, ancient polities employed both, as evidenced by Viking raiding and later settling across western, northern, and eastern Europe (8th–11th centuries CE) and later on a massive scale by the Mongolian 'Golden Horde' (1227–1291 CE).[140] If the goals are more modest and diverse, and are underpinned by limiting costs of actions, then raiding strategies can achieve political, economic, and military objectives.[141]

The cornerstone of effective raiding, to ensure surprise and damage to the maximum extent possible in the shortest possible time and create the greatest shock impact, is detailed in the section on information gathering and reconnaissance. Maurice of Byzantium suggested it was the elemental aspect for one of his identified conditions for conducting raids into hostile territory.[142] Viking raiders identified and prioritized for attack soft targets, such as monasteries and undefended settlements, to maximize benefits, guarantee shock, and limit costs.[143]

The effective combination of tactical intelligence and raiding could generate strategic effects in the ancient world. One example from ancient history stands out in this regard: the Roman raid on the camp of Syphax. Towards the end of the Second Punic War (218–201 BCE), and in a bid to force Hannibal to decamp from Europe and return to defend Carthage, Roman proconsul Publius Cornelius Scipio Africanus invaded northwest Africa and unsuccessfully attempted to besiege and seize Utica (204 BCE). The siege was broken by the arrival of the army of Carthage, under the command of Hasdrubal Gisco, and their allies, the Masaesyli, led by Syphax. Scipio remained in Africa and continued to threaten Utica, but the various armies camped for the winter and waited for spring before taking any major action.[144] Given the

[138] Gat, *War in Human Civilization*, p. 122.

[139] Lawrence Keeley, *Constant Battles: The Myth of the Peaceful, Noble Savage* (New York: St Martin's Press, 2003), p. 57.

[140] For details of Viking approaches and activities drawn from a wide range of sources, see F. Donald Logan, *The Vikings in History*, 3rd ed. (London: Routledge, 2005). An accessible work on the Mongols remains James Chambers, *The Devil's Horsemen: The Mongol Invasion of Europe* (New York: Atheneum, 1985).

[141] Archer Jones, *The Art of War in the Western World*, p. 55.

[142] Maurice, *Maurice's Strategikon*, p. 96.

[143] Neil Price, *Children of Ash and Elm: A History of the Vikings* (New York: Basic Books, 2020), pp. 304–305; 308; 346–351.

[144] Theodore Ayrault Dodge, *Hannibal: The History of the Art of War among the Carthaginians and the Romans, Down to the Battle of Pydna, 168 BC, with a Detailed Account of the Second Punic War* (London: Greenhill Books, 1994), pp. 579–585. More recent treatments include J.F. Lazenby, *Hannibal's War: A Military History of the Second Punic War* (Norman, OK: University of Oklahoma Press, 1978); Yann Le Bohec, *Histoire militaire des guerres puniques, 264–146 av. J.C.* (Paris: Editions Tallandier, 2014).

superiority of Carthaginian numbers, the length of the war, and the uncertainty Hannibal created through his actions, Syphax persuaded Scipio to begin negotiations for peace. Taking advantage of a truce to negotiate, Scipio selected a handful of legionnaires to visit Syphax's and Hasdrubal's camps, located in the hills some 6 miles south of Utica.[145] While those negotiating remained in armour and uniform, others accompanying them were disguised as servants or slaves. The servants wandered the camps at will, marking out distances and noting other information such as camp security and the construction of dwellings. Scipio ensured he sent different soldiers as the negotiations persisted, to collect different types of information and prevent the Carthaginian's suspicions from being aroused. After breaking off negotiations in the early spring, Scipio appeared to be moving to resume the siege of Utica.

This deception, however, masked his plan to attack both Hasdrubal's and Syphax's camps. The plan for the raids not only took account of the lax security measures at both camps but also the susceptibility of structures within them to being set afire. Theodore Ayrault Dodge concludes both raids had their intended effect. Fires created panic in both camps, and large numbers of the fleeing Carthaginians and Masaesyli were subsequently killed.[146] The raid, which has no name or date in the surviving sources of Livy and Frontinus, nevertheless evened the numerical odds in Scipio's favour and set the conditions within the Carthaginian Army for its subsequent defeat in the Battle of the Great Plains (203 BCE). Without the detailed survey of the camps conducted beforehand, it is questionable whether the raid would have been as effective as it was tactically and strategically.

The targets of ancient raids, besides those against leadership ones, fall into three broad categories. The first is soft military targets supporting fielded armies, such as baggage or encampments such as Syphax's. The second is supporting economic and military infrastructure, which includes essential crops, granaries and mills, livestock, and armouries. Although the common perception of the Middle Ages (1100–1550 CE) is dominated by knights strictly adhering to a code of chivalry asking for and giving quarter and placing honour and virtue above all other considerations, reflected in the idealized nature and aspirations of the authors in texts that survive the era, the realities of warfare during this time were different.[147] Harari, for example, explores in considerable detail the various types of raids conducted during this period in two of the three categories above. For economic and military infrastructure, he identifies primarily the successful or unsuccessful attempts at destruction, including bridges (Po bridge [1544]; Kelso bridge [1333]) as well as dams (Watten on the river Aa [1347]) and dikes (near Calais [1438]). He notes, however, that the scattered and distributed nature of industry, combined with the robust construction of much of the infrastructure, limited raids along these lines during this era.[148] Norman Housley

[145] B.H. Liddell Hart, *Scipio Africanus: Greater Than Napoleon* (London: Greenhill Books, 1992), pp. 126–127.

[146] Theodore Ayrault Dodge, *Hannibal*, pp. 589–590.

[147] The work by Christine de Pizan, *The Book of Deeds of Arms and of Chivalry*, is one such representative work.

[148] Harari, *Special Operations in the Age of Chivalry*, pp. 18–19.

suggests medieval raiding was 'a key feature of strategy' for two reasons, one that hurt the enemy and the other that enriched the raider, and presumably lightened the cost of campaigning on the sovereign.[149]

These targets could take on additional importance if one belligerent adopted a scorched-earth policy in the face of an invader. During Imperial Hapsburg King Charles V's invasion of Provence in 1536, marking the third war he fought with France, the French king François ordered the destruction of any sources of supply in the region as they retreated. A number of mills, however, remained standing. Several were destroyed in subsequent French raids but one in particular, in the town of Auriol, was still intact. Blaise de Monluc developed a plan to take 150 men, led by local guides, to attack the Auriol mill some 25 kilometres away. The plan was refined after De Monluc received specific information about the layout and garrisoning of the town. After evading Imperialist outposts and garrisons, the raiders split into two forces, a blocking and an assault force. The blocking force would prevent reinforcements from reaching the mill across town while the assault force, under Monluc's direct command, would seize and destroy it. Operating in darkness, Monluc and his men forced their way into the mill, dismantled some of its key components, and burned the rest, somehow managing to avoid Imperial cavalry sent to hunt them down. Although not decisive in and of itself, the raid on the mill of Auriol contributed to the supply difficulties that increasingly crippled the Imperial forces and eventually forced their retreat.[150] The combination of raiding and a Fabian strategy allowed the French, in Harari's estimation, to 'win a tremendous victory without fighting a single engagement'.[151] Other considerations weighing on Charles V's mind, such as wars on other fronts against the Turks and in Piedmont, however, must be factored into this outcome.[152]

The third type of target were symbolic targets, such as the Epidaurian olive wood statues seized by the Aeginetans mentioned previously. Symbolic targets could span everything from castles and keeps at one end to religious icons and relics at the other. Although Harari raises some intriguing possibilities, raids at the lower end of this scale seem rather limited.[153] The reasons so few examples in the category exist have to do with the more honourable, lucrative, shocking, and potentially decisive form of symbolic target: sovereigns, nobility, or other leaders.

Raids against symbolic people could take two primary forms recognizable to modern special operators: targeted killing or hostage rescue/capture. The reasons driving each are numerous and context specific; scores of works have been written particularly about the legality and morality of the former, impeded at times by terminology either antiseptic or value-laden, such as assassination. In addition, various periods of time are associated with domestic political intrigue and murder, as well as

[149] Norman Housley, 'European Warfare c.1200–1320', in Maurice Keen, ed., *Medieval Warfare: A History* (Oxford: Oxford University Press, 1999), p. 133.

[150] Harari, *Special Operations in the Age of Chivalry*, pp. 171–177.

[151] Ibid., p. 179.

[152] Charles Oman, *The History of the Art of War in the Sixteenth Century* (London: Greenhill Books, 1991), p. 213.

[153] Harari, *Special Operations in the Age of Chivalry*, pp. 32–34.

unscrupulous and often depraved behaviour, including the Roman Julio-Claudian dynasty of emperors as well as various houses of Renaissance Italy, the most notorious of which was the House of Borgia. Under scrutiny here, however, is the question of the instrumentality of these forms in the ancient world, as a means of achieving strategic and political effects and the precedents they established for contemporary special operations.

Setting aside religious texts and literary examples, one of the earliest identifiable groups employing targeted killing along the lines of special operations were the *sicarii* during the period of Roman occupation of Judea (63 BCE–13 CE). The specific details regarding their activities are difficult to tease out for a variety of reasons, not the least of which is our reliance on a single account by Josephus for the period. Other challenges include uneven scholarship that sometimes lumps together the *sicarii* with the Zealot movement, in part to the suicide of their members in Jerusalem who, fleeing a counterattack by other Jewish elements within the city, retreated to Masada. There they achieved notoriety by preferring death at their own hands rather than surrendering to a long and painful death at the hands of the Romans.[154]

The *sicarii* received their moniker based on their preferred instrument: a short, concealable sword or dagger, or *sicarius* in Latin. The individuals they selected and methods of dispatching they employed were both precise and calculated. They specifically targeted fellow Jews who chose to work with Roman governors in senior administrative positions or served in other official capacities. Other targets included the rural gentry and anyone from the local population who benefitted significantly from this association materially. Their initial wave of attacks, beginning with the assassination of Jonathan the High Priest, were conducted in a seemingly contradictory manner: public and clandestine at the same time. Moving within striking range of their target and in a location frequented by the public going about their secular or religious business, one or more *sicarii* would kill their victim and in the ensuing panic and chaos blend into the crowd and escape.[155] In a later phase, they attacked the owners of rural estates, killing them and setting their holdings alight.[156] Max Boot labels the *sicarii* as 'the antecedents of terrorism' to demonstrate 'that Muslims have no monopoly on the use of terror'.[157] Richard Horsely offers less dramatically that they fell into a more general pattern of resistance to occupation. Through their actions, the *sicarii* drew attention to the fragility of Roman rule, sought to break down cohesion both among Jewish ruling leaders and within society more broadly, and provoked wider actions, in terms of escalation of violence as well as forcing the population to take sides.[158] In the lead-up to the Jewish Revolt, the *sicarii* played a dramatic and

[154] Claims of which Morton Smith outlines and subsequently demolishes convincingly in 'Zealots and Sicarii: Their Origins and Relations', *Harvard Theological Review*, 64:1 (January 1971): 1–19.

[155] Richard Horsely, 'The Sicarii: Ancient Jewish "Terrorists"', *The Journal of Religion*, 59:4 (October 1979): 435.

[156] Flavius Josephus, *The Works of Flavius Josephus*, vol. II, trans. William Whiston (Philadelphia, PA: J. B. Lippencott & Co., 1878), p. 120.

[157] Max Boot, *Invisible Armies: An Epic History of Guerrilla Warfare from Ancient Times to the Present* (New York: Liveright, 2013), pp. 209–209.

[158] Horsely, 'The Sicarii', pp. 450–452.

outsized political role, which makes their subsequent passivity during it all the more puzzling. After Masada, *sicarii* activity was limited to efforts to stimulate revolt in neighbouring Egypt and Cyrene by repeating the type of actions they conducted initially in Jerusalem: public assassinations.[159]

One parallel that can be drawn to the *sicarii* is the 'Order of Assassins'. Even though they existed a millennia later, there are striking similarities between them. Like the *sicarii*, the Assassins held fundamentalist beliefs. In addition, they were just as 'patient and clever in carrying out their plots' and most frequently used concealed knives or daggers to dispatch their targets, singly or in groups. Their targets served a broader purpose: reprisals for wrongs committed against them or to eliminate and deter those who sought to conquer their territory. Some of their later assassinations, after the deaths of both their founder, Hasan-i Sabbah and Rashid al-Din Sinan, labelled 'the Old Man of the Mountain', included some of the first recorded contract killings. The vast majority of their assassinations, however, were forms of instrumental violence that served defensive or responsive purposes using the minimum amount of force possible. They differed from the *sicarii* in their longevity, operating across almost two centuries (1090–1273 CE), as well as success. The actions of the Assassins helped expand a Shi'a Ismaili state for the Nizari sect across Persia and Syria.[160] Their methods, although similar on the surface, diverged from the *sicarii*. The Assassins also hid in plain sight but often adopted elaborate costumes and covers, including posing as dervishes or Christians, to both gain access to and trust of their targets, who could be difficult to access. The individual Assassins, or *fida'i* (*fedayeen*), were willing to give their lives in the successful conduct of their assaults, which has led to subsequent creations of a series of myths surrounding the source of the motivation and indoctrination, further perpetuated and embellished in the eighteenth and nineteenth centuries.[161] Their targets included Nizem al-Mulik (Seljuq vizier and ruler, 1092), Conrad of Montferrat (the king of Jerusalem, 1192), Philip of Montfort (lord of Tyre, 1270) and Lord Edward (duke of Gascony, 1271).[162]

If the Trojan Horse is considered by some as the embodiment of an economy of force operation militarily, those of the Assassins should be considered economy of force politically and strategically. Their calculated killings, of between fifty and seventy-five religious or secular leaders, had a profound effect that contributed to both a mystique and widespread fear of Ismailis among Sunnis, Shi'as, Crusading Christians, and Mongols.[163] In the words of one author, 'The panic concerning them recorded in [contemporary] sources is tangible.'[164] Max Boot cites one manifestation

[159] Ibid., p. 456.

[160] Farhad Daftary, *The Assassin Legends: Myths of the Isma'ilis* (London: I.B. Taurus, 1994), pp. 1–7.

[161] Fahrad Daftary, 'The "Order of the Assassins": J. von Hammer and the Orientalist Misrepresentations of the Nizari Ismaelis', *Iranian Studies*, 39:1 (March 2006): 74–77.

[162] Bernard Lewis compiles an impressive list of Assassin targets, dates, and results in *The Assassins: A Radical Sect in Islam* (New York: Basic Books, 2003), pp. 125–140.

[163] James Wasserman, *The Templars and Assassins: The Militia of Heaven* (New York: Simon & Schuster, 2001), p. 109.

[164] David Cook, 'Were the Ismāʿīlī Assassins the First Suicide Attackers? An Examination of Their Recorded Assassinations', in Paul Cobb, ed., *The Lineaments of Islam: Essays in Honor of Fred McGraw Donner* (Leiden: Brill, 2012), p. 97.

of this panic, at least for those who were potential targets: 'No commander or officer dared to leave his house unprotected; they wore armor under the clothes.'[165]

The targeted killing of leaders was a fact of life in the ancient and medieval world. There are at least fourteen targeted or political reprisal or honour killings listed in Herodotus' *Histories* and seven in Thucydides' *Peloponnesian Wars.*[166] Harari surmises that during the Middle Ages, 'assassinating enemy leaders was probably the most cost-effective way of waging war in the age of chivalry', but despite the range of examples he provides, from Count Geoffrey Martel of Anjou (1106) to William the Silent (1584), a norm developed that generally moderated behaviour against it.[167] Maintaining this norm led to a relative degree of stability in western Europe, at least up until the 'Wars of Religion' began in earnest, and stood in contrast to its prevalent use and subsequent upheavals in 'medieval Middle East and Renaissance'.[168] Fear of reprisal in kind and moral qualms may have limited the use of assassination as a tool in medieval Europe. Those fears and qualms, however, appeared non-existent when it came to profiting from another form of targeting leadership: hostage taking and rescue.

The taking and keeping of hostages has a long history. Hostages were a cornerstone of ancient diplomacy. As a practice, the different Chinese dynasties took or exchanged hostages for 2,300 years.[169] They served as a guarantor of whatever terms were agreed to between polities, or between emperor or king and their tributaries. There is no single history of hostage taking but the process is most often attributed first to the Ancient Egyptians.[170] The tradition of voluntary or tributary hostages was continued by the Ancient Greeks and Romans. Should one violate the agreement, the hostages could be executed. To increase their bargaining value, hostages obtained were nobles, royals, and relatives. In Ancient China, sons were especially valued as hostages.[171] In Ancient Rome, hostages were treated as guests according to custom; the current term has its roots in the Latin '*hospes*'. Hostages could be male or female, child or adult and could number from individuals to several hundreds.[172] More recently, Joel Allen opines hostages in Ancient Rome could fulfil a number of roles, from social and family structure to the cultural and sexual.[173] Perhaps the most famous ancient hostage was the Athenian Themistocles, leader of the

[165] Boot, *Invisible Armies*, p. 207.

[166] Strassler, ed., *The Landmark Herodotus*, p. 912; Strassler, ed., *The Landmark Thucydides*, p. 644.

[167] Harari, *Special Operations in the Age of Chivalry*, pp. 25–27.

[168] Ibid., p. 29.

[169] Lien-Shang Yang, 'Hostages in Chinese History', *Harvard Journal of Asiatic Studies* 15:3/4 (December 1952): 507.

[170] Irene Hermann and Daniel Palmieri, 'A Haunting Figure: The Hostage through the Ages', *International Review of the Red Cross*, 87:857 (March 2005): 136–137.

[171] Yang, 'Hostages in Ancient China', p. 508.

[172] The sex, age, number, and social standing of hostages taken in Ancient Rome have been compiled and analysed in Cheryl Walker, 'Hostages in Republican Rome', PhD dissertation, 1980, published online through Creative Commons, pp. 44–67, https://chs.harvard.edu/read/walker-cheryl-hostages-in-republican-rome.

[173] Joel Allen, *Hostages and Hostage-Taking in the Roman Empire* (Cambridge: Cambridge University Press, 2006).

victorious fleet that defeated a Persian one at the Battle of Salamis (480 BCE).[174] A half-century later, prior to the beginning of the Peloponnesian War, Themistocles offered himself as a hostage to the Spartans so they could send a delegation to Athens to observe the status of fortifications they were building. According to Thucydides and later Polyaenus, Themistocles planned the hostage exchange as a ruse all along, to buy sufficient time for the Athenians to complete their walls around city and port of Piraeus, which suggests hostage taking could serve honourable or dishonourable purposes.[175] In addition to serving as an incentive to uphold an agreement, hostages could also be acculturated in the values and traditions of those retaining them, as the Egyptians and the Romans did, as well as increasing the prestige and fame of the leader who negotiated for their release.[176] The Greeks and later Romans distinguished another category of hostages, involuntary or seized ones, where their value in bargaining was profit as opposed to upholding a contract.[177] It is this second category of hostage taking that prompted actions related to modern special operations.

Hostage taking expanded during the medieval period in the West and East for several reasons. A normal form of hostage diplomacy continued throughout. Adam Kosto divides these into two categories: true or conditional hostageship.[178] The nature of feudal governance, with its distributed power and authority consolidated in the hands of a few, meant the relative value of individual rulers, princes, and their retainers went up considerably. In short, monarchs or relatives were willing and able to pay the considerable sums for their release. While slaying a noble in battle brought prestige and honour and some compensation, capturing them paid. Harari suggests that ransoms, which, depending on the social rank of the hostage, were 'often larger than a kingdom's yearly income', and honour dictated every effort should be made to secure their return.[179] Hostage taking could occur away from the battlefield or military campaign domestically, as a means of securing more favourable transition of power or to seize territory. Hostage rescues, therefore, could save costs or restore political fortunes. The numbers of hostages could be staggering depending on the stability and level of intrigue within a kingdom, especially if smooth transitions of power did not occur. For example, in the house of Valois of Burgundy in France from 1407 to 1483, there were no fewer than eighteen abduction or assassination attempts, real or imagined.[180] The operations could take a variety of forms. Some, such as the rescue of King Charles the Bad of Navarre in 1357, resemble a stealthy infiltration raid involving more than two dozen knights.[181] Others were more indirect 'shock' raids or involved treachery and betrayal. Unfortunately, there is no systemic study of hostage rescue successes versus attempts in the Middle Ages.

[174] Strassler, ed., *The Landmark Herodotus*, 8.80–8.96, pp. 633–641.

[175] Ibid., 1.90–1.93, pp. 49–51; Polyaenus, *Strategems of War*, trans. R. Shepherd (Chicago, IL: Ares Publishers Inc., 1974), p. 28.

[176] Simon Thijs, 'Hostages of Rome', *Athens Journal of History*, 2:3 (July 2016): 205–208.

[177] Herrmann and Palmieri, 'A Haunting Figure', p. 140.

[178] Adam Kosto, *Hostages in the Middle Ages* (Oxford: Oxford University Press, 2012), pp. 24–33.

[179] Harari, *Special Operations in the Age of Chivalry*, p. 30.

[180] Ibid., pp. 125–159.

[181] Ibid., p. 31.

Even unsuccessful hostage rescue attempts, however, could have significant political consequences. The Frankish King Baldwin II, king of Jerusalem and regent of the Crusader territories of Antioch and Edessa in modern-day Syria and Iraq, was ironically captured in the midst of a raid to free his cousin, Joscelin of Courtenay. Both Baldwin and Joscelin were held hostage at the fortress of Khartpert in 1123 CE. As evidence of the prolific nature of hostage taking during this period, both had previously been captured at the Battle of Harran in 1104 BCE and subsequently ransomed.[182] A group of Armenians garrisoning Behesni volunteered for a seemingly suicidal mission against a mountain fortress. Relying on local information from other Armenians and taking advantage of complacency given how far Khartpert was away from Frank-held territory, fifty Armenians of Behesni entered the fortress by means of a ruse and seized it, killing several guards in the process.[183] In a common flaw identified with a number of high-profile special operations failures throughout the years, more thought was given to the ruse, however, than to getting the king and other prisoners back to Frankish territory. Khartpert was invested in a siege by Muslim forces diverted away from an upcoming planned campaign and retaken; Baldwin was eventually ransomed the following year and safely returned. Harari suggests the rescue attempt 'seized the imagination of the Franks and gave an invaluable boost to their morale', as well as diverting Muslim attention away from their planned siege of Kafartab.[184] The attempted hostage rescue at Khartpert, while unsuccessful, illustrates two more of Colin Gray's master claims for the strategic utility of special operations: raising (friendly or local) morale and reassuring a (domestic) constituency.[185]

Except for tribal raiders, the *sicarii*, and Order of Assassins, most direct and indirect shock actions were conducted in an ad hoc manner. The absence historically during the ancient and medieval period of a truly professional class capable of raiding, carefully selected and screened for a certain set of qualities and characteristics, relates most likely to the relationship of citizens to the military and the polity. Early agricultural societies had few resources to spare for permanent armies, much less specialized units, and raiding was driven by quasi-private economic incentives, taking advantage of the otherwise fallow season between planting and harvesting. Few Greek city-states, with some notable exceptions, maintained standing armies and even these tended to be small and comprised the elite formations for inflicting direct shock on the battlefield. Much the same is true of China's Warring States and early dynasty periods. Improved agriculture and expanding territory increased the size of armies and navies, allowing for the development of specializations within them such as artillerists and engineers. Mass, training, and discipline, combined with effective leadership, still determined the outcomes on most battlefields. Ancient writers were

[182] Christopher Tyerman, *God's War: A New History of the Crusades* (Cambridge, MA: Belknap Press, 2006), pp. 186–187; Steven Runciman, *A History of the Crusades, Volume II: The Kingdom of Jerusalem and the Frankish East, 1100–1187* (London: Folio Society, 1994), pp. 35–36.

[183] Runciman, *A History of the Crusades, Vol. II*, p. 132; Harari, *Special Operations in the Age of Chivalry*, pp. 78–80.

[184] Harari, *Special Operations in the Age of Chivalry*, p. 88.

[185] Gray, *Explorations in Strategy*, pp. 175; 177.

not blind to the potential of a specific class of indirect raiders. There is a noteworthy exception to this general rule. The rulers of several empires historically coopted some particularly effective warlike tribes with whom they had clashed and employed them for their skills as elite forces, often for light infantry or skirmishers, including Balearic slingers in the ancient period and Tiroleans in the Austro-Hungarian Empire.

Kautilya envisioned his kingdom needing a special cadre of indirect shock forces to conduct special operations. The rationale for them flows logically from his discussion of spies and covert action. He devotes most of this space to guaranteeing the stability, security, and survival of the state by identifying the range of various potential internal threats. It follows, however, that one should also prepare to have at hand the means to erode the stability, security, and survival of competitors. Kautilya saves the best for last in his category of spies. He labels them *tiksna*, which interpreters have translated either as 'assassins' or the more evocative 'fiery spies'. *Tiksna* are characterized by their potential range of acts of daring and creation of mayhem.

The attention Kautilya pays to this specific category of spy, far more than any other, suggests that special operations conducted in the most audacious and unorthodox manner are appealing by virtue of these attributes, or qualities and characteristics, of those who conduct them. Summarizing these with the term 'firebrand', *tiksna* can appear anywhere and everywhere, using snakes, daggers, poison, fire, or misinformation to kill, maim, or sow chaos and confusion. A modern Indian author characterizes *tiksna* as 'dare-devils of the first rank ... employed to carry out raids on the enemy's rear, clandestine operations behind the enemy lines like letting off cruel beasts, poisonous snakes, flooding the enemy defensive areas by busting dams or rivers; carrying out lightning raids on enemy capitals, and laying ambuscades'.[186] Kautilya has good reason for wanting to find and reward *tiksna*, based on the range of tasks they can perform and their potential strategic impact, particularly if they are used pre-emptively or preventatively: 'A single assassin can achieve, with weapons, fire or poison, more than a fully mobilized army.'[187] The seemingly endless potential of such spies, appearing where a competitor or usurper least expects them, is what captures Kautilya's imagination. Much the same is true of modern special operations, where their mystique can, at times, run well ahead of their capability. In addition, Kautilya's emphasis on 'fiery spies' raises an enduring question central to special operations: how does one manage and control such unorthodox personalities to accomplish the outcomes desired by the leader, ruler, or commander?

## Military assistance and advising

Military assistance and advising missions do not appear often in ancient or medieval history. While some of the reasons relate to the fractured nature of armed forces and

[186] Surinker Kumar Bhakari, *Indian Warfare* (New Delhi: Munshiram Manoharial Publishers, 1981), p. 88.

[187] Kautilya, *The Arthasastra*, 9.6.54,55, p. 498.

states, another dimension explains why they occurred so infrequently. Such missions have only taken place historically in times of competition between two major states whose power and influence rests, in part, in the alliances they maintain with weaker states which possess desirable or scarce resources or geographic locations. For much of ancient history, either an empire has dominated large swathes of territory or none has. In the latter case, power and influence has been distributed across a range of different actors and is limited in scope or influence. For dominant empires, sending direct military assistance, in the form of large fleets and armies, could tip the balance in local conflicts.

Beyond the broader geopolitical conditions working against military assistance and advising, there is a simpler explanation. As polities grew in size and wealth, and wars grew increasingly frequent, it was much easier for their rulers to hire expertise in the form of mercenaries. Rather than seeking military assistance, city-states or wealthy groups of Greek exiles hired mercenaries to serve their cause frequently, as numerous mentions in Thucydides' history suggest.[188] One of the earliest first-hand accounts we have of a mercenary is provided by Xenophon in his work the *Anabasis*. The upheavals and nearly constant warfare of the Peloponnesian Wars led to a class of military experts willing to fight for pay. In some cases, such as Xenophon and the Spartan Clearchus, banishment from their city-state for a range of offences provided the motivation. Others were in 'political difficulties at home with the opposing party'.[189] Opportunity for Xenophon and his 10,000 fellow Greeks occurred in the form of an Achaemenid campaign led by Cyrus the Younger to overthrow Artaxerxes II. Working through intermediaries, Cyrus paid leaders including Clearchus to recruit an army through their social networks. The subsequent fighting retreat of 'the Ten Thousand' to evacuation and safety after their paymaster and sponsor's death at the Battle of Cunaxa (401 BCE; near Baghdad) is the stuff of military legend.

Hiring mercenaries, then as now, is a proposition fraught with perils. They include mutiny of the paid forces, which Clearchus had to address early during the march, up to seizing power of the polity for themselves, which occurred on several noteworthy occasions in Renaissance Italy two millennia later.[190] Machiavelli suggests mercenaries are 'easier to corrupt' the citizens of a polity and warns against employing them in *The Art of War*.[191] If specialized skill was not to hand or could not be cultivated, it could always be rented through mercenaries who made a name for themselves, including the Apiru mentioned in Egyptian and Hittite writings, the Swedish Varangians in Byzantium, the White Company in Renaissance Italy, the Italian condottieri, and during the last century, the American Volunteer Group in China, better known as 'the Flying Tigers'.

[188] Strassler, ed., *The Landmark Thucydides*, p. 677.

[189] Xenophon, *The Persian Expedition*, trans. Rex Warner (Baltimore, MD: Penguin Books, 1969), p. 19.

[190] Xenophon, *The Persian Expedition*, pp. 26–30; William Caferro, 'Warfare and Economy in Renaissance Italy, 1350–1450', *The Journal of Interdisciplinary History*, 39:2 (August 2008): 168.

[191] Machiavelli, *The Art of War*, trans. Neville, First Book, https://oll.libertyfund.org/title/machiavelli-the-art-of-war-neville-trans.

One of the exceptions to the geopolitical conditions mentioned above was the Peloponnesian Wars between Athens and Greece. As with most activities outlined in this chapter, this one was ad hoc in nature. Its context was the ultimately ill-fated Athenian invasion of Sicily (415–413 BCE), whose outcome helped determine the end of the war. For this reason, this specific example is not only clear but also an exceptionally detailed one. The Spartan mission to assist and advise the forces of the city-state of Syracuse, on the eastern coast of Sicily, was prompted by a speech delivered to the Spartan assembly by none other than Alcibiades. Alcibiades had convinced the Athenian assembly the previous year to invade Sicily, promising great riches and few obstacles, but ultimately fled Athens after some of his political rivals brought charges against him. Convinced by Alcibiades of the threat the Sicilian Expedition posed, the Spartan assembly nominated Gylippus to sail to Syracuse with as many Corinthian ships and reinforcements as he could. His arrival, as well as his combined force of cavalry, light troops, marines, and hoplites, was instrumental in shifting the momentum of the campaign away from the invading Athenians and their allies.

Gylippus and his force bolstered Syracusan morale at a crucial juncture, helped them improve their defences and formulate a plan to trap the besieging Athenians and regain the initiative. In a theme that appears repeatedly in history, the commander sent to advise soon took control over management of the preparation and conduct of a number of battles that followed.[192] In addition to leading forces in the field, Gylippus was instrumental in rallying other city-states on the island of Sicily to provide reinforcements to Syracuse, a not insubstantial task given existing local rivalries. As many military advisors have experienced since 413 BCE, the proxy forces with whom he was working had ideas of their own. Despite negotiating the surrender of the Athenian commanders Nicias and Demosthenes, and guaranteeing their safety, Gylippus was unable to control the Syracusans and prevent them from executing both as well as most of the remaining Athenian personnel.[193]

While other military advising and assistance missions doubtless existed during the Peloponnesian Wars, like Greek intelligence they tend to hide behind vague language. For example, having detailed the events of the Corcyraen revolution in 427 BCE, Thucydides summarizes that 'the whole Hellenic world was convulsed; struggles being everywhere made by the popular leaders to bring in the Athenians, and by the oligarchs to introduce the Spartans'.[194] Most other cases within the Peloponnesian Wars consisted of military assistance to restore factions, suggesting perhaps the utility of future study of what is codified in US SOF doctrine as unconventional warfare, or within intelligence circles as regime change. The line between security assistance and specialized military advice and assistance, however, is difficult to determine. It continues to remain blurry, as evidenced between four different terms used by the Department of Defense to describe much the same activity: security assistance,

[192] Strassler, ed., *The Landmark Thucydides*, pp. 427–442; 445–478.
[193] Ibid., pp. 477–478.
[194] Ibid., 3.82, p. 199.

security force assistance, security cooperation, and foreign internal defence.[195] For these reasons, among others, a detailed exploration of ancient and medieval military advice and assistance lies outside of the scope of this study. Hopefully other scholars will identify examples of military advice and assistance, as well as unconventional warfare, here, in Ancient China, and elsewhere historically.

## Asymmetric response

When polities clash militarily, there are circumstances and situations that arise that cannot be foreseen. Often, they create a need to adapt to and develop innovative solutions. Modern special operations were developed as one such adaptation. In the ancient and medieval world, numerous such adaptations and innovations occurred. Some of its logic is reflected in the modern aphorism penned by Conrad Crane: 'There are two ways of waging war, asymmetric and stupid.'[196] An example of such an asymmetric adaptation or innovation includes Mehmut the Conqueror's novel solution to engineer a way to move ships across land from the Bosporus to the Golden Horn during the siege of Constantinople (1453 BCE). From this location his fleet could sink Byzantine ships and cut off their maritime supplies, cannonade the fortified walls of the city, and build pontoon bridges to ferry troops across the Golden Horn.[197] A month later, after a costly siege, Constantinople fell for the final time.

Not all battlefield innovations occurred on such a grand scale. Others developed historically bear characteristics of special operations due to their qualities and characteristics. One such approach was the use of combat swimmers, often to overcome formidable defences or tactical obstacles. Such obstacles could be natural but surmounting them required little more than pluck or desperation. For example, during an early engagement during the Peloponnesian Wars, a Spartan force on the island of Sphacteria in the Bay of Navarino found itself isolated. The supplies they needed to sustain their defences were delivered, according to Leebaert, by some of the first 'combat swimmers' moving from shore to island and back.[198] The supplies got through, but the Spartan defences were undone by perhaps a different asymmetric special operation. A Messenian commander, leading 'some archers and light troops', managed to work their way through 'where the Spartans, trusting the strength of the ground, kept no guard', outflanking them and seizing the high ground behind the Spartan defensive position.[199] At least one scholar has suggested the tactics used by Thracian peltasts here started a subsequent trend in Greece resembling guerrilla

[195] Taylor White, 'Security Cooperation: How It All Fits', *Joint Forces Quarterly*, 72 (1st Quarter 2014): 106–108, https://ndupress.ndu.edu/Portals/68/Documents/jfq/jfq-72/jfq-72_106-108_White.pdf? ver=2014-03-13-152,407-220.

[196] Conrad Crane, 'The Lure of Strike', *Parameters*, 43:2 (2013): 5.

[197] Franz Basinger, *Mehmed the Conquerer and His Time*, ed. William Hickman, trans. Ralph Manheim (Princeton, NJ: Princeton University Press, 1978), pp. 88–89.

[198] Leebaert, *To Dare and to Conquer*, p. 56.

[199] Strassler, ed., *The Landmark Thucydides*, 4.36, p. 242.

warfare, although others dispute this.[200] The depleted and demoralized Spartans, caught with threat in front of and behind them, retreated from their positions on the island to a fort and ultimately surrendered after a seventy-two-day siege and blockade (425 BCE).[201] Spartan surrender, previously thought inconceivable, caused shock waves that rippled throughout the city-states of Greece.

More frequently, special operations were needed to defeat manmade obstacles, including barriers, city walls, and fortifications or others. During the Sicilian Expedition and the siege of Syracuse (413 BCE), and in order to protect their ships from Athenian raids, the Syracusans drove pilings 'in front of the old docks'.[202] The Athenians responded by constructing a fortress on a raft from which they sent combat swimmers to saw down the pilings underwater repeatedly while under missile fire.[203] In the words of Thucydides, during the siege all sides sought advantages in skirmishing with each other 'and every artifice of war was successfully practiced'.[204]

The preceding examples, while responsive, are not particularly asymmetric in nature. These uses of combat swimmers necessarily drove the combatants to experiment and adapt. Although the term 'asymmetry' can be problematic for reasons already well described elsewhere, it is useful when trying to capture an offsetting approach to dominant technology or tactics.[205] One of the most common was attempts to undermine, in the most literal sense, the threat presented by an overwhelming military power. The distinction of the first naval special operation perhaps belongs to Skyllias, a renowned diver and recovered or sunken treasure, as well as his daughter Hydne.[206] As the army of Darius I invaded Greece in 480 BCE, it was supported by a fleet shadowing along the eastern coast. While anchored somewhere off Magnesia, likely near Cape Pori, the Persian fleet encountered a storm that lasted, according to Herodotus, for four days.[207] The fleet may have lost as much as one-third of its vessels during the conflagration.[208] In addition to the number and location of ships lost, the factor contributing to their destruction is unclear: divine intervention, poor seamanship, or human agency. According to Greek poetry, Skyllias, either alone or with Hydne, dove beneath the Persian ships and cut their anchor ropes.[209] Once

[200] Jan Best, *Thracian Peltasts and Their Influence on Greek Warfare* (Leiden: Brill, 1969), pp. 24–25; Mary Francis Williams, 'Philopoemen's Special Forces: Peltasts and a New Kind of Greek Light-Armed Warfare (Livy 35:27)', *Historia: Zeitscrifte für Alte Geschichte*, 53:3 (2004): 268.

[201] Strassler, ed., *The Landmark Thucydides*, 4.38, pp. 243–244.

[202] Ibid., 7.25, p. 440.

[203] Ibid., 7.25, p. 441.

[204] Thucydides, *The Peloponnesian Wars*, vol. III, trans. William Smith (London: A.J. Valpy, 1831), p. 88.

[205] See, for example, the responses of Chace and Milevski to a white paper on the subject from the US Army Asymmetric Warfare Office. Lukas Milevski, 'Asymmetry Is Strategy, Strategy Is Asymmetry', *Joint Forces Quarterly*, 75 (4th Quarter, 2014): 77–83; Jesse Chace, 'Defining Asymmetric Warfare: A Losing Proposition', *Joint Forces Quarterly* 61 (2nd Quarter, 2011): 115–120. An earlier exploration of the subject is Steven Lambakis, James D. Kiras, and Kris Kolet, *Understanding 'Asymmetric' Threats to the United States*, NIPP Report (Fairfax, VA: National Institute for Public Policy, October 2002).

[206] His skills are mentioned specifically by Herodotus in 8.8. Strassler, ed., *The Landmark Herodotus*, 8.8, p. 604.

[207] Ibid., 7.188–7.192, pp. 577–578; W. Kendrick Pritchett, 'Xerxes' Fleet at the "Ovens"', *American Journal of Archaeology*, 67:1 (January 1963): 5.

[208] Strassler, ed., *The Landmark Herodotus*, 7.190, p. 578.

[209] Frank Frost, 'Scyllias: Diving in Antiquity', *Greece & Rome*, 15:2 (October 1968): 180–181.

beset by the weather, they could have easily smashed into each another or have been dashed upon the coast in the storm surge. While impossible to prove, it is within the realm of the possible, given known diving duration and depths during the time period, not to mention a well-established Greek diving tradition.[210]

Other examples followed in this method. A century and a half later, in 332 BCE, the Tyrians employed a similar approach against Alexander's blockading ships and mole he was building to ferry across his siege engines. To combat the ships at anchor, the Tyrians sent divers to cut the anchor cables, to which 'Alexander replied by replacing them with chains'.[211] The Tyrians were also alleged to have used divers to erode the foundations of the mole, as well as employing fireships, although other accounts suggest natural causes did subsequent damage.[212] During the later Ottoman Empire siege of Malta (1565 CE), defended by the Knights Hospitaliers of St John and as well as numerous local Maltese, Ottoman divers set to work removing the various connected woodworks comprising the palisade near the heights of Corradino but were interrupted. Several Maltese entered the water and 'a terrible and extraordinary struggle ensued, frequently beneath the surface of the water', resulting in the retreat of the Turkish divers.[213] At least four of the Maltese were singled out for their bravery in a subsequent Spanish account of the siege.[214]

Another asymmetric approach against naval vessels used divers to 'hole' them, or otherwise compromise the integrity of their hulls. One author, citing 'the Early Middle Ages', attributes to an unnamed Swedish admiral an idea to prepare for victory against a fleet of Baltic pirates the following day. He 'sent in at night men ... to dive beneath them and make holes' whose results during the subsequent battle so distracted the pirates, in trying to stem the inrushing water, that 'the Swedes gained a complete victory and the power of the pirates was annihilated'.[215] A medieval chronicle, the *Chronicon Pictum* or *Képes Chronicle*, identifies a knight named Búvár Kund and documents his exploits in 1052 CE. Kund, who also went by the name Zotman(n) or Zot(h)mund, allegedly took decisive underwater action against a fleet of Henry III, the Holy Roman emperor. Even if the deeds are fictional, Zothmund existed given that his name appears in a later 1055 CE letter mentioning other Hungarian knights.[216] Seeking to impose his claim over southern possessions and to help consolidate his rule, Henry the Black sailed a fleet to besiege Bratislava in his fourth campaign to pacify Hungary. The fleet subsequently sank in the Danube and Henry was forced to return home. According to *Chronicon Pictum*, 'The Hungarians found a very good swimmer called Zothmund who was sent to the fleet of the

[210] Ibid., pp. 181–183.
[211] J.F.C. Fuller, *The Generalship of Alexander the Great* (London: Eyre & Spottiswoode, 1958), p. 213.
[212] Kern, *Ancient Siege Warfare*, p. 210.
[213] Cyprian Bridge, 'Sub-Aqueous Warfare, Ancient and Modern', *Fraser's Magazine*, 18 (October 1878): 462.
[214] Ernle Bradford, *The Great Siege, Malta 1565* (London: Hodder & Stoughton, 1961), p. 152.
[215] Bridge, 'Sub-Aqueous Warfare, Ancient and Modern', p. 461.
[216] Csaba Szaro, 'Die militärischen Aspekte der deutsch-ungarischen Beziehungen während der Salierzeit', in Georg Stadtmüller, ed., *Ungarn-Jahrbuch: Zeitschrift für die Kunde Ungarns und verwandte Gebiete*, Band 21, Jahrgagn 1993–1994 (Munich: Verlag Ungarisches Institut, 1995), p. 13.

emperor in the middle of the night. He dove under the fleet and drilled holes in all the ships so that they what filled up with water.'[217] The sinkings had an immediate effect on the surrounding locals: 'the Germans were attacked upon every side, and compelled to retreat'.[218]

A similar approach was used in Asia to offset overwhelming power. In Vietnamese folklore, the hero Yết Kiêu, loyal servant of Tran Hung Dao, allegedly swam underneath the invading Mongol, or Yuan Dynasty, fleet and drilled holes in their hulls while they sailed on the Bach Dang River in 1288 CE. More credible scholarly sources attribute the subsequent Dai Viet victory to the preparations made by Commander Tran Hung Dao, which included 'stake yards' implanted in the river bed to impale or rupture the hulls of the Mongol fleet, comprising some 400 ships.[219] Subsequent archaeological expeditions have found evidence of the stakes in the river.[220] As intriguing as these examples are, their cultural significance makes them important in thinking about asymmetric approaches and in particular the idea of human ingenuity and agency characteristic of special operations.

A final ancient example of asymmetric approach deserves mention: the military reforms undertaken by Philopoemen of Megalopolis. According to the Roman history, Philopoemen developed 'a new kind of fighting' in 192 BCE. Mary Francis Williams argues what he accomplished was nothing less than creating the first 'special forces' known to war. She bases this conclusion on the forces and tactics Philopoemen developed and subsequently employed, including amphibious raids, nights attacks, fire, and small-unit manoeuvre and harassing attacks of superior forces conducted by lightly armed citizen-soldiers, as opposed to mercenaries.[221] While most had been used elsewhere, save night attacks, what makes Philopoemen's efforts stand out was the integrated use together as a unique form of war, in an era where hoplite arms and armour were getting heavier, as opposed to lighter. In addition, the effective use of light arms, such as bows, slings, and javelins, required 'extensive specialized training', often the preserve of tribes who hunted and fought using them out of necessity to survive.[222] While some of his insights were doubtless inspired by his upbringing in the Arcadian, as well as his own careful study of previous use of light infantry and auxiliaries by Philip V, Demosthenes, and others, there was no substitute for his own experience and willingness to learn and adapt.[223]

If Plutarch is to be believed, Philopoemen's activities as a mercenary on the island of Crete both made use of his considerable experience and transformed in the process. Plutarch writes that:

[217] Quoted in Gyorgy Szabados, 'The Hungarian National Defense during the German Wars 1030–1052', *Chronica*, 6 (January 2006): 79.

[218] Edwin Lawrence Godkin, *The History of the Hungary and the Magyars: From the Earliest Period to the Close of the Late War* (New York: Alexander Montgomery, 1853), p. 43.

[219] Jun Kimura, Mark Staniforth, Le Thi Lien, and Randall Sasaki, 'Naval Battlefield Archaeology of the Lost Kublai Khan Fleets', *The International Journal of Nautical Archaeology*, 41:1 (2013): 3–4.

[220] Kimura et al., 'Naval Battlefield Archaeology of the Lost Kublai Khan Fleets', pp. 6–7.

[221] Williams, 'Philopoemen's Special Forces', pp. 271–272.

[222] Ibid., p. 272.

[223] Ibid., pp. 275–276.

> In Crete he waged war in the service of the Gortynians; not the straightforward and honourable warfare of a Peloponnesian and Arcadian, but one in which he adopted the Cretan practices, and turning their tricks and wiles and stolen marches and ambuscades against themselves, speedily showed them that they were children opposing foolish and vain mischievousness to genuine military experience.[224]

The net effect of Philopoemen's reforms gave the military of the Achaean Confederacy all of the attributes of a modern special operations force: 'specialized, rapid, and innovative troops that performed special missions', one that fought asymmetrically against the prevailing military norm of its day: the heavily armed, heavily armoured hoplite phalanx.[225]

## Conclusion

The preceding sections identified a range of activities conducted from ancient history up to the sixteenth century that functionally resemble, in broad terms, the net activities SOF perform today. Instead of finding direct historical anecdotes and connecting them to these roles and missions, the analysis grouped the activities by broader purpose or function. These functions served the needs of a range of polities from city-states to empires. Although they might have differed in complexity and scale, what is perhaps most striking is their perceived or real utility across time, geography, and culture. This observation may seem rather anodyne, but its establishment is a necessary precondition for creating a general theory of special operations.

Some might question the value of such a survey, arguing it blurs the line between exclusively military activities and those more traditionally considered statecraft. This perspective reflects a modernist approach; it suggests thinking constrained by 'what's in the box' in terms of current frames of reference and experiences.[226] It also ignores a fundamental reality of pre-modern warfare—only the most powerful empires had anything close to divisions we are inclined to make between 'military' and 'intelligence', as well as 'military' and 'political'. These are socially constructed distinctions that reflect the size and complexity of the workings of a modern nation-state. Often within polities in the ancient world, the sovereign and the military commander were one and the same, easing the direction of activities and force or their *control.*

The only difference between them was whether a state of active hostilities existed between their polity and another. Several of the functions related to special operations, as discussed and understood by earlier authors, appear to cross the division commonly prevalent in the West between peace and war. In addition, as the

[224] Plutarch, *The Parallel Lives of Plutarch*, vol. X., trans. Bernadotte Perrin (Cambridge, MA: Loeb, 1923), pp. 292–293.

[225] Williams, 'Philopoemen's Special Forces', pp. 267–268.

[226] The 'general theory' derived from this insight is contained in Thomas Searle, *Outside the Box: A New General Theory of Special Operations*, JSOU Report 17-4 (MacDill Air Force Base, FL: JSOU Press, 2017), pp. 17–21.

preceding suggests, the lines between the different functions were not always as clear and discrete as we might like. Intelligence could support raids and treachery and subversion led to the capture of fortifications that 'could then become a base from which further destructive raids could be launched'.[227] In other words, the confusion and uncertainty inherent in the debate and decisions surrounding 'strategic competition' today, specifically related to preparing for conflict in the future as opposed to engaging in a range of competitive activities economically, politically, and militarily, exist precisely because the lines between them were and always have been blurred.

The functions related to special operations appear timeless and have been intuited and codified theoretically, as the works of Sun Tzu, Kautilya, and others suggest. What differed was their specific application and conduct. Even then, there has been more continuity than change. Much of this has to do with the pace of military change and the underlying logic and appeal of special operations. Writing about the Middle Ages, Harari opines that,

> from a military viewpoint, in the field of special operations continuity far outweighed difference. Strategically and politically, the targets of special operations were largely the same in eleventh-century Syria and in sixteenth-century France. Similarly, since special operations always depended on ingenuity, surprise, and daring far more than on formal tactics, changes in technology and military organization, including those resulting from the gunpowder revolution, had little impact on them. Indeed, there was little fundamental change in either the targets or the methods of special operations between the Ancient World and the eighteenth century.[228]

To frame this observation in a different way, while the nature or rationale, appeal, and purpose of special operations has not changed, how they have been conducted, or their character, is constantly shifting. This observation is consistent with the predominant understanding of the nature and character of war, first identified by Carl von Clausewitz in 1834 and whose continued relevance has been argued by scholars such as Colin Gray, Lawrence Freedman, Hew Strachan, Richard Betts, Antulio Echevarria, and others.

Modern special operations have taken on a much clearer form, and different character, for three reasons. One is the creation of sizeable, stable polities, in the form of the nation-state, given the resources they have and can generate. Another is the growth of the complexity of these states, containing particularly since the Industrial Revolution a range of specialized but vulnerable infrastructure required to maintain essential services. A third is perhaps the most important. While there have always been elite military forces, primarily to conduct direct shock on the battlefield, protect the sovereign, and ensure domestic stability, the shift to special operations occurred primarily due to technological developments permitting vastly

[227] John Gillingham, 'An Age of Expansion, c.1020–1204', in Maurice Keen, ed., *Medieval Warfare: A History* (Oxford: Oxford University Press, 1999), p. 81.

[228] Harari, *Special Operations in the Age of Chivalry*, p. 40.

improved speed, reach, coordination, and lethality. They include: the mechanization of armies and navies and the harnessing of airpower, permitting rapid access to distant locations; the development and increasing portability and reach of communications technologies such as the telegraph, wireless, and radio, to coordinate activities between units and with higher headquarters; and the improved lethality and destructive power of small groups thanks to improvements in firepower and explosives, including lightweight machine guns and assault rifles and trinitrotoluene (TNT), high explosives, and plastic explosives.[229]

Aside from the use of specialized classes of individuals to meet persistent requirements, such as intelligence gathering and collection and the use of spies, as well as military elites and or tribes skilled in raiding and skirmishing incorporated into the military, the norm for conducting those functions related to special operations historically was the creation and subsequent disbandment of ad hoc units to meet an urgent need or battlefield requirement.[230] The combat swimmers used at Syracuse, Tyre, and Malta are one example. It should be noted this cycle continues, as the next chapter makes clear, but what has changed is the ability and requirement to develop and maintain specialists in a variety of skills, honed to a high level. Using the term that comprises the subject of the next chapter, special operations *capability* has existed theoretically but has been largely absent historically, with few exceptions, such as Philopoemen's military.

The survey of ancient and medieval history suggests a recurring theme related to special operations functions: the dual nature of the honourable and dishonourable. This theme transcends cultural time periods and geographies. It is most evident in the functions of intelligence and treachery and subversion. Early Western heroic literature identifies and distinguishes between two traditions. The first is the paragon of the strong warrior, who seeks out honourable battle with his foes and stands on the battlefield with spear and shield, oblivious to personal danger and ready to go into a combat frenzy, embodied by Achilles. The second represents a different ideal: weakness, cunning, guile, deception, and a 'skulking war of war', armed with a bow and embodied by Odysseus.[231] The traditions are not mutually exclusive: Achilles used deception and Odysseus was capable of honourable, or 'upstanding', actions.[232]

229 T.E. Lawrence recognized the value of available technologies in enhancing the mobility, lethality, and coordination of small units during the Arab Revolt (1916–1918). See Lawrence, *The Evolution of a Revolt*, CSI Reprint (Fort Leavenworth, KS: Combat Studies Institute, n.d.), p. 17, https://usacac.army.mil/sites/default/files/documents/cace/CSI/CSIPubs/lawrence.pdf. Two decades later, Orde Wingate envisioned how aircraft and radios could free his 'long range penetration' forces from ground-based supplies, greatly extending their mobility. Scott McMichael, *A Historical Perspective on Light Infantry*, Research Survey No. 6 (Fort Leavenworth, KS: Combat Studies Institute, 1987), p. 30. More recently, Audrey Kurth Cronin discusses the impact of 'disruptive' technologies for individuals and small groups in *Power to the People: How Open Technological Innovation Is Arming Tomorrow's Terrorists* (New York: Oxford University Press, 2020).

230 Harari, *Special Operations in the Age of Chivalry*, pp. 9–34; Dunnigan, *The Perfect Soldier*, p. 9; Leebaert, *To Dare and to Conquer*, pp. 583–585.

231 The phrase was used by American colonists to describe the type of warfare conducted by local native American tribes in New England. Patrick Malone, *Skulking Way of War: Technology and Tactics among the New England Indians* (Lanham, MD: Madison, 1991); Jeffrey Asher, 'An Unworthy Foe: Heroic Ἔθη, Trickery, and an Insult in Ephesians 6:11', *Journal of Biblical Literature*, 130:4 (Winter 2011): 744.

232 Asher, 'An Unworthy Foe', p. 738.

What has been translated throughout the ages is an ideal and its antithesis, interpreted through notions of what is virtuous and honourable, as opposed to what is treacherous and dishonourable. Then, as now, these notions have proven very useful rhetorically and politically, to help bolster individual and collective self-images and denigrate opponents and their cowardly ways.

In a time of increasing budgetary and organizational competition, these notions remain surprisingly resonant; recent inquiries into the unethical and unprofessional behaviour of special operations personnel only serve to confirm some preexisting beliefs. Traditions remain inseparable when it comes to military practice. In the words of Everett Wheeler, their relationship belies a fundamental truth: 'the duality of alternating means of warfare characterizes the western tradition.'[233] It is evident in the writings of the Ancient Greeks, Romans, and medieval Europeans, including Christine de Pizan, who struggles with the realities of the battlefield clashing with the ideals of her faith.

Eastern writers have also grappled with this truth. The writings of T'ai Kung and Sun Tzu place heavy emphasis on cunning, deception, and trickery to win as decisively and economically as possible, the former through sponsoring the long-term subversion and moral decay of an adversary and the latter in both statecraft and, in the shorter term, to set the conditions for battlefield success. Subsequent Ancient Chinese writers sought to either distance themselves from these practices, by stressing honour and virtue or calling into question the legitimacy of the texts that advocate them.[234] Kautilya links virtue and reward with the wealth and prosperity of the empire, rationalizing harsh and seemingly unethical instruments as a means to guarantee it, including the use of assassins. In his estimation, why would a ruler not employ means that could forestall a conflict and render moot the fielding of an expensive army? The question, then as now, is not whether a polity needs a capability but rather how to control it. Creating and sustaining a special operations capability, and directing and controlling it effectively, form two of the elements of the general theory explored in the subsequent chapters.

[233] Wheeler, *Stratagem and the Vocabulary of Military Trickery*, p. 3.
[234] Sawyer, trans., *The Seven Military Classics of Ancient China*, pp. 30–31.

# PART II

# 3
# Special Operations Capabilities

This chapter continues the exploration of special operations capabilities, but from a theoretical perspective. It begins by examining the complications inherent in the concept of 'capability', a commonly used word in the national security lexicon. This section identifies that capability is often framed in overly narrow or greatly expansive ways, problems which extend to all military forces, including special operations. This work defines capability first as the sum product of various special operations forces (SOF) to conduct a variety of special operations roles and missions in an integrated manner, in demanding conditions and environments, over a sustained period of time, excluding ad hoc or 'one-off' actions. The next section explores capability lifecycles according to two predominant ideas related to military elites or SOF. While each provide useful insights for theory, their explanatory power is limited to SOF creation and misuse, rather than sustainable and effective long-term use.

From there, the chapter explores the idea of integration, in the form of connections that exist between SOF and conventional military units, as well as with resources of other government departments and agencies units. Integration is related to but differs from interoperability and cohesion. Beyond identifying the need for integration between different SOF units and others, this chapter develops an analytic framework for assessing the level of integration. These levels, ranging from none to high, reflect technical, organizational, and social linkages and can assess, in a broad evaluation, the degree of horizontal integration present within special operations. The chapter concludes with a short evaluation of the level of horizontal integration discernible across the SOF capabilities of India to test the utility of this analytic framework and finds weak linkages technically, organizationally, and socially.

The linkage between special operations capability and theory may be recent but it is a powerful one. In 1986, a senior researcher with the Congressional Research Service, John Collins, was tasked with providing a comparative analysis of the SOF of the United States and the Soviet Union. The timing of the tasking was significant: it occurred during the ongoing discussion and legislative wrangling to develop an amendment to the act formalizing jointness between the United States armed services that year. Submitted as House Resolution 3622 of the 99th Congress, the Goldwater-Nichols Department of Defense Reorganization Act was signed as Public Law No. 99–433 in October of that year. Collins' report, later published commercially under the title *Green Berets, SEALs, and Spetsnaz: U.S. and Soviet Special Operations*, played an important supporting role in developing an amendment to Goldwater-Nichols, sponsored by Senators Samuel Nunn, Jr. (D-GA) and

*Special Operations Success*. James D. Kiras, Oxford University Press. © James D. Kiras (2024).
DOI: 10.1093/9780198902096.003.0003

William Cohen (R-ME). This amendment created a joint headquarters command, the United States Special Operations Command (USSOCOM), along with its own separate major force programme or budget funding line, for American SOF. The report's contribution to the creation of USSOCOM was surpassed by the undue influence of its theoretical contribution: the articulation of guiding maxims, or 'truths', about special operations.[1]

Since their initial publication, the 'SOF Truths' have served a range of purposes within and without of the United States. In the earliest years of USSOCOM, the SOF Truths were valuable organizationally in creating a shared lexicon and common cultural reference point among Army, Navy, and Air Force special operations units. Distributed on printed cards by commanders and on posters within units, the Truths provide a reminder of the unifying bonds shared by all SOF. Politically, the SOF Truths have acted as both shield and sword for the special operations against calls to reallocate resources, disband units, or paradoxically, expand units. Outside of the United States, the SOF Truths have been used in support of efforts by special operators to create their own joint headquarters in countries including Canada, the United Kingdom, and Denmark. In theoretical terms, the SOF Truths serve a foundational purpose: the 'first principles' from which all else flows conceptually.

The self-evident and foundational nature of the SOF Truths, combined with the purposes they perform, is not without issue. Institutionally, the SOF Truths have been elevated by some from guiding principle to holy writ, never to be questioned. Given the constantly evolving and changing character of conflict and competition, such dogmatic adherence can stymie or short-circuit efforts to explore the relationship special operations should have with new technologies and the degree to which they will influence or drive future operations.[2]

At their root conceptually, the SOF Truths are related to capability in one form or another. The five SOF Truths are:

- Humans are more important than hardware
- Quality is better than quantity
- SOF cannot be mass produced
- Competent SOF cannot be created after emergencies occur
- Most special operations require non-SOF support.[3]

The first and second address issues of capability at the individual level and the third through the fifth at the organizational. Most importantly, they provide a set of institutional guard rails against the modern tendencies to privilege technological solutions to problems, offers that more of a useful capability is better, and react to, rather than anticipate, changes in capabilities. This chapter will not interrogate all five SOF Truths. Instead, it primarily answers the question of what 'competent' means,

[1] John Collins, *Green Berets, SEALs, and Spetsnaz: U.S. and Soviet Special Military Operations* (London: Pergamon-Brassey's, 1987), p. xiii.

[2] A problem identified in James D. Kiras, 'Future Tasks: Threats and Missions for SOF', *Special Operations Journal*, 5 (2019): 15–16.

[3] 'SOF Truths', United States Special Operations Command, www.socom.mil/about/sof-truths.

in the context of special operations generally, as well as the integration requirement for 'non-SOF support'. This examination begins with the elusive but necessary functioning core of integration: capability.

## Defining capability

The word 'capability' in its simplest form is easy to define: an ability to do something. That something is most commonly associated with a weapons system or technology. For example, if a platoon containing a squad possesses and has trained in how to employ FGM-148 Javelin or 9M133-EM Kornet missiles and can engage and destroy tanks and other armoured vehicles, it has an 'anti-tank capability'. Likewise, a naval vessel with a helicopter able to detect and engage submarines has an 'anti-submarine capability' and a jet fighter with radar and air-to-air missiles has 'an air-to-air capability' to destroy other aircraft. 'Doing something' as a common definition is far too vague for the military, so the US Department of Defense defines capability as follows: 'The ability to complete a task or execute a course of action under specified conditions and level of performance.'[4]

For SOF, confusion in understanding what capability means increases the further one travels up levels of command, away from the point of tactical mission accomplishment. The tasks American SOF should be prepared to perform are outlined in the service-specific Task Lists as well as Universal Joint Task List, the latter comprising 60 different types of operations, broken down further into 265 strategic national (SN) and strategic theatre (ST) tasks down through the operational (OP) to the tactical (TA).[5] While at the tactical level the specified conditions and levels of performance are clear, such measurements are increasingly challenging to determine and unsatisfactory the higher in operating level one progresses. Measures of performance only provide information on how well units or groups of units are operate relative to themselves, or assessing current performance against previous benchmarks. What they do not provide, however, is information or assessment on what effects task performance has against an adversary or opponent in competition or conflict, within a foreign government partner or proxy force in terms of training, advising, or equipping, or within a population whose attitudes and perspectives one is trying to change or influence.

The problems associated with measuring performance as opposed to effectiveness as a means of evaluating capability are well known, including for SOF.[6] This discrepancy is most clearly visible in the number of works on or about more direct missions

[4] The Joint Staff, *Charter of the Joint Requirements Oversight Council (JROC) and Implementation of the Joint Capabilities Integration and Development System (JCIDS)*, CJCSI 5123.01H (Washington, DC: The Joint Staff, 31 August 2018), p. GL-7.

[5] Chairman of the Joint Chiefs of Staff, *Universal Joint Task List (UJTL)*, CJCSM 3500.04c (Washington, DC: The Joint Staff, 1 July 2002), appendix A to enclosure B, 'Tasks', B-A-1 to 15; annex A to appendix C to enclosure B, 'Strategic National Tasks, Measures, and Criteria', B-C-A-1 to 152.

[6] See, for example, the discussion in Barnett Koven and Katy Lindquist, *Barriers to Special Operations Forces-Led Counterterrorism Effectiveness*, JSOU Report 21-7 (MacDill Air Force Base, FL: JSOU Press, 2021), pp. 189–195.

such as raids, and their readily available number of measures of performance: numbers of raids conducted, the percentage of successfully completed missions, and the number of killed or captured, amount of material seized and collected, or absence of one's own or civilian casualties. The volume of data on these missions stands in stark contrast to that available on the effectiveness of those missions over a period of time, leading critics to charge for counterterrorism that all such missions accomplish is 'mowing the grass' of terrorist leaders and managers while doing little over the long term to alter the conditions which spawn them.[7] For longer-duration 'indirect' missions, working by, with, and through foreign governments and proxies, the problem of measuring both performance and effectiveness is particularly acute.[8] Advisors tend to report progress optimistically, as do foreign governments and proxies, given the incentive associated with doing so: favourable performance reports and promotion, as well as continued resources and aid.[9] Rigorous scholarly attempts at quantifying the success of specific forms of theatre security engagement, such as raising the health standards in different countries, acknowledge problems of 'endogeneity' in models as well as significant issues related to causality.[10] One proposed solution to assessing special operations effectiveness offers a seven-step approach that relies heavily on ongoing operational (and therefore likely classified) information.[11]

Exploring special operations capability analytically abuts an initial methodological challenge all scholars, researchers, and journalists face: secrecy. The secrecy that enables the success of special operations, that sustains their unorthodoxy and allows them to undertake politically sensitive missions, maintain a low profile, or achieve surprise, also prevents access to much of the information needed to assess them objectively.[12] In particular, assessing capability is difficult based on sparse historical documentation, or actively hindered by the reluctance to release contemporary information. Data on individual units or platforms, and how they allow tasks to be performed under specified conditions, can be limited exclusively to those with need to know.

The reticence to disclose information, while a source of continuous frustration for researchers, is not without good reason. Unintentional disclosures of special

[7] Michael Shkolnik, '"Mowing the Grass" and Operation Protective Edge: Israel's Strategy for Protracted Asymmetric Conflict with Hamas', *Canadian Foreign Policy Journal*, 23:2 (Summer 2017): 185–189, DOI: 10.1080/11,926,422.2016.1229684; Jenna Jordan, 'When Heads Roll: Assessing the Effectiveness of Leadership Decapitation', *Security Studies*, 18 (2009): 755.

[8] Koven and Linquist, *Barriers to Special Operations Forces-Led Counterterrorism Effectiveness*, pp. 190–191.

[9] Walter Ladwig III, *The Forgotten Front: Patron–Client Relationships in Counterinsurgency* (Cambridge: Cambridge University Press, 2017), pp. 41–47; 175; 193–194; Tyrone Groh, *Proxy War: The Least Bad Option* (Stanford, CA: Stanford University Press, 2019), pp. 212–214.

[10] Glen Diehl and Solomon Major, 'MOE vs. M&E: Considering the Difference between Measuring Strategic Effectiveness and Monitoring Tactical Evaluation', *Military Medicine*, 180:1 (January 2015): 80.

[11] Linda Robinson, Daniel Egel, and Ryan Andrew Brown, *Measuring the Effectiveness of Special Operations*, RAND Report RR2504 (Santa Monica, CA: RAND Corporation, 2019), p. 105. The assessment process steps are: determine objectives, identify activities, define measures of effectiveness, develop indicators and collect data, conduct line of effort-specific analysis, consolidate analysis, and commander review. Robinson, Egel, and Brown, *Measuring the Effectiveness of Special Operations*, pp. 49–84.

[12] Mark Moyar, *To Oppose Any Foe: The Rise of America's Special Operations Forces* (New York: Basic Books, 2017), pp. 342–343.

operations capabilities in the past have led to organizational realignment, as occurred after the publication of Steven Emerson's *Secret Warriors* in 1988 and Mark Bowden's *Killing Pablo* in 2001.[13] To satisfy the public curiosity and fascination with special operations, accounts by former operators often go into lurid or highly specific technical details at the tactical level. Specific technical information about special operations equipment here, which is capability in its narrowest sense, is interesting but irrelevant for the purpose of the theory. Individual tactics, techniques, and procedures reflect the ability to conduct tasks under conditions, not the ability of SOF institutionally to perform missions successfully in a maintainable manner. These intentional disclosures by former operators have led to additional restrictions.

Such information releases are a source of frustration for those still active within special operations organizations and can lead to even more restrictive rules regarding its release. Books by Eric Haney and Sean Naylor have had such effects in the US.[14] Other work have prompted more unprecedented steps. In 2012, the commander of USSOCOM, Admiral William McRaven, sent a memo throughout the special operations community issuing specific guidance about publishing memoirs. Written in advance of the release of *No Easy Day* by Mark Bissonette, the Navy SEAL who claimed to kill Osama bin Laden, McRaven had to remind special operators about the detrimental effect such disclosures could have on current and future operations.[15] Other reactions were more severe. The books published in the United Kingdom by Andy McNab and Cameron Spence after the Gulf War (1990–1991) led Her Majesty's Government to rethink the Official Secrets Act and force current operators to sign a legally binding non-disclosure agreement from 1996 onwards.[16] In addition to those works by SOF operators seeking to capitalize on their experiences, others are motivated solely by collecting and publishing information on special operations units, partly out of curiosity but mostly of personal conviction, financial or repetitional gain, or all of the above.[17]

Some countries go to extraordinary lengths to limit information on special operations. Most readers are familiar with the standard practice of pixelating the faces of operators or placing a redaction bar (a black line) over their eyes in photographs to prevent identification. The editors of a recent volume on Swedish perspectives on special operations took the unusual step of providing three-letter aliases for

[13] Steven Emerson, *Secret Warriors: Inside the Covert Military Operations of the Reagan Era* (New York: G.P. Putnam's, 1988); Mark Bowden, *Killing Pablo: The Hunt for the World's Greatest Outlaw* (New York: Atlantic Monthly Press, 2001).

[14] Eric Haney, *Inside Delta Force: The Story of America's Elite Counterterrorist Unit* (New York: Delacorte Press, 2002); Sean Naylor, *Relentless Strike: The Secret History of Joint Special Operations Command* (New York: St Martin's Griffin, 2015).

[15] Josh Rogin, 'McRaven to OPSEC: Zip It', *Foreign Policy* (24 August 2012), https://foreignpolicy.com/2012/08/24/mcraven-to-opsec-zip-it.

[16] 'SAS Men Are Ordered Never to Write Books', *The Independent* (4 October 1996), www.independent.co.uk/news/sas-men-are-ordered-never-to-write-books-1356609.html.

[17] See, for example, Jeremy Scahill, *Dirty Wars: The World Is a Battlefield* (New York: Nation Books, 2013) and the documentary with the same name. Also, William Arkin, *Code Names: Deciphering U.S. Military Plans, Programs, and Operations in the 9/11 World* (Hanover, NH: Steerforth Press, 2005).

the authors of half of the chapters in a recent edited volume, citing 'strict security regulations' to 'protect them and their families'.[18] Wikipedia, a transparent source of information globally without a vested interest in national security, rather curiously states the number of Chinese operators of the People's Liberation Army Special Operations Forces as 'classified' despite publishing sensitive information about unit strengths and locations for many Western nations.[19]

Protecting information, however, can lead to pathological and counterproductive organizational behaviour, within units as well as their parent organizations and headquarters. Such behaviour includes groupthink within different special operations units compartmentalized to preserve operational security and prevent intelligence penetration.[20] It can also stretch to reinforcing failure in the form of operational losses and casualties as part of the 'sunk cost fallacy', in the face of damning evidence that an opponent understands the vulnerabilities of your capability. Specific examples include: the ambushing of a contracted Central Intelligence Agency aircraft over Manchuria on 29 November 1952 due to effective Chinese Communist counterintelligence in support of a Nationalist resistance movement that existed in name only;[21] the low success rate of agents parachuted into North Vietnam as part of Operation Plan 34-Alpha and later Laos as part of the 'Leaping Lena' programme from 1964 to 1965;[22] and the quick North Vietnamese responses to, and steadily inflicting losses upon, Military Assistance Command Vietnam Studies and Observation Group (MACVSOG) teams inserted from South Vietnam across the border into Laos to conduct reconnaissance and other missions against the Ho Chi Minh Trail during the Vietnam War.[23]

Protecting information can also be taken to absurd extremes. In the year following the 11 September 2001 terrorist attacks, officials in USSOCOM delayed the release and publication of its biannual 'posture statement' for almost three years, until all specific information related to its individual platforms, beyond an opaque mission statement, was removed. The stated reason from USSOCOM was that terrorists could use this information to counter SOF capabilities. I drafted a response to USSOCOM from the office representing special operations policy within the Pentagon, in which I argued such information had been published in the preceding three posture statements. In addition, I added the document was an important public information and relations instrument. Staffers and members of Congress review posture

[18] Gunilla Eriksson and Ulrica Petterson, eds., *Special Operations from a Small State Perspective; Future Security Challenges* (New York: Palgrave Macmillan, 2017), p. xi.

[19] 'People's Liberation Army Special Operations Forces', Wikipedia.org, https://en.wikipedia.org/wiki/People%27s_Liberation_Army_Special_Operations_Forces.

[20] The classic interpretation from a behavioural psychology perspective of the failure of the attempt to rescue American hostages held in Tehran in 1980 is Steve Smith, 'Groupthink and the Hostage Rescue Mission', *British Journal of Political Science*, 15:1 (January 1985): 117–123.

[21] William Leary, *Perilous Missions: Civil Air Transport and CIA Covert Operations in Asia* (Tuscaloosa, AL: University of Alabama Press, 1984), p. 140.

[22] Sedgwick Tourison, *Secret Army, Secret War: Washington's Tragic Special Operation in North Vietnam* (Annapolis, MD: Naval Institute Press, 1995), pp. 110–128.

[23] Richard Shultz Jr, *The Secret War against Hanoi: Kennedy's and Johnson's Use of Spies, Saboteurs, and Covert Warriors in North Vietnam* (New York: HarperCollins, 1999).

statements when considering budget appropriations. These arguments fell on deaf ears, however, and the document was published in a less substantial form.[24]

If accessing details related to tactical capability is both unlikely and unnecessary for deducing a general theory of special operations, a separate set of issues occur at the other end of the spectrum. At the national level, assessments of capability tend to be over-aggregated, vague, and opaque. Much of this reflects a deliberate lack of information, creating uncertainty. Uncertainty is often greatest in the national security arena when trying to assess opponent or adversary units and their intentions, leading to faulty assumptions and analysis. A few writers have provided a wealth of explanations for why this is the case, from competitors hiding information or actively trying to deceive one another.[25] One such example includes the exaggerated estimates of Soviet nuclear capabilities during the early Cold War, which were fuelled by elementary efforts at deception.[26] In other cases, cognitive shortcuts and behaviours can lead us to divine what we prefer through uncertainty. Such shortcuts and behaviours influence how we process information individually and collectively, and include anchoring, confirmation bias, and groupthink.[27] In a more general sense, signals sent from one adversary to another are prone to be misinterpreted, regardless of how clearly they are perceived and communicated.[28] During the Cold War, 'capability' was associated with attempts to estimate Soviet and Western gross military power strategically and combat effectiveness operationally.[29] In lieu of accurate, or accurately processed, information, those tasked with assessing adversary forces and intentions are likely to stick with verifiable facts and draw inferences and conclusions from there.

Two examples related to SOF illustrate these points. The People's Republic of China (PRC) takes considerable steps to limit the public release of information about its SOF and their capabilities. As a result, analysts can fall into the trap of lumping together elements into broad categories labelled as 'capability'. Darryl Lavender, for example, assesses China has 'a three-dimensional, all-weather infiltration approach capability' but an '[in]adequate deployment capability', limiting the threat they pose

[24] USSOCOM, *United States Special Operations Forces Posture Statement, 2003–2004: Transforming the Force at the Forefront of the War on Terrorism* (Washington, DC: USSOCOM, 2004).

[25] J. Bowyer Bell and Barton Whaley, *Cheating and Deception* (London: Routledge, 1991), Chapter 8.

[26] The two best-known cases are the collective set of estimates leading to an assessment of both a bomber and later missile gap. For reproduced, declassified contemporary assessments, see Donald Steury, ed., *Intentions and Capabilities: Estimates on Soviet Strategic Forces, 1950–1983* (Washington, DC: Center for the Study of Intelligence, 1996), pp. 5–138.

[27] An exceptionally clear survey on heuristics and biases is contained in Daniel Kahneman, *Thinking, Fast and Slow* (New York: Farrar, Straus and Giroux, 2013), pp. 119–198. On groupthink specifically, see Irving Janis, *Groupthink: Psychological Studies of Policy Decisions and Fiascoes*, 2nd ed. (Boston, MA: Wadsworth, 1982), Chapter 8.

[28] Robert Jervis, *Perception and Misperception in International Politics*, new ed. (Princeton, NJ: Princeton University Press, 1976), pp. 48–57.

[29] For a specific critique of one influential and controversial analysis of Soviet capability, see Kenton White, 'Mearsheimer's Folly: NATO's Cold War Capability and Credibility', *Infinity Journal*, 6:4 (Summer 2017): 22–31. John Mearsheimer doubled down on this earlier, concluding in 2014 that 'Russia lacks the capability to easily conquer and annex eastern Ukraine, much less the entire country', a line he has repeated after Russia's failed attempt to do so in 2022. John Mearsheimer, 'Why the Ukraine Crisis Is the West's Fault', *Foreign Affairs* (September/October 2014): 8.

to US forces in the Pacific theatre for the present.[30] In an ironic twist, Lavender, who works for the Defense Intelligence Agency, likely has access to much more detailed information about PRC forces but cannot disclose it to protect sources and methods of intelligence collection.

Lack of more specific information can tempt analysts to draw inferences from what data is available. For example, one could conclude based on numbers of units and operators alone that the most capable SOF in the world belong not to the United States but Democratic People's Republic of Korea (DPRK), more commonly known as North Korea. Estimates of their number of SOF ranges from between 100,000 to 200,000 'special purpose forces' personnel, compared to the United States' 72,000.[31] More astute observers will note that numbers signify quantity, not quality. In the words of one veteran national security analyst, 'A small fraction of that figure may be truly elite, but a lot must be trash.'[32] Such figures, however, can lead to wild overestimation of capability or an underestimation of what those forces can accomplish.

Limiting or obscuring information can also serve another purpose for special operations: preserving or enhancing the mystique surrounding them. Certain special operations units have obtained a near mythical status: the British Special Air Service (SAS), the American Navy SEALs, and Russian Spetsnaz, among others.[33] Some of this status reflects aspects of uncertainty outlined above: the deliberate withholding of information or deception of opponents. Cultivating and maintaining a mystique, however, is often done consciously and deliberately, to enhance public awareness, raise an institutional profile, or expand influence institutionally and politically. One early example was the 1968 movie *The Green Berets*, starring John Wayne and based loosely on a book (1965) by Robin Moore, who was given privileged access to the Special Forces community. More recent examples include the book (1999), movie (2001), and video game (2003) titled *Black Hawk Down*, the movie (2012) *Zero Dark Thirty*, the British television series (2022) *Rogue Heroes*, based on Ben Macintyre's book (2016), as well as the Chinese television series (2022) *Glory of the Special Forces*. A more recent trend in the United States attempts to rebalance the books in terms of institutional influence between the self-identified 'surgical strike' (translation: counterterrorism) and 'special warfare' (translation: unconventional warfare) elements within the American special operations community.[34]

[30] Darryl Lavender, *China's Special Operations Forces Modernization, Professionalization, and Regional Implications*, US Army War College Strategy Research Project paper (March 2013), pp. 9–10.

[31] Joseph Bermudez Jr, *North Korean Special Forces*, Second Edition (Annapolis, MD: Naval Institute Press, 1998), p. 8; Defense Intelligence Agency, *North Korea Military Power: A Growing Regional and Global Threat* (Washington, DC: Government Printing Office, 2021), appendix E: Special Operations Forces, pp. 52–54.

[32] John Collins, 'Why Special Operations Forces Are Special', unpublished presentation script (15 June 2004), p. 9.

[33] An article exploring this aspect of the SAS specifically is Paul Winter, 'Who Cares Who Wins? The Mythology of the SAS', *The Critic* (15 December 2022), https://thecritic.co.uk/who-cares-who-wins/?mc_cid=1d64a186cc&mc_eid=1606f88ecf.

[34] Charles Cleveland and Daniel Egel, *The American Way of Irregular War: An Analytical Memoir*, RAND Publication PEA301-1 (Santa Monica, CA: RAND, 2020), pp. 178, 182.

Uncertainty and myth perpetuation reflect political considerations that inhibit understanding of special operations capability. Between SOF units, politics most often manifests in the form competition and rivalries over which ones are the best at performing certain missions. Within US SOF, SEALs are referred to by others as 'squiddies' and Special Forces as 'green beanies'. In the UK, members of the SAS until recently disparagingly referred to their maritime colleagues of the Special Boat Squadron (SBS) after the Falklands as 'the shakies', an adaptation of their acronym into the 'Shaky Boat Squadron'.[35] At the level of service headquarters and above, however, competition between SOF component elements, and SOF and its conventional counterparts, can become increasingly fierce. Such competition reflects aspects of bureaucratic politics over organizational leadership, resourcing, and institutional influence.

During internal and inter-community competition, the SOF community collectively or individually are likely to use the term 'capability' in both senses of the word: in an opaque manner to describe ability in very generic terms, as well as highly specific one that means function or facility within a system or piece of equipment. In his 2003 Congressional testimony, one USSOCOM Commander referred to capability nineteen times in twenty-three pages in the following ways: the overall qualities of the forces; their functional ability in general terms; enemy forces, or as a statement of threat driving the need to develop new systems; planning functionality; personnel and personnel development; and, specific technologies and delivery platforms such as the CV-22 and the problem-riddled and subsequently cancelled Advanced SEAL Delivery System (ASDS) submersible.[36] A more recent academic assessment of 'the special capabilities of SOF' wanders between budgetary considerations and remarkably rough calculations of costs of operators to individual qualities and collective skills development, including close-quarter battle.[37] These illustrate how diverse and profligate the use of the term 'capability' can be in the military lexicon generally, and within SOF specifically.

As with definitions of special operations and who can and should conduct them, discussions of capability most often default to units and their prescribed roles and missions, or to specific platforms and equipment. The Air Force Special Operations

[35] Isobel Dickinson, 'SAS and SBS at War of "Shaky Boats" Banter That's "Undermining Morale"', *The Daily Star* (3 April 2021), www.dailystar.co.uk/news/latest-news/sas-sbs-war-over-shaky-23846976. A former SBS operator characterizes the rivalry between the two units during this period as much deeper and more divisive; he uses the term 'animosity'. Duncan Falconer, *First into Action: A Dramatic Personal Account of Life in the SBS* (London: Little, Brown & Co., 1998), p. 35.

[36] 'Statement of Lietenant [sic] General Bryan D. Brown, U.S. Army, Deputy Commander, United States Special Operations Command, Before the Senate Committee on Armed Services, Subcommittee on Emerging Threats and Capabilities on the State of Special Operations Forces', Senate Committee on Armed Services, Subcommittee on Emerging Threats and Capabilities on the State of Special Operations Forces (9 April 2003).

[37] Anthony King, 'What Is So Special about Special Operations Forces?', in Jessica Turnley, Kobi Michael, and Eyal Ben-Ari, eds., *Special Operations Forces in the 21st Century: Perspectives from the Social Sciences* (London: Routledge, 2018), pp. 274–277. King attempts to summarize the findings regarding capabilities in a volume of collective essays and concludes there are two viewpoints: 'the potentially essentialist and reductive' and 'collective and institutional'. Capability, he concludes, is the result of 'competences' (pp. 276–277).

Command (AFSOC), for example, has platforms that serve different functions but are described as providing specific capabilities. MC-130 aircraft possess the capabilities of refuelling, low-level penetration of enemy air defence systems, and delivering personnel and equipment by airdrop or landing, depending on the aircraft model and its modifications. More recently, some 'J' model MC-130s were slated to be altered to provide an amphibious or water-landing capability through the addition of seaplane floats, to permit more distributed and survivable basing options at islands and atolls across the Pacific Ocean.[38] The modified aircraft from the now-defunct effort were to be designated as MC-130J Commando II Amphibious *Capability* (MAC; italics added). Of them, an AFSOC spokesperson said, 'this capability allows the Air Force to increase placement and access for infiltration, exfiltration, and personnel recovery, as well as providing enhanced logistical capabilities for future competition and conflict'.[39] U-28A Dracos, in contrast, possess the capability of providing intelligence, surveillance, and reconnaissance information to ground forces and commanders given their range of onboard equipment, which includes 'advanced radio-comms suite, IR suppression, missile, hostile fire, and laser warning, [electro-optical] EO sensors, remote [signals intelligence] SIGINT, and advanced navigation systems. The primary Multispectral Targeting System includes [full motion video] FMV, [electro-optical/infrared] EO-IR, IR real-time video, and co-aligned laser designator. U-28 EQ+ mods added high-definition, FMV to six EQ/PC-12 configured aircraft for extended standoff 'find, fix, finish' capabilities in support of counter-ISIS op[eration]s'.[40] AC-130U gunships are capable of providing overwatch over as well as precision and volume fire support to ground forces, with 30-mm and 105-mm cannons and precision-guided munitions ranging from missiles to bombs, leading those forces to label the platform colloquially and respectfully as 'the [overhead] death barge'. For military advising and assistance missions, Army Special Forces ('Green Berets'), the 160th Special Operations Aviation Regiment, and units within the 492nd Special Operations Wing have the capability to conduct foreign internal defence missions, or training partner nation SOF ground and air forces, including helicopters (rotary wing) and aircraft (fixed wing). In more general terms, UK special forces doctrine refers to capability in a mission statement in terms of the collective value of the units, as well as their specific qualities and attributes:

> The UK Special Forces (UKSF) mission, assigned by the Chief of the Defence Staff (CDS) through the Deputy Chief of the Defence Staff (DCDS(C), is to deliver the Special Forces (SF) capability required to support Government Defence, Security

[38] Stew Magnuson, 'AFA News: Tests Show Converting C-130s into Amphibious Aircraft "Viable"', *National Defense* (21 September 2022), www.nationaldefensemagazine.org/articles/2022/9/21/tests-show-converting-c-130s-into-amphibious-aircraft-viable; Howard Altman, 'C-130 Seaplane Should Fly in 2023 Says Air Force Special Ops Commander', The War Zone (21 September 2022), www.thedrive.com/the-war-zone/c-130-seaplane-should-fly-in-2023-says-air-force-specops-commander.

[39] Altman, 'C-130 Seaplane Should Fly in 2023'.

[40] 'U-28A Draco', *Air & Space Forces Magazine*, n.d., www.airandspaceforces.com/weapons-platforms/u-28a.

> and Foreign Policy objectives. That capability, and especially the versatility of SF, enables them to achieve significant effects at all levels of warfare.[41]

For the purpose of developing a theory, and to escape the some of the lexical challenges related to special operations, capability refers subsequently to the ability to conduct special operations in the collective sense of the word, across the individual unit or group level. Capability is the sum total of various SOF units to work with others, either in a supported or supporting manner, to conduct special operations. It defines special operations in two ways, first as a statement of their combined, collective ability to conduct both defined and emergent missions along the functional lines identified in the previous chapter: intelligence and reconnaissance, treachery and sabotage, direct and indirect shock, military advising and assistance, and asymmetric response. In addition, to further separate this analysis from other pitfalls identified in the Introduction, capability here is the sum product of various SOF to conduct a variety of special operations roles and missions, in demanding conditions and environments, with a high likelihood of success over a sustained period of time. In other words, the use of standing SOF permits a range of special operations activities well beyond the scope of ad hoc responses to existing crises whose problems, once addressed, go away—an extant as opposed to a manufactured or created capability. The sum product of various SOF reflects the level of integration that exists between them. Having scoped capability down and in order to serve the purpose of a more general theory, the next section surveys the lifecycle of special operations capabilities: why and how they are created, sustained, misused, and often destroyed.

## Models of the SOF capability lifecycle

If crisis provokes a need to innovate and adapt, failure is also a powerful motivator to reconsider activities and examine them from a broader conceptual perspective. Sweeping technological change, in the form of a perceptible or imminent 'revolution in military affairs', is another. For Prussian writer Carl von Clausewitz, both conditions sparked his philosophical and practical study on the nature and character of war, as well as the conduct of operations.[42] Within the last century, the development of airpower and atomic and thermonuclear weapons spurred a range of different conceptual enquiries given the impact of technological change in opening up another domain of war fighting and in dramatically increasing the destructive power of individual weapons.[43] In the United States, the deeply unsatisfying outcomes of limited wars in Korea but particularly Vietnam compelled a number of writers to find a

[41] Development, Concepts, and Doctrine Centre, *Special Forces Operations, JDP 3-05*, Restricted (Swindon: DCDC, 2008), p. I-1.

[42] Carl von Clausewitz, *On War*, trans. and ed. Peter Paret and Michael Howard (Princeton, NJ: Princeton University Press, 1984), pp. 20–21.

[43] Some of the first works speculating on the future impact of ballistic missiles and nuclear weapons characterized them as a revolution in warfare or strategy. See B.H. Liddell Hart, *The Revolution in Warfare* (New Haven, CT: Yale University Press, 1947); William Borden, *There Will Be No Time: The Revolution*

theoretical explanation for them. Such explanations included an asymmetry of wills, incorrect centre of gravity analysis, and others.[44]

After the United States ended its support for South Vietnam but prior to its fall, Roger Beaumont tackled the question of the role of military elites in conflict. He perceived an inexplicable gap in military and strategic studies. In his words, 'It is amazing how people have accepted the usefulness of corps d'elite without considering their existence as symptomatic or pathological.'[45] The timing of Beaumont's book, along with its subject, suggests several possible motivations for writing on this subject: the growth and expansion of numerous elites during the Vietnam War, as well as their association with the conflict. No elite unit was more associated with Vietnam than the US Army Special Forces, or Green Berets, which in 1961 comprised 1,800 personnel.[46] President John F. Kennedy's public endorsement of the unit that year and authorization of their distinctive headgear the following one raised the public profile of the units exponentially. The Green Berets would be at the forefront of combating limited 'wars of national liberation' sponsored by the Soviet Union. The public fascination with the Green Berets was evident in print, film, and song, including Staff Sergeant Barry Sadler's chart-topping hit in 1966, 'The Ballad of the Green Berets'.[47] Dozens of works written about them included Robin Moore's previously mentioned book as well as articles in publications ranging in quality from *True* magazine to *National Geographic*.[48]

The fall of the Green Berets was as precipitous as their rise. The public disclosure of details during the trial of eight members of 5th Special Forces Group, Project Gamma, detachment B-57, Company E (Special Operations), for the murder of suspected North Vietnamese double agent, Chu Van Thai Khac, soured public opinion. According to Jeff Stein, this case, including pervasive lying up the chain of command, as well as the subsequent acquittal of all eight on the grounds of national security, prompted Daniel Ellsberg to copy the documents later released as *The Pentagon Papers*, which had the effect of turning more Americans against the war.[49] Organizationally, the Green Berets were caught up in the drawdown and budgetary belt-tightening that occurred as a result of President Richard M. Nixon's decision to 'Vietnamize' the war and withdraw US forces. The numbers of Army Special Forces

*in Strategy* (New York: Macmillan, 1946). A longer-term historical perspective is contained in MacGregor Knox and Williamson Murray, eds., *The Dynamics of Military Revolution, 1300–2050* (Cambridge, MA: Cambridge University Press, 2001), Chapter 1.

[44] Andrew Mack, 'Why Big Nations Lose Small Wars: The Politics of Asymmetric Conflict', *World Politics*, 27:2 (January 1975): 181–183; Harry Summers Jr, *On Strategy: A Critical Analysis of the War* (Novato, CA: Presidio Press, 1982).

[45] Roger Beaumont, *Military Elites: Special Fighting Units in the Modern World* (New York: Bobbs-Merrill, 1974), p. ix.

[46] David Tucker and Christopher Lamb, *United States Special Operations Forces* (New York: Columbia University Press, 2007), p. 93.

[47] Alasdair Spark, 'The Soldier at the Heart of the War: The Myth of the Green Beret in the Popular Culture of the Vietnam Era', *Journal of American Studies*, 18:1 (April 1984): 37–38.

[48] *True* (March 1970); Howard Sochurek, 'American Special Forces in Action in Viet Nam', *National Geographic* (January 1965): 38–65.

[49] Jeff Stein, *A Murder in Wartime: The Untold Spy Story That Changed the Course of the Vietnam War* (New York: St Martin's Press, 1992), pp. 383–385.

dropped considerably; three Special Forces Groups (3rd, 6th, 8th) were stricken from the Army rosters in 1971 and another in 1974 (1st SFG) and their active duty personnel slashed from a peak of 10,500 to 3,600 by 1977.[50] There were more Special Forces operators 'in-country' in 1968 from a single Special Forces Group, MACVSOG, and in intelligence detachment than there were collectively in the three remaining groups in 1977.[51]

Beaumont explained the rise and fall of the Green Berets, as well as other military elites including SOF, in a two-stage model he termed a repetitive 'cycle'. In reality, Beaumont's model consists of four stages. The first involves the decision to establish a military elite, which he suggests is driven by one of two criteria: 'a response to enemy success or threat of advantage'.[52] Having made this decision, the next stage involves building the elite through a process of screening, or selection. Although some form of informal screening has occurred with military elites throughout history, the first formal processes were created by psychologists and psychiatrists working within the British Special Operations Executive (SOE) and the American Office of Strategic Services (OSS) during the Second World War.[53] These processes, designed in response to the failure of clandestine agents in the field, went beyond the strength, endurance, and teamwork tests employed by various military elites.[54] In using a 'whole of man' approach, involving both objective and projective measurements, the testing screened for seven general and three special qualifications through a series of tests and personal interviews.[55] Contemporary SOF selection programmes are known for the high rejection rates and strenuous, if seemingly brutal and occasionally fatal stress exercises that bring with them unwanted media attention.[56]

Beaumont suggests scientific psychological screening is problematic and implies the need to staff military elites with sufficient numbers, as well as their undemocratic nature, conspire to undermine their effectiveness.[57] The concern for sufficient

[50] Charles Simpson III, *Inside the Green Berets: The First Thirty Years—A History of U.S. Army Special Forces* (Novato, CA: Presidio Press, 1983), pp. 61; 211; Susan Marquis, *Unconventional Warfare: Rebuilding U.S. Special Operations Forces* (Washington, DC: Brookings Institution Press, 1997), p. 40.

[51] Shelby Stanton authoritatively provides a number of 3,386 officers and enlisted 'in-country' in March 1968 with 5th Special Forces Group, MACVSOG, and Detachment B-63. See Stanton, *Green Berets at War: U.S. Army Special Forces in Southeast Asis, 1956–1975* (Novato, CA: Presidio Press, 1985), p. 72, n.14.

[52] Beaumont, *Military Elites*, p. 4.

[53] Details on the background and effectiveness of the OSS selection process are available in Louie Banks III, 'The Office of Strategic Services Psychological Selection Program', MMAS thesis, Command and General Staff College (Fort Leavenworth, KS, 1995), www.ossreborn.com/files/OSSPsychologicalSelectionProgram.pdf.

[54] For a colourful first-hand account of the British process, known as 'assessment', see William Morgan, *The O.S.S. and I* (New York: W.W. Norton & Co., 1957).

[55] The criteria and testing approach developed and used by the OSS is contained in The OSS Assessment Staff, *Assessment of Men: Selection of Personnel for the Office of Strategic Services* (New York: Rinehart & Co., 1948; reprint, New York: Johnson Reprint Corp., 1978), pp. 264–281.

[56] Tim Collins, 'Tested to Destruction: An Ex-SAS Officer Reveals the Terrifying Selection Process for Special Forces And Says That, Despite This Weekend's Tragedy, Standards Must Not Slip' *The Daily Mail* (16 July 2013), www.dailymail.co.uk/debate/article-2364723/Tested-destruction-Ex-SAS-officer-reveals-terrifying-selection-process-special-forces.html; Andrew Jeong, 'Navy Sailor Who Died after Hell Week Didn't Get Critical Care, Report Shows', *The Washington Post* (13 October 2022), www.washingtonpost.com/national-security/2022/10/13/navy-seal-death-kyle-mullen-hell-week.

[57] Beaumont, *Military Elites*, pp. 173–174.

numbers forms the third stage of his model: expansion. The ranks of military elites are initially staffed with highly motivated and competent personnel. If successful in their initial operations, elites can be expanded in size and activities following the principle of 'more is better'. To meet the seemingly voracious demand for Green Berets in Vietnam, while offsetting considerable combat losses and casualties, enlistment quotas increased eightfold in 1962 and first-time enlistees, or direct accession enlisted and junior officers, were accepted. Expansion, combined with high publicity, attracted recruits to the Green Berets for the wrong reasons and contributed later to their perception, among conventional forces, of consisting of loose cannons or 'cowboys'.[58] Beaumont offers that continued success, or unsuccessful operations, of military elites such as SOF leads to excessively high casualties and losses, on the order or 50 per cent or greater. Successes or failures lead inevitably to the last stage of Beaumont's cycle: destruction. He cites a few grim examples, including the fate of German U-Boat crews and Japanese pilots throughout the Second World War, as well as German and American paratroopers in battles on Crete in the case of the former, and on Gavatu-Tanambogo and Bougainville for the latter.[59] In other cases, destruction takes a different form: disbandment, after the crisis prompting their formation has passed.

There is little in the way of optimism in Beaumont's model; it has an air of inevitability, with only one possible outcome. The reason for his pessimism is his conviction that military elites represent an imperfect solution to poorly defined problems. Decision makers seek solutions to offset the structural or systemic weaknesses of their own militaries in the face of ongoing change in the character of warfare. They become overly enamoured with the qualities but more importantly the potential of military elites, without understanding their inherent costs. These costs take three forms: one qualitative, one quantitative, and one moral. The qualitative aspect Beaumont identifies was articulated earlier and more clearly by the Viscount William Slim, based on his experience with military elites under his command of the British Fourteenth Army in Burma and India during the Second World War.[60] Slim, with limited resources available in a peripheral theatre and in command of a disorganized and demoralized force, viewed military elites such as Orde Wingate's Chindits or Merrill's Marauders as unnecessarily siphoning off talented individuals who either waited for appropriate employment or, when employed, suffered disproportionately high casualties.[61] The psychological victory won by Wingate's Chindits, or 77th Indian Infantry Division as it was officially known, proving that Allied soldiers could operate just as effectively in dense tropical jungle as the Japanese, came at a high cost. Of the 3,000 members of the 77th Indian Infantry Division committed to Operation Longcloth (February–April 1943), only 20 per cent were fit for further duty after their return,

[58] The negative consequences of expansion and publicity are presented in a scathing manner in Simpson, *Inside the Green Berets*, pp. 66–68.

[59] Beaumont, *Military Elites*, pp. 175; 179–180.

[60] Beaumont offers a hypothetical scenario depicting losses of talent in a manner dripping with sarcasm. Ibid., pp. 171–173.

[61] William Slim, *Defeat into Victory* (London: Cassell, 1956), pp. 546–9.

thanks to combat fatalities and effects of long operations in the disease-filled jungle with insufficient medication, sanitation, or means of evacuation.[62]

Quantitative costs among elite units include casualties based on individual traits as well as the conditions of modern war, according to Beaumont. Individually, those seeking out adventure tend to be overly aggressive, romantic, or misanthropic.[63] As a result, they take imprudent risks and routinely throw themselves into circumstances that lead to the levels of losses outlined above, thanks to a degree of delusional self-confidence. The increasingly technological character of war creates a hostile and lethal battlefield for humans, a claim that observers of the recent war in Ukraine support.[64] The moral costs are less visible but not insignificant. In Beaumont's words, 'The use of "special forces" is a virtual acceptance by their masters that they are outside the laws of war and bounds of civility. Most modern corps d'elite were expected to behave more like gangsters than soldiers.'[65] This evidence of moral decay in the profession of arms and the conduct of war has the potential to create a cadre of reactionaries whose qualities can lead them to threaten the very institutions from which they sprang, a theme identified in the previous chapter and articulated well in fiction by Jean Lartéguy in *The Centurions* and later academically by Eliot Cohen.[66] Beaumont and others argue that modern technocratic warfare created the conditions for both the creation and destruction of elite capabilities.

George Monroe, writing more than three decades after Beaumont, shares elements of his pessimism. In an article published in *Armed Forces Journal* in 2008, Monroe develops a model more detailed than Beaumont's. Dubbed 'the Phoenix Cycle' (see Figure 3.1), Monroe identifies a pattern within the US Armed Services, specifically the institutional unpreparedness for and hostility to SOF capabilities. The root of these problems is an institutional preference for conventional war, with its expensive, high-technology platforms, despite their infrequent occurrence.[67] Air Force acquisitions and operations are consumed with preparations for such war, overlooking those necessary for irregular conflicts. When these more frequent conflicts do occur, an eight-step cycle begins. The first four steps consist of the conflict beginning, followed by the institutional recognition that its existing capabilities are

[62] W. J. Officer, 'With Wingate's Chindits', in James Stone, ed., *Crisis Fleeting: Original Reports on Military Medicine in India and Burma in the Second World War* (Washington, DC: Office of the Surgeon General, Department of the Army, 1969), p. 243.

[63] Beaumont, *Military Elites*, pp. 192–193.

[64] Thomas Gibbons-Neff, 'The Battlefield: A Kaleidoscope of Death', *The New York Times* (18 June 2022), www.nytimes.com/2022/06/18/world/europe/ukraine-battlefield-deaths.html; Mykhaylo Zabrodskyi, Jack Watling, Oleksandr V. Danylyuk, and Nick Reynolds, *Preliminary Lessons in Conventional Warfighting from Russia's Invasion of Ukraine: February–July 2022* (London: Royal United Services Institute for Defence and Security Studies, 2022), pp. 53–64.

[65] Beaumont, *Military Elites*, p. 7.

[66] Jean Lartéguy, *The Centurions*, trans. Xan Fielding (Paris: Presses de la Cité, 1960; reprint London: Penguin Classics, 2015); Eliot Cohen, *Commandos and Politicians: Elite Military Units in Modern Democracies*, Harvard Studies in International Affairs No. 40 (Boston, MA: Center for International Affairs, 1978).

[67] George Monroe, 'The Rebirth of the Outback Air Force', *Armed Forces Journal*, 1 February 2008, http://armedforcesjournal.com/the-rebirth-of-the-outback-air-force.

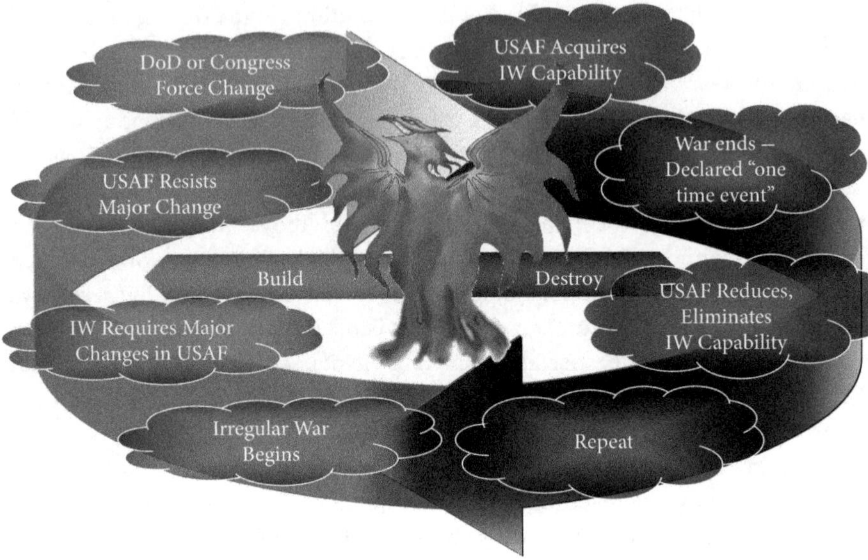


**Figure 3.1** Monroe's Phoenix Cycle.

*Source*: Headquarters Air Force briefing slide, 'Airpower in Irregular Warfare' briefing (unclassified), undated. Author's collection.

ill-suited to it. Recognition, however, does not translate into action due to institutional resistance.[68] In consequence, an external agent needs to force change to build the capabilities necessary to combat irregulars, either the executive or legislative branches of the US government: the president or Congress. Monroe's ideas are consistent with the conclusions of Barry Posen on how peacetime innovation occurs doctrinally within militaries but at odds with David Barno and Nora Bensahel on theatre-level technological adaptation.[69] The final four phases are marked by tactical adaptation and success but ultimately institutional failure. Although specific SOF capabilities are restored and employed successfully, thanks to accessible individual memory in the form of experienced retirees, institutional preferences ensure their dismantling in pursuit of more conventional ones at the 'end' of the conflict, thanks to irregular warfare being viewed as an aberration or 'one-time event'.[70] The next irregular conflict finds the United States having to repeat the process again, as it did in Vietnam, Afghanistan, and Iraq.

[68] Robert Komer refers to such preferences as 'institutional repertoires' in his scathing assessment of US military organizational culture in Vietnam. See Komer, *Bureaucracy Does Its Thing: Institutional Constraints on U.S.–GVN Performance in Vietnam*, RAND Report R-967-ARPA (Santa Monica, CA: Rand Corporation, August 1972), pp. 45–48.

[69] Barry Posen, *The Sources of Military Doctrine* (Ithaca, NY: Cornell University Press, 1984), p. 233; David Barno and Nora Bensahel, *Adaptation under Fire: How Militaries Change in Wartime* (New York: Oxford University Press, 2020), Chapter 6.

[70] Monroe, 'The Rebirth of the Outback Air Force'. A similar theme is at the heart of the article by Dennis M. Drew, 'U.S. Airpower Theory and the Insurgent Challenge: A Short Journey to Confusion', *The Journal of Military History*, 62:4 (October 1998): 809–833.

Monroe specifically argues for the creation of a suitable 'outback air force' within the US Air Force, one able to conduct missions with partner nations in remote locations using 'rugged, affordable' equipment.[71] The kernel of Monroe's 'outback air force' was the 6th Special Operation Squadron (SOS), which conducted the Air Force's Combat Aviation Advising (CAA) mission and was its sole special operations capability for aviation foreign internal defence (FID). Monroe's theory comports with the historical experience of this niche capability. From its roots as a special operations fighter squadron in 1944, the 6th SOS experienced several highs and lows, including efforts to expand the unit into a wing and being awarded its own distinctive brown beret in 2008, as well as being inactivated three times, ultimately and most recently on 6 October 2022.[72] In a rather cruel twist and one inconsistent with Monroe's notions, the decision to deactivate the Squadron came from senior special operations, as opposed to conventional Air Force quarters, in order to 'transition to the AFSOC We Will Need', which is institutional phrasing for platforms and personnel better suited for future conventional war.[73]

While both models of SOF capability lifecycle are useful, they have limitations. In common with the issues related to defining capability discussed at the beginning of the chapter, one focuses on capability too broadly and other too narrowly. Beaumont looks at military elites in the context of nations, societies, and polities fighting major conventional wars. His selection criteria for military elites are unnecessarily broad and ill-defined, ranging from specialized combat arms such as air forces and submarine and armoured services down to small and selective special operations units. His nine categories of corps d'elite, while interesting from an analytical perspective, accomplishes a goal of distinguishing different forms of military elites at the expense of understanding the form of SOF capability.[74] Monroe, in contrast, looks too narrowly at specific types of SOF capability in his model based on mission type. Initially he focuses on two based on existing SOF missions, direct action and FID, and zeroes in specifically on one within a single service, the Air Force: aviation FID.[75] Looking at capability through the lens of SOF roles and missions has the effect of focusing

[71] Monroe, 'The Rebirth of the Outback Air Force'.

[72] A RAND study provided quantitative analysis and evidence in support of expanding the 6th into an irregular warfare wing, or more than quadrupling its size, based on the stated desire of the current USSOCOM commander. Alan Vick et al., *Air Power in the New Counterinsurgency Era: The Strategic Importance of USAF Advisory and Assistance Missions*, RAND Report MG-509 (Santa Monica, CA: RAND Corporation, 2006), pp. 136–143; John Gresham, 'A Warrior's Life: An Interview with Gen. Bryan "Doug" Brown, USA (Ret.)', DefenseMediaNetwork.com (31 May 2011), www.defensemedianetwork.com/stories/a-warrior%E2%80%99s-life-an-interview-with-gen-bryan-%E2%80%9Cdoug%E2%80%9D-brown-usa-ret.

[73] Jonathan McCallum, 'Reassignment of 6th Special Operations Squadron Transitions a Storied Legacy', 492nd Special Operations Wing, www.492sow.af.mil/News/Article/3182721/reassignment-of-6th-special-operations-squadron-transitions-a-storied-legacy. AFSOC We Will Need is the guiding principle published in the following document: Air Force Special Operations Command, *AFSOC Strategic Guidance* (Hurlburt Field, FL: AFSOC, May 2020), pp. 1–2, https://media.defense.gov/2020/May/26/2,002,305,551/-1/-1/1/AFSOC%20STRATEGIC%20GUIDANCE.PDF.

[74] Beaumont's categories of corps d'elite are ceremonial/social, combat proven, praetorian, ethnic/cultural, politico-ideological, romanticist/atavistic, technological/proto-cybernetic, nihilistic, and functional/objective. Beaumont, *Military Elites*, p. 3.

[75] Monroe, 'The Rebirth of the Outback Air Force'.

unduly on specific organizations and units. In Monroe's case, this focus is on the 6th SOS, which limits the explanatory power of his model. Given their scope, neither provide insights on a more interesting question related to SOF capability: how to use it over a prolonged period of time to achieve the desired results without breaking it in the process.

To avoid the twin traps related to capability, casting the net too wide or too narrowly, the next section looks at special operations capability above the level of Monroe but below that of Beaumont. More specifically, it makes the case that special operations capability is best understood in its aggregate and relational form. In doing so, it elevates the subsequent analysis above the level of individual special operations actions, units, and roles and missions. While these remain fascinating and valuable to study and understand, the disproportionate attention they receive inhibits understanding of special operations in a broader, theoretical sense. Keeping the analysis below Beaumont's level avoids a problem well known in the study of irregular and civil wars, '"top-down" perspectives resulting in "teleologies" that can lead to slipshod analysis that does more to mystify than illuminate'.[76] An example illustrates this point. Beaumont tips his hand when he concludes that 'Elite forces are encapsulated delinquency in many instances ... Regression is regression, no matter how thick the disguise'.[77] The basis for this rather sweeping judgement, however, remains unclear.

## Integration: the heart of special operations capability

To further refine the definition of capability, this work uses the concept of integration. The overarching ideal to which a number of militaries subscribe all revolve around the idea of integration, from the American 'Multi-Domain or Joint All-Domain Operations' and the PRC's '*tixi*', or 'systems of systems warfare' to Russia's so-called 'Gerasimov doctrine'.[78] Integration, these various concepts suggest, is necessary given modern technological developments. Theorists and doctrine writers have intuited that the scale and tempo of warfare has increased dramatically, ensuring there are no safe havens or sanctuaries thanks to a proliferation of sensing systems connected to firepower, precision or otherwise, that can kill whatever it can see. In addition, new operating domains or environments will further increase the speed and reach of activities, including cyberspace and space. Even if wars have not been fought in them

[76] Orin Starn, quoted in Stathis Kalyvas, *The Logic of Violence in Civil War* (Cambridge: Cambridge University Press, 2006), p. 44.

[77] Beaumont, *Military Elites*, p. 192.

[78] 'Defense Primer: Army Multi-Domain Operations (MDO)', Congressional Research Service (21 November 2022), https://sgp.fas.org/crs/natsec/IF11409.pdf; Department of the Air Force/Space Force, *The Department of the Air Force Role in Joint All-Domains Operations*, Air Force/Space Doctrine Publication 3-99 (Washington, DC: Headquarters, Department of the Air Force/Space Force, 19 November 2021), www.doctrine.af.mil/Portals/61/documents/AFDP_3-99/AFDP%203-99%20DAF%20role%20in%20JADO.pdf; Valery Gerasimov, 'The Value of Science Is in the Foresight'. *Military Review* (January–February 2016): 23–29; Jeffrey Engstron, *Systems Confrontation and System Destruction Warfare: How the Chinese People's Liberation Army Seeks to Wage Modern Warfare*, RAND Report RR1708 (Santa Monica, CA: RAND Corporation, 2018), pp. ix–xiii.

yet, their importance to communications, control, and information sharing renders future conflict for control of them inevitable.

Despite these perceived realities, the bulk of research and sources on special operations continues to focus unduly on operations by specific individuals, units, or organizations on the land and sea. Studies and work need focus and scope. Consider the range of studies on or about the OSS and the SOE in the Second World War, which includes some excellent detailed academic works on the success or failure of operations in individual countries.[79] Most, however, focus on agents, teams or circuits, sections or regional headquarters, or organizations. There are also a number of excellent works on clandestine air and sea operations, including unit histories of and memoirs from the American 801st/492nd Bomb Group (better known as 'the Carpetbaggers'), British Special Duties squadrons, as well as a two-volume official history of the maritime infiltration operations.[80] Despite this wealth of material, however, there are no sources I am aware of that look at how the interactions of and interdependence between land, sea, and air during this global conflict contributed to their outcomes. The most comprehensive history of American and British special operations during the Second World War conspicuously omits any discussion of specialized air and maritime delivery units.[81] Any theory derived from these sources that continues to look at special operations exclusively from the perspective of the land or sea alone risks missing the whole in its focus on activities in a single domain.

Integration permits a level of qualitative assessment of kind and degree missing from most evaluations of capability. More specifically, integration can occur on different axes, such as horizontal across organizations and vertically, from the highest levels of command and direction to the lowest levels of execution. In addition, it breaks capability free from oversimplified, subjective, and binary assessments. Capability suggests, at the most basic level, either the ability to do something or not: able or unable to perform a task. Integration, through additional exploration, allows a better understanding of capability by providing analytic criteria necessary in answering the questions of special operations posed by Colin Gray: to what degree, or how well or poorly can they perform such missions, if at all.[82]

Moreover, there is an important precedent scholastically for integration's explanatory utility. Paul Staniland identifies the concept of integration as foundational to explain insurgent cohesion and collapse. He constructs a model consisting of two levels of integration set on horizontal and vertical axes to identify types of insurgent

[79] Examples include Roderick Bailey, *The Wildest Province: SOE in the Land of the Eagle* (London: Jonathan Cape, 2008) and Troy Sacquety, *The OSS in Burma: Jungle War against the Japanese* (Lawrence, KS: University Press of Kansas, 2013).

[80] Ben Parnell, *Air Commandos: The Saga of the Carpetbaggers of World War II* (New York: iBooks, 2004); Thomas Ensminger, *Spies, Supplies, and Moonlit Skies*, 2 vols. (Fairborn, OH: Hillborn Wildlife Refuge, 2001; Bloomington, IN: Author Solutions, 2001); K.A. Merrick, *Flights of the Forgotten: Special Duties Operations in World War Two* (London: Arms and Armour Press, 1989); Brooks Richard, *Secret Flotillas*, 2 vols. (London: Routledge, 2004).

[81] Andrew Hargreaves, *Special Operations in World War II: British and American Irregular Warfare* (Norman, OK: University of Oklahoma Press, 2013).

[82] Colin S. Gray, *Explorations in Strategy*, Contributions in Military Studies, No. 164 (Boulder, CO: Greenwood Press, 1996), p. 153.

organizations and their different degrees of cohesion. Staniland's model has power in explaining several cases of insurgent cohesion or collapse in the face of counterinsurgency, based on social coherence. Such coherence, particularly in the form of social ties or bonds formed and strengthened before conflict breaks out, are a key determinant of insurgent resilience. He argues that, 'Social bases represent a complex blend of agency, structure, and contingency that determines the ideational and social resources insurgent leaders can mobilize for war.'[83] The ability of insurgent leaders to mobilize social bases at the local level, while connecting them to other individuals and their networks, is critical. The ability to integrate these different resources together, vertically through command structure and horizontally across disparate elements, is highly desirable to offset the superior resources of the mobilized security apparatus of a state. The major innovation in American special operations over the past twenty years was the recognition that, in order to defeat integrated networks of terrorists and insurgents, SOF needed to develop its own.[84]

A methodological question related to integration must be addressed. Drawing on an aspect of a theory out its context and using it to explain a different phenomenon may seem problematic. The work addresses part of that concern in subsequent chapters, by introducing and developing the concept of vertical integration in Chapter 4 and connecting them together in a general theory of special operations in Chapter 5. As part of existing specialized units and members of a bureaucratic organization, SOF already possess all of the traits of an integrated organization. These traits include 'command and control, strategic assessment, and ideological production.'[85] In addition, they have other attributes of an integrated organization: 'bureaucratic specialization, standard operating procedures, a clear leadership structure, and a reasonably coherent ideology that is disseminated consistently throughout the organization.'[86] Integration begins early in special operations units, during selection processes designed to stress individuals and forge team cohesion, as well as through an innovative root organizational structure based on teamwork.[87]

A close connection exists between the social bonds central to Staniland's work and special operations. Special operations communities possess high levels of social cohesion because of the bonds forged during assessment, selection, training, and the creation of working teams. These bonds have been studied in some detail by anthropologists.[88] Horizontal integration is not just a function of bureaucracy and

[83] Paul Staniland, *Networks of Rebellion: Explaining Insurgent Cohesion and Collapse* (Ithaca, NY: Cornell University Press, 2014), p. 23.

[84] This principle was pithily summarized by General Stanley McChrystal as 'It takes a network to defeat a network.' Stanley McChrystal, 'It Takes a Network: The New Front Line of Modern Warfare', *Foreign Policy* (21 February 2011), https://foreignpolicy.com/2011/02/21/it-takes-a-network.

[85] Staniland, *Networks of Rebellion*, p. 27.

[86] Ibid.

[87] Anna Simons, *The Company They Keep: Life Inside the U.S. Army Special Forces* (New York: Free Press, 1997), pp. 128; 219–220.

[88] For such a study on American special forces, see Simons, *The Company They Keep*. Another, focused on the Norwegian version of the US Navy SEALs or British SBS, is Tone Danielson, *Making Warriors in a Global Era: An Ethnographic Study of the Norwegian Naval Special Operations Commando* (Lanham, MD: Lexington Books, 2018).

standardization. One of the attributes of special operations is offering unorthodox solutions to challenging problems due to organizational latitude and innovative, creative operators. The prevailing theory on SOF, as opposed to special operations, suggests three qualities of an unorthodox mindset allow operators to deal with battlefield friction more effectively than conventional forces: elite warriors, flexibility, and creativity.[89] There is a less visible social dimension, however, across different special operations units, which reflects social connections formed and strengthened as a result of mission focus, as well as extensive training, exercises, and operations. The term used within the special operations community for the phenomenon reflects this concept: 'habitual (working) relationships'.[90] Such relationships build confidence and trust, share knowledge to accomplish tasks more efficiently, and, through repetitive execution, lessen the amount of information required to make decisions, both within SOF and between SOF and conventional forces. The strength of these connections cannot be measured directly but can instead be inferred based on other evidence presented through case studies through process tracing.

To further connect the horizontal axis of integration as a method of assessing capability, it is necessary to look at how special operations are discussed in different works. Many works assess only the value of individual units, including Special Forces or the SEALs. Even those that consider wider special operations communities, for example USSOCOM or the Joint Special Operations Command, are apt to focus almost exclusively on the ground or sea combat elements: the 'trigger pullers' or 'shooters' in the vernacular, as well as the mission leaders. Consider an example: an individual raid or hostage rescue mission. Narratives almost always begin in a manner like Mark Moyar's description of the mission designed to capture several of Somali General Mohamed Farah Aideed's senior leaders in October 1993: 'Lifting off from a dilapidated airfield next to the Arabian Sea, the Black Hawks flew low and fast along the coast, where the Rangers see, between their dangling legs, the azure-blue waters splashing onto black rocks and sand.'[91]

In many narrative accounts of special operations, it is as if these units and their actions occur in a vacuum. Several questions go unanswered but more commonly, are not even asked, particularly those related to support functions such as aviation, intelligence, logistics, and basing, among others. Moyar at least mentions C-141 Starlifters transporting the members of Task Force Ranger to Mogadishu two months before the mission, code named Operation Gothic Serpent but more popularly known as 'Black Hawk Down', in addition to acknowledging the intelligence challenges the special operators faced.[92] Largely missing from many such gripping and evocative narratives is the information that theorist Carl von Clausewitz emphasizes is necessary to

[89] Robert Spulak Jr, *A Theory of Special Operations: The Origins, Qualities, and the Use of SOF*, JSOU Report 07-7 (Hurlburt Field, FL: JSOU Press, 2007), pp. 39–40.

[90] Jason Wesbrouck, Glenn Harned, and Preston Plous, 'Special Operations Forces and Conventional Forces Integration, Interoperability, and Interdependence', *Prism*, 6:3 (December 2016): 91–92; 95 n.26.

[91] Moyar, *To Oppose Any Foe*, p. 193.

[92] Ibid., pp. 199–200.

conduct critical analysis: how did these forces come to be here at this time? More specifically, what were the political discussions at the time, what other alternatives existed, how were forces based and sustained, and what information did they have? What other options were available, discussed, discarded, and why?[93] Such questions related to critical inquiry must necessarily be answered to determine the degree of capability, and more specifically, the level of horizontal integration between units conducting special operations.

Horizontal integration, as a means of evaluating capability in a general sense, also recognizes and accounts for a central but often unstated consideration. It is axiomatic that SOF often need support from elsewhere. This axiom is stated in USSOCOM's previously abandoned but rehabilitated fifth 'SOF Truth'.[94] Such support can include from others SOF units, including combat support and combat service support elements, as well as conventional forces. It can take a range of forms, including mobility and lift, supply, basing, fires or firepower, or information, surveillance, and reconnaissance (ISR).

The best way of understanding these points, and illustrating the connections between capability and integration, is through an historical example. The following narrative summarizes American SOF operations during the so-called 'Tanker War', a maritime expansion of the war between Iran and Iraq that occurred from 1987 until 1988. An edition of the USSOCOM history of special operations depicts the requirement, in spite of the presence of US Navy ships escorting tanker in the Arabian Gulf, that generated the need to send SOF:

> The protection offered by U.S. naval vessels, however, did not stop Iran, which used mines and small boats to harass the convoys steaming to and from Kuwait. To stop these attacks, the U.S. needed surveillance and patrol forces in the northern Persian Gulf and bases for these patrol forces. SOF, including Army helicopters and Navy SEALs and Special Boat Units, had the best trained personnel and most capable equipment for monitoring hostile activity, particularly at night when the Iranians conducted their missions. The Army's special operations helicopter crews trained to fly and fight at night. These helicopters were difficult to spot on radar and relatively quiet, allowing them to get close to a target. Shallow-draft [Naval Special Warfare] NSW patrol boats could ply waters that had not been swept for mines.
>
> In late July 1987, Rear Admiral Harold J. Bernsen, commander of the Middle East Force, requested NSW assets. Six Mark III patrol boats, other special boat assets, and two SEAL platoons deployed in August. At the same time, two MH-6 and four AH-6 Army special operations helicopters and 39 men received orders to the region in a deployment called Operation PRIME CHANCE I.[95]

[93] Clausewitz, *On War*, pp. 157–159; 163–167.

[94] To suggest SOF needed support from conventional forces ran contrary to the need to create an independent organizational culture from 1987 onwards. The fifth 'lost' SOF Truth was rediscovered in 2010. For a short history, see Stephen Fogarty and Jamie Nasi, 'Special Operations Forces Truth—Cyber Truths', *The Cyber Defense Review*, 1:2 (Fall 2016): 19–21.

[95] United States Special Operations Command History and Research Office, *United States Special Operations Command History*, 6th ed. (MacDill Air Force Base, FL: USSOCOM, 2008), p. 33.

In this instance, SOF forces possessed specific skills and equipment that could fill an operational need. They could do so in conditions that maximized surprise (at night, flying quiet helicopters at low level). In doing so, SOF forces could assume much less risk than other conventional forces. Its boats were unlikely to set off magnetic mines, and given that the Army helicopter pilots habitually trained for low-level night flying, there was little chance of accidents. In addition, SOF were able to deploy more rapidly than other potential options for the mission, including the Marine Corps. Finally, a force presence of around one hundred Americans and their equipment was less likely to draw Iranian attention or response.

But how did SOF come to be there at all? More specifically, where were they based and how did they continue to operate—to function? The USSOCOM history continues the narrative in remarkable detail:

> The Middle East Force decided to convert two oil servicing barges, Hercules and Wimbrown VII, into mobile sea bases. Besides obviating the need to ask for land bases, the mobile sea bases allowed SOF in the northern Persian Gulf to thwart clandestine Iranian mining and small boat attacks. Each mobile sea base housed ten small boats, 3 helicopters, fuel, ammunition, equipment, and workshops to support operations, and more than 150 men. In October, the mobile sea bases became operational.
>
> Soon SOF showed what they could do. On the evening of 21 September, one MH-6 and two AH-6 helicopters took off from the frigate Jarrett (FFG-33) to track an Iranian ship, the Iran Ajr. The helicopters observed the Iran Ajr extinguish its lights and begin laying mines. Receiving permission to attack, the helicopters fired guns and rockets, stopping the ship. As the Iran Ajr's crew began to push mines over the side, the helicopters resumed firing until the crew abandoned ship.
>
> Rear Admiral Bernsen then ordered the SEAL platoon from the [USS] Guadalcanal to board the Iran Ajr. Two patrol boats provided security. Shortly after first light, the SEALs boarded the ship and found nine mines and various arming mechanisms. The patrol boats rescued ten Iranians in a lifeboat and 13 in life vests floating nearby. Documents found aboard the ship showed where the Iranians had laid mines, implicating Iran in mining international waters. The Iran Ajr was sunk in deep water on 26 September.
>
> SOF next saw action in mid-October, three days after an Iranian Silkworm missile hit the reflagged tanker Sea Isle City near the oil terminal outside Kuwait City. Seventeen crewmen and the American captain were injured in the missile attack. In Operation NIMBLE ARCHER, four destroyers shelled two oil platforms in the Rostam oil field on 19 October. After the shelling, a SEAL platoon and a demolition unit planted explosive charges on one of the platforms to destroy it. The SEALs next boarded and searched a third platform two miles away. Documents and radios were taken for intelligence purposes.[96]

[96] United States Special Operations Command History and Research Office, *United States Special Operations Command History*, pp. 33–34.

Its narrative concludes with the following assessment of the utility of SOF during these actions in the Tanker War:

> Special Operations Forces provided the critical skills necessary to help CENTCOM gain control of the northern Persian Gulf and counter Iran's small boats and minelayers. Their ability to work at night proved vital, since Iranian units used darkness to hide their actions. The most important lessons to come out of Operation EARNEST WILL were the need to have highly trained Special Operations Forces capable of responding rapidly to crises anywhere around the globe and the vital need for interoperability between conventional and Special Operations Forces. Additionally, based on EARNEST WILL operational requirements, USSOCOM would acquire new weapons systems—the patrol coastal ships and the MARK V Special Operations Craft.[97]

In this example SOF units provided specialized equipment as well as highly skilled operators. Success or failure of SOF operations are often evaluated according to the skills and abilities of individual operators or units. While the skills of the individual SEAL or 160th Special Operations Aviation Regiment pilots were necessary, they alone could not guarantee mission success. By looking at the narrative from a slightly different perspective, that of integration, it widens the aperture of analysis and allows us to observe other explanations of success. More specifically, success was the sum product of SOF and conventional forces, including basing and other support functions, and their ability to work together. By possessing a high degree of horizontal integration within, reinforced through constant training and exercising, US SOF were able to respond rapidly and globally to a unique set of operational conditions. The USSOCOM authors also stress interoperability, and in particular, the ability of SOF to 'plug into' conventional support structures including basing and information-gathering assets. Interoperability, a common military term, captures only one facet of horizontal integration: the interaction between SOF and conventional forces.

The term 'interoperability', like 'joint' or 'all-domain', is problematic analytically for other reasons. It connotes, for example, not only the ability to work together but also a degree of substitution. In addition, 'interoperability' always conveys a sense of baseline functionality, rather than honed and refined elements combining to form something greater than the sum of its parts. Consider two examples. North Atlantic Treaty Organization (NATO) standardization agreements, or STANAGs, specify different ammunition rounds, including the 7.62mmx51 'heavy' cartridge and the 5.56mmx45 'light' one all the up to the heaviest artillery one through a Committee for Standardization, NATO Standardization Office, and Staff Group.[98] The general characteristics for rounds ensures a level of interoperability within and between the forces of all thirty NATO member states. Such interoperability assists in easing the

[97] Ibid., p. 35.
[98] 'Standardization', NATO (14 October 2022), www.nato.int/cps/en/natohq/topics_69269.htm.

logistics burden in times of crisis or conflict as forces can share ammunition regardless of nationality. Rounds can vary within calibre, however, to perform different functions. Such variation includes basic modifications, including phosphorus 'tracers' in the rear cavity of the bullet to aid in visually observing ballistic trajectory to adjust fire all the up to 'match grade' ammunition, carried by sniper units and produced with much higher tolerances to guarantee greater accuracy at long range. In addition, there is a wide range of propellant charges and bullet weights, measured in grains, all of which have different ballistic properties. The point is that 7.62mmx51 ammunition guarantees a basic level of interoperability between NATO forces, at the cost of specialized performance and function particularly for the most unique ammunition.

At the lowest level of implementation, SOF teams, there is a level of interoperability necessary under trying conditions and losses due to combat or other reasons. Team members are functionally cross-trained to ensure the unit can fulfil basic tasks: moving, shooting or calling fires, communicating, and providing medical assistance. Such cross-training, however, is a secondary skill set to individual specialization (communications, combat medicine, combat engineer, etc.).[99] At the broader organizational level, some of the same rationale is still true. Different service SOF components—land, sea, and air—are interoperable and interchangeable to a degree. Interoperability ensures a level of maximal use of available resources. Lost in the process is the optimal value that specialized units bring to the team, as the next example suggests.

Prior to 11 September 2001, US Army Forces Groups, and their subordinate units, were assigned to specific geographic regions. More specifically, 1st Group was assigned to the Pacific and Northeast Asia, 3rd Group Europe, 5th Group the Middle East, and 7th Group Latin and South America. The rationale for doing so was that by focusing on a specific region, each Group could better understand the specific socio-political dynamics within each. Prior to the establishment of USSOCOM, such expertise resided in 'area specialist teams' with whom incoming units would liaise.[100] A primary key to unlocking such understanding resides in linguistic specialization, which allows for cultural familiarity and the development of habitual working relationships with counterparts in foreign nations in the region. After 11 September, and especially after the overthrow of Saddam Hussein in Iraq in 2003, units from different Special Forces Groups were fed into rotation schedules to cover the mounting requirements levied on an increasingly stressed 5th Group, which was responsible for the region.[101] Pre-deployment training consisted mainly of optimizing weapons skills and drills, with language training largely ignored. As one Special

[99] For a history of the numbers of an Army Special Forces 'A-Team' and the rationale for its underlying composition, see Troy Sacquety, 'The Evolution of the Special Forces (SF) Operational Detachment-Alpha (ODA)', *Veritas*, 19:1 (2023): 23–27.

[100] Cleveland and Egel, *The American Way of Irregular War*, p. 22.

[101] US Government Accountability Office, *Special Operations Forces: Better Data Necessary to Improve Oversight and Address Command and Control Challenges*, GAO-23-105163 (Washington, DC: Government Accountability Office, October 2022), p. 22; Philip A. Buswell, 'Keeping Special Forces Special: Regional Proficiency in Special Forces', thesis, Monterey, CA: Naval Postgraduate School, 2011.

Forces battalion commander lamented, 'I only wish my operators treated language skills in the same manner they do weapons proficiency.'[102]

In breaking regional alignment to meet rotational requirements in Iraq and Afghanistan, Special Forces Groups and their subordinate units proved they were interchangeable with one another and interoperable to a degree with other units, at the loss of their primary value: language skills, cultural familiarity, and the development of a refined sense of the dynamics occurring within the country. In the words of an overly sympathetic assessment of the situation, in Afghanistan, by 2009:

> the essential fact of supporting the locals was forgotten as ideas of employing ever greater numbers of US troops gained traction. This flaw should have been noted by special operators, but the cold, hard reality was that they were unprepared to carry out this role except at the most tactical level. They were not employed correctly, and they did not have the opportunity—after [a] brief initial window—to employ themselves correctly.[103]

In other words, to meet ongoing and increasing requirements, SOF did not take control sufficiently early in Afghanistan of planning efforts, given their specific knowledge, skills, and abilities to serve as an integrating hub for capabilities, but rather allowed themselves to be used in an interoperable manner to meet increased general demand and rotational requirements.

The idea that interoperable does not mean interchangeable was further reinforced during the controversial first deployment of Marine Corps special operators to Afghanistan in 2007, after their recent incorporation into USSOCOM. The response of a Marines Special Operations Command (MARSOC) team to an ambush in Nangahar resulted in numerous Afghan casualties, the removal of the team and its rotation back to the US, and an investigation into the incident. The MARSOC company commander has argued that documentary evidence supports his claim of proportional response to the existing threat and that 'Fox Company's Marines were sent to war without the overarching direction and ground-level logistical assistance they would need to sustain themselves in theatre and prove effective on the battlefield' and his higher headquarters 'set us up for failure.'[104] Andrew Feickert, a professional researcher for the US Congress, reached an unusually candid and detached conclusion to explain the outcome: 'in an effort to relieve stress on special operations forces as well to demonstrate the effectiveness of newly-created MARSOC units ... this

[102] Lieutenant Colonel Mark Davis, conversation with author, Fort Campbell, KY (October 2008). Divorce rates among several 5th Group battalions, as well as other SOF units, skyrocketed during this time, the result of additional stress created on families by constant deployments. Gregg Zoroya, 'Special Forces' Marriages on Shaky Ground, Survey Shows', *USA Today* (9 July 2013), www.usatoday.com/story/news/nation/2013/07/09/marriages-military-families-elite-troops-stress/2432243.

[103] Linda Robinson, *One Hundred Victories: Special Ops and the Future of American Warfare* (New York: PublicAffairs, 2013), p. xviii.

[104] Andrew DeGrandpre, 'Task Force Violent: Part 2: The Unforgiven', *Military Times* (March 12, 2015), www.militarytimes.com/special-projects/task-force-violent/2015/03/12/task-force-violent-the-unforgiven-part-2.

unit was deployed into a situation that it may not have had the experience, cultural awareness, or maturity to handle'.[105]

Interoperability, as the preceding discussion suggests, is related to but not the same thing as integrated. The nature of the integrative relationship between SOF and conventional forces, when it exists, varies according to the stakes at risk and intensity of competition. In large-scale conflicts, including total wars and higher intensity limited wars such as the Korean Conflict, large numbers of conventional forces are often required. SOF have limited value based on their limited numbers and additional combat support and combat service support requirements. The primary value of SOF in such conflicts is to improve the strategic effectiveness of conventional forces over time.[106] Put simply, SOF should support conventional forces to the degree they can and avoid the tendency of being misused as a replacement for conventional forces because of their highly refined skills. The 2020 AFSOC Strategic Guidance recognizes this reality when it states, 'For most of its history, SOF has served as a key element of the joint force; often *supporting*, occasionally *supported*' (italics original).[107] In more limited conflicts, especially against irregular opponents, SOF are likely to play a more central role due to their honed skills and equipment. Conventional forces should not only integrate with, but also support SOF in such conflicts, as the consequences associated with their less discriminate use of force in such environments can be significant, such as influencing local populations to join or support insurgent or terrorist groups.[108]

A common error in some discussions of limited conflicts lies in the suggestion that more effective integration of military forces is not only necessary but sufficient to defeat insurgents, terrorists, or other armed groups. The irregular conflicts these actors engage in are highly political in nature and their central feature is the struggle for legitimacy.[109] As such, some of the most effective and indeed necessary responses include political, economic, and social reforms and programmes outside of the authority of the military.[110] For this reason, there is another aspect of special operations capability: integrating SOF horizontally with other government departments and agencies. In these conflicts, SOF cannot be decisive but their capabilities and actions, sufficiently integrated horizontally, can greatly improve the effectiveness of those other government department and agencies.[111]

The idea of integration, as it applies to special operations or SOF, is hardly novel. Authors almost a quarter century ago noted its significance. John Arquilla, for example, identified its importance in his anthology on special operations. He

[105] Andrew Feickert, *U.S. Special Operations Forces (SOF): Background and Issues for Congress*, RS21048 (28 June 2007), p. 6.

[106] James D. Kiras, *Special Operations and Strategy: From World War II to the War on Terrorism* (London: Routledge, 2006).

[107] AFSOC, *AFSOC Strategic Guidance*, p. 1.

[108] Emile Simpson, *War from the Ground Up: Twenty-First-Century Combat as Politics* (Oxford: Oxford University Press, 2013), pp. 7–8.

[109] Simpson, *War from the Ground Up*, pp. 26–29; 243–244.

[110] Cleveland and Egel, *The American Way of Irregular War*, p. 166.

[111] James Kiras, 'Special Operations and Strategies of Attrition', *Infinity Journal*, 2:4 (Fall 2012): 18–21.

concluded, 'The integration of special and conventional operations constitutes perhaps the most important analytical issue highlighted by these readings. Time and again, one sees success or failure hinging entirely upon the smooth coordination of regulars and raiders.'[112] The significance of integration is not limited historically, nor does it only apply to conventional military forces. Other authors have noted the need for greater integration between SOF and the Central Intelligence Agency (CIA) or observed it has already occurred to a high degree.[113] Thomas Henriksen notes the degree of integration between USSOCOM and the CIA has been part of a long-standing evolutionary process whose roots go all the way back to the founding of the United States.[114] Taking an argument related to SOF integration in a different direction, on the basis of outcomes in Afghanistan, Libya, and elsewhere, some authors have suggested the integration of SOF with airpower and local militias represents a paradigmatic shift in the conduct of war, or a new pattern for success.[115] In an extension of the argument to its limit, two journalists suggest this integration has created a 'new deterrence model' designed to dissuade terrorist attacks while a senior USSOCOM planner argues they form the basis of three 'new deterrence concepts' for strategic competition short of war: focused and unconventional deterrence, as well as unconventional resilience.[116]

As the preceding examples suggest, those works that identify integration, however, do so often for several narrowly focused purposes. Integration between special operations and either intelligence or conventional forces can make the former more effective operationally, by providing the type of information or support operators need to plan and execute missions. The reasons for integration include the limitations inherent to special operations and SOF, such as limited numbers, as well as the nature of the contemporary terrorist adversaries, which possess high levels of regional and global mobility. Works in favour of 'America's New Way of War' or 'the Afghan Model', for example, suggest integration occurred as a happy coincidence or convergence of several factors at the right time, permitting wars to be waged quickly and cheaply. In contrast, other authors emphasize the significant downsides of such integration.

[112] John Arquilla, ed., *From Troy to Entebbe: Special Operations in Ancient and Modern Times* (Lanham, MD: University Press of America, 1996), p. xviii.

[113] Colflesh Khoon Liat, 'Whither Special Forces? The Strategic Relevance of Special Operations', *Pointer*, 39:2 (2013): 38; Konrad Trautman, *Special Operations Intelligence: Guiding the Tip of the Spear*, JSOU Special Report, U//FOUO//REL to USA, FVEY (MacDill Air Force Base, FL: JSOU Press, November 2014), p. 1.

[114] Thomas Henriksen, *Eyes, Ears, and Daggers: Special Operations Forces and the Central Intelligence Agency in America's Evolving Struggle against Terrorism* (Stanford, CA: Hoover Institution Press, 2016), pp. 2–3.

[115] Norman Friedman, *Terrorism, Afghanistan, and America's New Way of War* (Annapolis, MD: Naval Institute Press, 2003); Craig Wills, Thomas Griffith Jr, and Richard Andres, 'Winning with Allies: The Strategic Value of the Afghan Model', *International Security*, 30:3 (December 2005); Richard Andres, 'The Afghan Model in Northern Iraq', *Journal of Strategic Studies*, 29:3 (2006): 395–422; Géraud Laborie, 'The Afghan Model More Than 10 Years Later', *ASPJ Africa & Francophonie* (3rd Quarter 2013): 49–60.

[116] Erik Schmitt and Thomas Shankar, *Counterstrike: The Untold Story of America's Secret Campaign against Al Qaeda* (New York: Times Books, 2011), pp. 46–64; Bob Jones, 'Deterring "Competition Short of War": Are Gray Zones the Ardennes of Our Modern Maginot Line of Traditional Deterrence?', in Aaron Bazin, ed., *On Competition: Adapting to the Contemporary Strategic Environment*, JSOU Report 21-5 (MacDill Air Force Base, FL: JSOU Press, 2021), pp. 42–50.

There are two identifiable themes in these works. The first suggests that a high degree of integration with conventional forces is detrimental to special operations. In essence, special operations and SOF lose their unique mindset, skills, and abilities by seeking to demonstrate their value and utility to conventional forces when integrating together for planning or missions.[117] The second theme notes the integration between SOF and the CIA but, recalling Beaumont's concern, suggests it is illegal, immoral, or both. Robert Chesney assesses the current level of integration, or convergence, between special operations and intelligence as 'remarkable' and points to burden and resource sharing, as well as legal flexibility: 'In pursuit of the same counterterrorism goals, they share information and personnel, and both construct and execute operations jointly—toggling between operating under CIA or military authorities as circumstances may dictate.'[118] The cost of such operational efficiency, however, is difficulties for members of Congress in distinguishing military from intelligence activities outlined in US Code and performing its oversight functions, as the fallout from the ambush of special forces soldiers killed in Niger in 2017 detailed in Chapter 4 suggests. Ineffective or weak oversight will inevitably lead to abuses of power, as has occurred throughout American history, or the creation of a de facto American empire globally and self-reinforcing system of 'forever wars' without end.[119]

These purposes and their themes are largely focused on, and limited to, the American experience in and with special operations. They reflect a long liberal democratic tradition, particularly in the United States, of suspicion of any forms of secret activity within open societies, either on general principle despite a history of covert action that goes back to the founding of the Republic.[120] As interesting and informative as it is, the American experience and its elements of practice, legal authority, and accountability are not universal. Other nations have different forms of government, openness, oversight, and requirements—in a word, different experiences. Given them, a general theory of special operations should rise above the specific functional and systemic limitations of any one country, especially the one whose experience so dominates the literature.

Integration between special operations and others occurs most frequently by necessity rather than design. In the Chindits example, the creation of the Project 9, later the 1st Air Commando Group, under the command of Philip Cochran, was mandated by President Franklin Roosevelt and Prime Minister Winston Churchill as a means of addressing the limitations identified during Operation Longcloth. Special air forces provided much-needed mobility, resupply, fire support, and casualty evacuation for the Chindits, expanded to brigade size, for the much more ambitious

[117] Thomas Adams, *US Special Operations Forces in Action: The Challenge of Unconventional Warfare* (London: Frank Cass, 1998), pp. 10–11.

[118] Robert Chesney, 'Military-Intelligence Convergence and the Law of the Title 10/Title 50 Debate', *Journal of National Security Law and Policy*, 5 (2012): 578–579.

[119] Jennifer Kibbe, 'The Rise of the Shadow Warriors', *Foreign Affairs*, 83:2 (March/April 2004): 102–115; Nick Turse, *The Changing Face of Empire: Special Ops, Drones, Spies* (Chicago, IL: Dispatch Books, 2012); Mark Danner, *Spiral: Trapped in the Forever War* (New York: Simon & Schuster, 2016).

[120] Stephen Knott, *Secret and Sanctioned: Covert Operations and the American Presidency* (New York: Oxford University Press, 1996).

Operation Thursday.[121] More broadly speaking, necessity is the result of several factors internal and external to special operations. Internally, integration is required to overcome capacity issues inherent to SOF, in the form of limitations, such as small numbers of operators and insufficient support functions. There are too few operators who can be used for any period, beyond a short independent raid, that do not require the services and support of outside forces, agencies, and departments. Consider SOF of many NATO member states deployed to Afghanistan. Several small states or medium powers, including Denmark, the Netherlands, Norway, and Canada, have numbers of operators ranging from the hundreds to just over one thousand. All have created overarching special operations organizations in order to better integrate within their SOF, as well as identify specific internal support functions. These support functions, however, cannot provide all the requirements necessary to sustain their national SOF to function effectively halfway around the world over a rotational period. As a result, the SOF of these forces rely on levels of integration with their own national and other NATO SOF and conventional forces, as well as the support of other agencies and departments. For example, Canada makes use of its close ties with the United States, as well as other partnerships and alliances, to receive intelligence information, in-country mobility, and other support functions.[122]

Externally, integration has been driven by factors including the misuse and destruction of SOF, the proliferation of SOF units, poor prior performance or failure in conflict, strategic shock and surprise, and novel adversary approaches. Perhaps the most well-known and frequently cited example is the conclusion reached by the members of the Special Operations Review Group, more commonly known as the Holloway Commission, in their report to the joint chiefs of staff and the secretary of defense after the failure of the SOF rescue operation conducted in Iran in 1980. Otherwise remarkable individual efforts, such as Richard Meadows' intelligence gathering activities in Iran, were subsumed by a combination of ill-defined command relationships, excessive compartmentalization of activities complicating the flow of information, ad hoc planning, and bad luck.[123] More importantly, shortly after the failure the Chief of Staff of the Army and legendary special operator, General Edward 'Shy' Meyer, 'recognized that special operations could not be effective without full cooperation and support among the services and other U.S. Government agencies'.[124] As Chapter 6 makes clear, the creation of British Directorate of Special Forces (DSF) in 1987 was driven by similar factors, including poor coordination

[121] R.D. Van Wagner, *Any Place, Any Time, Any Where: The 1st Air Commandos in World War II* (Atglen, PA: Schiffer Military, 1998), pp. 20–29.

[122] Much of this support is implicit in yet crucial to the success of the nascent Canadian counterterrorism SOF unit, Joint Task Force-2, to Afghanistan in 2002 and again in 2005, including intelligence, mobility, and fire support. Bernd Horn, *'We Will Find a Way': Understanding the Legacy of Canadian Special Operations Forces*, JSOU Report 12-2 (MacDill Air Force Base, FL: JSOU Press, February 2012), pp. 39–45.

[123] Special Operations Review Group, *Rescue Mission Report* (Washington, DC: Joint Chiefs of Staff, 1980), pp. 57–62.

[124] Marquis, *Unconventional Warfare*, p. 73.

between different special operations units during the 1982 Falklands War, as much as it was a recognition that the basis for British crisis decision making relied too much on informal personal relationships and ad hoc arrangements and not on standing processes of coordination between ministries.

Looking to other national experiences, limitations in the Dutch SOF ability to contribute in Afghanistan in 2009 were apparent when they were integrated into a broader International Stabilization Force (ISAF) special operations organization in Afghanistan and demands upon them increased. In the candid words of two official historians, 'The deployment [of Task Force 55] was plagued by a structural lack of necessary enablers, support, and resources. The fact that these assets had to be borrowed from Task Force Uruzgan and Regional Command South was far from practical for all units involved.'[125] Deriving insights from an ongoing conflict is fraught with peril. A group of authors from the Royal United Services Institute, however, assess that disparate levels of preparation and integration ensured the failure of Russian special operations during the first 72-hours of their attempt to seize Kiev, the capital of Ukraine.[126]

Novel adversary approaches, and the surprise and problems they generate, can also be a significant driver for integration. In January 1961, Nikita Khrushchev announced the Soviet Union's support for 'wars of national liberation', to compete with Western influence in the undeveloped or underdeveloped countries of the world without it escalating to possible nuclear war. In private, he stressed to President John F. Kennedy his belief that such wars conducted by 'the people', meaning guerrillas and insurgents, were 'sacred'.[127] At the time there were eight active 'national liberation' movements with associated insurgent forces. US Special Forces by themselves stood no chance of defeating such threats; to succeed they worked through an existing integrative structure—the Military Advisory Assistance Group, or MAAG.[128] Special Forces advisors not only benefitted from the existing relationships that existed within the MAAG but between it and the State Department, USAID, and the like. They brought value in training highly specialized and mobile counter-guerrilla units and provided first-hand information and assessments of the realities in the countryside, by virtue of their activities where the insurgents were operating.

One illustration of the success of special forces operating in an integrated manner, to combat a novel threat, occurred in Bolivia in 1967–1968. The details of Ernesto 'Che' Guevara's failed attempt to foment a *foco*-based insurgency in that country

[125] Arthur ten Cate and Martijn van der Vorm, *Callsign Nassau: Dutch Army Special Forces in Action in the 'New World Disorder'* (Leiden: Leiden University Press, 2016), pp. 286–287.

[126] Zabrodskyi et al., *Preliminary Lessons in Conventional Warfighting from Russia's Invasion of Ukraine*, pp. 26; 28.

[127] Office of the Historian, *Foreign Relations of the United States, 1961–1963, Volume V, Soviet Union*, Document 95. Paper Prepared in the Department of State, 12 June 1961 (Washington, DC: Department of State, 1998), pp. 245–246.

[128] MAAGs successfully operated until 1981, when Congress stipulated a requirement for prior (and explicit) approval. To better understand the scope of their use, see J. Kenneth Fasick, *Profiles of Military Assistance Advisory Groups in 15 Countries*, ID-78-51 (Washington, DC: General Accounting Office, 1 September 1978), www.gao.gov/assets/id-78-51.pdf.

are well known. One portion contributing to that failure resulted from the training of a specialized unit, the Bolivian Rangers, to respond to any internal guerrilla threat. A Special Forces Mobile Training Team from the 8th Special Forces Group, MTT-BL 404–67X, arrived in Bolivia in April 1967 under the command of Major Ralph 'Pappy' Shelton.[129] Shelton and his fifteen men set up a training programme and in October 1968, members of B Company, 2nd Ranger Battalion, Bolivian Rangers, tracked down and captured or killed the remaining column of 'El Che's' guerrillas.[130]

While the death of Che after his capture remains a source of controversy and conspiracy in some circles, the SOF effort should be placed in context. The Kennedy government tied military and economic aid to the country to efforts at expanding land reform started in 1952. SOF forces worked closely with CIA assets, including legendary paramilitary officer Felix Rodriguez, as well as the embassy country team.[131] US Agency for International Development and the Peace Corps conducted development efforts in impoverished rural areas, a pattern discerned elsewhere by *foco* theorist and ally of Guevara, Régis Debray.[132] Some focus on the coincidental and parallel efforts in Bolivia; Che's Guevara's attempt to foment insurgency and attract 'the people' to his revolution, and the MTT's efforts to create a unit to deny guerrillas their advantages in mobility. Yet the defeat of Che Guevara was necessary but not sufficient to quell the threat of insurgency in the country; it was one part of an integrated effort to do so.[133] By combining the unique properties of different departments and agencies together in a coordinated manner, with SOF efforts to create a specialized counter-guerrilla unit, the US had a level of temporary success in Bolivia. Twenty years later, and in the face of a more elusive threat in the form of cocaine production and shipping, a distinct lack of integration of SOF with other US government departments and agencies led to failure of the counternarcotics mission there.[134]

Threats become novel when they operate contrary to established norms. Whether it is terrorists enacting a global plot to fly aircraft into commercial buildings or states mobilizing a hodgepodge of political organizations, militias, and social organizations to create unrest, upheaval, and force a rapid vote on political succession, as in the case of Russia in the Crimea in 2014, such events reveal gaps and seams in existing capabilities, legal frameworks, organizational approaches, or a combination of

[129] Charles Briscoe, 'Welcome to Bolivia, MTT-BL 404-67X', *Veritas*, 4:4 (2008): 62–67.

[130] Kenneth Finlayson, 'The 2nd Ranger Battalion and the Capture of Che Guevara', *Veritas*, 4:4 (2008): 94–97. For a first-hand account from the Bolivian Ranger commander's perspective, see Gary Prado Salmón, *The Defeat of Che Guevara: Military Response to Guerrilla Challenge in Bolivia* (Westport, CT: Praeger, 1987), pp. 172–177.

[131] Rodriguez's role in Che's capture, including the last words the latter spoke before his execution, only became known with the publication of the former's autobiography twenty-one years after the event. See Felix Rodriguez and John Weisman, *Shadow Warrior: The CIA Hero of a Hundred Unknown Battles* (New York: Simon & Schuster, 1989).

[132] Régis Debray, *Revolution in the Revolution*, trans. Bobbye Ortiz (London: Penguin, 1967), p. 52.

[133] To understand the broader context of the SOF mission to Bolivia, see Mitch Weiss and Kevin Maurer, *Hunting Che: How a U.S. Special Forces Team Helped Capture the World's Most Famous Revolutionary* (New York: Berkeley, 2013), pp. 9–23.

[134] Cleveland and Egel, *The American Way of Irregular War*, pp. 41–42.

some or all of them.[135] When confronting a more powerful opponent one cannot match in terms of capabilities, rational actors actively seek out these vulnerabilities to exploit them to accomplish objectives.[136] As the definition in the Introduction suggests, special operations are extra normative when their activities seek to operate within and between these seams. From a position of weakness, special operations seek to overcome strengths either by attacking them directly, or more often, by exploiting vulnerabilities in an opponent's normative system. Special operations seek to overcome strengths, exploit weaknesses, or respond in a manner that attempts to fill gaps and seams in capabilities, normative frameworks, and operational approaches.

As a result, special operations capabilities are more than just individual units, their equipment, or their assigned roles and missions. Capability is a function of the level of integration existing between special operations units. SOF need to integrate due to limitations in their size, the complexity of their missions and requirements to be supported over extended periods of time, and their inherent enabling value strategically. Integration refers its connections within SOF units to one another and between SOF and conventional forces or other government agencies and departments. This connectivity, termed 'horizontal integration', can be assessed based on its level of robustness formed and maintained through habitual working relationships. The more interactions SOF have with others, the more robust the interactions and the greater the degree of horizontal integration. The fewer the interactions, the less the degree of horizontal integration. Horizontal integration is an important determinant of the effectiveness of special operations, operating in dynamic circumstances against an opponent over an extended period. The value proposition and utility of SOF capability is a function of the level or degree of horizontal integration within SOF, and between SOF and conventional forces and other government agencies and departments. Integrating SOF horizontally across the span of other government departments and agencies is crucial. Such horizontal integration allows SOF to tap into other resources and authorities, while providing useful enabling functions and information beyond their reach.

## Assessing SOF integration

Integration, as with capability, is not a novel concept when it comes to special operations. Its use, however, is often limited to its identification in successful cases or the need for more or greater integration to ensure more effective future operations but often stops there analytically. Various authors have pointed out its need

[135] Activities outlined clearly in Assessing Revolutionary and Insurgent Strategies (ARIS) Project, *'Little Green Men': A Primer on Modern Russian Unconventional Warfare, Ukraine 2013–2014* (Fort Bragg, NC: United States Army Special Operations Command, 2014), pp. 54–58.

[136] Conditions noted in the current strategic competition between the US with Russia, China, and Iran. See Seth Jones, *Three Dangerous Men: Russia, China, Iran, and the Rise of Irregular Warfare* (New York: W.W. Norton & Co., 2021), pp. 18–20.

in explicit or implicit ways, some negative and others positive. Most negative discussions focus on the misunderstanding and friction that arises culturally and organizationally between SOF and conventional forces. Authors including Beaumont, John Arquilla, Colin Gray, David Tucker and Christopher Lamb, Charles Cleveland, myself, and a host of others have written about the misuse of SOF, to their detriment or even destruction, given the inability of conventional military minds to understand their proper use in different contexts. Gray, for example, wrote in 2015 that:

> There is extraordinary need for SOF to seek to maintain what can be termed mission integrity, not least because SOF assets lend themselves all too easily to misuse and even abuse by a high command that may well not really understand the particular capabilities that SOF could bring as assets to a conflict.[137]

Tucker and Lamb conclude that while problems of misuse have been solved by 'SOF ... now in charge of SOF', they have morphed into overuse in ill-defined roles and missions.[138] In other words, military misunderstandings have been replaced by a civilian one. This notion of misuse was important culturally within American SOF during the political struggle for organizational independence in the mid-1980s. It continues to serve a purpose, along with the SOF Truths, in the sustaining of a collective SOF identity. Along the way, however, this notion of misuse has cemented itself within American SOF to the point of being conventional wisdom in the minds of unconventional warriors. In addition, I have heard this myth of misunderstanding and misuse invoked by Western SOF leaders during political battles over budgets, resources, and influence.

Ralph MacDonald's remarkable 1994 work about special operations cohesion provides a positive example of integration. He defines cohesion doctrinally but then introduces four different types that range in scope and scale: horizontal, vertical, organizational, and societal. He draws the different types from a study of military cohesion on the Falklands War published by Nora Kinser Stewart in 1988 for the US Army Research Institute for the Behavioral and Social Sciences.[139] While Kinser Stewart's study pursued lines of inquiry based on military sociology and military psychology, MacDonald focuses the bulk of his effort instead on what he terms as 'building blocks' of cohesion organizationally.[140] These building blocks are doctrine, education, and training. He then details various efforts within each to achieve cohesion and identifies a number of problems.[141] As proof

[137] Colin S. Gray, *Tactical Operations for Strategic Effect: The Challenge of Currency Conversion*, JSOU Special Report (MacDill Air Force Base, FL: JSOU Press, November 2015), p. 27.

[138] David Tucker and Christopher Lamb, *United States Special Operations Forces*, 2nd ed. (New York: Columbia University Press, 2020), pp. 46; 217.

[139] Ralph E. MacDonald, *Cohesion: The Key to Special Operations Teamwork*, Research Report No. AU-ARI-94-2 (Maxwell Air Force Base, AL: Air University Press, October 1994), p. 2.

[140] Nora Kinzer Stewart, *South Atlantic Conflict of 1982: A Case Study in Military Cohesion*, Research Report 1469 (Alexandria, VA: US Army Research Institute for the Behavioral and Social Sciences, April 1988), pp. v–vi.

[141] MacDonald, *Cohesion*, pp. 17–33.

perhaps of recurring cycles of interest within the military, MacDonald's proffered solution to enhance cohesion is one currently pursued across the US Department of Defense: wargaming.[142] The end result of his study is a number of recommendations designed to increase AFSOC and USSOCOM cohesion and improve preparation prior to conflict.[143] MacDonald's work introduces several interesting ideas but uses them for positive, prescriptive purposes. Reflecting a military inclination for problem solving, he identifies problems, offers concepts to explore them, and provides solutions in the form of actionable recommendations. MacDonald, however, employs the ideas related to cohesion pragmatically and functionally to improve preparations to make SOF more responsive and tactically effective.

To better understand integration (since 'cohesion' is better suited to military studies in psychology and sociology), this section develops a framework and means to identify its presence and assess its extent. While a quantitative approach might yield useful insights for integration, I use qualitative ones here for several reasons. Beyond design issues, access to useful data, in both quality and quantity, is unlikely for reasons outlined previously. Given them, I start with a general evaluation of SOF capability where data is available: the size and organization of SOF units and their stated missions.

As noted previously, numerical strength alone is problematic in determining capability. Surveying special operations units is necessary but insufficient. To develop a more refined understanding of capability, SOF resources are compared to missions they are expected to accomplish (tasking). These missions or taskings follow the categories established in the previous chapter: intelligence and reconnaissance, treachery and sabotage, direct and indirect shock, military advising and assistance, and asymmetric response. Doing so performs several functions. First, it provides continuity throughout the evaluation process across this work. Second, and crucially, it liberates the evaluation from American SOF 'core activities', a number of which are specified legally in Title 10 US Code § 167.[144] Works on SOF often repeat these core activities along the same lines as the SOF Truths and do not submit them to objective analysis, hindering more holistic efforts to assess actual capabilities.[145] Serious evaluation of capability can be stifled given the association of core activities with specific SOF units and communities. Lumping roles and missions into categories reflecting mission time horizons, profile, and the like, including 'direct and indirect' or 'special warfare and surgical strike', have been unsuccessful in separating activities from

[142] MacDonald, *Cohesion*, chapter 3. For a contemporary discussion of why and how to 'revamp wargaming' that bears a resemblance to similar calls during the 1980s and 1990s, see Jeff Appleget, Jeff Kline, and Rob Burks, 'Revamping Wargaming Education for the U.S. Department of Defense', CIMSEC (Center for International Maritime Security) (17 November 2020), https://cimsec.org/revamping-wargaming-education-for-the-u-s-department-of-defense.

[143] McDonald, *Cohesion*, pp. 59–63.

[144] '10 U.S. Code § 167—Unified Combatant Command for Special Operations Forces', Cornell Law School, www.law.cornell.edu/uscode/text/10/167.

[145] For example, see Lloyd Newton and Richard Corson, 'USSOCOM View on Capabilities', in Richard Shultz Jr, Robert Pfaltzgraff Jr, and W. Bradley Stock, eds., *Special Operations Forces: Roles and Missions in the Aftermath of the Cold War* (Washington, DC: Department of Defense, 1995), pp. 245–250.

SOF identities.[146] Using the mission categories outlined above permits assessment of SOF capabilities without relying on problematic US categories, definitions, or standards.

A general evaluation of capability is possible by comparing numbers, units, and missions against the conditions and standards to which SOF tasks are performed.[147] These include:

- the commitments or operating distances at which SOF will notionally operate (near, medium, and far)
- the time horizon for activities (short to long)
- the operating environments (ranging from friendly and permissive to hostile and denied), or their survivability
- the scope and complexity of activities (from simple and narrow to compound and highly technical)
- their sustainability, or regenerative ability, to conduct ongoing operations rather than single missions.

Most SOF are capable of conducting a single indirect-shock mission, or raid. Are they able to mount an operation quickly, at far distances, in denied environments? If so, can they repeat the feat? How quickly, or for what period of time?

The British conducted numerous maritime special operations during the Second World War for intelligence and reconnaissance purposes as well as indirect shock. The ability to do so sustainably, however, dramatically decreased the farther the targets were away from the coastline. One of their most famous indirect-shock missions was against shipping targets in the German-occupied French port of Bordeaux from 7 to 12 December 1942. The mission, codenamed Operation Frankton and employing operators using folding canvas kayaks, was successful in damaging six ships, a remarkable feat given the distance and conditions involved. The courage and persistence demonstrated by the operators is undeniable. Too much was asked of them and their leadership, particularly commanding officer Herbert 'Blondie' Hasler, were too eager to demonstrate their value and utility.[148] As a capability, however, the Royal Marines Boom Patrol Detachment (RMBPD) was not sustainable: 80 per cent of the operators involved were either captured or killed by the Germans.[149] The British mounted only three such operations, in much less drastic conditions, in the remaining three years of the war: two in Asia, with mixed results, and one

[146] Linda Robinson, *The Future of U.S. Special Operations Forces*, Council Special Report No. 66 (New York: Council on Foreign Relations, April 2013), pp. 10–11; US Army Special Operations Command (USASOC), *ARSOF 2022: Part 1* (Fort Bragg, NC: US Army Special Operations Command, 2013), p. 10.

[147] James D. Kiras, 'The Role of Special Operations Forces: Past, Present, and Future', *Pointer*, 37:2 (Winter 2011): 80–95.

[148] Ewen Southby-Tailyour, *Blondie: A Biography of Lieutenant Colonel H.G. Hasler, DSO, OBE, Croix de Guerre, Royal Marines* (London: Leo Cooper, 1998), pp. 78–84.

[149] C.E. Lucas Phillips, *Cockleshell Heroes: Epic Exploit of the War* (London: Pan, 1956), pp. 4; 234–235; 244.

in the Mediterranean.[150] Andrew Hargreaves suggests 'it is debatable whether [its] operations alone justified the time, equipment, and personnel taken up in forming the unit'.[151]

Returning to the analytic framework, two overarching categories assist in this evaluation of horizontal integration, processes and connections. Processes comprise observable elements of preparation including a coordinating body and its level of management of various SOF units (from non-existent or loose to tight), joint or interoperable training exercise scope and intensity, doctrine and professional development and distribution, and the level of inclusiveness of non-SOF organizations in SOF preparation activities. Non-SOF organizations include conventional military units, as well as governmental departments, ministries, and agencies, and non-governmental organizations. Connections consist of the level and degree of exchanges that occur, in terms of personnel and information, between SOF as well as non-SOF elements. Exchanges include working groups, education, wargaming activities, and liaising.[152] To overcome the problem of insufficient information, processes and connections are evaluated along a spectrum of formality, from none through informal to developed, formal ones. Operational use, if it occurs, provides unique insights into both capability and integration as different dimensions exposed during crisis and conflict. From these, a qualitative judgement of horizontal integration within a SOF capability is possible, in four degrees: high, medium, low, or none.

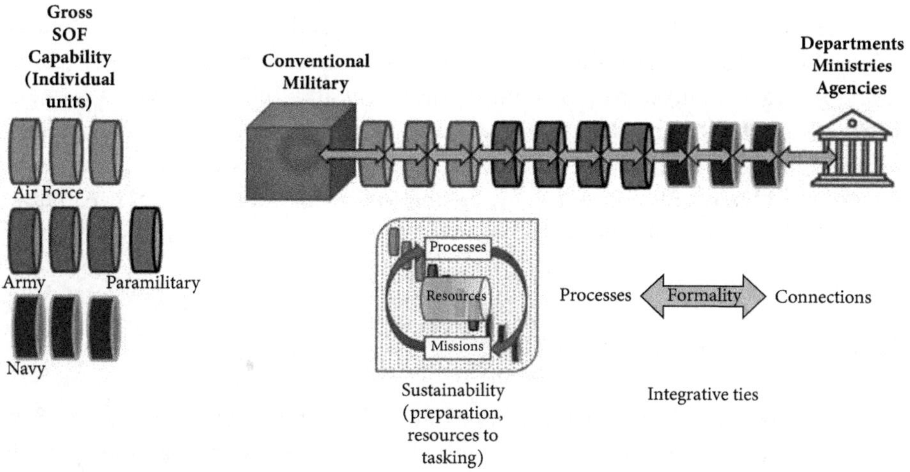


**Figure 3.2** Framework to assess SOF capability and horizontal integration.
*Source*: Author's work.

[150] Paddy Ashdown, *A Brilliant Little Operation: The Cockleshell Heroes and the Most Courageous Raid of WW2* (London: Aurum, 2012), p. 346.

[151] Hargreaves, *Special Operations in World War II*, pp. 255–256.

[152] Wesbrock, Harned, and Plous, 'Special Operations Forces Integration, Interoperability, and Interdependence', p. 91.

The goal of this analytic process, whose framework is depicted in Figure 3.2, is not to develop a baroque or self-contained theory of SOF capability. Instead, the purpose is to provide the first and foundational pillar for the development of a general theory of special operations. In addition, this approach permits a more objective assessment of SOF capability and integration beyond mythmaking and mystique, national prejudices, personal preferences, or institutional agendas. This analytic framework can supply two distinct images of capability integration. The first, based on sufficiently detailed information, offers a snapshot of integration at a given point in time. The second, resulting from a lack of such information, provides an evaluation of integration over time based on observable trends and patterns. Although this framework for capability and integration is designed with an organizational or institutional analysis in mind, it is general and flexible enough to scale up or down to other levels of analysis. The examination of control in SOF in the next chapter develops a framework encompassing vertical integration.

Theoretical frameworks and processes should be more than just rigorous. They ultimately must have explanatory power.[153] The ability of such concepts to illuminate a phenomenon, or provide a deeper understanding of them, is what determines their utility. To test the explanatory power of capability and horizontal integration, as part of a general theory of special operations applicable universally and not just to the 'great powers', the next section examines SOF capabilities of a South Asian regional power, India. India provides a useful example for several reasons. Its SOF have developed over time in a non-Western democracy, have expanded considerably over the past dozen years, and have been involved in a range of different activities including subversion and sabotage, indirect-shock missions, and asymmetric response (countering insurgents and terrorists) both domestically and outside of India's borders. India possesses SOF across the different branches of the armed services, in considerable numbers. But how capable are they? Assessing the level of horizontal integration tests the validity of this concept in evaluating India's special operations capabilities and also answers this question.

## Capability example: India

India's special operations capability has evolved dramatically over six decades. Several drivers led to its initial development. They include the size and geographic diversity of the country, a continuing territorial dispute with Pakistan over the territory of Jammu and Kashmir, hostilities flaring into intermittent conflict with Pakistan, recent military humiliation at the hands of the PRC in a separate territorial dispute that erupted into a month-long conflict (the Sino-Indian War, 1962), and the geopolitics of the Cold War. Regarding the latter, representatives from the

[153] Harold Winton, 'An Imperfect Jewel: Military Theory and the Military Profession', *The Journal of Strategic Studies*, 34:6 (December 2011): 855.

CIA contacted leaders in India in 1954, and again in 1962, to help them establish a 'stay behind', or unconventional warfare force, along the lines of others established in Europe.[154] This first force, trained by American special forces and designated as the Special Frontier Force (SFF), was succeeded by the Meghdoot Force after the Sino-Indian War. This unit was the brainchild of Major Megh Singh who, in a manner resembling David Stirling's efforts to establishing the SAS, lobbied his military superiors repeatedly for its establishment.[155] This force was renamed and expanded into two battalions that were used for raids in the short but decisive Indo-Pakistan War in 1971. The two battalions were themselves expanded into eight Parachute (Special Forces) battalions under Army control.[156] Other SOF units were created in response to the growth in international terrorism and the dramatic hostage rescue conducted in Britain at Princes Gate in 1980 (see Chapter 6), as well as domestic crises such as the ill-conceived Operation Blue Star to recover control of the Golden Temple in 1984. India's foreign intelligence agency, the Research and Analysis Wing (RAW), established the Special Group, or 22 SF, in 1983 to conduct a range of overt and covert operations.[157] The National Security Guard (NSG), or 'Black Cats' as they are more popularly known, were formed as a unit of India's national police in 1986. The NSG, under the control of the Ministry of Home Affairs, are designed to respond to domestic crises. The Indian Navy developed its own SOF unit in 1987, first the Marine Special Force and later renamed the Marine Commandos (MARCOS), to retake oil drilling platforms seized by terrorists in the Bombay East quadrant in the Indian Ocean.[158] The Indian Air Force, spurred on by attacks on air bases in Jammu and Kashmir by Pakistani-supported proxy forces, created its own SOF unit, the Garuds, in 2004.[159]

These various SOF units overlap in some mission areas but these have changed in scope or focus over time according to changing institutional and national priorities. In an overt manner, the Parachute (Special Forces), along with MARCOS, are tasked with performing raids or indirect-shock missions into enemy territory up to a depth of 20 kilometres or at sea. These raids are against facilities, such as bases and staging areas, or military platforms such as aircraft, naval vessels, or submarines. Most raids are envisioned against Pakistan and in support of other conventional operations, largely during wartime. A development within the last decade are punitive raids in peacetime, dubbed 'surgical strikes', as part of an evolution in India's national security doctrine. The extent to which some of these forward-deployed forces contribute

[154] P.C. Katoch and Saikat Datta, *India's Special Forces: History and Future of Indian Special Forces* (New Delhi: Vij Books, 2013), pp. 3–4; Otto Fiala, *Resistance Operating Concept (ROC)* (MacDill Air Force Base, FL: JSOU Press, 2020), pp. 167–169.

[155] Katoch and Datta, *India's Special Forces*, pp. 5–6.

[156] Ibid., p. 92.

[157] Sandeep Unnithan, 'Operation Bluestar [sic]: The League of Shadows', *India Today* (10 February 2014), www.indiatoday.in/magazine/the-big-story/story/20140210-operation-bluestar-indira-gandhi-singh-bhindranwale-army-raw-paramilitary-unit-800036-1999-11-29.

[158] G.M. Hiranandani, *Transition to Guardianship: The Indian Navy, 1991–2000* (New Delhi: Lancer, 2013), Chapter 12.

[159] Kaloch and Datta, *India's Special Forces*, p. 92.

to the doctrine's specified deterrence mission is unclear. The new doctrine came as a result of changes to Indian national security policy, which moved it away from a defensive to more responsive and offensive approach following the testing and formal acknowledgement of nuclear weapons in both India and Pakistan, the Kargil Conflict in 1999 which raised the spectre of nuclear escalation, and raids by Pakistani proxy forces on Indian Air Force and Army bases in Jammu and Kashmir in 2004 and 2016.[160] The 'surgical strike' doctrine was also used to justify an Indian cross-border raid into Myanmar in 2015.[161] Other overt tasks for military SOF include quick reaction in defence of military facilities, as well as targeted killing and counterinsurgency missions against the Liberation Tigers of Tamil Eelam (LTTE) as part of India Peace Keeping Force (IPKF) operations in Sri Lanka (1987–1990).[162] To counter terrorism and other threats domestically, the NSG is responsible for self-described hostage rescue, direct action, and personal protection missions. Special Group have planned to conduct overt hostage rescues as well as raid high-value facilities such as the Kahuta nuclear facility in Pakistan.[163] This unit also performs high-profile event security and crisis response. SFF acts as a commando unit, spearheading attacks and counterattacks and bolstering conventional infantry in disputed high-mountain areas in which India has repeatedly clashed with Pakistan and China. Overt aerial intelligence, surveillance, and reconnaissance missions are performed through a variety of collection methods and platforms in the RAW's Aviation Research Centre.[164]

The scope of covert missions of Indian SOF, much less their details, are difficult to determine given the paucity of available information. From reputable sources and media reports, though, a general picture emerges. It includes cross-border special reconnaissance and intelligence missions conducted by the SFF into China, the Parachute (Special Forces) into Pakistan, and the Special Group into Pakistan, China, and Bangladesh.[165] In addition, as part of its doctrine, Indian SOF have provided military advice and assistance in the raising, training, and supporting of proxy forces in East Pakistan (Baluchistan), Pakistan, and Burma, for intelligence gathering, treachery, and subversion.[166] Other covert missions include so-called 'pseudo' missions for

[160] Vivek Raghuvashni, 'India Embraces New War Doctrine', *Defense News* (8 November 2004), p. 14; Vinay Kaura, 'India's Pakistan Policy: From 2016 "surgical strike" to 2019 Balakot "airstrike"', *The Round Table*, 109:3 (2020): 277–287. For a Pakistani response suggesting India's doctrine is both nefarious and needlessly escalatory, see Ikram ul-Majeed, 'The Farce of the "Surgical Strike"', *Defence Journal*, 20:3 (October 2016): 76–77.

[161] Rashi Iyengar, 'Indian Special Forces Cross Border into Myanmar in Pursuit of Insurgents', *Time.com* (12 June 2015).

[162] Kaloch and Datta, *India's Special Forces*, pp. 40–43.

[163] Ajit Dubey and Namrata Biji Ahuja, 'Close Encounters of the Covert Kind', *The Week* (9 October 2016), www.theweek.in/theweek/cover/covert-operations.html.

[164] Praveen Swami, 'RAW to Shut Down Its Covert Air Wing, Assets Will Go to NTRO and IAF', *The Indian Express* (18 September 2005), https://indianexpress.com/article/india/india-others/raw-to-fold-its-covert-air-wing.

[165] Dubay and Ahuja, 'Close Encounters of the Covert Kind'.

[166] Amarjit Singh, 'How to Make Proxy War Succeed in Baluchistan', *Indian Defence Review* (20 December 2014), www.indiandefencereview.com/news/how-to-make-proxy-war-succeed-in-baluchistan/2; Subir Bhaumik, 'Guns, Drugs, and Rebels', *Gateway to the East*, India Seminar #550 (June 2005), www.india-seminar.com/2005/550/550%20subir%20bhaumik.htm.

asymmetric response, such as counterterrorism and counterinsurgency. These missions, conducted by the Tiger Troop of the Para Commandos as part of the IPKF, followed a model established by the British first in Malaya and later in Kenya in the fight against the Mau-Mau.[167] Operators dressed as guerrillas, patrolled in contested areas, gathered information, conducted ambushes, and spread disinformation to increase distrust among their opponents.[168] In keeping with the principle of plausible deniability, Special Group and RAW are suspected of conducting indirect-shock missions, such as the overseas targeted killings of individuals hostile or threatening to the state, as the recent case of Hardup Singh Nijjar in Canada and others elsewhere suggest.[169]

Exact data about Indian SOF remain elusive but general figures are available. They include fifteen battalions (13,500 personnel) of Parachute (Special Forces), six battalions (4,800) of SFF, approximately two battalions (1,500 personnel) each of Garuds and MARCOS, four squadrons within the Special Group (240), a like number in the Aviation Centre (240), and somewhere between 7,500 and 10,000 NSG, divided into two main groups the Special Action Group (SAG) and Special Ranger Group (SCG). Other divisions of the NSG include five Special Composite Groups (SCG) and an Electronic Support Group (ESG).[170] The very rough estimates contained here suggest an overall SOF strength of 33,000 personnel.

By all accounts selection processes and individual training of Indian SOF operators are formalized and excellent but increasingly opaque as one approaches the institutional level. Some initial training of cadres was done by foreign SOF and several selection processes are modelled along Western or Israeli lines. MARCOS commandos, for example, undergo at least fifteen weeks of basic training followed by advanced skills training for those who pass through initial selection.[171] Online sources suggest training is rigorous with a 'pass rate' of candidates as low as 2 per cent.[172] This rate for NSG commandos is higher, at between 30–50 per cent but advanced training involves expending volumes of ammunition to guarantee accuracy levels of 85 per

[167] Kaloch and Datta, *India's Special Forces*, p. 39; Lawrence Cline, *Pseudo Operations and Counterinsurgency: Lessons from Other Countries* (Carlisle Barracks, PA: Strategic Studies Institute, June 2005), pp. 2–6.

[168] James D. Kiras, 'Terrorism and Irregular Warfare', in John Baylis, James Wirtz, Eliot Cohen, and Colin S. Gray, eds., *Strategy in the Contemporary World: An Introduction to Strategic Studies* (Oxford: Oxford University Press, 2002), p. 225.

[169] Smriti Chaudhary, 'Mossad or RAW: Indian Spy Agency Is as Lethal as Israeli Mossad But Remains behind the Curtains: Former RAW Official', *The EurAsian Times* (7 December 2020), https://eurasiantimes.com/mossad-or-raw-indian-spy-agency-is-as-lethal-as-israeli-mossad-but-remains-behind-the-curtains-raw-official; Greg Miller, Gerry Shih, and Ellen Nakashima, 'An Assassination Plot on American Soil Reveals a Darker Side of Modi's India', *The Washington Post* (29 April 2024), www.washingtonpost.com/world/2024/04/29/india-assassination-raw-sikhs-modi.

[170] Organization and figures compiled from a range of sources, including Katoch and Datta, *India's Special Forces*, pp. 92–95; Ewen Southby-Tailyour, ed., *Jane's Amphibious and Special Forces*, 17 (Coulsdon, Surrey: Jane's Information Group, January 2007), pp. 288–289; and various online sources, including Wikipedia.

[171] Hiranandani, *Transition to Guardianship*, p. 82.

[172] 'Indian Navy Marine Commandos (MARCOS)', Boot Camp and Military Fitness Institute, n.d., https://bootcampmilitaryfitnessinstitute.com/elite-special-forces/indian-elite-special-forces/indian-navy-marine-commandos-marcos.

cent in close quarter battle situations such as hostage rescues.[173] The rapid expansion of both the NSG and Parachute (Special Forces) in the wake of the Mumbai terrorist attack for the former and almost tripling and proposed quadrupling of the latter between 2004 and 2019 have strained limited training facilities and personnel.[174] In line with the practice of SOF of other nations, some Indian candidates have attended and passed foreign SOF selection programmes. Above the individual to the unit level, Indian SOF participate in an unspecified level of joint training and have conducted exercises over the past twenty years with Russia, Kyrgyzstan, Israel, Australia, the United Kingdom, and the United States, among others.[175] While training exercises have led, in some cases, to the acquisition of new unit capabilities such as MARCOS' ability to conduct airborne operations at sea, their number and distribution appear transactional in nature, with informal linkages at best.[176] Skills and capabilities, however, are tempered by a 'lack of quality manpower and officers' overall, particularly a professional, empowered, non-commissioned officer SOF cadre.[177]

A recurring pattern in the history of Indian SOF, however, points to several significant barriers to a high degree of horizontal integration. For example, although intelligence cues from RAW provided information to enable a decapitation strike against LTTE leaders meeting at Jaffna University in October 1987, the Parachute (Special Forces)'s lack of dedicated lift or habitual working relationships with Indian Air Force pilots led to a complete lack of surprise in subsequent missions. SOF were inserted piecemeal over twenty sorties using a limited number of available helicopters.[178] During the response to the terrorist attack in Mumbai in late November 2008, the NSG took almost 10 hours to have forces in place and ready to rescue hostages. While some of the delays were political and bureaucratic in nature, a key factor related to capability was a lack of dedicated airlift standing by on alert. In this specific case, responding NSG units were based 165 miles away from a suitable aircraft, which had to be crewed, fuelled, and readied for take-off.[179] An Il-76 was

[173] Kaushik Sircar, *Guns and a Butterfly: The Black Cats' Strike—Operation Ashwamedh* (Mumbai: Himalaya Books Pty. Ltd., 2009), pp. 10–11.

[174] Vivek Raghuvashni, 'Indian Army Special Ops Gain Troops, Command', *Defense News* (24 January 2005), pp. 9; 78–79; 85; Devansh Shah, 'Here's How the NSG Black Cat Commandos Train to Face the Worst', *MensXP* (1 March 2019), www.mensxp.com/health/motivation/50743-here-rsquo-s-how-the-nsg-black-cat-commandos-train-to-face-the-worst.html.

[175] 'Press Time: Into-Russia Games Slated', *Defense News* (29 August 2005), p. 3; 'India-Kyrgyzstan Joint Special Forces Exercise Concludes at Bakloh', *SP's Land Forces* (6 April 2022); Vivek Raghuvashni, 'India to Create Strike Force', *Defense News* (12 May 2003), p. 34; 'Army Chief Extends Invitation to Australian Special Forces to Train with India', *SP's Land Forces* (14 July 2016); Vivek Raghuvashni, 'India Seeks U.S. Spec Ops Arms', *Defense News* (19 January 2004), p. 17; 'For the Record: U.S.–India Counterterror Pact', *Defense News* (26 July 2010), p. 3.

[176] Hiranandani, *Transition to Guardianship*, p. 82.

[177] Kaloch and Datta, *India's Special Forces*, p. 96; Arjun Subramaniam, *A Military History of India since 1972: Full Spectrum Operations and the Changing Contours of Modern Conflict* (Lawrence, KS: University Press of Kansas, 2021), p. 119.

[178] Kaloch and Datta, *India's Special Forces*, p. 40.

[179] Angela Rabasa et al., *The Lessons of Mumbai*, Occasional Paper OP249 (Santa Monica, CA: The RAND Corporation, 2009), p. 10.

required to transport the twenty tonnes of equipment the NSG deemed necessary to resolve the hostage crisis.[180]

In India, as in other countries, the political or bureaucratic response to crises or failures often consists of reorganizing SOF or proposing consolidation. Too often, these changes made do not address the underlying issues related to integration. For the NSG, distributing forces into five regional SCGs answered the question of speed of response to crises. Some observers characterize the central issue of post-crisis reforms as one of 'good intentions being poorly implemented, or else the original vision and intent getting lost'.[181] Other measures designed to improve integration take inordinately long periods to implement from their initial proposal. The creation of a combined headquarters to coordinate the activities of military special operations, first proposed in 2005 and identified as essential by the members of the Naresh Chandra Committee on Defence Reforms in India in 2011, only occurred in 2018.[182] The resulting organization, the tri-service Armed Forces Special Operations Division (AFSOD), which announced its first joint, combined special operations exercise the following year in the Andaman and Nicobar Islands and operationally deployed to Jammu and Kashmir one month later, exerts administrative but no operational control over Indian SOF.[183]

AFSOD will have an uphill struggle in the coming years. In an unusually candid and critical assessment, former Special Forces Lieutenant General P.C. Katoch details a catalogue of SOF maladies, including: insufficient, inoperable equipment, a lack of dedicated support assets; poor facilities; duplication of effort in terms of missions; high annual personnel turnover; and haphazard systems acquisitions.[184] The latter is a function of India's courting of equipment from a host of different nations for political rather than functional reasons, not to mention issues with domestic major programme development of assault weapons and tanks.[185] The maladies Katoch mentions, combined with an insufficient understanding of their value and limitations politically and militarily, reduces Indian SOF to 'one tenth' of the capability of a single component of American SOF, the special forces, despite having

[180] Sandeep Unnithan, *Black Tornado: The Three Sieges of Mumbai, 26/11* (Noida: HarperCollins India, 2020), p. 80. It is noteworthy that Unnithan's book has become the basis for a television series, *State of Siege 26/11*, serving to bolster the image of the NSG.

[181] Vinayak Dalmia, Vindra Kapoor and Saikat Datta, 'India's Enduring Challenge of Intelligence Reforms', ORF Issue Brief No. 428, December 2020, Observer Research Foundation, www.orfonline.org/research/indias-enduring-challenge-of-intelligence-reforms.

[182] Grummet Kanwal, 'Restructuring Defence Reforms for National Security', *Indian Defence Review*, 30:3 (July–September 2015), www.indiandefencereview.com/news/restructuring-defence-reforms-for-national-security/2.

[183] Vishal Thapar, 'Tri-Services Special Forces Division Fielded for First Time in Large Scale Exercise', *SP's Naval Forces* (18 October 2019); Manjeet Negi, 'Army, Navy, IAF Special Forces Deployed in Kashmir to Hunt Terrorists', *India Today* (7 April 2022), www.indiatoday.in/india/story/army-navy-iaf-special-forces-deployed-in-kashmir-to-hunt-terrorists-1622243-2019-11-24.

[184] Katoch and Datta, *India's Special Forces*, p. 96.

[185] For one example, see the following assessment of Indian small arms acquisition and development: Javin Aryan, 'India's Quest for a New Assault Rifle Is Almost Over', Observer Research Foundation (26 February 2021).

equal overall numbers.[186] Writing two years after publishing his initial assessment, Katoch implies the situation facing Indian SOF is even more daunting, citing basic equipment, navigation and targeting, and mobility shortfalls.[187]

Indian joint doctrine specifies how special operations should be applied as a form of military power, following the lead of the aspirational joint doctrine for special operations published by the HQ Integrated Defence Staffs in 2008.[188] In addition, the joint doctrine outlines theoretical 'integrated and joint structures' but the reality for special operations is that better integration is inhibited by persistent and bitter infighting, 'lack of coordination', 'rampant turf wars', as well as levels of distrust, suggesting informal processes and connections.[189]

Amit Kumar assesses that the Indian government is taking some of the 'right steps' to address the issues inhibiting SOF integration but categorizes the range of structural and institutional problems as: doctrinal challenges, lack of centralized command structure, internal security duties, political will, intelligence apparatus synchronization, equipment for special forces, and human resource issues.[190] Other observers are more circumspect in their assessment of Indian SOF capabilities. Given an inability to bridge special operations with precision airpower in neighbouring Pakistan, they conclude special operations have 'an even more limited capability' in any form of regional military power projection.[191] Thus, Indian SOF are likely to remain as instruments of reprisals against Pakistan, in tit-for-tat exchanges, designed for little above '*identifying the enemy's reversible strengths, and innovatively striking at them*' (italics original), as well as symbolic extra-regional prestige missions, as with the deployment of a SOF contingent as part of an Indian contribution to the United Nations' peacekeeping mission in the Congo.[192] Based on the information presented above, manning appears to outweigh equipping, units possess mixed levels of competence, and processes and connections within the Indian system, while formalizing, are largely informal in nature. In addition, while the creation of an integrating headquarters is welcome, it appears to have far too little in the way of resources or authorities to pull Indian SOF away from their armed service parents. Therefore, India possesses a '*low*' degree of SOF capability based on its assessed level of horizontal integration.

[186] Katoch and Datta, *India's Special Forces*, p. 95.

[187] P.C. Katoch, 'Equipping Infantry and SF Units: Current Weaknesses', *SP's Land Forces* (April 2015): 6.

[188] Ministry of Defence, *Joint Doctrine Indian Armed Forces*, JP-01/2017 (New Delhi: Ministry of Defence, 2007), p. 26; 'Joint Doctrine for Special Forces Operations Unveiled', Livefist, Defence.com (30 September 2008).

[189] Ministry of Defence, *Joint Doctrine Indian Armed Forces*, JP-01/2017, p. 50; 'How Special Forces of India, MARCOS and Para SF Compete against US Navy SEALs?' *The EurAsian Times* (28 July 2019), https://eurasiantimes.com/why-special-forces-of-india-marcos-para-sf-are-no-match-to-us-navy-seals.

[190] Amit Kumar, 'India's Special Forces: An Appraisal', *Journal of Defence Studies*, 9:2 (April–June 2015): 165–173.

[191] Joshi Shashank, *Indian Power Projection: Ambitions, Arms, and Influence*, Whitehall Papers, 85:1 (2015): 112.

[192] Prem Mahadevan, a senior researcher with the Center for Security Studies at the ETH Zurich, has Pakistan in his sights exclusively and sees little value for Indian SOF elsewhere, including China. Prem Mahadevan, *An Eye for an Eye: Decoding Global Special Operations and Irregular Warfare—A Vision for India* (New Delhi: Wisdom Tree, 2014), p. xi.

Despite excellence in the operational ranks, and in addition to overheated expansion and retention issues, the primary impediments to integration occur just where it is most necessary: the institutional level. Unsatisfactory operational experiences driving external and internal calls for reform and restructuring have largely gone unheeded; there seems to be no driving force or imperative capable of breaking the institutional hold that different Indian armed services hold over military SOF or institutional barriers that exist with other domestic security and intelligence agencies. One reason for continued independent institutional intransigence is bureaucratic competition brought on by a lack of overarching guidance or vision for employment from the executive level, either through misunderstanding, disinterest, or some other explanation.

## Conclusion

This chapter made the case for the first of two conceptual elements of a general theory of special operations. The starting point is the notion of 'capability', which can be difficult to pin down intellectually. Having identified the twin challenges to defining capability, overly general and vague and too platform or equipment specific, this chapter surveyed some of the barriers related to special operations specifically. These barriers include secrecy and lack of information and profligate use of the term without specifying its level or providing context. 'Capability' becomes a catch-all term, interpreted however the recipient cares. To provide some initial grounding to the inquiry, this chapter next explored two general descriptive models for the lifecycle of SOF capability: Roger Beaumont's 'selection-destruction' cycle for military elites and George Monroe's 'Phoenix Cycle'. Beaumont overgeneralizes and is overly determinative, based in part on his preformed opinion regarding the appeal of military elites and the danger they pose to democracies. Monroe, in contrast, overgeneralizes based on his goal of ensuring a very finite capability, embodied in a specific aviation SOF unit, is preserved, nurtured, and grown.

Capability is best understood at the institutional level, or across SOF. Integration is a net result, the product of investment in personnel and equipment, as well as training and exercises, that leverage the strengths of each to offset their vulnerabilities. This section also distinguished integration from jointness and interoperability and identified the most common internal and external stimuli for it. In addition, the chapter explored several examples whose outcomes are better understood from the perspective of integration. Too often with SOF, monocausal explanations that favour one community over another are employed. The chapter next looked at integration analytically. Although numerous authors have pointed out the value or need for integration, including a study on special operations cohesion, its precise meaning and scope is often unclear, left to the beholder, or held hostage by institutional imperatives. This specific section offered an analytic framework for assessing SOF capability institutionally in latitudinal terms: the level or degree of horizontal integration occurring between SOF units and with conventional forces and other government ministries, departments, and agencies. The framework begins with a gross estimate of special operations capability based on numbers or units; a 'bean count' of SOF. From there, examining functional expectations of SOF against these resources can identify any initial mismatches. Processes and connections developed between units are observable in the form of joint and combined training exercises, as well as the influence of a higher integrating headquarters.

The case study of Indian SOF provided the means to test the value of the analytic model. Indian SOF perform admirably in a range of incredibly demanding environments. Their units possess some of the finest and most experienced high-altitude mountain operators in the world. In addition, MARCOS is routinely mentioned among the list of the most capable maritime SOF. Despite this individual excellence and tactical acumen, Indian SOF collectively are beset by a range of institutional and structural issues that inhibit more effective horizontally integrated SOF capability. If Indian SOF remain focused specifically on cross-border reprisal actions against Pakistan and domestic stability and security, this will not remain a problem. Successive Indian political and military leaders, however, argue for Indian regional hegemony necessitating power projection. Given these stated aspirations, Indian SOF do not possess the tight, formalized connections or processes to turn them into reality. The mismatch of political aspirations and lack of will to resources form one form of control over special operations. The next chapter explores political and military control of SOF as the next element in a general theory of special operations.

# 4
# Control of Special Operations and Vertical Integration

If capabilities identify what is possible for special operations to accomplish under a set of operational or functional conditions, the next dimension for a theory of special operations is control. Control is a foundational concept in several military theories, and central to more recent ones on insurgent behaviour and the use of violence in civil war. Military theories tend to view control in terms of either command of forces or of an operating domain. Control of special operations is a product of the type of government or political system directing them. In addition, it captures the relationship that exists between the collective agent, or special operations capability and the directive principal of that government or political system whose goals and objectives they are working towards. Within the academic disciplines of politics, political science, and sociology there is a healthy literature that explores civil–military relations, as well as the direction and oversight of clandestine or covert activities within political systems. Unfortunately, that literature tends to view 'the military' in monolithic terms, not allowing for greater or lesser levels of agency among different military components, as well as privileging a top-down view of decision making. In addition, the literature is dominated by American perspectives and focus, which limits its transferability to and explanatory value for other national contexts.

The principal and agent relationship lays the foundation for a different element of a special operations theory, once suitably expanded and modified to account for their unique aspects. This chapter examines the relationships that exist between different principals and agents, from the decision-making executive authority through levels of command and control, to special operations capabilities. What makes special operations unique in this hierarchy of relationships is the concept of risk, an elusive concept but one central to special operations. This chapter develops a model of risk assessment and prioritization that works jointly from the decision maker down and from operational units up. Challenges in risk translation and communication threaten the stability and coherence of control. Control up and down the chain of command, and its type or degree, is labelled here as 'vertical integration', in contrast to the horizontal integration discussed in Chapter 3.

Vertical integration is a central concept to special operations theory in two ways. First, control and risk assist in breaking the analysis out of its current prison of discussions focused unnecessarily on special operations missions and their time horizons, or so-called 'direct' and 'indirect' mission sets. This functional distinction inhibits a more holistic, theoretical understanding of special operations given their association

*Special Operations Success*. James D. Kiras, Oxford University Press. © James D. Kiras (2024).
DOI: 10.1093/9780198902096.003.0004

and ownership by specific communities or units, as well as their identities. To break free from this, Chapter 2 identified special operations functions that have occurred historically. Second, political and strategic risks differ from operational and tactical ones, leaving room from misinterpretation as well as room for special operations forces (SOF) to exert agency. Left with opaque guidance or none at all, special operators have developed processes and methods of communicating risk to obtain as much agency and as little oversight as possible, leading to a condition where the tactical or operational tail wags the dog of strategy and policy. High levels of vertical integration can contribute to success but are difficult to sustain over time, as conditions change. Low levels of vertical integration are not necessarily disastrous but require very different control arrangement and risk communication to succeed.

## Military control: theory and doctrine

Prior to the twentieth century, the concept of control, as it relates to military matters, was inextricably connected to the idea of generalship, or command. Regardless of changes in the scope and scale of war, Martin van Creveld argues command has certain needs, responsibilities, and functions that are eternal: collecting information and making sense of information, defining objectives, making decisions, and monitoring them to ensure subordinate units are operating in concert or a degree of unity of purpose and effort.[1] Exerting control over army units and fleet vessels operating at increasingly long distances from other and well out of sight of commander required technological solutions to transmit and receive information, as well as the development and expansion of supporting staffs to process, interpret, and manage it. Although individuals could still exercise command, their contact with and influence over their subordinates was increasingly diminished by the early twentieth century.

Pining for a previous era when commanders were in direct, personal contact with their troops and sailors, J.F.C. Fuller lamented that:

> It was, I think, ever increasing size, with its concomitant complexity of control, which more than any factors created this change both in industrial and military organizations. The more management, or command, became methodized, the more dehumanized each grew; the worker, or the soldier, becoming a cog in a vast soulless machine was de-spiritualized, the glamour of work, or of war, fading from before his eyes, until working, or fighting, became drudgery.[2]

Other scholars noted a tension at the very heart of command and control. Command, embodied in a single individual, the commander, was the realm where genius and individual agency, in the form of 'leadership, originality and inventiveness', could

[1] Martin van Creveld, *Command in War* (Cambridge, MA: Harvard University Press, 1985), pp. 6–7.
[2] J.F.C. Fuller, *Generalship: Its Diseases and Their Cure* (Harrisburg, PA: Military Service Publishing Co., 1936), p. 14.

continue to be exercised.[3] Control, however, is the active bureaucratic suppression of agency to ensure subordinate elements worked to fulfil the promise of that singular genius.

The term control took on a different meaning militarily with the development of theoretical works reconceptualizing historically useful operating environments, or domains. Technological advancements in naval forces, for example, provided the stimulus for theories on strategy, tactics, and the use of maritime power by authors including Alfred Thayer Mahan, Julian Corbett, Raoul Castex, and Stepan Makarov.[4] These advancements included revolutionary changes in propulsion, fuel, armour, armament, optics, and communications.[5] In addition to conceptualizing how future naval battles would be fought, Mahan and Corbett opined about the purpose they should serve. Mahan, drawing upon the ideas of Antoine Henri de Jomini, contributed to the confusion surrounding command and control in his writings. For him, virtually uncontested command of the sea is obtainable by decisively defeating a major naval power's fleet of capital ships through offensive action.[6] Mahanians view control and command of the sea as inextricably linked; the former is the exploitation of the domain once the latter is secured.

Corbett and others took issue with Mahan's overly simplistic causal logic undergirding his theory. Castex, for example, labelled Mahan's notion 'utopian' and 'giv[ing] the impression that its beneficiary enjoys the marvelous privilege of having to himself the immense expanse of the oceans, of building a sort of barricade whose key he puts in his pocket, thus totally banning its peacetime users'.[7] Corbett argued command and control were separable phenomena, the latter consisting of four general states resulting from a dyad of conditions when the domain was contested: general and local, and permanent and temporary. Command of the maritime domain, which Corbett expanded beyond naval forces to comprise commerce and shipping, was unlikely given its size and uninhabitability for humans. For Corbett, the state of control of the domain was also a function of both the political and military objective. For example, a predominant maritime power operating in a limited war context may need only to exert temporary, local control of the sea to land expeditionary forces while sustaining its maritime exports and imports. Command of

[3] Maurice Tugwell, 'Adapt or Perish: The Forms of Evolution in Warfare', in David Charters and Maurice Tugwell, eds., *Armies in Low-Intensity Conflict: A Comparative Study of Institutional Adaptation to New Forms of Warfare*, ORAE Extra-Mural Paper No. 38 (Ottawa, ON: Operational Research and Analysis Establishment, December 1985), p. 4.

[4] Alfred Thayer Mahan, *The Influence of Sea Power Upon History, 1660–1783* (Boston, MA: Little, Brown and Co., 1890); Julian Corbett, *Some Principles of Maritime Strategy* (Annapolis, MD: Naval Institute Press, 1988); Raoul Castex, *Strategic Theories*, trans. and ed. Eugenia Kiesling (Annapolis, MD: Naval Institute Press, 1994); Stepan Makarov, *Discussion of Questions in Naval Tactics*, trans. John Bernadou (Annapolis, MD: Naval Institute Press, 1990).

[5] Karl Lautenschläger's article from 1983 remains an indispensable resource for understanding the nature and impact of these changes. See Karl Lautenschläger, 'Technology and the Evolution of Naval Warfare', *International Security*, 8:2 (Fall): 3–51.

[6] Mahan, *The Influence of Sea Power Upon History*, pp. 46–47; Harlan Ullman, 'Mahan's Illusory Command of the Seas', *Naval Institute Proceedings*, 147:6 (June 2021), www.usni.org/magazines/proceedings/2021/june/mahans-illusory-command-seas.

[7] Castex, *Strategic Theories*, p. 53.

the domain only exists to the extent the maritime lines of communication can be controlled effectively.[8]

Technology, specifically powered flight, opened possibilities to exploit other domains including the air, leading to the development of additional theories. In lieu of long historical experience to draw upon, as in the case of the sea, theorists here sought to mine the most recent ones or speculate on future command and control of the domain based on theories developed for other domains. Guilio Douhet framed 'conquest of the command of the air' in very Mahanian terms: decisive battle to seize command that, once obtained, conveys one-sided advantages in terms of control to its possessor. Similarly, Douhet eased the control of commanded air forces through their concentration. John Slessor, in contrast, argued the nature of the domain and modern industrialised adversaries would mean that command, which he termed 'air superiority', was dynamic, localized, and relative. Its control would likely remain contested throughout a conflict while localized control and exploitation was possible.

Command of a domain is separate from that exercised over forces operating within it. For Mahan and Douhet, command over air or sea forces is assured through the simplest control arrangement: concentration. Concentration is always necessary to guarantee sufficient mass at a decisive point, in huge operating spaces, without the technical means of coordinating forces. Providing mass ensures opportunities to engage in decisive battle are not missed as forces search for each other. Slessor and Corbett, in contrast, argue that technology permits concentration in a relatively short period of time. Control over those forces is a function of the system put in place to command them and, more specifically, the efficiency of its processes.[9] Such process efficiency lies at the heart of the theory of operational command and control developed by John Boyd in the 1980s. Boyd simplified the command process in a four-step, iterative model, known as the 'OODA' or observe-orient-decide-act loop.[10] Process efficiency against a competitor, translated as tempo as opposed to speed, determines outcomes because command becomes increasingly difficult to exert as control breaks down with each iteration.[11]

Other theories more directly relatable to special operations have control at their centre but exclude command. They explain complex human phenomena: insurgency and civil war, involving asymmetric threats that SOF were designed to emulate or counter. The first wave of theories was written by imperial soldiers looking to

[8] Ibid., p. 56; Corbett, *Some Principles of Maritime Strategy*, pp. 94; 316–317.

[9] Van Creveld, *Command in War*, p. 262.

[10] Boyd's core ideas are available and evolve in seven presentations he gave, beginning with 'A Discourse on Winning and Losing' and ending with 'Revelation'. His presentations, originally on acetate slides, have been collected together and made publicly available recently. Grant Hammond, ed., *John Boyd: A Discourse on Winning and Losing* (Maxwell Air Force Base, AL: Air University Press, March 2018), www.coljohnboyd.com/static/documents/2018-03__Boyd_John_R__edited_Hammond_Grant_T__A_Discourse_on_Winning_and_Losing.pdf.

[11] Boyd's ideas have been championed by a small group of dedicated exponents, which was necessary as he never translated his presentation into written form. The best although hardly objective explanation of Boyd's ideas is contained in Frans Osinga, *Science, Strategy and War: The Strategic Theory of John Boyd* (New York: Routledge, 2007), Chapters 5 and 6.

put down colonial insurrections and rebellions, or pacify regions such as Charles Callwell, Charles Gwynn, Hubert Lyautey, and Thomas Bugeaud.[12] As colonial forces were often at a numerical disadvantage relative to indigenous populations, the authors of such works advocated approaches designed to overawe, punish, offset technologically, or progressively occupy terrain. Control was synonymous with the absence of rebellion or disorder, an absolute state not unlike Mahan and Douhet's advocacy for command of their respective domains.

Notions of control along these lines continued through the post-colonial period after the Second World War, in slightly modified and softened form. Control in limited areas in the form of security for the local population was a necessary precondition for governance measures to take root. Theorists including Robert Thompson, Frank Kitson, David Galula, Roger Trinquier, John McCuen, and Abdul Haris Nasution placed different emphasis on control measures necessary to win over the support of the local population and isolate the insurgents in return.[13] A common feature related to control among them is an acknowledgement that insurgent approaches are designed to compete and 'out-administer' the incumbent regime with a more effective form of shadow government. Ideas from such writers about territorial and population control have been translated into military doctrine, as a defining attribute of insurgency and a central feature of countering terrorism, as well as theories on the use of violence selectively in civil wars to achieve political outcomes.[14] Most recently, David Kilcullen suggests 'competitive control' within dense urban ecosystems sensitive to harsh measures, given the breakdown of state authority and increasing capabilities and interconnectedness of populations, will be the defining characteristic of future insurgencies.[15]

In addition to control in the broader conceptual sense, works also devote considerable space to the subject in its most basic functional one identified by Maurice Tugwell: the process by which commander's intent is translated into uniform action on the ground. Doctrine, or codified best practices, provides an initial framework for commanders to respond to a range of different situations with proven structures and processes. In counterterrorism (CT) operations, for example, the aspirational command and control (C2) structure is 'flexible and responsive ... one that

[12] Charles Callwell, *Small Wars: Their Principles and Practice*, 3rd ed. (London: Parnell Book Services, Limited, 1976; reprint 1906); Charles Gwynn, *Imperial Policing* (London: Macmillan and Co., Ltd, 1934). For French theorists, see Douglas Porch, 'Bugeaud, Galliéni, Lyautey', in Peter Paret, Gordon Craig, and Felix Gilbert, eds., *Makers of Modern Strategy: From Machiavelli to the Nuclear Age* (Princeton, NJ: Princeton University Press, 1986), pp. 376–407.

[13] Thompson, *Defeating Communist Insurgency*; Kitson, *Low Intensity Conflict*; David Galula, *Counterinsurgency*; Roger Trinquier, *Modern Warfare*; John McCuen, *Countering Revolutionary Warfare*; Abdul Haris Nasution, *Fundamentals of Guerrilla Warfare*.

[14] 'Insurgency in the most basic form is a struggle for control and influence ... An *insurgency* is the organized use of subversion and violence to seize, nullify, or challenge political control of a region' (italics original). Headquarters, Department of the Army, *FM 3-24/MCWP 3-33.5, Insurgencies and Countering Insurgencies* (Washington, DC: HQ, Department of the Army, May 2014), pp. 1-1 to 1-2; The Joint Staff, *Joint Publication 3-26: Counterterrorism* (Washington, DC: The Joint Staff, October 2014), p. I-2; Stathis Kalyvas, *The Logic of Violence in Civil War* (Cambridge: Cambridge University Press, 2006).

[15] David Kilcullen, *Out of the Mountains: The Coming Age of the Urban Guerrilla* (New York: Oxford University Press, 2013), pp. 114–138.

enables coordination at all appropriate levels, rapid decision making and approval, and timely action, and maintains unity of effort. In complex operational environments, the [Joint Force Commander] may use the support command relationship to provide the organizational agility necessary for CT operations'.[16] The C2 structure for counterinsurgency, while preserving flexibility, places much more emphasis on the need for decentralization. Decentralization is necessary given the geographic scope of operations, range of tasks, and their simultaneity, while also acknowledging that the best information for action exists at the point of contact with the population and not the headquarters.[17] Based on the specific contextual circumstances, doctrine and its specified control processes serve as a hedge against conflicting and incomplete information. Through a series of validated processes, commanders identify and prioritize objectives, select courses of action, and translate these down the chain of command into orders. To preserve flexibility and allow subordinate leaders to exercise local judgment and initiative, orders specify a commander's intent but not his or her specific method of achieving the objective. As Michael Kometer points out, control processes functionally consist of two elements: the method of decision as well as interaction of various decisions made within a command system.[18] Anthony King observes that in the twenty-first century, generals give up control to command resources spread across different organizations as well as geographic space, with headquarters staff becoming increasingly professionalized in the process.[19] He concludes cohesion 'has increasingly become more difficult' as a result.[20]

The military translation of control, then, consists of two dimensions: one conceptual and the other functional. The conceptual dimension refers to dominance of a particular operating environment or domain. Theories divide on the question of dominance and more specifically, whether it is absolute or relative in time and space. The functional dimension covers the processes and procedures designed to ensure that a leader's objectives are accomplished organizationally through orders and feedback. Central to the functional dimension is communication and monitoring. Communication is necessary to clarify orders, report status, and request support. Monitoring, which can take the form of loose or rigid control measures, is necessary to ensure compliance of orders to guarantee unity of purpose and unity of effort. Relinquishing control and delegating command to lower levels of authority is desirable in theory but difficult to do in practice, given the desire of senior leaders to exhibit their leadership in command. The issues inherent to control also appear at the level of strategy, where the military, policy, and politics collide.

[16] The Joint Staff, *Joint Publication 3-26: Counterterrorism*, p. IV-5.

[17] Headquarters, Department of the Army, *FM 3-24/MCWP 3-33.5, Insurgencies and Countering Insurgencies*, pp. 6.3 to 6.4.

[18] Michael Kometer, *Command in Air War: Centralized Versus Decentralized Control of Combat Airpower* (Maxwell Air Force Base, AL: Air University Press, June 2007). p. 3.

[19] Anthony King, *Command: The Twenty-First-Century General* (Cambridge: Cambridge University Press, 2019), pp. 440–442.

[20] King, *Command*, p. 71.

## Political control: civil–military relations

The study of civil–military relations in the United States is recent relative to the age of the Republic. Founded on the notion of a citizen militia mobilized in times of crisis, both the standing Army and the Navy were kept purposefully weak to prevent their use as a means of state coercion of the citizenry. As the founder of modern civil–military relations scholarship, Samuel Huntington, notes, the Founding Fathers thought that:

> The distance of the United States from Europe meant that it required no permanent military force except for small frontier garrisons to deal with the Indians. Consequently, it was generally agreed that primary reliance must be put on a citizen military composed of part-time officers and enlisted men. This was the only form of military force suitable for the new republic.[21]

This notion was so important as to be codified in the greatest source of legal authority in the country, the Constitution, in Article 1, Section 8. While a weak military served the purpose of the Founding Fathers well in times of peace, in times of war the necessity of training, arming, equipping, fielding, employing, and sustaining rapidly expanding numbers of citizen-soldiers created numerous, and not insubstantial growing pains, depending on the scale and intensity of the conflict.

Scrutiny of civil–military relations grew out of the recognition that the existing military model in the United States was inadequate to confront the challenges of the international system after the Second World War. Such inadequacies were not apparent during the First World War, especially given the substantial commitment of American forces only in 1918—the last year of that conflict. The United States rapidly developed and fielded armed forces numbering in the millions of men and women during the Second World War and almost as quickly demobilized them afterwards. The small size of the peacetime American armed forces was complemented by foreign policy largely based on strict neutrality and reflective of the American belief in the creation of a new world order. President Harry Truman found himself leading the United States in the creation of this order in 1945 following the unconditional surrender of the leaders of Nazi Germany and Imperial Japan. Not only was the United States the leader of the Western world by dint of its economic strength but also its military might. The advantages conveyed by being the sole nuclear power gave the United States the unprecedented ability to shape international economic and legal systems.[22] The detonation of an atomic weapon by the Soviet Union in 1949, combined with the tightening of its political grip on countries liberated during the Second World War, translated into questions about the sufficiency of

[21] Samuel Huntington, *The Soldier and the State: The Theory and Politics of Civil—Military Relations* (Cambridge, MA: Belknap, 1959), pp. 166—167.

[22] John Ikenberry, *Liberal Leviathan: The Origins, Crisis, and Transformation of the American World Order* (Princeton, NJ: Princeton University Press, 2011), p. 2.

American citizen-soldier model to meet this new, and potentially existential threat to democracy.

Samuel Huntington's 1959 work *The Soldier and the State* is an outgrowth of then contemporary security debates, most notably the creation of the 'New Look' national security policy of the presidential administration of Dwight Eisenhower. The 'New Look' emphasized nuclear weapons as a cheaper alternative to excessive spending on large standing military forces to guarantee national security. Huntington concluded 'One of the more basic and obvious facts of our time is that changes in technology and international politics have combined to make security the final goal of policy rather than its starting assumption.'[23] The focus of Huntington's work was on military officers, their level of professionalism and technical competence, and the control relationship they should have towards their civilian leaders.

Huntington concluded there are two methods of control for professional military officers: subjective and objective. Under the subjective control model, civilian control is designed to civilianize the military, through such mechanisms as government institutions, social class, and constitutional form.[24] Objective control, in contrast, reflects a bargain between civilian leaders and military officers. In exchange for not interfering in political matters, civilian leaders grant military officers a degree of autonomy given their technical proficiency and military professionalism, in essence rendering the military 'politically sterile and neutral.'[25] The popularization of Huntington's model of objective control has fostered a myth of military detachment and isolation from politics unreflective of reality over the past half-century.[26]

Morris Janowitz, looking at the phenomenon of civil–military relations from a sociological perspective, took issue with Huntington's assumptions. Janowitz concludes that the American military preserves security best by upholding the existing international order through its frequent use in constabulary operations, termed 'irregular warfare' today. Such operations require professionalism less characterized in technical terms specified by Huntington than by military self-restraint. With less emphasis on winning wars, and a greater emphasis on maintaining the normative status quo, military officers subscribe willingly to the value system they are sworn to preserve. Shared values mean military officers cooperate willingly with civilian leaders towards a common security objective, ensuring civilian control.[27] Neither Huntington nor Janowitz adequately explore civilian control and its mechanisms in sufficient depth, preferring to focus instead on professionalism as their dependent variable.[28]

Peter Feaver's *Armed Servants* moves beyond characterizing the type of control civilians impose upon the military, assessing instead how control occurs using

[23] Huntington, *The Soldier and the State*, p. 3.
[24] Ibid., pp. 80–83.
[25] Ibid., p. 84.
[26] Eliot Cohen, *Supreme Command: Soldiers, Statesmen, and Leadership in Wartime* (New York: Simon & Schuster, 2002).
[27] Morris Janowitz, *The Professional Soldier: A Social and Political Portrait* (New York: Free Press, 1960).
[28] Peter Feaver, 'The Civil–Military Problematique: Huntington, Janowitz, and the Question of Civilian Control', *Armed Forces & Society*, 23:2 (Winter 1996): 167.

a principal/agent model. In this model, Feaver's assumptions include a minimal standard of rationality, that civilian decision makers and senior military leaders embrace the concept of principal and agent, and that there are only two players in this civil–military 'game'—the civilian principal and the military agent.[29] Within the game the military agent has two choices: carry out their orders as directed by the civilian principal, or exercise a degree of agency. According to this model, exercising agency involves 'shirking' and following the agent's preferred courses of action instead. As Feaver suggests, while the principal and agent agree on the overall objective, such as security for the state, they often disagree over exactly how to achieve it with the resources they have at their disposal. Activities that qualify as shirking can comprise a 'broader range of deviant behavior' beyond mere preference and can include passive forms of resistance through insubordination all the way to conducting a coup in pursuit of its institutional preferences.[30]

To minimize the likelihood of shirking, civilian principals must monitor their military agents to ensure compliance. Monitoring, however, is only as effective as its ability to detect and punish, or impose costs, for perceived military shirking. Civilian agents seek to align military actions with their preferences of state security outcomes. Where preferences diverge, however, the military is likely to shirk if civilian monitoring is weak and uneven and the punishment for shirking is low.[31] Special operations, or SOF, are rarely mentioned in his work given its scope and focus, in which the military is a unitary actor and not comprising separate agents with unique outlooks, modes of performance, and methods of behaviour. Where SOF do appear in Feaver's work, such as his discussion of the manhunt for Mohammed Farah Aideed in Somalia, they are largely inseparable from the military as a monolithic entity.[32] Feaver does, however, make an intriguing statement that hints at possible cleavages within the monolith: 'The rest of the U.S. military strongly opposed the mission.'[33] In other words, he confers agency to the US military collectively as a single entity in which SOF are subsumed.

The question of agency relates to a critique common to these three major theories of civil–military relations. Each tends to view civil–military relations as a top-down process involving a single agent, 'the military', which minimizes the rich and fractious interactions that occur within and between the various Armed Services. Another recent approach suggests the unit of focus for civil–military relations after 1973's shift to the All Volunteer Force concept should not be the officer but rather the professional enlisted ranks. Such analysis emphasizes the degrees of specialization within the military, resulting in different 'forces' within, such as convergent, divergent, and segmented.[34] Culture, as a number of security studies scholars argue, cannot be

[29] Peter Feaver, *Armed Servants: Agency, Oversight, and Civil–Military Relations* (Cambridge, MA: Harvard University Press, 2003), pp. 97–98.

[30] Feaver, *Armed Servants*, p. 60.

[31] Ibid., pp. 58–62.

[32] Ibid., pp. 243–248.

[33] Ibid., p. 244.

[34] David Segal and Karin De Angelis, 'Changing Conceptions of the Military as a Profession', in Suzanne Neilsen and Don Snider, eds., *American Civil–Military Relations: The Soldier and the State in a New Era* (Baltimore, MD: Johns Hopkins University Press, 2009), pp. 204–206.

overlooked as it is the primary driving force behind Service choices and preferences. According to this critique, culture drives and informs preferences, whether it be choice of doctrine[35] or force development and employment.[36] Other scholars remain unconvinced of the role of culture in interest or preference formation and suggest it is nothing more than the context within which decisions are made and is subordinate to rational choice.[37]

One of the limits of the top-down approach is apparent in the single scholarly offering attempting to link together the subject of civil–military relations with SOF. Published shortly after the Vietnam War, Eliot Cohen's *Commandos and Politicians* concludes that elite forces pose a threat to the civil–military balance within democratic states disproportionate to their military effectiveness. In the foreword to the book, Samuel Huntington is at a loss to explain where elite units such as SOF fit within his construct of a professional military.

In Huntington's estimation, elite forces present his conceptualization of an objective-controlled military with a paradox. They are highly professional and possessed of military virtue, and yet are also highly political in nature and often are closely connected to civilian leaders who find elite forces appealing. Cohen explores this paradox and concludes that, 'By subverting the chain of command, courting favor with politicians, and distorting perceptions they disturb the professional basis of healthy civil-military relations—what Huntington terms "objective civilian control" of the military.'[38] In addition to being overly wedded to Huntington's objective civilian control model, Cohen's work does not systematically treat his subject but conducts instead a broad and rather thin qualitative analysis of elite forces. How his interpretation of the objective civilian control model would account for the adoption of the All-Volunteer Force mentioned above, in which the entirety of the force is professionalized and places tactical acumen above the technically proficient officer, much less the organizational entrenchment of elite forces in the form of the United States Special Operations Command in 1987, is unclear.

This cursory review of the major works in civil–military relations suggests there is a gap in the literature. The existing literature subsumes anecdotes on SOF specifically into the more generalized conceptualization of 'the military'. General Stanley McChrystal's public resignation in 2013, in the wake of reports about disparaging remarks he and his staff made to a reporter about the president and vice-president, can be explained as civilians reasserting objective control in Huntingtonian terms, or evidence of the mechanisms of civilian control 'punishing' shirking according

[35] Elizabeth Kier, *Imagining War: French and British Military Doctrine between the Wars* (Princeton, NJ: Princeton University Press, 1997).

[36] Jeffrey Legro, 'Military Culture and Inadvertent Escalation in World War II', *International Security*, 18:4 (Spring 1994): 108–142; Jeffrey Donnithorne, *Four Guardians: A Principled Agent View of American Civil–Military Relations* (Baltimore, MD: Johns Hopkins University Press, 2018).

[37] Patrick Porter, *Military Orientalism: Eastern War through Western Eyes* (New York: Columbia University Press, 2009); Thomas Mahnken, *Secrecy and Stratagem: Understanding Chinese Strategic Culture* (Sydney: Lowy Institute for International Policy, February 2011).

[38] Eliot Cohen, *Commandos and Politicians: Elite Military Units in Modern Democracies*, Harvard Studies in International Affairs No. 40 (Cambridge, MA: Harvard Center for International Affairs, 1978), p. 79.

to Feaver's model.[39] General Michael Flynn, who served as McChrystal's director of intelligence, has become active in politics. He has done so since his retirement, making a number of statements before Congress and being incarcerated briefly for not disclosing foreign contacts while serving as President Donald Trump's national security advisor.[40] While Cohen might view this evidence with modest alarm, his solution would be a simple one: consign SOF back to the shadows from whence they came by restricting their missions, denying them publicity, restricting the size of such units, and inducing self-restraint on the part of civilian leaders to use them.[41] Such conclusions, however, do little to advance our resolution of the paradox of the 'quiet professionals' who can be, at the same time, highly political and possessed of a high level of agency separate from 'the military' as a whole.

## American special operations and civil–military relations

Peter Feaver further fleshes out the details of civilian control in one of his earliest works on civil–military relations. In *Guarding the Guardians*, he explores the two traditional Huntingtonian models of civilian control: delegative, which he prefers to Huntington's objective control; and subjective control. To these Feaver adds assertive control, which involves 'direct civilian supervision over the military' foreshadowing his subsequent development of the principal/agent approach to civil–military relations.[42] He goes further to discuss what he classifies as the 'always/never problem' associated with the use of nuclear weapons, given the stakes associated with accidental launch or the inability to use, in the face of pre-emptive strikes, due to overly restrictive control measures.[43]

Nuclear forces provide a departure point for the study of civil–military relations associated with American SOF, as both can be high-risk enterprises where the stakes of failure are significant. Among the first generation of scholars exploring special operations, David Charters and Maurice Tugwell, Colin Gray, and Lucien Vandenbroucke consider special operations as 'of an unorthodox and frequently high risk nature' that require 'oversight at the highest level'.[44] Although Gray excludes the question of oversight as unnecessary to the definition, he makes a connection between special operations and nuclear war based on shared, elevated risk: 'planners

[39] Barack Obama, *A Promised Land* (New York: Crown, 2020), pp. 577–580.

[40] Pamela Engel, 'Michael Flynn's Former Colleagues Describe Him as Volatile and "Unbelievably Arrogant"', *Business Insider* (11 December 2016), www.businessinsider.com/michael-flynn-former-colleagues-arrogant-2016-12; Dook Snyder, 'Smoke Signals from the Swamp: The Fall of Flynn', *The Berkshire Edge* (12 November 2017), https://theberkshireedge.com/smoke-signals-from-the-swamp-the-fall-of-flynn.

[41] Cohen, *Commandos and Politicians*, pp. 97–100.

[42] Peter Feaver, *Guarding the Guardians: Civilian Control of Nuclear Weapons in the United States* (Ithaca, NY: Cornell University Press, 1992), p. 9.

[43] Feaver, *Guarding the Guardians,* pp 12-20.

[44] David Charters and Maurice Tugwell, 'Special Operations and the Threats to United States Interests in the 1980s', in Frank R. Barnett, B. Hugh Tovar, and Richard H. Schultz, eds., *Special Operations in U.S. Strategy* (Washington, DC: National Defense University Press, 1984), p. 35; Colin S. Gray, *Explorations in Strategy* (Westport, CT: Praeger, 1996), p. 145; Lucien Vandenbroucke, *Perilous Options: Special Operations as an Instrument of U.S. Foreign Policy* (New York: Oxford University Press, 1993), p. 187.

approach special operations and "strategic" nuclear war in much the same way, for many of the same reasons. The risks are extraordinary at both ends of the conflict spectrum'.[45] For Gray, risk is the element that connects political consequences to military failure. He implies a high degree of civilian control over SOF, which translates into little opportunity for the latter to exercise agency.

The type of civilian control over SOF, in turn, is a function of the degree of political risk associated with their activities. Individual raids, such as hostage rescues or kill/capture missions, tend to have a high degree of assertive control based on their episodic nature and their political stakes. The failed attempt to rescue the fifty-three Americans held hostage in Iran in April 1980, Operation Eagle Claw, is one such raid. President Jimmy Carter was connected, via satellite radio, with the joint task force commander for the operation, Major General James Vaught. Another is the raid to kill or capture Osama bin Laden in Abbottabad, Pakistan, in May 2011, code-named Operation NEPTUNE SPEAR. The degree of assertive control for this raid appears to be vividly and visually captured by a picture of senior civilian principals, including the President of United States, watching a video feed in the White House Situation Room of the raid in progress.

For both Gray and Feaver, risk is a key independent variable. Risk was originally discussed in social science literature in the context of nuclear deterrence and escalation, reflecting rational cost-benefit calculations inherent in game theory.[46] For Feaver, risk underpins the 'always/never problem' of nuclear weapons use. Gray also views it as a shared element between nuclear war and special operations. Although he points to SOF risk being 'extraordinary', Gray conflates military with political risk. Military risk involves the risk to SOF should they lose the tactical initiative in overt engagement or be compromised or discovered in covert or clandestine missions.

Political risk differs from the military one in degree and kind. The contemporary literature about political risk focuses largely on two areas: socio-cognitive considerations connected to the decision-making calculus behind large-scale, overt state military interventions[47] or as a function of the collective context influencing leader preferences.[48] In special operations, political risk is never fatal to those authorizing and overseeing the missions even though it may be the case to those carrying them out militarily. At worst, civilian leaders may lose their jobs through firing, impeachment, or election in states with democratic elections and accountable, representative government.

Unlike raids or clandestine and covert actions, civilian control over American SOF has been largely delegative for longer-term activities with a low degree of political risk, even after publicized mission failures or politically embarrassing incidents. One of the best-known mission failures occurred during Operation RED WINGS, from

[45] Gray, *Explorations in Strategy*, pp. 147; 158 n.11.

[46] Thomas Schelling, *Arms and Influence* (New Haven, CT: Yale University Press, 1966), pp. 92–125; Herman Kahn, *On Escalation: Metaphors and Scenarios* (New York: Hudson Institute, 1965), pp. 11–14.

[47] Yakob Vertzberger, *Risk Taking and Decisionmaking: Foreign Military Intervention Decisions* (Stanford, CA: Stanford University Press, 1998).

[48] Mikkel Rasmussen, *The Risk Society at War: Terror, Technology and Strategy in the Twenty-First Century* (Cambridge: Cambridge University Press, 2006).

late June until early July 2005, during which nineteen Navy SEALs and Army special operations aviators were killed and one survivor badly wounded. Details of the mission only became public two years later after the lone survivor published his account of the ordeal.[49] In this case, civilian control was delegated through the normal chain of command to a combined joint special operations task force for Afghanistan, or CJSOTF-A. Other factors contributing to delegative control included the nature of the activity, as one mission within a rotational tour of a SOF unit. The mission was in support of conventional forces operating in Kunar province. In this case, the SEALs were conducting reconnaissance and pathfinding in advance of a raid and subsequent Marine Corps and Afghan National Army battalion-sized sweep of the Korengal Valley.[50] The publication of some details about the mission after the fact had little impact politically on missions of SOF in Afghanistan. Special operations missions only obtained a degree of political attention and subsequent discussion of risk from 2010 onwards, after Afghan President Hamid Karzai called for controversial 'night raids' to end in November of that year.[51]

Delegative control of SOF is not problematic provided civil–military relations operate along the principal/agent model outlined by Feaver. Notions of working and shirking are useful in abstracting the nature of relations between civilian decision makers and the Armed Services, whose hierarchical organizational structures and bureaucratic processes are designed to maintain discipline and good conduct, uniformity of behaviour, continuity, and rigor in handling large-scale equipment and complex systems acquisition and employment in joint, combined arms warfare. There is good reason to question the validity of this civil–military relations construct as American SOF already have a greater degree of agency than conventional military forces given the nature of the sub-state adversaries and state proxies they face and higher levels they are apparently seeking.

As Feaver points out, decreased monitoring becomes a civil–military concern when military agents are willing and able to shirk the orders civilian principals give them.[52] Yet the qualities and characteristics of the operators within SOF, or what some might term 'the special operations mindset', suggest the opposite is the case. The special operations mindset, or approach to problem solving, reflects the selection processes and organizational culture within SOF.

The selection process is both physical and psychological in nature, prizing an extraordinary drive and determination to successfully complete the mission in the face of friction, adversity, and uncertainty.[53] Commitment to completing the mission

[49] Marcus Luttrell and Patrick Robinson, *Lone Survivor: The Eyewitness Account of Operation Redwing and the Lost Heroes of SEAL Team 10* (New York: Back Bay Books, 2007).

[50] Ed Darack, *Victory Point: Operations Red Wings and Whalers—The Marine Corps' Battle for Freedom in Afghanistan* (New York: Berkley Books, 2009).

[51] Thom Shankar et al., 'Despite Gains, Night Raids Split U.S. and Karzai', *The New York Times* (15 November 2010), www.nytimes.com/2010/11/16/world/asia/16night.html?_r=0.

[52] Feaver, *Armed Servants*, pp. 54–58.

[53] Robert Spulak Jr, *A Theory of Special Operations: The Origins, Qualities, and Use of SOF* (Hurlburt Field, FL: Joint Special Operations University Press, 2007), pp. 18–20; Tony Balasevicius, 'Finding the Right Stuff: Special Operations Forces Selection', in Bernd Horn and Tony Balasevicius, eds., *Casting Shadows on the Light: Canadian Perspectives of Special Operations Forces* (Kingston, ON: Canadian Defence

is not unique to SOF; the Marine Corps motto 'every Marine a rifleman' is an example. What separates SOF from their conventional colleagues, however, is the different emphasis inherent in the mindset that operators place on two associated aspects of the mission: the conditions under which the mission is undertaken and the standards to which it will be performed. For example, while

> any force can be trained to capture a high value target with a great likelihood of success ... SOF, however, are able to conduct multiple missions over the same period of time across a wide variety of space with almost no reduction in their standard of execution—in other words, in conditions such as at night or in adverse weather while still carrying out missions beyond the ability of other units.[54]

Marines, therefore, could have conducted the raid to capture Islamic State leader Abu Bakr al-Baghdadi in Syria in October 2019, but their likelihood of successful completion might be lower and the potential costs in casualties and escalation much higher.

The conditions and standards to which SOF train and execute is a function of their numerical disadvantage against their enemies. To overcome their finite numbers, SOF rely on a combination of surprise, speed of action, skills proficiency, and a host of other combat multipliers to achieve a relative tactical advantage.[55] Stated simply, SOF seek to limit tactical and political risk and guarantee positive battlefield effects with minimal losses. Individuals who are task-driven and -orientated to achieve meaningful results, or generate strategic effects, by using their limited resources and exceptional training, skills, and equipment in innovative ways populate SOF. SOF derive high levels of satisfaction and accomplishment from challenging missions.

Shirking, in the operational sense, is almost unheard of within SOF. If anything, seemingly impossible missions provide SOF with the opportunity to draw upon their unique mindset and flexibility to solve problems in innovative ways.[56] SOF prefer to be used as often as possible and not remain in garrison constantly training for missions they never execute. This preference continues when operators become leaders and senior representatives of SOF. Indeed, such leaders see SOF as the most effective and efficient instruments of policy to address the nation's greatest challenges and form the basis of 'a strategy like no other—a strategy for success in today's uncertain world'.[57] As former United States Special Operations Command (USSOCOM) commander Peter Schoomaker was fond of saying when SOF were not employed as often

Academy Press, 2007), pp. 49–50; Tone Danielson, *Making Warriors in a Global Era: An Ethnographic Study of the Norwegian Naval Special Operations Commando* (Lanham, MD: Lexington Books, 2018), Chapter 5.

[54] James D. Kiras, 'The Role of Special Operations Forces: Past, Present, and Future', *Pointer*, 37:2 (Winter 2011): 83.

[55] William McRaven, *Spec Ops, Case Studies in Special Operations Warfare: Theory and Practice* (Novato, CA: Presidio Press, 1996), p. 1.

[56] Danielson, *Making Warriors in a Global Era*, pp. 87–88.

[57] William McRaven, 'Special Operations: The Perfect Grand Strategy?', in Bernd Horn, John Paul de B. Taillon, and David Last, eds., *Force of Choice: Perspectives on Special Operations* (Kingston, ON: McGill-Queen's University Press, 2004), p. 77.

during the Clinton administration, 'It was like having a brand-new Ferrari in the garage, and nobody wants to race it because you might dent the fender.'[58]

SOF prefer working to a fault. Once civilian or military leaders make the decision to employ them, operators favour conducting the mission with as little interference as possible. Over time, this narrow mission focus creates a cultural 'bias for action', which in turn leads to high tempos and the potential for reductions in effectiveness. In 2017, USSOCOM took on executive authority for other mission areas, including countering weapons of mass destruction. The net effect, according to USSOCOM's own admission, was a force 'fatigued, worn and frayed around the edges' as well as 'selective implementation of ... recommended actions ... [that only] temporarily alleviated some symptoms'.[59] American SOF are more likely to see monitoring actions, such as orders to 'stand down' or not work, as needless and unwanted interference.[60] Such frustration is evident in the attitudes prevalent in several SOF works.

Paradoxically, the set of behaviours that could be considered shirking, in a normative sense, is one of the qualities that make SOF so appealing to policy makers. When confronted with obstacles to mission accomplishment, special operators seek to overcome them in unexpected, unorthodox, extranormative, and less costly manners. Extranormative does not mean that SOF violate or skirt rules as a matter of course. They are more than capable, and are indeed characterized by their ability to operate on the margins or stretch the limits of existing rules or constraints. Such margins include legislative ones regarding defined roles and missions, in Title 10, Title 50, and Title 22 of the US Code, among others.[61] By pushing the boundaries through their unorthodox solutions, and operating on a small scale relative to conventional forces, SOF provide additional, flexible, and seemingly low-cost, low-risk options to civilian leaders.

Yet, given their ability to create unorthodox solutions to knotty/difficult problems, SOF sometimes propose and conduct actions that exceed civilian leaders' preferences. Although a rarity, such possibilities can include unauthorized missions or unpalatable actions.[62] At other times, American SOF have been given considerable agency by civilians provided they obtain results. During the hunt for Abu Musab Zarqawi in Iraq in 2006, SOF special missions units 'could raid geographical [areas of responsibility] without consulting higher-ranking officials, a circumvention of the chain of command that enabled Special Forces to act on intelligence gathered by interrogators earlier that day'.[63] A decade and a half earlier, by advocating for and seeking an update of Title 10 of the US Code (Section 2011), the leaders of

[58] Shultz Jr, 'Showstoppers', p. 26.

[59] USSOCOM, 'Comprehensive Review', unclassified (23 January 2020), p. 13, https://sof.news/pubs/USSOCOM-Comprehensive-Ethics-Review-Report-January-2020.pdf.

[60] Vandenbroucke, *Perilous Options*, pp. 166–169.

[61] W. Hays Parks, 'Special Forces' Wear of Non-Standard Uniforms', *Chicago Journal of International Law*, 4:2 (Fall 2003): 493–560.

[62] Sean Naylor, *Relentless Strike: The Secret History of Joint Special Operations Command* (New York: St Martin's Press, 2015), pp. 62; 89; Matthew Aid, *Intel Wars: The Secret History of the Fight against Terror* (New York: Bloomsbury Press, 2012), pp. 74–75.

[63] Aki Peritz and Eric Rosenbach, *Find, Fix, Finish: Inside the Counterterrorism Campaigns the Killed bin Laden and Devastated Al Qaeda* (New York: PublicAffairs, 2012), p. 128.

US 'Special Operation Command had essentially created its own, separate foreign policy track' for the training of forces in other countries.[64] SOF activities had limited or denied such training to countries identified for their human rights abuses under the strictures of the Leahy Amendment, passed in 1994. The result of these and other excesses, advocacy of more aggressive or risky courses of action, and resistance to more effective inter-service integration, has led some commentators to label SOF 'pariah cowboys'.[65] The problem of exceeding civilian policy makers' preferences becomes more acute the more SOF are operating clandestinely and covertly to combat threats in the 'Gray Zone', with the relaxed monitoring necessary to maintain a level of plausible deniability for civilian leaders. Such missions are more likely to exceed policy makers' desires, if the short history of American covert action serves as a guide.

The character of ongoing and future threats has driven a desire within the special operations community for greater independence and agency with low levels of civilian monitoring. The logic behind the assessment of these threats is clear enough. American SOF are responsible for combatting violent extremist organizations (VEOs) overseas, which continue to adapt and spread to uncontrolled areas or challenge weak and fragile regimes allied to or partnered with the United States. Modern VEOs' ability to network and share information and resources translates into the capability to achieve surprise, maintain the initiative, and conduct increasingly complex and lethal violent attacks. Given their ideological beliefs, VEOs and state-sponsored proxies will not hesitate to use a weapon of mass destruction; their access to information and resources makes such an acquisition increasingly likely. This convergence between sub-state groups and proxies and their ambitions led USSOCOM, in part, to lobby to become DoD's lead agent for countering weapons of mass destruction (WMD) in 2017.[66] The net effect of these current and future developments is the creation and exploitation by China, Russia, Iran, and VEOs of a 'Gray Zone' that blurs distinctions between military and other activities for which traditional American responses, including uses of force, may be ill suited.[67]

Addressing Gray Zone phenomena requires the use of limited special operations manpower and resources to permit greater freedom of action. One way is through the creation first of a 'Global SOF Network', which will allow 'persistent engagement' to 'win population-centric conflicts, oftentimes, and preferably, before they start', in a 'Global SOF Posture' designed to 'enhance deterrence and better posture the nation's

[64] Dana Priest, *The Mission: Waging War and Keeping Peace with America's Military* (New York: W.W. Norton & Co., 2003), pp. 110–111.

[65] Shultz Jr, 'Showstoppers', p. 31.

[66] Timothy Szymanski, 'Statement of Vice Admiral Timothy Szymanski, Deputy Commander, United Special Operations Command before the 117th Congress House Armed Services Committee Subcommittee on Intelligence and Special Operations', pp. 2–3.

[67] Richard Clarke, 'Statement of General Richard Clarke, USA, Commander, United Special Operations Command before the 117th Congress Senate Armed Services Committee', p. 1, www.armed-services.senate.gov/imo/media/doc/2022%20USSOCOM%20Posture%20-%20Clarke%20-%20SASC%20(5APR22)%20(FINAL).pdf.

response to emerging threats'.[68] One manifestation of this network is Operation GALLANT PHOENIX, discussed in detail in Chapter 8. As part of 'rebalancing the force' to maintain relevance in strategic competition, USSOCOM is emphasizing its less kinetic authorities, including as the 'designated joint [DoD] component for [military information support operations, or MISO]'.[69] As the USSOCOM commander testified in 2022, 'our MISO activities to counter strategic competitors have also more than doubled over the past three years—comprising over 40% of our MISO activities worldwide in FY21'.[70] The key enablers of the Global SOF Network are increased numbers of support and liaison personnel, funding, and expanded authorities such as the 2013 'Forces For Unified Commands Memorandum which 'gives USSOCOM Combatant Command authority over the Theater Special Operations Commands (TSOCs)—units assigned to each of the seven Combatant Commands'.[71] In essence, USSOCOM continues to evolve from a sub-unified command with responsibilities to equip the force, oversee force development, and manage its budget, to a combatant command with global responsibilities for force employment and synchronization of efforts against VEOs, WMD, and information influence operations.

The net effect of increased funding and authorities is the ability to 'achieve these strategic ends with a small footprint, while not constituting an irreversible foreign policy decision'.[72] In essence, SOF leaders are assuring civilian decision makers they have an ability to address vexing foreign policy problems, at a marginal cost in resources and manpower, while preserving flexibility of options and choice. Colin Gray summarizes this argument in terms of two special operations claims to strategic utility, namely 'economy of force' and 'expansion of choice'.[73] The larger potential cost to ensuring such strategic utility, however, lies in granting SOF expanded authority. The operational rationale for this is old as modern SOF themselves: decentralized control offers such forces greater flexibility and responsiveness in conducting their missions.[74] In civil–military terms, however, expanded authority increases the military effectiveness of SOF at the expense of monitoring.

Alice Hunt Friend, for example, concludes that American SOF conducted a long-term influence campaign, conducted in three distinct phases over thirty years to shape civilian decision makers' preferences to ensure they adopted SOF's as if they were their own.[75] The major issue with such frameworks, however, is their

[68] William McRaven, 'Posture Statement of Admiral William McRaven, USN Commander, United Special Operations Command before the 113th Congress House Armed Services Committee', p. 3, https://docs.house.gov/meetings/AP/AP02/20140408/102072/HHRG-113-AP02-Wstate-McRavenA-20140408.pdf.

[69] The Joint Staff, *Joint Publication 3-13.2, Military Information Support Operations* (Washington, DC: The Joint Staff, 21 November 2014), p. viii.

[70] Clarke, 'Statement of General Richard Clarke', p. 4.

[71] McRaven, 'Posture Statement of Admiral William McRaven', p. 5.

[72] Ibid., p. 4.

[73] Colin Gray, *Explorations in Strategy* (Westport, CT: Greenwood Press, 1996), pp. 168–174.

[74] Andrew Hargreaves, *Special Operations in World War II: British and American Irregular Warfare* (Norman, OK: University of Oklahoma Press, 2013), p. 142.

[75] Alice Hunt Friend, 'Creating Requirements: Emerging Military Capabilities, Civilian Preferences, and Civil–Military Relations', unpublished PhD dissertation, American University (2020), pp. 110–115; 117–120.

transferability and therefore utility in other national contexts. The dominant civil–military relations frameworks reflect unique American organizational, structural, and cultural conditions that, while in theory apply generally to other democracies, cannot account for those existing elsewhere. These observations go back to the publication of Huntington's *The Soldier and the State*.[76] Douglas Bland, for example, points out that for Britain the notion of control is misleading. Based on a long historical tradition, as well as a small standing army, what has involved there is a more collaborative model or what he terms 'shared responsibility'.[77] Reaching Bland's stated goal of a unified theory of civil–military relations requires addressing four problems he identifies—the praetorian problem, maintaining good order and discipline, subordination to a political body, and the relation of the expert to the minister—as well as extend the model to include other forms of government.[78] In non-democratic states, particularly autocracies, the American-centric schools of thought offer little insight. Amos Perlmutter attempted to develop a taxonomy differentiating the professional soldier from the praetorian and professional revolutionary soldier in support of his argument about military corporatism in the armed forces of post-revolutionary and developing states.[79] Two reviewers concluded his work 'is flawed by historical misinterpretations, loose theoretical perceptions and definitions, and outright errors of fact and citation', leaving little in the way of successor theories.[80] Although other scholars have touched upon civil–military relations in developing states, they often focus on cultural or technological dimensions rather than structural and organizational arguments.[81]

To account for these differences in developing a general theory of special operations and not a new one for civil–military relations, this work proposes additional models of political control based on government type. These models were derived from observation and interaction with various SOF in their national contexts, as well as extensive study. They accept Feaver's foundational premise of preference divergence between a principal and an agent, in addition to his delegative and assertive, or what I prefer to call 'directive control'. I have also included Bland's collaborative to account for the peculiarities of parliamentary democracies based as much on uncodified and informal arrangements as they are specified, formal ones.

At least two other possibilities beyond delegative, directive, and collaborative control are identifiable intuitively: competitive as well as co-optive control. Competitive control in civil–military terms addresses the problem of agency by keeping

[76] Samuel Finer, *The Man on Horseback: The Role of the Military in Politics* (London: Pall Mall Press, 1962).

[77] Douglas Bland, 'A Unified Theory of Civil–Military Relations', *Armed Forces and Society*, 26:1 (Fall 1999): 10–13.

[78] Bland, 'A Unified Theory of Civil–Military Relations', pp. 12–13.

[79] Amos Perlmutter, *The Military and Politics in Modern Times: On Professionals, Praetorians, and Revolutionary Soldiers* (New Haven, CT: Yale University Press, 1977), pp. 9–17.

[80] Daniel Beaver and Lowell Wenger, 'Military and Politics in Modern Times', review of *The Military and Politics in Modern Times: On Professionals, Praetorians, and Revolutionary Soldiers* by Amos Perlmutter, *Armed Forces and Society*, 5:1 (Fall 1978): 138.

[81] See, for example, Stephen Biddle and Robert Zirkle, 'Technology, Civil–Military Relations, and Warfare in the Developing World', *Journal of Strategic Studies*, 19:2 (June 1996): 171–212.

agents busy in two ways: both through the principal assigning tasks and, more importantly, doling out rewards in a piecemeal manner to keep agents vying for influence and favour. Nazi Germany and Imperial Japan provide two well-known and stark historical examples of this form of control.[82] While efficiency suffers due to bureaucratic competition, it preserves the principal's power by focusing energy away from potential plotting or coup attempts. In its simplest terms, autocratic leaders employ a strategy of 'divide and conquer', through inter-service and interpersonal rivalry. Competitive control modifies and elevates one of Graham Allison and Philip Zelikow's lenses of the analysis, the government politics model, to a civil–military strategy.[83]

Co-optive control operates on similar lines as competitive but features its own twist. Rather than parcelling out inducements piecemeal to keep agents vying for attention, co-optive control binds agents directly to the principal using not only inducements but also social connections including kinship and tribal or ethnic affiliations. As an example of co-optive control, Indonesian President Suharto assured the loyalty of Prabowo Subianto through a mixture of kinship ties, including marriage to his daughter Titiek, as well as inducements. Prabowo received choice commands in and served for three years as commander and restructurer of Indonesia's SOF: *Kopassus*.[84] Strengthening social ties as well as enrichment prevented Prabowo, a potential rival to Suharto's power, from threatening it.

Defining different types or models is necessary for civil–military relations theory but insufficient for a general theory of special operations. As the preceding discussion argues, while Feaver's principal-agent theory of civil–military relations still has considerable explanatory utility, it contains four problems for more general theory. First, by treating 'the military' as a single entity, it precludes the possibility of exploring subordinate units or communities of practice such as SOF. While necessary to streamline theory, cracking open the monolithic 'black box' of 'the military' is necessary. Second, while principal/agent problems are universal in nature, most of civil–military theory focuses on and is shaped by the American example that accepts the assumption of civilian primacy and military subordination based on clear oversight, transparency, and open monitoring processes and communications. As Bland and others have pointed out, this normal theory insufficiently accounts for variations in government, culture, or historical patterns outside of the United States. A general theory of special operations requires expanded categories to account for these differences. Third, the current normal theory of civil–military by Feaver suggests militaries have two choices in the face of civilian orders: working or shirking. Given the qualities and characteristics specifically recruited, selected, and refined, SOF

[82] Internal 'squabbles' and internecine fighting characterized the early Nazi Party; Hitler capitalized on them institutionally. Richard Evans, *The Coming of the Third Reich* (New York: The Penguin Press, 2003), p. 199; Geoffrey Megargee, *Inside Hitler's High Command* (Lawrence, KS: University Press of Kansas, 2000), pp. 233–236. He also used direct inducements. For details, see Norman Goda, 'Black Marks: Hitler's Bribery of His Senior Officers during World War II', *Journal of Modern History*, 72:2 (June 2000): 413–452.

[83] Graham Allison and Phillip Zelikow, *The Essence of Decision: Explaining the Cuban Missile Crisis*, 2nd ed. (New York: Longman, 1999), p. 264.

[84] K. Joy Sihotang et al., *Pengabdian Korps Baret Merah Abad XX* (Jakarta: Kopassus, 2000), pp. 106–107.

are unlikely to shirk. Instead, they are prone to hyper-working or, should they disagree with the civil authorities, overthrow them in a coup, as several authors have warned in works suggesting SOF are inherently anti-democratic elements. Fourth, and finally, the principal/agent approach necessitates fixing the analysis at a single level: the strategic, or at the level where policy is created by the principal and turned into strategy by the agent. In this model, dynamic processes and interactions between the strategic, operational, and the tactical level tend to be glossed over or ignored. Arguably, it is in this rich dialogue up and down and across the chain of command, and the resulting tensions, frictions, and compromises that arise, that lead to both a more nuanced and valuable understanding of highly political military agents such as SOF in both senses of the term: the effects they generate as well as engagement with their superiors.

To break control free from the specific and current military and political paradigms and to explore how it relates more specifically to SOF, the next section provides the linkage that both defines how civilian leaders and military commanders at the strategic level frame the utility of special operations from the top down to how SOF view tactical mission parameters and communicate success criteria up the chain of command. This linkage, the conceptualization and communication of risk, is at once a common reference point for discussion but also a source of considerable confusion and misunderstanding.

## Risk associated with special operations

Special operations are defined by several criteria including the visibility of the missions (covert, clandestine, or overt), the use of specialized tactics or equipment, and the operating environment, or actors within it. One attribute frequently associated with special operations is risk. American doctrine suggests risk within special operations is relative. In this case, the risk inherent in special operations is higher relative to comparable conventional military options. Such risk occurs in three categories: physical, strategic, and political/diplomatic.[85]

Physical risk within special operations is the easiest to comprehend and illustrate. David Charters and Maurice Tugwell define special operations according to high levels of risk associated with the activities themselves.[86] Hostage rescue missions, for example, often have high levels of inherent risk and their outcomes are anything but assured. Recent examples include the successful rescue of Captain Richard Phillips from a lifeboat of the *Maersk Alabama* at sea as well as Jessica Buchanan and Poul Hagen Thisted who were held captive in Somalia.[87] Unsuccessful attempts include

[85] The Joint Staff, *Joint Publication (JP) 3-05: Special Operations* (Washington, DC: The Joint Staff, 2014), p. I-1.

[86] David Charters and Maurice Tugwell, 'Special Operations and the Threats to United States Interests in the 1980s', in Frank R. Barnett, B. Hugh Tovar, and Richard H. Schultz, eds., *Special Operations in U.S. Strategy* (Washington, DC: National Defense University Press, 1984), p. 35.

[87] Leon Panetta, *Worthy Fights: A Memoir of Leadership in War and Peace* (New York: Penguin Press, 2014), p. 402.

Linda Norgrove, who was killed during the clearing actions to rescue her from a hillside compound in Afghanistan.[88] Others connect risk in special operations to the forces that conduct them, emphasizing the drastic outcomes for those conducting special operations should the mission fail.[89] Such risks are well understood and accepted by the operators conducting the missions.[90]

Strategic risks associated with special operations are connected to the expected outcome. Most frequently, the outcome is justified in terms of a highly disproportionate result relative to the investment made in forces. Emily Spencer and her fellow contributors characterize outcomes as a function of special operations comprising an 'economy of force' or 'economy of effort' option for decision makers.[91] Colin Gray acknowledges such economy, which is one of his overarching claims for the strategic utility of special operations, writing 'in conflict with a competent foe the achievement of great strategic returns will require the expenditure of great effort. Special operations represent an attempt to circumvent this principle'.[92] The high risks of special operations, in other words, are appealing because of their potentially high payoff. The weighing of costs versus benefits serves as the theoretical departure point for Malvesti's explorations of risk and special operations in counterterrorism. From a rational choice theoretical starting point, she uses prospect theory to conclude that decision makers will authorize special operations when operating from a position of losses, particularly internationally, as opposed to gains.[93]

The realm of political risks is the most elusive of the three categories related to special operations. Physical risk is intuitively relatable to personal danger, and the missions themselves are well understood as unorthodox and inherently risky ventures. Strategic risk is translatable into the rational weighing of costs versus benefits, an action most conduct daily. Political risk, however, is more problematic for several reasons. First, political risk can be equated incorrectly with strategic risk, which is problematic given their different purposes. Second, political risk can be confused for politics, which can have a highly negative connotation for military action. Americans in particular, but Westerners in general, view politics as unnecessary meddling with military courses of action. Such popular perceptions are often reinforced in the memoirs of senior special operators. The assumption behind such perceptions is that political leaders are uninformed at best, or ignorant and dismissive of military advice at worst. A causal logic chain connecting this assumption to the politics of decision-making and different types of risks associated with special operations is best illustrated in the following quote from David Tucker and Christopher Lamb:

[88] Naylor, *Relentless Strike*, pp. 371–373; 403–407; 425–427.

[89] Gray, *Explorations in Strategy*, pp. 146–147.

[90] Joseph Votel, 'Advance Policy Questions for Lieutenant General Joseph L. Votel, USA Nominee for Commander, United States Special Operations Command', Headquarters, USSOCOM (10 July 2014), p. 38.

[91] Emily Spencer, ed., *Special Operations Forces: A National Capability* (Kingston, ON: Canadian Defence Academy Press, 2011), especially Chapters 1, 3, 5–7.

[92] Gray, *Explorations in Strategy*, p. 147.

[93] Michelle Malvesti, 'Risk-Taking in Countering Terrorism: A Study of United States Presidential Decisions to Use Special Operations and Covert Action', unpublished PhD dissertation, June 2002, pp. 269–270.

> Politicians will thus have a tendency to discount the risks inherent in special operations in hopes of achieving their policy objectives on the political cheap ... The 'cheapness' of SOF appears to lower the political risk of using them. Yet, an inability to understand the operational risks involved in their use leads politicians to underestimate the chance of operational failure and hence to underestimate the political risk they run by using special forces. In other words, when civilians direct the use of SOF, civilian preferences and prejudices paradoxically may create precisely the political risks that politicians think they are avoiding.[94]

If Tucker and Lamb's assertion holds true, one would expect misunderstandings of risk and high-profile failures given the almost continuous use of special operations over the past decade. In addition, the unintended and paradoxical creation of political risk should threaten the sustained use of special operations. Yet there is little evidence of the fallout of such political risk when special operations fail. To search for possible answers for why this is not the case, this chapter explores characterizations of risk beyond the realm of special operations.

## Risk in a broader context

The challenges associated with deciphering different types of risk stem from the central idea itself, which is frequently confused with related concepts of danger and uncertainty. In its most common definition, risk is best understood as an understanding or judgment of the likely outcome of a series of events, after weighing probabilities, means, the environment, and desired outcomes.[95] A recent formulation suggests risk is the product of calculations of threat as well as assessments of one's own vulnerability.[96] Risk assessment, however, is more than just a best guess. Within different fields, including life insurance and finance, understandings of levels and degrees of risk result from a careful weighing of all available information and probabilities within a closed system. For example, in actuarial calculations, the risk of paying out a life insurance policy is balanced against demographic factors including lifestyle, life expectancy, family health history and geographic location, among others. In this case, the level of risk is offset by the premium charged to the policyholder: the greater the risk, the higher the premium.

Assessing risk is further complicated when assigning probabilities to outcomes in an open system such as foreign policy or national security decision making. Problems in this realm are dynamic and complex, and dependent on inadequate and incomplete information, the impact of human and organizational preferences and

[94] David Tucker and Christopher Lamb, *United States Special Operations Forces* (New York: Columbia University Press, 2007), p. 230.

[95] Vertzberger, *Risk Taking and Decisionmaking*, p. 19.

[96] Stanley McChrystal and Anna Burtico, *Risk: A User's Guide* (New York: Portfolio Penguin, 2021), p. 10.

behaviour within the decision-making process, as well as actions against a thinking, active, and wilful opponent.[97] Yaakov Vertzberger suggests different types of risk inherent in national security decisions must be parsed out and understood individually. He identifies three types of risk in the national security decision-making process: real, perceived, and acceptable [98] Perceptions of risk, in his estimation, are 'a scientific question that is affected by non-scientific considerations (e.g., cognitive biases)' whereas acceptability of risk 'is a political question that is affected by non-political considerations (e.g., the nature of risk)'.[99] To evaluate the range of factors associated with risk in decision-making related specifically to military interventions, Vertzberger develops a complex socio-cognitive approach that includes contextual factors, problem framing, social and group impacts on decision making, and organizational factors' impact on risk assessment and acceptability preferences.[100]

Complex models of risk in decision making such as Vertzberger's provide detail at the expense of simplicity and clarity. Critics of this approach counter that the key insight of such models, why a risky decision was made, is better understood through the lens of rational choice.[101] Rational choice examinations of decision making privilege the decision of the ultimate arbiter, the executive, at the expense of the rich detail inherent in debates advisors have about risk. To anchor this point in a special operations context, the alleged divergent opinions represented by Vice-President Joseph Biden on one hand, and Secretaries Robert Gates and Hillary Clinton on the other, are largely irrelevant to President Barack Obama's approval for the special operation against Osama bin Laden in 2011.[102] Prospect theory, employed by Rose McDermott, Michelle Malvesti, and others, offers 'a theory of decision-making under risk that permits a systematic explanation and prediction of risk propensity' based on expected gains or losses.[103]

Other scholars conclude models that explore risk in decision-making mistake the symptom for the cause. In their estimation, risk is not simply a consideration or part of a process, but the greater context shaping all decision making. Scholarship espousing this view of risk has sociologist Ulrich Beck's 'risk society' as its conceptual starting point. Beck, originally writing in 1986, suggests 'the social production of wealth is systematically accompanied by the social production of risks'.[104] The nature of modern economies and technologies, based are on globalization, creates problems, or risks, that can and should be mitigated. He labels this interactive process

[97] Carl von Clausewitz, *On War*, ed. and trans. Michael Howard and Peter Paret (Princeton, NJ: Princeton University Press, 1976), pp. 90–121.
[98] Vertzberger, *Risk Taking and Decisionmaking*, pp. 18–19.
[99] Ibid., p. 25.
[100] Ibid., p. 110.
[101] Rose McDermott and Jonathan Cowden, 'Hit or Miss: Sociocognitive Approach to Decisionmaking', Review of Yaakov Vertzberger, *Risk Taking and Decisionmaking, International Studies Review*, 1:1 (Spring 1999): 125.
[102] David Sanger, *Confront and Conceal: Obama's Secret Wars and Surprising Use of American Power* (New York: Crown, 2012), pp. 77–80; 85–100; Nicholas Schmidle, 'Getting Bin Laden: What Happened That Night in Abbottabad', *The New Yorker* (8 August 2011), www.newyorker.com/magazine/2011/08/08/getting-bin-laden.
[103] Malvesti, 'Risk-Taking in Countering Terrorism', p. 47.
[104] Ulrich Beck, *Risk Society: Towards a New Modernity* (London: Sage, 1992), p. 3.

of risk creation and mitigation 'reflexive modernity'. The net effect is that 'politics must catch up with the self-limitation that has been carried out historically'.[105] Politics will inevitably give way to sub-politics—a new openness—as groups within societies are increasingly aware and empowered. Writing in advance of the widespread adoption of the Internet, Beck's theory appeared to many scholars to be highly prophetic and his ideas applicable to a range of socially significant issues related to emerging technologies, including genetics and cybernetics.[106]

Social scientists have embraced the 'risk society' as an explanation for characteristics of modern war. War between states has lost its former utility and appeal. The interconnectedness of states also means they have more to lose through going to war than they have to gain.[107] Stability, in other words, has trumped conquest. Technology and interconnectedness that provides benefits also breeds its own challenges or risks beyond war, such as refugee migration, pandemics, and violent extremism. In other words, risks come from non-state groups at the level of sub-politics, which draws upon anxiety and individual insecurities. Traditional military force has limited utility against such a diffuse and kaleidoscopic variety of risks. In consequence, the risks created by the environment must be managed instead. As Christopher Coker suggests, the very concept of war is changing from threats to risks and from war as defence against threats to war as security against risks.[108] War is not an exercise in decisively changing outcomes so much as it is limiting the costs of responses while managing risks. The analogy to describe contemporary risk management is policing, Yee-Kuang Heng argues, where areas are surveilled, monitored, and patrolled with only intermittent and preventative or pre-emptive intervention.[109]

The 'risk society' does more than provide a macro-level explanation of the security environment. Mikkel Rasmussen has connected the 'risk society' directly to special operations. Beginning from a similar set of assumptions and conditions as Coker and Heng, he departs in his analysis, however, by looking at war through conceptual, as opposed to contextual or case study lenses. The three lenses Rasmussen chooses are technology, doctrine, and agents that comprise the realm of strategy.[110] While the first two are self-explanatory, the third is not. Agents, in this case, refer to the United Nations and violent extremism. The former has bureaucratized and constrained state-based aggression and war to the point of disutility. Violent extremism, embodied by al-Qaeda, is the sub-state reaction to contemporary technology, society, and culture, or the risk states seek to manage.

[105] Beck, *Risk Society*, p. 233.

[106] Barbara Adam, Ulrich Beck, and Joost van Loon, *The Risk Society and Beyond: Critical Issues for Social Theory* (London: Sage, 2000).

[107] For an informed yet accessible explanation of the reasoning for not going to war, see Christopher Blattman, *Why We Fight: The Roots of War and the Paths to Peace* (New York: Viking, 2022), especially Chapters 1–3.

[108] Christopher Coker, *War in an Age of Risk* (New York: Polity, 2009), p. 63.

[109] Yee-Kuang Heng, *War as Risk Management: Strategy and Conflict in an Age of Globalised Risks* (London: Routledge, 2006), pp. 38–39.

[110] Mikkel Rasmussen, *The Risk Society at War: Terror, Technology and Strategy in the Twenty-First Century* (Cambridge: Cambridge University Press, 2006), p. 6.

Special operations, and more specifically SOF, are an integral part of Rasmussen's risk management approach. SOF represent an empowered, sub-political response to the non-state threat posed by al-Qaeda and, more recently, Daesh. Rasmussen argues SOF are not part of the traditional bureaucratic hierarchy of the military and therefore are not beholden to its ways of thinking or its rule sets. The political and systemic risk of non-state threats is offset by the personal risk special operators are willing to assume for a greater purpose. Rasmussen's characterization of the motivation of other sub-political response to non-state threats, members of non-governmental organizations (NGOs), could equally apply to special operators: 'They felt that they could influence world events and do the right thing, and only the risks that they took proved worthwhile'.[111] He deduces that SOF and NGOs will increasingly work together to manage the risks posed by non-state actors based on a shared understand of personal risk as well as their ability to perform 'edgework', the 'skilled performance of the dangerous activity, involving the ability to maintain control over a situation that verges on complete chaos that requires, above all, 'mental toughness', the ability not to give in to fear'.[112] Calls within the Department of Defense (DoD) to shift to so-called 'Phase 0' or 'shaping' operations against violent extremists or state-based 'SOF campaigning' against strategic competitors, with SOF providing monitoring, early warning, and a means of preemption, suggest Rasmussen's theoretical strategy of risk management may become a reality.[113]

At first glance, the concepts of the 'risk society' and risk management either completely or tenuously explicate the political risk associated with the contemporary use of special operations. Both concepts suggest political risk is the product of the current global security environment. As is the case with strategic culture, context is an explanation for behaviour and outcome but cannot be the only explanation. Other explanations likewise fall short. Prospect theory abstracts away too many variables within the risk assessment and decision-making process and is best suited for discrete, episodic events in response to crises and not the sustained use of special operations. While prospect theory may be too abstract, the socio-cognitive approach demands a high fidelity of information relative to the power of its explanatory return. Obtaining sufficient information to satisfy this approach is unlikely for reasons well known to researchers of special operations, given their sensitive nature. To connect together physical, strategic, and political/diplomatic risk, a different approach is needed.

[111] Rasmussen, *The Risk Society at War*, p. 180.

[112] Deborah Lupton, quoted in Ibid., p. 179.

[113] Richard Clarke, 'Statement of General Richard Clarke, USA, Commander, United Special Operations Command before the 117th Congress Senate Armed Services Committee', pp. 4–5. www.armed-services.senate.gov/imo/media/doc/2022%20USSOCOM%20Posture%20-%20Clarke%20-%20SASC%20(5APR22)%20(FINAL).pdf; Brian Petit, *Going Big by Getting Small: The Application of Operational Art by Special Operations in Phase Zero* (Parker, CO: Outskirts Press, 2013); Malcolm Brailey, 'The Transformation of Special Operations Forces in Contemporary Conflict', in Bernard Loo, ed., *Military Transformation and Strategy: Revolutions in Military Affairs and Small States* (Abingdon: Routledge, 2009), pp. 80–100.

## Risk and control in special operations

There are almost as many frameworks for assessing risk, with an eye towards managing it, as there are definitions. According to risk management training for its personnel, the US DoD considers any possible contingency, from a flat tire to a lack of supplies, as a risk to the mission that must be identified and mitigated, no matter how seemingly trivial or banal. At the programmatic level, however, the DoD perspective on risk shifts away from the local and environmental, or daily hazards and events, to broader concerns. They include the ability of the Department to prepare and maintain its readiness to address military contingencies to its ability to acquire and maintain sufficient resources in the future. Risk perceptions are driven by different considerations of what is crucial at different operating levels, from the political down to the individual. In other words, no one set of risk variables is sufficient for risk in special operations from the tactical to the political, making the concept a 'nebulous ... and subjective' one.[114]

A framework for examining risk for special operations need not be created out of whole cloth. Scholarship on and official documents describing special operations contain useful insights. Bernd Horn suggests three types of risk: preventable, strategic, and external.[115] US Joint doctrine, in contrast, offers three broad categories of risk: physical, strategic, and political/diplomatic. It is possible to equate physical risk to the tactical level, or special operations at the point of execution. The leap from physical to strategic risk, however, is too large to make in a single bound. Doctrine identifies actors and command levels, and specifies command relationships between the tactical and strategic but does not specify risk considerations. In cases of sustained use of special operations, which has been the norm for the past twenty years and appears to be the future of Western SOF, the intervening operational or campaign level should have its own risk considerations. The next level in the hierarchy is strategic risk. Risk at this level can shutter the 'guiding light' that ensures adequate means are used to generate strategic effects in service of policy goals.[116] The last level in the hierarchy of risks for special operations is the political/diplomatic. The following analysis starts at the point of execution for SOF and works its way upwards to the strategic level.

## Physical or tactical risks

The small scale of special operations relative to comparable conventional ones, combined with selective manning and specialized equipment and training, grants SOF greater flexibility at the cost of increased physical or tactical risks. Special

[114] Bernd Horn, 'Risk: A Nebulous Concept', in Bernd Horn, ed., *Risk and Decision-Making* (Ottawa, ON: Canadian Special Operations Forces Command, 2019), pp. 1; 3.

[115] Horn, 'Risk', pp. 3–6.

[116] Colin S. Gray, *The Strategy Bridge: Theory for Practice* (Oxford: Oxford University Press, 2010), p. 124.

operations are often conducted with small forces at long ranges, in hostile or contested territory, without the security of large numbers or inherent combat support functions. Should forces lose surprise, mobility, and/or initiative, the personnel conducting special operations are vulnerable to capture or death, as several historical and more recent actions such as Roberts Ridge in 2002 suggest.[117] Malvesti, in a report on the subject of risk in counterterrorism, terms its negative consequences as 'casualties'.[118] Given the relatively small size of the forces committed for special operations, the first set of tactical risks focus on the sufficiency of the resources assigned to conduct the mission according to the task as well as the anticipated resistance in or around the specific objective.[119] In other words, the specific threats in the operating area allow for the identification of fairly concrete risk to forces. Some authors charge that the past twenty years of counterterrorism operations have instilled in US SOF 'seven deadly norms of risk avoidance' at the tactical level, based on assured communications and overly cautious force protection measures, that will inhibit future special operations in future conflicts against state competitors.[120]

The next dimension of physical risk beyond the forces themselves deals with the mission they have been assigned: risk to mission. Mission can be understood in two different ways. The first refers to the type of activity being undertaken. For example, indirect-shock missions and working through proxy forces, or treachery and subversion, both occur in hostile or denied areas and incur higher levels of mission risk due to the overall qualities of the operating environment.[121] Military advising and assistance mission, including civil affairs activities, are missions that happen in relative permissive environments and therefore incur lower levels of risk as a mission set.[122] The second way risk can be characterized, in terms of mission, is through the specific environmental and support factors that could adversely affect mission success at the point of enemy contact. Such factors include tangibles like weather, enemy reinforcements, and the terrain alongside intangibles such as enemy morale, resistance, or courses of action. At the tactical level, functional experts from various organizations provide information that manifests itself in the development of plans

[117] William McRaven, *Spec Ops, Case Studies in Special Operations Warfare: Theory and Practice* (Novato, CA: Presidio Press, 1996), p. 6; Sean Naylor and Christopher Drew, 'SEAL Team 6 and a Man Left for Dead: A Grainy Picture of Valor', *New York Times* (27 August 2016), www.nytimes.com/2016/08/28/world/asia/seal-team-6-afghanistan-man-left-for-dead.html.

[118] Michele Malvesti, *Policy Considerations in Combating Terrorism: Decision-Making Under Conditions of Risk and Uncertainty* (West Point, NY: Combating Terrorism Center, January 2012), pp. 2; 5.

[119] Threat evaluation guidelines tactically, that feed into a risk calculation matrix, are available in Headquarters, Department of the Army, *Special Forces Detachment Mission Planning Guide*, GTA 31-01-003 (Washington, DC: HQ, Department of the Army, January 2020), pp. 41–43; 48–49, https://irp.fas.org/doddir/army/gta31_01_003.pdf.

[120] Spencer Reed, 'Recalibrating Special Operations Risk Tolerance for the Future Fight', War on the Rocks (31 January 2023), https://warontherocks.com/2023/01/recalibrating-special-operations-risk-tolerance-for-the-future-fight.

[121] The Joint Staff, *Joint Publication 3-05: Doctrine for Joint Special Operations* (Washington, DC: The Joint Staff, 2014), pp. II-5; II-9; Mark Grdovic, *A Leader's Handbook to Unconventional Warfare*, SWCS Pub 09-1 (Fort Bragg, NC: John F. Kennedy Special Warfare Center and School, November 2009), pp. 21; 25.

[122] The Joint Staff, *Joint Publication 3-22: Foreign Internal Defense* (Washington, DC: The Joint Staff, 2021), pp. V-5 to V-6.

that contain requested levels of support from other special operations or conventional forces passed up to the operational level. Different levels of force visibility for some or all the forces committed to the mission, which range from overt to clandestine and covert, can mitigate some risk. Missions at the tactical level are limited in scope, duration, and investment of forces. A setback tactically can be disastrous or fatal for those involved but its impacts, particularly when the tactical mission is part of an ongoing campaign and not a response to a national-level crisis, are unlikely to have significant strategic or political consequences. Two examples include the downing of one MH-47 helicopter, Turbine 33 (July 2005), and a CH-47, Extortion 17 (August 2011), the latter 'the largest single loss of life of the entire Afghanistan war'.[123] Despite these losses, there was little discernible political or strategical consequence for the conduct of SOF campaigns in Afghanistan, aside from Congressional hearings.[124]

To mitigate physical risk and improve the odds of mission success, forces conducting special operations can conduct extensive and realistic pre-mission rehearsals of concepts that exploit the experience and training of the operators if time allows. Problems, however, can result from overlapping risk calculations. One is repeating previously successful mission templates with inadequate alteration, or modified risk calculation, of changing circumstances such as an adaptive enemy. These circumstances allegedly contributed to the casualties suffered in the 'Blackhawk Down' mission mentioned previously.[125] Another is the growth of the forces assigned to the mission, including operators and support forces, to reduce risk to negligible levels. Such forces including intelligence, surveillance, and reconnaissance aircraft, fire support assets, and perimeter security forces can mitigate physical risk while jeopardizing low visibility. As one operator depicts the problem, 'One way we tend to think we mitigate risk is by adding more capabilities for this contingency and that contingency' because, from the standpoint of decreasing physical risk, 'more is better'.[126] The size of such force packages as a hedge against risk and uncertainty have grown to the point that they have been evocatively named 'gorilla packages', including prolific air support in the form of 'fighter planes and AC-130 (gunships)'.[127]

## Operational or campaign risks

At the operational or campaign level there is an increased scope and scale of special operations activities, accompanied by an exponential growth in the range of factors

[123] Panetta, *Worthy Fights*, p. 357.

[124] Marcus Luttrell and Patrick Robinson. *Lone Survivor: The Eyewitness Account of Operation Redwing and the Lost Heroes of SEAL Team 10* (New York: Back Bay Books, 2007); Ed Darack, *Victory Point: Operations Red Wings and Whalers—The Marine Corps' Battle for Freedom in Afghanistan* (New York: Berkley Books, 2009), pp. 137–145; Don Brown, *Call Sign Extortion 17: The Shoot-Down of SEAL Team Six* (Guilford, CT: Lyons Press, 2015), pp. 225–241.

[125] Jonathan Stevenson, *Losing Mogadishu: Testing U.S. Policy in Somalia* (Annapolis, MD: Naval Institute Press, 1995), p. 94; Mark Bowden, *Blackhawk Down: A Story of Modern War* (New York: Atlantic Monthly Press, 1999), p. 22. Bernd Horn terms this as 'contempt for the enemy', 'Risk', pp. 19–20.

[126] Shultz Jr, 'Showstoppers', p. 32.

[127] Chris Wallace with Mitch Weiss, *Countdown Bin Laden: The Untold Story of the 247-Day Hunt to Bring the Mastermind of 9/11 to Justice* (New York: Avid Reader Press, 2021), p. 244.

to account for, which increases overall complexity and places emphasis on a different kind of risk. At the operational level, risks are a function of the need to conduct numerous actions over prolonged periods of time cumulatively and/or sequentially to achieve objectives as part of a campaign.[128] In Western practice a combined joint special operations task force commander (CJSOTF), combined forces special operations component commander, or joint task force commander is responsible for identifying and managing risks at the operational or campaign level.

Risk to forces is present at the operational or campaign level but reflects different considerations such as supply, logistics, and basing. Risks at this level, once identified, are mitigated through integration with conventional and host nation chains of command. To assist tactical forces in mitigating risk, operational level commanders extensively aid in the planning, integration, and coordination of internal processes and resources. For example, a single tactical operation in Afghanistan in 2015, Operation ZUMMARI 241, had the following requirements:

> communications, protective air cap (Coalition Air Force fighters overhead), fire support coordination (Coalition Air Force close air support [CAS] and Army attack helicopters), anti-aircraft artillery (AAA), communications jamming during ingress, holding and egress along pre-planned routes, selecting alternate landing zones and "catch up" loiter positions, combat search and rescue (CSAR), casualty evacuation (CASEVAC), downed aircraft situation and subsequent "bump" plans for ground forces and aircrews, and refueling—aerial and ground.[129]

Such risks also include potential 'friendly fire incidents', from failing to integrate with conventional forces, and they can be addressed at the operational level through coordination or separation, in the form of unique geographic area designated as a Joint Special Operations Area.[130]

These considerations of risk at the operational level relate to overall campaign objectives as insufficient familiarity with the mission, method of operations, or operating environment all create risk to mission. To ensure a degree of 'stability and cohesion', and to alleviate organizational risk, several commanders and senior enlisted advisors repeated tours in CJSOTF-Afghanistan from 2002 to 2014.[131] Other risk to mission at the campaign level includes 'operational failure'.[132] For example, operational failure throughout 2004 and 2005 at the campaign level for special operations in Iraq would be the expansion of Abu Musab al-Zarqawi's networked Sunni insurgency, al-Qaeda in Iraq (AQI). AQI leaders developed a highly effective and

[128] J.C. Wylie, *Military Strategy: A General Theory of Power Control* (Annapolis, MD: Naval Institute Press, 1989), pp. 24–25.

[129] Charles Briscoe, 'Operation ZUMMARI 241: Braving Fire Ambushes in Afghanistan', *Veritas*, 12:2 (October 2016): 56.

[130] The Joint Staff, *Joint Publication 3-05*, p. III-11.

[131] Michael Krivdo, 'CJSOTF-A (Combined Joint Special Operations Task Force-Afghanistan): A Short History, 2002–2014', *Veritas*, 12:2 (October 2016): 9, 13.

[132] 'Policymaking during Risk and Uncertainty: The Role of Special Operations', *Yale Journal of International Affairs* (March 2012), p. 95, www.yalejournal.org/publications/policymaking-during-risk-and-uncertainty-the-role-of-special-operations.

distributed organization that exploited ethnic, religious, and tribal ties in a cohesive network. The strength and cohesion of its network also contained the source of its downfall, its inability to adapt to change.[133] To accomplish the campaign objective, General Stanley McChrystal struck a compromise. To decrease risk to mission, he radically altered his special operations organization, Task Force 714, as well as its method of planning and executing actions.[134] Innovation and adaptation occurred because of leadership, allocation of sufficient resources and personnel, as well as tactical latitude and functional authorities. In the process of innovating, the risk to force for Task Force 714 increased while the risk to mission decreased. The level of physical risk went up tactically with increased contact with the enemy while the development of a special operations network conducting relentless actions against an insurgent network, decreased risk to mission.[135] In this example, discussed in detail in Chapter 8, the primary risk to mission was not the competence of AQI, but rather the inability of Task Force 714 to innovate and adapt.

Operational-level SOF solutions, however, can also generate wider risk to mission. One such problem existed in Afghanistan. Local village security was a problem for more than a decade. Tribal leaders and villages were vulnerable to Taliban forces on one hand, or warlord militias on the other. One approach to help solve this problem was the embedding of special operations personnel in specific villages in Konar province, as Jim Gant advocated.[136] In Iraq, US forces mapped out tribal relations to better understand the complex and occasionally overlapping power and influence of leaders at the family, lineage, clan, tribe, and confederation levels.[137] While this approach had some tactical success, it needed to be conducted on a much wider operational scale to achieve a campaign goal. The creation of the Village Stability Operations/Afghan Local Police concept in 2010 was a widespread special operations operational effort to create a 'local defense programme aimed at increasing its chance of success and decreasing the risk that it would spawn out-of-control militias'.[138] More clearly defined variables for operational-level risk assessment, along with specific measures of effectiveness, would assist SOF in better understanding their impact during campaigns as opposed to the measures of performance often used.

A final risk associated with SOF missions at the operational level relates to the inflation of the mission set to the level of strategy. Put simply, the repetitive authorization and use of special operations, either out of routine or towards a definable goal, can confuse the means and ways of strategy for the ends.[139] For example,

[133] Chad Serena, *It Takes More Than a Network: The Iraqi Insurgency and Organizational Adaptation* (Stanford, CA: Stanford University Press, 2014), pp. 16–22.

[134] Richard Shultz Jr, *Transforming U.S. Intelligence for Irregular War: Task Force 714 in Iraq* (Washington, DC: Georgetown University Press, 2020), pp. 6–7.

[135] McChrystal, *My Share of the Task*, pp. 147–149.

[136] Jim Gant, 'One Tribe at a Time: A Strategy for Success in Afghanistan', unpublished white paper (2009), www.globalsecurity.org/military/library/report/2009/2009_one_tribe_at_a_time.pdf.

[137] Michael Eisenstadt, 'Tribal Engagement Lessons Learned', *Military Review*, 88:5 (September–October 2007): 17.

[138] Linda Robinson, *One Hundred Victories: Special Ops and the Future of American Warfare* (New York: PublicAffairs, 2013), p. 24.

[139] Peritz and Rosenbach, *Find Fix Finish*, pp. 231–233.

direct action missions that characterize counterterrorism missions run the risk of becoming an endless series of 'tit-for-tat' exchanges between SOF and terrorist groups, which may or may not lead to success, as the Israeli experience in several campaigns suggests.[140] Potentially worse, losses among their own units can drive SOF to conduct immediate, substantial reprisals, leading their adversaries to escalate their response in what Boaz Ganor and Mia Bloom have both termed a 'boomerang effect', that can occur tactically, operationally, or even strategically.[141] Counterterrorism and asymmetric responses have become, at times, the strategy against terrorism instead of an operational approach to combat it.[142] These missions are only one of the means available to address terrorist threats but they are appealing politically due to their responsiveness, flexibility, and tactical effectiveness. Devoid of meaningful direction and dialogue, special operations at the campaign level run the risk of performing the function demanded by risk societies: episodic actions designed to respond to circumstances rather than addressing systemic causes.

## Strategic-level risks

Strategy is '[t]he direction and use made of means by chosen ways in order to achieve desired ends'.[143] More than just a plan, strategy should seek continuing advantage and not focus on the current challenge but in anticipation of subsequent challenges.[144] In theory, actors at the political level sets the objectives, or desired ends, while the strategic directs how resources are managed and used through plans. The reality can be very different.

Determining specifically where in the national security apparatus strategy is developed and managed can be challenging. Commonly, strategy is 'owned' at the national level by advisors who work for the senior executive, in the case of the United States the president or in Great Britain the prime minister. One dimension of strategy is military strategy, which is the responsibility of the military and civilian leaders within a department or ministry of defence or war. Within the United States DoD, USSOCOM, and its policy arm, the Office of the Assistant Secretary of Defense for Special Operations and Low Intensity Conflict (OASD SO/LIC), identify and mitigate the military strategic-level risks related to special operations. At this level, the risks for special operations become more opaque, are primarily organizational, and include bureaucratic politics.[145]

[140] Daniel Byman, *A High Price: The Triumphs and Failure of Israeli Counterterrorism* (New York: Oxford University Press, 2011), p. 365.

[141] Boaz Ganor and Liram Koblentz-Stenzler, *Israel's Targeted Killing Policy: Moral, Ethical, and Operational Dilemmas* (New York: Palgrave Macmillan, 2022), Chapter 7; Mia Bloom, *Dying to Kill: The Allure of Suicide Terror* (New York: Columbia University Press, 2005), p. 20.

[142] Simpson, *War from the Ground Up*, pp. 131–132.

[143] Gray, *The Strategy Bridge*, p. 18.

[144] Everett Dolman, *Pure Strategy: Power and Principle in the Space and Information Age* (London: Frank Cass, 2005), pp. 4–6.

[145] Allison and Zelikow, *The Essence of Decision*.

Risks at the level of military strategy for special operations reflect two different sets of responsibilities. The first is the current use of military resources to achieve national objectives. The second are institutional responsibilities to guarantee the long-term health and welfare of the organization and its members. In both cases, those developing and implementing military strategy should look beyond the use of force with an eye towards the future. Michael Mazarr outlines a DoD framework that accomplishes this, consisting of four elements:

1. Force management risk, defined as threats to 'the ability to recruit, retain, train, equip sufficient numbers of quality personnel and sustain the readiness of the force while accomplishing its many operational tasks'.
2. Operational risk, dangers to the capacity to achieve military objectives in conflict.
3. Future challenges risk, threats to an ability to 'invest in new capabilities and develop new operational concepts needed to dissuade or defeat mid- to long-term challenges'.
4. Institutional risk, threats to the ability to manage defense resources efficiently.[146]

Strategic-level risks reflect the collision of two different responsibilities in pursuit of national objectives. Current use of military resources drives organizations such as USSOCOM and OASD SO/LIC to connect policy to operations and tactics through administration, advocacy, and lobbying. Evidence of such connections includes the introduction of specific authorities into Congress, such as Section 1208 and later 127e funding authorities for special operations to assist proxy forces, to classified plans. The second set of responsibilities reflects organizational risks that relate to the management, stewardship, and long-term health and readiness of the special operations community. USSOCOM and OASD SO/LIC staff and leaders identify the specific force management, future challenges, and institutional risks, along with plans to mitigate them. Organizational risk mitigation, however, creates additional risk, primarily as a reflection of institutional self-image, identity, and history, as well as relative rank and prestige with comparable, competing organizations such as the Armed Services (the Army, Navy, Air Force, etc.) or other agencies (the Central Intelligence Agency). Risk within this set of responsibilities mirrors special operations organizational preferences and culture.

Special operations leaders at the strategic level address risk to current military operations and institutional challenges by shaping political calculations through their lobbying and advocacy. How risk is communicated to decision makers is central to such shaping, as the risk generated here is not strategic but political. In preventing the recurrence of past misuse or non-use, extending the current political preference to rely on special operations, and expanding special operations resources, authorities, prestige, and influence, special operations leaders and staffs may view instruments

[146] Michael Mazarr, *Rethinking Risk in National Security: Lessons of the Financial Crisis for Risk Management* (New York: Palgrave, 2016), p. 36.

of government, both civilian and military, as a threat to the growth, primacy, and long-term health of the organization. Much like at the operational and tactical levels, special operations address threats at the strategic level through extensive training, preparation, and planning, using unorthodox approaches and unconventional tactics. In short, the line between operational, strategic, and political can be compressed or obscured depending on the type of special operation conducted and the character of the political authority.[147] Hostage rescues and indirect-shock missions, where opportunities are both golden and fleeting, such as the strikes against Anwar al-Awlaki and Osama bin Laden, render the dividing lines between political, strategic, operational, and tactical moot. The danger for special operations at the strategic level is in the conceptualizing of risks not as strategic challenges but rather political ones.

## Political risks

Given the sizable literature on decision making, there is a robust set of factors already identified on political risks connected to special operations. Yakob Vertzberger separates out two crucial aspects of political risk: risk texture and risk taste. He further offers eight textures of risk: risk transparency, risk severity, risk certainty, risk horizon, risk complexity, risk reversibility, risk controllability and containability, and risk accountability.[148] These textures of risk offer a useful departure point for interview questions for civilian national security leaders, and in particular, how textures may shape perceptions of the effectiveness and utility of one military response over another, for example, special operations as opposed to airpower or specific forms of precision strike.

Risk calculation, as a reflection of risk texture, shapes risk taking at the political level. Vertzberger identifies three patterns. The first is deliberative risk taking, which most closely aligns with Malvesti's model of relative losses and gains from prospect theory. The second pattern, dispositional risk taking, reflects the impact of personality on the likelihood of the decision maker to embrace risk or not. Some individuals are more inclined to take risk than others. The third pattern, socially driven risk taking, is a function of group preferences and dynamics, includes collective psychological and behavioural pathologies such as groupthink and cognitive closure.[149]

Both the texture of risk and response patterns raise more questions than they answer at this stage in terms of the sustained use of special operations. Previous studies of special operations responses to specific crises or incidents have favoured one pattern over another, for example the adverse impact of groupthink on risk calculations and decision making during the ultimately unsuccessful special

[147] Andrew Cockburn, *Kill Chain: The Risk of High-Tech Assassins* (New York: Henry Holt, 2015), p. 191.
[148] Vertzberger, *Risk Taking and Decisionmaking*, pp. 26–27.
[149] Ibid., pp. 29–30.

operations hostage rescue mission in Iran in 1980.[150] Others favour explanations of a relative position of losses internationally[151] or other deliberative criteria, such as non-compensatory decision dimensions, or concerns about re-election.[152] Singular explanations relating to preferences, particularly those that focus heavily on individual events, are insufficient to answer questions about a series of decisions. For example, President Bill Clinton was notoriously reticent to commit to certain types of special operations, such as counterterrorism missions, but readily approved others for humanitarian missions in Haiti or enforcing international norms, such as seizing Persons Indicted for War Crimes (PIFWCs), during his tenure.[153]

Political risk calculations are influenced by risk communication from the tactical, operational, and strategic levels. Such communication is noticeably absent in studies that discuss decision making about special operations. Although the dialogue between civil and military leaders, or principles and agents, may be an unequal one, most of the considerations of risk occur because of such communication of options and their risks.[154] Risks are often reformulated or repackaged when they are communicated at the political level. I recall Secretary of Defense Donald Rumsfeld had a map in his office in the Pentagon in 2002 with pins indicating the locations of special operations units. Special operations staff officers were reticent to discuss unnecessary or unprofitable missions. They preferred to let the secretary believe detachments were actively engaged everywhere and mitigating the risk of future terrorist actions and never suggested removing a pin. Better understanding how and why special operations leaders frame and communicate risk to political leaders, or avoid it, might ameliorate some of the problems identified in the previous section.

To summarize the preceding discussion and address the shortfalls of other models, the author presents a risk model that spans the tactical to political, or national decision-making level (see Figure 4.1). At the tactical level, special operators primarily focus on risks to forces as well as risks to mission. Risk at this level is readily understood as problems, including specific threats, that are bounded in space and time in a localized area. At the operational, more uncertainty exists as the scale and time horizon of actions grow. Risk here shifts to risk to mission for the sustainment of missions in the field and their connection to campaign objectives. At the strategic level, risk calculations start to diverge primarily as a function of organizational or bureaucratic considerations. Finally, at the political level, risk calculations reflect political survival and national interest, rational cost/benefit calculations and prestige concerns, among others.

[150] Steve Smith, 'Groupthink and the Hostage Rescue Mission', *British Journal of Political Science*, 15:1 (January 1985): 117–123.

[151] Malvesti, 'Risk-Taking in Countering Terrorism'.

[152] David Brulé, 'Explaining and Forecasting Leaders' Decisions: A Poliheuristic Analysis of the Iran Hostage Rescue Decision', *International Studies Perspectives*, 6:1 (February 2005): 99–113.

[153] Shultz Jr, 'Showstoppers', p. 26; Walter Kretchik, Robert Baumann, and John Fishel, *Invasion, Intervention, 'Intervasion': A Concise History of the U.S. Army in Operation Uphold Democracy* (Fort Leavenworth, KS: US Army Command and General Staff College Press, 1998), pp. 33–42, https://apps.dtic.mil/sti/pdfs/ADA528265.pdf; Pete Blaber, *The Mission, the Men, and Me* (New York: Berkeley, 2008), pp. 43–46; Alice Friend Hunt, 'Creating Requirements', pp. 42–43.

[154] Cohen, *Supreme Command*; Feaver, *Armed Servants*.

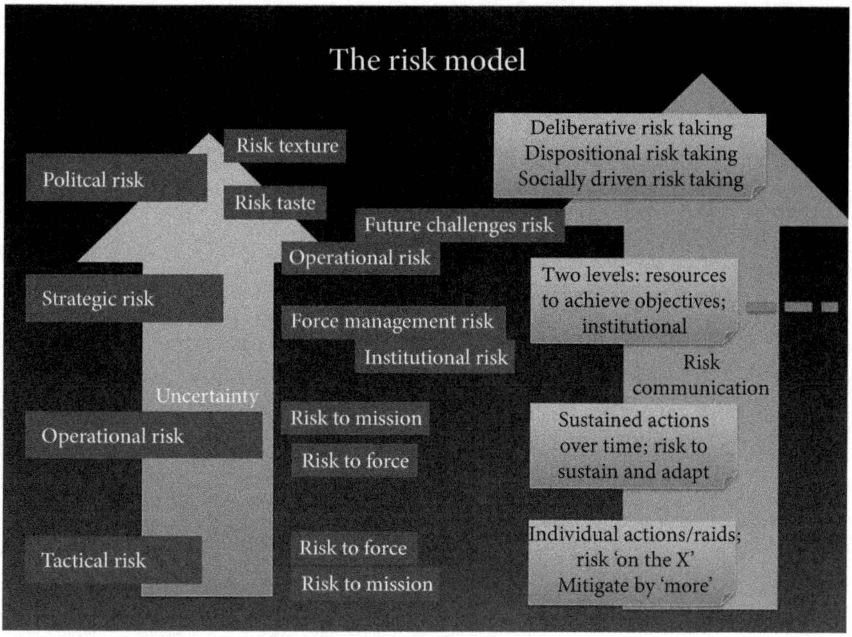


**Figure 4.1** The risk model.
*Source:* Author's work.

Three elements of risk from this model are relevant to the specific subject of risk evolution. The first is that risk communications change, due to different frameworks of understanding by political and military leaders. Different frames of reference diverge the higher the level of decision making, due to considerably increased uncertainty and paucity of accurate information. The nature of risk communication changes due to a tension inherent in linking resources to accomplish specific objectives against threats and institutional considerations. Such considerations, the second element, are a reflection of bureaucratic self-interest and include institutional longevity and health. Scholars have explored in great depth the frequent divergence of military and policy considerations, especially in works examining organizational behaviour. The third and last element is that considerations of risk change at the political level—the level of national security decision making—for several reasons reflecting response patterns to uncertain threats and adversary intentions. One is deliberative risk taking, explored in prospect theory as the calculation of relative risks and gains. Another is dispositional risk taking, which includes individual psychology and personality, and translates as a senior leader's threshold or 'taste' for risk. A final reason is socially driven risk taking, the result of group preferences and dynamics, or collective psychology. For the remainder of this work, and to bound it sufficiently, I focus primarily on bureaucratic considerations at the operational and strategic levels and socially driven risk taking at the political. Sufficient detail on mission planning at the tactical level is unavailable and likely remains classified for special operations conducted over the past half-century, if at all. For these reasons, this work ignores tactical level risk calculations but instead focuses on risk calculation and communication up the chain of command to senior military and/or political decision makers, as well as how the latter communicate risk tolerance down. This iterative dialogue framing and communicating risk in different ways is central to understanding the control measures designed to mitigate it.

## Vertical integration: control and risk

The analysis thus far has identified the two overarching forms of control of SOF, as well as the five types it can take: delegative, assertive or directive, collaborative, competitive, and co-optive. In addition, it has presented the different forms risk calculation and communication take from the political to the tactical level both up and down a chain of command. An element missing from the discussion, however, is the methods or means of control and their relationship to risk. These methods or means of control fall along three general lines: resource constraints or capabilities, authorities or oversight measures, and political or bureaucratic influence. Political and military leaders can control SOF by limiting the means or capabilities they can employ, the ways in which those means or capabilities can be employed, their power, or any combination of the three.

Chapter 3 examined capabilities from the standpoint of how and why they are developed, sustained, and integrated horizontally. Given the political and military concerns associated with highly motivated, restless, and unorthodox elite SOF outlined previously, the simplest control solution is to cut capabilities, in terms of mission sets and personnel, or place limitations on the resources they receive. For example, in the wake of the reported murder and torture of two Somali men in 1993, the Canadian Airborne Regiment was disbanded in 1995 following evidence of systemic, uncontrolled immoral practices and behaviour. In the words of the authors of the final report investigating the incident for the Canadian Ministry of Defence, their findings demonstrate 'the extent of the morass into which our military has fallen', particularly in terms of leadership failures, necessitating drastic action.[155] More recently, the Australian Special Air Service Regiment (SASR) came under political and military scrutiny for similar excesses in its operations in Afghanistan.[156] The scrutiny, which led to the disbanding in 2020 of one of its squadrons, 2 Squadron, resulted largely from an inability of the Australian Special Operations Command to investigate and take effective action in response.[157]

[155] Ministry of National Defence, *Report of the Somalia Commission of Inquiry*, Executive Summary, Conclusion (Ottawa, ON: Ministry of National Defence, 1997), https://web.archive.org/web/20070404053127/http://www.dnd.ca/somalia/vol0/v0s27e.htm.

[156] Christopher Knaus, 'War Crimes: Former Minister Reveals Why Canada Disbanded Its Special Airborne Force after Scandal', *The Guardian* (2 December 2020), www.theguardian.com/australia-news/2020/dec/03/war-crimes-former-minister-reveals-why-canada-disbanded-its-special-forces-after-scandal.

[157] 'Chief of Army Disbands 2 Squadron SASR', *Australian Defence Magazine* (20 November 2020), www.australiandefence.com.au/defence/land/chief-of-army-disbands-2-squadron-sasr;

Other control measures need not be so drastic. Under some political regimes, SOF can be controlled by buying them off. In exchange for additional resources or engaging in economic entrepreneurship, SOF offer loyalty and protect the regime and can serve as a hedge against other powerful players in weak democracies or autocracies.[158] Resource controls can include reductions to personnel, equipment, facilities, or funding. These reductions can be a reaction to years or decades of growth and expansion as the case of USSOCOM suggests; a natural market reaction, metaphorically speaking. After more than doubling its budget and dramatically increasingly its numbers of personnel, President Donald Trump directed a 12 per cent budget cut to USSOCOM in his 2021 budget request to Congress.[159] Critics, for example, argue downsizing is necessary as the growth in the investment in SOF personnel is not commensurate with their return, relatively to conventional forces. Richard Hooker Jr opines that SOF

> do not, whatever their proponents may say, deliver decisive strategic results. Neither are they true economy-of-force assets ... they come at a price in funding and manpower that does not square with their actual contributions to campaign success. Soldier for soldier, they are far more expensive to recruit, train, equip, and retain.[160]

At other times reductions are driven for similar reasons as disbandments, through an accumulation of evidence of unethical behaviour or a perceived sense of entitlement. The latter sense can be particularly dangerous, as Nora Bensahel and David Barno point out, stemming initially from an 'unhealthy sense of privilege from [an SOF operator's] very first days in the [special operations] community. And the pressures of constant deployments and casualties make it all too easy for that privilege to transform into entitlement and a sense that normal military standards and discipline no longer apply.'[161]

Political leaders looking to control SOF can also legally constrain the actions SOF are permitted to take. Within the American system, for example, Congress often places time restrictions on SOF legal authorities, making them temporary in nature. At the strategic level, these included what were known as the '1200' series of

Inspector-General of the Australian Defence Force (ADF), *Report of Inquiry under Division 4A of Part 4 of the Inspector-General of the Australian Defence Force Regulation 2016 into Questions of Unlawful Conduct Concerning the Special Operations Task Group in Afghanistan, Part 1—The Inquiry, Part 3—Operational, Organisation and Cultural Issues*, Official (Redacted) (Canberra: ADF HQ, 2020), pp. 26–27.

[158] Roya Izadi, 'State Security or Exploitation: A Theory of Military Involvement in the Economy', *Journal of Conflict Resolution*, 66:4–5 (2022): 729–754; Mwangi Kimenyi and John Mukum Mbaku, 'Rents, Military Elites, and Political Democracy', *European Journal of Political Economy*, 11:4 (April 1996): 699–708.

[159] US General Accounting Office, *Special Operations Forces: Additional Actions Are Needed to Effectively Expand Management Oversight*, GAO-19-386 (May 2019), p. 1; Jon Harper, 'Special Ops Command Faces Funding Cuts', *National Defense* (23 April 2020), www.nationaldefensemagazine.org/articles/2020/4/23/special-ops-command-faces-funding-cuts.

[160] R.D. Hooker Jr, 'America's Special Operations Problem', *Joint Forces Quarterly*, 108 (1st Quarter 2023): 53.

[161] David Barno and Nora Bensahel, 'How to Fix U.S. Special Operations Forces', War on the Rocks (25 February 2010), https://warontherocks.com/2020/02/how-to-fix-u-s-special-operations-forces.

authorities, named for their respective sections of the National Defense Authorization Act of 2006, which were designed to give SOF greater flexibility in providing security assistance and funding support to partner nations and proxies largely for counterterrorism missions. These authorities were designed to last for a year, after which time Congress would review, modify, renew, or allow them to lapse. Section (§) 1208 initially began as a means to provide SOF the ability to fund unconventional warfare activities but was expanded in 2006, reviewed annually, and formally added to US Law as §127e of Title 10 in 2016.[162] This authority, actions under which are overseen by OASD/SOLIC, has led to considerable discussion and debate over the ease with which SOF can arm and employ proxies against terrorist groups well beyond the original scope of the 2002 Authorization to Use Military Force, as well as Article II of the US Constitution.[163] Media reports suggest in 2020 there were nineteen 127e programmes operating in at least twelve different countries.[164] After successive years of review and modification, Section 1206 became more formalized as § 2282, Title 10 United States Code in 2012 but was eventually repealed by Congress in 2016.[165] The law was redrafted in 2017, as § 333 authority, to improve the flexibility of 'capacity building' efforts with partner nation forces and came with a series of streamlined, but still onerous processes.[166] This funding authority came to national attention and scrutiny after the Tongo Tongo ambush in Niger in 2017 discussed below.

A counterterrorism example illustrates the dynamic process at work in controlling special operations actions through authorities. To establish a level of control from the political level through the strategic to the operational, authorities such as 127e are further specified as execute orders (EXORDs) from the chairman of the joint chiefs of staff, which translates an order from the president or secretary of defense into a military action through Global Combatant Command (GCC) headquarters. When SOF have used a bottom-up process for targeting terrorists based on actionable intelligence, their request for action, in the form of a concept of operation (CONOP) stating the specific authority being used (along with other information) goes through a seven-step process for review and authorization, under a limited sixty-day window.[167] At times, presidents have relaxed control and delegated mission authorization to combatant commanders. They in turn have issued their own

[162] Title 10, Subtitle A, Part 1, Chapter 3, §127e. Support of special operations to combat terrorism. https://uscode.house.gov/view.xhtml?req=granuleid:USC-prelim-title10-section127e&num=0&edition=prelim.

[163] Concerns about 127e are outlined in an accessible manner in Katherine Yon Ebright, 'What Can a Secretive Funding Authority Tell Us about the Pentagon's Use of Force Interpretations?', Lawfare (11 October 2022), www.lawfareblog.com/what-can-secretive-funding-authority-tell-us-about-pentagons-use-force-interpretations.

[164] Nick Turse and Alice Speri, 'How the Pentagon Uses a Secretive Program to Wage Proxy Wars', The Intercept (1 July 2022), https://theintercept.com/2022/07/01/pentagon-127e-proxy-wars.

[165] 1303–1307; US Congress, Public Law 114-328, 'Subchapter VII—Administrative and Miscellaneous Matters' (23 December 2016), https://uscode.house.gov/statviewer.htm?volume=130&page=2504.

[166] Melissa Dalton et al., *Shifting the Burden Responsibly: Oversight and Accountability in U.S. Security Sector Assistance* (Washington, DC: Center for Strategic and International Studies, April 2019), p. 8.

[167] Cora Currier, 'The Kill Chain: The Lethal Bureaucracy behind Obama's Drone War', The Intercept (15 October 2015), https://theintercept.com/drone-papers/the-kill-chain.

EXORDs, interpreting presidential guidance or specific executive orders to expand the authority of SOF actions, and take more direct control of special operations missions, as Central Command (CENTCOM) commander David Petraeus is alleged to have done in 2009.[168]

At the operational level, control of SOF is exercised militarily through intricate command relationships (COMRELs). These relationships are necessary to spell out in order 'to achieve unity of command of the force and unity of effort with partners' for functional task forces such as those employed by SOF.[169] As Joint doctrine specifies, for US SOF this starts with command authority (COCOM) vested in the USSOCOM commander, who oversees the management of the force through a process known as 'Global Special Operations Synchronization'.[170] GCCs exercise operational control (OPCON) for their respective areas through their theatre special operations component headquarters.[171] Furthermore, relationships are identified to avoid potential confusion, such as which commanders or headquarters have what operational, tactical, and administrative control over individual units, identifying who is supported and supporting.[172] Working out these complex and sometimes confusing command relationships is not only necessary for clarifying control but is also an integral part of mitigating both mission and force risk.[173]

Political leaders in the United States also maintain control through reporting and notification requirements to Congress as well as study and reporting mechanisms outside of SOF influence. These include reports commissioned for members of Congress conducted by Congressional Research Service staff, studies and recommendations produced by the General Accounting Office, and analyses and reviews conducted by the Office of Management and Budget. Such studies have identified gaps in SOF command and control data, as well as problems relating to creating, 'right sizing', and ending SOF C2 structures, training and preparation of staff, and an inability to properly staff increasing numbers of subordinate operational-level headquarters.[174]

The last method of control is curtailing SOF's bureaucratic influence relative to other arms of the government. The birth of modern joint SOF in the in United States, with the formation of USSOCOM in 1987, saw an early version of this within DoD.

[168] Marc Ambinder, 'Obama Gives Commanders Wide Berth for Secret Warfare', *The Atlantic* (25 May 2010), www.theatlantic.com/politics/archive/2010/05/obama-gives-commanders-wide-berth-for-secret-warfare/57202.

[169] Deployable Training Division, Joint Staff J-7, *JTF C2 and Organization* (Washington, DC: The Joint Staff, April 2017), p. 1.

[170] This synchronization occurs every five years and its steps and processes are outlined in more than thirty pages of instructions and terms of reference. See, for example, USSOCOM, 'USSOCOM Global Special Operations Synchronization (GSOS), FY2018–2019, Preparation Instructions', unclassified//for official use only, 18 June 2015.

[171] The Joint Staff, *Joint Publication (JP) 3-0: Operations* (Washington, DC: The Joint Staff, 2018), p. IV-8.

[172] These relationships are summarized clearly and graphically in The Joint Staff, *JP 3-0: Operations*, p. III-3.

[173] Deployable Training Divising, Joint Staff J-7, *JTF C2 and Organization*, p. 14.

[174] Government Accounting Office, *Special Operations Forces: Better Data Necessary to Improve Oversight and Address Command and Control Challenges*, GAO-23-105,163 (Washington, DC: Government Accounting Office, October 2022), pp. 19–24.

When Congress forced DoD and the Armed Services to accept the new unified command for SOF, a question naturally arose about where to locate it. According to SOF lore, DoD chose Tampa for two reasons: first, there was an available headquarters there already, Readiness Command (REDCOM), modifiable into a SOF one, and second, that the headquarters would be as far away as one could get from Washington, DC, limiting its presence and therefore its influence and power.[175] As Susan Marquis notes, one influential office on the Joint Staff 'was pushing for anywhere *outside* of the Washington area' (italics original).[176] Over the past decade, however, Congress and successive presidents have granted numerous SOF resource and authorities requests, including expansion of personnel. While agreeing to other requests designed to enhance interagency coordination, such as the creation of a Joint Interagency Task Force for the National Capital Region (JIATF-NCR), they have denied others that would expand SOF prestige, bureaucratic power, and influence.[177] Rejected requests included making SOF its own armed service, the creation of a separate department of irregular warfare or unconventional warfare command, and allowing SOF to have its own degree-granting professional military education.[178]

Political leaders can limit SOF power and influence through institutional competition, distributing or redistributing SOF units within competing government ministries, agencies, and departments, or pitting a consolidated SOF organization against others. While indicative of a broader approach to power designed to guarantee his absolute hold upon it from before the war, Adolf Hitler tacitly approved the transfer of the Brandenburgers from the control of the intelligence service (*Abwehr*) to the army (*Wehrmacht*) during it. He increasingly relied on his ideological paramilitary (*Schutzstaffeln*, or SS; more specifically, Amt VI-S of the *Reichssicherheitshaupamt*) for high-risk special operations.[179] Dennis Showalter concludes a combination of ideology and technological developments 'favored developing elites' within Nazi Germany and many were born out of the 'vanity' of various leaders.[180] In other cases, SOF units have been punished for perceived disloyalty. In the Russian Federation, President Boris Yeltsin inflicted 'an insult' on KGB Vympel in 1993 by transferring the organization first to the Ministry of the Interior and then to its law enforcement branch, for backing the losing faction in a coup attempt.[181] The peerless example of

[175] Susan Marquis, *Unconventional Warfare: Rebuilding U.S. Special Operations Forces* (Washington, DC: Brookings Institution Press, 1997), p. 154.

[176] Marquis, *Unconventional Warfare*, p. 152. Aside from Tampa, other contenders included Fort Bragg in North Carolina and Fort Belvoir in Virginia.

[177] Naylor, *Relentless Strike*, pp. 428–429.

[178] John Mulholland, 'Retying the Gordian Knot: US Special Operations Command as a Service', *Military Times* (7 March 2021), www.militarytimes.com/opinion/commentary/2021/03/07/retying-the-gordian-knot-us-special-operations-command-as-a-service; David Tucker and Christopher Lamb, *United States Special Operations Forces* (New York: Columbia University Press, 2007), pp. 218–219; Charles Cleveland and Daniel Egel, *The American Way of Irregular War: An Analytic Memoir*, PEA301-1 (Santa Monica, CA: RAND, 2020), pp. 219–220; JSOU, *Special Operations Forces Reference Manual*, 4th ed. (MacDill Air Force Base, FL: JSOU Press, June 2015), pp. 2–9.

[179] Franz Kurowski, *The Brandenburgers: Global Mission*, trans. David Johnston (Winnipeg, MB: J.J. Fedorowicz, 1997), p. 184; Kenneth Campbell, 'Otto Skorzeny: The Most Dangerous Man in Europe', *American Intelligence Journal*, 30:1 (2012): 143.

[180] Dennis Showalter, 'German Army Elites in World Wars I and II', in A. Hamish Ion and Keith Nelson, eds., *Elite Military Formations in War and Peace* (Westport, CT: Praeger, 1996), pp. 152; 154.

[181] Plater-Zyberk, *Russia's Special Forces*, p. 6.

competition is the Iranian Revolutionary Guard Corps, including its Quds Force and Saberin battalions, which serve to offset army and navy forces, although an element of co-option by the regime also cannot be ignored.[182]

The different methods of control are easy enough to understand in practice, particularly when events or processes are viewed in isolation. Much the same is true of risk. The previous chapter introduced the concept of horizontal integration related to capabilities. This one introduces a different concept related to control and risk: vertical integration. Such integration is not a new concept. Military doctrine discussing command and control aspires to achieve it. Scholarship on civil–military relations either explicitly or implicitly identifies the conditions under which it can occur, directed from the policy maker or principal down to the agent, who should work to achieve the desired outcome. Rarely are considerations of this relationship seen as a joint process where the agent can in turn influence or exert a level of control on the principal.[183] Theoretical works on grand strategy and strategy also privilege top-down, unidirectional ideas of integration. Those lamenting a specified grand strategy for the United States often assume that one artfully crafted and sufficiently bounded by the senior executive or his counsel will solve the problems of inefficient government meandering aimlessly between different crises.[184] A similar false notion occurs in literature on strategy, particularly those who advance the idea of 'the master strategist', one who is an omniscient, all-controlling genius.[185]

The concept of control and vertical integration was inspired by two sources. The first was David Tucker and Christopher Lamb's discussion of vertical and horizontal integration as necessary attributes for effective future American special operations organization.[186] Many of their prescriptions for more effective function coordination were adopted, as Chapter 8 outlines. The second, which sparked a transformation in my thinking, is Edward Luttwak's work *Strategy.* In it, Luttwak synthesizes a number of Clausewitz's ideas into a general principle of strategy, based on its paradoxical logic of struggle between opposing wills or adversaries, in what Clausewitz termed 'a continuous interaction of opposites'.[187] Luttwak also introduced a broader Western audience to the idea of an intervening 'middle' level of competition and conflict

[182] Frederic Wehrey et al., *The Rise of the Pasdaran: Assessing the Domestic Roles of Iran's Islamic Revolutionary Guard Corps*, MG-821-OSD (Santa Monica, CA: RAND, 2009), especially Chapter 2 and 5; Afshon Ostovar, *Vanguard of the Imam: Religion, Politics, and Iran's Revolutionary Guards* (Oxford: Oxford University Press, 2016), pp. 54–61.

[183] Stathis Kalyvas' theory of selective violence concludes it is based on a joint process between an incumbent or insurgent and the local population, whose agency is often overlooked. Stathis Kalyvas, *The Logic of Violence in Civil War* (Cambridge: Cambridge University Press, 2006), p. 209.

[184] Stephen Walt, *The Hell of Good Intentions: America's Foreign Policy Elite and the Decline of U.S. Primacy* (New York: Farrar, Straus and Giroux, 2018); Robert Wilkie, 'Why America Needs a Grand Strategy', *National Review* (19 July 2021), www.nationalreview.com/2021/07/why-america-needs-a-grand-strategy. For a useful corrective to the two primary schools of thought on US grand strategy, see Hal Brands, 'Rethinking America's Grand Strategy: Insights from the Cold War', *Parameters*, 45:4 (Winter 2015): 7–16.

[185] See, for example, Dolman, *Pure Strategy*, p. 194; Ketan Patel, *The Master Strategist: Power, Purpose, and Principle* (New York: Hutchinson, 2005). Lawrence Freedman addresses this 'myth' without offering a notion of his own, beyond the dominant civilian, in *Strategy: A History* (Oxford: Oxford University Press, 2013), pp. 236–244.

[186] Tucker and Lamb, *United States Special Operations Forces*, pp. 224–233.

[187] Carl von Clausewitz, *On War*, ed. and trans. Michael Howard and Peter Paret (Princeton, NJ: Princeton University Press, 1976), p. 136.

between grand strategy, strategy, and tactics: the operational, or theatre level.[188] In the process of exploring the paradoxical logic across different levels of activity Luttwak introduces the idea of harmonization, across vertical and horizontal layers, in what he termed 'interpenetrations'.[189] For Luttwak, tactical actions through the operational and strategic militarily comprised the vertical layer and grand strategy the horizontal. What he was attempting to do for grand strategy, and later Colin Gray for special operations, was fleshing out more fully Clausewitz's intriguing but elusive metaphor of 'currency conversion', turning tactical actions into meaningful strategic outcomes.[190]

While drawing from Tucker, Lamb, and Luttwak, my discussion of vertical integration differs in one central respect. The focus here is not on grand strategy but rather special operations working together in an integrated manner over time. The interaction of control measures, as well as risk communication up and down a chain of command from policy intention to special operations execution creates dynamics that complicate matters well beyond Luttwak's categories of 'harmony' or 'disharmony'.[191] Weak or insufficient control, as well as vague or impenetrable guidance, can lead SOF to game processes to avoid what operators perceive as unnecessary or counterproductive scrutiny and oversight. In contrast, strong or overbearing control or that insufficiently delegated can limit SOF creativity and flexibility, denying them the chance to exploit emerging opportunities, as Chinese lectures on special operations suggest.[192] Although based on Luttwak's concept, vertical integration extends these ideas to special operations and incorporates missing dimensions from his discussion: military and political control and risk.

As with the previous chapter, the level of integration can be in four degrees: high, medium, low, or none. Integration in the vertical dimension, and control and risk in particular, is much more sensitive to contextual changes than are capabilities. Risk calculations and communication can change suddenly. The terrorist attacks on 11 September 2001 changed American risk calculations in a single day. For these reasons, a general theory should seek to evaluate integration between SOF, military commands, and political leaders based on the type of control, the specific measures being used to implement it, and the communication of risk up and down the chain of command.

To assess the level of vertical integration, the next section examines the ambush and killing of four SOF operators near Tongo Tongo in Niger in 2017 and its subsequent consequences. The facts of the ambush are unequivocal. On 4 October, four American special forces senior enlisted officers from Team Ouallam, Operational

[188] Edward Luttwak, *Strategy: The Logic of War and Peace*, rev. ed. (Cambridge, MA: Belknap Press, 2001), p. 112.

[189] Luttwak, *Strategy*, pp. 236–242.

[190] Colin Gray, *Tactical Operations for Strategic Effect: The Challenges of Currency Conversion*, JSOU Special Report (MacDill Air Force Base, FL: JSOU Press, November 2015).

[191] Luttwak, *Strategy*, p. 234.

[192] Guo Ming. *Lectures on the Science of Special Operations*. Academy of Military Sciences Master's Postgraduate Series Teaching Materials. 2nd ed. U//FOUO translation, 2016. SAIC Selected Book Translations CVI (Beijing: Military Science Press, July 2013), p. 203.

Detachment Alpha 3212, 3rd Special Forces Group—Staff Sergeants (SSG) Jeremiah Johnson, Bryan Black, and Dustin Wright, along with Sergeant (SGT) LaDavid Johnson—were patrolling with Nigerian partner forces they had been training and advising. The object of the mounted ground patrol was a local Islamic State in Iraq and Syria—Greater Sahara (ISIS-GS) leader and the team was to be in support of another scheduled to fly in by helicopter. Ultimately the intelligence driving the operation was faulty and team returned with their Nigerian partners to their advanced operations base (AOB). Stopping at the village of Tongo Tongo to resupply, the team talked with locals and departed. Shortly thereafter, the team and the Nigerian partner forces advanced to contact with a sizable ISIS-GC force. In the ensuing firefight, the team and its partners realized they were outnumbered and in danger of being surrounded. Three members of the team failed to break contact and steadily increasing, massed fire overwhelmed SSGs Black, J. Johnson, and Wright. SGT L. Johnson, along with two Nigerian soldiers in his vehicle, were shot and killed subsequently.[193]

Subsequent media coverage after the ambush reflected sparring characteristic of a charged political environment and oversized personalities and egos. Normal processes of control and accountability, however, were working behind the scenes. These included a DoD instigated, Combatant Command-led investigation as well as briefings delivered to members of the Senate Armed Services Committee two weeks after the ambush, following a meeting between the secretary of defense, James Mattis, and the Committee's ranking member, John McCain, the previous week. Secretary Mattis and the chairman of the joint chiefs of staff also issued separate statements to the press, although some confusion resulted from how they characterized the nature of the mission. Another contributing factor was insufficient information, in degree and kind, exchanged between members of President Trump's administration and members of Congress.[194] Congressional interest in the subject is evident in a report by its research arm, the Congressional Research Service, answering ten questions to better place these events in their geopolitical, strategic, operational, and legal context.[195]

Subsequent discussion of control measures supporting the mission focused on the series of legal authorities underpinning it. Members of Congress voted to consolidate capacity building authorities under §333 in the month preceding the ambush. They also, however, codified their unease about 'reporting requirements' and 'monitoring' functions related to capacity building and its effectiveness.[196] Analysts outside government noted difficulties integrating these functions 'with existing combatant command processes and concepts', leading staff to 'check the box' instead.[197] Katherine Yon Ebright concluded reporting for §333 activities were transparent and robust,

[193] The preceding was summarized from the DoD's official investigative report. US Africa Command, 'Oct 2017 Niger Ambush, Summary of Investigation' (Stuttgart: Africa Command, 10 May 2018), pp. 2–4, https://dod.defense.gov/portals/1/features/2018/0418_niger/img/Oct-2017-Niger-Ambush-Summary-of-Investigation.pdf.

[194] Rebecca Kheel, 'Mattis, McCain Meet a Day after Subpoena Threat', The Hill (20 October 2017), https://thehill.com/policy/defense/356450-mattis-mccain-meet-a-day-after-subpoena-threat.

[195] Congressional Research Service, *Niger: Frequently Asked Questions about the October 2017 Attack on U.S. Soldiers*, CRS Report R44995 (Washington, DC: Congressional Research Service, 27 October 2017).

[196] Congressional Research Service, *Niger*, p. 7.

[197] Dalton et al., *Shifting the Burden Responsibly*, p. 8.

they were being used to stretch and extend the provisions of the Constitution on the use of force in several of the fifty-two countries in which they were being run, including Niger and Somalia, constituting a 'risk'.[198] She also charged that SOF exploited specious 'overlapping operational authorities' to authorize the Team Ouallam mission, while Alice Hunt Friend suggested they and others interpret a central aspect of self-defence as 'looking for a fight'.[199] Secretary Mattis testified before Congress that SOF and the DoD were fulfilling the reporting requirements to ensure sufficient oversight and monitoring. More specifically, he stated that for global SOF counterterrorism operations DoD was submitting to Congress between six and seven reports per day on average.[200] While factually correct, Congress had neither the resources nor resident expertise for adequate monitoring, particularly given the volume of reports submitted, virtually guaranteed that control actually increased political risk.[201] Claims by its members to be unaware of scope and scale of SOF missions at the time were politically self-serving, given the reporting requirements it put in place.[202]

Africa Command's (AFRICOM's) subsequent investigation identified several shortfalls specifically relating to operational and tactical risk and its communication. Its investigation determined Team Ouallam was insufficiently prepared, possessing inadequate pre-mission training, a lack of airborne intelligence collection to provide overwatch for the mission, and no assigned quick reaction forces in the event the team made contact with the enemy. These risks were communicated clearly up the chain of command but sloppy staff work including a 'cut and paste from previous CONOPS' and issues with detail and notification 'contributed to a general lack of situational awareness and command oversight at every echelon'—meaning there was insufficient control in the military chain of command.[203] In addition, the distribution of control between operating elements in Niger, the coordination element in Chad, and higher headquarters in Baumholder and Stuttgart, Germany, likely also played a role in the breakdown.[204]

Subsequent statements by the special operations component commander for AFRICOM (SOCAFRICA), Major General Mark Hicks, provide additional insights into how tactical and operational risks would be mitigated moving forward. According to a memo Hicks sent his command in the aftermath of the incident: 'going forward … we must reduce our risk exposure and build trust in our ability to exercise sound judgment and disciplined planning and execution'. He added:

[198] Katherine Yon Ebright, *Secret War: How the U.S. Uses Partnerships and Proxy Forces to Wage War under the Radar* (New York: Brennan Center for Justice, 2022), pp. 8–10.

[199] Yon Ebright, *Secret War*, p. 11; Alice Hunt Friend, 'The Accompany They Keep: What Niger Tells Us about Accompany Missions, Combat, and Operations Other Than War', War on the Rocks (11 May 2018).

[200] *The Authorizations for the Use of Military Force: Administration Perspective, Hearing Before the S. Comm. on Foreign Relations*, 115th Cong. (2017) (Statement of James Mattis, Secretary of Defense), www.govinfo.gov/content/pkg/CHRG-115shrg37142/html/CHRG-115shrg37142.htm.

[201] Yon Ebright, *Secret War*, pp. 13–14.

[202] Rukmini Callimachi et al., '"An Endless War": Why 4 U.S. Soldiers Died in a Remote African Desert', *The New York Times* (20 February 2018).

[203] US Africa Command, 'Oct 2017 Niger Ambush, Summary of Investigation', p. 6.

[204] Ibid., p. 8.

> I expect you to modify your assumptions about the level of risk you can accept. I expect you to plan and conduct operations with an increased margin of safety ... Exercising a capability is not a sufficient reason to plan or execute a mission. Activity does not equal progress. I want us to aggressively pursue strategic objectives which at least for the near term means less aggressively pursuing tactical objectives.[205]

Operational headquarters, in other words, was attempting to buy down risk to mission at the campaign level by mitigating risk to mission and force at the tactical level.

Subsequently, Maj Gen Hicks was reprimanded by the USSOCOM commander, as were a handful of junior officers.[206] These actions suggest sufficiency in control measures, specifically punishment when an agent makes missteps. Others disagree, including a former SOCAFRICA commander, Donald Bolduc. He points out that while Hicks was both inexperienced and relatively recent to a very diverse, geographically distributed, and complex command, punishment should occur at higher levels of command. In particular, Bolduc suggests specific recommendations to prepare address inexperience and recency 'fell on deaf ears at SOCOM and AFRICOM'.[207] Translating his argument into risk terms, force management risks ballooned into operational ones. After the action and in response, USSOCOM mitigated institutional risks without addressing the deep-seated, force management ones that translate into additional ones operationally.

On its surface, the events leading up to, during, and after the Tongo Tongo ambush suggest a *high* degree of vertical integration in US SOF. Investigations and inquiries revealed shortfalls in control mechanisms and risk calculations, particularly from the operational level down. Punitive action was taken to punish those responsible for the shortfalls and corrective actions implemented at the operational and strategic levels. Yet the control and risk model suggests vertical integration may not be as high at first glance. Tactically, Team Ouallam exerted considerable agency and assumed considerable risk to force and mission. The chain of command from Team Ouallam to SOCAFRICA exerted its own agency in interpreting authorities and insufficiently filling out proper approval processes so it could conduct the mission. They did so based on the dynamic nature of the ISIS-GC threat and an emergent opportunity: the chance to capture or kill one of its senior leaders. SOF did not shirk in this case but risk to force tactically expanded into political risk for the Trump administration, which had entrusted expert DoD and SOF agents to address terrorism threats globally. This short analysis suggests the control and risk model in this chapter allows for

[205] M.G. Hicks memo cited in Christina Goldbaum, 'U.S. Gen. Reins in Special Operations Forces in Africa after Niger Deaths and *Daily Beast* Investigations', *The Daily Beast* (11 February 2018), www.thedailybeast.com/us-gen-reins-in-special-operations-forces-in-africa-after-niger-deaths-and-daily-beast-investigations.

[206] 'General, Five Others Reprimanded in Connection with 2017 Niger Ambush That Left Four Dead', *Stars and Stripes* (5 November 2018), www.stripes.com/news/general-five-others-reprimanded-in-connection-with-2017-niger-ambush-that-left-four-dead-1.555288.

[207] Donald Bolduc, 'One Major General Was the Fall Guy for the Niger Ambush. The Real Fault Lies Elsewhere'. Task and Purpose (8 November 2018), https://taskandpurpose.com/news/marcus-hicks-niger-ambush-africom.

more nuanced and insightful analysis of SOF activities and therefore forms part of a general theory of special operations.

## Conclusion

Unquestionably, SOF organizations and activities have long appealed to civilian leaders. John F. Kennedy held the special forces in high regard, and thanks to him, the organization has its eponymous headgear, the Green Beret. British Prime Minister Margaret Thatcher was a frequent guest of the British Special Air Service and held its operators in high esteem, particularly after their performance in rescuing hostages held at the Iranian Embassy in London in 1980.[208] The reason for this appeal is simple. SOF provide civilian decision makers with additional, low-cost options. Moreover, few tasks are too daunting for them to tackle and the value of positive options during crises cannot be understated. They offer the possibility of addressing problems judiciously, as forcefully or quietly as decision makers would like, while minimizing risk to the highest degree possible, especially in comparison to conventional force options that may result in unnecessary entanglements, large numbers of casualties, and adverse publicity.

The key to understanding why control of SOF is not a simple prospect reflects their agency, chains of command from the national down to the tactical, and calculations and communication of political and military risk. When civilian leaders are faced with crises where the political and military risks are high, they often turn to surgical strike units optimized to conduct missions to very high, 'no fail' standards. Where political and military risks are low, civilian leaders can employ sabotage and subversion units to maintain access, build relationships and capacity, and provide more refined, detailed information. Where the political and military risks are mixed, civilian decision makers have the option of employing sabotage and subversion, military advising and assistance, and asymmetric response units in proportion and duration to the level of threat and policy outcomes desired.

Understanding control and risk issues related to special operations is essential not just for the purpose of developing a general theory but also to challenge and put to rest several persistent myths and canards regarding SOF. As the previous chapter discusses, several influential works on SOF lament their misuse historically. The contemporary manifestation of this argument is embodied in three words: 'the easy button'.[209] The culprit most often identified behind misuse is usually conventional military minds, but has included political leaders, who lack understanding, clear goals, or imagination. While disbandment seems highly unlikely given the high stock political leaders place in SOF, the misuse myth endures and appears, in one form or another, in almost every recent work on the subject. The establishment of

[208] Alastair MacKenzie, *Special Force: The Untold Story of 22nd Special Air Service Regiment (SAS)* (London: I.B. Tauris, 2011), p. 174; John Newsinger, *Dangerous Men: The SAS in Popular Culture* (London: Pluto, 1997), pp. 28–29.

[209] Russell A. Burgos, 'Pushing the Easy Button: Special Operations Forces, International Security and the Use of Force', *Special Operations Journal*, 4:2 (2018): 109–128.

USSOCOM, and its policy advocate in the Pentagon, OASD SO/LIC, offers a high degree of institutional and bureaucratic insulation against current or future misuse. The prevention of misuse, and a degree of institutional stability established in 1987, created a new problem: SOF have a level of agency, based on their ability to connect more directly to civilian leaders, generate options, and provide solutions. If misuse does occur, one question worthy of answering is the degree to which those forces contribute to the 'unintended acceleration' of growth in scope and duration of global American commitment of forces, the size of the SOF community, and their authorities or operating constraints.

Agency is not a problem provided principals have the desire and means to ensure agents are working effectively, as well as punish them when they shirk. While this model is both a necessary and useful departure point for studying the civil–military relations of SOF, it is insufficient for two primary, interrelated reasons. The first is the special operations mindset, which is an inherent trait deliberately screened for and cultivated in the operators themselves. SOF live to work as effectively and efficiently as possible and provide solutions to challenging national security problems as often as possible. Without sufficient monitoring and control, SOF have two discrete modes of operation that differ from the working/shirking choices outlined by Feaver that contribute to the 'unintended acceleration' phenomenon. SOF either *accede* and follow closely the directions they have been given or, when the opportunity to exercise agency is available, *exceed* the mandate given to them by civilian decision makers. The goal of this chapter is not to create a new theory of civil–military relations. Rather it points out the limitations of current theories as they relate to special operations. This chapter takes existing civil–military relations theory and modifies it, by adding in other control types, and introducing risk into the conversation.

The second reason relates to the willingness and ability of civilian principals to sufficiently monitor the activities of their agent. For SOF, the political and strategic value the agent provides the principal, combined with the ways the agent conducts the mission, mitigates against effective monitoring. In over simplified terms, the more the agent—SOF—succeed, the less likely the principal—civilian leaders—are to oversee or monitor their activities. In addition, the principal is likely to approve the requests of SOF leaders for more resources, less restrictive authorities, and greater support at the expense of monitoring, given their responsiveness and seeming economy compared to conventional force requests. SOF are inclined to take on more than they can handle reasonably, when additional resources and increased bureaucratic influence and prestige are available.

The results of reviews or inquiries in the US, Australia, and Britain suggest SOF have great difficulty in viewing or grading themselves and what they do objectively. Their commitment to mission success is laudable but masks deeper institutional problems that only external monitoring can diagnose and resolve. A contributing factor to this condition are different frames of reference for risk, with unique aspects at various levels, that often get lost in translation. The risk model introduced here suggests control of SOF is complicated by three factors: differing risk considerations and frames of reference, translation or communication of risk issues and concerns up and down the chain of command, and the nature of SOF to want to be used and useful, or 'unleashed' to do what they do best with as little in the way of restrictions as possible. These restrictions including sufficient monitoring and oversight, which special operators can view as unhelpful at best or unnecessary at worst in the performance of their missions. All were apparent in and contributed to the outcome at Tongo Tongo in 2017.

Both reasons highlight a paradox of SOF for civilian leaders. To have the greatest success tactically and operationally and resolve risks at those levels, SOF require a high degree of agency and autonomy and little intrusive monitoring. Yet a level of intrusive yet sustainable oversight is necessary to ensure SOF do not create greater risks politically or strategically by continuously exceeding what was asked of them. Little evidence suggests the metaphorical SOF 'Ferrari' is likely to be garaged any time soon. It has the potential to accelerate out of control, or careen off the track, if this paradox is not understood and addressed. This tendency is also observable in other countries, democratic or otherwise, and the interplay between control and risk forms a crucial part of the general theory of special operations developed in the next chapter.

# 5
# The Theory of Special Operations Coherence

## Introduction

The previous two chapters laid the groundwork for a theory of special operations by examining first their horizontal and next their vertical dimensions. The unifying idea linking these dimensions is integration. The integration of capabilities, the horizontal, determines the scope and complexity of tasks that special operations forces (SOF) have the potential to accomplish. Control and risk considerations and their integration vertically outline political and military expectations for SOF and determine the degree of autonomy they possess. Each allows a snapshot in time of capabilities and control relationships; individually both have explanatory value. Yet they are insufficient in providing a deeper understanding of special operations success and failure. To accomplish this, the theoretical models must be combined. Together they comprise a special operations system that connects special operations capabilities institutionally in aggregate and with other military and government resources, as well as from the tactical level of execution up to the political level. The only missing element from this system, however, is the dynamic nature of competition and conflict.

The goal of this chapter is a conceptual framework for academics, decision makers, and special operators to understand better special operations performance in times of competition and conflict, against multiple opponents. This theory tackles the problem of 'currency conversion' or exchanging special operations effects into meaningful political outcomes. It answers an important question for special operations of 'why strategy and tactics seemed often to be sailing past each other as ships in the night, passing unseen and therefore unacknowledged in the diversity of their navigation'.[1] Previous theoretical explorations of special operations likewise have sailed past each other, focusing on either tactical explanations of roles and missions or exclusively political and strategic ones. The theory presented in this chapter links these two together, the bottom-up and top-down perspectives, to ensure a more holistic understanding of special operations.

This chapter accomplishes this goal by setting into motion or animating capabilities and control by identifying how effectively this special operations 'system'

[1] Colin S. Gray, *Tactical Operations for Strategic Effect: The Challenge of Currency Conversion*, JSOU Special Report (MacDill Air Force Base, FL: JSOU Press, November 2015), p. 7.

*Special Operations Success*. James D. Kiras, Oxford University Press. © James D. Kiras (2024).
DOI: 10.1093/9780198902096.003.0005

responds to the input of four primary factors. The first is the impact of time as capabilities are developed, used, expanded, or disbanded to meet a range of different requirements and contingencies. Too often special operations are discussed in a single operation or contingency. This theory accounts for responses to a range of activities and contingencies, which can occur simultaneously. The second, the political stakes involved, directly reflects those requirements by shaping control mechanisms and risk calculations. The third is perhaps the most important factor, one that continues to make competition and conflict a dynamic process. This factor consists of the actions, reactions, and consequences of adversaries, competitors, partners, and allies. The fourth and last is more operational or functional in nature: the bearing and calibration of special operations against an opponent. Special operations can be either offensive or defensive in bearing, during competition, in anticipation of or reflecting conflict. This orientation, and its shift, influences how SOF are calibrated: overtly, covertly, clandestinely, or a proportionate mix.

The theory grapples with a central issue related to special operations: how to utilize unorthodox forces that are inherently dynamically unstable institutionally to achieve desired political outcomes, while maintaining stability in terms of their political influence and preparation for future competition and conflict. For reasons that will become clear, the logical yet overly simplistic answers to maintaining stability—increased capabilities or tighter control—are problematic in their own right. At its core, this theory attempts to resolve the tension inherent in the last three 'SOF Truths' identified in Chapter 3: that SOF cannot be mass produced, created after emergencies occur, and require non-SOF support.

## Four elements creating a dynamic theory

### Time

Few elements within special operations are as misunderstood or mischaracterized as time. No expression better captures this misunderstanding than the old saw repeated in numerous works about Afghanistan and, in particular, the differences in cultural frames of reference between the Taliban and Afghanistan: 'You have the watches but we have the time.'[2] This statement has become enshrined in popular lore as a self-evident explanation for why a coalition of states could never achieve victory in Afghanistan. Time could only work to the Taliban's advantage as the coalition states would eventually withdraw. No special operations could prevent this inevitable outcome, no matter how proficient they were or how successful tactically. For the

[2] Evan Solomon, 'Fighting in Afghanistan: "You have the watches. We have the time"', *Macleans Magazine* (2 September 2017), https://macleans.ca/news/fighting-in-afghanistan-you-have-the-watches-we-have-the-time.

current generation of operators and decision makers, this statement has become the equivalent of Harry Summers Jr's pithy exchange with a North Vietnamese counterpart after the conclusion of the Vietnam War.[3]

Time is an acknowledged but rarely explained element in theoretical works on special operations. This occurs given inadequacies of existing institutional definitions or a narrow focus on SOF tactical mission execution. Harry Yarger, drawing on the Department of Defense *Dictionary of Military Terms*, emphasizes 'time and political sensitivity' for his American theory of special operations.[4] Time is central to William McRaven's theory of relative superiority. He links it to initiative and surprise; the longer a tactical action goes on, the greater the likelihood that raiders will find themselves on the losing end of a 'correlation of forces' and be overwhelmed by numerically superior defenders.[5] Other works imply extended time works against SOF for surgical or Commando operations.[6] Finally, Robert Spulak Jr suggests time and space are enduring dimensions of strategy and SOF provide 'great strategic value'. They do so by enhancing and focusing the concentrated mass of others, working 'by, with, and through' these proxies, to overmatch enemy forces in space.[7] Most theories of special operations or SOF time view time in a limited, functional way. Time is related to mission execution at the tactical or task-completion level.

At the operational or campaigning level, time is also acknowledged but conceptualized in military terms differently. Much of this reflects the growing scale and complexity of conflict in campaigning, introducing what Waldermar Erfurth describes as a 'time lag'.[8] Some authors, such as Spulak Jr, acknowledge this reality. A more frequent error equates time with tempo, or pace of activity. In its simplest terms, speed is linked to initiative, whose possession leads inexorably to success. There were two drivers of these ideas and writings: overcoming the numerical inferiority of North Atlantic Treaty Organization (NATO) forces in the face of superior Warsaw Pact ones who had the presumed advantage of the initiative, and technological developments in battlefield sensors and precision-guided weapons. By slowing Warsaw Pact forces throughout their considerable depth with deep and accurate strikes, NATO could seize the initiative and conduct large-scale counter-manoeuvres

[3] '"You know you never defeated us on the battlefield", said the American colonel. The North Vietnamese colonel pondered this remark a moment. "That may be so", he replied, "but it is also irrelevant"'. Repeated in 2014 by James Dubik, 'Winning Battles, Losing Wars', *Army* (December 2014): 16.

[4] Harry Yarger, *21st Century SOF: Toward and American Theory of Special Operations*, JSOU Report 13-1 (MacDill Air Force Base, FL: JSOU Press, April 2013), p. 21.

[5] William McRaven, *Spec Ops, Case Studies in Special Operations Warfare: Theory and Practice* (Novato, CA: Presidio Press, 1995), pp. 382–388.

[6] Edward Luttwak, Steven Canby, and David Thomas, 'A Systematic Review of "Commando" (Special) Operations, 1939–1980, Part I', unpublished report, C & L Associates (24 May 1982), p. S-4; David Charters et al., 'Surgical Operations: Military Lessons from Selected Case Studies', unpublished report, Centre for Conflict Studies, University of New Brunswick (Fall 1982), pp. 259–260.

[7] Robert Spulak Jr, *A Theory of Special Operations: The Origin, Qualities, and Use of SOF*, JSOU Report 07-7 (Hurlburt Field, FL: JSOU Press, October 2007), p. 31; Linda Robinson, 'SOF's Evolving Role: Warfare "By, With, and Through" Local Forces', RAND Blog (9 May 2017), www.rand.org/blog/2017/05/sofs-evolving-role-warfare-by-with-and-through-local.html.

[8] Waldermar Erfurth, *Surprise*, Sean Possony and Daniel Vilfroy, trans. (Harrisburg, PA: Military Service Publishing Company, 1943), pp. 40–41.

offensively and end a future conflict quickly.[9] Coordination of forces and strikes across widespread geographic space can take considerable time, as well as investment in adequate communications, but too often the pace of operations, or their tempo against an opponent, is confused with one's own synchronization and speed of execution.

The ideas of John Boyd, as well as writings surrounding operational art that dominated military theory focused on warfighting in the last decades of the Cold War, focus specifically on time and tempo. Boyd pointedly criticized the army's focus on synchronization in its doctrine in the following way: 'You synchronize watches not people ... This idea of synchronization will ruin the Army.'[10] He based his inductive conclusion regarding time and tempo narrowly from his tactical skill and experience in fighter combat. More broadly and more problematically, he based his observe-orient-decide-act (OODA) loop theory, which has been extended to the operational and strategic level, on the second law of thermodynamics from physics, and more specifically its idea of entropy, or energy transfer and loss.[11] A new generation of theorists, including Antoine Bousquet, suggest theories based on thermodynamics were sufficient for industrialized or mechanized war but not the emerging one: cybernetic or chaoplexic, based instead on self-synchronization and coordination, itself based on automated information flows.[12]

At the level of strategy, however, time functions much more slowly. Everett Dolman offers the idea that time is a constant dimension of strategy and contains an 'event horizon', ensuring that the strategist is enmeshed in the present reality yet incapable of predicting the future[13]. Time favours neither competitor nor belligerent. The reason behind this conclusion is simple. Within competition and conflict, the meaning of and therefore value placed on time is different depending on the distances involved and level of activity. The geographic location of the United States, combined with a policy of basing the bulk of its military units there, contributes to one of its biggest challenges related to its interests: projecting force quickly enough to prevent a fait accompli. In the prescient words of Robert Leonard, the challenge the United States faces in the Pacific is as follows: 'the issue in any future conflict overseas boils down to getting there with enough power *in time* to attain our strategic goals'.[14] Tactical actions in conflict can be decided in seconds or minutes, operational ones in hours, days, or weeks, and strategic ones in weeks, months, or years. Colin Gray notes that 'the strategic time frame is longer than the operational, and the operational is longer than the tactical. This is a significant reason why the strategist is in

[9] John Romjue, *From Active Defense to AirLand Battle: The Development of Army Doctrine, 1973–1982* (Fort Monroe, VA: Historical Office, TRADOC, June 1984).

[10] Quoted in Stephen Robinson, *The Blind Strategist: John Boyd and the American Art of War* (Dunedin, New Zealand: Exisle Publishing Pty Ltd, 2021), p. 55.

[11] For greater detail on why this extension is problematic, see James D. Kiras, *Special Operations and Strategy: From World War II to the War on Terrorism* (London: Routledge, 2006), pp. 30–34.

[12] Antoine Bousquet, *The Scientific War of Warfare: Order and Chaos on the Battlefields of Modernity* (New York: Columbia University Press, 2009), pp. 236–237.

[13] Everett Dolman, *Pure Strategy: Power and Principle in the Space and Information Age* (London: Frank Cass, 2005), p. 9.

[14] Robert Leonhard, *Fighting by Minutes: Time and the Art of War* (Westport, CT: Praeger, 1994), p. 7 (italics original).

need of greater mental and moral fortitude than is the operational-level commander'. More time is not always beneficial, as Gray suggests. He adds that, 'The strategist has ample time to have the kind of second and third thoughts that are likely to shake his self-confidence.'[15] SOF operators have noted an individual variation of the strategist's dilemma related to time: adrenaline and trained reaction during most tactical engagements. Only later, or when there is more time to consider different outcomes as often occurs with moral choice, do such operators second-guess themselves or overthink their decisions, particularly after a team's hide site is discovered by a civilian (often a child).[16]

Special operations appear to break this temporal mould given their ranging scope, effect, and duration. Some special operations, such as raids, appear to resolve crises in hours or minutes, such as hostage rescues. Others, including military advising and assistance missions, can take years to develop competent proxy forces. In addition to transcending time, they appear to compress or expand the levels of competition and conflict: tactical actions with the potential to generate immediate strategic effects or actions in a longer duration campaign that steadily turn the tide against an opponent in cumulative operational approach of associated individual wins and losses that add up to eventual victory. This conceptualization, driven largely by the net successes of submarine and anti-submarine warfare during the Second World War, was identified initially by Naval War College professor Herbert Rosinski and found form in a book by one of his students, J.C. Wylie.[17]

These ideas seemingly connected seamlessly for special operations with those of 'strategic emergence', or the self-generated patterns that result from unguided activity driven by local conditions or 'way finding' in a strategy without design.[18] The connection and its subsequent approach appear to be the means of military transformation towards a design-based approach, one for which SOF appear to be tailor-made. Design approaches to strategy, like SOF, are meant to bend, rewrite, or obviate existing rules of conflict.[19]

As appealing as they may be, claims of the 'extra-normative' nature of SOF, however, do not apply to the laws of time and space. They do not possess mechanics of their own to warrant a quantum theory wholly different from their parent phenomenon. Special operations, it is worth recalling, comprise military and intelligence activities designed to generate meaningful operational and strategic effects within a larger context of competition and conflict, in pursuit of political goals. As such, they

[15] Colin S. Gray, *The Strategy Bridge: Theory for Practice* (Oxford: Oxford University Press, 2010), p. 75.

[16] Marcus Luttrell and Patrick Robinson, *Lone Survivor: The Eyewitness Account of Operation Redwing and the Lost Heroes of SEAL Team 10* (New York: Little, Brown & Co., 2007), pp. 202–203; Leonard Cheshire, *The Face of Victory* (London: Hutchinson, 1961); Barry Davies, *Shadow of the Dove* (London: Bloomsbury, 1996); Josh Vriens, '"My only regret is not doing more"—Why I Reported Eddie Gallagher for War Crimes', Task & Purpose blog (25 January 2022), https://taskandpurpose.com/news/eddie-gallagher-war-crimes.

[17] J.C. Wylie, *Military Strategy: A Theory of Power Control* (Annapolis, MD: Naval Institute Press, 1989), pp. 22–23.

[18] Robert Chia and Robin Holt, *Strategy without Design: The Silent Efficacy of Indirect Action* (Cambridge: Cambridge University Press, 2009), chapters 1, 6.

[19] Ben Zweibelson, *Understanding the Military Design Movement: War, Change, and Innovation* (London: Routledge, 2023), pp. 4–5.

are bound by laws governing the nature of war, even if they contribute to and can change its character. This nature suggests special operations serve an instrumental purpose; they are but one of a range of options, or another means, to achieve an outcome. In any event, their high cost to develop and sustain and small scale relative to other means or actions governments or non-state actors can use limit the effects they can generate. They are enablers and additive to those means and are not decisive on their own.

Indecisive should not be mistaken for inconsequential and this is especially the case with special operations and their relationship to time. The greatest impact SOF have is at the operational or campaign level. SOF can be tremendously useful in achieving military and political outcomes by serving to increase or decrease the tempo of a campaign. Special operations accomplish this by targeting or rendering unusable critical vulnerabilities or capabilities in an opponent's system. There are several examples historically of both forms in plan and practice, including delaying operations. Two special operations in Norway in 1943 and 1944 targeting heavy water production and distribution, Operation Gunnerside and later the sinking of the ferry SF *Hydro*, slowed down German attempts to developing a working fission weapon.[20] Multiple special operations over the past thirty years, including sabotage, subversion, and assassination using physical and cyber means by Israel and the United States, have similarly slowed Iranian efforts to develop fissile material for its own nuclear weapons.

During the Cold War, 10th Special Forces Group as well as 22nd Special Air Service (SAS) teams had plans and exercises to link up with existing teams of 'stay behind' agents in countries overrun by Warsaw Pact forces. These SOF units would conduct sabotage of crucial infrastructure, such as rail and road tunnels, junctions, bridges, and other manoeuvre 'bottlenecks', in an attempt to reduce the tempo through 'retardation operations' of enemy offensives.[21] Specialized two-man teams, designated 'Green Light', would use transportable Special Atomic Demolition Munitions (SADMs) to destroy targets in various western and eastern European countries, as well as others in the Middle East and Southwest Asia.[22] In the wake of the Russian annexation of Crimea, NATO has released an operating manual for resistance forces to help them prepare and equip to conduct 'resistance operations' designed to slow down the tempo of future Russian offensives. As an example of this 'resistance operating concept' in practice, the most recent conflict between Russia and Ukraine saw Ukrainian SOF conduct specialized attacks against fuel depots and logistics convoys

[20] For details of both operations, see James D. Kiras, 'The Perfect Instance: Strategic Sabotage of German Heavy Water Production, Operation Gunnerside, February 1943', in Bernd Horn, James D. Kiras, and Emily Spencer, eds., *The (In)Visible Hand: Strategic Sabotage* (Ottawa, ON: Canadian Special Operations Forces Command, 2021), pp. 43–54.

[21] Tamir Sinai, 'Eyes on Target: "Stay-Behind" Forces during the Cold War', *War in History*, 23:3 (2021): 696–698.

[22] William Arkin, 'Nuclear Backpacks', *Bulletin of the Atomic Scientists*, 41:4 (April 1985): 4–5; Adam Rawnsley and David Brown, 'The Littlest Boy', *Foreign Policy* (30 January 2014), https://web.archive.org/web/20210203203101/https://foreignpolicy.com/2014/01/30/the-littlest-boy.

to slow down Russian offensives in the northern suburbs of Kiev during the first weeks of the war.[23]

Special operations can also speed up the tempo by overcoming critical obstacles, seizing key terrain, or inducing chaos and confusion in adversary leadership and command systems. German SOF experienced a mixed record of these types of operations during the Second World War in the Netherlands in 1940 and Belgium in 1944 in Operation *Greif*. The former drew British and French forces north, opening a gap subsequently exploited by German forces in their drive through the Ardennes. In the latter example, English-speaking German operators in the Ardennes four years later wore American uniforms and had enemy or visibly modified equipment to look like American equipment to seize key locations, including crossroads and bridges, or sow confusion among retreating or advancing forces.[24]

SOF activities can also increase tempo by negating the basis of an adversary's defences. The German 'elastic defence' concept on the Eastern Front during the Second World War hinged on the ability to shunt rapidly heavily armoured, mobile 'firefighting' forces useful in blunting serious Soviet breakthroughs of the defensive front. Soviet SOF, working with partisan forces, supplied by air, and controlled by Stavka headquarters, launched two massive, coordinated campaigns to destroy railways and bridges in advance of Operation Bagration in July 1944.[25] On a single evening, partisans controlled in this manner 'blew up about 40,000 rails and derailed 147 trains', threatening the estimated 100,000 tons of daily supplies the German forces in this sector required.[26] Tying down an additional 100,000 troops for rear-area security and rail repair and unable to swiftly move their mobile panzer reserves behind the front lines, much of German Army Group Centre was encircled and Soviet forces annihilated twenty-five divisions during the offensive.[27] This offensive, conducted on the level of army 'fronts', finally realized the ambitions of their pre-war 'deep battle' doctrine and accomplished one of Josef Stalin's military goals since 1942.[28]

Not all efforts need to be physical to speed up the tempo of an offensive. The combination of an Allied air campaign in advance of Operation Desert Storm, along

[23] Yaroslav Trofimov, 'Ukraine's Special Forces Hold Off Russian Offensive on Kyiv's Front Lines', *Wall Street Journal* (4 March 2022), www.wsj.com/articles/ukraines-special-forces-hold-off-russian-offensive-on-kyivs-front-lines-11646417190?st=qrzrh5leq99gqft&reflink=share_mobilewebshare.

[24] Peter Caddick-Adams, *Snow and Steel: The Battle of the Bulge, 1944–1945* (Oxford: Oxford University Press, 2015), pp. 361–367.

[25] Kenneth Slepyan, *Stalin's Guerrillas: Soviet Partisans in World War II* (Lawrence, KS: University Press of Kansas, 2006), pp. 98–99; David Glantz, *The Military Strategy of the Soviet Union: A History* (London: Frank Cass, 1992), p. 142.

[26] Leonid Grenkevich, *The Soviet Partisan Movement, 1941–1944: A Critical Historiographical Analysis* (London: Frank Cass, 1999), pp. 241; 259.

[27] Heinz Guderian, *Panzer Leader* (New York: E.P. Dutton, 1957), p. 336; Grenkevich, *The Soviet Partisan Movement, 1941–1944*, pp. 239–240; 248.

[28] David Glantz, *Soviet Military Operational Art: In Pursuit of Deep Battle* (London: Frank Cass, 1991), pp. 138–142. Partisan control measures are outlined in detail in John Armstrong, ed., *Soviet Partisans in World War II* (Madison, WI: University of Wisconsin Press, 1964), pp. 46–55; 129–133.

with psychological operations including broadcasts and leaflet drops to bombed Iraqi troops, assisted the eventual advance of ground forces in a one-hundred-hour campaign.[29] By altering the tempo of a campaign by speeding it up or slowing it down, SOF can not only increase or decrease costs but also 'buy' time, in the process realizing several of the 'master claims' for the strategic utility of special operations: expand choice, showcase competence, raise friendly moral will, sustain political will, reassure a polity, and humiliate an enemy.[30]

Metaphors, particularly for time, are necessary to grasp given its abstract nature but they must also be approached with caution. Colin Gray notes, for example, 'Clocks tick for all belligerents politically, strategically, operationally, and tactically. The strategic time frame is longer than the operational, and the operational is longer than the tactical.'[31] It is tempting to think of these time frames as different-size gears of a clock, intermeshed but moving at different rates. Applied literally, this metaphor is problematic for reasons scholars and practitioners have identified, including casting war as a closed, finite system operating in a repetitive, predictive, cyclical fashion.[32] Heuristically, however, the different gears at different levels have a certain value in understanding how special operations impact time relatively, in a dynamic struggle. When special operations increase the operational tempo of a campaign, they put an enemy strategy and system at risk by increasing the friction within it. In contrast, slowing down the tempo can buy time operationally and strategically—important breathing space—to recover from shock, surprise, or misfortune. The previous chapter offered another possibility: special operations exceeding their original mandate due to insufficient control. This excess can lead to special operations following a logic somewhat of their own, in terms of tempo, working at an intensity which alters the strategy or becomes one.

These observations regarding time, despite their heuristic value, are abstract and theoretical. As such, they are *ceteris paribus*, or all other things being equal. Conflict and competition, however, are not equal and involve a dynamic resulting from several factors. These factors are expanded below and include most notably the issue that is at stake, how much each side values it, the costs or burdens they are willing to pay, and the adversary's reaction. All three affect the bearing and calibration of special operations.

[29] Details of the psychological operations used to 'prepare the battlespace' are available in Jared Tracy, '"Building the Airplane in Flight": PSYOP in Operation DESERT SHIELD, Part 2', *Veritas*, 19:1 (2022): 49–71.

[30] Colin S. Gray, *Explorations in Strategy* (Westport, CT: Greenwood Press, 1996), pp. 174–179.

[31] Gray, *The Strategy Bridge*, p. 75; Carl von Clausewitz, *On War*, Peter Paret and Michael Howard, trans. and eds. (Princeton, NJ: Princeton University Press, 1984), p. 178.

[32] See, for example, Glenn Cunningham and Charles Allen, 'Applying Clausewitz and Systems Thinking to Design', in J. Boone Bartholomees, Jr., ed., *U.S. Army War College Guide to National Security Issues, Volume I: Theory of War and Strategy* (Carlisle Barracks, PA: Strategic Studies Institute, 2012), pp. 246–247; Amos Fox and Thomas Kopsch, 'Moving beyond Mechanical Metaphors: Debunking the Applicability of Centers of Gravity in 21st Century Warfare', Strategy Bridge blog (2 June 2017), https://thestrategybridge.org/the-bridge/2017/6/2/moving-beyond-mechanical-metaphors-debunking-the-applicability-of-centers-of-gravity-in-21st-century-warfare.

## Stakes

Much like risk, the stakes of competition and conflict are intuitively grasped but poorly understood analytically. The reason is relatively simple. Stakes, like risk, are subjective and based on perceptions. Unlike risk, which involves calculations of various threats in an environment, stakes are based on perceptions of the significance of different physical or virtual ones. The continued existence of a country or collective polity physically is a vital interest, while others based on ideas, such as reputation or norms, are much more difficult to attach meaning or consequence to. Articulating stakes gives meaning to our various interests in a relative manner, by communicating them to opponents, yet they are open to misinterpretation or misunderstanding for any one of a few different reasons, including diverse cultural frameworks or overstating the value of an interest. As an example of the former, Saddam Hussein considered remaining in power a vital interest in the lead-up to the invasion of Iraq in 2003, for which he was willing to give up territory, a calculation unimaginable to coalition strategists. For an example of the latter, Russia routinely has overstated a number of its different 'vital interests' in its most recent conflict with Ukraine, giving life to both debate in Western circles about potential responses and the adage that if everything is a vital interest, nothing is.[33] As Bernard Brodie noted, 'The phrase *vital interests* rolls portentously and somewhat granatically off the tongue, pregnant with meaning but nevertheless obscure.'[34]

Stakes became central to theories of deterrence developed during the Cold War and subsequently, as threatening a state's vital interest(s) could lead to mutual nuclear annihilation. Most often, international relations literature focuses on stakes as they relate to power, which is based on gross calculations of capability and the value leaders place on the use of force to secure their interests, as well as a state's reputation. Below the threshold, however, authors have explored how a state's interests could be pursued without threatening vital ones and triggering a major conflict. States can pursue their interests through a range of different types of competition with one another, through limited conflict or in third-party states, or more indirectly through proxies or protégés.[35] Stakes have external and internal dimensions, and as Ivan Arreguin-Toft points out, major powers lose asymmetric conflicts against non-state actors when domestic perceptions of stakes increasingly diverge from those articulated by leaders.[36]

Stakes are also crucial to theories looking at ending conflicts, or war termination. They play a central role in these theories primarily due to a stated assumption regarding deterrence, statecraft, and war: all fundamentally rest on economic assessments

[33] Alexander Motyl, 'Putin's Imperial Aspirations: How Ukraine Went from Being Ignored by America to a Vital National Interest', *Milwaukee Independent* (28 April 2022), www.milwaukeeindependent.com/syndicated/putins-imperialist-aspirations-ukraine-went-ignored-america-vital-national-interest.

[34] Bernard Brodie, *War and Politics* (London: Cassell, 1973), p. 341 (italics original).

[35] Vesna Danilovic, *When the Stakes Are High: Deterrence and Conflict among Major Powers* (Ann Arbor, MI: University of Michigan Press, 2002).

[36] Ivan Arreguin-Toft, *How the Weak Win Wars: A Theory of Asymmetric Conflict* (Cambridge: Cambridge University Press, 2005), chapter 2.

of value subject to negotiation and bargaining.[37] This has led to a dichotomous understanding of stakes, framed in terms of 'total' or 'limited' war or 'brute force' and 'coercion'.[38] In a theory exploring limited war, one which distinguished between limited, unlimited, and total war, Robert Osgood was cautious. He noted that while limited wars are easy to identify historically, they 'are not a uniform phenomenon' and a dynamic can exist within them, including limited geography but unlimited weapons, which also applies to belligerents and stakes.[39] For coercion, the challenges of effectively coercing an opponent have as much to do with the practical difficulties of putting ideas into practice as they do with accomplishing anything other than the military's desire for complete victory. Daniel Byman and Matthew Waxman point out that 'coercive strategy often represents the ideal rather than the reality of policy making. At times, policy makers have no strategy for coercion at all'.[40]

Stakes play an additional role. Byman and Waxman add that 'leaders may see their country's interests as limited and not meriting a serious and integrated overall strategy', in which 'the means-end chain used to coerce may not be internally consistent and thus will be unlikely to produce measurable success'.[41] For terrorists to succeed in coercing a much more powerful state opponent, they must conduct 'the armed struggle in the gray area that lies between reactionary and effective coercion by the threatened and assured concession from the [political] center'.[42]

While some authors have linked nuclear weapons and special operations in terms of their control, doing the same in terms of stakes is highly problematic. Simply put, the consequences of their actions, and the responses they potentially trigger, are entirely different save for scenarios where SOF attempt to seize or deny a state's nuclear weapons, or the 'countering weapons of mass destruction' role, as it is known legally and doctrinally. Notions of independent SOF deterrence, including focused or unconventional deterrence, rest on tenuous assumptions including the fragility of regime control and access to and depth of the latent hostility resident in a suppressed population.[43] More muted claims outline SOF's ability to support or contribute to integrated deterrence emphasize the needs to integrate existing capabilities operationally, as the US Army's 'cyber-space-SOF triad' is designed to do, or to fill a gap in competition comprising so-called 'gray zone activities' through overt escalation,

[37] Stated most simply and directly by Thomas Schelling in *Arms and Influence* (New Haven, CT: Yale University Press, 1966), p. 1. See also more recent discussions in James Fearon, 'Rationalist Explanations for War', *International Organization*, 49:3 (Summer 1995): 379–414; R. Harrison Wagner, 'Bargaining and War', *American Journal of Political Science*, 44:3 (July 2000): 469–484.

[38] Brian Brennan, 'Limited vs. Total War', *Armor*, 11:5 (September–October 2002): 8–11; Schelling, *Arms and Influence*, pp. 2–6.

[39] Robert Endicott Osgood, *Limited War: The Challenge to American Strategy* (Chicago, IL: University of Chicago Press, 1957), pp. 2; 237–250.

[40] Daniel Byman and Matthew Waxman, *The Dynamics of Coercion: American Foreign Policy and the Limits of Military Might* (Cambridge: Cambridge University Press, 2002), p. 28.

[41] Byman and Waxman, *The Dynamics of Coercion*, p. 29.

[42] J. Bowyer Bell, *The Dynamics of the Armed Struggle* (London: Frank Cass, 1998), p. 15.

[43] Bob Jones, 'Deterring "Competition Short of War": Are Gray Zones the Ardennes of Our Modern Maginot Line of Traditional Deterrence?', in Aaron Bazin, ed., *On Competition: Adapting to the Contemporary Strategic Environment*, JSOU Report 21-5 (MacDill Air Force Base, FL: JSOU Press, 2021), pp. 42–47.

denial actions, or punishment ones.[44] These contributions and courses of action, however, are based entirely 'on the stakes at risk'.[45]

Stakes and risk are inextricably intertwined. They follow the same pattern as the military and politics in conflict: 'the relationship between the worlds of politics and of military force is, or should be, one of reciprocal influence'.[46] Because strategy is both purposeful and consequential in nature, the stakes involved can reciprocally influence or even overwhelm considerations of risk. Harkening back to the previous chapter and its different levels and types of risk, as well as postures, reframing the concept of a 'risk society', with its range of environmental factors to be managed, is important. The risk society is premised on a lack of existential threats to a polity as well as relative stability in the international system. The lack of such a threat, and the episodic nature of others such as terrorism, climate change, transnational crime, and immigration, rarely call for decisive, coordinated action, much less maximum effort. They rarely threaten a state's survival but, paradoxically, receive greater political attention in their management, increasing the likelihood of friction and miscommunication between organizations responsible for addressing specific parts of them. Alex Schmid illustrates this point in the twenty different and changing definitions of terrorism between various federal departments and agencies of the United States government.[47]

High stakes change the texture of risk politically and can overcome risk-taking postures, including deliberative, dispositional, or socially driven risk taking. At the strategic level, risk regarding future challenges, force management, and institutions may be significantly limited. From the operational down to the tactical, risk to mission can also subsume considerations of risk to force. Translated back into terms specific to special operations and its capabilities, high stakes to the state explain the seeming indifference of political and military leaders to the health of SOF. Those calculations do not rise above the level of noise when the very future of the state or a polity is at risk. Stakes and risks, therefore, are related to one another through the common medium of consequences and perceptions. The consequences of perceived success and failure, rather than simple gains and losses, are the basis of calculating stakes. Perceptions related to threats, and the consequences of not addressing them adequately, frame our understanding of risk.

A less complex, esoteric, and jargon-laden understanding of stakes was articulated almost two centuries ago. Carl von Clausewitz expounded briefly on the 'value of the object' in his discussion of the reasons for which conflicts end. The first two relate to unachievable goals, brought on either as a result of insufficient means, miscalculation, or both. His last reason deals with a different logical conclusion: the price

[44] Will Beaurpere and Ned Marsh, 'Space, Cyber, and Special Operations: An Influence Triad for Global Campaigning', Modern War Institute at West Point (6 September 2022), https://mwi.usma.edu/space-cyber-and-special-operations-an-influence-triad-for-global-campaigning; Robert Haddick, *How Do SOF Contribute to Comprehensive Deterrence?*, JSOU Report 17-11 (MacDill Air Force Base, FL: JSOU Press, 2017), pp. 57–65.

[45] Haddick, *How Do SOF Contribute to Comprehensive Deterrence?*, p. 60.

[46] Gray, *The Strategy Bridge*, p. 31.

[47] Alex Schmid, ed., *The Routledge Handbook of Terrorism Research* (London: Routledge, 2011), p. 44.

of victory outweighs the value you ascribe to it. He writes, 'Since war is not an act of senseless passion but is controlled by its political object, the value of the object must determine the sacrifices to be made for it in magnitude and also in duration. Once the expenditure of the effort exceeds the value of the political object, the object must be renounced and peace must follow.'[48] As Michael Handel notes, though, this observation is a 'normative recommendation', based on Clausewitz's exploration of a purely theoretical form of war, and 'not as a description of reality'.[49] Stakes can exist anywhere along a spectrum of possibilities, as Clausewitz suggests: 'wars can have all degrees of importance and intensity, ranging from a war of extermination down to simple armed observation.'[50]

The value of the object can be at one end absolute and non-negotiable, as was the case for the Allies with the unconditional capitulation of the Nazi German, Fascist Italian, and Imperial Japanese totalitarian regimes, or the utter annihilation of an opponent, as Rome annihilated Carthage in the Third Punic War. At the other end is exceptionally limited value: contributing military forces to peacekeeping missions, Clausewitz's 'armed observation', to ensure belligerents are upholding ceasefire agreements in support of global norms regarding the cessation of armed conflict. The value of the object in these cases is apparent in which nations choose to contribute SOF to United Nations and which ones do not. Often, small or medium powers such as Denmark value the object of supporting such missions as an extremely economical way of increasing their profile, prestige, influence, and reputation internationally to offset their 'lack of relative power' and as a means of generating or creating it.[51] Rising powers such as China also view peacekeeping missions outside of Asia as a low-cost, low-risk way to increase their international prestige and power.[52]

Rather than looking at stakes in isolation, a general theory of special operations should consider three other possibilities. The first is the potential for stakes to change over time. The value of an object may wane as time advances and the costs increase. For example, no number of special operations, particularly those conducted by 'special squads' conceived by William Nicol Gray and conducted by former SAS operator Roy Farran, could have convinced British leaders or the public that continuing control of Palestine was worth the effort.[53] The second is the prospect of the consideration of multiple stakes at any one time. Crises and contingencies do not arise in a discrete sequence but often overlap and occur simultaneously. Great Britain, for example, was dealing with covert support to mujaheddin in Afghanistan and overt

[48] Clausewitz, *On War*, p. 92.

[49] Michael Handel, 'Who Is Afraid of Carl von Clausewitz? A Guide to the Perplexed', unpublished monograph, 6th ed. (Summer 1997), www.clausewitz.com/readings/Handel/Handlart.htm.

[50] Clausewitz, *On War*, p. 81.

[51] Troels Burchall Henningsen, 'Small State Strategy in Practice: The Formulation and Execution of Danish Theories of Success in the United Nations Security Council, 2005–2006', *Scandinavian Journal of Military Studies*, 5:1 (2022): 129–130.

[52] Xiao Tianliang, ed., *The Science of Military Strategy* (2020), China Aerospace Studies Institute, trans. (Maxwell Air Force Base, AL: CASI, January 2022), p. 289.

[53] David Cesarani, *Major Farran's Hat: Murder, Scandal, and Britain's War Against Jewish Terrorism, 1945–1948* (London: William Heinemann, 2009), pp. 185; 216–218.

and ongoing 'aid of the civil power' operations in Northern Ireland in 1982, based on evaluation of their respective stakes, when a crisis in the Falklands escalated into war.[54] For the United States, the year between 1967 and 1968 was both monumental and busy. Reflecting the different calculations of stakes, most of its SOF were committed to Southeast Asia, specifically Vietnam, Cambodia, and Laos, followed by the Middle East, Latin America and the pursuit of Che Guevara and his lieutenants, and Europe. The third and final possibility is the underlying rationality of decision making. Decision making is inherently subjective and never occurs in a vacuum or in isolation. It is subject to the influence of the changes in the environment, prior experience and cognitive shortcuts, and inherent bias resulting from neurological as well as cultural or normative sources.[55] Notions of rationality underpinning deterrence theories have been further challenged by identifying sources of cognitive distortion, including competing political imperatives, ideology, narcotics and alcohol (ab)use, sleep deprivation, and stress.[56]

Stakes, therefore, are important to include in a general theory of special operations but difficult to understand or assess accurately, much less precisely. They exist along a spectrum from absolute or complete, requiring the mobilizing of all resources, through conflict of increasingly diminishing scale and intensity, down to individual deployments designed to monitor a situation, or reassure, support, or sustain a partner or proxy. Like risk, stakes are not invariable but are subject to change as circumstances are. Consideration of multiple stakes can occur simultaneously, as the result of crises as well as ongoing commitments. In addition to their mutability, stakes are also subject to a host of internal and external influences, which raises questions about their rationality in a pure theoretical sense. A lack of pure rationality should not be confused with a lack of coherence or cogency. Rather, decision-making calculations, including stakes and risk, are subject to various pressures including emotions and bias. The span of stakes, combined with their mutability and additive and subjective nature, helps animate a general theory of special operations, even if they are not calculable using 'the seductive simplicity of the scientific method'.[57]

The decision to employ force purposefully is a consequential one. We like to think these decisions are made coldly and in a calculated manner in support of interests, one of Thucydides' three identified causes of war. Stakes, however, are more likely to reflect the other two, less instrumental ones: fear and honour. Consider the following example. On 12 October 2011, President Barack Obama authorized the dispatching of a Joint Special Operations Task Force (JSOTF) to Uganda, the

[54] Rory Cormac, *Disrupt and Deny: Spies, Special Forces, and the Secret Pursuit of British Foreign Policy* (Oxford: Oxford University Press, 2018), p. 223.

[55] Chiara Acciarini, Federica Brunett, and Paolo Boccardeill, 'Cognitive Biases and Decision-Making Strategies in Times of Change: A Systematic Literature Review', *Management Decision*, 59:3 (2021): 639–641.

[56] Keith Payne, *The Fallacies of Cold War Deterrence and a New Direction* (Lexington, KY: University Press of Kentucky, 2001), pp. 39–61.

[57] Thomas Cutler, 'Review of Carl von Calusevitz [sic], *On War*', *Proceedings*, 135:1 (January 2009), www.usni.org/magazines/proceedings/2009/january/further-review.

Central African Republic, the Democratic Republic of the Congo, and South Sudan.[58] JSOTF forces were to conduct military advice and assistance missions, designed to assist the security forces of four states to combat the threat posed by Joseph Kony and his armed group, the Lord's Resistance Army (LRA). The commitment of forces was not prompted by rational calculations of American interest in the region. Previous engagement since 2008 had accomplished little against the group. Thanks to the media campaign of a non-governmental organization, Invisible Children, and a social media clip about LRA activities that went viral with the hashtag #Kony2012, public interest in the region was stimulated, leading members of Congress to call for action.[59] While upholding US commitments in the region, one might argue that honour, in the form of upholding stated values by making good on promises to take action against human trafficking, more so than any definitive interest in the region, prompted the dispatch of SOF. In his notification letter to Congress, President Obama messaged both the risks he was willing to assume, as a reflection of the stakes involved, 'toward the removal of Joseph Kony from the battlefield' given the public commitment of such a small SOF contingent.[60] Kony continues to elude capture more than a decade later, interest in the issue domestically has all but been forgotten, and there is limited news coverage, suggesting SOF efforts contributed to little more than a passing policy fad involving limited risk and even less in terms of stakes for the United States.[61]

## Opponents and allies

Military theories, aspects of which get translated into doctrine, seek to provide answers, insights, and even guidance and direction in the pursuit of victory more generally in war as well as in specific domains. The prescriptive insights many provide, by authors ranging from Sun Tzu to Bleddyn Bowen, reflect the consequentialist nature of conflict and strategy, as well as their pragmatic nature. Conflict is consequential given how the means within it interact, primarily through death and destruction. Given the potentially high costs and uncertain outcomes, war between states is a relatively rare occurrence. War between major powers is rarer and even more consequential, not only in terms of costs but also for its ability to include third

[58] Office of the Press Secretary, The White House, 'Letter from the President to the Speaker of the House of Representatives and the President Pro Tempore of the Senate Regarding the Lord's Resistance Army', 14 October 2011, https://obamawhitehouse.archives.gov/the-press-office/2011/10/14/letter-president-speaker-house-representatives-and-president-pro-tempore.

[59] Erica Orden and Emily Steel, 'How "Kony" Clip Caught Fire Online', *The Wall Street Journal* (9 March 2012), www.wsj.com/articles/SB10001424052970204781804577271692294533870?mod=article_inline.

[60] Office of the Press Secretary, The White House, 'Letter from the President to the Speaker of the House of Representatives'.

[61] Michael Phillips, 'Pizzas, Loudspeakers, and Moms: The U.S. Military's Unorthodox Mission against Joseph Kony', *The Wall Street Journal* (10 March 2017), www.wsj.com/articles/pizzas-loudspeakers-and-moms-inside-the-u-s-militarys-most-unorthodox-mission-1489158880; Emma Madden, '"Kony 2012", 10 Years Later', *The New York Times* (8 March 2022), www.nytimes.com/2022/03/08/style/kony-2012-invisible-children.html.

parties and other actors within them. Wars within states, in struggles over political control, are deadly but their consequences are largely confined with their borders. Strategy's nature is pragmatic in the sense that it must be done to have any practical value, otherwise it is nothing more than an academic and unvalidated theory of victory. Strategy, the purposeful application of available and developed means in operational ways to achieve the ends set by policy, remains difficult. Theories and doctrine provide us with answers, in the form of principles, sequences, and rules based on concepts and best practices, yet victory often eludes us and competition and conflict morph into something we did not anticipate. Why does this occur?

Scholars have answered this question in both similar and different ways. Some argue the issue is with our perception and understanding of victory. According to them, its conceptualization in the West is quaint and outdated; victory has changed, thanks largely to the impact of information technologies, changing demographics, and shift away from war as a useful instrument of policy for states.[62] Emile Simpson argues that the combination of non-state actors, including terrorists, insurgents, and local populations, has an entirely different frame of reference: they are often unwilling to concede defeat as they value a different object to the intervening forces. Not struggling over the same object forces intervening powers to align their localized actions with continuing rhetoric.[63] The perceptual is more important than physical action. In other words, competition and conflict is primarily a battle of influence; force only has value in the sense that it reinforces messaging and convinces popular opinion in a range of different constituents. A combination of 'strategic atrophy' in the West with a state of 'durable disorder' globally enabled by new technologies requires different methods as well as new 'rules' at most or reconceptualized principles and near-principles of war at the very least.[64]

The quest to explain failure to apply overwhelming power meaningfully against non-state actors, or to anchor new theories, overlooks or redefines a central truth of war and strategy: the impact of the interaction of opponents on each other. Chinese military authors from their National Defense University conclude that strategy is necessarily 'confrontational' and filled with 'antagonism', given its political nature and existence of 'opposing parties'.[65] Some of this is attributable to the scientific socialist belief in an inevitable clash of incompatible ruling systems: capitalism versus socialism. Using a different form of dialectical reasoning from the one Karl Marx employed to reach this conclusion, Carl von Clausewitz explored the truth of war and strategy.

[62] See, for example, Thomas Hammes, *The Sling and the Stone: On War in the 21st Century* (New York: Zenith Press, 2006); M.L. Cavanaugh, 'Military Victory Is Dead', Modern War Institute blog (11 September 2016), https://mwi.usma.edu/defeat-military-victory.

[63] Emile Simpson, *War from the Ground Up: Twenty-First-Century Combat as Politics* (Oxford: Oxford University Press, 2013), pp. 54–61.

[64] Sean McFate, *The New Rules of War: Victory in the Age of Durable Disorder* (New York: HarperCollins, 2019), pp. 1–10; Marvin Pokrant, *New Principles of War: Enduring Truths and Timeless Examples* (Lincoln, NE: Potomac Books, 2021), pp. 286–291.

[65] Xiao Tianliang, ed., *The Science of Military Strategy* (2020), China Aerospace Studies Institute, trans. (Maxwell Air Force Base, AL: CASI, January 2022), pp. 22–23.

In its purest, conceptual logical form, conflict is a struggle not between two systems but rather between two opposing wills. Each attempts to impose their will on the other, while simultaneously resisting the effort of their opponent, which Clausewitz traces deductively through the ideal notion of polarity, the relationship of the offence to the defence, uncertainty and chance, and the subjective nature of war that is a source of friction within an organization.[66] Both the objective and subjective aspects of war conspire to make strategy difficult. Only 'knowledge ... becom[ing] capability' in the sense of informed judgement, or the intuitive ability of the educated and prepared mind to pierce through the fog of war to see a potential path to victory, as well as the courage of conviction to 'follow that faint light wherever it might lead', can overcome these difficulties.[67] Whether this synthetic form of genius identified by Clausewitz exists in a single individual, collectively within a staff or its processes, or can exist at all is a philosophical question that will be debated for some time.[68] Chinese military writers imply a hybrid type of genius in future war, resulting from the combination of an overall leader's judgement informed by a five-step iterative process—an expanded version of Boyd's OODA loop—based on machine-assisted analysis and simulating courses of action.[69] The interaction of the two opponents, which Clausewitz compares metaphorically to wrestlers grappling with one another, leads to a paradox within strategy that accounts for the effects opponents and allies have.[70]

This paradox, inherent in the logic of strategy, was first identified and explored by Edward Luttwak in 1985. In essence, he merged Clausewitz's idea of interaction of opposites with his arguments about the inseparable, dynamic relationship that exists between offence and defence. Upending the conventional wisdom of the day and seemingly flying in the face of Napoleon's advice and activity, Clausewitz concludes that the defence is the stronger of the two but is passive and has a negative object. The offence has a positive object and retains the initiative but increasingly loses its strength the longer it goes on. The defence either gains or preserves its strength, absorbing and dispersing the energy and resources of the offence.[71] At a select point, the offence reaches its culminating point where it can no longer continue or consolidate or hold what it has taken. This point is the optimal one for the defence to go on the offence and the see-saw struggle of wills continues until one is either physically destroyed or morally collapses. Clausewitz extends this abstract depiction from the campaign level to the strategic one, suggesting there is a culminating point of victory as well.[72]

[66] Clausewitz, *On War*, pp. 75–77; 85–87.

[67] Ibid., pp. 102; 127.

[68] See, for example, Richard Betts, 'Is Strategy an Illusion?', *International Security*, 25:2 (Fall 2000): 5–50; Clifford Rogers, 'Clausewitz, Genius and the Rules', *Journal of Military History*, 66:4 (October 2002): 1167–1176; Lawrence Freedman, *Strategy: A History* (Oxford: Oxford University Press, 2013), pp. 237–244; Hew Strachan, 'Strategy in Theory; Strategy in Practice', *Journal of Strategic Studies*, 42:2 (2019): 171–190.

[69] Tianliang, *The Science of Military Strategy*, p. 10.

[70] Clausewitz, *On War*, p. 75.

[71] Ibid., pp. 357–369.

[72] Ibid., pp. 566–573.

Luttwak expands this discussion of interaction and culmination to include the very logic of strategy itself. The most direct path to victory is also the most heavily defended as the opponent is aware of this vulnerability. In a nod to Basil Liddell Hart's notion of successful military approaches taking an indirect approach, Luttwak takes the point a step further. According to his idea of the reversal of opposites, the paradoxical logic of strategy suggests the appealing indirect path becomes an unsurprising, direct one, as the case of 1940 and 1944 offensives in the Ardennes suggests.[73] The failure to adapt in peace or war means that 'the logical will induce a self-negating evolution, which may reach the extreme of a full reversal, undoing war and peace, victory and defeat, and all they include'.[74] Interaction, over time, has an evolutionary effect on opponents, visible technologically and organizationally in 'reciprocal force-development effects', more commonly known as the measure/countermeasure cycle, in the quest for greater efficiency and effectiveness.[75] Those incapable of adapting to the particular circumstances of peace or war will find themselves at supreme disadvantage or even defeated. Attempts to achieve positive outcomes are likely to be thwarted by a combination of inherent adaptation and systemic entropy.

The paradoxical logic acknowledges the interaction of opposites, including their means or capabilities, but overlooks a vital element: the stakes. These stakes, or the value of the object discussed previously, determine how much energy and effort each opponent is willing to expend. The will of an opponent is even more difficult to divine than the stakes. Many military theories, when subsequently put into practice, collapsed for this reason. Notions of strategic bombing, premised largely on the ideas and assumptions of Guilio Douhet and Hugh Trenchard, failed to produce the civilian panic or social unrest necessary for capitulation.[76] Despite failing to achieve its intended effect, paradoxically, strategic bombing in the Second World War became, in Phillips O'Brien's evaluation, nevertheless the 'war-winning failure' as conflict was primarily a struggle based on industrial production.[77] Thomas Schelling's concept of graduated escalation, put into practice in the form of two air campaigns, Flaming Dart and Rolling Thunder, made little impact on North Vietnamese will or their determination to unify their country. The message signalled behind the actions was either misunderstood or ignored, or perhaps paradoxically hardened North Vietnamese commitment and resolve.[78] When underestimations of will are combined

[73] Allied surprise in 1944 centred not on where a German counteroffensive would occur but instead on faulty intelligence assessments that they were capable of one at all and command biases. Harold Winton, *Corps Commanders of the Bulge: Six American Generals and Victory in the Ardennes* (Lawrence, KS: University Press of Kansas, 2007), pp. 79–84.

[74] Edward Luttwak, *Strategy: The Logic of War and Peace*, rev. ed. (Cambridge, MA: Belknap Press, 2001), p. 16.

[75] Luttwak, *Strategy*, p. 35.

[76] Tami Davis Biddle, *Rhetoric and Reality in Air Warfare: The Evolution of British and American Ideas about Strategic Bombing, 1914–1945* (Princeton, NJ: Princeton University Press, 2002), pp. 106–107; 289–291.

[77] Phillips Payson O'Brien, *How the War Was Won: Air-Sea Power and Allied Victory in World War II* (Cambridge: Cambridge University Press, 2015), p. 479.

[78] Richard New Lebow, 'Thomas Schelling and Strategic Bargaining', *International Journal*, 51:3 (Summer 1996): 563–567; Alexander Field, 'Schelling, von Neumann, and the Event That Didn't Occur', *Games*, 5:1 (2014): 71.

with overestimations of capability, the results can be deadly. Adolf Hitler's supreme gamble and fundamental strategic error of the Second World War, the invasion of the Soviet Union in June 1941, was premised on the fragility of the Soviet state and its inferiority in all categories, including will, of its people, a view shared by most of the German General Staff.[79] Although just how deadly the errors of calculation will be to Vladimir Putin and Russia, its invasion of Ukraine in February 2022 expected victory within 76 hours, yet after a year it has suffered costs estimated at one-quarter of a million casualties and close to 40 per cent of its pre-war fleet of tanks.[80] The Ukrainian will to resist the Russian invasion, given the stakes involved and consequences of defeat, surprised not only the Russians but also well-informed conflict analysts.[81]

A general theory of special operations needs to account for the dynamic created by opposing wills in conflict and competition, as well as different agendas pursued by allies. Colin Gray points out that while strategy does not need an enemy, it requires an adversary to limit study to 'contexts wherein there is politically motivated competition between and among polities. It is assumed that any or all instruments of coercion may be used or threatened'.[82] Given these considerations, the intensity of the dynamic is shaped by the perceived stakes that the use of force, or its threat, is determined to accomplish. A threat to use force invokes a less intense dynamic against an adversary than does its direct use against them. It reaches its peak intensity in a conflict where both sides value the object equally. This point is explored specifically for special operations further in the section on orientation.

The dynamic of competing wills assumes two players, or opponents. It becomes much more unpredictable and complicated when others are included, such as allies, partners, and proxies. The benefits of having allies and partners are clear. They can increase legitimacy for a series of actions, improve overall capability and capacity, spread costs through burden sharing, and expand the number of options available by opening different possibilities. Proxies provide the benefit of indirect action against an adversary in pursuit of an interest, limit the risk of escalation with a major power, and bear the brunt of costs in terms of casualties and scrutiny.[83] Yet allies, partners, and proxies come with their own costs, which reflect their assessment of stakes, political agendas, and agency.[84] They are capable of exercising their own will in pursuit

[79] Megargee, *Inside Hitler's High Command*, pp. 103–116.

[80] Caitlin McFall, 'UK Claims Some 220,000 Russian Casualties in War as Ukraine Prepares Counteroffensive', Fox News (29 March 2023), www.foxnews.com/world/uk-claims-some-220000-russian-casualties-war-ukraine-prepares-counteroffensive; Dan Sabbagh, 'Russian Army Has Lost Up to Half of Key Battle Tanks, Analysts Estimate', *The Guardian* (15 February 2023), www.theguardian.com/world/2023/feb/15/russian-army-has-lost-up-to-half-of-key-battle-tanks-analysts-estimate-ukraine.

[81] Erik Schmitt, Helene Cooper, and Julian Barnes, 'How Ukraine's Military Has Resisted Russia So Far', *The New York Times* (3 March 2022), www.nytimes.com/2022/03/03/us/politics/russia-ukraine-military.html.

[82] Colin S. Gray, *Theory of Strategy* (Oxford: Oxford University Press, 2018), p. 65.

[83] Jonathan Wilkenfield et al., 'Escalation Management in Gray Zone Crises: The Proxy Factor', *International Studies Quarterly*, 66:3 (September 2022).

[84] Alan Doig et al., 'Marching in Time: Alliance Politics, Synchrony, and the Case for War in Iraq, 2002–2003', *Australian Journal of International Affairs*, 61:1 (2007): 23–40.

of their political goals, may not be as committed to a specific outcome (or have a different one), and may attempt to 'free ride' by not paying their share of the costs.[85]

In addition to control issues, or allies, partners, and proxies exercising their own will, the decision-making process is often slowed down due to a need to reach consensus, often across different languages. Byman and Waxman identify three other challenges to coalitions, specifically when it comes to trying to coerce collectively. These challenges include a lack of common agenda, the diminished credibility of threats, and an inability to achieve rapid and focused escalation dominance.[86] In summary, allies, partners, and proxies can solve problems while creating others. They are a desirable means of accomplishing goals provided commitments, trust, and capabilities are sufficient and maintained. The commitment of various NATO partners, while prone to inevitable friction and disagreement and a degree of 'organizational routine', nevertheless sustained the attempt to create a stable, function democracy for almost twenty years in Afghanistan, even though the experiment ultimately failed.[87]

During the Second World War, Adolf Hitler and Nazi Germany could not have made worse choices for allies among the Axis powers—Fascist Italy and Imperial Japan—and Hitler's choice of partners, including Hungary, Romania, and Finland, was no better.[88] Several became liabilities in terms of their commitments and capabilities and Italy involved the Third Reich in tertiary campaigns in Greece, North Africa, and the Mediterranean. Consensus in collective decision making, in addition to being slow, can introduce a different dimension to the dynamic, becoming 'more like a business deal ... clogged with diplomatic reservations' in which' negotiators are reluctant to commit all their forces to a common cause'.[89]

In other words, the relationship between allies, partners, and proxies can become transactional in substance, requiring inducements or other pledges in exchange for support. In these cases, unless bound by a formal treaty as well as the willingness to uphold it, allies, partners, and proxies may use the circumstances to extort concessions, particularly if they know their resources are crucial in assuring the outcome.[90] Should their extortion fail, they can concede support or exercise other options, including seeking an arrangement with the competitor. As a means of accomplishing goals, special operations are subject to the dynamic and paradoxical logic inherent within strategy, as well as the influences allies and partners have on coalition efforts. Special operations, in sum, are influenced by and at the mercy of strategy and politics.

[85] Thomas Plümper and Eric Neumayer, 'Free-Riding in Alliances: Testing an Old Theory with a New Method', *Conflict Management and Peace Science*, 32:3 (2015): 247–268.

[86] Byman and Waxman, *The Dynamics of Coercion*, pp. 159–163; 167–171.

[87] Sten Rynning, *NATO in Afghanistan: The Liberal Disconnect* (Stanford, CA: Stanford University Press, 2012), p. 6.

[88] For details, see Burkhart Muller-Hillebrand, *Germany and Its Allies in World War II: A Record of Axis Collaboration Problems* (Lanham, MD: University Publications of America, 1980).

[89] Hugh Smith, 'The Womb of War: Clausewitz and International Politics', *Review of International Studies*, 16:1 (January 1990): 48.

[90] Jesse Johnson, 'The Cost of Security: Foreign Policy Concessions and Military Alliances', *Journal of Peace Research*, 52:5 (September 2015): 677.

## Bearing and calibration

The previous dynamic elements are rather ethereal compared to the final ones. They result from the first three and determine whether special operations benefit from the initiative or are forced to respond to the actions of an opponent. Bearing refers to how SOF are postured: whether they are on the offence or defence. As the preceding discussion suggests, offence and defence are relative and subject to change. As the section on time outlines, however, the rate of that change varies drastically and slows much more considerably from the tactical level up to the grand strategic. Stated differently, change occurs much more slowly at the grand strategic level, specifically in terms of the offence and defence, than it does from the strategic level down. A theory of special operations accounts for differences in time scales and relates them to offence and defence as a method of interpretation of how and why decisions are made to employ them.

Understanding the relationship between offence and defence conceptually is often impeded by issues of currency and relevancy. The desire to extract meaningful lessons learned from infrequent contemporary conflicts has led to errors of judgement and analysis. Much of this is related to advocacy for or against specific battlefield technologies or platforms. Two of the Western apostles of operational manoeuvre restoring the power of the offence prior to the Second World War, J.F.C. Fuller and Basil Henry Liddell Hart, extracted lessons from the Spanish Civil War. Both concluded the defence was predominant, thanks to development in armaments, specifically aircraft and anti-tank and anti-aircraft and guns.[91] Virtually every major conflict of over the past half-century has produced analysis suggesting the main battle tank is an anachronism, including most recently the Nagorno–Karabakh War (2020) and the war between Russia and Ukraine (2022–present).[92]

The offence is inextricably linked to the defence; one cannot exist without the other. Rather counterintuitively, and in a logical conclusion that is too clever by half, Clausewitz notes that the defender starts the war. After all, only by choosing to defend does a conflict begin by 'bringing the two parties into existence'.[93] The defender has the choice to not resist at all. This observation aside, Clausewitz and later Luttwak noted the relationship between offence and defence across different levels of competition and conflict. In particular, both stressed that being on the defence strategically does not condemn one to operational or tactical defence automatically.

The opponent on the strategic defence can and should go on the operational and tactical offence as soon as possible, to buy additional time and space strategically to marshal other resources, including equipment and will. British prime minister

[91] Brian Bond, *Liddell Hart: A Study of His Military Thought* (London: Cassell, 1977), p. 97; J.F.C. Fuller, 'Tanks Won't Do Much', *The War Weekly*, issue 3 (10 November 1939), p. 64.

[92] David Johnson provides a useful historical survey in 'The Tank Is Dead: Long Live the Javelin, the Switchblade, the ...?' War on the Rocks blog (18 April 2022), https://warontherocks.com/2022/04/the-tank-is-dead-long-live-the-javelin-the-switchblade-the.

[93] Clausewitz, *On War*, p. 377.

Winston Churchill had this idea in mind when he sanctioned 'butcher and bolt' Commando raids, established and commissioned the Special Operations Executive to 'set Europe ablaze', and authorized strategic bombing raids against targets in France and Germany conducted by the Royal Air Force.[94] While the British were in the midst and eventually won a hard-fought operational defensive victory during the Battle of Britain in the summer of 1940, forestalling an invasion of the islands by the Germans, nothing would restore popular morale than offensive actions taking the war to the continent. Three early British Commando raids in the Second World War, in particular Operations Collar (France, June 1940), Ambassador (German-occupied Guernsey, July 1940), and Claymore (German-occupied Norwegian Lofoten Islands, March 1941), accomplished this political objective even though their military results were negligible.[95] The latter mission provided an unintended benefit: the capture and exploitation of cryptographic material that provided the crucial first link in the chain that broke the 'uncrackable' Enigma cypher.[96]

A more current example further illustrates the linkages between offence and defence across different levels of war. In the shock that accompanied the terrorist attacks on New York and Washington in 2001 were several effects, including the very real belief among senior decision makers that more and even worse was yet to follow. Some members of Congress called for an overwhelming reprisal in response, including using nuclear weapons. The Bush Administration needed to respond and elected to use a mixture of paramilitary and military special operations as some of the first actions of Operation Enduring Freedom. The subsequent offensive campaign that unfolded operationally and tactically, discussed in much greater detail in Chapter 6, did more than just drive the Taliban and al-Qaeda out of power in Afghanistan. Strategically, it bought the United States precious time—breathing space—to better comprehend the global threat al-Qaeda posed and to reassure the American public and the country's allies and partners.

Special operators prefer to be on the offence tactically, operationally, and strategically, if possible. Possessing the initiative confers numerous advantages, including striking at the time and place of your choosing, achieving surprise, and possessing a positive aim, even when you are outnumbered. In words that are just as insightful a century later, having and maintaining the initiative provides '[t]he power of making our adversary's movement conform to our own'.[97] More recently, and from the perspective of naval special operations against Russia, Kevin Stringer argues that in addition to the traditional role of discombobulating an opponent, possessing the initiative in competition can impose 'greater costs on Russian actions' in the contested regions of the Baltic, Black, and Caspian Seas in an operational campaign,

[94] Glen St. J. Barclay, '"Butcher and Bolt": Admiral Roger Keyes and British Combined Operations, 1940–1941', *Naval War College Review*, 35:4 (July–August 1982): 99–101; E.H. Cookridge, *Inside S.O.E.: The First Full Story of Special Operations Executive in Western Europe, 1940–1945* (London: Arther Barker, 1966), p. 3.

[95] Andrew Hargreaves, *Special Operations in World War II: British and American Irregular Warfare* (Norman, OK: University of Oklahoma Press, 2013), pp. 23–24.

[96] Gray, *Explorations in Strategy*, p. 142; Hargreaves, *Special Operations in World War II*, pp. 216–217.

[97] Robert Cherry, 'The Initiative in War', *Royal United Services Institution Journal*, 66:461 (1921): 87.

including preparation of the environment discussed below.[98] Chinese authors link the initiative, which reflects 'the offensive nature of special operations', with concentration of forces in time to achieve decisive victory, flexibility, and the notion of 'subjective dynamism': the combination of personality traits possessed by special operators and intelligence that tips the balance over objective factors.[99]

Calibration, in contrast, refers to operational of functional orientation of special operations. This orientation has three different types: overt, covert, and clandestine. Each is defined and explored below. The reason for placing calibration in a broader context analytically and theoretically is to break free from a tautology inherent in the discussion of special operations capability. Too often, in narratives about special operations, the strategist's questions of 'why' and 'so what' when SOF are used overtly, covertly, or clandestinely default to one of three explanations. The first is how a capability is designed to operate. Clandestine sensors for intelligence and reconnaissance missions, for example, are developed to be employed and operate clandestinely. The second is a function of how special operations or SOF are defined. Just as some definitions apply 'high risk' to all special operations, they also employ 'political sensitivity' in an all-encompassing way, without delineating how their calibration relates to it, much less explain why they occur simultaneously or change over time. The third is a function of bureaucratic competition and institutional rivalries between SOF calibrated differently for prestige and influence, particularly within democracies.

Although special operators deserve a certain amount of credit for largely remaining 'quiet professionals', at the institutional level who gets the recognition matters within SOF organizationally. For obvious reasons, clandestine SOF usually avoid attention and scrutiny as much as possible, yet this yearning for acknowledgement remains. Works by authors including Erik Schmitt, Thom Shankar, and Sean Naylor could not be written if numerous sources representing different institutional equities recounted episodes and provided information in the volume and to the level of detail they did.[100] There is no other explanation for why so many otherwise responsible professionals committed actions that could be interpreted as misguided at best or treasonable at worst. A related aspect of this issue is leaders more often wittingly than not revealing details of clandestine activities or confirming their existence, through public pronouncements or sanctioned leaks to journalists, for near-term political advantage, as information related to the Stuxnet virus suggests.[101]

[98] Kevin Stringer, 'Jomini and Naval Special Operations Forces: An Applied-Competition Approach to Russia', *Naval War College Review*, 74:4 (Autumn 2021): 84–85, 88.

[99] Guo Ming, *Lectures on the Science of Special Operations*, Academy of Military Sciences Master's Post-graduate Series Teaching Materials, 2nd ed. (U//FOUO) translation, 2016, Selected Book Translations CVI (Beijing: Military Science Press, July 2013), pp. 103; 106–108.

[100] Erik Schmitt and Thom Shankar, *Counterstrike: The Untold Story of America's Secret Campaign against Al Qaeda* (New York: Times Books, 2011); Sean Naylor, *Relentless Strike: The Secret History of Joint Special Operations Command* (New York: St Martin's Press, 2015).

[101] David Sanger, *Confront and Conceal: Obama's Secret Wars and Surprising Use of American Power* (New York: Crown, 2012), pp. 188–209.

Of the three forms of special operations calibration, overt is the simplest to understand and easiest to identify. Overt orientation is almost indistinguishable from any other form of military activity or statecraft. Operators are recognizable by several identifiable characteristics: uniforms of one sort or another, national or unit crests, equipment, or liveries, such as camouflage painting schemes and the like. In other words, there is little to no attempt to disguise who operators represent. Political sensitivities or the risk of escalation, however, may dictate that operators not broadcast their arrival or presence in another country, and take appropriate measures to conceal. This is especially true when such measures are agreed upon by both countries in advance, as occurred during advising and assistance missions in the Philippines to combat Abu Sayyaf Group. Operators often flew by commercial carriers into civilian airports dressed in civilian clothes first as part of Task Force 510 and later with Joint Special Operations Task Force-Philippines (JSOTF-P) in keeping with domestic political sensitivities.[102] Once there and while working with their Philippine counterparts, operators were in uniform, as they are during most overseas training missions and exercises.[103] JSOTF-P had its own blog page with a fact sheet that also included a picture of its assembled headquarters staff, reinforcing its overt presence.[104] Personnel wearing civilian clothing while off-duty was done there as elsewhere to respect host nation or regional power political sensitivities, not irritate the local population, and serve as a force protection measure by offering personnel a measure of anonymity.[105]

Operators and activities for overt missions are attributable and linked; there is little attempt to conceal their presence or activities. In fact, missions are conducted overtly to ensure an opponent recognizes their presence and activities. Overt calibration more broadly serves purposes related to strategic utility through a form of signalling and acknowledgement, in support of broader coercive goals in both competition and conflict. These goals include reinforcing friendly morale, bolstering the credibility of commitments, humiliating an opponent, denying an opponent its capabilities, and serving as a potential tripwire for conflict escalation, among others.[106] The allegedly 'secret' deployments of US Green Berets and Marine Corps Raiders to Taiwan in 2020, for example, were conducted in such a manner so as not to provoke the People's Republic of China (PRC). The military advice and assistance missions they conducted, including countering internal irregular threats from the PRC as well as

[102] Linda Robinson, 'The SOF Experience in the Philippines and the Implications for Future Defense Strategy', *Prism*, 6:3 (December 2016): 151–167.

[103] See, for example, the picture published with the following news story: T.D. Flack, 'U.S. Aid Agency Focuses Its Efforts on Mindanao', *Stars and Stripes* (10 March 2007).

[104] 'Joint Special Operations Task Force: Philippines Fact Sheet', JSOTF-P blog (1 July 2011), http://jsotf-p.blogspot.com/2011/09/type-your-summary-here_20.html.

[105] The legal rationale and functional reasons for this decision are spelled out in W. Hays Parks, 'Special Forces' Wear of Non-Standard Uniforms', *Chicago Journal of International Law*, 4:2 (Fall 2003): 498–504.

[106] Schelling, *Arms and Influence*, p. 22. For good reasons scholars more recently have questioned the logic underpinning tripwires. See, for example, Dan Reiter and Paul Poast, 'The Truth about Tripwires: Why Small Force Deployments Do Not Deter Aggression', *Texas National Security Review*, 4:3 (Summer 2021): 33–53.

preparing the Taiwanese military to conduct resistance activities should it invade, have been discussed openly by journalists and politicians.[107] This suggests a form of indirect signalling to Beijing, likely to reduce ambiguity in US policy and intentions as well as to respond to PRC overflights of the island, but in a manner unlikely to provoke Beijing, while increasing the level of risk associated with future actions.[108]

To further complicate adversary risk calculations strategically and operationally by increasing uncertainty through ambiguity, special operations can be conducted covertly or clandestinely, through attempts to disguise who is sponsoring or conducting them. Few concepts related to special operations, however, are as misunderstood as covert and clandestine. The terms are often confused for each other—or worse, used interchangeably or used together. The Central Intelligence Agency's (CIA's) Directorate of Operations, known for a time as the National Clandestine Service, is its arm for conducting covert action. Unpicking how to accurately label Operation NEPTUNE SPEAR, the raid in Abbottabad, Pakistan that killed Osama bin Laden, is a journey to confusion between the different legal authorities, chains of command, and means used, a point developed further in Chapter 8.[109]

The difference between covert and clandestine is a subtle but significant one. Covert actions are designed to obscure both the hand conducting the mission and the act itself strategically. Clandestine activities, in contrast, only seek to do so tactically. The act, but not the actor, is temporarily concealed.[110] Within the United States, covert action is defined statutorily with a clear chain of authorization: the president of the United States signs a finding, often classified, to authorize the mission. Clandestine actions, in contrast, are not defined legally but rather doctrinally, to separate out their hidden activities reported to Congress intermittently from traditional military activities in support of intelligence activities, including covert action. Both have common ground of interest to policy and decision makers: 'a serious risk of exposure of US involvement, compromise of information, or loss of life'.[111]

If conducted effectively, the authorization and conduct of covert action should rarely come to light, or if it does, a sufficient period should have elapsed, particularly if it is conducted on a modest scale. The size and scale of the CIA's Operation CYCLONE, the covert effort of advice and assistance to Afghan mujahideen fighting against occupying Soviet forces, started modestly at under three-quarters of a million dollars.[112] At its height, it involved more than five countries with single-year funds

[107] Gordon Lubold, 'U.S. Troops Have Been Deployed in Taiwan for at Least a Year', *Wall Street Journal* (7 October 2021); Seth Cropsey and Harry Halem, 'Taiwan's Special Operations Forces', Real Clear Defense (20 October 2021), www.realcleardefense.com/articles/2021/10/20/taiwans_special_operations_forces_799768.html; Keoni Everington, 'Taiwan Confirms Presence of US Green Berets on Outer Islands', *Taiwan News* (15 March 2024), www.taiwannews.com.tw/news/5115290.

[108] Demetri Sevastopulo, 'U.S. Special Forces Secretly Train Taiwan's Military', *Financial Times* (8 October 2021), www.ft.com/content/46794116-7355-4669-b947-ce32bc5aff1a.

[109] Joseph Berger III, 'Covert Action: Title 10, Title 50, and the Chain of Command', *Joint Forces Quarterly*, 67 (4th quarter 2012): 32–39.

[110] Michael DeVine, *Covert Action and Clandestine Activities of the Intelligence Community: Selected Definitions*, CRS Report R45195 (Washington, DC: Congressional Research Service, 29 November 2022), pp. 5–7.

[111] DeVine, *Covert Action and Clandestine Activities of the Intelligence Community*, p. i.

[112] 'Document 76, Summary of Conclusions of a Special Coordination Committee Meeting, Subject: Covert Action', 23 October 1979, in David Zierler, ed., *Foreign Relations of the United States, 1977–1980, Volume XII: Afghanistan* (Washington, DC: Government Printing Office, 2018), p. 208.

peaking at $700 million—the largest covert action in history.[113] While great care was taken early to provide the mujahideen with equipment from third-party sources, in an effort to maintain a level of plausible deniability for the United States, the introduction of American FIM-92 Stinger and British Blowpipe man-portable air defence missiles from 1986 onwards, the former thanks largely to the efforts of Congressional staffer Michael Pillsbury, removed any possible doubts over who was behind the activities.[114] In contrast, the range of propaganda activities against North Vietnam, conducted by the Military Assistance Command, Vietnam (MACV), as part of its Studies and Observation Group's (SOG's) Psychological Operations Group (POG) in support of Operations Plan 34-A, included white (overt), grey (clandestine), and black (covert) propaganda efforts. The most overt were leaflet drops; the most covert, the so-called 'black letters' sent in an effort to discredit North Vietnamese leaders, remained undisclosed for almost a quarter-century after the end of the war.[115] Countries other than the United States, including democracies such as Great Britain, France, and Israel, as well as autocracies including Iran, the PRC, and Russia, offer little in the way of disclosure of details of covert and clandestine activities, much less acknowledge them, for considerably longer periods of time. This concealment, unfortunately, leads some to speculate irresponsibly and sometimes wildly about the scope of such activities.

A more recent example illustrates why clandestine and covert activities are difficult to identify, particularly while they are unfolding. On 26 September 2022, three of the four Russian-owned Nord Stream 1 and 2 natural gas pipelines ruptured. These pipelines, which traverse most of their length underwater in the Baltic Sea, provide almost half of the European Union's supply. The ruptures occurred in the lengths of pipeline nearest the coast of the Danish island of Bornheim. Suspicions of sabotage arose after Swedish reports that the seismic profile of ruptures appeared more humanmade than any naturally occurring phenomenon.[116] Since then, attribution of the alleged sabotage actions has been placed by various official and media sources on Russian, American, British, and Ukrainian operatives. Seymour Hersh identified the American unit he claims was responsible for and capable of the attack: divers of the US Navy's Diving and Salvage Center.[117] Hersh's claims rest on dubious sources and were rejected officially by the Biden Administration and refuted by open-source investigation.[118] Other allegations in favour of a pro-Ukrainian group conducting

[113] Congress appropriated $700 million in 1987, thanks largely to the efforts of Texas House of Representatives member Charlie Wilson. The government of Saudi Arabia pledged to match any funds provided by the US. For details, see Steve Coll, *Ghost Wars: The Secret History of the CIA, Afghanistan, and Bin Laden, from the Soviet Invasion to September 10, 2001* (New York: Penguin, 2004), pp. 65–67.

[114] Alan Kuperman, 'The Stinger Missile and U.S. Intervention in Afghanistan', *Political Science Quarterly*, 114:2 (Summer 1999): 225–228.

[115] Kenneth Conboy and Dale Andradé, *Spies and Commandos: How America Lost the Secret War in North Vietnam* (Lawrence, KS: University Press of Kansas, 2000), pp. 204–212.

[116] 'Blasts Near North Stream Were Explosions, Not Earthquakes, Swedish Seismologist Says', Reuters News (27 September 2022), www.reuters.com/world/europe/blasts-near-north-stream-were-explosions-not-earthquakes-swedish-seismologist-2022-09-27.

[117] Seymour Hersh, 'How America Took Out the Nord Stream Pipeline', substack post (8 February 2023), https://seymourhersh.substack.com/p/how-america-took-out-the-nord-stream.

[118] Matt Burgess, 'Online Sleuths Untangle the Mystery of the Nord Stream Sabotage', *Wired* (28 March 2023), www.wired.com/story/nord-steam-explosions-mystery-osint.

the mission have been debated ferociously online, with some claiming Russia conducted a 'false flag' operation in an attempt to depict Ukraine as a state conducting terrorism. They point out, not unreasonably, that Russia has sabotaged pipelines previously, in 2006, which provided a prelude to invading Georgia two years later.[119] The depth of the pipelines and their relative inaccessibility make confirming any of the details of the incident challenging. Until more definitive proof or an official admission is made, the who and why behind this alleged sabotage will remain open to speculation.

Clandestine military special operations conducted by the United States are often termed 'preparation of the environment', a form of shaping activities prior to the start of larger operations. A useful, non-doctrinal translation of the 'overuse[d]' term is 'an amalgam of lesser activities combined to minimize surprise and manage uncertainty by leveraging the capabilities and assets at the disposal of SOF to shape the environment. The "environment" that is being shaped ... includes both the physical environment and "human terrain"'.[120] The concern within the United States over clandestine preparation activities, particularly among members of Congress based on a perceived need to keep military and intelligence activities separate functionally and legally, is inordinate blurring or 'convergence' that has occurred, leading to problems of oversight and control.[121]

Covert and clandestine special operations can be conducted in support of each other, in combination with overt activities, or as a prelude to them. In addition, they can be conducted singly or together. The Israeli Mossad, for example, has conducted different covert actions against Syrian and Iranian nuclear programmes, some individually and others together. Although the strike against the Al Kibar nuclear site in September 2007, Operation Out of the Box or Orchard, was conducted by the Israeli Air Force, its precision was enhanced by a clandestinely inserted team of *Shaldag* or *Sayeret Matkal* operators.[122] In addition, Mossad is alleged to have engaged in a campaign targeting first Iraqi and later Iranian scientists, in the latter case employing remotely operated and controlled machine guns.[123] In theory, activities preparing the environment during competition should smooth the way for subsequent actions during conflict. Russian and Chinese analysts have suggested this pattern is characteristic of modern conflict. The first step of the so-called 'Gerasimov doctrine' in

[119] 'Nord Stream Could Be "False Flag" Operation: Germany', Yahoo News (8 March 2023); C.J. Chivers, 'Explosions in Russia Cut Gas Pipelines to Georgia', *The New York Times* (22 January 2006), www.nytimes.com/2006/01/22/international/europe/explosions-in-russia-cut-gas-pipelines-to-georgia.html.

[120] Joshua Kuyers, '"Operational Preparation of the Environment": "Intelligence Activity" or "Covert Action" by Any Other Name?', *National Security Law Brief*, 4:1 (2013): 21, 24.

[121] Robert Chesney, 'Military-Intelligence Convergence and the Law of the Title 10/Title 50 Debate', *Journal of National Security Law and Policy*, 5 (2012): 578–579.

[122] Uzi Mahnaimi, 'Snatched: Israeli Commandos "Nuclear" Raid', *Sunday Times* (18 November 2007); Yaakov Katz, 'Israel's Covert Mission to Destroy a Secret Syrian Reactor', *Tablet* (15 May 2019), www.tabletmag.com/sections/israel-middle-east/articles/israel-mission-syrian-nuclear-reactor.

[123] Ronen Bergman and Farnaz Fassihi, 'The Scientist and the A.I.-Assisted, Remote-Control Killing Machine', *The New York Times* (18 September 2021), www.nytimes.com/2021/09/18/world/middleeast/iran-nuclear-fakhrizadeh-assassination-israel.html.

transitioning from political to military action is a covert action aimed at 'The transformation of differences into contradictions and their recognition by the political military leadership', leading to tension, crisis, localized conflict, and its swift resolution.[124] Among the changing 'rules of war' is the use of 'military means of a concealed character, including carrying out actions of informational conflict and the actions of special operations forces'.[125] Some Chinese analysts stress 'covertness' as one of four central attributes of special operations.[126]

Coordinating covert and clandestine special operations presents its own challenges, given alternative control procedures and information-sharing limitations. Keeping operations covert or clandestine requires limiting access to information about them strictly to those who need to know, as well as special access procedures. Host nation police and others may have little to no knowledge of clandestine activities in their areas, as may political leaders. Such was the case during the Basra prison incident in September 2005, when two SAS operators, conducting clandestine surveillance and tradecraft better suited to Belfast than Iraq, were captured and detained by Iraqi police forces.[127] The intercession of a combined special operations and conventional forces task force storming the prison in which they were held, in violation of orders from the Permanent Joint Headquarters in Northwood, caused numerous ripples politically in Iraq as well as the United Kingdom.[128] The Raymond Davis incident in Pakistan in January 2011 highlights the political fallout that can occur as a result of clandestine activities being revealed. A retired operator, Davis was working as a CIA contractor doing a routine reconnaissance when he was in the process of being either robbed or kidnapped in heavy traffic in Lahore. Davis responded to what he perceived as a threat by killing his two would-be assailants.[129] His seizure, detention, and release by Pakistani police 'set US-Pakistan relations back years' and perhaps contributed to the decision of Obama Administration officials, including CIA director Leon Panetta, not to provide their Pakistani counterparts with advanced notification of their decision to conduct a raid in Pakistan targeting Osama bin Laden 10 months later.[130]

[124] Valeri Gerasimov, 'The Value of Science Is in the Foresight: New Challenges Demand Rethinking the Forms and Methods of Carrying out Combat Operations', *Military Review* (January–February 2016): 28.

[125] Gerasimov, 'The Value of Science Is in the Foresight', p. 24.

[126] Ming, *Lectures on the Science of Special Operations*, p. 96.

[127] Mark Urban, *Task Force Black: The Explosive True Story of the Secret Special Forces War in Iraq* (London: St Martin's Griffin, 2012), pp. 95–96; 99–106.

[128] 'SAS Hero Was Rescued as He Was about to Be Shot', *The Scotsman* (1 May 2009), www.scotsman.com/news/sas-hero-was-rescued-he-was-about-be-shot-2443884; Stavros Atlamazoglou, 'How British Special Operators Defied Their Bosses to Rescue 2 Comrades Captured on a Secretive Mission in Iraq', Business Insider (21 July 2021), www.businessinsider.com/british-sas-troops-defied-orders-to-save-comrades-basra-iraq-2021-7.

[129] Mark Mazzetti, *The Way of the Knife: The CIA, a Secret Army, and a War at the Ends of the Earth* (New York: Penguin, 2013), pp. 261–265; 272–290.

[130] Ryan Parry and Emma Forst, 'Exclusive: CIA Contractor Who Shot Two Pakistani Robbers Then Feared Death at the Hand of Mob Breaks Silence to Tell of Real-Life Homeland Plot Which Became Diplomatic Crisis', *The Daily Mail* (30 June 2017), www.dailymail.co.uk/news/article-4652448/CIA-agent-speaks-held-prisoner-Pakistan.html.

Ideally, the bearing and calibration of special operations would entirely reflect a leader's preferences and choices. While there is choice in how the means of special operations are employed in different ways, the dynamic imposed by Clausewitz's interaction of opposites across different levels limits options, influences choices, and denies certain possibilities. These possibilities include strategical paralysis in its most recently articulated forms, reflexive control, and anaphylaxis, which suggest activities of special operations can occur in such a manner as to either paralyze an opponent cognitively in advance by manipulating their perceptions or 'disrupt what Clausewitz described as the trinity between the government, the people, and the armed forces'.[131] Individual context, or rather its seven different forms that combine into a collective identity, shapes what is possible within a polity. These seven forms include the socio-cultural, economic, technological, military, geographical, historical, and political.[132] Canadian concerns over its country's political tradition, historical reputation as an honest broker and good global citizen, geographic location, and the like conspire to limit its choices for the bearing to be defensive strategically and operationally and the calibration of its SOF to be overt. Clandestine activities, while possible, heighten political risks internally as well as externally, should they be revealed. Covert action is simply beyond the Canadian scope in the current context unless the stakes and threat are incredibly high, as they were during the Second World War.

## Two trinities: merging horizontal with vertical integration

The discussion of integration relative to the two primary axes, horizontal and vertical, is a necessary theoretical foundation but insufficient on its own to consist of a general theory of special operations. Capabilities, the horizontal axis, provides a theory of the means of special operations. Control and risk, the vertical, describes how the ways of special operations influence a specific, limited end: guaranteed military subordination of special operations to civil 'work' requirements. What is missing is how the capabilities connect with control and risk, as well as the historical special operations functions identified in Chapter 1. Connecting these together, and introducing how the four dynamic elements of competition and conflict impact them, establishes a general theory of special operations.

Clausewitz logically derived his trinity to capture the elements for which any theory of war should account. These included primordial violence, an expression of the passions of the people, chance embodied in the commander and their interactions with the military instrument, and reason, or the rational, instrument purpose for which violence is employed.[133] The three exist in dynamic tension. A theory, in

[131] Alastair Finlan, 'A Dangerous Pathway? Toward a Theory of Special Forces', *Comparative Strategy*, 38:4 (Summer 2019): 265. For a primer on Russian theories of reflexive control, see Timothy Thomas, 'Russia's Reflexive Control Theory and the Military', *Journal of Slavic Military Studies*, 17 (2004): 237–256.

[132] Gray, *The Strategy Bridge*, pp. 38–41.

[133] Thomas Waldman, *War, Clausewitz, and the Trinity* (Burlington, VT: Ashgate, 2013), pp. 6–7.

Clausewitz's estimation, 'maintains a balance between these three tendencies, like an object suspended between three magnets'.[134] The theory presented below addresses these three tendencies in different ways. Risks and stakes reflect the amount of effort decision makers and a polity are going to exert in the face of different threats and contingencies that arise as part of competition and conflict. The ongoing interplay of control, political and military, with capabilities is the realm of chance.

This general theory of special operations starts with an initial trinity of its own based on the instrumental, institutional level of execution. The first point of the trinity is SOF capabilities, in their collective, institutional sense outlined previously. These can be military or paramilitary in nature, ensuring that different national organizational constructs can be included. American SOF make a sharp division based on legal codification between military (Title 10) and paramilitary (Title 50), as well as law enforcement (Title 22) special operations. In other national contexts, such as the United Kingdom or France, the convergence between them or blurring of legal lines that so worries some observers is a matter of preference, policy, or both. One of the primary challenges for Americans in deciphering British special operations is the intentional blurring between military and intelligence activities, which has as much to do with historical tradition as it does necessity based on limited resources.[135]

Capabilities are connected to a second vertex of this conceptual trinity, the functions that SOF perform. These enduring functions from Chapter 2 on historical precedents are intelligence gathering and reconnaissance, treachery and subversion, direct and indirect shock, military assistance and advising, and asymmetric response. The last vertex of this notional theoretical triangle is the effects that SOF and special operations generate. At the centre of this trinity is effectiveness. Effectiveness derives from the interplay between capabilities, functions, and effects. SOF capabilities insufficiently integrated to perform their assigned functions, or generate sufficient effects, are ineffective. Introducing the dynamic elements of opponents, allies, and partners, and bearing and calibration as influences on this trinity, impacts special operations effectiveness. To a greater or lesser extent, the first two dynamic elements shape whether SOF are used overtly, covertly, or clandestinely as well as whether special operations will spearhead efforts against opponents and in support of allies and partners, and in turn act independently or be supported by conventional military forces and other government agencies. Alternatively, they can shape whether SOF will support these forces and agencies.

Effectiveness is a product of capabilities, functions, and effects and a reflection of their specific context. The scope of the functions and capabilities of American SOF, and the effects they can generate, is staggering to SOF from partner nations. Given their breadth of functions and range of American interests, however, they are not necessarily effective. SOF within smaller countries, focused on specific functions generating specific effects, may in fact be more highly effective. Belgium's military

[134] Clausewitz, *On War*, p. 89.

[135] Cormac, *Disrupt and Deny*, p. 9; Ken Connor, *Ghost Force: The Secret History of the SAS* (London: Weidenfeld & Nicolson, 1998), pp. 334–335; 341; 347.

and police SOF, including its Special Forces Group, Paracommando Brigade, and *Escadron Spécial d'Intervention*, or *Le Group Diane*, are sufficient to conduct its assigned functions: intelligence gathering and reconnaissance, military advising and assistance, and asymmetric response to terrorism domestically and abroad, as well as direct and indirect shock overseas.[136] They stand in contrast to the capabilities of the Bahamian SOF, whose personnel comprise a Commando squadron and Drug Enforcement Unit, whose mandates are expansive and overwhelm their numbers. The functions they are assigned include intelligence gathering and reconnaissance (maritime surveillance) as well as asymmetric response in the form of narcotics and immigration interdiction. Their capabilities are insufficient to meet all of these requirements, even when augmented by unmanned aerial systems and integrated with other Bahamian military and police functions, Caribbean Community, and other forces from other nations, including the United States.[137]

While effectiveness is not quantifiable, it can be determined, like horizontal and vertical integration, qualitatively over time. A qualitative judgement can be formed by observing several criteria aside from those mentioned above. In addition, the sufficiency of SOF capabilities relative to functions and effectiveness can be determined by looking at three other criteria: the suitability of capabilities against threats and contingencies, their adaptability or innovation in the face of the unexpected and unanticipated related to them, and their responsiveness or ability to anticipate changes related to threats and contingencies and prepare to meet them. Conceptually, this theoretical trinity, with effectiveness at its core, avoids a common error in professional military education of interpreting Clausewitz's construct heuristically: seeking an idealized form in which the core is balanced statically and the vertices equal, as in an equilateral triangle. Balance is relative to the size and shape of a triangle, and it can be disturbed, or oscillate, under change or dynamic pressure. This construct, subject to these observable and reportable changes over a period, is the basis for the interaction of special operations *means* and *ways*. This portion of the theory is depicted in Figure 5.1.

This conceptual trinity of special operations means and ways does not invalidate previous theoretical enquiries. William McRaven's theory of relative superiority nests within it as an explanation for SOF outcomes tactically in two specific functions, direct or indirect shock and asymmetric response. Success or failure determines the effects they generate as well as their effectiveness. Robert Spulak's theory of SOF also relates to effectiveness. He evaluates the link between SOF attributes, selected and trained, and how these specifically generate effects. Their effectiveness, he would

[136] Ewen Southby-Taylour, ed., *Jane's Amphibious and Special Forces*, 17 (Coulsdon: Jane's Information Group, January 2007), p. 59. See also 'Primary Tasks of the Unit', Special Forces Group website, n.d., www.sfg.be/page/the-primary-tasks-of-the-unit.

[137] Royal Bahamas Defence Force, 'Benefits of the Bahamas Unmanned Aerial System, *The Bahamas Weekly* (31 December 2019), www.thebahamasweekly.com/publish/local/Benefits_of_The_Bahamas_Unmanned_Aerial_System_printer.shtml; Southby-Taylour, *Jane's Amphibious and Special Forces*, p. 47; 'RBDF Commando Squadron Infantry Training', Royal Bahamian Defence Force blog (2 September 2022), http://rbdf.gov.bs/rbdf-commando-squadron-infantry-training.

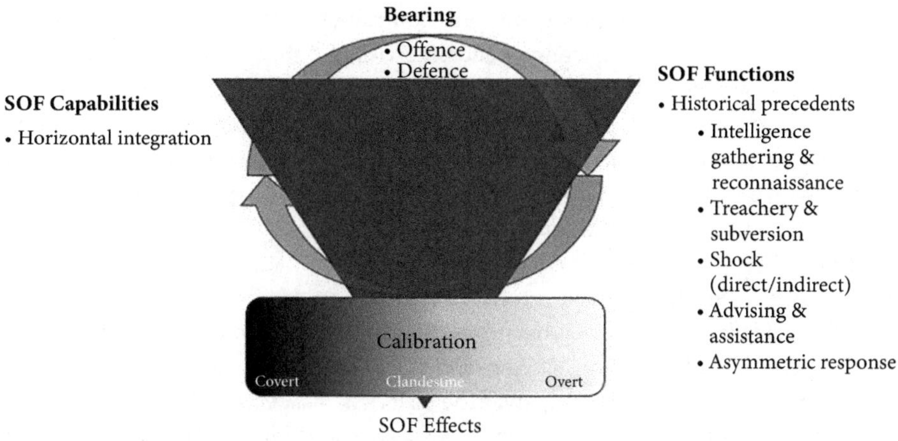


**Figure 5.1** Special operations effectiveness trinity.
*Source:* Author's work.

argue, results from these attributes, which include creativity and flexibility, by overcoming friction. Work in a similar vein that explores SOF adaptation and innovation, individually and collectively, has been explored in theoretical terms but often in an exclusively national context by Richard Shultz Jr, Tone Danielsen, Rikke Haugegaard, Rebecca Jensen, and others.[138] While context specific, they all relate to and are compatible with special operations effectiveness, particularly its criteria of adaptiveness and innovation.

The theory of special operations coherence also subsumes the general theory of special operations offered by Tom Searle. Searle's conception of special operations capabilities and functions being 'outside of the box' of conventional operations performs a useful but limited function.[139] While it separates and differentiates and can therefore be included in a broader general theory, it fails to perform a crucial function of theory: explaining how special operations capabilities integrate within other overlapping departments and agencies, much less when they are most effective and when they are not. His specific focus on the American authorities and organizations also limits its generalizability. These same shortcomings are also evident in Harry Yarger's theoretical foray. While identifying twenty-six premises and fourteen principles useful in considering the subject of special operations, some are specific to the United

[138] Richard Shultz Jr, *Military Innovation in War: It Takes a Learning Organization—A Case Study of Task Force 714 in Iraq*, JSOU Report 16-6 (MacDill Air Force Base, FL: JSOU Press, 2016); Tone Danielsen, *Making Warriors in a Global Era: An Ethnographic Study of the Norwegian Naval Special Operations Commando* (Lanham, MD: Lexington Books, 2018); Rikke Haugegaard, Karina Mayland, and Allan Shapiro, 'Thinkers at the Cutting Edge: Innovation in the Danish Special Forces', in Gitte Hjostrup Christensen, ed., *Interdisciplinary Perspectives on Special Operations Forces*, Conference Proceedings No. 4 (a) (Copenhagen: Royal Danish Defence College, 2017), pp. 88–107; Rebecca Jensen, 'The Diffusion of Innovation: SOF in Iraq', in Jessica Turnley, Kobi Michael, and Eyal Ben-Ari, eds., *Special Operations Forces in the 21st Century: Perspectives from the Social Sciences* (London: Routledge, 2018), pp. 226–240.

[139] Tom Searle, *Outside the Box: A New General Theory of Special Operations*, JSOU Report 17-4 (MacDill Air Force Base, FL: JSOU Press, 2017), pp. 17–21.

States and a few, as Searle points out, are contradictory.[140] Instead of invalidating these previous theoretical or conceptual enquiries, the theory of special operations coherence is capable of accommodating each of their ideas within its framework.

The second trinity, connected to the first, has goals or desired outcomes at its apex. Although some might prefer the term 'end state' here instead, the ones selected here were done carefully. The value of this general theory lies in explaining the success and failure of special operations contributing to them over time. While tactical missions and campaigns should have clear and precise end states, when one moves to the level of strategy, they become more elusive and opaque. End states connote finality; as Clausewitz reminds the unwary military officer, 'the outcome of a war is not always to be regarded as final'.[141] Consider, for example, the strategic end state in Afghanistan and against terrorism over the past two decades. Goals and desired outcomes strategically change over time, as successes and failures accumulate, which puts pressure on SOF but also provides them with the opportunity to exceed the mechanisms designed to control them discussed in Chapter 4. This is especially true in competition between opponents, as opposed to more direct conflict with them, as continuing successes and low costs make special operations an appealing option, which opens the door for increased influence and prestige.

The second vertex of this additional trinity is a direct result of perceived threats against which special operations will be used. The nature of the opponent and an alliance with partners and proxies shape decision makers' understanding of two particular aspects related to threat: stakes and contingencies. How threatening that opponent is, or how valuable the alliance or partnership, drives a perception of the stakes in play. The value of the object will determine how much effort will be used in what particular manner. Competing with the PRC and Russia is important to the United States, given the stakes at risk, but they are not enough to escalate to conflict given the consequences that might occur. One reason why some voices emphasize the importance of competition between the United States and the PRC is the ultimate prize or stake: leadership and control over the global economic and normative system, with all the benefits it provides. As John Ikenberry summarizes, the United States has managed 'to organize and maintain an open economic system, providing security and access to markets for other states while enjoying a steady stream of economic benefits for itself'.[142] For the US, supporting Ukraine as partner or proxy against Russia realizes several objectives, each with their own stakes. These include imposing costs on Russia, weakening it as a potential threat, enhancing the credibility of the US commitment to its NATO allies (particularly those that border Ukraine and fear a Russian return), and indirect conflict with Russia through a partner to limit its escalation, among others.

[140] Yarger, *21st Century SOF*, pp. 47–67; Searle, *Outside the Box*, pp. 53–55.

[141] Clausewitz, *On War*, p. 80.

[142] G. John Ikenberry, Review of *America's Global Advantage: U.S. Hegemony and International Cooperation* by Carla Norrlof, Cambridge University Press, 2010, *Foreign Affairs* (September/October 2010), www.foreignaffairs.com/reviews/capsule-review/2010-09-01/americas-global-advantage-us-hegemony-and-international.

Defined threats and ongoing competition can lead to crises short of conflict that nevertheless constitute contingencies requiring a response. These contingencies, such as the invasion of a partner or a regional crisis, are an acknowledgement within the theory of the central idea of the paradoxical logic: opponents can think, are reactive, and have a high degree of agency. Each of the case study chapters features different contingencies, including the Falklands and First Gulf War for Great Britain, al-Qaeda's global operations and regional operations recurrently in Iraq and later in Syria for the United States, and South Africa's activities in Angola and Mozambique. Countries can take actions which force the need to respond or back down, which can threaten reputation and influence. States seek continuing advantage dictated as much by their goals as their opponent's, reflected in their strategy. In Everett Dolman's words, actors seek to continue to compete 'because continuation not culmination is the goal'.[143] The exception to Dolman's observation, however, is the case where the stakes are so high, or the consequences so marginal, that annihilating an opponent, or demanding their submission, is preferable. The twenty-year period for the case studies illustrates efforts in using special operations for the purposes of continuation or completion. For special operations specifically, pursuing continuing national political objectives can put considerable pressure on SOF capabilities in several ways. As the case studies suggest, SOF are appealing given they deliver results at much lower costs in terms of resources than other available options. SOF are happy to reciprocate as they prefer to be used and useful, justifying the investment in their capabilities. Sustaining SOF capabilities over time, in the face of different contingencies with varying stakes, can be challenging for these reasons.

Completing this trinity is risk, discussed at length in the preceding chapter. Risk is pervasive; it is attached to all human activity, including competition and conflict. Threats, real or potential, drive calculations of risk. These calculations, like time, manifest themselves and are understood and communicated differently across the different levels of organization. The primary focus of this general theory of special operations lies in the interaction between the political, strategic, and operational level, and the effects resulting from the expansion or contraction that occurs as this system responds to different inputs, including external pressures. Risk miscommunication and misunderstanding open the potential for SOF to respond in a manner consistent with policy. Alternatively, they provide SOF with the opportunity to go beyond the limits of policy. The previous chapter identified these operating modes as acceding to or exceeding policy. Conformity, just as much independence, can threaten outcomes.

At the centre of this second trinity is control. Some form of control over SOF is necessary to ensure outcomes are achieved, goals are reached, and they do not become a threat to the regime or political leadership. Control can occur in different modes as well as several forms. These forms include delegative, directive, collaborative, competitive, and co-optive. The primary differences between them lie in the nature of the relationship between the political principal and the collective special operations

[143] Dolman, *Pure Strategy*, p. 4

agent, as well as the primary level at which control over them occurs. This level can be political, strategic, or operational. The stakes that are involved, contingencies that arise, threats addressed, and risks perceived create a spectrum of control integration ranging from tight, or high, to loose and distributed to military organizations or even operational commands. Control is the critical connection between the *ends* being pursued to *means* through the *ways* they are being employed, with *risk* shaping all three in the process. Figure 5.2 depicts this trinity centred around control.

Connecting the concept of effectiveness with control results in a general theory of special operations coherence, depicted in Figure 5.3. The subordinate elements and dynamic influences of control and effectiveness result in a theory that attempts to explain the coherence of special operations. Coherence occurs in three ways: special operations development, use, and sustainment. As a result, the theory of special operations coherence moves beyond previous ones focused on capabilities that define or categorize but do not explain, as well as those that insist that their creation must inevitably result in special operations misuse or destruction in the cycle discussed in Chapter 3. In addition, the theory of special operations coherence incorporates, or connects to, theories of civil–military relations, broadening them beyond the dominant American or Western democratic states focus by defining a range of control possibilities that account for less transparent, representative forms of government, including autocracies. Of particular interest to civil–military relations scholars is the degree to which special operations pose a threat to the systems that create them. Much of this work assumes special operations are anti-systemic, in both its positive and negative forms. This notion, first identified by Anton Asklund Johnsen and Gitte Højstrup Christensen in 2016, was limited to a discussion of SOF ability to move beyond prevailing boundaries and norms within the system that created them.[144] Here it is expanded, revealing a duality. SOF can serve as a form of responsive anti-systemic antibody (at best) as well as a contagion (at worst) within their own system

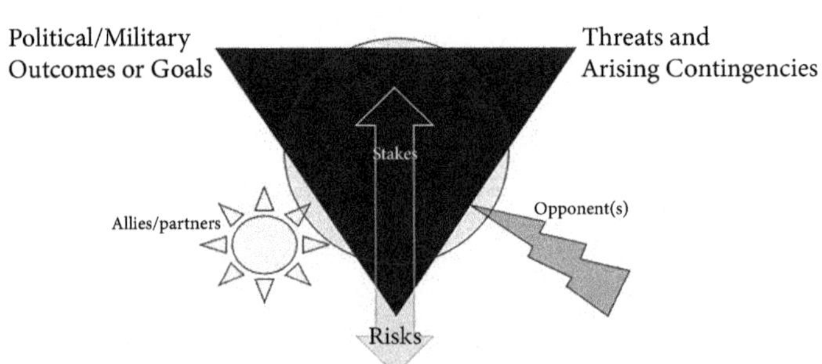


**Figure 5.2** Special operations ends trinity.
*Source:* Author's work.

[144] Anton Asklund Johnsen and Gitte Højstrup Christensen, 'Clarifying the Antisystemic Elements of Special Operations: A Conceptual Inquiry', *Special Operations Journal*, 2:2 (Fall 2016): 106–123.

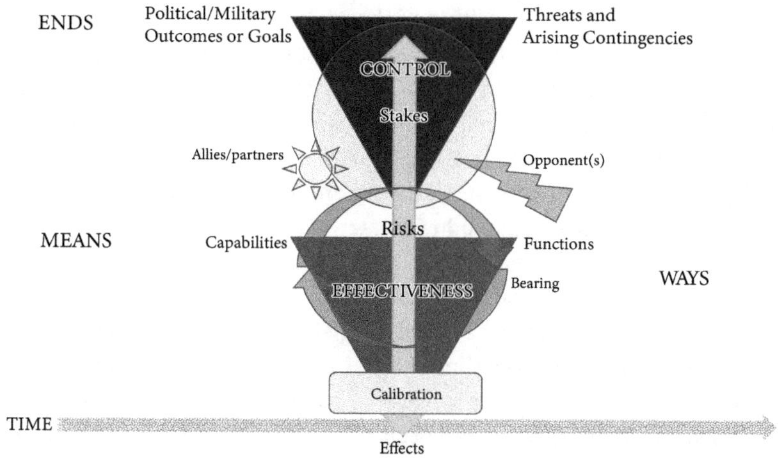


**Figure 5.3** Theory of special operations control and effectiveness.
*Source:* Author's work.

or against that of an opponent. In this way, Johnsen and Christensen's ideas connect with those contained within Alastair Finlan's theory of special operations use against an opponent, as well as Anthony King's observations about the rising influence of British SOF politically in the domestic context.[145]

In moving the understanding of special operations beyond these specific conceptual or agenda-filled silos, this general theory fulfils its primary purpose. Theory rarely informs policy, but it can serve an educational function. If this general theory helps inform the dialogue between decision and policy maker, what American SOF have called characteristically 'educating the consumer', it will have served an additional purpose. To assist in that educational process the range of concepts within this general theory connect to and account for Art Lykke's influential and pervasive model of strategy: the relationship that exists between ends, ways, means, and risk.[146] The ends equate to the desired outcomes or goals, against various threats or for different contingencies. The means are the product of special operations capabilities and the functions SOF are asked to fulfil. Finally, the ways reflect how those means are utilized, specifically the extent to which the bearing and calibration of special operations capabilities contribute to those goals and outcomes—how effective they are. Balancing them over time is the key to special operations success, this theory holds, which is challenged by changing capabilities and contingencies, shifting political goals and desired outcomes, and greater or lesser threats and different contingencies.

[145] Finlan, *passim*; Anthony King, 'The Special Air Service and the Concentration of Military Power', *Armed Forces and Society*, 35:4 (July 2009): 646–666.

[146] For two interestingly similar yet different accounts of this model by his former students and later peers, see Gary Guertner, 'Why Is Strategy Difficult?', in Gary Guertner, ed., *The Search for Strategy: Politics and Strategic Vision* (Westport, CT: Greenwood Press, 1993), chapter 1; H. Richard Yarger, 'Toward a Theory of Strategy: Art Lykke and the U.S. Army War College Strategy Model', in J. Boone Bartholomees Jr, ed., *U.S. Army War College Guide to National Security Issues, Volume I: Theory of War and Strategy* (Carlisle Barracks, PA: Strategic Studies Institute, 2012), chapter 3.

Lykke's framework, pervasive within the special operations community, allows for a common starting point for dialogue and education by articulating and mapping out the relationships that exist in special operations sustainment and use. There is a degree of distrust between special operators and others, which includes their conventional peers and civilian decision makers, as much a result of what they do as it is who they are culturally. This general theory and its elements incorporate issues of trust, agency, and control discussed in Chapter 4 up and down the different levels of command: strategic, operational, and tactical.

This general theory serves another specific purpose for the special operations community, providing a framework for analysing special operations results, from the operational to the political over time. Rather than merely describing the primary challenges for special operators or suggesting converting their activities into strategic effects in vague theoretical terms, special operators have in this theory a means to understand the points at which frictions or pressures threaten to undercut the ability of special operations to accomplish goals and objectives operationally or strategically. The British use of special operations outlined in Chapter 6, for example, suggests loose control mechanisms politically and operationally worked perfectly well in limited situations where political leaders sought to limit costs and risks. The pressures and demands of costlier, higher-risk activities in a conventional war, however, tested these mechanisms and led to overuse and under-delivery of results. Alternatively, the theory of special operations coherence allows those seeking to understand the history of special operations various tools to reinterpret organizations, operations, and outcomes, as the case study chapters suggest.

The theory of special operations coherence accounts for critical factors and develops links that others have not. First, it accounts for special operations roles and missions and overcomes the primary barrier to evaluating their effectiveness: their different approaches (direct and indirect) and outcome time frames (short versus long term). To escape issues of organizational ownership and institutional identify and break free from the dogma that doctrine sometimes imposes, this theory recasts special operations roles with reference to the observable past. Second, a theory of special operations must address and explain how they occur across the completion–conflict continuum. It is worth pointing out that the dividing line on this continuum for when competition becomes conflict is increasingly difficult to determine, a point some Russian and Chinese authors stress as well. Third, the stakes at risk in any activities along the competition and conflict continuum shape control mechanisms, including authorities and rules of engagement, elements included within this framework. Fourth, this theory frames SOF capabilities as the sum of individual units working together and accounts for their independent use as well as their integration with military and other government resources. After all, while special operations are primarily enabling activities, they are capable of independent actions. This theory accounts for this apparent conundrum by linking capabilities and their functions to the stakes involved, their control, and operational bearing and calibration. Fifth, this theory accounts for shifts that can occur between the tactical, operational, strategic, and political levels when it comes to control mechanisms and risk calculations and communication. Sixth, this theory of special operations accounts for the different orientation of special operations, in overall terms such as offence and defence, as well as their operational one: overt, covert, and clandestine. Seventh, competition and conflict are interactive and adversarial in nature and this theory accounts for several dynamic factors, including the actions of one or more opponents, allies, and partners.

Harold Winton concludes theory has five tasks or functions to fulfil: defining, categorizing, explaining, connecting, and anticipating.[147] The theory of special operations coherence defines them and the capabilities required to conduct them, and sheds light on their comprehensive utility. Some elements of the theory were extracted from previous works while others were developed from a combination of observation, experience, and reflection. It categorized different special operations functions observable historically and connected them to capabilities. Other connections made explicit here are between discussions of special operations functions and capabilities and their relationship to control issues central to scholarship on civil–military relations. The critical explanations within this theory, however, are of the relationships that occur within and between the different elements of this theory, such as capability and function, risk and control, and stakes and orientation. On a more functional level, this theory links special operations to decision making, organizational behaviour, military operations, and statecraft. Winton's last task for theory is to anticipate. While this theory cannot predict outcomes, it does anticipate special operations activities in space and cyberspace, and suggests under what conditions they might be useful or not. Theories should also have a central proposition, preferably stated as clearly and concisely as possible. The central proposition of the theory of special operations coherence is that successful special operations use lies in balancing the tension between the political and military appetite for their use against SOF's own ambitions against their need to be effective and adaptive in response to circumstances over time.

What remains for this general theory of special operations is to set it in motion to see what range of possibilities it can explain. The previous chapter examined control relationships between principal and agent in the American context. What about other political systems, specifically autocracies? China, Russia, Iran, and North Korea all possess sizeable SOF. Autocracies have an antipathy towards military elites, including SOF, constituting as they do a potential revolutionary threat to power based on their capabilities. All but Iran have subordinated much of their SOF to conventional military formations. In the case of the PRC, SOF brigades are subordinated to army groups or specialized counterterrorism and domestic security forces allocated to disparate military regions. Subordination and allocation in this manner suggests tight control and specific assigned functions. Tight vertical integration comes at the expense of horizontal integration, within and between different special operations capabilities.

Where stakes are high, and in support of a conventional military invasion of Taiwan without it being able to count on outside support, PRC SOF can perform their assigned supporting functions well: intelligence gathering, reconnaissance, and

[147] Winton, 'Imperfect Jewel', p. 855.

indirect shock. The risks are likely to be limited to operational ones. Risks for the PRC should Taiwan's partners and allies become involved would increase the stakes dramatically, leading to even tighter control over SOF, to prevent unnecessary escalation, given the deliberative and dispositional risk taking exhibited by Chairman Xi Jinping. To limit time for response at increased risk and lower stakes, Chairman Xi could authorize PRC SOF to conduct a *coup de main* of the island. Unless they achieve compete surprise, PRC SOF would likely suffer heavy losses due to the control restrictions placed on them and their inability to adapt, much less reap the benefits from horizontal integration. The product of this application of the theory is purely speculative, but it suggests the potential to work through a range of different scenarios and contingencies depending on PRC choices. This general theory of special operations, when used in a heuristic manner, can also assist in analysing allied and partner SOF.

The various elements within this theoretical construct are malleable: they are intended to expand and contract, stress and strain as they interact with conditions in their environment. To illustrate these points, SOF capabilities can be limited in size and integration. Provided their functions are limited, they face few significant threats or unexceptional contingencies, have little need to operate covertly, and are used in pursuit of modest political goals, few difficulties will arise (beyond the normal tensions). Should those limited SOF be used for multiple functions against more daunting threats in a constrained and socially driven risk-taking environment, with unclearly specified outcomes, the result is likely to be considerable tension within the system and eventual institutional collapse—or an attempted coup, depending on the type of leadership and government. The animating elements, particularly when time is combined with SOF (in)effectiveness, especially during adaptation, suggest that the dynamic tension within the system employing SOF can result in strain or compression if not managed adequately politically and militarily. The central problem is understanding the conditions sufficiently to know when one is in danger of strain or compression and having the strength of conviction to alter a chosen course of action. A general theory can aid the former but not the latter.

Determining the explanatory value of this model in its abstract form requires 'testing' it empirically. To avoid problems of looking at individual episodes in isolation and ripping events out of their context, the case studies that follow have a common frame of reference: a twenty-year period. This duration allows for the inclusion of a range of different dimensions, including changes in leadership or administration as well as policy in democracies, competition and conflict involving a range of opponents and allies state and non-state, conducted over various stakes, and employing SOF capabilities in different functions, sometimes simultaneously. They include actions of various form and duration, occasionally overlapping, that allow the incorporation of different aspects of the theory. Depending on what information is available, the case studies suggest that the theory of special operations coherence is useful in providing meaningful insights about capabilities and control, across successes and failures, when they have used SOF over an extended period.

# PART III

# 6
# Stretched beyond Footholds
## The United Kingdom, 1971–1991

### Introduction

This chapter assesses British special operations from 1971 to 1991. Structurally, it does so differently than the subsequent chapter on South Africa for a simple reason. Unlike the South African case, where the full spectrum of special operations activities could be examined in depth thanks to the collapse of the apartheid regime and a subsequent release of information, the details of British special operations in the more recent past remain elusive and uneven. In this regard it shares more in common with the United States case study in Chapter 7, in terms of unevenness of information, for different reasons. For example, much is available to the scholar on the use of British special forces[1] in the Falklands War (1982), ranging from personal accounts by Special Air Service (SAS) commanders down to the level of individual troopers. The British official history of the campaign mentions or discusses special forces and their role and impact on the campaign. Surprisingly, some specific records, including operations orders, have been declassified and are available for public review at the National Archives at Kew.

In contrast, little official is available on the use of British special operations during the lengthy period of 'The Troubles' in Northern Ireland or during the First Gulf War (1990–1991) in Kuwait and Iraq. A series of memoirs related to the latter led Her Majesty's Government to take the unprecedented step of instituting an enhanced form of gag order in 1996 to prevent further releases, an approach relatively but not entirely successful. Sensitivities relating to 'the dirty war' in Northern Ireland, political, institutional, and individual, keep records related to special operations valuable for this enquiry firmly behind the wall of Section 3.4 of the Public Records Act protection on the basis of 'security or other specified grounds'. These including Ministry of Defence periodic and technical reporting, as well as Cabinet Office and specific memos and papers of the prime minister.[2] Other potential avenues of information previously available to the author, such as the SAS Regimental Association, are closed

[1] In keeping with the convention nationally in the United Kingdom, 'special forces' here is the collective noun for all British units where special operations forces (SOF) is used elsewhere.

[2] For example, see the listings for DEFE 24/3037, 'Northern Ireland' (1971–1983) and PREM 15/1027, 'Special Operations in Northern Ireland' (1971–1972) on Discovery, the National Archives' online catalogue, https://discovery.nationalarchives.gov.uk/results/r?_ps=60&_q=%22special+operations%22+AND+%22Northern+Ireland%22.

*Special Operations Success*. James D. Kiras, Oxford University Press. © James D. Kiras (2024).
DOI: 10.1093/9780198902096.003.0006

as a series of inquests are underway surrounding the conduct and potential excesses of British special forces during their operations in Afghanistan.[3]

Given significant gaps in useful specifics to the researcher, this chapter employs aspects of the theory of special operations coherence as a framework for analysis. After a brief background that outlines the development and use of British special operations up to 1971, the chapter explores British capabilities and control mechanisms for special operations. The former are relatively well known, thanks to a number of sanctioned and unofficial histories and journalistic accounts. Control mechanisms are somewhat more difficult to chart, thanks to the opaque, relational nature of British government. Existing literature on British civil–military relations sheds valuable light on them and the overall system of approval and monitoring. The remainder of the chapter, the case study analysis, proceeds chronologically and is divided in overlapping time periods as follows: the transition from operations overseas in Oman to Northern Ireland from 1971 until 1977, the period from 1975 until 1981 that saw the internally driven development a counterterrorism capability whose use shot the SAS into fame and recognition, the 1982 Falklands War, the return in focus to Northern Ireland from 1982 through to 1991, and the First Gulf War in several short weeks from 1990 until 1991. This case study highlights several aspects of special operations central to the theory absent in other cases, including the United States. These include shifting from overseas to domestic internal missions, the identification and filling of capability shortfalls, the move from low-intensity activities at home to high-intensity operations overseas not once but twice, the conduct of special operations in unilateral as well as coalition uses of force, a change in activities from supporting to supported activities in terms of command relationships, and the creation of the first post-Second World War headquarters ostensibly to control special operations.

This inquiry serves several purposes related to the theory of special operations. First, it interprets a series of well-known, and in certain cases infamous, episodes in British special operations history to derive novel and unique insights about them. Doing so through the framework of the theory is designed to overcome the substantial issues surrounding British special forces and their sources of information identified by Alastair Finlan: 'by stifling informed debate it encourages ignorance (myths replace reality) that can affect, not just the general public, but more worryingly, conventional soldiers who may end up working with/commanding these units in war'.[4] Second, the overall period overlaps with that of the South African case study, with which it shares a number of similarities yet there are also significant differences. The South African case study looks at special operations primarily from the perspective of historical mission sets and their bearing and orientation, including

[3] Dan Sabbagh, 'SAS Had "Gone Too Far" in Afghanistan, Inquiry into Claims of Unlawful Killings Told', *The Guardian* (15 October 2023), www.theguardian.com/uk-news/2023/oct/15/sas-had-gone-too-far-in-afghanistan-inquiry-into-claims-of-unlawful-killings-told; Hannah O'Grady and Joel Gunter, 'SAS Unit Repeatedly Killed Afghan Detainees, BBC finds', BBC News, 12 July 2022, www.bbc.co.uk/news/uk-62083196.

[4] Alastair Finlan, 'The (Arrested) Development of UK Special Forces and the Global War on Terror', *Review of International Studies*, 35 (2009): 971.

whether they were conducted overtly, covertly, or clandestinely. This case study, in contrast, looks at British special operations chronologically. While their similarities and differences are compared in Chapter 9, the same time period can be examined using different approaches of the theory of special operations to arrive at the same destination: an evaluation of horizontal and vertical integration and how the system of control and capabilities responds to various changes over time. Third and finally, the British case study from 1971 until 1991 assesses, through the lenses of the theory, a medium power's employment, use, and expansion of capabilities in circumstances of global power projection, unilaterally and as a coalition partner, unlike the South African case, which was limited to domestic, local, and regional use. To set the context for British special operations capabilities in 1971, the next section looks at their historical creation, expansion, disbandment, and use from 1940 onwards.

## Background

British military and paramilitary special operations have their modern roots in the Second World War. The special operations took the form of an eventual constellation of 'private armies' by war's end, conducting activities along two primary lines. The first, and most numerous, were the range of British Army and Royal Navy and Air Force overt military units. These units, the brainchildren of various military entrepreneurs, ranged in scale and impact from the small Sea Reconnaissance Unit to the brigade of operators from the SAS, composed of two regiments, which took part in the invasion and liberation of Europe. For clandestine paramilitary operations, its various functions were split between the Secret Intelligence Service (SIS), the Special Operations Executive (SOE), the Ministry of Economic Warfare (MEW), and the Political Warfare Executive (PWE). SIS primarily undertook intelligence-gathering missions in Nazi-occupied Europe and elsewhere. SOE, which also conducted intelligence gathering and reconnaissance missions, focused on providing military advice and assistance, as well as conducting treachery and subversion and indirect-shock missions using clandestine methods such as industrial or military sabotage.

Wartime exigencies, particularly given the desperate circumstances of national survival facing the United Kingdom in 1940, brought out various experts, entrepreneurs, and misfits for special operations. As the case of T.E. Lawrence during the First World War suggests, the United Kingdom produces skilled specialists, adventurers, romantics, and eccentrics, separately or together in one person. These traits, P.A. Tobin notes, form part of 'a British characteristic, nurtured by centuries of maritime supremacy, to make use of small parties of adventurers to discomfort our enemies by attacking them in the most improbable places and by devious means'.[5] Experts such as Jock 'Blondie' Hasler, a small boating enthusiast, skilled marine

[5] P.A. Tobin, 'The Bertrand Stewart Prize Essay, 1952: British Special Forces', *Army Quarterly*, 65 (October 1952/January 1953): 161.

navigator, and equipment innovator, found a willing audience and trial for his idea of using small, collapsible canoes to conduct long-distance, clandestine insertion maritime sabotage raids in the Royal Marine Boom Patrol Detachment (RMBPD), discussed in more detail in Chapter 3. Ralph Bagnold, a noted desert explorer, navigator, and innovator before the Second World War, offered his services in North Africa to General Archibald Wavell to conduct effective reconnaissance and intelligence gathering behind Italian lines denied by air.[6] Wavell authorized the formation of a unit, the Long Range Desert Group (LRDG), to do so by circumnavigating areas of the Sahara Desert thought impassible by the Italians.[7] David Stirling, a member of the Scottish landed class as well as of the Guards Commando Regiment, used a combination of bluff and bluster to sell his concept for an airborne desert raiding force to attack *Luftwaffe* and *Regio Aeronautica* aircraft at the most vulnerable point, on the ground at airbases behind the lines.[8] Orde Wingate, the father of long-range penetration special operations, was known as much for his ideological zealotry and eccentricities as he was for his brilliance.[9] Misfit characters, whose bravery and ferocity were legendary but who were also beset by deep character flaws, included Michael 'Mad Mike' Calvert, Blair 'Paddy' Mayne, and a host of others.

British special forces, paramilitary and military, proliferated during the war to such a degree that some have questioned their utility and cost-effectiveness. Nineteen numbered Commando units were created for amphibious raiding with a mixed record of success. The value and utility of Wingate's Chindits, which grew in numbers of personnel to the size of an understrength army division, have been the subject of considerable debate since their initial creation, which continues to the present.[10] The SOE comprised around 13,000 personnel, the size of a 1939 British Army division. Its reputation, and assessments of its value, waned given its wartime competition and feuding with SIS and the British military establishment over resources and missions and subsequent disbandment almost immediately after the war.[11] Arguments against special forces included a drain on military talent away from conventional forces, underutilization or underperformance relate to the resources they consumed, and

[6] For Bagnold's pre-war explorations, see Ralph Bagnold, *Libyan Sands: Travel in a Dead World* (London: Eland, 2010, repr. 1935). For an account of Bagnold's meeting with Wavell, see Trevor Constable, 'Bagnold's Bluff', *The Journal of Historical Review*, 18:2 (March/April 1999): 6.

[7] For Owen's account, see David Lloyd Owen, *The Desert My Dwelling Place* (London: Cassell and Co., 1957).

[8] Gavin Mortimer makes a revisionist case that Bill Stirling, not David, was the instrumental driving force in bringing the 'L Detachment' concept to General Claude Auchinleck's attention. Auchinleck replaced Wavell in June 1941 after the latter lost Churchill's confidence. Gavin Mortimer, *David Stirling: The Phoney Major* (London: Constable, 2022), pp. 92–93.

[9] Simon Anglim, 'Orde Wingate and the Theory behind the Chindit Operations: Some Recent Findings', *The RUSI Journal*, 147:2 (2002): 92–98.

[10] The most recent assessment is Simon Anglim, *Orde Wingate, Unconventional Warrior: From the 1920s to the Twenty-First Century* (London: Pen & Sword, 2015).

[11] Michael Foot, 'Was SOE Any Good?', *Journal of Contemporary History*, 16:1 (January 1981): 171. Discussions about the future activities and postwar countries of interest are available in CAB 301/53, 'The Future of Special Operations Executive (S.O.E.)'. Anthony Eden, the foreign minister, raised the issue of control with the prime minister on 23 November 1944 after Lord Selbourne's resignation. FO 1093/108, 'Top Secret: Prime Minister', PM/44/716. Keeping two organizations for covert action and intelligence gathering was ultimately 'uneconomical': CAB 301/53, 'Future of SOE', undated revised draft, p. 2.

their inevitable growth and decline in skill, observations which commentators echo about modern SOF.[12] In weighing these, Andrew Hargreaves concludes that despite an uneven record, British special forces were cost-effective given their results.[13]

The dramatic rise of British paramilitary and military special forces was matched by their fall, in the form of disbandment. Several factors contributed to this but the two most significant were the defeat of Winston Churchill, whose inspiration and patronage sustained their development and growth, and the economic requirement to move from a wartime footing to peacetime recovery while reasserting control over the various colonies. These drove the need to demobilize and return to the economy those drafted into uniform.[14] The RMPBD, given its demonstrated expertise in beach reconnaissance work during the war, survived the cuts to become the basis for revived Special Boat Sections (SBS). The SAS, expanded into a brigade during the war, was pared down to a rump reserve unit, the Artists Rifles, before becoming part of the Royal Air Force in 1947 as the 21st Special Air Service Regiment (Artists Rifles). Close to 300 personnel of the SOE skilled in paramilitary special operations found a home in the SIS's Special Operations Branch after their organization dissolved early in January 1946, despite postwar plans to preserve its capabilities within the Ministry of Defence.[15] The strains of the period on British special forces are evident in the objections raised by commanders of 21 SAS Regiment (Artists) to the proposed renaming RMPBD units as 'special boat sections' over a two-year period, from 1949 through 1951.[16] They relented, becoming the Special Boat Company in 1951.[17]

Colonial needs and countering Soviet influence drove the employment of British special operations from 1945 until 1971. The needs were not evenly spread out as different requirements existed for special operations covertly and overtly. Paramilitary British special operations through covert action had a mixed record, tipped substantially towards failure. Efforts to provide military advice and advice to proxies in Ukraine (1945–1946), the Baltic States (1946–1952), Albania (1949–1952), and Indonesia (1957–1959), either Soviet satellite regimes or leaning towards the Soviet Union, were unsuccessful, despite joint efforts between the SIS and CIA in some cases. American confidence in the British paramilitary special operations capabilities was instilled not only as a result of coordination efforts during the Second World War but also because of the efforts of senior SIS leaders to depict themselves to their

[12] Jan Kallberg, 'Time to Radically Downsize the West's Special Forces', Center for European Policy Analysis, 29 November 2023, https://cepa.org/article/time-to-radically-downsize-the-wests-special-forces. Viscount Slim provides an eloquent articulation of this argument. William Slim, *Defeat into Victory: The Magnificent Account of a Great Campaign in the Second World War* (London: Cassell, 1956), pp. 455–458.

[13] Andrew Hargreaves, *Special Operations in World War II: British and American Irregular Warfare* (Norman, OK: University of Oklahoma Press, 2013), p. 269.

[14] Daniel Todman, *Britain's War: A New World, 1942–1947* (Oxford: Oxford University Press, 2020), pp. 777–832. Voices from this time period are captured admirably in David Kynaston, *Austerity Britain, 1945–1951* (London: Bloomsbury, 2008).

[15] Philip Davies, 'From Special Operations to Special Political Action: The "Rump SOE" and SIS Post-War Covert Action Capability, 1945–1977', *Intelligence and National Security*, 15:3 (2000): 55–76; HS 8/821, 'Policy and Planning: Future of SOE'.

[16] ADM 201/102, 'Use of Title "Special Boat Sections": Objections by 21 SAS Regiment (Artists) TA'.

[17] Saul David, *SBS: Silent Warriors* (London: William Collins, 2021), p. 419.

American counterparts as 'an integrated organisation for intelligence and special operations', an amalgamation of the SIS and SOE.[18] The lone successes occurred in Greece, within a larger civil war between communist and royalist forces, and almost two decades later in Yemen (1962–1968).[19]

Coordination between various administrative bodies, or integration between them, missing in a number of prior efforts, was a contributing factor towards the latter's success, according to Rory Cormac.[20] SIS capabilities to conduct paramilitary covert action were limited by small numbers of personnel, fiscal constraints, a lack of supporting assets necessary to support and sustain proxy armies, increased demands on its strategic intelligence collection and analysis functions, and significant disagreements within Whitehall, particularly between the Cabinet Office, the Ministry of Defence, the Colonial Office, and the Foreign Office over its utility and necessity, in part resulting from a lack of policy clarity, strategic focus, and risk aversion. SIS gained far more influence politically, in Whitehall and with the CIA, for its access and placement of its agents behind the Iron Curtain, as well as the quality of its individual operators and abilities to stoke treachery and subversion.[21] These qualities and abilities were firmly on display in Operation BOOT, the British codename for the early exemplar of Cold War treachery and subversion special operations, the overthrow of Iranian prime minister Mohammad Mossadeq in 1953.[22]

The Special Boat Companies and Sections, in contrast, did not live up to their immediate postwar potential. Blondie Hasler argued in 1946 for two distinct categories of missions: tactical and strategic.[23] All of their missions, save perhaps for indirect-shock missions conducted as part of a wider special operations campaign to interdict North Korean supplies during the Korean War, were tactical in nature. Their 'underemployment' from 1946 until 1971 can be attributed to two factors. First, the unit and its capabilities were virtually unknown in the Ministry of Defence, much less the rest of Whitehall.[24] Second, its highly specialized skill sets centred around specific reconnaissance and indirect-shock missions, which meant opportunities for their application were limited and infrequent.[25]

The rising special operations star from 1947 until 1971, however, was the SAS. The reasons for that rise were related to but different from those during the Second

[18] Jack Easton, assistant chief of SIS, quoted in Keith Jeffrey, *The Secret History of MI6: 1909–1949* (New York: Penguin, 2010), p. 719.

[19] Clive Jones, *Britain and the Yemen Civil War, 1962–1965: Ministers, Mercenaries, and Mandarins: Foreign Policy and the Limits of Covert Action* (Brighton, UK: Sussex Academic Press, 2004); Duff Hart-Davis, *The War That Never Was: The True Story of the Men Who Fought Britain's Most Secret Battle* (London: Arrow, 2012).

[20] Rory Cormac, 'Coordinating Covert Action: The Case of the Yemen Civil War and the South Arabian Insurgency', *Journal of Strategic Studies*, 36:5 (2013): 692–717.

[21] Rory Cormac, *Disrupt and Deny: Spies, Special Forces, and the Secret Pursuit of British Foreign Policy* (Oxford: Oxford University Press, 2018), pp. 158–165.

[22] Christopher Andrew, *Her Majesty's Secret Service: The Making of the British Intelligence Community* (New York: Viking, 1986), p. 494; Ervand Abrahamian, 'The 1953 Coup in Iran', *Science & Society*, 65:2 (Summer 2001): 182–215.

[23] David, *SBS*, pp. 418–419.

[24] Cormac, *Disrupt and Deny*, p. 165.

[25] Ibid.

World War. More specifically, SAS operators not only demonstrated their inherent qualities of flexibility and adaptability individually and organizationally, but they also proved to be an effective economy of force option for successive Whitehall cabinets seeking to limits costs and preserve British interests in a global South increasingly swept up in revolutionary, anti-colonial movements.[26] Special operations conducted by the SAS rarely provided to be the decisive factor against such movements, but they were invaluable in one regard. The SAS were able to match and counter asymmetric threats on their own terms.

British SAS efforts in its initial campaigns in Greece and Palestine were a mixture of clandestine military advice and assistance and covert action. In Greece specifically, the informal network of SAS personnel tapped their social connections and their patrons lobbied for an SAS capability to be retained and then employed sub rosa as trainers to the Greek National Army (GNA). Given the mountainous terrain that favoured the insurgents, as well as the GNA's propensity to employ artillery and air power against them, the senior leadership of the British Military Mission (Greece) recognized the need for specialized force to combat them.[27] Rather than employ SAS forces directly, several of its officers were employed as trainers and advisors to proposed Greek Commando units. The missions were not a coincidence but reflected the close connections 1 SAS had during the war with Free Greek Sacred Squadron.[28] In contrast to the American- and British-trained conventional forces, the three SAS-trained Greek Raider Squadrons matched guerrillas in their mobility in the mountains and aggressive small-unit actions, including ambushes.[29] A combination of conventional offensives, backed by precision air power, winnowed *Kommounistikó Kómma Elládas* (KKE), or Greek Communist Party, to their strongholds in northern Greece after their failed 1947 conventional offensive and denial of sanctuary by Yugoslavia in 1949.

In Palestine, SAS activities between 1945 and 1948 were informal and, although sanctioned, proved controversial. To combat the asymmetric threat posed by the *Irgun Zvai Leumi* (Irgun) the *Lochmei He'rut Israel* (Lehi, or the Stern Gang) on equal terms, former Chindit Bernard Fergusson approved the use of military advising and assistance officers with special operations experience to 'lead 10-man covert squads' of police constables to locate and ambush Irgun and Lehi members while operating in plain clothes.[30] The three officers chosen all had SAS, SOE, or SIS experience and included Roy Farran, who in turn recruited many of his former 2

[26] For details, see the various chapters in Erez Manela and Heather Streets-Salter, eds., *The Anticolonial Transnational: Imaginaries, Mobilities, and Networks in the Struggle against Empire* (Cambridge: Cambridge University Press, 2023).

[27] Tim Jones, *Postwar Counterinsurgency and the SAS, 1945–1952: A Special Type of Warfare* (London: Frank Cass, 2001), pp. 45–53.

[28] Tim Jones, *SAS: The First Secret Wars* (London: I.B. Tauris, 2005), pp. 93–94.

[29] David Sutherland, an operator who served with the SAS and SBS in the Second World War, was one of the Raiders' trainers. He summarizes the training programme in his memoir, Sutherland, *He Who Dares: Recollections of Service in the SAS, SBS, and MI5* (Annapolis, MD: Naval Institute Press, 1999), pp. 181–182.

[30] Jones, *Postwar Counterinsurgency and the SAS, 1945–1952*, pp. 35–36.

SAS colleagues for his squad.[31] He later achieved notoriety in the disappearance of Alexander Rubowitz, a member of the Stern Gang. Farran's escape from custody, and subsequent trial and acquittal, are representative of perceived heavy-handed, morally questionable, and possibly illegal British anti-colonial tactics.[32] The case led directly to the disbandment of the covert squads. Although not SAS activities per se, this example illustrates the workings of informal social networks between British special operators.

The SAS was elevated from a reserve or Territorial Army unit back into the active-duty British Army as a result of its performance during the Malayan Emergency (1948–1960). In September 1943, during the Second World War, an SOE unit, Force 136, contacted the Malayan People's Anti-Japanese Army (MPAJA). The MPAJA conducted Maoist-inspired guerrilla warfare against occupying Japanese forces. Its liaison with Force 136 was Chin Peng, who later led the Malayan Communist Party efforts against the British after the war. British support, in the form of special operations military advice and assistance and supplies, only began in earnest in 1945 for a variety of reasons, too late to have any effect on the war. The MPAJA formed the nucleus of its successor, the MCP's Malayan National Liberation Army (MNLA), which in turn was a disciplined, well-led, and well-supplied guerrilla army. Socio-economic turmoil in Malaya at the end of the Second World War, along with the MCP's thwarted political ambitions and waves of political unrest and low-level street violence, led to increased tension within the colony. The assassination of several plantation owners in June 1948 provided a pretext to crack down on the MCP and its related organizations. Those who survived, including Chin Peng, formed the MNLA in February 1949.

The key vulnerabilities of the British in Malaya were its isolated police detachments, large conventional military units, and resource infrastructure, specifically rubber plantations and tin mines. The MNLA conducted guerrilla ambushes of the first two and evaded their subsequent sweeps, while sabotaging or destroying infrastructure, to convince British authorities the costs were not worth continued colonial rule. Initial heavy-handed responses by the British characteristic of colonial policing designed to overawe and overwhelm by force played to the MNLA's advantage in terms of recruiting.[33] The British responses to the worsening crisis—a counterinsurgency (COIN) plan conceived by Harold Briggs (1950), a permanent secretary of defence in the form of Robert Thompson, a former Chindit, and successive High Commissioners in the form of Henry Gurney and Gerald Templar—were instrumental in turning the tide against the MNLA. The subsequent interpretation of events by former participants and COIN enthusiasts created several myths surrounding the British response as a model for emulation.[34] Elements of this model were

[31] Roy Farran, *Winged Dagger* (London: Fontana Books, 1956), p. 347.

[32] David Cesarani, *Major Farran's Hat: Murder, Scandal, and Britain's War Against Jewish Terrorism, 1945–1948* (London: William Heinemann, 2008), pp. 168–174; 186–191.

[33] Charles Callwell, *Small Wars: Their Principles and Practice*, 3rd ed. (London: Parnell Book Services, Limited, 1976), p. 24.

[34] See especially John Nagl, *Learning to Eat Soup with a Knife: Counterinsurgency Lessons from Malaya and Vietnam* (Chicago, IL: University of Chicago Press, 2005), pp. 87–114. For a corrective, see Karl Hack,

tried unsuccessfully in Vietnam, in the form of the Strategic Hamlet programme: crop destruction and jungle defoliation using aerially sprayed chemicals in Operation RANCH HAND, and in Afghanistan in the Village Stability Operations effort discussed in Chapter 8.[35]

The problem for the British in Malaya, after the Briggs Plan started up in earnest, was not isolating the population from the MNLA but rather coming to grips effectively with the enemy in their jungle sanctuaries. 'Mad Mike' Calvert proposed a solution in the form of a special force employing tactics along the lines of the Chindits, in which he served as a company and later brigade commander. Calvert's concept, which drew on his own experiences as well as a short-lived unit dubbed Ferret Force, saw the formation of the Malayan Scouts (SAS), initially comprising a single squadron.[36] That unit, A Squadron, would conduct long-range jungle patrols seeking to make contact with or track MNLA guerrilla bands. To assist in the latter, Malayan Scout patrols often made use of indigenous Ibans as trackers. Its initial operations in 1951 provided proof of the concept even though numerous tactical mistakes were made.[37]

The concept was initially successful enough to prompt the expansion of SAS involvement and units. It occurred first through a B Squadron from 21 SAS (Artists) expecting to be deployed to the Korean War, and later a C Squadron of recruited Rhodesians as well as D Squadron of regular volunteers. Discipline was problematic within the Scouts, particularly within A Squadron, who followed the lead of their incredibly brave but unstable commanding officer, Calvert. The problems were such that the B Squadron commander, having had his requests to transfer his unit elsewhere denied, took the unusual step of resigning his command in July 1951.[38] James Ladd puts the differences between commanders down to the issue of laying the foundation for permanent special forces against Calvert's for an impromptu, ad hoc one disestablished once the need passed.[39]

Calvert's concept rested on the ability of the Malayan Scouts to remain in the jungle for extended periods of time. Key to it was linking up with and gaining the support of the Orang Asli, the indigenous population of 40,000 people comprising three main ethnic groups. Two of the three, the Semang and Senoi, would be the focus of the Malayan Scouts' attention, the former residing in hilly, jungle-clad northern sections of the Malayan Peninsula and the more numerous latter in the main

'The Malayan Emergency as a Counter-Insurgency Paradigm', *The Journal of Strategic Studies*, 32:3 (June 2009): 384–414.

[35] William Buckingham Jr, *Operation Ranch Hand: The Air Force and Herbicides in Southeast Asia, 1961–1971* (Washington, DC: Office of Air Force History, 1982), pp. 4–5.

[36] For a contemporary report, see '"Ferret Force" Is Answer to Reds', *Townsville Daily Bulletin* (25 October 1948), p. 2, https://trove.nla.gov.au/newspaper/article/63369806.

[37] Mistakes included the loss of equipment; a trooper went missing and his body was later recovered; and an indigenous tribesman was accidentally killed. For details, see John Leary, 'Searching for a Role: The Special Air Service (SAS) Regiment in the Malayan Emergency', *Journal of the Society for Army Historical Research*, 73:296 (Winter 1995): 256–257.

[38] Alan Hoe and Eric Morris, *Re-Enter the SAS: The Special Air Service and the Malayan Emergency* (London: Leo Cooper, 1994), pp. 71–73.

[39] James Ladd, *SAS Operations* (London: Robert Hale, 1986), pp. 106–107.

spinal mountain range.[40] In addition to providing additional manpower and serving as guides, the Orang Asli provided a source of invaluable support for the MNLA and MCP as British population control measures were put in place, including food rationing: growing nutrient-rich root vegetables such as manioc and sweet potatoes in cleared-out sections of the jungle.[41]

After Calvert's departure, the unit was transformed thanks to rest, refitting, and imposing standards, and took a leaf from their opponent's handbook. In particular, the Scouts adopted self-criticism sessions with the purpose of identifying lessons to improve tactical performance. The net effect of these changes was the weeding out of unsuitable candidates, optimizing training for the mission at hand, and professionalizing the unit to a common standard, increasing its overall efficiency and effectiveness.[42] Self-criticism led to the development of another practice loosening control: tactical planning and decision making by committee, with little regard for rank, subsequently known as a 'Chinese parliament'.[43] Operational control procedures included a regimental headquarters within the country coordinating via radio, as far away as 150 miles, with squadron headquarters distributed where they were most needed: 'in the jungle'.[44] At this juncture, in early 1952, the Malayan Scouts (SAS) were renamed as 22 SAS Regiment.

To extend the patrolling time, range, and operating depth, 22 SAS embraced historical and developed new approaches for operations. The former was established by the Chindits for long-range penetration patrols: the use of aerial resupply to established bases in the jungle. The bases were little more than temporary camps. Drops to bases often alerted the MNLA to the SAS's presence and location. For this reason, and based on the frequency of the air drops, SAS patrols relocated to new bases every three or four weeks for a period of three months, to maintain pressure on the MNLA in their sanctuaries.[45] The new approach involved a modification of the SAS's airborne delivery by parachute. Traditional methods of mobility required the SAS to be delivered by helicopter or dropped from an aircraft. The former was in short supply, expensive, had short range relative to aircraft, required a clearing in order to land, and, lastly, telegraphed their location to an opponent due to the tremendous noise they generated.[46]

Airborne delivery by aircraft solved a number of these problems but the jungle created others. Parachutists do not penetrate the thick jungle ceiling as their parachute canopies often snag on branches. In Malaya, this created issues as the ceiling was often

[40] John Leary, 'Searching for a Role: The Special Air Service (SAS) Regiment in the Malayan Emergency', *Journal of the Society for Army Historical Research*, 73:296 (Winter 1995): 254.

[41] J.M. Woodhouse, 'Some Personal Observations on the Employment of Special Forces in Malaya', *Army Quarterly*, 66 (April and July 1953): 69; Johnny Cooper, *One of the Originals: The Story of a Founder Member of the SAS* (London: Pan Books, 1991), p. 125.

[42] Hoe and Morris, *Re-Enter the SAS*, pp. 102–105.

[43] The practice is described in detail in Tony Jeapes, *SAS: Secret War* (London: HarperCollins, 1996), pp. 65–66.

[44] Woodhouse, 'Some Personal Observations on the Employment of Special Forces in Malaya', p. 70.

[45] Hoe and Morris, *Re-Enter the SAS*, pp. 107–109.

[46] Arthur Barondes, *The Accomplishments of Airpower in the Malayan Emergency (1948–1960)* (Maxwell Air Force Base, AL: Aerospace Studies Institute, 1963), pp. 19–22; 35–36.

70 feet high, the equivalent of a six-storey building. By including a length of webbing strap to the parachute harness, once released from the parachute, an operator could then abseil down from the jungle canopy safely to its floor.[47] For its initial operations, however, such as Operation HELSBY (1951), 22 SAS employed more conventional airborne drops and helicopter insertions, along multiple axes of approach, to destroy an MNLA base near the Malay–Thai border.[48]

The period from 1952 through 1954 was one of experimentation, adaptation, and organizational change for 22 SAS. Beyond continued tactical experimentation, the Regiment was to operate in and among the Orang Asli for two primary purposes. One was to serve an integrating function between the Orang Asli and the Malay police force. Having built up trust with some of them by providing basic medical services through the patrols, even if for theatrical purposes, SAS troopers could convince them that a more permanent police presence, in the form of local forts, was beneficial to the tribes.[49] Maintaining trust was not assured, especially given the SAS's role in early, ill-designed attempts to resettle Orang Asli groups.[50] The other, comprising intelligence-gathering and reconnaissance missions, would improve British understanding of tribes, which was scant, as well as on the MNLA in the jungle hideouts (Operation GALWAY).[51] Contacts within the MNLA, if they were with significantly important MCP individuals, resulted in operations shifting to more traditional cordon and search tactics: surrounding an area with several battalions of troops, whose advances would be spearheaded by SAS patrols. This occurred in 1953 when GALWAY was supplanted by Operation VALIANT.[52] Organizationally, A Squadron was inactivated in 1952, its members having served their enlistment period.

From 1954 until 1957, when the SAS was withdrawn, MNLA fortunes steadily declined thanks to a comprehensive COIN approach, employing a range of coercive measures, in which special operations played a valuable, supporting role. That role continued to be as an interface between the Malayan government and the Orang Asli, while sustaining pressure on the MNLA through continuous, aggressive patrolling to deny the insurgents sanctuary. In addition to searching for MCP members, SAS troopers specifically targeted ethnic Chinese members of the Asal Protection Corps. This Corps consisted of a cadre vital in proselytizing, recruiting, training, and leading Orang Asli against the Malayan government and provided a crucial link with

[47] The equipment and procedures are described in specific detail in D.D. Ranft, 'Parachuting in Malaya', *Army Quarterly*, 66 (April and July 1953): 205–207.

[48] Hoe and Morris, *Re-Enter the SAS*, pp. 114–118; Malcolm Postgate, *Operation Firedog: Air Support in the Malayan Emergency, 1948–1960* (London: Her Majesty's Stationery Office, 1992), pp. 81–82.

[49] Lord Guthrie, serving in an additional capacity as a medic on SAS patrols, notes the benefits of this theatricality: 'It astonished me what aspirin could do; it was the perfect palliative for local tribespeople. That was all to the good in winning "hearts and minds"'. Charles Guthrie, *Peace, War and Whitehall: A Memoir* (Oxford: Osprey, 2021), p. 66.

[50] John Leary, 'Searching for a Role: The Special Air Service (SAS) Regiment in the Malayan Emergency', *Journal of the Society for Army Historical Research*, 73:296 (Winter 1995): 259–260.

[51] Hoe and Morris, *Re-Enter the SAS*, p. 164.

[52] John Leary, 'Searching for a Role: The Special Air Service (SAS) Regiment in the Malayan Emergency', *Journal of the Society for Army Historical Research*, 73:296 (Winter 1995): 261–262.

the MCP. SAS troopers did their utmost to target Asal members during contacts, but not the Orang Asli, to sever it. These efforts had tactical and operational effects. They influenced hostile Orang Asli to switch allegiances while the collective COIN efforts deprived the MNLA of the initiative, putting the movement on the defensive for the remainder of the Emergency.

SAS efforts in Malaya, while important, were not decisive. The nature of the conflict, with its primary emphasis on security, stability, and asserting Malayan government control leading to independence, guaranteed this. In terms of contacts or kills, SAS operations were small: 108 out of a total of 6,400.[53] Kills, however, are a poor measure of effectiveness. The SAS took part in fifty-three named operations, gathering useful information on virtually unknown elements of the Orang Asli and the MNLA, while keeping the MCP and Asal Corps on the defensive.[54] More importantly, the SAS provided an interface, previously missing, between the Orang Asli and the police.[55] In addition, the SAS offensive operations bought time for the development of a more effective and, in the long run, lasting solution and interface with the Orang Asli. This solution was the fielding of the Senoi Praaq, a unit of the Royal Malaysia Police. The unit, composed mainly of Orang Asli, conducted patrol and engagement missions along the lines of SAS from forts and initially served as an auxiliary to the Regiment, allowing for the cross-pollination of skills and best practices. Although they did not have to deploy for extended periods like the SAS, their language skills and cultural understanding were more valuable in winning over hostile or ambivalent Orang Asli.[56]

Beyond the establishment of 22 SAS Regiment on the roster of the British Army in 1957, Malaya had an important impact on British special operations for the next four decades. The conflict established in the mind of senior military and civil leaders the ability of the SAS to work discretely and semi-independently, with minimal support, in the most forbidding terrain. These attributes would prove useful in future conflicts. In addition, the Malayan experience cemented the tactical utility of self-criticism on the individual level, through Chinese parliaments, as well as the value of creating standard operating procedures (SOPs). Both contributed to an evolution in culture within the SAS, but the latter served as a means of knowledge transfer for incoming operators to replace not insignificant losses and the rotation in of new, non-specialist units. Prior to the development of a selection and assessment programme, SOPs shortened the learning curve and evolved to reflect best practices. They further professionalized British special operators and by extension, special operations. These reflected a broader institutional embracing of codifying best practices, as evidenced by the writing, publishing, and distribution of a pocket manual, Conduct of Anti-Terrorism Operations in Malaya, which was revised periodically.[57]

[53] Hoe and Morris, *Re-Enter the SAS*, p. 205.
[54] The named operations are listed in appendix A in ibid., pp. 210–211.
[55] See James Ladd's patriarchal and colonial assessment in Ladd, *SAS Operations*, p. 113.
[56] Roy Jumper, *Death Waits in the 'Dark': The Senoi Praaq, Malaysia's Killer Elite* (Westport, CT: Greenwood Pres, 2001), pp. 53–70.
[57] Jones, *Postwar Counterinsurgency and the SAS, 1945–1952*, pp. 134–137.

Historically, best practices were transferred informally but within a generation most skilled British special operators experienced in the jungle had retired or moved on.

The generational dimension and professionalization of the SAS is underreported. Malaya, however, saw the passing of the torch in this regard, between the Second World War Chindits and SAS 'Originals', such as Mike Calvert and Johnnie Cooper, through an interregnum of conventional soldiers experienced in the war but new to special operations such as John Woodhouse, to the next generation of lifetime special operators including Tony Jeapes and Peter de la Billère, who served as a Captain in the latter stage of the conflict.[58]

The SAS during the case study period was also shaped by its activities in other jungles, this time in Borneo. Borneo is the largest of the Sunda Island chain and its territory was split between Indonesia, Malaysia, and Brunei. The Indonesian label their territory, as well as the entire island, Kalimantan. The Malaysian portions were North Borneo and Sarawak, with Brunei in between. At issue in 1963 was the creation of country of Malaysia, which united Malaya with North Borneo and Sarawak, both formerly Crown colonies. Indonesian President Sukarno objected to it and, prompted by a revolt in Brunei the previous year and perhaps inspired by unsuccessful British and American covert actions to destabilize his country five years prior, dispatched his own detachments of small groups of Indonesian Army soldiers to do the same in the former colonies.[59] Those soldiers trained, equipped, advised, and often led detachments of proxies labelled by Commonwealth forces as either the Clandestine Communist Organisation (CCO) or Indonesian Border Terrorists (IBTs).[60]

The subsequent actions, initially and subsequently, were never a declared war between the countries involved but termed by the Indonesians *Konfrontasi*, or Confrontation. They comprised small-units actions designed to raise proxy forces and destabilize the colonies on the Indonesian side, prevent infiltration across the border, counter those detected that had done so, and conduct cross-border raids to pre-empt future attempts. Emile Simpson uses the *Konfrontasi* as the singular historical case study for his theory of modern warfare, to illustrate points about pragmatic decision making and narrative coherence, made possible by the British 'understanding the language of force'.[61] The *Konfrontasi* lasted three years, until 1966, when Sukarno was deposed in a military coup.

The cross-border operations authorized in 1964, in which 22 SAS played a role, were code-named 'Claret'. At the operational or campaign level, they followed the pattern set in Malaya: respond defensively to surprise in an uncoordinated manner,

[58] Ladd, *SAS Operations*, pp. 111–112; Anthony Kemp, *The SAS: Savage Wars of Peace, 1947 to the Present* (London: John Murray, 1994), pp. 23–25; 30; Cooper, *One of the Originals*, pp. 120–122.

[59] For details of the 1962 Brunei Revolt, see Alexander Shaw, 'British Counterinsurgency in Brunei and Sarawak, 1962–1963: Developing Best Practices in the Shadow of Malaya', *Small Wars & Insurgencies*, 27:4 (2016): 702–725.

[60] Raffi Gregorian, 'CLARET Operations and Confrontation, 1964–1966', *Conflict Quarterly* (Winter 1991): 48–49.

[61] Emile Simpson, *War from the Ground Up: Twenty-First Century Combat as Politics* (Oxford: Oxford University Press, 2013), p. 177.

regroup and develop a comprehensive plan, then regain the initiative and put the enemy on the defensive through offensive actions aimed at their sanctuaries. The primary difference between Malaya and Borneo was those sanctuaries existed in the sovereign territory of another country: Indonesia, and more specifically, in Kalimantan. The raids were a natural evolution of the established military practice of countering small-unit raids by cordoning off areas, killing or capturing those surrounding, or should they evade or escape the cordon, following and harrying until they were destroyed. Such activities often took place across the border, in accordance with the principle of 'hot pursuit'.[62]

Claret operations continued for the remainder of the conflict, ending in 1966. To ensure cross-border incursions did not escalate the conflict, the director of Borneo Operations, Walter Walker, outlined seven conditional 'golden rules' guiding their conduct. The rules outline the basic principles for clandestine military special operations, including selected forces, senior-level approval, pre-mission rehearsals, specific conditions to limit casualties, and sanitized equipment and cover stories.[63] Escalation was less likely if objectives were limited, force used in a selective manner, and plausible deniability maintained.

Special operations conducted by 22 SAS under Claret were designed to improve the performance of conventional forces at the campaign level. They did so primarily by conducting reconnaissance missions to locate CCO units, so that conventional forces could engage them, or locate training and support bases to be assaulted or attacked from the air. Thanks to declassified official records and campaign summaries, Christopher Tuck assessed SAS activities at the operational level. In total some 545 missions by type were conducted and of these, more than half, almost two-thirds, consisted of reconnaissance missions.[64] Of those 332 reconnaissance missions, 29 per cent were conducted by the SAS, greatly outpacing the number of ambushes and 'hearts and minds', or psychological operations missions.[65] Those missions, however, achieved mixed results, due in part to the nature of terrain and geography, environment, equipment breakdowns (many caused by the latter), a lack of air support, and insufficient cultural understanding. Scholars debate the effectiveness of Claret operations, and more specifically the SAS's role within them, with claims ranging from 'one of the winning elements' to tactical effectiveness in conventional tasks characteristic of strategic misuse.[66] Tuck concludes 'no amount of artistry at the operational level in the use of Special Forces can compensate for weaknesses in strategy'.[67]

[62] Christopher Tuck, 'Shaping Hearts and Minds: Claret Operations in Borneo, 1965–1966', *Small Wars & Insurgencies*, 34:4 (2023): 805.

[63] Tom Pocock, *Fighting General: The Public and Private Campaigns of General Sir Walter Walker* (London: Collins, 1973), pp. 196–197.

[64] Tuck, 'Shaping Hearts and Minds', p. 808.

[65] Christopher Tuck, 'The Limits of Covert Action: SAS Operations during "Confrontation", 1964–66', *Small Wars & Insurgencies*, 27:6 (2016): 1005.

[66] Gregorian, 'CLARET Operations and Confrontation, 1964–1966', pp. 48–49; Alastair MacKenzie, *Special Force: The Untold Story of 22nd Special Air Service Regiment (SAS)* (London: I.B. Tauris, 2011), pp. 126–130.

The desire in Whitehall to limit the scope and costs of the conflict, while denying Sukarno the means to fulfil his strategy, suggests political risk calculations drove strategic and operational choices along pragmatic lines, given the modest stakes involved.

The operational and tactical realities of 22 SAS operations were irrelevant in the political and institutional sense. The political perception in Whitehall was that 22 SAS could be counted on to conduct operations skilfully, with discretion, at little cost, to maintain British influence in former or current Crown colonies. At the same time as Claret operations were initiated, the reconstituted A Squadron was committed to Aden on a training mission in 1964, after recent terrorist attacks the previous year. While the tactics it used in Malaya proved problematic, resulting in casualties and public acknowledgement of the Regiment's presence in the media, its clandestine and covert activities had a mixed record. Within Aden, SAS teams operated clandestinely in small teams, specifically in the contested district within the main town, as part of Operation NINA in a manner almost identical to the type of patrols conducted by Roy Farran in Palestine.

In contrast to Ken Connor's positive tactical evaluation of their effectiveness, Alastair MacKenzie assesses their performance as poor, as they were 'poorly coordinated' with other clandestine teams, suggesting low levels of horizontal integration.[68] In neighbouring Yemen, 'sheep-dipped' operators 'on leave' from the Regiment conducting covert activities were a source of alarm in the Foreign Office and nearly caused a major diplomatic flap.[69] Others worked through David Stirling's company, Watchhouse International Incorporated. Operators such as Cooper and Woodhouse did so ostensibly while retired and in a private military capacity but maintained strong, informal ties to Whitehall.[70] The nature of the ties is illustrated in a sardonic yet accurate quote about how these operators came together informally: 'With the tacit support of SIS's Chief ... Most were approached over whiskey and sodas at White's and the Special Forces Club in London.'[71] The result of their efforts, for various reasons, is unfortunately unknown but they were increasingly involved in more robust covert action, coordinated within Whitehall within the newly established Joint Action Committee in 1964, followed two years later by the South Arabia Action Group.[72]

Both Aden and Borneo and Aden saw the emergence of behaviour last seen in the Second World War: special operations commanders lobbying commanders for immediate employment. In Borneo, Woodhouse had approached Walker and, given

[68] Ken Connor, *Ghost Force: The Secret History of the SAS* (London: Weidenfeld & Nicolson, 1998), p. 129; MacKenzie, *Special Force*, p. 102.

[69] Cormac, *Disrupt and Deny*, p. 156.

[70] Anthony Kemp, *The SAS: Savage Wars of Peace, 1947 to the Present* (London: John Murray, 1994), pp. 88–89; Connor, *Ghost Force*, p. 128.

[71] Calder Walton, *Empire of Secrets: British Intelligence, the Cold War, and the Twilight of Empire* (New York: Overlook Press, 2013), p. 325.

[72] Rory Cormac, 'Coordinating Covert Action: The Case of the Yemen Civil War and the South Arabian Insurgency', *The Journal of Strategic Studies*, 36:5 (2013): 704–713.

the Regiment's recent experience in Malaya, made an easy transition into Claret operations. In Aden, the A Squadron commander, Peter de la Billère, sought out Charles Harrington, commander-in-chief of Middle Eastern Command, to alter his unit's mission from a training mission to an operational one—in the former's words, 'a live operation'.[73] MacKenzie opines the subsequent issues with special operations, including insufficient preparation and coordination, reflected undue haste in finding opportunities for employment rather than taking time to adapt to a very different set of operating conditions.[74] Operational and tactical coordination may have been lacking in its overt and clandestine activities but its willingness to be employed in lost causes in colonies soon to be independent anyway, as a means of countering asymmetric threats on their own terms, suggests Whitehall's perceptions of 22 SAS's political and strategic value greatly outweighed the realities on the ground.

## Capabilities

In contrast to the opacity surrounding the processes and control mechanisms associated with British special operations discussed in the next section, the structure of its special forces circa 1971 is clearly demarcated functionally and organizationally. In terms of function, overt special operations are the realm of British military special forces and covert of the SIS, more commonly known as MI6. Except for its activities related to Northern Ireland, MI5's role in intelligence gathering against internal threats is excluded here due to lack of information. British special operations make little distinction between covert and clandestine, reflecting the relative size of their units and the informal nature of British governance historically in a nuanced manner more in common with their South African counterparts than their litigious American cousins.

In 1971, British overt military special operations capabilities resided in two units: the SAS and the SBS. The SAS was divided into three regiments, an active-duty and fully operational one, 22 SAS, as well as two Territorial Army regiments split geographically between the south, 21 SAS (Reserve), and the north, 23 SAS (Reserve). Within 22 SAS, there were four 'sabre' squadrons comprising between sixty and sixty-five operators designated alphabetically: A, B, D, and G. Each of the squadrons is divided into four troops, based on a specialized function, including Air, Boat, Mobility, and Mountain, as well as a headquarters or command one. The troops can then further be subdivided down into teams ranging from between two and four operators, depending on circumstances. In Malaya, three-men teams maintained redundancy and tactical effectiveness in close contact firefights in the deep jungle

[73] Peter de la Billère. *Looking for Trouble: SAS to Gulf Command, the Autobiography* (New York: HarperCollins, 1994), p. 213.
[74] MacKenzie, *Special Force*, pp. 98–100.

while preserving stealth.[75] The reserve regiments comprise three alphabetically designated squadrons. These included A, C, and E for 21 SAS and B, D, and G for 23 SAS.[76] When activated during a war, the reserve regiments would either conduct long-range reconnaissance missions for the British Army of the Rhine in Europe against the Warsaw Pact, in a direct supporting role for conventional forces, or backfill combat losses for 22 SAS in other, ongoing conflicts should the need arise.

For historical reasons outlined above, 22 SAS often conducted independent operations in a clandestine manner in support of British foreign policy and national security objectives, employed as individual troops and infrequently as entire squadrons. The primary missions of the SAS remained countering asymmetric threats, its *raison d'être* after the Second World War, along with intelligence gathering and reconnaissance, military advice and assistance, and indirect-shock missions behind enemy lines driving its original creation.

Unlike 22 SAS which had an independent role, the SBS were attached directly to the Royal Marine Corps and, by extension, served as an element of the Royal Navy. There were six SBS, numbered accordingly. They were based out of different locations depending on need and interest, including Singapore and Malta. In addition, 1 and 2 SBS supported war plan requirements to delay Warsaw Pact forces through sabotage missions, primarily bridge destruction.[77] Working in direct support of the Amphibious Warfare Group, SBS units were often tasked by either the Joint Intelligence Bureau (JIB), MI6, or both, to fill gaps in knowledge through intelligence collection missions such as clandestine beach and other nautical survey missions.[78] Other missions for the various SBS detachments included reconnaissance, indirect-shock missions including maritime raids, and fleet support missions, including demining. SBS units were able to conduct intelligence-gathering missions for JIB and MI6 thanks to their clandestine approach, either in detached small teams or transported by Royal Navy submarines, launching and recovering from the same.

British covert special operations capability resided largely in one shadowy element of SIS, about which little is known and even less published. Only its existence has been alluded to or revealed inadvertently, and there are no details about its size, structure, location, or composition.[79] Identified in different sources as 'E Squadron', or 'the Increment', its role is covert action through the conduct of paramilitary special operations. The E Squadron designation makes sense in one regard: as a unit within or extension of the SAS.

[75] Woodhouse, 'Some Personal Observations on the Employment of Special Forces in Malaya', pp. 70–71; Cooper, *One of the Originals*, p. 139.

[76] Philip Warner, *The Special Air Service* (London: William Kimber, 1972), p. 239; William Seymour, *British Special Forces* (London: Sidgwick & Jackson, 1985), p. 303.

[77] James Ladd, *SBS: The Invisible Raiders* (London: Arms and Armour Press, 1983), pp. 213–215.

[78] John Parker, *SBS: The Inside Story of the Special Boat Service* (London: Headline Book Publishing, 1997), pp. 215–216.

[79] Mark Urban, 'Inside Story of the UK's Secret Mission to Beat Gaddafi', BBC News, 19 January 2012, www.bbc.com/news/magazine-16573516.

The designation implies much about British special operations capabilities beyond a possible connection to the SAS. First, it is an example of a different form of the paramilitary and military convergence discussed in Chapter 8 based on practical considerations. Given the small size and secrecy of the British special operations community, the only practical method of increasing capacity for certain mission sets when crises arise is through cross-designating personnel already trained and expert in them. The small size of British SOF entails their deployment in much smaller elements with little in the way of external support, requiring them to operate clandestinely and avoiding contact at all costs as a means of mitigating risk and survival. SIS personnel, who can travel as individuals or in small numbers, require security detachments in higher-risk zones, a function that cross-designated personnel can conduct discretely. Finally, 22 SAS and to a lesser extent SBS were employed routinely clandestinely in 'bush wars' throughout the Empire and former colonies, meaning that operators were conducting intelligence gathering and reconnaissance missions, as well as countering asymmetric threats, using clandestine tradecraft. Therefore, assertions that E Squadron operators have been 'sheep dipped', or conducted activities as part of E Squadron as paramilitary only to return to the uniformed military fold, or have conducted high-risk covert reconnaissance missions in denied territories, are plausible and unsurprising.[80]

In terms of capabilities as defined in Chapter 3, British military special operations were optimized for countering asymmetric threats, indirect-shock missions, and intelligence gathering. Their forté were such missions on a small-scale, yet global level. These characteristics are attributable to the individual units and operators. Collectively, each occupied a niche and often conducted activities in isolation from each other. There is little evidence of cross-training for missions despite some for the skills sharing. Both were products of their parent services, the British Army and the Royal Navy, adopting along the way some of their culture but largely looked at with suspicion and contempt by them. For example, John Parker notes that even among the Royal Marines, under whom the SBS fell and itself an elite element of the Royal Navy, viewed the special operators as 'overbearing eccentrics diffident to authority, and always asking for expensive new toys'.[81]

From their parent organizations, as well as their historical predecessors, they gained a unique, distinct unit collective identity, one that values tradition, history, heritage, and independence that characterizes British military units, and the Army and Royal Marines in particular. Their different service orbits, combined with their different mission focus areas, translated into theoretically compatible but mutually exclusive, domain and environment-specific capabilities despite a degree of overlap in some tactical missions, such as those involving watercraft between the various boat troops and SBS. Although they each operated clandestinely, they did so independent

[80] William Arkin, 'Inside the Military's Secret Undercover Army', *Newsweek* (17 May 2021), www.newsweek.com/exclusive-inside-militarys-secret-undercover-army-1591881.

[81] Parker, *SBS*, p. 204.

of each other. For these reasons, at the start of the case study period the degree of horizontal integration among capabilities was medium, achieving higher levels with their parent services but suffering from a lack of common training, equipping, and mission guidance. Figure 6.1 at the end of the chapter graphically depicts these capabilities, their functions, and their effectiveness, along with their calibration and informal linkages (the vertical dashed line), according to the theory of special operations coherence.

## Control

Five different forms of control were identified in Chapter 4. They are delegative, directive, collaborative, competitive, and co-optive. Douglas Bland offered collaborative control as a different model that reflects the qualities, attributes, and quirks of the parliamentary democracy tinged with residual class distinctions. This section will not repeat Bland's argument but rather places it within a body of scholarship that has control as its central element: civil–military relations theory. There have been three successive waves of British scholarship on the subject, the most recent stimulated by British decision making for the war in Iraq and concerns over prime ministerial overreach and executive agency. The first provides unique insights on the specific form of collaborative control in the United Kingdom for special operations.

Scholastic work on or related to civil–military relations in the UK is a relatively recent phenomenon. This recency reflects the nature of British government and scholarship. The impact of British government on civil–military relations begins with its different conceptualization of 'executive' and how power is wielded. James Cable provides a concise summary of the executive authority and power during the time period covered by this paper:

> In Britain, therefore, the Executive is the Cabinet, dominated but not entirely controlled by the Prime Minister. This executive is not directly elected and has no intrinsic legitimacy, but derives its authority and its powers, in unequal proportions, from Crown, Parliament and Party. Of these three the Crown legally confers substantial powers, but can not, in the person of the Monarch, effectively restrict their exercise. Parliament confers greater powers, both in theory and in practice. Parliament also enjoys over the government a power which is real, but seldom used, being held in check by the ability of the Prime Minister to threaten the House of Commons with dissolution and a general election. Legally, Parties can neither give power nor control it, but in practice they are taking growing share in both functions.[82]

Other scholars have characterized the system of a small number of Cabinet offices, headed by politically appointed ministers and administered by agnostic permanent

[82] James Cable, *Political Institutions and Issues in Britain* (London: Macmillan, 1987), p. 32.

civil servants, as a 'core executive' that 'stand[s] at the apex of the policy process'.[83] The executive responsible for military affairs is the Minister for Defence, who receives professional advice from the senior ranking uniformed member, the Chief of the Defence Staff (CDS), whose nomination has political but not process undertones. Members of the UK Defence Forces are volunteers, who owe their allegiance to the person of the monarch and not Parliament or the prime minister, for long-standing historical reasons.[84] Policy matters, including those of the military, are influenced and guided by Cabinet Committees, comprising representatives from different Cabinet offices, lending what in the United States would be classified as an 'interagency' dimension to decision making. The only constant feature of the British government is change, in terms of power distribution between ministers and some prime ministers' 'instinctive sense of what he [sic] can get away with', making it comparable metaphorically to 'a living organism'.[85] The depiction of the British system by Cable as a reflection of its society suggests a level of cultural deterministic and intuitive accuracy: 'It also suits the British temperament, which distrusts theory, logic, and lawyers, preferring improvisation to planning. The favorite slogan of the British governing class is that every problem should be treated on its own merits.'[86]

The British distrust of theory, as it relates to civil–military relations, is evident in the initial wave of scholarship on the subject. Three years after the publication of Samuel Huntington's *The Soldier and the State*, British military historian Michael Howard commented in *Political Science Quarterly* that focus on the subject was not only 'unprecedented' but 'perhaps ... disproportionate'.[87] The overall point of Howard's article is that fears of 'the man on horseback', or praetorianism threatening democracy, were overexaggerated and advice offered by senior military leaders can never be strictly military in content, or at least should not be if the system of government is functioning normally.[88] Samuel Finer expanded upon a number of these themes a decade and a half later in *The Man on Horseback* and Howard edited a volume offering nine chapters, each exploring in a focused manner a historical case of the relationship between generals and decision makers.[89]

Huntington's work, a structural explanation of the power 'choices' between civil and military authorities, was challenged by Morris Janowitz's social one. Janowitz focused on the social dimensions of militaries as they relate to their functions. British political activity was moderated by its use overseas in constabulary for limited objectives. A handful of British scholars, including Gwyn Harries-Jenkins and

[83] Martin Burch and Ian Holliday, 'The Prime Minister's and Cabinet Offices: An Executive Office in All but Name', *Parliamentary Affairs*, 52:1 (January 1999): 32.

[84] The reasons are summarized admirably and generally in Corelli Barnett, *Britain and Her Army, 1509–1970: A Military, Political, and Social Survey* (London: Allen Lane, 1970), pp. xviii–xx.

[85] Cable, *Political Institutions and Issues in Britain*, pp. 14, 21.

[86] Ibid., pp. 20–21.

[87] Michael Howard, 'Civil–Military Relations in Great Britain and the United States, 1945–1958', *Political Science Quarterly*, 71:1 (March 1960): 35.

[88] Howard, 'Civil–Military Relations in Great Britain and the United States', p. 35.

[89] Samuel Finer, *The Man on Horseback: The Role of the Military in Politics* (London: Routledge, 2002, 1974 repr.); Michael Howard, *Soldiers and Governments: Nine Studies in Civil–Military Relations* (Bloomington, IN: Indiana University Press, 1959).

C.B. Otley, adopted a 'Janowitzian' approach to the study of the social backgrounds of British officer cadres or tranches over different years. Military historian Hew Strachan responded to these theories by demonstrating that soldiers in the British Army have been professional and political throughout its history, despite the thresholds and criteria outlined by Huntington and Janowitz.[90]

Empirical evidence, particularly when marshalled by those a broad and deep an understanding of history as Howard and Strachan, had far greater explanatory power than social science models for the British case. Howard and Strachan also masterfully, if at times indelicately, pointed out issues with a model-based approach, including contradictions within the underlying assumptions, problems with limited explanations, and reductionism and even incorrect representation of historical events and facts.[91] Underpinning British scepticism of civil–military relations theory was the logic of Sir Arthur Conan Doyle's 'dog that didn't bark' historically. In Douglas Bland's words:

> The British have a long and, generally, satisfactory (though at times turbulent) history of civil military relations. Among other things, this results from a high degree of harmony between political leaders and officers based on the assumption that they have complementary roles to play in deciding and managing national defense.[92]

Easing that harmony was the fact that for a significant period in modern British history, political leaders and officers came from a network of 'Old Boys': members of the same elite classes, including the aristocracy or landed gentry, who attended the same or similar public schools that instilled nearly identical values.[93]

If the public discussion of civil–military relations in the UK has been about overreach and under delivery, scholastically the dominant themes have been harmonization and integration.[94] Robert Egnell builds upon Bland's unified theory, drawing as it does upon regime theory, and he looks at the impact of the organization of the civil–military interface on military effectiveness. He concludes that the in the UK setting there is a high degree of integration between civil leaders and the military given the former dictates the function of the latter than is the case in the United States, leading to more effective outcomes in complex peace operations such as Iraq.[95] In a more critical assessment of British campaigns in 'small wars' from 1945 onwards,

[90] Hew Strachan, *The Politics of the British Army* (Oxford: Clarendon Press, 1997), p. 263.

[91] Strachan, *The Politics of the British Army*, pp. 10–19.

[92] Douglas Bland, 'A Unified Theory of Civil–Military Relations', *Armed Forces and Society*, 26:1 (Fall 1999): 12.

[93] On its influence politically at the height of the British Empire, see Andrew Adonis, *Making the Aristocracy Work: The Peerage and the Political System in Britain, 1884–1914* (Oxford: Oxford University Press, 1993), pp. 164–204. On public schools for elite and their influence on military even in the mid-twentieth century, see David Walsh and Anthony Seldon, *Public Schools and the Second World War* (London: Pen & Sword, 2020).

[94] Daniel Korski, 'British Civil–Military Integration: The History and Next Steps', *The RUSI Journal*, 154:6 (2009): 14–24.

[95] Robert Egnell, *Complex Peace Operations and Civil–Military Relations: Winning the Peace* (London: Routledge, 2011), p. 5.

Aaron Edwards concludes 'there has frequently been a disconnection in civil and military relations, so much so that failure—and the risk of failure—has only been averted once these relations have been harmonized and unity of effort restored between Britain's political and military establishment'.[96] A final, noteworthy contribution to the discussion is David Charters' work on the state of British civil–military relations in the initial years of the British Army's involvement in 'The Troubles' in Northern Ireland. He concludes harmonization in civil–military relations was accomplished specifically between 1968 and 1974. The price paid for it, however, comes in the time to it takes to coordinate various committees and offices, especially in the face of limited political decision-making manoeuvring space, and in transparency by limiting access of individuals to information and denying alternatives analysis.[97]

Scholarship on British civil–military relations deviate from 'normal', uniquely American theories. Britain has a longer, more diverse, and in consequence richer political tradition than the context spawning the American experience. In addition, the attributes and characteristics of British government suggest less emphasis on control in a normative sense than direction, which results from more informal mechanism of government. If existing theory cannot be sufficiently modified for the British case, then new theory is perhaps required.

One quality of British special forces makes their control aspects different from other national contexts. It is related to the idea of direction in a more informal, integrated decision-making environment. The study of American special operations forces, and civil–military relations in general, is uncomplicated given clear chains of command, hierarchical organization, published authorities, and well-defined checks and balances, in the form of principals monitoring and punishing agents, in the system of government. The British system, in comparison, can only be characterized as 'opaque', particularly as it relates to authorization, authorities, and monitoring. The American authorization chain for the raid against Osama bin Laden in 2011 can be disentangled given an understanding of the terminology and lines of authority, as Joseph Berger III does in an article, cleared for public release, in an official government journal.[98] The very existence of an article publicly discussing a covert action a year after the event would be looked upon with horror by those in Whitehall, in which secrecy is the norm and 'need to know' is as much a means of protecting the system as it is for exclusion, for the purpose of wielding power.[99]

The British system involves fewer levels of hierarchy and therefore 'control' in the sense of the normal American theories on civil–military relations. Authorization exists but according to precedent and not an established legal code separating out functions. Decisions are made informally, often without consulting Parliament,

[96] Aaron Edwards, *Defending the Realm: The Politics of Britain's Small Wars since 1945* (Manchester: Manchester University Press, 2012), pp. 287–288.

[97] David Charters, *Whose Mission, Whose Orders? British Civil–Military Command and Control in Northern Ireland, 1968–1974* (Montreal: McGill-Queen's University Press, 2017), pp. 37–38.

[98] Joseph Berger III, 'Covert Action: Title 10, Title 50, and the Chain of Command', *Joint Forces Quarterly*, 67, 4th Quarter (2012): 32–39.

[99] Cable, *Political Institutions and Issues in Britain*, p. 61.

by ministers at the behest of the prime minister, and enacted through committees or various appointed executives, such as the secretary of state for Northern Ireland. SAS and other personnel are used in a uniformed capacity in support of traditional military missions but also in paramilitary capacity for covert action under the direction of the SIS. Rory Cormac explains a crucial difference between the American and British conceptualization of special forces: 'Unlike their American counterparts, which are fundamentally a military asset, Britain's special forces are a national instrument, paid for by the Ministry of Defence, but working closely with SIS [government headquarters, the rough equivalent to the National Security Agency and US Cyber Command], and the Foreign Office.'[100] These close working relationships in the British system, based on informal ties, make it much more likely for the 'agent' in Feaver's theory, in this case special forces, to engage with the 'principal', in the form of ministers as well as the prime minister directly in some cases.

The benefits of this type of system for those in the American one—less monitoring and oversight, along with sanctioning methods—are apparent but occur at a cost. Unlike covert action, which is subject to authorization by the Joint Intelligence Council and monitored by parliamentary committees on intelligence, military special operations are exempt from oversight. A 2023 report summarizes control mechanisms for British SOF, or rather the lack of them, as follows:

> Strategic decisions regarding their deployment and operationalisation, however, are not subject to the same extent of consultation with the House of Commons as deployment authorisations for the regular Armed Forces. In contrast to other branches of the Armed Forces, any strategical decision related to Special Forces is taken alone by the Prime Minister, the Defence Secretary, and the Director of Special Forces. This appears to be at odds with what is clearly stated in the authoritative statement of how the MoD operates.
>
> The Special Forces are the only branch of the UK military not subject to external oversight mechanisms. In addition, a lack of transparency over internal oversight and accountability mechanisms, including investigations conducted by the Royal Military Police, raises concerns over their ability to operate independently and effectively.[101]

In summary, clearly mapping control aspects related to special forces in the UK is more challenging than the American context for structural and cultural reasons. Therefore, this section derives insights from successive waves of scholarship exploring civil–military relations in the UK context. Existing theories of civil–military relations, with their emphasis on model development and abstraction of details in pursuit of wider explanatory power and focused on the unique American case, with

[100] Cormac, *Disrupt and Deny*, p. 9.

[101] Hamza Bozkurt et al., *Accountability and Oversight of UK Special Forces* (Colchester, UK: Human Rights Centre Clinic, June 2023), p. 5.

its greater transparency and clear delineations, are likely to overlook or miss completely the subtleties and nuance contained in the more organic system based on relationships and personalities. Figure 6.1 depicts collaborative control. Not always clear and direct, reflecting the opacity of the British system, its informality is depicted vertically as a dashed line, not always in alignment in capabilities.

## Case study, 1971–1991

The time period is significant for the theory of special operations coherence for a number of reasons. First, there are a wide range of special forces activities in a number of different environments in conditions: from independent actions to integration with police and conventional forces in two very different campaigns, irregular or counterinsurgency in the case of Northern Ireland and conventional during the Falklands War. That integration occurred to varying degrees, with visible effects on the outcomes. Special operations activities include offensive, indirect-shock actions (raids or direct-action missions), intelligence gathering and reconnaissance including special reconnaissance and clandestine surveillance, and countering asymmetric threats including insurgents and terrorists. Some activities took place within Britain itself, others in British territory (Northern Ireland and Gibraltar), and yet more outside the United Kingdom in locations as far away as the Falkland Islands in the South Atlantic to Oman, Dhofar, and Iraq in the Arabian Peninsula. During this time period the organizational structure of British special forces, specifically its command-and-control structure, was more formalized through the creation of the Directorate of Special Forces (DSF), with impacts on the relationships that existed between special forces units as well as the machinery of Whitehall.

The case study is divided into four subordinate time periods: from 1971 until 1976, with SAS forces committed to Oman and ending with Prime Minister Harold Wilson's public commitment of the Regiment to Northern Ireland, a break with previous government policy. The next section explores the development of new units to fill capability gaps, including one for hostage rescue, employed dramatically and very publicly in London at the Iranian Embassy in 1980. This period was marked by withdrawal from Oman and increased coordination in Northern Ireland. In the wake of the hostage rescue, the third section explores the consequences of this action for vertical control with a new 'principal', Prime Minister Margaret Thatcher, and her ministers, including the activities of the SAS and SBS in the Falklands War in 1982. The final section, covering from the end of the Falklands War until John Major's election as prime minister, concludes with the end of the Gulf War. Significant events during this time include increased use of SAS forces in ambushes in Northern Ireland, the creation of DSF, the pre-emptive action taken against an Irish Republican Army Active Service unit in Gibraltar in 1988, and the First Gulf War, which ended in 1991.

## Period 1: 1971–1976

Regular training and exercising, as well as new and old forms of preparation, occupied British special operations capabilities. The continuing mission involved 21 and 23 SAS preparations to support the British Army of the Rhine (BAOR) in the defence of West Germany against a Warsaw Pact invasion. These preparations consisted largely of training and exercises largely for the purpose of tactical and operational reconnaissance, by allowing their observation positions to be overrun and providing situation updates on the direction, strength, and identity of successive Soviet echelons of armoured, mechanized, and motorized units.[102] One operator describes the frenetic pace of activities, a mixture of operations, training, and exercises as a 'mad rush' to ensure skills were 'aired and sharpened'.[103]

Another composite unit, the British Military Exchange Mission (Brixmis) stationed in East Berlin, used its diplomatic access to conduct technical and operational intelligence gathering on new Soviet equipment and units, including taking photographs as well as stealing what components they could. The unit was an amalgamation of specialists from the different armed services, including RAF air reconnaissance and photo interpretation analysts and SAS personnel, as well as Foreign Office personnel. To integrate the team and develop it into an intelligence-gathering capability, all personnel seconded to Brixmis attended a specially designed two-and-a-half-week course that began in 1972. Training included conducted recce missions on US and British bases in the United States using specialized equipment they would be issued subsequently, including night-vision goggles and high-resolution cameras.[104] The new mission, countering the asymmetric threat posed by transnational terrorism, is discussed subsequently.

British SOF were committed operationally abroad and in a more recent development domestically over this five-year period. Abroad, the predominant focus was on the remaining Crown colonies in the southern Arabian Peninsula increasingly threatened by Yemen, which was backed by the Soviet Bloc. Understanding of the SAS's role in Oman and Dhofar has been coloured unfortunately by a single event during this period: the defence of Mirbat on 19 July 1972. Mirbat features dramatic elements of a desperate colonial siege: vastly outnumbered defenders prevailing despite the odds due to determination, pluck, and courage. In addition, it has received specific treatment by popular military non-fiction author Rowland White, guaranteeing wider recognition.[105] To make the story more dramatic, the odds are often depicted

[102] Described more fully in de la Billère, *Looking for Trouble*, pp. 161–165; Tony Geraghty, *Beyond the Front Line: The Untold Exploits of Britain's Most Daring Cold War Spy Mission* (London: HarperCollins, 1996), p. 204.

[103] Lofty Large, *Soldier against the Odds: From Infantry to SAS* (Edinburgh: Mainstream Publishing, 1999), p. 259.

[104] Geraghty, *Beyond the Front Line*, pp. 316–318.

[105] Rowland White, *Storm Front* (London: Corgi Books, 2012).

in overwhelming terms: a mere nine members from the SAS held off between 200 and 300 Adoo guerrillas and inflicted a crushing defeat upon them. In addition, many narratives focus on the actions of Sergeant Talaiasi Labalaba, who single-handedly manned the lone crew-served artillery piece and was subsequently killed. Mirbat has been equated with the defence of the Alamo (1836) or Rorke's Drift during the Zulu War (1879) in romanticized terms that call to mind the Victorian colonial adventure novels *Beau Geste* or *The Four Feathers*.[106] Critics charge that such narratives represent little more than latent British militarist fantasizing and myth-perpetuation.[107] Left out of many accounts was the presence of just over one hundred Omani and Pakistan Balochi policemen or soldiers—if mentioned, their contributions are denigrated or dismissed.[108]

While British special operations in Oman and Dhofar predated 1971, stretching back to 1958, they gathered momentum from 1970 onward. A change in leadership of the Sultanate, by coup conducted by family members installing Sultan Qaboos, opened the door to more comprehensive security measures. Qaboos favoured a British-designed strategy to sway popular support to his regime and away from the guerrillas of the Population Front for the Liberation of the Arabian Gulf (PFLOAG) or Oman (PFLO), known collectively and informally as *adoos*, or foes. The accession of Qaboos to power in 1970 provided the SAS its initial entree, as he requested its support first in providing for his personal security and training his bodyguard forces.[109] Johnny Watts, the officer commanding 22 SAS, outlined the strategy termed 'the Five Fronts'. They included intelligence gathering, propaganda or information, medical, veterinary, and the raising of the *firqat*.[110] As Alexander MacKenzie points out, while comprising a comprehensive approach, Watts' strategy was not novel but rather reflected best practices in COIN and special operations in Malaya, Borneo, and Vietnam.[111]

After initial operations in 1971 and 1972 to secure terrain, involving two squadrons of 22 SAS, Watts shifted special operations to proxy forces, and in particular the *firqat*, as the locus of local security measures.[112] These forces were backed by Sultan's Armed Forces (SAF) reconnaissance and air strikes and led by SAS teams, suggesting the approach used by the Americans in 2001 and later in 2014 discussed in Chapter 8, dubbed 'the Afghan model', is not unique. Military control was exerted over attempts to import a special operations practice from Aden and Borneo, clandestine cross-border raids, given the risks they posed politically in escalating the conflict with

[106] MacKenzie, *Special Force*, p. 145; Stavros Atlamazoglou, 'Battle of Mirbat: The Near Alamo of the Special Air Service', SOFREP, 5 March 2020, https://sofrep.com/news/special-air-services-secret-war-oman-part-5.

[107] John Newsigner, 'The Cult of the SAS Is about Britain's Imperial Delusions', Jacobin, 12 October 2022, https://jacobin.com/2022/10/sas-cult-britain-imperial-delusions-military-history.

[108] Seymour, *British Special Forces*, p. 299.

[109] J.F. Peterson, *Oman's Insurgencies: The Sultanate's Struggle for Supremacy* (London: Saqi, 2007), p. 328.

[110] John Strawson, *A History of the S.A.S. Regiment* (London: Guild Publishing, 1985), p. 202.

[111] Alastair MacKenzie, *Special Force*, p. 138.

[112] Strawson, *A History of the S.A.S. Regiment*, p. 209.

Yemen's Soviet Bloc sponsors and drawing unwanted attention to Britain's increasing role in the conflict. The British General Office Commanding (GOC) in Dhofar, Timothy Creasey, walked a fine line between appeasing Sultan Qaboos, who wanted widespread covert raids, and Whitehall ministers anxious about the wider political implications should the British role be revealed.[113] In order to train the proxy Mahra tribesmen, the Sultan personally 'requested an additional presence of SAS personnel', suggesting continuing close relations between the sovereign and British special forces.[114]

It was in the role of military advising and assistance of the *firqats* that British special operations came into their own in the Dhofar War (1971–1976; see Map 6.1). To counter an estimated 5,000 *adoo*, divided almost in half between guerrilla regulars and auxiliaries, 22 SAS squadrons on a rotating basis trained, advised, and led some 1,600 personnel of 21 different *firqat* units.[115] The common practice was for roughly two teams of seven SAS operators, operating officially as British Army Training Teams (BATTs), to train, equip, and lead detachments of roughly a platoon of local personnel, some 30-odd men. Recruiting from among various tribes, *firqats* were bolstered from another colonial practice in Kenya, Malaya, and elsewhere: the employment of 'turned' guerrillas, former members of the PFLOAG, given '[t]heir knowledge of the ground and their influence with the civilians' that John Akehurst, commander of the Dhofar Brigade who had attended Staff College with SAS Squadron Commander Tony Jeapes, assessed as 'indispensable'.[116]

*Firqat* were part of a layered defence approach to PFLOAG infiltration from neighbouring Yemen. Defensive barrier lines comprised barbed wire and mines, patrolled by aircraft with security posts at intervals in between. The *firqat* provided the tactical offensive element of this defence, patrolling intrusively into enemy-held or contested territory in an effort to interdict *adoo* supply lines.[117] They deployed primarily in mountains of the Eastern Qara range, territory difficult to patrol or cordon off, operating from 'mountain-top, Government-held bases' that sought to contest PFLOAG influence and sanctuary in the rugged terrain.[118] To win suspicious hill tribes of *jebalis* and convince them of the benefits of siding with the government after security had been established, the Sultanate established in 1975 a Civil Aid Department (CAD) to extend welfare and social services. A SAS officer, Martin Robb, served as CAD's first coordinator, suggesting the Regiment's growing influence at a critical stage of the conflict.[119]

[113] Geraint Hughes, 'A Proxy War in Arabia: The Dhofar Insurgency and Cross-Border Raids into South Yemen', *Middle East Journal*, 69:1 (Winter 2015): 97–100.

[114] Hughes, 'A Proxy War in Arabia', p. 97.

[115] Philip Towle, *Pilots and Rebels: The Use of Aircraft in Unconventional Warfare, 1918–1988* (London: Brassey's, 1989), p. 152; MacKenzie, *Special Force*, p. 141.

[116] John Akehurst, *We Won a War: The Campaign in Oman, 1965–1975* (Wilton, UK: Michael Russell [Publishing] Ltd, 1982), p. 43.

[117] MacKenzie, *Special Force*, p. 142.

[118] Ranulph Fiennes, *Where Soldiers Fear to Tread* (London: Hodder and Stoughton, 1975), p. 242.

[119] Peterson, *Oman's Insurgencies*, p. 393.

**Map 6.1** Map of Oman, with Dhofar highlighted.

*Source:* United States Central Intelligence Agency. *Oman* (Washington, DC: Central Intelligence Agency, 1996). Map. www.loc.gov/item/96,681,131.

Military advising and assistance special operations in Dhofar were fraught with control-related issues. Based on his experiences working with them, John Akehurst characterized the *firqat* as 'unpredictable charges' and offered the following assessment: 'Properly motivated, and with the prospect of financial or other gains, they could be splendid fighters, as good for us as they had been against us; but equally they could, if the mood took them, be intransigent and uncooperative, sometimes aggressively so.'[120] While it is easy to explain such qualities culturally, they reflected rational self-interest and a keen self-awareness of where the locus of agency truly lay: with the locals and not their advisors. The *firqat* perfectly understood that advisors 'had no executive authority; he had to persuade the man with power to issue the order', leading to delays in action and no small measure of exasperation.[121] Patience was one virtue assisted by another: communication. Unlike in Malaya and Borneo, where operators spoke local languages and dialects, increasing numbers of SAS operators from G, B, and other squadrons rotating through Dhofar spoke Arabic, easing coordination and preventing miscues or misunderstanding through inept translation. Despite this, few operators were fully integrated with their proxies: 'they were with the firqa but not of them' and 'lived and ate apart', as Jeapes concludes, to avoid subsuming their identities as well as firewalling the frustrations they experienced trying to control and direct them.[122]

Control-related issues were not limited to barriers between operator and proxy. They extended vertically to include control of operations and operators. Radio communication between 22 SAS Headquarters in the United Kingdom and forward-deployed squadrons was tenuous. To prevent communications from being intercepted, they needed to be encrypted. The only practical method of transmission was via Morse code. The net effect of these limitations of and impositions on radio communication meant only short messages were sent, usually once a day, updating the commander of the local situation. To gain a better understanding of evolving conditions, the 22 SAS commander and deputy flew forward in alternating rotation every six weeks, a journey that could take several days.[123] While comprehensible for the maintenance of operational control of forces under command halfway around the world, the rotation of commanders led to friction with at least one squadron commander. In a peculiarly British form of conflict resolution characteristic of collaborative control, Peter de la Billère, the deputy commander, had the G Squadron commander removed for resisting an administrative control order from the director of the SAS. He did so by leveraging the political influence of a retired patron of the Regiment, Dare Newell.[124] Collaborative control is evident in another episode later in the campaign, a contretemps between the director of the SAS and the 22 SAS commander. Watts, promoted to director, countermanded the order by Jeapes to have a junior officer removed from theatre as well as the Regiment. After considerable disagreement and a series of orders and counterorders between the two, eventually 'a face saving compromise' was reached, transferring the officer to a different squadron.[125]

[120] Akehurst, *We Won a War*, p. 43.
[121] Jeapes, *SAS*, p. 60.
[122] Tony Geraghty, *Who Dares Wins: The Story of the Special Air Service, 1950–1980* (London: Arms & Armour Press, 1980), p. 133.
[123] De la Billère, *Looking for Trouble*, pp. 268–269.
[124] Ibid., pp. 272–273.
[125] Connor, *Ghost Force*, pp. 157–158; MacKenzie, *Special Force*, p. 143.

Further down the chain of command, the distribution of SAS teams among the various *firqat* created other problems, suggesting weak vertical control links. The loyalty of *firqat*, as with most proxies, depended on a steady supply of rent. Individual SAS teams dispersed the funds with little in the way of oversight. Given the scope of the war, which Prime Minister Harold Wilson sought to keep small by design, normal control mechanisms for monies, in the form of Royal Army Pay Corps personnel, were scarce.[126] A whistleblower from the Regiment's ranks alerted higher headquarters of dubious practices taking place in the field: claims of *firqat* requiring payment that existed on paper only, whose monies were pocketed by the operators reporting them. MacKenzie assesses the practice was not isolated but endemic, with members of every squadron engaging in it during their four-month rotations.[127]

Addressing the problem posed a challenge for 22 SAS Commander de la Billère: follow standard procedure and the matter would bring unwanted attention to a campaign almost unknown among the British and threaten relations with the Sultanate. By not following procedure, inevitably some details would find their way into the press, leading to claims of a cover-up. In the end, the matter was referred to the representative of another informal means of collaborative control: the colonel commandant of the SAS, a symbolic and not executive position, who in 1974 was Antony Head, the 1st Viscount Head. Head ensured vertical reporting up through his informal connection and assumed personal responsibility for the incident, using his political capital and influence as collateral. In this way, 'the matter was brought under control' politically while a board of inquiry within the Regiment set upon the task of housekeeping.[128] A few were censured officially and some monies paid back, while others were sent back to their former conventional regiments, a form of social sanctioning known as an RTU (returned to unit).[129]

The fate of the whistleblowing officer points to a different aspect of weak vertical control. MacKenzie relates that he was 'ostracised by the soldiers for publicising and curtailing what was, obviously, a very lucrative income supplement'.[130] The term 'ostracise' brilliantly illustrates a point within the Regiment at this time. Informal social power and influence existed among the non-commissioned officers and low-level ranks. Officers in charge of troops as well as squadrons were with decreasing frequency promoted from the ranks as had occurred after the end of the Second World War through Borneo. For example, Johnny Cooper, an SAS 'original' trooper who served with David Stirling, rose to become 22 SAS commander before retiring. Increasingly, a new generation of officers without special operations were put in charge of troops or squadrons. In garrison such officers could learn their trade over time and gain the respect and confidence of those under their command. In combat

126 De la Billère, *Looking for Trouble*, pp. 286–288.
127 MacKenzie, *Special Force*, p. 146.
128 De la Billère, *Looking for Trouble*, pp. 287–288.
129 Ibid.
130 MacKenzie, *Special Force*, p. 146.

conditions in Dhofar, however, the lack of combat experience and knowledge led to more than episode of troops taking action first without waiting for orders.[131]

Success in Dhofar at the end of a six-year war was achieved at minimal cost for a variety of reasons. British special operations played a role, particularly in their ability to serve an integrating function with other capabilities. While other technical advisors and British officers and non-commissioned officers (NCOs) provided a professional core to the Omani forces fighting in Dhofar, including its modest but effectively wielded fleet of strike and mobility aircraft, British special operators provided an effective link between those forces, the *firqat*, and the population. Close working relationships existed between *firqat* and other Omani forces, which often worked in conjunction with each other, taking advantage of their mutual strengths: 'The firqat's understanding of the ground and their speed of manoeuvre were superior to SAF troops', but when it came to straight military tactics, the SAF's military discipline told every time. The two forces were complementary; neither could have won the war alone'.[132]

Military strength alone could not defeat the PFLOAG. Central to it was a comprehensive approach to COIN that considered methods of influencing and attracting the various tribes of *jebali* to have confidence that the Sultanate took their interests seriously. The Five Fronts strategy, developed by Watts without prompting, provided just such a comprehensive approach in theory. Putting that theory into practice, however, required British special operators to play outsized roles in psychological operations and civic action programmes. A striking element about special operations in the Dhofar War is the influence of informal social ties serving an integrating function. Such informality included the assigning of SAS personnel to key positions, from the Sultan's initial bodyguard to the head of CAD.

These informal ties were also evident as BATT was stood down to be replaced by Firquat Forces, whose leadership ranks were increasingly filled with former SAS operators.[133] PFLOAG tactics also played a role in alienating the population. In line with requirements to report *jebali* whose political loyalties were questionable to higher headquarters, local guerrilla commanders duly provided lists of names, many of which reflected personal vendettas or ambivalence towards accuracy, resulting in the application of seemingly indiscriminate violence.[134] Success was further enabled by unity of command characteristic of an autocratic monarchy, albeit one influenced by informal social ties. A final and not insignificant factor was resources and more importantly the bonanza the global energy crisis in 1973, driven by war between Egypt and Israel, resulting in vastly increased oil revenues for the Sultanate that not only funded the war but also a series of infrastructure and social welfare

[131] Connor, *Ghost Force*, pp. 168–169.

[132] Jeapes, *SAS*, p. 237.

[133] Ibid., p. 236; David Arkless, *The Secret War: Dhofar, 1971/1972* (London: William Kimber, 1988), p. 92.

[134] Fiennes, *Where Soldiers Fear to Tread*, p. 167.

investments.[135] The campaign in Dhofar influenced British special operations when their focus shifted closer to home across the Irish Sea.

The historical context preceding the last wave of 'The Troubles' in Northern Ireland, as well as that of the creation and employment of British special forces, fill volumes. Paul Gill and John Horgan, quoting fellow scholar John Whyte, note Northern Ireland 'is the most heavily researched area on earth.'[136] For the sake of analytic focus and space, the discussion of British special forces capabilities is limited to four aspects related to control. The first is the nature of British military policy towards Ireland and after 1922, Northern Ireland. Despite stationing and rotating military units through Ireland since the Act of Union in 1798, successive British governments have 'steadfastly avoided using it to the control the country' and were 'almost as nervous about its army in Ireland as it was about Ireland itself'.[137] This aversion and nervousness contributed significantly to historical civil–military crises such as the 'Curragh Affair' in 1914, in which British officers threatened to resign when they translated opaque instructions as orders to prepare to use force against Loyalist Protestants.[138] It also help explain the decisions, subject to much scrutiny and criticism by commentators later, not to use units of the British Army more forcefully when the Troubles reignited in 1968 following civil rights marches that subsequently escalated into sectarian violence the Royal Ulster Constabulary (RUC) was unwilling or unable to control.

British reticence to use force is further explained by the dual nature of governance in Northern Ireland, which had a parallel system of ministers, cabinets, and its own indigenous, Protestant-dominated police force, the RUC. The informal, ad hoc responsive nature of the British style of governance meant early opportunities to restore order and gain a level of confidence or at least impartiality of the Catholic-minority population in Northern Ireland were squandered. The integration and smooth functioning characteristic of this form of governance in the face of crisis only occurred after missteps, confusion, and escalation of the conflict in 1972. They include the events of Bloody Sunday, subsequent reprisal bombings, the implementation of Direct Rule from Britain and institution of a series of emergency powers, and the conduct of focused offensive, Operation MOTORMAN (1972), designed to take back control of traditional Catholic-held sections, considered 'no-go' areas, of Londonderry and Belfast (see Map 6.2).[139] From 1972 onwards the British placed significant limits on the use of force and the Wilson Cabinet thought the security situation was stable enough in 1977 to announce a

135 Ibid., p. 242; Peterson, *Oman's Insurgencies*, p. 447.

136 Paul Gill and John Horgan, 'Who Were the Volunteers? The Shifting Sociological and Operational Profile of 1240 Provisional Irish Republican Army Members', *Terrorism and Political Violence*, 25:3 (2013): 435.

137 Elizabeth Muenger, *The British Military Dilemma in Ireland: Occupation Politics, 1886–1914* (Lawrence, KS: University Press of Kansas, 1991), p. 1.

138 M.L. Connelly, 'The Army, the Press, and the "Curragh Incident", March 1914', *Historical Research*, 84:225 (August 2011): 535–557.

139 Andrew Sanders, 'Operation Motorman (1972) and the Search for a Coherent British Counterinsurgency Strategy in Northern Ireland', *Small Wars & Insurgencies*, 24:3 (2013): 465–492; Charters, *Whose Mission, Whose Orders?*.

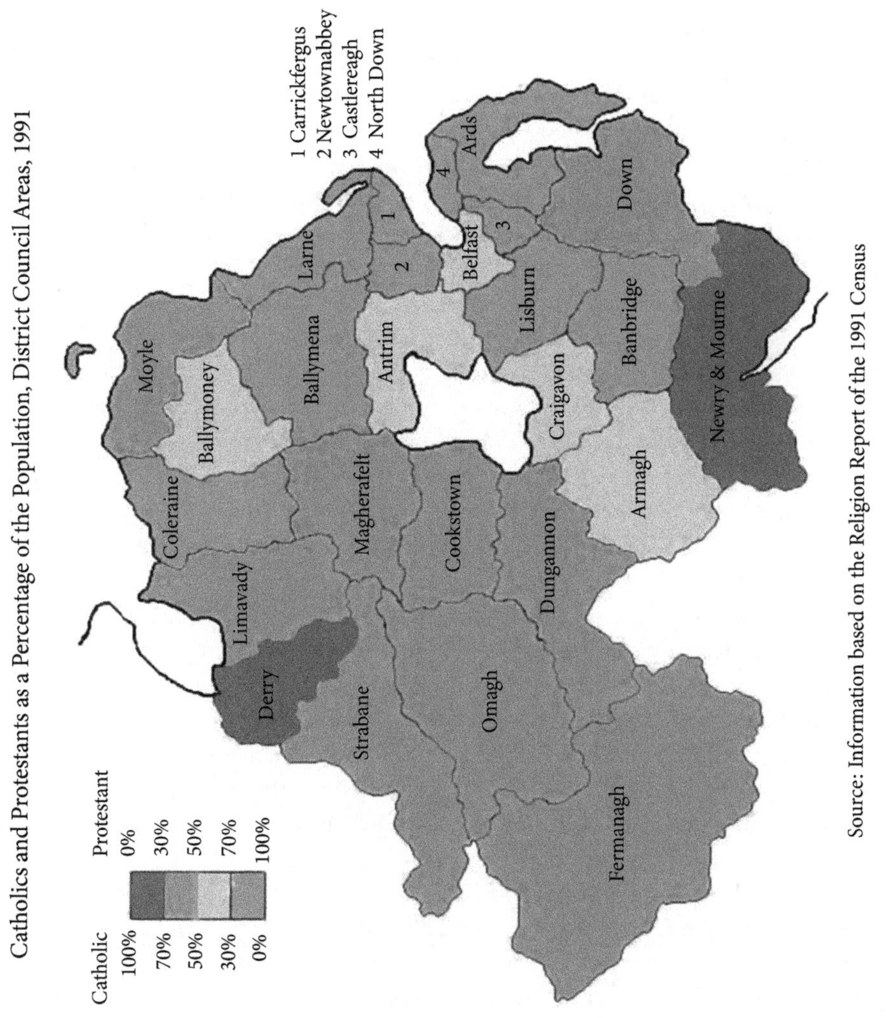


**Map 6.2** Northern Ireland, map of religious affiliation by council area, 1991.
*Source:* CAIN (https://cain.ulster.ac.uk/images/maps), permission to reprint granted.

policy of 'police primacy'.[140] This policy and events leading up to it shifted towards a law enforcement approach to countering terrorism, with the RUC as the lead agency, that shaped the character of, and imposed constraints upon, the collection of intelligence and the use of military force in Northern Ireland in aid of the civil power.

The second element was the changing character of irregular violence in Britain and globally. In Northern Ireland, this character is associated with the Irish Republican

[140] Bradley Bamford, 'The Role and Effectiveness of Intelligence in Northern Ireland', *Intelligence and National Security*, 20:4 (2005): 584.

Army (IRA). Within the IRA, however, there was a struggle between an older generation waiting for the right circumstances to rise up against British control along more traditional revolutionary lines and a younger generation, inspired first by the successes of the civil rights movement in the United States as well as by events globally in 1968, 'the year of the heroic guerrilla', to conduct more forceful and ruthless urban guerrilla warfare.[141] The split within the IRA occurred the following year between the Official IRA and the more radical, socialist-inspired Provisional IRA (PIRA), to peak radically for Britain in Northern Ireland, including dreams of a 'year of victory' in 1972 and much later an Irish 'Tet Offensive'.[142] As a result of police and military pressure, as well as developments in other urban guerrilla movements elsewhere and taking advantage of an almost two-year ceasefire, the PIRA shifted from a more formal, hierarchical structure to networked one based on cells and small groups, or Active Service Units (ASUs), much more difficult to authorities to identify, isolate, and combat.[143] Technologies, in the form of accessible air travel, televised news covered, and the miniaturization of electronics, gave groups like the PIRA unprecedented mobility and access, coordination, and influence, further complicating British government efforts.[144]

A final aspect related to irregular violence, particularly in Northern Ireland, was the array of different violent and non-violent non-state actors, which included student and labour protest movements, political and front groups, Protestant paramilitary organizations and armed groups, human rights groups, criminal gangs, and media organizations. The net effect on British security operations was increased capability and reach of terrorist groups, a complex operating environment, and considerable media coverage of, scrutiny of, and interest in actions taking place Northern Ireland.

The third element was the use of British special forces within and without prior to 1971. Within the United Kingdom and Northern Ireland specifically, one source puts the first deployment of SAS forces to Belfast, in a limited and covert, plain-clothes capacity, in 1966 and training exercises in 1969.[145] MacKenzie, however, suggests overt 22 SAS foot patrols were conducted the same year as the exercises. The patrols, Operation CUFF, served as a visible presence of security forces to dissuade

[141] Robert Daniels, *Year of the Heroic Guerrilla: World Revolution and Counterrevolution in 1968* (Cambridge, MA: Harvard University Press, 1987); Michael MacDonald, *Children of Wrath: Political Violence in Northern Ireland* (Cambridge: Polity Press, 1986), pp. 85–86; Richard Clutterbuck, *Protest and the Urban Guerrilla* (London: Abelard-Schuman Ltd, 1973).

[142] Alan Parkinson, *1972 and the Ulster Troubles* (Dublin: Four Courts Press, 2010); Dieter Reinisch, 'Dreaming of an "Irish Tet Offensive": Irish Republican Prisoners and the Origins of the Peace Process', The Irish Story.com, 6 October 2018, www.theirishstory.com/2018/08/06/dreaming-of-an-irish-tet-offensive-irish-republican-prisoners-and-the-origins-of-the-peace-process; Ed Moloney, *A Secret History of the IRA* (New York: W.W. Norton & Company, 2002), p. 255.

[143] Jacob Shapiro, *The Terrorist's Dilemma: Managing Violent Covert Organizations* (Princeton, NJ: Princeton University Press, 2013), pp. 182–185.

[144] James Kiras, 'Terrorism and Globalization', in John Baylis and Steve Smith, eds., *The Globalization of World Politics: An Introduction to International Relations*, 3rd ed. (Oxford: Oxford University Press, 2005), pp. 482–483.

[145] Connor, *Ghost Force*, p. 177.

more sectarian violence flaring in Ulster. They were carried out with little fanfare or publicity, and nothing was done to distinguish or disguise 22 SAS from other British regiments taking part.[146] A combination of limited resources and more urgent foreign policy priorities, particularly commitments in Aden and Dhofar, limited the availability and desirability of sending teams from 22 SAS prior to 1976.

So too did the behaviour of operators in the city. Their conduct suggested that private enterprise along criminal lines in Dhofar was still a problem, leading to official RUC complaints and colouring subsequent decisions to employ them to Northern Ireland.[147] The 22 SAS operators brought with them practices and SOPs from Dhofar, Borneo, and Malaya suitable for operations in colonies, protectorates, and possessions overseas. How much of what subsequently happened in Northern Ireland is attributable to colonial legacy, or the nature of special operations more generally, is unclear. Tony Jeapes summarizes the core of the issue with special operators synoptically: 'SAS soldiers are fighting men: you do not train tigers and expect them to sing like canaries', in explaining why BATT forces took casualties earlier than commanders anticipated.[148] The word commonly used to describe 22 SAS operations in Northern Ireland from 1971 until 1980, given their practices, is 'aggressive'.

This aggression contributed to the fourth and final element: the development of new British special forces capabilities. Some of the tactical skills and experience the operators of 22 SAS had honed and refined in combating irregulars overseas was transferrable to activities in Northern Ireland, including mobile patrolling and reconnaissance from static observation posts or heavily camouflaged observation posts, or OPs, also known as hide sides. Aggression was one contributing factor. Another far more important one, however, was related to gaps in intelligence regarding the PIRA. OPs operated overtly but the 'most effective' were clandestine ones conducted in close proximity to their targets, at distances of 20 metres or less.[149] Conventional British units fielded the majority of such OPs in their newly created 'close observations platoons', or COPs.

Urgent operational necessity in new and very different conditions in Northern Ireland, particularly densely populated urban areas divided along sectarian lines. The dual nature of governance, and law enforcement focus of counterinsurgency activities, meant British Army units were reliant initially on information from Northern Ireland Special Branch. This civil agency was understaffed, ill-coordinated, and 'staggeringly inefficient', qualities shared with Special Branch organizations elsewhere in British colonies when crises erupted.[150] Informants, casual or otherwise, proved

[146] MacKenzie, *Special Force*, p. 198.
[147] Ibid., pp. 199; 318, n568.
[148] Jeapes, *SAS*, p. 61.
[149] Alastair MacKenzie, *Pilgrim Days: A Lifetime of Soldiering from Vietnam to the SAS* (London: Osprey, 2019), p. 150; A.F.N. Clarke, *Contact* (London: Secker & Warburg, 1983), p. 30.
[150] Connor, *Ghost Force*, p. 178; David Charters, 'Professionalizing Clandestine Military Intelligence in Northern Ireland: Creating the Special Reconnaissance Unit', *Intelligence and National Security*, 33:1 (2017): 131; Martin Thomas, *Empires of Intelligence: Security Services and Colonial Disorder after 1914* (Berkeley, CA: University of California Press, 2008), pp. 109–117; 244–246.

difficult to obtain from the Catholic-minority population after British Army forces did little to halt militant Protestant actions from 1969 onwards.

An initial experiment to coordinate covert intelligence gathering by conventional forces, the Military Reaction Force (MRF), was designed to fill some of the information vacuum. The experiment, the brainchild of Frank Kitson who developed the concept of pseudo-guerrilla operations used in Africa (see Chapter 7), backfired publicly and dramatically when IRA forces raided the front companies and assets of the MRF in 1972.[151] Senior officers within the British Army, however, were less concerned about the MRF's blown cover and more with what they perceived as 'a lack of professionalism and accountability'.[152] Two innovation solutions were developed, after much coordination and discussion at the Ministry of Defence level, to provide better intelligence clandestinely: a plain-clothes mobile, close and technical reconnaissance unit, the Special Reconnaissance Unit (SRU), in 1973, and more controversially, a plain-clothes human intelligence source development and agent-running unit, the Force Research Unit (FRU), in 1981. To get the first new unit and capability up and running as quickly as possible in 1973, thirty operators, or just under half of a Sabre Squadron, withdrew from the Regiment on a temporary basis to fill fully one-quarter of SRU's ranks through a process known as 'de-badging', or no longer appearing on the SAS's roster.[153] Both SRU and FRU could operate in areas denied to other British special forces and alleviated the political and military concerns about deploying 22 SAS, whose reputation for action was well established.

SRU also was developed based on additional special operations failure. Early attempts by 22 SAS operators to conduct close surveillance in urban environments were unsuccessful. James Rennie, an SRU operator, outlines several reasons why. The residents of neighbourhoods in which close clandestine surveillance would be conducted were both suspicious of outsiders and observant of changes in their environment: 'families that had lived in the same house and street for years, maybe two or even three generations ... They were interrelated or had known one another from when they were children. Strangers were noticed'.[154] Compromise of small clandestine teams could have fatal consequences, particularly if locals identified anything that raised their suspicions, as the case of Robert Nairac suggests.

[151] Kitson's seminal work on pseudo-operations remains *Gangs and Counter-Gangs* (London: Barrie and Rockliff, 1960). Due in part to Kitson's reputation as a COIN expert and given his hand in creating the MRF, he has become in the minds of some a spider at the centre of a web of clandestine activities in Northern Ireland equivalent in conspiracy to Edward Lansdale in the United States. For example, see David Burke, *Frank Kitson's Irish War: Mastermind of the Dirty War in Ireland* (Cork, Ireland: Mercier Press, 2021).

[152] Geraint Hughes, 'The Use of Undercover Military Units in Counter-Terrorist Operations: A Historical Analysis with Reference to Contemporary Anti-Terrorism', *Small Wars & Insurgencies*, 21:4 (2010): 572.

[153] Charters, 'Professionalizing Clandestine Military Intelligence in Northern Ireland', p. 134; Harry McCallion, *Undercover War: Britain's Special Forces and Their Secret Battle against the IRA* (London: John Blake, 2020), p. 20.

[154] Robert Brown, *Special Forces Operator: Serving with the SAS and MRF* (Barnsley: Pen & Sword, 2021), p. 123.

Nairac, an early SRU member, was taken from a pub after trying to pass himself off as a Republican sympathizer. He was interrogated and killed, his body never recovered.[155] The 22 SAS operators on clandestine mobile surveillance were unable to disguise the fact they were the 'tigers' Jeapes described. Their attempts to blend in to urban environments, by wearing plain clothes, were problematic—operators tended to dress in civil society in a suspiciously uniform way.[156] In addition, some could not mask their disdain for Republicans and glared at them 'like angry dogs on the leash' and 'lacked subtlety'.[157] A subsequent 22 SAS attempt to conduct a mobile reconnaissance outside of the cities, where less subtlety was required, only seemed to confirm this assessment. In May 1976, eight operators were stopped, detained, and arrested over the border in Eire (the Republic of Ireland), leading to diplomatic tensions and public questions about the Regiment's value in Parliament.[158]

The SAS's reputation was used politically and publicly by Prime Minister Harold Wilson in a dramatic manner. By all accounts, Wilson's declaration on 7 January 1976 that Britain was sending 22 SAS to Northern Ireland came as a surprise to most, including both the IRA and members of the Regiment. His decision was in response to sectarian violence violating the 1975 ceasefire, including the Kingsmill Massacre of 10 Protestant building workers by PIRA gunman two days prior.[159] It also highlights an important dimension of British civil–military relations and control of special forces. From a strictly military perspective, committing 22 SAS made little sense operationally. A quarter of the Regiment was still involved to operations in Oman and others were resting, refitting, retraining, or preparing to deploy there. In addition, most of the Regiment's operators had not conducted integrated operations with conventional British Army regiments. When they had, it occurred overseas in virtually unknown wars, not in a developed country where they would face media scrutiny and public inquiry. A prevailing perception within the ranks of British Army units was that special forces were ill-prepared 'cowboys' at best or 'trigger happy hoodlums' at worst.[160] A brigadier's response to a proposed covert action in his locale elicited a telling response: 'I'll have no dirty tricks in my area. This is bloody disgraceful. You call yourself an officer?'.[161] Even more problematically for the Regiment, public commitment potentially put at risk the ability to deploy 22 SAS personnel covertly as members of the media would already be alerted to their presence.[162]

From a political standpoint, Wilson's action made sense. In attempting to force the PIRA to deescalate its violence after the Kingsmill Massacre, Wilson escalated with

[155] For contrasting views on Nairac, see John Parker, *Death of a Hero: Captain Robert Nairac, GC, and the Undercover War in Northern Ireland* (London: Metro Books, 1999), pp. 259–261; Raymond Murray, *The SAS in Ireland* (Dublin: Mercier Press, 1990), pp. 147–154.

[156] Michael Asher, *The Regiment: The Definitive Story of the SAS* (London: Penguin, 2008), p. 433.

[157] James Rennie, *The Operators: Inside 14 Intelligence Company—The Army's Top Secret Elite* (London: Century, 1996), pp. 166; 200–201.

[158] Geraghty, *Who Dares Wins*, pp. 153–154; MacKenzie, *Special Force*, pp. 202–204.

[159] Ed Moloney, *A Secret History of the IRA* (New York: W.W. Norton & Company, 2002), pp. 320–321.

[160] A.F.N. Clarke quoted in Murray, *The SAS in Ireland*, p. 182; MacKenzie, *Special Force*, pp. 199–200.

[161] Geraghty, *Who Dares Wins*, p. 150.

[162] Connor, *Ghost Force*, p. 179.

one of the only instruments at his disposal for the threat to remain credible. It also upheld Britain's policy of Direct Rule by suggesting the country and its leaders were willing and had the resources not to leave Northern Ireland any time soon. Peter Neumann concludes that the political theatrics of the Wilson's declaration aside, 'the fact that the SAS—with its reputation for ruthlessness and bravery—was soon blamed for every suspicious incident in the province seemed to confirm its value [as a deterrent].'[163] Wilson's confidence in the SAS may also have been bolstered by the tactical displays put on for 'VIPs' in skills such as freefall parachuting conducted intermittently.[164]

Despite operators' concerns about jeopardizing future operations, the deterrent value of Wilson's declaration may have increased because of the Regiment's prior clandestine activities—suspected or otherwise. Either reflecting their revolutionary ideology, innate suspicion, or for some other reason, senior PIRA leaders believed the SAS had been committed to Northern Ireland well in advance of Bloody Sunday, as a vanguard of future political action, agents provocateurs, or both.[165] Some of this belief could have been informed reasonably by PIRA counterintelligence and countersurveillance activities of the movements of SAS personnel to and within the province, observations by the local population, and inference. It also might have been the product of the paranoia and suspicion associated with survival in clandestine revolutionary groups.[166]

After the initial public and rocky start, 22 SAS activities settled into a steady routine of surveillance and patrolling. SAS operations were expanded in the province, with a troop from each of the Regiment's squadrons operating from Bessbrook, in South Armagh, as well as Londonderry and Belfast, in support of each of the three Army Brigades stationed there.[167] According to former minister of defence and then secretary of state for Northern Ireland, Roy Mason, 'the effect of the SAS in Armagh proved worthwhile and I decided therefore to take the decision that the SAS should operate province-wide.'[168] Tours of duty for troops were eighteen months.[169] Contributing to this expansion was perception among leaders such as Mason that 22 SAS operations, specifically ambushes like the one resulting in the death of Seamus Harvey one year after Wilson's declaration, had a deterrent effect on PIRA units if only locally.[170]

[163] Peter Neumann, *Britain's Long War: British Strategy in the Northern Ireland Conflict, 1969–1998* (New York: Palgrave Macmillan, 2003), p. 107.

[164] AIR 29/3994, 'RAF Detachment to Special Forces' (1966–1970) and AIR 29/4268, 'RAF Detachment to Special Forces' (1971–1975).

[165] Richard English, *Armed Struggle: The History of the IRA* (Oxford: Oxford University Press, 2003), p. 172.

[166] Nathan Bos, ed., *Human Factors Considerations of Undergrounds in Insurgencies*, 2nd ed. (Fort Bragg, NC: United States Army Special Operations Command, 2013), pp. 99–100.

[167] Jack Ramsay, *SAS: The Soldier's Story* (London: Macmillan, 1996), p. 170.

[168] Quoted in Murray, *The SAS in Ireland*, p. 198.

[169] Ramsay, *SAS*, p. 176.

[170] Geraghty, *Who Dares Wins*, pp. 152–153; James Adams, Robin Morgan, and Anthony Bambridge, *Ambush: The War between the SAS and the IRA* (London: Pan, 1988), pp. 82–83.

Other British special forces, including SBS and SRU, maintained a low profile operationally and politically in the provinces as well as abroad with a single exception in 1972. Political concerns associated with countering international terrorism, discussed in greater detail below, extended to fears in maritime realm in two ways: commercial shipping, especially passenger liners, could be bombed or hijacked; and Britain's emerging source of revenue, oil from the North Sea, could be held for ransom should an oil rig be seized. The first concern was raised by a mariner working for the Cunard line about an upcoming chartered cruise on the *Queen Elizabeth II* (*QE2*). A subsequent bomb threat led a troop from the SBS parachuting and boarding her in 1972, accompanied by individual SAS and Royal Army Ordinance Corps disposal technicians along with unwanted media reporting of the incident.[171] This proof of concept was expanded and in future SBS personnel accompanied additional sailings operating clandestinely, either as passengers or as uniformed Cunard employees.[172] With Ministry of Defence approval, and capitalizing on the media coverage to become renamed as the Special Boat Service in 1973, it formed a standing Maritime Counter-Terrorism Force (MCTF) in 1975.[173] The MCTF was short-lived, however, and the SBS troop operating on alert for oil rigs was replaced by another from its same parent organization, the Royal Marines.[174] Three reasons explain why SBS and SRU operations did not receive greater political attention: their smaller size and fewer number of operatives compared to the SAS; the nature of their missions, conducting maritime reconnaissance and clandestine close urban surveillance or running agents, which necessitated high degrees of operational security to protect the lives of operators; and their supporting role relative to much larger organizations, working as part of an integrated intelligence gathering machine in Northern Ireland or as part of the Royal Marines and Royal Navy. The relative newness of SRU cannot be discounted.

The aggressiveness attributed to SAS officers which drew attention, specifically in ambushes, was a product of several individual and collective sources. They include experience outside Britain in Dhofar and Borneo, where there were fewer restrictions. Individually, senior operators wanted to fight the enemy, and newer operators wanted to prove their 'bottle' and pass their probationary period. In the words of one operator, R.F.,

> We knew what jobs we had to do and we did them … On that first tour I was really under scrutiny. When you join the regiment, you go back down in rank to trooper, and my boss gave us a real chance to prove ourselves. Instead of just letting us look like patrol members, he gave us patrols to run.[175]

[171] Peter Mercer, *Not by Strength, by Guile* (London: Blake, 2000), pp. 103–110.

[172] John Parker, *SBS: The Inside Story of the Special Boat Service* (London: Headline Book Publishing, 1997), pp. 252–258.

[173] Parker, *SBS*, p. 270.

[174] Ibid., pp. 272–273.

[175] Ramsay, *SAS*, p. 173.

Such responsibility further tests the suitability of newer operators and builds their experience and confidence. It also exposes them to circumstances where their individual judgement is tried under exceptionally trying circumstances, captured vividly by senior SAS NCO Ken Connor.[176] They include the killing of seven known PIRA members and the arrest of an unknown number of others. Unfortunately, in such circumstances, mistakes occurred, including killing innocent victims.[177] The shooting death of teenager John Boyle on 11 July 1978, and the subsequent public trial and acquittal of SAS Sergeant Alan Bohan and Lance Corporal Temperly, had significant political and military consequences.[178]

Subsequent SAS operations, from 1978 until 1984, were largely restricted to intelligence-gathering missions only. The effect on operators only added to their growing frustration. First was a seeming breach of trust by having their individual weapons registered forensically when the Regiment initially deployed, then additional legal constraints, compounded by unclear political goals, tedious and challenging reconnaissance missions watching movements and individuals, and law enforcement tasks such as evidence gathering.[179] All contributed to a collective sense of futility and frustration with special operations in Northern Ireland, and within British special forces ranks, including the SRU, particularly after operators were killed but their targets prosecuted and jailed.[180]

This futility and frustration, while understandable operationally, suggests, however, that vertical control measures depicted in Figure 6.1 were in place and working. They included formal military chains of command as well as dual civilian investigative and due processes mechanisms. Perceived excesses, specifically in the use of force, led to parliamentary inquiries and trials of operators on criminal charges. Little evidence exists to suggest that British special operations in Northern Ireland exceeded the vertical control mechanisms in place. Each 22 Squadron notionally operated under the control of General Officer Command, Headquarters Northern Ireland, through one of three brigade headquarters. Complicating matters, however, was the fact that the director of the SAS had formal access in London civil and military authorities, and informally through the colonel commandant. This access could be the source of tension. While serving as director in 1976, Watts was ordered by the chief of defence staff, the senior-most member of the Ministry of Defence, to explain the circumstances of the Regiment's commitment to Northern Ireland after Wilson's declaration, on the presumption he had gone to the prime minister directly.[181] Watts demurred on this occasion. On another, he threatened to resign. At issue was the prosecution and criminal charges facing the eight operators detained

[176] Connor, *Ghost Force*, pp. 195–196.

[177] MacKenzie, *Special Force*, p. 204.

[178] Murray, *The SAS in Ireland*, pp. 225–235; Geraghty, *Who Dares Wins*, pp. 155–161, Adrian Weale, *Secret Warfare: Special Operations Forces from the Great Game to the SAS* (London: Hodder and Stoughton, 1997), pp. 222–224.

[179] MacKenzie, *Special Force*, p. 200; Ramsay, *SAS*, pp. 170–173.

[180] Duncan Falconer, *First into Action: A Dramatic Personal Account of Life in the SBS* (London: Little, Brown and Company, 1998), pp. 107–109; McCallion, *Undercover War*, pp. 113–114; 131–133.

[181] Asher, *The Regiment*, p. 430.

by the Irish Republic *Gardai*. A sense of futility turned to fury within SAS ranks at possible sentencing and jailing, leading to a near mutiny requiring a visit by Watts from London.[182] This state of affairs within the ranks, as well as Watts' and the colonel commanding 22 SAS's subsequent threat to resign, suggested a civil–military crisis as consequential as the Curragh Affair. Their bluff called, both failed to do so, suggesting political and military controls were robust and healthy.[183]

The greatest potential source for British special forces to exceed their authority occurred with the commitment and creation of various units. In addition to 22 SAS, the SBS appears to have a role, if a very limited one, in conducting individual, specific intelligence-gathering and reconnaissance missions conducted exclusively from the sea and largely focused on interdicting historical gun-running operations into Northern Ireland.[184] A few of its operators cross-decked over to the SRU, after passing the latter's selection process which not all did, in order to take part in the active special operations missions available.[185] The growth and development of new units such as SRU and FRU contributed to a perennial problem within special operations, and between special forces and conventional units: competition for scarce suitable human capital and in the process creating 'a shortage of operatives'.[186]

To ensure sufficient tactical coordination and operational control of special operations, including reconnaissance and the few indirect-shock missions that did occur, the Intelligence and System Group (Int & Sy Group) was created in 1980. Int & Sy Group performed two critical functions. It coordinated activities within and between various special operations and conventional elements and placed special forces in Northern Ireland under the single control of a single leader.[187] Its creation might also have been prompted by the diminishing viability of SRU's cover name, the Northern Ireland Training and Advisory Team.[188]

Barriers to more effective integration were social in nature. Cultural traditions and the tribal identity of different regiments created between elites over who was more qualified, or special, to conduct activities in Northern Ireland. A.F.N. Clarke, from the Parachute Regiment, provides a quintessential example. In discussing publicized SAS errors, he concludes, 'Cowboys, the lot of them; there are some guys I've recognised who have failed [Para] selection, so how did they get into the S.A.S.? I wouldn't give them the time of day ... They are a joke.'[189] These feelings were reciprocated. In his assessment of army liaison officers to Special Branch, MacKenzie suggests they 'were of mixed ability and more than one was an SAS "wannabe"'.[190]

[182] Connor, *Ghost Force*, p. 181; Asher, *The Regiment*, p. 436.
[183] MacKenzie, *Special Force*, p. 203.
[184] Peter Hennessy and James Jinks, *The Silent Deep: The Royal Navy Submarine Service Since 1945* (London: Allen Lane, 2015).
[185] Falconer, *First into Action*, pp. 136–146.
[186] Ibid., p. 167.
[187] Kemp, *The SAS*, p. 129.
[188] Charters, 'Professionalizing Clandestine Military Intelligence in Northern Ireland', p. 135.
[189] Clarke, *Contact*, p. 117.
[190] MacKenzie, *Pilgrim Days*, p. 151.

Initially such coordination occurred informally and out of necessity, given a paucity of informant and agent handlers. To fill the gap:

> more SAS officers were drafted in to control agents in the field. This had the advantage that, without any overt structural changes in Intelligence-gathering; [sic] there would be a cadre of people accustomed to working together who could act as a sort of cement to bond an organization that seemed, at times, to be falling apart.[191]

Other individual units, including SRU, and later Int & Sy Grp dispatched liaison officers to coordinate activities with civilian police and special operations units. Robert Nairac performed such duties well after passing SRU selection, serving as a vital link and network connection between his own unit, the SAS, and RUC Special Branch.[192] He did so largely based on informal social and bureaucratic connections he already had within the latter. What started as an interim measure, spearheaded by the SAS out of necessity on an informal basis, was formalized in Int & Sy Grp.

With the establishment of police primacy, RUC Special Branch undertook other efforts to coordinate special operations and intelligence gathering and sharing between its own units and British ones. Starting initially in 1978 on a trial basis, its Tasking and Co-ordination Group (TCG) in Belfast was so successful in integrating efforts horizontally at the operational level, by 'concentrat[ing] the products of intelligence gathering in one place' via a 'centralised command structure' that it developed two more, one in Armagh for the south of Northern Ireland and another in Derry for the north the following year.[193] Rory Finegan concludes the TCGs linked together military and civilian intelligence and fused this with the means to capture or kill PIRA members: '"executive action"—locking together intelligence from informers with the surveillance and ambushing activities of undercover units'.[194] It was a variation of the 'find, fix, finish' methodology for manhunting used by American special operators discussed in detail in Chapter 8, with notable differences. Police primacy ensured civil, vice military objectives were met as the goal was arrest and prosecution. This goal, in turn, meant much stricter potential human rights, media, and legal scrutiny on the process lest the individual walk free from prosecution over an oversight or technicality. Finally, it occurred far more infrequently because of both.

Control of special operations, as well as other military and police activities, was necessary given the political goal at hand. Rhetoric supporting the image of a strong national leader, as a means of bolstering public morale and confidence, differed from the reality of British objectives in Northern Ireland. Margaret Thatcher's famous public declaration about never negotiating with terrorists belied the fact she was doing so covertly through SIS intermediaries, the fact of which even some of her close staff

[191] Geraghty, *Who Dares Wins*, p. 147.
[192] Asher, *The Regiment*, p. 434.
[193] McCallion, *Undercover War*, p. 80.
[194] Rory Finegan, 'Shadowboxing in the Dark: Intelligence and Counter-Terrorism in Northern Ireland', *Terrorism and Political Violence*, 28:3 (2016): 507.

were unaware.[195] Thatcher was far from the first prime minister to do so. Serious talks with the IRA, and between the IRA and loyalists, began in earnest as early as 1972 in pursuit of a 'balanced settlement' made possible by appealing to the reason of a 'decent majority' over the opposition of 'men of violence'.[196]

## Period 2: 1972–1980

PIRA terrorism was only one form concerning to British leaders. As mentioned above, the nature of transnational terrorism was driven home internationally with the hijacking of El Al Flight 426 by members of the Popular Front for the Liberation of Palestine on 22 July 1968 (PFLP). Bruce Hoffman considers this incident the first act of 'modern, international terrorism', although for Britain perhaps a stronger case can be made for the maritime hijacking of the SS Santa Maria in 1961, given the SBS response to the threat to the *QE2*.[197] The successful negotiation of the hostages' release encouraged more hijackings, some of whose power was magnified by live television coverage. Sources and details vary on who, how, and when 22 SAS developed its initial idea for counterterrorism response. Ken Conner suggests the original idea was the brainchild occurred roughly around the time of the El Al hijacking, at a time when the Regiment was under-committed and, according to him, 'scratching around for gainful employment'.[198] This anticipation and experimentation resulted in a staffing paper 'in the spring of 1972', ultimately rejected within the Ministry of Defence on the grounds of greater government concerns at the time.[199] Nevertheless, a few individuals within the Regiment persisted and gained informal approval, and considerable latitude and resources but no formal authority. They experimented in a Counter-Revolutionary Wing (CRW), with new techniques and equipment for close-quarter battle in confined spaces that became the genesis for the capability.[200]

The need for such a capability was driven home by dramatic televised hostage taking of Israeli athletes in Munich in September 1972. The subsequent botched rescue attempt at Fürstenfeldbruck airport, conducted by cobbled-together police units, resulted in the deaths of the hostages, terrorists, and helicopter pilots charged with flying them out. The shock of the event and the outcome prompted British Prime Minister Edward Heath to direct the Ministry of Defence to develop a counterterrorism rescue force.[201] The gravity and seriousness of the request is evident in the

[195] Harmonie Toros, '"We Don't Negotiate with Terrorists!": Legitimacy and Complexity in Terrorist Conflicts', *Security Dialogue*, 39:4 (August 2008): 407–408; 'Margaret Thatcher "Negotiated with IRA"', BBC News, 30 December 2011, www.bbc.com/news/uk-northern-ireland-16366413.

[196] Neumann, *Britain's Long War*, pp. 85–88.

[197] Bruce Hoffman, *Inside Terrorism*, 3rd ed. (New York: Columbia University Press. 2017), p. 65; Henry Zeiger, *The Seizing of the Santa Maria* (New York: Popular Library, 1961).

[198] Connor, *Ghost Force*, p. 203.

[199] MacKenzie, *Special Force*, p. 213.

[200] Connor, *Ghost Force*, pp. 204–205; Seymour, *British Special Forces*, p. 305.

[201] De la Billère, *Looking for Trouble*, pp. 280–281; Paul de B. Taillon, *The Evolution of Special Forces in Counter-Terrorism: The British and American Experiences* (Westport, CT: Greenwood Press, 2000), p. 38.

response to the 22 SAS commander's request about the funding available for the unit: 'on one of the few occasions in my life I heard the glad answer, "Money's no problem—as much as you like"'.[202] The length of the siege dominated the headlines and airways and Heath wanted to ensure resources were sufficient to prevent such a tragedy from occurring on British soil.

Resources are important but SOF also attribute success to access and placement of their operators. The profile and access of SAS grew politically with the demonstrated utility of Special Projects Team (SPT), the official name given to Operation Pagoda, the on-call counterterrorism response force, consisting originally of between sixteen and twenty operators drawn from across the SAS's four squadrons.[203] Subsequently every Sabre Squadron committed a team to 'Pagoda duty' on a rotating basis, adding to the missions but diminishing specific capability given the need to spin up, retrain, and stand on alert.[204] SPT was allegedly mobilized for the first time operationally in the UK in December 1975 during the Balcombe Street siege in Westminster, London. According to BBC News, the four hostage takers surrendered after six days 'when they heard on the radio the SAS was to be sent in'.[205] Two years later, SPT provided technical advice, specialized equipment, and planning support to its West German counterpart, *Grenzschutzgruppe* (GSG) 9, in the successful storming of Lufthansa Flight 181, hijacked by members of the PFLP in Mogadishu Somalia.[206] The technical advice and equipment occurred first as request for support, followed by conversations between German Chancellor Helmut Schmidt, British Prime Minister James Callaghan, and two SPT members: Major Alastair Morrison and Sergeant Barry Davies.[207] Davies points out the British technical special forces support washed away a diplomatic logjam between the West German and British governments over more than £112 million in foreign exchange costs related to the BAOR.[208] More broadly, Britain's assistance to West Germany was 'good for Britain's relations with Europe', leading to increased counterterrorism collaboration.[209]

British SOF access politically increased thanks to two developments. One was communications, in the form of a direct links between the SPT in Hereford and the director of the SAS in London, as well as between the latter and the Ministry of Defence. Liaising contributed to vertical integration. In de la Billère's words: 'we had immediate access to the Ministry of Defence—and indeed kept a representative permanently stationed there. By this means we generally managed to pick up the first

[202] De la Billère, *Looking for Trouble*, p. 281.
[203] Connor, *Ghost Force*, p. 208; Asher, *The Regiment*, p. 421.
[204] Asher, *The Regiment*, p. 430.
[205] 'UK Balcombe Street Gang's Reign of Terror', BBC News, 9 April 1999, http://news.bbc.co.uk/2/hi/uk_news/315216.stm.
[206] Rolf Tophoven, *GSG 9: German Response to Terrorism* (Koblenz: Bernard & Graefe Verlag, 1985), pp. 67–73.
[207] Barry Davies, *Fire Magic: Hijack at Mogadishu* (London: Bloomsbury, 1994), pp. 1–7; Paul de B. Taillon, *Hijacking and Hostages: Government Responses to Terrorism* (Westport, CT: Greenwood Press, 2002), pp. 128–138.
[208] Davies, *Fire Magic*, p. 168.
[209] Geraghty, *Who Dares Wins*, p. 173.

hints of impending trouble'.[210] The second was the establishment of Britain's crisis response facility, the Cabinet Office Briefing Rooms, or COBR (also COBRA, for unknown reasons), in 1972.[211] Domestic civil unrest, in the form of the crisis resulting from striking coal miners, and government's uncoordinated response to it drove the creation of COBR.

Both developments would feature prominently in the British special operations' coming of age. The publicity surrounding the storming of the Iranian Embassy on Princes Gate in South Kensington, London on 5 May 1980 marked a new era of prestige, influence, and political access for British special forces. The Embassy was seized by six members of the Democratic Revolutionary Front for the Liberation of Arabistan (DRFLA), seeking the release of ninety-one fellow Arabs imprisoned in Iran. With the hostage taking turning into a standoff, COBR was activated to coordinate various ministries at the highest levels. Through very informal channels, the 22 SAS commander, Michael Rose, was notified of the incident. Almost immediately thereafter, the special forces director sought 'official permission to deploy our team' from the Ministry of Defence while also actively 'angling for a place in' COBR.[212]

Although the prime minister, Margaret Thatcher, was absent from COBR meetings, she worked through the home secretary, William Whitelaw, to ensure her intentions were clearly communicated and she remained up to date.[213] 'Police primacy' remained the rule throughout most of the siege but tensions increased throughout, although circumstances changed on the sixth day with the execution of one of the hostages, Abbas Lavansani.[214] In a quick and informal sequence of events, authority shifted from the Metropolitan Police to the Army under Military Aid to the Civil Power, in part due to the assertion of the on-scene special forces commander, Rose, for 'written authority'.[215] More than twenty-four troops from B Squadron, then serving on Pagoda duty, abseiled into the Embassy from different directions, relying on speed, surprise, and confusion to rescue the hostages and neutralize the threat of terrorist violence during Operation NIMROD in 34 minutes.[216] One of the operators involved states the prime minister personally thanked them, allegedly saying, 'Gentlemen, there is nothing sweeter than success, and you boys have got it', as well as watching the news rebroadcast of the rescue with them at Regent's Barracks.[217] News coverage of the siege was televised to captive audiences in Britain and elsewhere. Coincidentally, live broadcasts of the siege

[210] De la Billière, *Looking for Trouble*, p. 314.

[211] COBR and its functions and processes are outlined in The Cabinet Office, *Responding to Emergencies: The UK Central Government Response—Concept of Operations* (London: HMSO, 2010), pp. 21–43.

[212] De la Billière, *Looking for Trouble*, pp. 319–320.

[213] As an aside but also as an illustration of how intimate social connections are in the United Kingdom, Whitelaw had drilled and instructed a young David Stirling as a cadet and found him wanting. Asher, *The Regiment*, p. 21.

[214] George Brock et al., *Siege: Six Days at the Iranian Embassy* (London: Macmillan, 1980), pp. 108–109.

[215] De la Billière, *Looking for Trouble*, p. 331.

[216] Michael Paul Kennedy, *Soldier 'I' SAS* (London: Bloomsbury, 1989), pp. 194–197.

[217] Quoted in Kennedy, *Soldier 'I' SAS*, pp. 199–200.

occurred over a six-day period and Operation NIMROD was conducted early in the evening of a Bank Holiday Monday, ensuring a domestic audience of 'millions' and creating an image that subsequently worked for and against British special operations.[218]

## Period 3: 1980–1982

Public attention raised the profile of British special forces, especially the SAS, in positive and negative ways politically and militarily in the months and years to follow. The increase in credibility and political access was accompanied by two floods. One was foreign government requests for SAS expertise now that Britain had joined the exclusive club of countries successfully carrying out risky hostage rescue missions.[219] The other occurred in the form of new applicants for SOF, mostly for the wrong reasons.[220] In the words of one operator, the influx of those who were 'ambitious' and 'hungry for money ... can have a big effect in a small unit'.[221] In common with the normalization and broader acceptance of special forces by conventional forces, officers saw SAS tours as helpful and not detrimental to their career. The public profile of British special forces was sustained with an inquest into the shooting deaths of five of the DRFLA hostage takers. While steps were taken to protect the identities of the four SAS troopers who testified at the inquest, and the shootings ruled justified, several operators were uneasy with this development. As one said acidly, 'We now had to be seen to be accountable. Democracy in action.'[222] Events involving the SAS, should they receive media attention, would from this point forward receive much greater, and more critical, scrutiny.

Such was not the case in the lead-up the Falklands War. SAS troops continued their rotations to Northern Ireland, to conduct reconnaissance and surveillance missions, while other teams were dispatched for low-profile training missions in support of proxies or governments in Afghanistan, Cambodia, and Gambia.[223] MacKenzie offers SAS missions working with proxies resulted from 'operational cuts in the Security Services'.[224] These cuts were part of a longer-term trend affecting British special operations: a slow and steady decline and dispersal of SIS covert action capabilities.[225] SAS mission expansion also included absorbing other special forces. Cormac concludes, '[SRU] and the SAS had been combined under one commanding officer',

[218] Paul Hastie, 'John McAleese: The SAS Soldier Who Was the Man behind the Mask', BBC Scotland, 26 August 2021, www.bbc.com/news/uk-scotland-tayside-central-58328164.
[219] Connor, *Ghost Force*, p. 235.
[220] Parker, *SBS*, p. 166.
[221] Rusty Firmin, *The Regiment: 15 Years in the SAS* (Oxford: Osprey, 2015), p. 191.
[222] Kennedy, *Soldier 'I' SAS*, p. 204.
[223] Connor, *Ghost Force*, pp. 267–268; MacKenzie, *Special Force*, p. 217.
[224] MacKenzie, *Special Force*, p. 217.
[225] Cormac, *Disrupt and Deny*, p. 215.

suggesting the director of the SAS, which is confirmed by at least one SRU member's recollection.[226] Rotations of SAS units, working through Int & Sy Grp, continued apace.

The only new development in Northern Ireland was an addition to the Group's capabilities in 1981. It consisted of a plain-clothes unit, the FRU, designed to run informers and other agent networks penetrating PIRA as well as loyalist ranks. A unit of the Intelligence Corps, FRU joined the ranks of other special forces, undertaking a special operations capability that completed the 'find, fix, finish' puzzle. Through its informer networks and compartmented from RUC Special Branch, whose loyalist leanings increasingly came under suspicion, FRU provided the information that found PIRA, Ulster Defence Association (UDA), and Ulster Volunteer Force (UVF) members. Once found, SRU conducted surveillance and tailing, mobile and static, to narrow down their habits and routines or 'fix' them to specific locations.[227] At an appropriate time and location, 22 SAS troops or teams would complete or 'finish' the mission by capturing or killing individuals. By this time, four separate SRU detachments orientated geographically operated in Northern Ireland: one in the north, another in the east which covered Belfast, and two in the south, split between interdiction and PIRA cells.[228] FRU detachments collocated and operated with these and so too, presumably, did the SAS.[229] All were designed to prevent, to the extent possible, PIRA surprise attacks.

Surprise did in fact occur but thousands of miles away. By most accounts, the Argentine seizure of the contested Falkland Islands caught British leaders unawares, including Prime Minister Thatcher, leading some to conclude there was a failure of intelligence. As with any crisis, tensions increased and decreased, signals were either misinterpreted or missed, and other intelligence biases crept into British understanding of events.[230] Based on differing claims of sovereignty dating back to the 1830s, the conflict began when Argentine forces seized the Falkland Islands through an amphibious assault on 2 April 1982, followed by South Georgia and the South Sandwich Islands (see Map 6.3). Three days later, a British naval task force, comprising 127 ships ranging from aircraft carriers to trawlers, sailed on an 8,000-mile voyage to retake the islands.

The war, Operation CORPORATE, consisted of British Vulcan bomber airstrikes comprising an epic 7,600-mile round-trip flight, a hotly contested battle for sea and air control in and around the islands in which eight British and two Argentine ships were sunk, and a month-long (21 May to 14 June) ground campaign to retake key terrain, beginning with Port San Carlos in the west and ending with the surrender of Argentine forces at the capital in the east, Stanley. For Britain, the Falklands War

[226] Ibid., p. 209; Rennie, *The Operators*, p. 157.

[227] Weale, *Secret Warfare*, p. 222.

[228] Sarah Ford, *One Up: A Woman in Action with the S.A.S.* (London: HarperCollins, 1997), p. 139.

[229] Rob Lewis, *Fishers of Men* (London: Hodder and Stoughton, 1999), p. xvi.

[230] Lawrence Freedman, *The Official History of the Falklands Campaign. Volume I: The Origins of the Falklands War* (London: Routledge, 2005), p. 219.

was the first major projection of conventional military power in a conflict since the Suez Crisis in 1956.[231]

British special forces played an outsized role in the campaign in four ways, marked by success, failure, and tragedy. The first, comprising the initial British military actions in the Falklands War, were via intelligence-gathering and reconnaissance missions. Three different special forces detachments conducted such missions up to a month prior to the fleet's arrival. One involved an unspecified number of G Squadron teams three weeks before. The teams, delivered at night by helicopter some distance, would move by foot some distance on the sparsely populated and construct hide sites, or OPs, close to their targets.[232] Collecting information by observation during the day, the teams were forced at night to hike back to the landing zones, at distances up to 6 miles, to transmit their reports via unencrypted radios to prevent location.[233] Finally, other efforts much farther to the east in the South Atlantic had mixed success. An effort by Mountain Troop, D Squadron to set up an OP on Fortuna Glacier overlooking Leith Harbour on South Georgia Island, some 860 miles east of the Falklands, ended in disaster. Weather conditions threatened the survival of the troop, leading to its commander to call for an extraction, but extreme winds and snow caused two helicopters attempting it to crash. A third attempt by a dangerously overloaded remaining helicopter from HMS *Antrim* barely managed the feat.[234] In the prime minister's words, 'It was a terrible and disturbing start to the campaign.'[235] For members of the British Antarctic Survey, the results were not surprising. Their efforts to dissuade the Mountain Troop commander, given their intimate knowledge of the extreme conditions there, were dismissed in pursuit of the mission.[236]

Additional reconnaissance missions tried to develop better understanding of potential threats where other intelligence was unavailable. SBS operators and M troop, 42 Commando, surveilled and reconnoitred Argentine positions in the remote, former whaling station at Gryviken, reporting a detachment of one hundred Argentine marines as well as a Second World War-era submarine, the ARA *Santa Fe*.[237] The *Santa Fe* was subsequently damaged by depth charges and returned back to port, alerting the Argentines that British forces were in the area.[238] Given the size

[231] Alexander Haig Jr, 'Military Intervention: A Case Study of Britain's Use of Force in the 1956 Suez Crisis', unpublished student thesis, 8 April 1966, Carlisle Barracks, PA, US Army War College. Alexander Haig, secretary of state for the Reagan Administration, attempted twice to mediate a solution between Argentina and the United Kingdom. For details, see Lawrence Freedman, *The Official History of the Falklands Campaign. Volume II: War and Diplomacy* (London: Routledge, 2005), chapters 10–12.

[232] A sense of the nature and tension of these missions is conveyed evocatively in Kennedy, *Soldier 'I' SAS*, pp. 222–224.

[233] Barry Davies and Nigel McCrery, *The Complete History of the SAS* (London: Andre Deutsch, 2011), pp. 162–163.

[234] Asher, *The Regiment*, p. 459.

[235] Margaret Thatcher, *The Downing Street Years* (New York: HarperCollins, 1993), p. 205.

[236] Alan Bell, 'On a Wing and a Prayer: Task Force 317 and the Recapture of South Georgia Island', in Bernd Horn, ed., *Risk: SOF Case Studies* (Ottawa, ON: CANSOFCOM, 2020), p. 142.

[237] Roger Perkins, *Operation Paraquat: The Battle for South Georgia* (Chippenham: Picton, 1986), pp. 162–173.

[238] Jeffrey Ethell and Alfred Price, *Air War South Atlantic* (London: Sidgwick & Jackson, 1983), pp. 39–40.

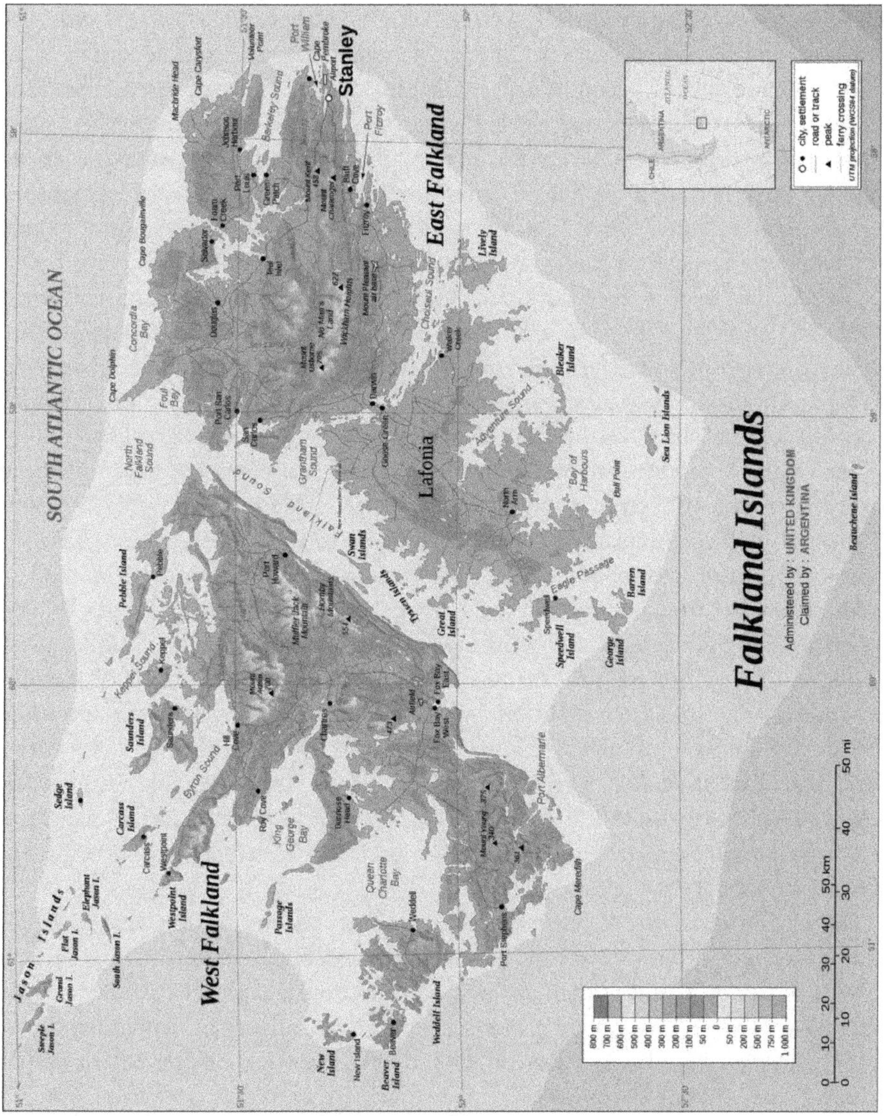


**Map 6.3** Map of the Falklands Islands.

*Source*: Wikimedia Commons, creator Eric Gaba (Sting). Permission is granted to copy, distribute and/or modify this document under the terms of the GNU Free Documentation License provided creator is attributed.

of the detachment, and with the help of gunfire support from HMS *Antrim* and *Plymouth*, the seventy-five-man force comprising two troops from D Squadron, along with the SBS and M troop, 42 Commando, advanced on the British Antarctic Survey station to have its garrison surrender, with barely a shot fired.[239] The assertion that

[239] Cedric Delves, *Across an Angry Sea: The SAS in the Falklands War* (London: Hurst and Company, 2020), pp. 84–94.

'The SAS observation party on Harbour Point could have polished [the garrison] off without any additional support' indicates the mindset within D Squadron at the time.[240] Fortunately it was not tested, as the four different lines of command within this aggregate task force 'had all the hallmarks of a disaster in the making'.[241]

Another intelligence-gathering mission sought to fill gaps in operational intelligence for the task force by calling on favours diplomatically. The 'special relationship' with the United States led to sharing of satellite and other imagery information, as well as specific special forces support discussed below.[242] Within South America, Britain maintained close working relations with Argentina's western neighbour, Chile. The Ministry of Defence dispatched RAF officer Sidney Edwards there clandestinely, specifically to determine how much support the Chileans would provide. Edwards almost immediately connected with his Chilean counterparts, gaining access to some of their intelligence information. Because of a geographical oddity, Chilean radars based in the mountains had excellent coverage of Argentinian air bases to the east. While not a special operator per se, Edwards did provide a vital link in the intelligence capability chain. That link was subsequently filled by British special forces. The director of the SAS, de la Billère, called Edwards to his office prior to his departure; they had met on the diplomatic circuit some years prior.[243] This meeting, taking advantage of an informal social connection, turned into a special operation. Once it became clear that Chilean radar could provide advanced warning of Argentinian sorties from their four southernmost airfields against the task force, Edwards arranged for an additional portable British radar to be sent, along with a clandestine four-man SAS team to relay warnings via tactical satellite communications (SATCOM) to the fleet.[244] In this case, British special forces played a crucial role in integrating capabilities at the operational level, thanks largely to informal social ties at the national level.

The second consisted of indirect shock, or independent raiding actions. In the Squadron commander's words, the 'time felt right for a spot of derring-do, we were more than up for it, and might have found the appropriate objective to dare to win against'—all that was required was one confirmed by intelligence and planning.[245] An appealing target, based on available maps and imagery, was the Argentine airfield on Pebble Island. After a three-day reconnaissance of it by a team from Boat Troop, a combined 'party of 45 men of the SAS and a Royal Navy forward observer landed' on 14 May, a week before the main amphibious invasion.[246] Supported by naval gunfire from the HMS *Glamorgan* and recalling David Stirling's original concept for the SAS, the D Squadron party, led by its commanding officer Cedric Delves, suddenly

[240] Perkins, *Operation Paraquat*, p. 180.

[241] Bell, 'On a Wing and a Prayer', p. 136.

[242] Nigel West, *The Secret War for the Falklands: The SAS, MI6, and the War Whitehall Nearly Lost* (London: Little, Brown & Co., 1997), pp. 78–88.

[243] Sidney Edwards, *My Secret Falklands War* (Lewes, Sussex: Book Guild Publishing, 2014), pp. 8–10.

[244] Edwards, *My Secret Falklands War*, pp. 41–43.

[245] Delves, *Across an Angry Sea*, pp. 132–133.

[246] Freedman, *The Official History of the Falklands Campaign. Volume II*, p. 431; Asher, *The Regiment*, pp. 459–460.

and unexpectedly wreaked havoc on the Argentinian airfield there. At a cost of no casualties to themselves, the D Squadron raiders destroyed eleven close-support aircraft, fuel supplies, and ammunition before safely withdrawing.[247] The report of the results noted the operation was 'very cost effective' and 'removed as many aircraft in 45 minutes as had been destroyed by other means in 14 [days]'.[248] The *Fuerza Aerea Argentina* (Argentine Air Force), however, 'more than replaced the losses' of aircraft five days later.[249] An unintended consequence of the raid was that Argentine air defences on the Falklands improved in one respect as a result, as the pilots whose aircraft were destroyed were repurposed as air observers on surrounding mountain tops, feeding reports back by radio.[250] Another raid, less reported but no less dramatic the previous week, saw an SBS team conduct an air assault to seize and sink the ARA *Narwhal*, which had been shadowing and collecting intelligence on the task force as it sailed towards the Falklands.[251]

Direct-shock missions, or raids designed as shows of force, served tactical purposes. A military deception mission, conducted by the remaining forty members of D Squadron on 20 May, sought to convince an Argentine garrison that the main landing was near Darwin and not San Carlos. Using mostly small arms, D Squadron fired copiously and conspicuously in the night but supporting naval gunfire only occurred later the following morning, thanks to limitations in communications.[252] Despite subsequent claims of success by the 22 SAS Commander, there is little to suggest the deception succeeded in large part due to horizontal integration issues thanks to outdated radios and encryption codes.[253] Far more successful was an integrated action on Fanning Head the same evening, in which a composite unit of 2 and 3 SBS, with a Spanish-speaking interpreter, used new technology in the form of a thermal imager, naval gunfire, and a successful appeal to the Argentine unit on observation duties to surrender.[254]

The third was operating in support of British grounds forces on Falkland Island. Teams from 2 and 6 SBS conducted invasion support activities for Commando and Para forces, including beach and close reconnaissance as well as pathfinding as forces when ashore on 21 May.[255] SBS forces also directed naval gunfire at targets ashore, specifically in Fox Bay and Port Howard.[256] Much later in the campaign, the ubiquitous D Squadron, in conjunction with 3 Commando Brigade, manoeuvred

[247] Department of the Navy, *Lessons of the Falklands: Summary Report* (Washington, DC: Department of the Navy, February 1983), p. 61.

[248] WO 305/5052, FM CTG 317.8 to CTF 317.17 May 1982.

[249] Martin Middlebrook, *The Fight for the Malvinas: The Argentine Forces in the Falklands War* (London: Penguin, 1989), p. 142.

[250] Middlebrook, *The Fight for the Malvinas*, p. 136.

[251] Falconer, *First into Action*, pp. 282–290.

[252] Middlebrook, *The Fight for the Malvinas*, pp. 145–146.

[253] Matthew Miller, 'Special Operations in the 21st Century: Revisiting the Falklands War', *Inter Populum: The Journal of Irregular Warfare and Special Operations*, 1:1 (Spring 2023): 63.

[254] Parker, *SBS*, p. 304; Middlebrook, *The Fight for the Malvinas*, p. 146.

[255] Lawrence Freedman, *The Official History of the Falklands Campaign. Volume II*, pp. 544–545; Parker, *SBS*, p. 305.

[256] Parker, *SBS*, p. 306; Mercer, *Not by Strength, by Guile*, p. 64.

and seized key terrain at Top Malo and Mount Kent.[257] The SAS seizure of Mount Kent in particular, although delayed, led to the only SOF versus SOF engagement of the war, when a section of Argentina's 602 Company tried unsuccessfully on 29 May to dislodge the British operators, costing them two dead and thirteen wounded.[258] A follow-up operation to seize Port Stanley was unnecessary as the Argentine garrison subsequently surrendered.

The fourth and final way British special forces played a role in the campaign was to meet an emergency requirement resulting from the first week of the conflict. The combined intelligence-gathering and reconnaissance mission would address the primary and most dangerous threat to the task force: limited numbers of Argentine Exocet anti-ship missiles and aircraft capable of launching them from Rio Grande in the Tierra del Fuego. The threat became very real on 4 May when an Exocet struck and severely disabled HMS *Sheffield*, which eventually sank. Three weeks later, the SS *Atlantic Conveyor* was hit by two Exocets and sank after two days.[259] The loss of HMS *Sheffield* was initially shocking to the task force but the *Atlantic Conveyor*'s loss acutely felt by 3 Commando, as eleven of the twelve helicopters planned to support its airmobile went down with her.[260] To address the Exocet threat, three related special operations were planned but only one conducted. Disaster struck the preliminary operation, code-named PLUM DUFF, when the two helicopters transporting a nine-member SAS reconnaissance team ultimately crashed on 18 May.[261]

A proposed subsequent raid, Operation MIKADO, was cancelled. The mission's extreme range and desperation tactically mirrored the *Giretsu* special operation, conducted by Japanese forces on 24/25 May 1945 to destroy enemy aircraft on Kadena and Yomitan airfields on Okinawa.[262] MIKADO envisioned two C-130s landing on Rio Grande airfield, with fuel only to make the trip there but not back to Ascension Island.[263] The aircraft would disgorge 45 B Squadron operators who, after destroying the Exocets and aircraft would attempt to reload and fly to neighbouring Chile. The more likely outcome was the C-130s would crash-land, forcing the operators to escape on foot to Chile and seek asylum there.[264]

B Squadron operators were uncomfortable with the plan because of the lack of specific intelligence information about the target, including the number of aircraft and location of missiles. That unease grew considerably after news of PLUM DUFF's failure and the loss of surprise. Discomfort led to hostility among the ranks of

257 Julian Thompson, *No Picnic: 3 Commando Brigade in the South Atlantic, 1982* (London: Guild, 1985), pp. 107–109.

258 Miller, 'Special Operations in the 21st Century', pp. 63–64.

259 Department of the Navy, *Lessons of the Falklands*, p. 57.

260 David Brown, *The Royal Navy and the Falklands War* (London: Arrow Books, 1989), pp. 228–230.

261 For a first-hand account of the operation by the helicopter pilot who flew the mission and escaped with the SAS team, see Richard Hutchings, *Special Forces Pilot: A Flying Memoir of the Falklands War* (Barnsley: Pen & Sword, 2008), pp. 143–173.

262 James Belote and William Belote, *Typhoon of Steel: The Battle for Okinawa* (New York: Harper & Row, 1970), pp. 272–273.

263 Hutchings, *Special Forces Pilot*, pp. 123–124.

264 West, *The Secret War for the Falklands*, pp. 137–141.

experienced senior enlisted operators, the resignation of one, and the Squadron's commander relief from duty by the director of the SAS after declining the mission in response.[265] These events constituted the worst crisis in SAS internal control in the Regiment's history and further exacerbated the tensions between officers and enlisted.[266] A third mission, to be conducted by the SBS focused on the Argentine airfield at Puerto Deseado code-named Operation KETTLEDRUM, was also cancelled due to lack of accurate operational information that threatened the loss not only of the special operators but also the submarine transporting them.[267]

British special operations owed their successes largely to integration of capabilities. Tactically, SAS and SBS joint training missions, while occasionally vexed by personality clashes between operators, nevertheless facilitated integration as occurred during the assault on Mount Kent. Joint training developed relationships between operators. SAS boat troops within the squadrons had a natural connection in combining training and sharing tactical lessons learned, including the best equipment to scale heights in a maritime environment. In addition, SAS sometimes acted as opposing forces for SBS in tactical exercises.[268] The colocation of D Squadron and 2 SBS on the Royal Fleet Auxiliary *Fort Austin* during the sailing of the task force southward doubtless also strengthened relationships and informal connections.[269]

Other informal connections filled equipment and capability gaps within British special forces. Close, fraternal working relationships between the SAS and American special mission units provided not only AIM-92 Stinger portable air defence missiles that were much lighter and more capable than the British-issued Blowpipe, and emergency training on them in a 'private arrangement', but also, even more importantly, secure, man-portable satellite communications on loan.[270] The SAS were initially issued with high-frequency radios without encryption and of limited range. Tactical SATCOM provided connection between deployed SAS squadron headquarters aboard the task force with the director of the SAS in London, as well as between specific teams relaying vital intelligence such as early warning of Argentine sorties mentioned above. In Ken Connor's words, the SATCOM 'revolutionised our communications on long-range ops, enabling the CO in the Falklands to talk to Hereford in real time'.[271] Unfortunately, there was at least two downsides to immediate communications: higher headquarters imposing itself on previously largely autonomous local special forces commanders, and politically and bureaucratically within the

[265] Firmin, *The Regiment*, pp. 224–226; Asher, *The Regiment*, pp. 454–457.

[266] Connor, *Ghost Force*, p. 249.

[267] Ewen Southby-Tailyour, *Exocet Falklands: The Untold Story of Special Forces Operations* (Barnsley: Pen & Sword, 2014), pp. 248–257.

[268] Falconer, *First into Action*, pp. 240–245.

[269] Parker, *SBS*, p. 279.

[270] Asher, *The Regiment*, pp. 462–463; Connor, *Ghost Force*, pp. 250–251; Documents 111, 'Telegram from the Department of State to Secretary of State Haig', 13 April 1982 and 69, 'Action Memorandum From the Acting Assistant Secretary of State for European Affairs (Scanlan), the Assistant Secretary of State for Inter-American Affairs (Enders), and the Assistant Secretary of State for Economic and Business Affairs (Hormats) to Secretary of State Haig', in *Foreign Relations of the United States, 1981–1988, Volume XIII: Conflict in the South Atlantic, 1981–1984* (Washington, DC: Government Printing Office, 2015), pp. 129; 232–233.

[271] Connor, *Ghost Force*, p. 253.

Ministry of Defence and Whitehall, including allegedly the prime minister herself, over the director of the SAS's communication channel that kept him better informed first than others.[272]

Lack of integration, as well as chance in the form of bad luck, contributed to the operational losses British special forces suffered in the conflict. Half of the two British combat losses were caused by a preventable 'own goal', when an SBS team was ambushed by a SAS patrol on 4 June 1982, resulting in the death of Sergeant Ian Hunt. Subsequent discussion of the incident in print has descended into invective and censure but two facts related to integration make the cause clear: incompatible radios between the SAS and SBS 'made it impossible for them to coordinate' and poorly delineated and separate and distinct areas of responsibility both played a role in the tragedy.[273] More formal as opposed to informal ties could have reasonably made it less likely to occur.

Little, however, could have prevented the single event that caused the greatest number of British special forces casualties, the downing of a Sea King helicopter on 19 May. The helicopter was allegedly struck by a bird while cross-decking, or lifting from one ship to another, around twenty-four operators from D and G Squadrons, as well as support operators from 264 (SAS) Signal Squadron.[274] Twenty operators lost their lives, including one of the two operators trained to use the Stingers.[275] In consequence, B Squadron operators had to be airlifted and dropped by parachute near the fleet in freezing South Atlantic waters to make up the losses, although only thirty arrived thanks to a mechanical issue with one of the two RAF C-130s transporting them.[276] Air accidents claimed special forces casualties indiscriminately. Constituting just over half of Argentina's special forces fatalities, six operators from the 601st National Guard Special Forces Company died when their Puma helicopter crashed in a low-level flight near Mount Kent.[277]

Despite some success, including the Pebble Island raid and support to 3 Commando Brigade, British special forces experienced their fair share of failures and false starts. In the finger-pointing that followed, several aspects of vertical and horizontal control are evident. First, senior civilian ministers as well as some uniformed commanders had undue expectations for what special forces could accomplish. For example, the minister of defence, John Nott, was convinced 'SAS/SBS deployed from [nuclear submarines] will be the answer to winning back the Falklands' the day after they were invaded.[278] Michael Asher suggests some of this is due to the

[272] Ibid.; Falconer, *First into Action*, p. 297.

[273] Falconer, *First into Action*, p. 291; Parker, *SBS*, pp. 305–306; Connor, *Ghost Force*, p. 257.

[274] Mark Aston, *SAS: Sea King Down* (London: Penguin, 2022), p. 224; Tony Shaw, *SAS: South Georgia Boating Club* (Barnsley: Pen & Sword, 2022), pp. 111–112.

[275] Asher suggests survivors heard a 'bang' while crash survivor Mark Aston reported 'a distinct thump and the immediate winding-down of the engines', possibly indicating a mechanical failure, which is not unlikely given the age and overuse of Royal Navy helicopters during the campaign. Asher, *The Regiment*, p. 458; Aston, *SAS*, p. 224.

[276] Firmin, *The Regiment*, pp. 230–235.

[277] Middlebrook, *The Fight for the Malvinas*, pp. 200, 303.

[278] Freedman, *The Official History of the Falklands Campaign. Volume II*, p. 223.

publicity surrounding Princes Gate and creation of two images of British special forces: 'the "real" SAS, and the SAS of public imagination'.[279] To these another should be added: the SAS politically cultivated by the Regiment, or an SAS of political expectation. For example, the units and commanders tasked with KETTLEDRUM, already aboard HMS *Onyx*, reached conclusions about those planning them in London. They considered the plan 'a Boys Own Paper Scheme ... I could not believe that anyone sane would have considered mounting a military operation on the basis of the information that they had been given'.[280] A senior SBS officer added: 'As with Plum Duff and Mikado, no one in the United Kingdom had done their sums'.[281] The impact of the cultivated image is discussed subsequently.

The second is that the climate of secrecy surrounding the planning for special operations in the Falklands was confined to limited numbers in Whitehall, contributing to the development of plans divorced from operational realities and prone to several of the problems identified by the members of the Holloway Commission in the wake of the failed American hostage rescue in Iran.[282] The small numbers and secrecy are evident in the planning for Operation PARAQUET. During it, 'discussions were confined to the Prime Minister, Defense Secretary, and the most senior commanders', according to the official historian of the campaign.[283] While the overall success of the war played a part, the lack of official inquiry into the missteps, accidents, and mistakes related to special operations is curious. Intuitively, the political capital accompanying the lauding of its premier special operations capability would likely be undercut by any public scrutiny of the failures.

The third, and perhaps most important, fact is that British special forces sought to capitalize on their access by cultivating civilian and military principals by aggressively demonstrating their value in all circumstances. Special forces leaders did little to dispel the 'myths' surrounding them, much less the 'fatuous expectations' of what they could, or could not, accomplish.[284] The director of the SAS and 22 SAS commander used their access and influence to 'lobby hard' for an expanded role in the campaign and engage in 'a marketing expedition' by courting the chief of the defence staff, the minister of defence, as well as the prime minister to ensure special forces were represented at the highest level of planning prior to and during the campaign.[285] The director went so far as to convince them, in his own words, that the 'SAS could make a significant contribution to the plan for recovering the islands'.[286] Such representing, as well as haste, led to flights of fancy in planning

[279] Asher, *The Regiment*, p. 455.
[280] Southby-Tailyour, *Exocet Falklands*, p. 249.
[281] Ibid., p. 252.
[282] Special Operations Review Group, Rescue Mission Report (Washington, DC: Joint Chiefs of Staff, 1980), pp. 13–29.
[283] Freedman, *The Official History of the Falklands Campaign. Volume II*, p. 222.
[284] Ibid., p. 729; Southby-Tailyour, *Exocet Falklands*, p. 249.
[285] Ramsay, *SAS*, p. 195; MacKenzie, *Special Force*, p. 157.
[286] De la Billière, *Looking for Trouble*, p. 344.

including PARAQUET, PLUM DUFF, MIKADO, and KETTLEDRUM, in spite of insufficient intelligence necessary to plan and conduct special operations at the operational and environmental extreme limits of capability. Connor observes, with some relish, that another contributing factor was the temptation of increasingly careerist special forces staff officers to 'accentuate the positive and eliminate the negative altogether' in their reporting up the chain of command.[287]

Further down the chain of command, at the campaign level, conventional commanders were frustrated by special forces possessing better communications than even the task force commander, provided by their American counterparts. In addition, their 'readiness to act outside the standard command structures', and excess capacity, in terms of numbers deployed relative to tasks, suggest special forces leaders were looking for opportunities to justify their relevance and utility, presumably to expand their influence bureaucratically and politically.[288] Although demanding secrecy to preserve operational security, which created its own challenges, conventional British commanders were incensed when the SAS broke it for its own political gain. For example, shortly after Mount Kent was seized, after almost a week's delay, 3 Commando commander Julian Thompson was 'furious when I was told that boastful remarks were being made back in England by those who should have known better to the effect that the British were now holding the high ground overlooking Stanley'.[289]

Such claims could only have been made by the SAS, transmitted to the director of the SAS in London by the 22 SAS commander, Michael Rose. Rose had attached to him on Mount Kent a correspondent from *The Daily Telegraph*, Max Hastings. Hastings, who cut an odd figure with the operators, 'made a deal' to Rose to 'use [his satellite] phone to talk directly with the Brigadier back in the UK'.[290] The net effect of these realities for British special forces was mixed: the SBS acceded to their control and supported the Royal Marines and the Royal Navy as initially ordered.[291] Their focus on such fleet duties, as well as combined exercising and training, integrated seamlessly with conventional forces. That did not, however, prevent sources within the SBS talking to journalists as communications note with some irritation.[292] The SAS exceeded their control, which should have rested with the operational naval and ground commanders.[293] At points during the campaign, their actions led to legitimate queries about their elements acting under their own initiative, uncoordinated with other units in the area.[294] Senior SAS personnel leveraged their access to Whitehall to engage more than necessary in a mixed range of special operations that attempted to be all things to all operational requirements.

[287] Connor, *Ghost Force*, p. 248.

[288] De la Billière, *Looking for Trouble*, p. 342; Freedman, *The Official History of the Falklands Campaign. Volume II*, p. 729.

[289] Thompson, *No Picnic*, p. 109.

[290] Tony Hoare, *Born for War* (London: Welbeck, 2022), p. 215.

[291] WO 305/5052, FM CINCFLT to RBDNPS/MGRM TRF, 3 April 1982.

[292] WO 305/5052, FM HMS FEARLESS to RM POOLE, 28 April 1982.

[293] WO 305/5052, FM CTF 317 to RBDFNR/CTG 317.0, CTG 317.1, CTG 317.9, 8 May 1982.

[294] WO 305/5052, FM CTU 317.1.1 to CTG 317.1, 4 June 1982.

The conclusion of the war with Argentina was the not the end of British special forces' involvement in the Falkland Islands. CORPORATE's commander recommended B Squadron remain on the island, while others returned home, as they were 'acclimatised' and 'would make idea troops to counter any possible insertion of ARG special forces'.[295] In subsequent years a team of operators continued to travel to them in a clandestine manner complete with cover stories. Their purpose was planned 'retaliatory action against Argentina, in the event of attacks against our forces in the South Atlantic'. Those plans were reassessed by members of the visiting teams. Given the challenges British special forces faced during the war, based on weak horizontal integration, how such plans for retaliation would be 'workable' in any sense is unclear.[296]

## Period 4: 1982–1991

While the Falklands War had little impact on the activities of the SRU and FRU, whose undercover activities continued, the SAS returned to operations in Northern Ireland and its refresher training and exercising domestically and in the Mediterranean, Kenya, and elsewhere.[297] SBS personnel, meanwhile, conducted clandestine operations to gather intelligence against Soviet forces, test new equipment and experiment with new procedures, participate in training and exercises to maintain their skills, and conduct liaison missions with their American counterparts.[298] Although returning to 'Military Aid to the Civil Power' missions ostensibly meant continuing restrictions, or direction and tighter control of British special forces, that proved not to be the case. This period is significant for three reasons: the escalation in the number of SAS ambushes within Northern Ireland, the use of special forces in a pre-emptive manner outside of Britain, and the formalization of control arrangements designed to enhance special forces integration and seemingly curb evidence of excesses.

Various political events define this period in Northern Ireland, including the PIRA bombing of the Conservative Party conference at the Grand Hotel, Brighton (1984), the signing of the Anglo-Irish Agreement and creation of a governmental vehicle for cooperation between the two (1985), disagreements and dissention with PIRA ranks over 'abstentionism' or not occupying Republican elected seats in the south, and signals from both the leader of the political arm of IRA, as well as the secretary of state for Northern Ireland, for a negotiated end to the conflict (1989). The Brighton bombing reflected the desire of PIRA leaders for a shocking 'spectacular' forcing a crisis, as much as it did target Margaret Thatcher personally and derail upcoming talks

[295] WO 305/5052, FM CTG 317.1 to RBDEC/CTF 317, 16 June 1982.

[296] DEFE 23/236, 'Special Forces in the Falklands Islands', 30 October 1984.

[297] AIR 29/5056, 'RAF Detachment to Special Forces' (1983–1984), ledger entries May 1983 and July 1983.

[298] Falconer, *First into Action*, pp. 299; 305; 310–311; 315.

between her and her Republic of Ireland equivalent, Taoiseach Garrett FitzGerald.[299] It also was part of a growing trend in the sophistication and innovation of PIRA military capabilities and tactics, exposing the influence of training and equipment provided by Libya and support and resources from Irish sympathizers in and supply networks from the United States. These capabilities included machine guns, rocket propelled grenades, surface-to-air missiles, and Semtex plastic explosives, as well as homemade mortars and increasingly sophisticated bombs, detonators, and timers.[300]

The escalation of the number of SAS ambushes during this period, and the number of fatalities, has been explained in various ways. One suggests there was a 'shoot-to-kill' policy tacitly understood between the military agent and the civil principal.[301] The theme is consistent within the narratives of human rights activists, socialists, and critical theorists. Mark Urban suggests an external source is to blame, reflected by change in leadership within the SAS, whose officers were drawn increasingly from the ranks of the aggressive Paras. He cites that more than 52 per cent of 22 SAS officer cadre came from this source during this decade.[302] Ken Connor, offering an experienced operator's perspective, dismisses the idea the SAS went 'out of control' for internal reasons. Instead, a functional, procedural one led to the tacit understanding: 'The rising kill rate coincided with a change in terminology which, while a long way short of orders to kill, did give SAS men more leeway of interpretation on operations.'[303] Other operators and soldiers echo Connor's point about the restrictive nature of the rules of engagement, embodied in the infamous 'Yellow Card' soldiers often affixed to their rifle stocks, and suggest the implications of violating them were well understood among the ranks.[304] The statistics from the era are unequivocal, as MacKenzie points out: twenty PIRA killed by the SAS, compared to three by the vastly more numerous other regiments of the British Army, from 1983 to 1987.[305] Bradley Bamford suggests another explanation—the policy of prevention running directly up against the 'judiciary's reluctance to convict based on intelligence and informer testimony that did not meet the high threshold of evidence required in a court of law'.[306] In other words, the results were a confluence of loosened restrictions, oversight, and control and the inability, or unwillingness to adapt legal frameworks to

[299] Gary McGladdery, *The Provisional IRA in England: The Bombing Campaign, 1973–1997* (Dublin: Irish Academic Press, 2006), pp. 125–129; M.L.R. Smith, *Fighting for Ireland? The Military Strategy of the Irish Republican Movement* (London: Routledge, 1995), p. 177. See also Rory Carroll, *There Will Be Fire: Margaret Thatcher, the IRA, and Two Minutes That Changed History* (New York: Putnam, 2023).

[300] Oppenheimer, *IRA Bullets and Bombs*, p. 141.

[301] John Newsinger, *Dangerous Men: The SAS in Popular Culture* (London: Pluto, 1997), p. 33; Murray, *The SAS in Ireland*, p. 454.

[302] Mark Urban, *Big Boys' Rules: The Secret Struggle against the IRA* (London: BCA, 1992), pp. 171–172.

[303] Connor, *Ghost Force*, pp. 190–191.

[304] Ford, *One Up*, p. 152; Michael Dewar, *The British Army in Northern Ireland* (London: Arms and Armour Press, 1985), p. 220.

[305] MacKenzie, *Special Force*, p. 206.

[306] Bamford, 'The Role and Effectiveness of Intelligence in Northern Ireland', p. 595.

match PIRA lawfare, or mobilizing the law as a weapons of war, whose insufficiency was noted by the GOC in Northern Ireland in 1976.[307]

The Loughgall ambush in 1987 exemplifies the 'shoot-to-kill' policy for some but can be explained in less dramatic ways. Better information, and improved information sharing, ensured advanced warning of an impending PIRA attack against an RUC station located in Loughgall in north Armagh. Urban suggests the perpetrators of the attack, from the notorious East Tyrone Brigade ASU, were under surveillance and an informer provided details of the plot, whom Harry McCallion suggests was 'a high ranking PIRA member'.[308] More recent accounts cast doubt on the idea of an informer and suggest instead the aggregation and analysis of smaller pieces of informational tips and cues, such as PIRA men scouting for a digger, provided a clear intelligence picture.[309] The plot involved using a 300-pound bomb, carried in the bucket of the digger, which would circumvent the wall surrounding the RUC station.[310] Eight heavily armed men from the ASU would also assault the station. Near-constant surveillance of the East Tyrone confirmed the attack was proceeding and a plain-clothes team shadowed and reported on the movements of the ASU members to their objective.

At least twenty-four SAS operators from G Squadron, participating in Operation JUDY, took up positions inside the station and around the area.[311] The ASU and its support arrived around 1915 on 8 May and commenced their assault, firing on the station.[312] The SAS returned fire, cutting down the eight who were armed mostly with assault rifles and wearing bulletproof vests. Also killed in the hail of fire was an unarmed member supporting the ASU, as well as an unfortunate citizen passing by the station in a car.[313] Despite some residual bureaucratic rivalry and friction noted in a number of accounts, the intelligence processes of the Intelligence and Security Group, Northern Ireland (ISG), combined with smoother integration with the TCG, was making better use of field resources such as SRU and FRU troops, leading to better and more detailed information about PIRA ASU activities.

In contrast, there was a lack of coordinated and confirmed information on overseas targets, which contributed to the outcome of the Gibraltar ambush in 1988. Surveillance of two PIRA operatives, Daniel McCann and Sean Savage, included their travel overseas. Their trip to Spain in 1987, combined with information about a new PIRA bombing technique and countersurveillance, suggested a bomb plot in

[307] Orde Kittrie, *Lawfare: Law as a Weapon of War* (Oxford: Oxford University Press, 2016); Aaron Edwards, 'Misapplying Lessons Learned? Analysis of British Counterinsurgency Strategy in Northern Ireland, 1971–1976', *Small Wars & Insurgencies*, 21:2 (2010): 310–311.

[308] Urban, *Big Boys' Rules*, p. 227; McCallion, *Undercover War*, pp. 172–173.

[309] English, *Armed Struggle*, p. 254; Connla Young, 'IRA Man Tells the Inside Story of the Loughgall Attack and the SAS Ambush', *The Irish News* (8 May 2017), www.irishnews.com/news/2017/05/08/news/ira-man-tells-the-inside-story-of-the-loughgall-attack-and-the-sas-ambush-1019349.

[310] Adams et al., *Ambush*, p. 111.

[311] Ibid., pp. 112–113; Urban, *Big Boys' Rules*, pp. 227–230.

[312] Connor, *Ghost Force*, p. 190.

[313] MacKenzie, *Special Force*, p. 205.

the British territory of Gibraltar.[314] Details of a potential plot were assembled by the Joint Intelligence Committee and briefed to the prime minister. A troop on SPT duty was dispatched to shadow McCann, Savage, and Mairead Farrell and prevent them from carrying out a bomb attack outside the British governor's residence. The troop members conducting Operation FLAVIUS were provided with an updated version of the 'Yellow Card' rules dictating the conditions, which included the Use of Force, Opening Fire, Firing Without Warning, and Warning Before Firing.[315]

On 6 March 1988, the vehicle allegedly containing the bomb was seized and cordoned off, with police ceding control of the pursuit of McCann, Savage, and Farrell to the SPT. In the confusion that accompanies tactical action details vary but the SPT team, split up to cover McCann and Farrell and also Savage, perceived threatening behaviour in the form of weapons. The SPT troopers following Savage presumed he had a detonator and intended to use it as he rapidly turned around. All three were cut down by gunfire at close range.[316] Subsequent examination revealed the three were unarmed and the vehicle contained no bomb, although one linked to McCann was discovered subsequently across the border in Spain.[317] In addition to errors in operator judgement in the heat of the moment, significant intelligence gaps contributed to the event, most notably a lack of coordination and coverage between British and Spanish authorities.[318] The absence of a public inquiry, and only a coroner's inquest, contributed to perceptions the killings were sanctioned.

Loughgall, Gibraltar, and more aggressive SAS actions in Northern Ireland were contributing to the British strategy of attrition against the PIRA. Ambushes removed from the PIRA experienced, difficult-to-replace cadres, and in the case of Loughgall, much of a regional ASU at once.[319] The net effect was designed to be a deterrent one through denial by convincing IRA leaders armed struggle would not crack British will, and they were losing the means to do so anyway.[320] Individual ambushes were rarely followed up. Although Loughgall prevented bombings of a similar scale for more than two years, the shock on the PIRA cadres lasted less than two months, according to the official British Army history.[321] Following up with more immediate action, the analysis implies, including special operations, might have tipped PIRA leaders and cadres to quit. The strategy of attrition had two apparent, connected prongs discerned by PIRA members, at least in Tyrone: 'the SAS went for IRA members while the UVF went for the families'.[322] Beyond that, special operations served

[314] Connor, *Ghost Force*, pp. 191–192.

[315] Nicholas Eckert, *Fatal Encounter: The Story of the Gibraltar Killings* (Dublin: Poolbeg, 1999), pp. 67–69.

[316] Adams et al., *Ambush*, pp. 158–167.

[317] June Tweedie and Tony Ward, 'The Gibraltar Shootings and the Politics of Inquests', *Journal of Law and Society*, 16:4 (Winter 1989): 464.

[318] Moloney, *A Secret History of the IRA*, pp. 330–331.

[319] Ibid., p. 305.

[320] Neumann, *Britain's Long War*, p. 132; M.L.R. Smith and Peter Neumann, 'Motorman's Long Journey: Changing the Strategic Setting in Northern Ireland', *Contemporary British History*, 19:4 (December 2005): 429.

[321] Chief of the General Staff, *Operation Banner: An Analysis of Military Operations in Northern Ireland*, Army Code (London: Chief of the General Staff, July 2006), pp. 8–10.

[322] Quoted in Moloney, *A Secret History of the IRA*, p. 322.

an additional function: eliminated several barriers to peace, in the form of hardcore militant members, paving the way for the Good Friday Agreement.[323] Peter Taylor assesses the attrition conducted by special forces worked in a complementary, if radically different manner: the SAS in 'a vital and often spectacular rale ... at infrequent intervals ... [and] a blaze of publicity' at one end, and SRU 'ground away at the IRA and its loyalist counterparts day after day under their very noses'.[324]

MacKenzie's claim that SAS actions were damaging to British relations with the Republic of Ireland, the United States, and internationally rings only partly true.[325] The relations with the first two were governed by more pragmatic considerations, including peace and stability for the former and the 'special relationship' in the case of the latter. The United States had political embarrassments of its own related to Northern Ireland, including the contradiction in policy between a hard line against terrorism internationally but a seeming ambivalence towards overt support for the PIRA with sections of its northeast. Politically, British special forces could operate in ways useful to ministers, including the prime minister, in a manner that guaranteed them a level of deniability, or holding at arm's length, in a complex domestic COIN campaign.

SAS ambushes sent the message, and had the effect for which they were intended. They had unintentional consequences as well. Especially in the case of Loughgall, incorrect information initially distributed from official sources, combined with an inquiry that exonerated the eight SPT operators, damaged the credibility of the government and served to some as further evidence that British special forces were playing by a different set of rules and were above the law, according to Father Raymond Murray.[326] Rory Cormac reaches an even more pointed conclusion. In addition to the political victories such incidents provided the PIRA, they amounted to a *de facto* policy of sanctioned, targeted killing, despite claims to the contrary.[327]

Prior to the Loughgall and Gibraltar ambushes, British special forces were undergoing an organizational change designed to implement better integration and direction with other Ministries and Committees. To little fanfare, the position of director of special forces replaced director of the SAS in March 1987. Much of the impetus for this had occurred because of the reviews conducted after the Falklands War highlighting command-and-control problems. If nothing else, in normalizing the position, DSF would be more likely to follow committee guidance such as the Special Operations Group and Special Operations Coordinator Teams in future.[328] The position consolidated British special forces under the command of a brigadier. For SBS and SRU, whose named changed during this period to 14 Field Security and Intelligence Company and later 14 Intelligence Company, this change offered the

[323] McCallion, *Undercover War*, pp. 169; 218–219.
[324] Peter Taylor, *Brits: The War against the IRA* (London: Bloomsbury, 2001), p. 143.
[325] MacKenzie, *Special Force*, p. 209.
[326] Murray, *The SAS in Ireland*, pp. 398–422.
[327] Cormac, *Disrupt and Deny*, p. 209.
[328] Southby-Tailyour, *Exocet Falkland*, p. 275.

promise of improved coordination, integration within the Ministry of Defence, and normalization in terms of rank and promotion. For critics, the price to be paid was high and the bill came due during the Gulf War: expertise needed within units was sacrificed to fulfil staff functions with DSF headquarters.[329] The creation of DSF was a harbinger of things to come: with the electoral defeat of Margaret Thatcher and her patronage, the end of the Cold War, and the quest for efficiencies, the era of special access for British special forces, or at least the SAS, was coming to a close.

The next major commitment of British special forces, now combined under DSF, was the First Gulf War (1990–1991). By all accounts, Iraqi President Saddam Hussein's invasion and occupation of neighbouring Kuwait in August 1990 caught the British and Americans unprepared, focused as they were on an overwhelming number of issues vying for attention in the wake of the collapse of the Soviet Union and Warsaw Pact. As a result, neither was well positioned proactively to subvert or overthrow his regime. British covert action immediately following the invasion centred on Kuwait and military advising and assistance missions to Kuwaiti exiles who formed a nascent resistance movement. The mission soon 'fizzled out'.[330] Covert propaganda efforts to stimulate treachery and subversion inside Iraq were no more successful, due in part to Saudi concerns about using their territory to house facilities for broadcasts.[331] Paramilitary special operations, in other words, were not an option in response, especially during the buildup of coalition forces to retake Kuwait, dubbed by the Americans as Operation DESERT SHIELD.

One possible avenue for special operations reflected a coalition political concern: the rescue of its citizens seized and held in Iraq. Although SPT on Pagoda duty could notionally conduct the mission, its capacity was largely limited to rescue missions conducted in a single location. Saddam ordered these hostages distributed throughout the country and had them placed at locations he thought might be targeted by coalition aircraft. Specific detailed information was wanting and the volume of potential missions, or requirements, overwhelmed the ability to coalition special forces to find and rescue hostages.[332] Sources disagree on the number of British special operators assembled by mid-November 1990 to plan and conduct pre-mission rehearsals to rescue the hostages, but 500 seems to be the most credible figure.[333] Those preparations, however, were rendered moot. In a concession to buy time, or perhaps forestall the inevitable, Saddam conceded to releasing the 2,000-odd hostages in early December.[334]

British special operations played a limited role during subsequent military operations, backed by a United Nations mandate, which began on 17 January 1991. Several

[329] Connor, *Ghost Force*, p. 327.
[330] Cormac, *Disrupt and Deny*, p. 243.
[331] Mark Urban: *UK Eyes Alpha: Inside British Intelligence* (London: Faber & Faber, 1997), p. 163.
[332] Davies and McCrery, *The Complete History of the SAS*, p. 179; Parker, *SBS*, pp. 314–315.
[333] Asher, *The Regiment*, p. 484; Parker cites 800, an implausible figure. John Parker, *SBS*, pp. 315–316.
[334] Dana Priest, 'Saddam Orders the Release of All Hostages', *The Washington Post* (7 December 1990), www.washingtonpost.com/archive/politics/1990/12/07/saddam-orders-the-release-of-all-hostages/a48a7129-b97b-4de3-aa56-8e1c0f443b7c.

factors account for this. Coalition forces, under the command of American General Norman Schwarzkopf Jr, expected considerable resistance from the Iraqi Army, then the third largest in the world and one well equipped with Soviet material, including complex air defences. There was little doubt within the coalition as to which country was leading and controlling it. Schwarzkopf initially limited special operations during the buildup to military advising and assistance missions of Saudi and Kuwaiti military forces, treachery and subversion through tactical psychological operations, and direct support of conventional forces in reconnaissance missions.[335]

His unwillingness to commit special operations for more substantial missions, although he was eventually convinced to accept an expanded British role, had a number of sources. They included his experience as chief army advisor to the commander in Grenada in 1983 as well as his view that coalition air superiority could perform the functions of special operations more effectively, and with less risk, than teams of operators behind enemy lines, particularly in open desert.[336] It happened that the commander of Schwarzkopf's British contingent was de la Billère, who lobbied on several occasions for a greater special forces role in the conflict. Schwarzkopf relented in mid-January, after de la Billère identified a mission tertiary to the main effort within Iraq and reinforced the value of special operations to his commander with an SAS demonstration showcasing their abilities.[337]

Coalition planners were rightly concerned with Iraq's ballistic missiles, particularly its mobile units capable of moving, erecting the missile, and launching it. These Scud missiles, combined with chemical or biological weapons, threatened coalition marshalling areas, departure points, and supply dumps. Rather than trying to find the launchers, coalition special forces would focus on severing their command-and-control links. These links, buried underground, could not be jammed. The first British special operations mission against one of these targets occurred at the start of the second week of the war, on 23 January 1991. An SBS team of thirty-six operators from an unspecified section, requiring two RAF special duties Chinooks from 7 RAF Squadron to transport them, flew within 36 miles of Baghdad to a junction site for the cables to conduct Operation MAUDE.[338] Accompanying them was a small detachment of American Army and Air Force operators 'to coordinate close air support'.[339] In an operation lasting 90 minutes on the ground, the links were blown up with explosives.[340] SAS teams from A, B, and D Squadrons launched on the same evening to conduct two types of missions centred around Scud missiles: intelligence-gathering road watches as the LRDG had done during the Second World War but this time to

[335] USSOCOM, *United States Special Operations Command History, 15th Anniversary* (MacDill Air Force Base, FL: USSOCOM, 2002), pp. 38–41; 45–46.

[336] Peter de la Billière, *Storm Command: A Personal Account of the Gulf War* (London: HarperCollins, 1992), p. 191.

[337] De la Billière, *Storm Command*, p. 192.

[338] Parker, *SBS*, p. 320; Urban: *UK Eyes Alpha*, pp. 168–169.

[339] United States Special Operations Command History and Research Office, *United States Special Operations Command History, 15th Anniversary*, p. 43.

[340] United States Special Operations Command History and Research Office, *United States Special Operations Command History*, p. 43.

identify mobile launchers, or indirect-shock missions against other command-and-control links.[341] The units were deployed as mobile half-squadron fighting columns or much smaller eight-man foot patrols.[342]

The experiences of several of these patrols are sufficiently well known to avoid recounting their details here. Of three B Squadron patrols, one foot patrol was compromised and all but one of its operators captured or killed (Bravo Two Zero), another made to the tactical decision at a landing zone that patrolling was futile in the terrain and returned to base (Bravo One Zero), and a third returned after passing along radio message about Scud locations and after being detected and chased by Iraqi security forces (Bravo Three Zero). What the successful patrols managed to accomplish in preventing Scud launches remains contentious.[343]

The cost to British special operations tells another story: A Squadron suffered one dead and one captured and B Squadron three dead and four captured, with likely the same number wounded as captured.[344] While nowhere near as catastrophic as the losses the Regiment incurred in the Falklands, one important difference is worth pointing out: the losses were, in large part, preventable. In the Falklands the SAS and SBS suffered two fatalities in combat and one of these was from friendly fire. In the First Gulf War, the casualties were much higher despite being under the umbrella of coalition air supremacy. In other words, integration gaps existed in British special forces capabilities.

These gaps represent a recurring theme in operator narratives in relating their experiences about the war. A recurring one is the appalling state of equipment available.[345] Unsurprisingly, several compare the state and quantity of their 'kit' unfavourably with that provided to their American counterparts, a product of cultural differences as well as successive decades of British defence budget cuts. Too few global positioning satellite receivers, or Magellans, were available to outfit all units, driving fighting columns to larger sizes, up to half-squadrons or more than thirty personnel.[346] One-third of the SAS deaths resulted from hypothermia due to the unavailability of cold-weather gear or it not being issued.[347] Meteorological reports said they would not need such gear and a somewhat cavalier attitude prevailed within the Sabre Squadrons but prior British special operations experience in deserts in January suggested otherwise.[348] For the two fielded A Squadron teams he was able to locate, the Regimental Sergeant Major (RSM) managed to deliver locally procured fleece-lined goatskin overcoats.[349]

[341] Des Powell and Damien Lewis, *SAS: Bravo Three Zero* (London: Quercus Editions Ltd, 2021), pp. 184–185.

[342] Davies and McCrery, *The Complete History of the SAS*, p. 180.

[343] The arguments are summarized well in two paragraphs in MacKenzie, *Special Force*, p. 186.

[344] Powell and Lewis, *SAS*, p. 329.

[345] MacKenzie, *Special Force*, p. 185.

[346] Ibid.

[347] Chris Ryan, pseudo. Colin Armstrong, *The One That Got Away* (London: BCA, 1995), p. 7.

[348] Asher, *The Regiment*, p. 518.

[349] Peter Radcliffe, *Eye of the Storm: Twenty-Five Years in Action with the SAS* (Miami, FL: Lewis International Inc., 2000), pp. 213–214.

Personal survival on the teams depended on more than just warmth. One leg of the infanteer's tactical triad of move-shoot-communicate was less than reliable. Mobile teams carried two radio sets: satellite and high-frequency burst transmission radios. The former was unencrypted, and the latter encrypted, or the most used for important tactical and operational transmissions.[350] Issues with the burst transmitters included the lengthy time of set-up and incomplete communications between deployed operators and headquarters.[351] Emergency distress beacons carried by the teams, known as tactical beacons (TACBEs), proved fickle. Bravo Two Zero's, activated when they were in contact and under fire from Iraqi forces, failed to elicit a reassuring response.[352] A Regimental 'hot debrief' at the end of the campaign, conducted over two days, revealed two unpleasant truths. Neither the TACBEs nor the man-portable Clansman radios used by B Squadron patrols had been set to the correct frequencies by 264 (SAS) Signals.[353] In addition, even when communications had been available, the 22 SAS commander would not have dispatched combat search and rescue aircraft and personnel to risk their loss as well.[354] That D Squadron did not have some of the same mechanical or functional issues as either A or B leads to possible inferences about differences in maintenance cultures between them at the time.

Vertical and horizontal integration issues evident in British special operations during the Falklands War persisted during the First Gulf War. Although combined under a command headquarters, the SAS and SBS fought separate wars. Rather than combine capabilities together and take advantage of their mutual strengths, British special operations for the remainder of the short First Gulf War were separated geographically along an arbitrary dividing line in southern Iraq: SAS in the western portion and to the east, the SBS.[355] Separate areas of responsibility ease command-and-control issues but connote an inability or unwillingness to integrate capabilities together. In addition, the unusual step of relieving a SAS squadron commander and replacing them with an RSM, Peter Radcliffe, looks different through the lens of control of integration. The A Squadron commander, a transfer officer from the SBS called 'Graham' in the accounts in command of Alpha One Zero and 'not one of us' as one of the troopers put it, lost the confidence of his commanding officer for not being aggressive enough.[356] Little accommodation was made to bridge the different unit operating cultures and the fact that the RSM, the senior enlisted soldier in the Regiment, had to be sent in suggests two other serious matters in terms of command and control: impatience in not wanting to wait an extra 48 hours for a replacement officer to arrive, or a crisis of confidence among the enlisted under Graham's command with a potential loss of tactical control.[357]

[350] Cameron Spence, *Sabre Squadron* (London: Michael Joseph, 1997), p. 70.
[351] Powell and Lewis, *SAS*, p. 329.
[352] Ryan, pseudo, *The One That Got Away*, pp. 54–55.
[353] Powell and Lewis, *SAS*, p. 323.
[354] Ibid., p. 322.
[355] Parker, *SBS*, p. 319.
[356] Spence, *Sabre Squadron*, pp. 31–32.
[357] Ibid., pp. 118–124; Asher, *The Regiment*, pp. 486–487; Radcliffe, *Eye of the Storm*, p. 205.

That Radcliffe says he had to challenge the men in his new command, and undercut the agency conveyed them by 'Chinese parliaments', suggests the latter was a very real possibility.[358]

## Conclusion

This chapter explored the British experience with special operations from 1971 until 1991. This period saw special operations used in a wide range of contingencies, including indirect shock and intelligence gathering and reconnaissance conducted domestically under tight restrictions and control and abroad in two major conventional conflicts. In addition, military special operations were engaged early in post-colonial military advising and assistance missions in Oman, specifically Dhofar, and assumed responsibility for covert action by serving as SIS's action arm. The sheer range of activities in which British special operators engaged is astonishing, according to the few sources that discuss them in any significant detail. While access to official records will provide a more complete picture upon their release, analysing and assessing British special operations during this period will remain hostage to official secrecy and uneven information by former participants whose works often conceal biases and agendas, personal or otherwise. One claim of the theory of special operations coherence is that it can provide different analytic insights despite a problem inherent in the study of special operations: incomplete, uncritical, and biased accounts. This chapter puts that claim to the test.

The foundational elements of the theory are control and capabilities. This case study suggests something different from normal American theories of civil–military control at work due to the idiosyncrasies of British government. Informal mechanisms and connections, combined with responses that privilege flexibility over deliberate advanced planning, means that control in the sense envisioned by normal theory does not exist—it is undesired or even antithetical. The working of British government stresses direction, as opposed to control, and prizes secrecy, familiarity, and flexibility at the expense of agility and, at times, consistency. Many changes have occurred since the period in question related to great public transparency and accountability, prompted by a number of stressful national events and subsequent parliamentary inquiries. The degree to which those changes have influenced British civil–military relations, however, is for other scholars to determine.

Several insights related to control are evident from 1971 until 1991. First, academic concerns regarding praetorianism among military elites such as special forces are unwarranted in the British case. At the apogee of their political influence senior British special forces leaders were not interested in removing or replacing leadership. In part, this is explained by the second insight, which is that civil authorities continued to hold British special forces in high esteem, even if it did not always appear to be the case. Criticisms based on operational factors of Harold Wilson's

[358] Radcliffe, *Eye of the Storm*, pp. 222–223.

decision to commit the SAS to Northern Ireland misses an overall point. 'Strategic' deployment is as much political as it is military, if not more so. The restrictions placed on British SOF during the first five years were a function of a fluid political situation as much as they were integration and normalization into a broader counterinsurgency approach following a policy of restraint. Not committing forces specialized for 'low-intensity conflict', to preserve them for more 'strategic' missions as some former operators suggest, was simply not an option for civil authorities. In keeping its curious form of continuity in terms of some policy, successive British prime ministers until John Major viewed special forces and their capabilities highly favourably.

Third, direction vice control of British special operations worked for and against British political leaders and SOF. They permitted Wilson to maintain plausible deniability in Dhofar and provided him and Thatcher with immediate response options to signal and coerce the PIRA at crucial points in the conflict. For special forces leaders, specifically those of the SAS, informal social connections proved useful in gaining access to the prime minister, expanding both its influence and resources, particularly from 1975 until 1982. These connections, however, inhibited more intrusive monitoring and control by Parliament and were instrumental in making the problems of SAS operators engaged in personal enterprise in Oman disappear. Such enterprise resulted from insufficient control of operators at the campaign level as they moved seamlessly between overt, covert, and clandestine activities once the SIS gutted its covert action capability. At the tactical level, insufficient control over career enlisted operators by temporary, careerist officers, enabled by norms and practices such as the 'Chinese parliament', was a friction point that only got worse and flared up most often in conventional conflicts due to increased stakes and pressures. This finding directly refutes Nora Kinzer Stewart's claims about horizontal and vertical bonding and loyalty in specialized elites.[359] They also inhibited more effective integration between the SAS and SBS, leading to friendly fire deaths and separate special operations campaigns during the Falklands War.

Informal links benefitted both and worked well until they did not. When under tight operational control with significant political restrictions, as police primacy in Northern Ireland, British special operations performed well based on integration with conventional forces and local constabulary. Friction points remained, specifically regarding information sharing and duplication of effort, but they were inevitable and manageable. When restrictions were eased and British special operations fell into more of a supporting role they butted up against the limitations of informal social ties or moved well beyond them, particularly in the Falklands War. Those ties provided the SAS, given their close connections to American special mission units, with communications capabilities more immediate and responsive than conventional commanders had.

[359] Nora Kinzer Stewart, *South Atlantic Conflict of 1982: A Case Study in Military Cohesion*, Research Report 1469 (Alexandria, VA: US Army Research Institute for the Behavioral and Social Sciences, April 1988), p. 33.

Informality became a source of tension when SAS commanders used them to circumvent traditional command lines and lobby for greater use in missions increasingly beyond their capabilities, such as MIKADO. The increasingly absurd plans had the potential to squander scarce resources. Two authors coldly reached the following conclusion about special operations during the Falklands campaign: 'The point is often forgotten that elite forces like the SAS and SBS are designed to be expendable and take high risks when the objective merits such risks. The objective seems to be valid in every case.'[360] Subsequent revelations suggest this is so. Much heavier planned casualties had MIKADO and other 'good ideas' been launched, in combination with the unexpected loss of roughly one-tenth of SAS Regiment in a cross-decking accident, would have caused irreparable harm. That harm to vertical integration within the Regiment and the SBS, and between the SAS and political decision makers at a much higher and consequential level politically, would have shattered trust and subsequent efforts at control. As it stood, at least one near mutiny was only narrowly averted at the cost of weakened ties between SAS officers and leaders.

Finally, for control connected back to the idea that British civil–military relations privilege informality and deniability, British special operations provided civil leaders with a unique foreign policy and military instrument to address political challenges. British special forces, precisely due to their lack of formal specified missions or authorities, could undertake actions along a considerable spectrum, from overt to covert, policing to intelligence, among many others. Their flexibility and responsiveness provided decision makers with a range of Gray's expansion of choice not only strategically but also politically. Such demonstrable political utility included reinforcing deterrence, assisting allies and partners for pragmatic reasons, and forming the basis for an informal coalition against collective threats such as international terrorism. British special forces operated in areas, and conducted missions in politically sensitive environments, designed both to conceal the hand guiding them but also to introduce deliberate ambiguity about policy, such as occurred with continuing assessments about 'shoot to kill'.

The billpayer for informal control in Britain was special operations capabilities. The spectrum of these capabilities increased dramatically from 1971 until 1991 to include intelligence-gathering and reconnaissance missions, indirect shock, asymmetric response, and military advice and assistance. New units, such as SRU and FRU, were created specifically for covert reconnaissance in modern urban environments, given shortfalls in existing clandestine ones. In addition, the SPT evolved in response to changing threats and responded effectively against an asymmetric one domestically in 1980, propelling the SAS to stardom. Training to maintain currency and expertise in specific skills, along with small-scale deployments, stretched the SAS too far as it tried to maintain footholds simultaneously in support to conventional warfighting missions, small team operations overtly and clandestinely, asymmetric

[360] Anthony Cordesman and Abraham Wagner, *The Lessons of Modern War, Volume III: The Afghan and Falklands Conflicts* (Boulder, CO: Westview Press, 1990), p. 349.

responses for terrorist events, as well as covert action missions on behalf of the SIS. In contrast, SBS teams focused on supported of the fleet possessed a highly level of integration with them but not their SAS counterparts. Fully one-quarter of the SAS was tied down on Pagoda duties, as G Squadron was during the Falklands War, ensuring the losses were felt even more acutely in other units. It also led D Squadron, in part, to become a special operations jack of all trades throughout the campaign.

When conventional conflicts did occur, integration between special operations capabilities was extremely limited, leading to individual tactical successes but little in the way of coherent operational or strategic focus in the Falklands or the First Gulf War. In the former, the crucial success occurred in Chile by accident, providing additional advanced warning of Argentinian air attack. Equipment shortfalls, such as incompatible radios between the SAS and SBS, led to friction and losses in the Falklands. Some shortfalls were overcome by leveraging relationships with other SOF. Such a system could work so long at their American counterparts were not similarly engaged and had no equipment to spare. Consequently, during the First Gulf War, SAS patrols were forced into one of two mission approaches: full Sabre Squadron patrols to limit risk to force and to mission, at the expense of stealth and concealment, or small teams to preserve it. The net effect was uneven tactical accomplishments, increased intra-unit tensions, and preventable losses, particularly as an integrating special operations headquarters was created four years before the war.

The discussion of capabilities and control mechanisms at the beginning and end of the case study can be framed using terms from the theory of special operations coherence. In adding, they can be represented visually as Chapter 5 suggests. Early in this chapter there are references to Figure 6.1, which is depicted here. Doing so permits comparison with analysis with Figure 6.2, explained and appearing subsequently. To summarize, British special operations in 1971 were a state of relative coherence, based on their small size and relatively limited functions. As the deployments to Oman and Northern Ireland suggest, the calibration of British special forces was weighted heavily to clandestine or covert operations. British special operations capabilities to conduct independent small teams operations globally, in asymmetric responses, intelligence gathering and reconnaissance, and military advice and assistance against relatively insignificant opponents, were a bright spot throughout this period. This capability remained high. Publicity from 1980 onward shifted the calibration of British special forces more overtly (Figure 6.2).

Capability integration relative to functions, however, was medium where political stakes were higher but declined to low, as Figure 6.2 suggests. As the intensity of conflict increased and greater integration was necessary, support from allies and partners filled the void as occurred during the Falklands War. Absent that support and in the face of increasing responsibilities and functions, against moderate stakes, those shortfalls were laid bare during the First Gulf War, also depicted in Figure 6.2.

Control measures in Northern Ireland influenced integration domestically but efforts to impose it across British special forces occurred too late. Although subordinated to British Army control, the SAS, SRU, and FRU operated based on informal coordination, rather than tight integration. The tight integration between the SBS

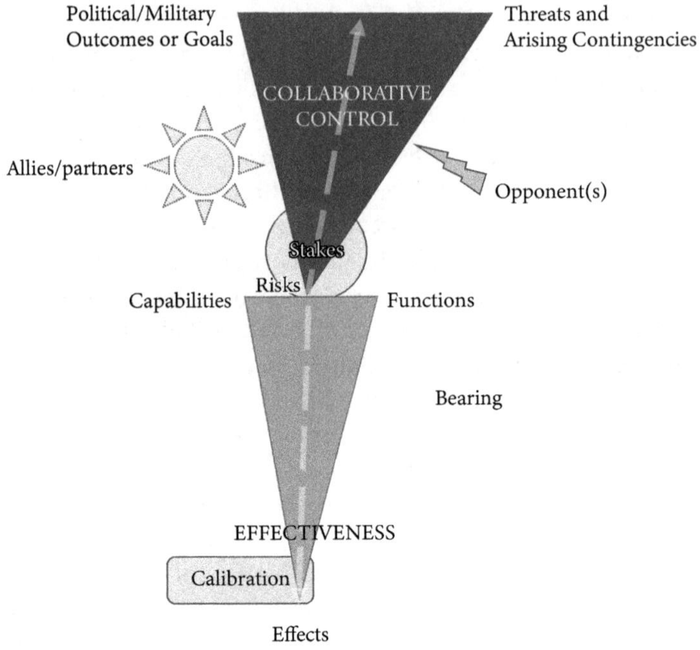


**Figure 6.1** UK special operations congruence, 1971.
*Source:* Author's work.

and the Royal Navy served the needs of the fleet well at the expense of tighter special operations integration. The Falkland and First Gulf Wars demonstrated what occurred when individual and institutional ambitions ran ahead of limitations in capability, due to insufficient preparation and integration in advance. The establishment of DSF in 1987 did little to overcome gaps in capability, relative to functions, much less integration, as Figure 6.2 suggests. 'Handfuls of heroes' could conduct great deeds, based on finesse and guile, but they struggled in situations where the strength resulting from capabilities integration and coherence, as opposed to divergence (Figure 6.2), was required.[361]

Peter de la Billière's fortunes serve as an imperfect analogue for those of British special operations, while also pointing to their central problem. His ascent, with its roots in Malaya, was accompanied by increased profile, ambition, and influence. His informal social connections, specifically with the commandant colonel, avoided a scandal and increased parliamentary scrutiny and control. His lobbying efforts during the Falklands and First Gulf War, within the Ministry of Defence and with General Schwarzkopf, ensured an oversized role for special operations. The successes of his first book on his experiences in the First Gulf War prompted a second about his special operations experiences that revealed operational details and embarrassing

[361] Colin S. Gray, 'Handfuls of Heroes on Desperate Ventures: When Do Special Operations Succeed?', *Parameters*, 29:1 (Spring 1999): 2–24.

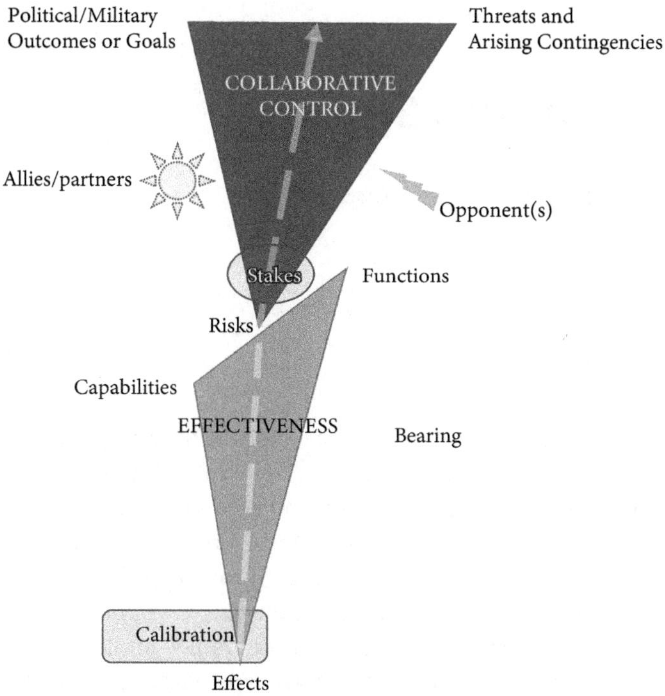


**Figure 6.2** UK special operations divergence, 1991.
*Source:* Author's work.

political ones and also triggered an avalanche of former operator memoirs also seeking to cash in. Yet his subsequent sanctioning by Guthrie was not a formal one; de la Billière was never charged with violating the Official Secrets Act. Instead, Guthrie imposed a potentially worse one: the tribe of operators turning its collective back to him and ostracizing de la Billière from its social functions and clubs, which speaks to the power and resonance of informal social ties within the community.[362]

[362] Guthrie, *Peace, War and Whitehall*, p. 206.

# 7
# System within a System

## South African Special Forces, 1972–1992

### Introduction

In the twentieth century, few countries have witnessed the rising and falling fortunes of special operations more than South Africa. With no tradition of organized special operations, South Africa created, developed, and expanded capabilities in short order to tackle a range of domestic and regional threats. From a wide-ranging and feared capability in the 1970s and 1980s able to strike throughout the region, by 1993 the rebranded units were hidden from public scrutiny as conventional parachute forces. Despite the considerable span of this shift, there are few academic studies and little interest in exploring the special operations experience in the southern African continent. There are several possible explanations: limited Western scholarship on African military history generally and an uneven weighting in what does exists on colonial and post-colonial history and political economy subjects. Observers must also navigate source barrier issues related to South Africa's eleven official languages.

The dearth of interest in South African military operations is also largely a function of South Africa's racist past, specifically the period of apartheid or white-minority rule from 1948 until 1994. Apartheid, an institutionalized system of racist anti-black marginalization and oppression, is the anathema of liberal democratic values. Aided by Western popular culture, charismatic leaders such as Nelson Mandela and Bishop Desmond Tutu provided eloquent condemnations of the evils of apartheid. A jaundiced view of South Africa continued after its end with the creation of the first post-Cold War private military companies led and employing former South African special operators. Given these, few scholars outside the region engage with South African military history and those who do are looked upon warily.

Yet the apartheid era provides an intriguing special operations experience to examine for several reasons. An African case study, South Africa represents an 'outlier' in special operations research. If the general theory of special operations proves sufficient for a non-Western case such as South Africa, it should also be able to explain others. Source material also favours this case study. Although archival access is erratic and the material in Afrikaans, there is a range of accessible, English-language source material. The dissolution of the apartheid system in 1994 prompted the publishing of numerous personal memoirs, political and military, some apologetic and others uncompromising, in English. With appropriate care, the aggregation of these

*Special Operations Success*. James D. Kiras, Oxford University Press. © James D. Kiras (2024).
DOI: 10.1093/9780198902096.003.0007

sources permits a clear portrait of South African special operations activities across all historical mission sets. In addition, the post-apartheid governments of Nelson Mandela and Thabo Mbeki established a Commission to study and report on the state security apparatus. The Truth and Reconciliation Commission published its findings in five volumes, the second of which offers valuable insights into the use of special operations across the range of orientation, offensive and defence, as well as bearing—overt, covert, and clandestine activities. This case is unique in that South Africa had neither a long-standing tradition nor existing special forces[1] prior to 1972. In under a decade, the country created highly capable and diverse special force, leading to questions about how and why this occurred. Finally, South Africa provides a case of different form of government from the other case studies: an oligarchic and exclusive one with an apparent high degree of directive civilian control and possessed a wide range of special operations capabilities in a highly integrated manner—yet ultimately, one that failed politically. The theory of special operations coherence provides insights into why this occurred.

Moreover, while most contemporary special operators and decision makers accept the premise underpinning the national security concept of strategic competition, they struggle in balancing new capabilities with old ones. For special forces, this problem is particularly acute. Current resourcing choices focus on shedding capabilities used successfully over the past two decades in pursuit of ones that may never be used in the future. The South African case study saw a wide range of special operations activities, involving military and paramilitary special forces and conventional forces in an integrated manner termed 'composite warfare' by Eeben Barlow.[2] This case study also saw the widespread use of sponsored proxy forces, including armed militias and insurgent groups, as well as third-party actors, such as East Germany and Cuba. Thanks to the publication of the KGB Mitrokhin Archive and new volumes of the American Foreign Relations of the United States series, as well as scholarly studies on Cuban foreign policy and intelligence activities, the South African case study sheds important new light on how special operations contribute to local, regional, and global competition and conflict. South Africa conducted the full range of historical special operations missions within its borders, offensively and defensively, in an overt, covert, and clandestine manner, often simultaneously.

This chapter begins with a brief history of South African special forces development prior to 1976. The control of special operations capabilities, the subject of the next section, was transformed through the creation of the National Security Management System (NSMS) implemented by P.W. Botha. This system provided South African special forces with a high degree of agency through delegative control. These forces covered the full spectrum of activities, from individual covert indirect-shock missions to support to conventional forces later in Angola. Separating and exploring them by size, orientation, and bearing allows each to be understood in a method

[1] As with the previous chapter, the accepted term for the collective, special forces (vice special operations forces, SOF), is used here. SOF has a different connotation in this context as subsequent discussion reveals.

[2] Eeben Barlow, *Composite Warfare: The Conduct of Successful Ground Force Operations in Africa* (Pinetown, South Africa: 30° South Publishers [Pty] Ltd, 2016), p. 424.

less onerous than a chronological narrative covering parallel operations across seven countries. Doing so also puts to the test the elements of the theory of special operations coherence. The concluding section ties the various operational strands together, looking to the theory to explain events, with graphic representations of it at its beginning and ending points for comparison.

## Background

Conditions in the decades prior to 1972 shaped the development of special forces in South Africa. Early South African settler citizen militias were initially retained for local self-defence, but the Union of South Africa pledged forces for the defence of British imperial possessions a handful of years after the Second Anglo-Boer War (1899–1902). These forces, trained and equipped along British lines, took part in campaigns in eastern and western Africa during the First World War and North Africa and Italy during the Second World War, as well as the abortive Suez Expedition. Following a referendum in 1960 driven partially by anti-imperialism but mostly by race-based Afrikaner nationalism, South Africa broke ties with the British monarchy and formed a Republic in 1961. Freed from the obligations of maintaining interoperability with British forces, white South African political and military leaders focused primarily on rising threats on the South African sub-continent.

The post-colonialism wave that brought the Afrikaner National Party to power contained a powerful threat to white-minority rule: black African independence movements, inspired and supported by communist powers such as the Soviet Union (USSR) and the People's Republic of China (PRC). Most worrisome for South African political and military leaders, however, was the upwelling of insurgencies in neighbouring Rhodesia, Mozambique, and Angola, as well as the independence of Botswana in 1966 (see Map 7.1), and the potential ramifications for the South African-controlled mandate territory of (formerly German) Southwest Africa, or Namibia. For South Africa leaders, these five countries and territories provided a buffer or breakwater against any communist-backed insurgencies arising elsewhere on the continent. Given Rhodesia's proximity to Johannesburg and Pretoria, the economic and political centres of South Africa, as well as increasing calls for Namibia's independence in the United Nations and the newly formed Organization of African Unity, these two territories formed a particular focus for South African security attention and concern.

The threat of communist-inspired political violence was brought home to Afrikaners in 1960 and 1966. Near the police station at Sharpeville, in the province of Transvaal, police opened fire on a crowd of black South Africans protesting racially restrictive travel laws, killing 69 and wounding almost 200 on 21 March 1960. In response, the African National Congress (ANC) and the South African Communist Party (SACP) formed an armed wing, uMkhonto we Sizwe (MK; see Table 7.1), conducting a wave of sabotage attacks and bombings in reprisal. More dramatic were a series of assassination attempts on Prime Minister Hendrik Verwoerd, leading

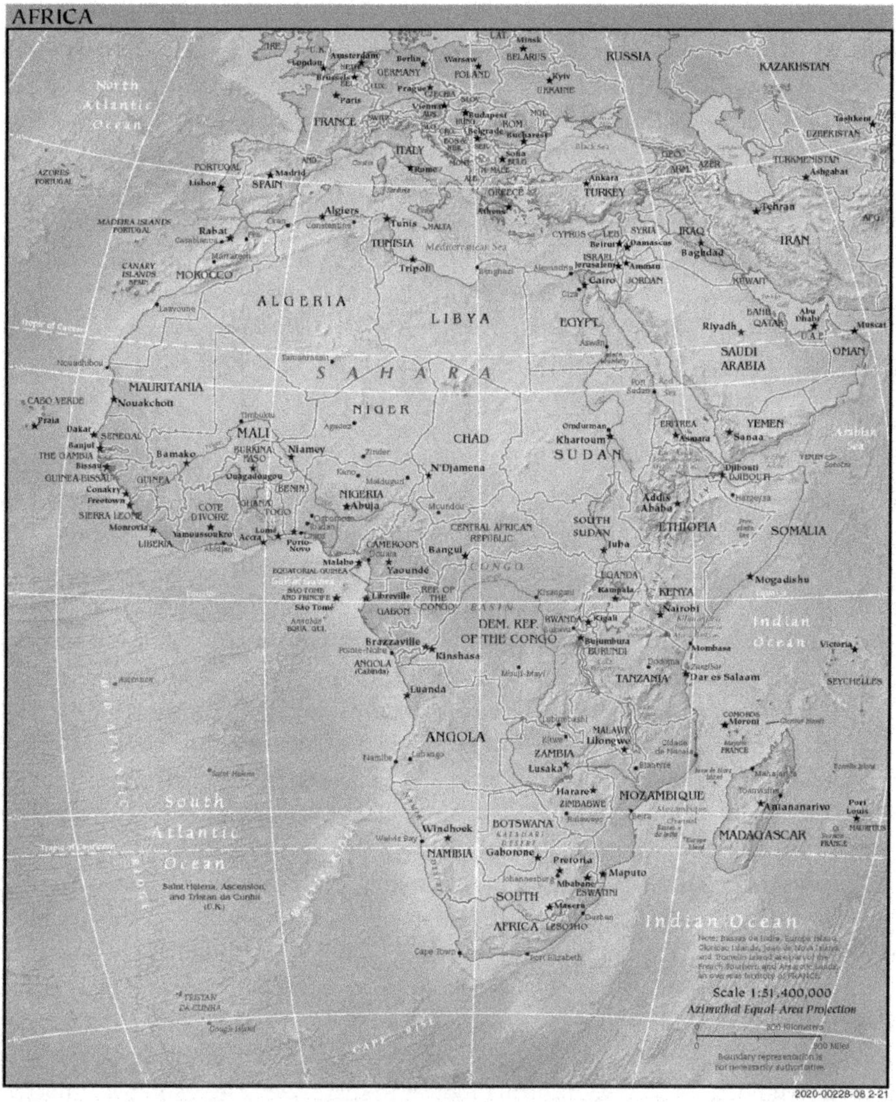


**Map 7.1** Topographical and political map of southern Africa.

*Source: The World Factbook 2021*. Washington, DC: Central Intelligence Agency, 2021. www.cia.gov/the-world-factbook/about/archives/2021/static/1649a3e16a86dcf2f65e53fa39b4a1e7/africa_phy.pdf.

implementer of apartheid and member of the *Broederbund*, the secret society espousing the notion of a distinctive Afrikaner *volk*, two weeks later. Although struck in the head by one bullet and grazed by another, Verwoerd survived until 6 September 1966, when he was stabbed in Parliament. His assassin, Dmitri Tsafendas, had links to the armed element of the Communist Party of Greece and a record of agitation in neighbouring Mozambique. Connecting the dots between subversion, sabotage, and assassination led anti-communist South African political and military leaders to link the assassination with Moscow's Politburo.

Despite the perceived threat of communist revolutionary warfare at home and abroad in the Afrikaner ruling classes, South African Defence Forces (SADF) remained complacent. British officers were purged from the senior ranks of the SADF after the formation of the Republic, taking with them their experience and administrative skills. Far harder to purge was the British military cultural legacy, including doctrine and traditions, which placed greater emphasis on sporting enthusiasm and prowess than on innovations in the profession of arms. A 'rugby mafia' within SADF controlled career advancement and promotion into the 1970s and early 1980s, reflecting a culture characterized by 'brute strength and ignorance'.[3]

In this context, the development of South Africa's initial special operations capability was uneven, given the lack of a tradition of their use. Commandant (later Lieutenant General) Willem Louw, entrusted with the development of Union and later Republic of South Africa's elite infantry formation, 1 Parachute Battalion, or 'the Parabats', questioned their value in future wars in southern Africa. His scepticism was likely based on both the necessity of large-scale vertical envelopment operations in Africa and their viability, given the relative few numbers of capable airframes in the South African Air Force (SAAF).[4] The Rhodesian experience using helicopters to lift, transport, support, and control distributed ground forces, under their 'Fireforce' concept, could respond effectively against insurgent threats.[5] A contemporary of Louw, Jan Breytenbach, went on an exchange tour to the United Kingdom, where he read extensively about British special operations during the Second World War, especially those in North Africa.[6] From this he concluded small-team unit operations, operating deep and independently in areas the enemy felt secure, better suited the conditions in southern Africa, were more easily supported logistically, and had the potential for operational or strategic impact.

Breytenbach, Louw, and others understood the realities of contemporary warfare but had no South African special operations precedent on which to draw. Unlike several other British Commonwealth countries, South Africa did not provide volunteers in support of British efforts to defeat the communist guerrillas in Malaya from 1948 onwards (see Chapter 6 for details). At the time, it was experiencing considerable difficulties reaching its own personnel recruiting targets for the Permanent and Citizen Force—there was little manpower to spare overseas.[7] Moreover, this reticence was not challenged by the British government, as training the South Africans in special operations ran a not inconsiderable risk of a Commonwealth gamekeeper reverting into a much better-prepared and equipped poacher in the future. These concerns echoed but did not completely mirror those expressed by the head of MI5

[3] Les Rudman interview, Pretoria, 8 June 2023; Louis Bester interview, Langebaan, South Africa, 12 June 2023.

[4] Paul Els, *We Fear Naught but God: The Pictorial Edition of the South African Special Forces*, 4th rev. ed. (Cape Town, South Africa: PelsA Books, 2014), pp. 19–20.

[5] This concept is outlined succinctly in Chris Cocks, *Fireforce: One Man's War in the Rhodesian Light Infantry*, 2nd ed. (Roodepoort, South Africa: Covos Books, 1997), pp. 31–33.

[6] Els, *We Fear Naught but God*, p. 18.

[7] Ian van der Waag, *A Military History of Modern South Africa* (Johannesburg: Jonathan Cape, 2015), pp. 228–230.

against the creation of a South African security service.[8] Another suggested that future might not be too far off, given that nationalist Afrikaners were vocal and popular in their criticisms of and increasing opposition to the UK/Commonwealth. Their efforts increased after the electoral defeat of pro-British Prime Minister Jan Smuts in 1948.[9] The policies of apartheid, instituted by Daniel Malan after Smut's defeat, was the fulfilment of more extreme Afrikaner nationalist ideals and further drove a wedge between the United Kingdom and South Africa. A side effect of this nationalism was a form of 'Afrikanization' of the Union Defence Forces, purging British officers from its ranks in favour of loyal and politically reliable Afrikaners.[10]

Louw and Breytenbach built an initial capability as South Africa was increasingly being drawn into an asymmetric conflict in its protectorate of Southwest Africa. The South West Africa People's Organization (SWAPO), a black African nationalist political organization, was committed to achieving independence from South Africa. In keeping with other Maoist-inspired movements, SWAPO as a political organization had a parallel one for the armed struggle, the People's Liberation Army of Namibia (PLAN). SWAPO attempted to build base areas in Southwest Africa to conduct guerrilla war, as revolutionary movements were doing elsewhere in southern Africa, including Angola, Mozambique, and Rhodesia (see Table 7.1).[11] The dissolution of the Federation of Rhodesia and Nyasaland in 1964, part of a continent-wide colonial withdrawal gaining momentum, led to Rhodesia's Unilateral Declaration of Independence from the United Kingdom in 1965 and the establishment of socialist government in Zambia sympathetic and supportive of revolutionary movements in its neighbours. SWAPO and other revolutionary movement leaders were given additional support by the Organization for African Unity, established in 1963, one of whose principles was 'Absolute dedication to the total emancipation of the African territories', including Angola, Mozambique, Rhodesia, and South African-ruled Southwest Africa.[12] From their bases in Tanzania, an initial cadre of SWAPO leaders and fighters established a footprint in Omgulumbashe, a remote settlement area in northern Southwest Africa, in 1966.

At this time, the South African Police (SAP) were responsible for the security and internal stability of Southwest Africa. Personnel shortages dogged the SAP just as they did the SADF. Working on a model of one-third South Africans and two-thirds Southwest African recruits, SAP personnel were hard-pressed to patrol the territory, much less fight a SWAPO insurgency. Information gathered from local sources tipped off the SAP to the SWAPO base at Omgulumbashe.[13] In response, the SAP

[8] Van der Waag, *A Military History of Modern South Africa*, p. 225.

[9] J.C. Smuts, *Jan Christian Smuts* (London: Cassell & Company, 1952), pp. 510–511.

[10] Leopold Scholtz, *The SADF in the Border War, 1966–1989* (Solihull, UK: Helion, 2015), p. 7; van der Waag, *A Military History of Modern South Africa*, pp. 222–224.

[11] Mao Tse-Tung, 'On Protracted Warfare', in *Selected Military Writings of Mao Tse-Tung* (Beijing: Foreign Languages Press, 1963), pp. 219–221.

[12] 'OAU Charter', OAU (25 May 1963), p. 4, https://au.int/sites/default/files/treaties/7759-file-oau_charter_1963.pdf.

[13] Els, *We Fear Naught but God*, p. 22.

**Table 7.1** Southern African guerrilla organizations by country

| | | | |
|---|---|---|---|
| South Africa | uMkhonto we Sizwe (**MK**) <<br>• African National Congress (**ANC**) | | |
| Namibia | People's Liberation Army of Namibia (**PLAN**) <<br>• South West Africa People's Organization (**SWAPO**) | | |
| Angola | *Frente Nacional de Libertação de Angola* (**FNLA**) +} | *Forças Armadas Populares de Libertação de Angola* (**FAPLA**) ʌ<br>• *Movimento Popular de Libertação de Angola* (**MPLA**) + * < | *União Nacional para a Independência Total de Angola* (**UNITA**) + >} { |
| Rhodesia | Zimbabwe People's Revolutionary Army (**ZIPRA**) <<br>• Zimbabwe African People's Union (**ZAPU**) | Zimbabwe African National Liberation Army (**ZANLA**) ><br>• Zimbabwe African National Union (**ZANU**) | |
| Mozambique | Frente de Libertação de Moçambique (**FRELIMO**) + ~ < | Resistência Nacional Moçambicana (**RENAMO**) +-} | |

Symbols: +, Combined political party and military organization;
*, Became ruling political party in 1977 after three-year civil war;
ʌ, Armed forces of MPLA, elevated to national army;
~, Became ruling party in 1975;
<, Marxist-Leninist (USSR-supported proxy);
>, Maoist (PRC-supported proxy);
{, Western-supported proxy;
}, South African-supported proxy;
-, Rhodesian-supported proxy.

requested assistance from the Parabats in creating its own rapid reaction force. Jan Breytenbach and a handful of others trained 44 SAP officers in air assault tactics from aircraft and helicopters. The Parabats not only provided military assistance and advising to the SAP platoon but also integrated into the units, accompanying them on the operation that took place on 26 August 1966.[14] Integration occurred

[14] Paul Els, *Ongulumbashe: Where the Bushwar Began* (Wandsbeck, South Africa: Reach Publishers, 2007), pp. 113–127.

not only tactically but administratively: Breytenbach and the other seven Parabats were deputized as law enforcement officers to accompany the raid.[15] SWAPO suffered twenty-nine casualties; all but two were captured and subsequently detained and served prison sentences in South Africa.[16]

The raid on Omgulumbashe is important for three reasons. First, it marked the start the subsequent 'Border War', between South Africa and SWAPO, which continues to celebrate the date as one of three national holidays related to the war.[17] Second, the raid established a pattern of tactical integration between special operations capabilities and military ones that became the hallmark of subsequent operations during the Border War. Finally, the raid was the first special operations experience of then-Captain Jan Breytenbach, who subsequent South African operators refer to as 'the father of South African special operations'.[18]

The period between Omgulumbasche and 1972, the start of the case study, saw developments related to special operations capabilities. Officers other than Breytenbach, including Major Eddie Webb, advocated in the SADF for the creation of 'SAS-type' training, an organization, or both. With the blessing of Louw, now promoted to chief of the Army, Webb, Breytenbach, and nine other SADF officers and non-commissioned officers attended the selection course for the Rhodesian SAS in 1967.[19] Just over half completed it, which included segments on land and water navigation, escape and evasion, survival, and bushcraft.[20]

Another effort involved a Citizen Force specialized unit, the Hunter Group. This unit was the brainchild of Commandant G. Van Kerckhoeven. Van Kerckhoeven noted counterinsurgency actions in neighbouring countries and sought in 1968 to develop these skills among South Africa's more numerous National Servicemen, should the need for to deploy them arise.[21] The training curriculum of the Hunters centred on individual and group immediate action drills, familiarity with national and foreign weapons, abseiling, demolitions, and house clearing.[22]

The activities of Webb, Breytenbach, and others paid dividends in 1969 as individual efforts to develop a special operations capability coalesced. At a meeting in Pretoria, Charles Allen 'Pops' Fraser briefed the concept of a special forces unit to the SADF senior commanders, which included footage of Hunter Group training.[23] Perhaps most importantly for South African special operations capabilities and control was the presence at that meeting of P.W. Botha, then minister for defence.

[15] Dan Lamprecht interview, Pretoria, South Africa, 8 June 2023.

[16] Randoph Vigne, 'SWAPO of Namibia: A Movement in Exile', *Third World Quarterly*, 9:1 (January 1987): 90.

[17] Annually, 26 August is celebrated as 'Heroes Day'. The two other holidays are Independence Day (21 March) and Cassinga Day (4 May).

[18] Frans Fourie interview, Pretoria, South Africa, 9 June 2023.

[19] Dan Lamprecht interview, Pretoria, South Africa, 8 June 2023.

[20] Els, *We Fear Naught but God*, p. 24.

[21] Peter Stiff, *The Silent War: South African Recce Operations, 1969–1994* (Alberton, South Africa: Galago Press [Pty], 1999), pp. 62–63.

[22] Curriculum detailed by one of the first nine Hunter Group instructors, Stanley Botten. Another instructor was unit commander, Lt Col Kerchhoeven. For details, see Stanley Botten, *We Dare: History of the 2nd Reconnaissance Regiment, South African Special Forces* (Scotts Valley, CA: CreateSpace, 2017), pp. 13–25.

[23] Els, *We Fear Naught but God*, p. 8.

Following the approval of the concept, Breytenbach was tasked by Louw with developing a special operations cadre in 1970, comprising himself and eleven others. They nicknamed themselves 'the Dirty Dozen'.[24] The cadre developed specific training expertise and divided along lines necessitated by different insertion methods, labelled the clandestine air and sea groups. The first South African special operations unit of sixteen personnel, 1 Reconnaissance Commando (1 Recce), was formally established on 30 August 1972 and publicly revealed four weeks later.[25]

Further skills transfer and experience development in countering asymmetric threats occurred among the SAP in 1967. An insurgent incursion into Rhodesia was comprised not only Zimbabwe People's Revolutionary Army (ZIPRA) fighters but also South African revolutionaries from uMwS. Exchanges between national intelligence agencies led the government of Prime Minister John Vorster (1966–1978) to offer assistance in the form of SAP personnel and SAAF helicopters, to Ian Smith's Rhodesia.[26] The Rhodesians viewed the arrangement as a 'mixed blessing, politically and militarily' but SAP personnel benefitted immensely.[27] They learned lessons regarding rural tracking and bushcraft as well as coordinated operations between military and civilian agencies characteristic of Rhodesian counterinsurgency operations.[28]

## SADF takes control in Namibia

The SWAPO threat to South African rule in Namibia crystallized South African national security thinking, as Vorster and his Cabinet came under increasing pressure from various quarters to take more drastic action in response. This was despite SWAPO's logistical challenges in moving fighters on foot from their bases in Zambia through Angola or via the Caprivi Strip, targeted by aggressive Portuguese and South Africa efforts to locate, isolate, and eradicate them. SWAPO's leaders, however, were encouraged to ratchet up the armed struggle by the January 1970 passage of United Nations Security Council Resolution 276 declaring South Africa's continuing presence in and governance of Namibia 'illegal and invalid', a decision upheld by the International Court of Justice in July 1971.[29]

The rulings also spurred public and private organizations in Namibia to rebel. A series of strikes, including in the major port and mining cities in 1971–1972 designed

[24] Interview Dan Lamprecht, Pretoria, 8 June 2023.

[25] Els, *We Fear Naught but God*, p. 49.

[26] Stiff, *The Silent War*, p. 240.

[27] Ken Flower, *Serving Secretly: Rhodesia's CIO Chief on Record* (Alberton, South Africa: Galago, 1987), p. 108.

[28] Legacy Conversations, 'Lt Colonel Stanley F. Botten, Hunter Group & 2 Recce 2/Ic', YouTube video, 1:30:22, 7 May 2022, www.youtube.com/watch?v=BCQgZII3_G4.

[29] United Nations Security Council Resolution 276 (1970), 'The Situation in Namibia' (30 January 1970), p. 2; International Court of Justice, 'Legal Consequences for States of the Continued Presence of South Africa in Namibia (South West Africa) Notwithstanding Security Council Resolution 276 (1970)', Advisory Opinion (21 June 1971), p. 53.

to force a UN plebiscite, paralyzed the country but were suppressed. Civil disobedience actions occurred simultaneously with armed struggle, including the placement of landmines throughout four 'operational zones' designated by SWAPO spokesman Andres Shepanga.[30] The scale of SWAPO infiltrations into Namibia expanded exponentially, increasing more than tenfold from 1966 to 1974. The net effect of these actions overwhelmed SAP forces despite their experience in countering insurgency in Rhodesia and expanded policing powers in the territory.[31] In his assessment of the 1972 strikes, Leopold Scholtz concluded the problem was not one of SAP capability but rather capacity: 'the police found it impossible to cope with internal security as well as the insurgency'.[32]

One exception to Vorster's characteristic caution was the use of special operations personnel, who were more experienced than SADF in countering revolutionary warfare. Although 1 Recce comprised just over two dozen operators in 1974, they had experience conducting intelligence-gathering and reconnaissance missions in southern Angola to familiarize operators with the terrain as well as conducting clandestine surveys to identify scarce watering holes.[33] After Breytenbach lobbied the senior staff officer for special operations to gain combat experience, five-man teams also conducted long-duration bush patrols in Mozambique.[34] These patrols were designed to hone scouting and tracking skills and apply combat skills should the groups make contact with or ambush *Frente de Libertação de Moçambique* (FRELIMO) guerrillas, an extension of ongoing training operations with the Rhodesian SAS.[35] These operations expanded dramatically over the next two decades.

Not all of 1 Recce's missions were successful. A small team of operators, led by Jan Breytenbach, transited by a South African Navy (SAN) submarine, the SAS *Emily Hobhouse*, to Dar es Salaam, Tanzania for Operation STARLIGHT (1972). Their mission, conducted in conjunction with the Portuguese, was to topple President Julius Nyerere for his support to FRELIMO.[36] Although the charges planted by the kayak-borne operators detonated, the attack had little impact strategically.[37] Plans to raise and employ Zambian subversion forces in 1971, led by Breytenbach, failed miserably. This failure led to expanded use of Recces in cross-border actions against SWAPO columns in southern Angola.[38] In addition, Recces were tasked to mine and severely damage or sink a ship alighting in Durban suspected of carrying munitions

[30] Al Venter, *The Zambezi Salient: Conflict in Southern Africa* (Old Greenwich, CT: The Devin-Adair Company, 1974), pp. 54; 56.

[31] Annette Seegers, *The Making of the Military in Modern South Africa* (London: I.B. Tauris, 1996), pp. 137–138.

[32] Leopold Scholtz, *The SADF in the Border War, 1966–1989* (Solihull, UK: Helion, 2015), p. 12.

[33] Alexander Strachan, *1 Recce: The Night Belongs to Us* (Cape Town: Tafelburg, 2018), p. 56.

[34] Strachan, *1 Recce*, pp. 344–345.

[35] Els, *We Fear Naught but God*, p. 88.

[36] Dan Lamprecht interview, Pretoria, South Africa, 8 June 2023.

[37] Arne Søderland and Douw Steyn, *Iron Fist from the Sea: Top Secret Seaborne Recce Operations (1978–1988)* (Johannesburg, South Africa: Delta Books, 2018), pp. 3–4.

[38] Els, *We Fear Naught but God*, p. 52.

for FRELIMO. Several mines attached to the ship failed to detonate.[39] During a clandestine mounted patrol in southern Angola in June 1974, Breytenbach's unit found tracks indicating that a force of SWAPO was nearby.[40] A close-range firefight ensued and resulted in the first Recce combat fatality, Lieutenant Freddie Zeelie.[41] Deniability was only maintained by acknowledging Zeelie's death but obfuscating his unit and location.[42] The clandestine nature of the patrols and ambushes, while deferring attacks on some local police stations, 'did not do much to hinder SWAPO's build-up.'[43]

Events in April 1974 decisively changed South Africa's geopolitical and military circumstances. After years of lobbying by senior officers, Vorster and his Cabinet turned over primary authority for security issues in Namibia to the SADF on 1 April. Vorster's overarching policy with other African nations, given his characteristic caution as one of 'live and let live', led him to circumscribe or reject proposed plans that would upset other African leaders including cross-border raids against SWAPO camps in Zambia and Angola.[44] Senior South African officers, perceiving an approaching wave of revolutionary warfare coming from Angola, Mozambique, Rhodesia, and Zambia as early as 1968, were anxious to gain experience for SADF personnel before the storm arrived.

Soon South African political and military leaders faced their worst geopolitical nightmare. A military coup and civil uprising in Portugal on 25 April 1974, dubbed the Carnation Revolution, toppled more than the government; it upended South African strategic plans in several interconnected ways.[45] South African policy and strategy was based on maintaining strategic depth, or neighbouring country buffers, against insurgencies to prevent them from spreading—keeping them at arm's length. This strategic depth had been sustained by providing material counterinsurgency support to Portugal and Rhodesia. The rapid disengagement of Portuguese military forces from Angola and Mozambique ended 'four and a half centuries' of presence on the African continent.[46] Almost overnight, South Africa's strategic depth evaporated and groups sympathetic to SWAPO provided territory and support to communist-influenced insurgencies in neighbouring countries.

The collapse on 1 January 1975 of the Alvor Agreement, a half-hearted attempt at power sharing between the different insurgent groups in Angola, set in motion a chain of events related to local, regional, and global geopolitical competition between the Soviet Union, the PRC, and the United States. Angola, rather than

[39] Ibid., p. 48.

[40] Stiff, *The Silent War*, pp. 53–55.

[41] Strachan, *1 Recce*, pp. 74–78; Dan Lamprecht interview, Pretoria, South Africa, 8 June 2023.

[42] Stiff, *The Silent War*, p. 55.

[43] Scholtz, *The SADF in the Border War, 1966–1989*, p. 14.

[44] Al Venter, *Vorster's Africa: Friendship and Frustration* (Johannesburg, South Africa: Ernest Stanton Publishers, 1977), p. 22.

[45] For details, see Ronald Chilcote, *The Portuguese Revolution: State and Class in the Transition to Democracy* (Lanham, MD: Rowman & Littlefield, 2010).

[46] John Cann, *Counterinsurgency in Africa: The Portuguese Way of War, 1961–1974* (Westport, CT: Greenwood, 1997), p. 13.

Mozambique, emerged as the focus of competition for four reasons: capable and apparently malleable proxy there in the *Movimento Popular de Libertação de Angola* (MPLA), its geographic location on the Atlantic coast easing air and sea transportation, its position to the next potential front of Namibia, and the relative success of Rhodesia in containing insurgent incursions into the country. Rhodesia weakened FRELIMO in Mozambique, creating and supporting its own proxy insurgency, the *Resistência Nacional Moçambicana* (RENAMO). Within Angola, the *Frente Nacional de Libertação de Angola* (FNLA), *União Nacional para a Independência Total de Angola* (UNITA), and MPLA jockeyed for position, seeking competitive advantage against one another through outside military assistance and sponsorship.

At first glance, assistance appeared divided along geopolitical lines between Western and Soviet Bloc supported groups but this is misleading. Insurgents and counterinsurgents alike sought aid from a variety of sources. In the early 1970s, for example, MPLA leader Agostinho Neto received aid and support from the Soviet Union and the PRC, ideological rivals, as well as socialist outcast Yugoslavia.[47] Each group recruited and largely held enclaves within different tribal areas of the country but of the three, the Soviet and Cuban-backed MPLA was the strongest militarily. Competing geopolitical and national agendas made for strange bedfellows in the provision of aid and support. For example, in 1975 PRC advisors were providing military training to FNLA forces, supported with American covert aid under programme IA FEATURE.[48] At the same time this aid also funded British and French ill-fated efforts to recruit and field a mercenary contingent to bolster FNLA numbers and capabilities.[49] Chinese and American goals converged in Angola in trying to undercut further Soviet expansion of influence, the result of the rapprochement between the two countries against a shared opponent.

The Alvor Agreement disintegrated as FNLA, UNITA, and MPLA sought predominance through a series of escalating reprisal actions, heightening South African concerns. Vorster and his Cabinet envisioned MPLA control of the country, allowing it to provide a home for SWAPO bases. By July 1975, civil war resumed in Angola, leading to South African efforts to prevent the SWAPO and MPLA gaining control of the country, particularly after the latter seized the capital, Luanda. Of two other groups, Holden Roberto's FNLA had been the best-supported and appeared to be strongest militarily. The MPLA appeared to be on the brink of defeat when Soviet military aid and assistance, alongside Cuban military advisors, helped turn the tide in its favour. On 4 September 1975 the Vorster government authorized military advising and assistance missions to stabilize both the reeling FNLA as well as UNITA, which also included equipment and logistical support.

[47] Piero Gleijens, *Conflicting Missions: Havana, Washington, and Africa, 1959–1976* (Chapel Hill, NC: University of North Carolina Press, 2002), pp. 242–243.

[48] Detailed in John Stockwell, *In Search of Enemies: A CIA Story* (New York: W.W. Norton & Co., 1984).

[49] Odd Arne Westad, *The Global Cold War: Third World Interventions and the Making of Our Times* (Cambridge: Cambridge University Press, 2007), p. 227; Anthony Mockler, *The New Mercenaries: The History of the Mercenary from the Congo to the Seychelles* (London: Guild Publishing, 1985), pp. 167–169.

Given his experience with Zambian forces, Breytenbach was a natural choice to assess UNITA and FNLA and lead the training mission. Accompanying him were two 1 Recce officers and two enlisted operators, augmented by Citizen Force conscripts, as well as 'helpful ... US training manuals'.[50] Breytenbach soon discovered FNLA appearances were deceiving; their well-equipped officers stood in stark contrast to the soldiers, whom he described as 'the scruffiest, most underfed, worst armed, and most unwarlike body of troops I had ever seen in my life'.[51]

A training regime focused on weapons employment and tactics, combined with harsh discipline, forged the company-sized element of FNLA in Rundu into a reasonably competent combat force. The unit, known as Bravo Group, followed Alpha Group established the year prior by Commandant Delville Linford. Linford's task was almost the exact opposite of Breytenbach's. Rather than special operators training and leading a conventional force of recruits into battle, Linford used liaison experience with the Portuguese in Angola to organize combat-hardened and experienced former Portuguese Army Ovambo trackers, known as *fletchas* (arrows), whose skilled and accurate tracking abilities were unparalleled.[52] Both groups proved their worth as part of Task Force Zulu, one of two combat and manoeuvre elements participating in South Africa's first sizable conventional military operation since Korea.

The operation, SAVANNAH, began on 14 October 1975 and ended ten weeks later. SAVANNAH covered 3,300 kilometres in four to six weeks and involved hastily cobbled together battle groups of different elements, featuring small-scale battles of varying intensity, ranging from pitched battles to ambushes and meeting engagements.[53] Task Force Zulu, roughly battalion-sized, was later joined by three other battle groups augmented by South African personnel, armoured cars, armoured personnel carriers, artillery, and Zairean Army troops. The operation was commanded by 'A handful of senior officers ... in the manner of a special forces operation', delegating control to them.[54]

Alarmed at the speed of the South African-led advances, Fidel Castro quickly deployed almost 600 hundred special operations advisors, augmented by other reinforcements, to stiffen FAPLA's resistance. His efforts proved decisive at the Battle of Kifangondo (8 November) and outside Luanda (mid-November to early December).[55] The combination of positions defended in depth and increasingly heavy and accurate artillery fire left Vorster with an unpalatable choice: commit more sizable and less deniable SADF forces overtly, putting at risk the clandestine nature of South African participation in the war, or withdraw. His decision to commit

[50] Jan Breytenbach, *The Live by the Sword: 32 'Buffalo' Battalion—South Africa's Foreign Legion* (Alberton, South Africa: Lemur, 1990), pp. 12–13.

[51] Jan Breytenbach, *Forged in Battle* (Pretoria, South Africa: Protea, 2014), p. 18.

[52] Delville Linford, *As the Crow Flies: My Bushman Experience with 31 Battalion* (Pretoria, South Africa: Protea, 2014); Tim Stapleton, 'Bush Tracking and Warfare in Late Twentieth-Century East and Southern Africa', *Historia*, 59:2 (November 2014): 243.

[53] Task Force Zulu's advance rate in combat averaged a staggering 100 kilometres per day. Strachan, *1 Recce*, p. 140.

[54] Seegers, *The Making of the Military in Modern South Africa*, p. 213.

[55] Edward George, *The Cuban Intervention in Angola: 1965–1991* (London: Frank Cass, 2005), pp. 77–78.

South African Citizen Forces, and specifically the recently established reserve and training 2 Reconnaissance Commando, as well as the attention received by the decision of his government to extend the term of national service from twelve to twenty-four months, forced his hand.[56]

International media reports, including an article in the *Washington Post*, identified South African forces in the operation. This timing of the revelations was disastrous, occurring as they did almost immediately after the fall of Saigon. Regional political and military support for South Africa evaporated, as did American covert aid and assistance as Congress had no interest in foreign anti-communist operations that could become another Vietnam.[57] South African gains were erased after it withdrew its forces by March 1976. The space was filled by SWAPO camps in southern Angola and 'the worst fears of the military planners—the fears that had led in such large measure to Operation "Savannah" being launched—started coming true'.[58]

Operation SAVANNAH was a watershed for South African special operations. SWAPO camps in Angola led to a drastic increase its insurgent activities in Namibia, prompting South Africa to respond by creating new special operations capabilities. SAVANNAH also established a pattern of recurring 'externals', or conventional force projection forays, into Angola to destroy SWAPO bases, disrupt SWAPO operations, and punish the MPLA-led government of Angola that continued into the 1980s. Despite some early successes, SAVANNAH highlighted glaring shortcomings in South African doctrine and military equipment against Soviet-supplied and Cuban supported proxies, particularly artillery, armoured personnel carriers, and tanks. In addition, anticipated MPLA and Cuban reaction to future South African operations lent weight to expanding South African special forces to conduct covert and clandestine 'externals' in Angola and other states providing support to SWAPO and the ANC.[59]

The outcome of SAVANNAH also fed Afrikaner suspicions that they could count on little outside support. Soviet support to the ANC increased significantly after 1976, including the training of a new generation of cadres and foot soldiers who left South Africa after the police crackdown of the 'spontaneous' Soweto uprising, in which several hundred protestors were killed.[60] For National Party members and many senior government leaders, including in the SADF, the uprising and Soviet support to the ANC were proof of an active, ongoing revolutionary conspiracy. This outlook subsequently became the basis of a policy justifying executive authority and control over all military activity, including special operations.

Moreover, the deployment of Citizen Forces or other SADF personnel abroad technically contravened the Defence Act (1957) and its provision to seek Cabinet

[56] Kenneth Grundy, *The Militarization of South African Politics* (Bloomington, IN: Indiana University Press, 1987), p. 22; Strachan, *1 Recce*, p. 140.

[57] Center for National Security Studies, 'CIA's Secret War in Angola', *Intelligence Report*, 1:1 (1975): 5; 8–10, www.cia.gov/readingroom/docs/CIA-RDP88-01314R000100660020-1.pdf.

[58] Willem Steenkamp, quoted in John Turner, *Continent Ablaze: The Insurgency War in Africa, 1960 to the Present* (London: Arms and Armour Press, 1998), p. 37.

[59] Søderland and Steyn, *Iron Fist from the Sea* (2018), p. 16.

[60] Westad, *The Global Cold War*, p. 243.

approval first for named operations, as well as Parliamentary approval. Vorster's decision to submit an amendment to the Act, making the sending of SADF troops legal retroactively, set the conditions for a dramatic change in the political control of special operations. With the benefit and perspective of hindsight, Annette Seegers concludes: 'The Cabinet, the Department of Foreign Affairs, and Parliament never recovered from the twin blows of Operation Savannah and the Defence Amendment Act of 1976.'[61]

South African special operations capabilities underwent a transformation after SAVANNAH. That transformation began conceptually. The creation of SWAPO bases in southeastern Angola cut their infiltration distances from Zambia by two-thirds. Therefore, the military focus for special operations from 1976 onwards related specifically to locating and destroying those bases. A secondary objective was preventing the infiltration of SWAPO units into Namibia and South Africa by locating and destroying them en route. A third and final objective was the destruction of logistical support, in the form of supplies and munitions, from supporting countries including Angola and Zambia.

## Capabilities

The dramatic growth in responsibilities and potential targets for special operations, however, came up against a cold reality: a lack of capacity. Given the small size of the Recces in 1976—no more than four dozen operators in both units—concentrating them came at the expense of abandoning other lucrative targets. The solution to this quantitative problem came in the form of expansion of degree, rather than of kind. The clandestine sea capability originally developed in 1 Recce was split during the reorganization and formed into 4 Recce in the summer of 1978. This Reconnaissance Commando occupied former SAN facilities in Langebaan, with operational training facilities across Saldanha Bay.[62] In addition to training for repeat operations along the lines of STARLIGHT, 4 Recce personnel also took combat diving courses, facilitated by French and Israeli advisors.[63] 4 Recce's purview was intelligence gathering and reconnaissance, including beach surveys, as well as indirect-shock and asymmetric missions through the littorals and into riverine environments. Limited numbers reflected the limited functions or purposes South African special forces were designed to fulfil at the early stage, which is captured visually in Figure 7.1.

No special operations capability transformation was more dramatic than the composition of the new unit, 5 Reconnaissance Commando. This unit was transformational given the apartheid context, consisting of mostly black African soldiers, many refugees from Angola who served in 31 or 31 Battalion. While the SAP had black Africans as well, SADF persisted in segregating them into non-combat,

[61] Seegers, *The Making of the Military in Modern South Africa*, p. 212.

[62] Arne Søderland and Douw Steyn, *Iron Fist from the Sea: South Africa's Seaborne Raiders, 1978–1988* (Solihull, UK: Helion Books, 2014), pp. 51–54.

[63] Irené 'Tuffy' Joubert interview, Langebaan, 12 June 2023.

auxiliary support units. The Permanent Force elevated its first primarily black unit, the Colored Corps, to its order of battle in 1963, but racial barriers to employment and command persisted.[64] Kenneth Grundy also points out the desire of South African apartheid regime to demonstrate to the world, primarily the West, progress on inclusion and representation even if the gestures were superficial.[65] Parliament passed legislation allowing blacks to serve in Permanent Force units in 1975 but the results were minimal. Five years later, only 490 did so.[66]

The anticipated tasks assigned to 5 Recce were dangerous and risky but consistent with operations conducted elsewhere. SADF officers noted the success the Rhodesians had with units of Africans officered by whites, the Selous Scouts. Their operations were in turn inspired by British ones in Kenya against the Mau Mau known as 'pseudo' or 'counter-gang' operations.[67] Their underlying logic is simple: the best response to an asymmetric threat is one of your own, in this case guerrillas who 'turned' against their former comrades. Pseudos involve an element of deception and subversion. Taking advantage of trust, pseudos were more likely to get local populations to provide information on nearby guerrilla groups. Those groups, in turn, seeing their apparent comrades-in-arms, would lower their guard, making them vulnerable to ambush. Over the longer term, distrust could sour the local population against the guerrillas or guerrilla groups fighting among themselves.

The Selous Scouts had an unenviable operational record of success. This record was matched only by jealousy and loathing among other Rhodesian units. Along with their record, the Scouts also developed a reputation for indiscipline and ruthlessness, a home for 'vainglorious extroverts and a few psychopathic killers'.[68] To learn from the Scouts, an advance team of the initial cadre of more than dozen 5 Recce operators trained with them before developing their own training and selection course of over forty black African recruits.[69] Later, the 5 Recce pseudo experiment would provide a bridge for some black African operators to cross some of apartheid's racial boundaries.

Prizing operator quality over quantity had its advantages and disadvantages. The output of operators was fractional and acute, given that in some selection periods only a few, and in some cases, none succeeded.[70] One solution lay in taking advantage of existing human capital, which the SADF possessed in Alpha and Bravo Groups: a pool of Ovimbundu, Choke, and Ovambo tribesmen with peerless local knowledge

[64] Kenneth Grundy, *The Use of Blacks in the South African Armed Forces*, Collected Seminar Papers, Institute of Commonwealth Studies (1981), pp. 139–140, https://sas-space.sas.ac.uk/4094/1/Kenneth_W_Grundy_-_The_use_of_Blacks_in_the_South_African_armed_forces.pdf.

[65] Grundy, *The Use of Blacks in the South African Armed Forces*.

[66] Kenneth Grundy, 'A Black Foreign Legion in South Africa?', *African Affairs*, 80:318 (January 1981): 108.

[67] The development and use of the concept of 'pseudo' operations or 'counter-gangs' is most frequently attributed to Frank Kitson. His inspiration and account of their genesis is contained in Frank Kitson, *Bunch of Five* (London: Faber and Faber, 1977), pp. 34–35.

[68] Flowers, *Serving Secretly*, p. 124.

[69] Els, *We Fear Naught but God*, pp. 256–257.

[70] Dan Lamprecht interview, Pretoria, South Africa, 8 June 2023; Frans Fourie interview, Pretoria, South Africa, 9 June 2023.

of terrain, language, and culture. A number had lost land or family to MPLA or UNITA, making them motivated to fight. For these reasons Alpha and Bravo Groups subsequently expanded and became 31 and 32 Battalion respectively.

Although not 'special forces' as the SADF defined them, 31 'Bushmen' and 32 'Buffalo' Battalion are SOF instead: not designed as such but capable of conducting some special operations missions. What set 31 and 32 Battalion apart from special forces was the frequency of their employment and their calibration: apart from the Recces, no other SADF units employed in the field using enemy equipment and standard uniforms. Although officered and commanded by white South Africans, 31 and 32 Battalion operated clandestinely in a manner other SADF units could not. Recce small teams were used to locate SWAPO camps and radio the information back to tactical and sector headquarters. They could stay in place and observe until local reinforcements in the form of nearby 32 Battalion companies or platoons arrived, after which the camps would be assaulted. Alternatively, if the camps were larger and better defended, the Recce teams could withdraw and assemble a larger force from 32 Battalion to assault it later.

Even so, more special operators were needed to address state requirements. Immediately following the end of SAVANNAH, SADF took three actions: increasing the visibility of the Recces within its ranks through a promotion campaign, conducting multiple selection courses in the same year, and contracting special operations personnel from outside of the country. All three had their intended effect. The overall number of Recces doubled, and the annual qualification number jumped almost three-fold, from sixteen to thirty-nine operators.[71] Personnel shortages, however, remained acute, leading to an ineffectual and short-lived effort directed by SADF Headquarters in 1981 of direct accession. Rather than following the traditional pathway, via the Permanent or Citizen Force and a period of military service as well as selection and qualification, males called up for their period of national service could be trained as special operators.[72] This controversial program only lasted eighteen months.[73] Contracted special operations or elite personnel were solicited and vetted from retired operators, most frequently from other Commonwealth countries—the wider fraternity of SAS and 'Para' veterans.[74]

An increase in operators was accompanied by organizational and functional changes in South African military special operations units, authorized in 1976. 1 Recce and 2 Recce remained but their functions were distributed across various Reconnaissance Commando units. 1 Recce now assumed responsibility for special operations in urban areas, likely in the wake of concerns following the Soweto uprising and the SAP's indiscriminate reaction to it. In addition, the responsibility for selection and training resided in 1 Recce. 2 Recce remained a Citizen Force

[71] Els, *We Fear Naught but God*, pp. 118–119.
[72] Chris Serfontein interview, Pretoria, South Africa, 7 June 2023.
[73] Els, *We Fear Naught but God*, p. 127.
[74] Peter McAleese, *No Mean Soldier: The Autobiography of a Professional Fighting Man* (London: Orion Books, 1993); Alastair MacKenzie, *Pilgrim Days: A Lifetime of Soldiering from Vietnam to the SAS* (Oxford: Osprey, 2019), chapter 6.

unit, serving both to retain skilled operators on the roster who had completed their national service obligations and to maintain their skills. In extreme circumstances, 2 Recce also provided a ready reserve of trained personnel for either specific special operations, to feed ongoing operations, or *in extremis* circumstances including national emergencies. To these established units, as well as the headquarters established in Pretoria, 3, 4, and 5 Recce were added on 1 May 1976.[75]

Anecdotal evidence suggests the expansion of the Recces from 1975 onward had an adverse impact on the quality of the force as well as the overall capability of South African special operations.[76] Some also confirmed the prevailing opinion within conventional forces of special operators as undisciplined renegades. For example, prior to going on selection in 1981, Major Hermann van Niekerk advised 4 Recce schemes to mount and fire artillery from skiffs and jet-skis, which he considered 'stupid and absurd ... an idiotic idea'.[77] Personnel shortages also led to employing operators without necessary qualifications. A well-known example included a combat jump by Sergeant Neville Clack despite him not 'having undergone a military parachute training course'.[78]

One casualty of expansion was integration between South African special forces, SOF units, and conventional forces. These were evident during Operation PIGEON (1977). Because of a combination of unspecified 'problems with [Recce] pathfinders' and low-altitude, nighttime air navigation, airdropped Recces found themselves 15 kilometres from their objective, a SWAPO base in Angola known as Eheke.[79] Eighty-three operators and headquarters personnel, comprising a significant percentage of 1 and 2 Recce personnel as well as most of 5 Recce's black operators at the time, parachuted in. Once landed and rallied up, the Recce task force was to assault the base from the opposite direction, northeast, after three companies of 32 Battalion fixed the defenders' attention from the southwest on 28 October 1977.[80] Communication problems plagued coordination efforts between the two separated assault elements of Task Force 101, despite the use of communications relay aircraft equipped with a Telstar satellite communications system.[81] The 32 Battalion companies assaulted one SWAPO camp en route and while marching north to conduct another had its reserves attacked by a SWAPO column. Further complicating matters was the fact that the tired and thirsty Recce operators were navigating parallel to the main base, Eheke 1, encircling it.

Control issues also played a part. The ground force special operations commander, Major Hennie Blouw, tried to call off the drop, and failing that, the assault as

[75] Els, *We Fear Naught but God*, pp. 172–173; Chris Serfontein interview, Pretoria, South Africa, 7 June 2023.

[76] Christo Roelfse interview, Pretoria, South Africa, 8 June 2023.

[77] Hermann van Niekerk, *Born to Storm: A Courageous Boy Became a Fearless Soldier* (Pretoria, South Africa: Group 7 Drukkers en Uitgewers [Pty] Ltd, 2019), p. 227.

[78] Stiff, *The Silent War*, p. 199.

[79] Botten, *We Dare*, p. 158; Els, *We Fear Naught but God*, p. 120.

[80] Specific number, based on names of two-thirds of the participants, from Strachan, *1 Recce*, p. 312. Els vaguely refers '80 to 90 members' of the Recces being involved in PIGEON. Els, *We Fear Naught but God*, p. 120.

[81] Strachan, *1 Recce*, p. 323.

well given the tactical muddle on the ground, which was well within his authority. In both cases, however, the sector commander, Brigadier Ian Gleeson, overrode the requests.[82] Gleeson forced the issue due to political pressure he was under from higher headquarters, and presumably the State Security Council (SSC).[83] Instead of the planned assault from the rear, the composite special operations force attacked directly into the SWAPO defenders. Meanwhile, the 32 Battalion companies, hearing the firefight in which the Recces were engaged, moved to contact what proved to be an abandoned SWAPO satellite camp southwest of the base.[84] The Recce assault on the Eheke 1 came at a cost the special operators and SADF could ill afford: seven dead and eight wounded, or almost 20 per cent of the assaulting force.[85]

## Control

Eleven months after PIGEON, political changes altered the landscape of South African special operations integration, fortunes, and command and control. John Vorster resigned in 1978, paving the way for the National Party to promote P.W. Botha to prime minister. Botha served as minister of defence for fifteen years and was a 'hawk', with a favourable view of the role of the military. While defence minister, he tripled SADF's share of the national budget: from less than two to five per cent of the GDP, almost 20 per cent of the Republic of South Africa's government spending.[86] In addition, he played an instrumental role in approving of SADF's proposal to establish a military special operations capability in 1969.

Botha's appointment to prime minister affected South African strategy. Its basis rested on two concepts: Total Onslaught and Total Strategy. The former encapsulated all nationalist Afrikaner fears and perceptions about a unified global communist movement. In this view, by mobilizing all economic and military means available and sponsoring and directing waves of revolutionaries, the Soviet Union sought to overwhelm the only remaining bulwarks of white civilization on the continent: Rhodesia and South Africa. To combat such an onslaught, a total strategy was required to preserve Afrikaner power, wealth, and even civilization but only by mobilizing state resources in an integrated 'grand strategy', which included addressing domestic political and economic grievances.[87]

On the conceptual level, South African thinking owed a debt to French General André Beaufre. Revered in the mid-1960s on the same plane as Thomas Schelling and Bernard Brodie, Beaufre's ideas resonated with SADF General Allan Fraser in 1965. Allan noted 'the absence of the study of the theory of strategy was a serious gap in

[82] Stiff, *The Silent War*, p. 196; Christo Roelfse interview, Pretoria, South Africa, 8 June 2023.
[83] Cornelius Bornman interview, Pretoria, South Africa, 9 June 2023.
[84] Piet Nortje, *32 Battalion: The Inside Story of South Africa's Elite Fighting Unit* (Cape Town: Zebra Press, 2004), p. 124.
[85] Els, *We Fear Naught but God*, p. 51; Strachan, *1 Recce*, p. 339.
[86] William Gutteridge, 'South Africa's Defence Posture', *The World Today*, 36:1 (January 1980): 26.
[87] C.J. Jacobs, 'The Forward Defence Strategy of the South African Defence Force (SADF), 1978–1989', *The Southern Journal for Contemporary History*, 31:3 (2006): 30–31.

South African military training' and commissioned two professors, Dean Fourie and Ben Cockram, to develop lectures on the subject for senior military officers.[88] Successive year groups of senior officers were exposed to Beaufre's ideas through reading, lectures, or conversations with the author. Beaufre also provided 'encouragement and valuable suggestions' to Fraser when he drafted his original doctrine document and was invited to visit the Defence Staff College in 1974.[89]

Beaufre sought to fill a gap he identified in contemporary military thought, noting that ideas failed to keep pace with social and technological developments in revolutionary warfare and nuclear arms. What was needed was a 'praxeology', or 'science of taking action'.[90] Western societies were being confronted by political, economic, social, and technological challenges simultaneously. The only logical response was a coordinated effort emanating from 'the top of the pyramid and under direct control of the Government ... *total strategy*' (italics original).[91] The result of his analysis was a 'universal Einstein-type formula' suggesting strategy is the product of 'any specific factor applicable to the case concerned ... military force ... the psychological factor ... and ... time'.[92]

SADF officers 'borrowed' Beaufre's ideas on Total Strategy and counterinsurgency.[93] Total Strategy needed as its foundation a rational, holistic evaluation. '[A] full-blown, systematic and detailed analysis of the multivariate aspect of the South African political system, the culture and the economy' never occurred, largely for reasons of circumstance, culture, and preference.[94] The ideology underpinning apartheid, with its simplistic divide as the solution to complex and growing social, political, and economic problems, had its own influence on national security policy and strategy in South Africa. Through concepts such as Total Onslaught and Total Strategy, domestic policy reflecting party ideology transformed into a means of interpreting and responding to changing continental and geopolitical circumstances.

This ideology impacted political organizations and more importantly the relationship of civil authorities to the South African military, especially its mechanisms of control, during P.W. Botha's tenure first as prime minister (1978–1984) and later as state president (1984–1989). South Africa transformed from a Parliamentary democracy into a securocracy, a form of government involving reporting chains and authorities dominated by the military.

[88] Dean Fourie, Review of *A Military History of Modern South Africa* by Ian van der Waag, Jonathan Ball Publishers, 2015, in *Scientia Militaria, South African Journal of Military Studies*, 43:2 (2015): 202; van der Waag, *A Military History of Modern South Africa*, p. 252.

[89] C.A. Fraser, 'Revolutionary Warfare: Basic Principles of Counterinsurgency (Being a Review of "Lessons Learnt from Past Revolutionary Wars")', n.d., RESTRICTED, p. iii; van der Waag, *A Military History of Modern South Africa*, p. 252.

[90] André Beaufre, *An Introduction to Strategy, with Particular Reference to Problems of Defense, Politics, Economics, and Diplomacy in the Nuclear Age*, trans. R.H. Barry (New York: Frederick Praeger, 1965), p. 14.

[91] Beaufre, *An Introduction to Strategy*, p. 14.

[92] Ibid., p. 129.

[93] Daniel Douek, *Insurgency and Counterinsurgency in South Africa* (London: Hurst, 2020), p. 55.

[94] Philip Frankel, *Pretoria's Praetorians: Civil–Military Relations in South Africa* (Cambridge: Cambridge University Press, 1984), p. 53.

## National Security Management System

South Africa's transition into a securocracy began when Botha assumed office and created the Joint Management System, better known as the National Security Management System or NSMS, in August 1979. The goal of NSMS was 'to integrate all government department in the day-to-day running of the country so that all facets of government would have an effect on the overall security picture'.[95] Integration was necessary as previous attempts to coordinate government more effectively, such as the creation of a SSC in 1972 and the development of a Joint Management Centre (JMC) for Namibia, failed to prevent debilitating competition between departments or weed out those who owed their positions to National Party and family patronage but were marginally competent at best. Botha was looking for a 'clean' government, free from corruption and nepotism, with 'teamwork' his guiding tenet.[96] Teamwork ensured that the primary goal of national security, 'protecting all South Africans from any security threat', was accomplished along five policy lines of effort, including pre-emptive and preventative self-defence, preserving the Republic against external and internal threats, mobilizing the population, and domestic arms manufacturing self-sufficiency.[97]

Structurally, the NSMS pursued state security by creating a chain of coordinating units linking the state executive to the local level, where Local Management Centres were responsible for security. They connected vertically upwards to Mini-JMCs that coordinated activities through two mechanisms: Joint Liaison Forums and sub-committees for intelligence, operations, communications, politics, social and economic development, and security. These 450 Mini-JMCs reported up to 60 Sub-Joint Management Centres to 11 Regional Joint Management Centres up to the National Joint Management Centre (NJMC), each containing the same 5 sub-committees and hosting Joint Liaison Forums as well.[98] The system would function provided all departments made good-faith efforts to work together. One participant suggests other departments merely 'paid lip service' to coordination and resource sharing, however.[99]

Parallelling the NJMC were numerous Working Groups and thirteen Interdepartmental Committees for all aspects of national security, including critical infrastructure, strategic supplies, manpower, and cultural activities. Both the NJMC and Working Groups were chaired by the deputy minister for law and order and represented by the directors general of all other government departments. Their activities were overseen by a Secretariat comprising four branches: administration, national intelligence, strategic communication, and strategy. The Secretariat reported to the

[95] Hilton Hamann, *Days of the Generals* (Cape Town, South Africa: Zebra Press, 2001), p. 58.

[96] Seegers, *The Making of the Military in Modern South Africa*, p. 162.

[97] Magnus Malan, *My Life with the SA Defence Force* (Pretoria, South Africa: Protea Books, 2006), pp. 150–151; van der Waag, *A Military History of Modern South Africa*, p. 253.

[98] Seegers, *The Making of the Military in Modern South Africa*, p. 163. Peter Stiff graphically represents this system concisely in *Warfare by Other Means: South Africa in the 1980s and 1990s* (Alberton, South Africa: Galago, 2001), p. 80.

[99] Les Rudman interview, Pretoria, South Africa, 8 June 2023.

SSC, chaired first by the prime minister and later the state president, with only a select group such as the ministers of foreign of affairs, defence, law and order, and finance and their most senior advisors, including the chief of SADF.

In line with precepts of Parliamentary democracy, the Cabinet was notionally 'in formal command and control' of the system of government.[100] During the Botha years, however, the president and SSC increasingly came to dominate it. The SSC oversaw and authorized the execution of strategy but did not develop it. To assist SADF planners for higher-risk cross-border raids, and to prevent the political fallout that occurred after SAVANNAH, SSC began issuing guidelines in 1979 specifically for 'planned' as opposed to 'hot-pursuit' operations. Such planned operations 'required the assent of the SSC'.[101] Strategy, planning, and operations were driven largely by SADF which 'dominated' the NSMS through the security committees or *VEIKOMS*, and 'each of the joint management centers coincided with SADF commands and overall the NSMS reflected a military approach towards threat perceptions' for much of the period.[102]

In essence, the NSMS, particularly the SSC, became 'a government within a government' run by the military, as it 'was available and [had] better knowledge and better know-how' meaning that for all intents and purposes it 'drove the system'.[103] To shield their activities from interference or unwanted scrutiny, including external oversight, *VEIKOMS* invoked national security as grounds for classifying or withholding information and often only notified other departments after they had taken actions, including special operations authorizations.[104] At the executive level, by rewriting South Africa's Constitution in 1983 and merging the powers of prime minister and president into a single authority of state president, Botha could dominate Cabinet and largely ignore the concerns of members of Parliament. In essence, Botha became the epitome of the strong unitary executive authority or, in Brian Pottinger's less-flattering depiction, an 'imperial presidency'.[105]

Creating a bureaucracy to implement Beaufre's idealized 'Total Strategy' had its advantages and disadvantages in countering revolutionary warfare in southern Africa. Total Strategy implied almost total control; NSMS suggested not only control but also tight integration between different elements of governments. To guarantee a shared understanding of what was at stake, Botha's staff reproduced and distributed a truncated version of Pops Fraser's doctrine on countering revolutionary warfare. In terms of tight control and integration, however, the reality for South Africa within its borders and elsewhere was anything but. NSMS did little to mitigate the rivalries

[100] James Selfe, 'South Africa's National Management System', in Jacklyn Cock and Laurie Nathan, eds., *War and Society: The Militarisation of South Africa* (Claremont, South Africa: David Phillip, Publisher [Pty] Ltd, 1989), p. 151.

[101] Truth and Reconciliation Commission, *The Truth and Reconciliation Commission of South Africa Report*, volume 2 (Pretoria: Truth and Reconciliation Commission, 1998), p. 144.

[102] Robert Griffiths, 'South African Civil–Military Relations in Transition: Issues and Influences', *Armed Forces & Society*, 21:3 (Spring 1995): 398.

[103] Hamann, *Days of the Generals*, p. 59.

[104] Seegers, *The Making of the Military in Modern South Africa*, p. 167.

[105] Brian Pottinger, *The Imperial Presidency: P.W. Botha in the First 10 Years* (Johannesburg, South Africa: Southern Book Publishers [Pty] Ltd, 1988).

that existed between the SADF, SAP, and the National Intelligence Service. In addition, the leaders and members of these three organizations largely denigrated or ignored other government agencies whose responsibilities included addressing the underlying political, social, and economic conditions in Namibia and South Africa to undercut ANC and SWAPO recruiting. Those within government who did not fully support NSMS security objectives, or were known to be critical of them, were circumvented or excluded.[106]

There were also few bureaucratic incentives to engage too energetically within the NSMS. In Seegers' cutting assessment, instead of promotions or rewards, officials should have contented themselves with 'the ecstasy of knowing you were efficient', secure in their 'quasi-missionary' calling of 'undoing the wrongs and correcting ... injustices'.[107] Frankel concludes that rather than a 'tightly integrated design' that implemented Total Strategy, what resulted within the South African military in particular was 'actually far less coherent, less internally integrated and considerably less conspiratorial' in action.[108] Stability became the norm through security actions and normality remained a never-attained goal. The national policy of preventative self-defence, combined with the bias for action particularly within SADF ranks, led to counterproductive approaches to bureaucratic coordination and integration. SADF leaders assumed control of much of the running of the higher levels of the NSMS, not only given their expertise in planning and execution of tasks but also because the SAP 'did not have the capabilities' to do so themselves.[109] The SAP, in turn, were in charge of the Sub-JMC level and below, which it managed as best it could with its strained personnel resources.[110]

The small size of SADF Permanent Forces, and their high cohesion given strong informal social ties, led to tight integration at the level of mission execution. Within special forces, ties were even tighter, given successive General Officer Commanding (GOC) Special Forces Headquarters (*SPESKOP*) knew the different Recce unit mission commanders well, as they had worked together previously.[111] Such tight social integration, however, also conspired to create a degree of praetorianism within its ranks, serving as a barrier to more effective coordination at the operational and strategic levels. This praetorianism not only resulted from leading efforts to counter-revolutionary warfare in Namibia and abroad but was also a function of trust and confidence Botha routinely expressed in their abilities. Botha's special regard for special forces largely resulted from his familiarity with SADF given his previous position as minister of defence, leading to the establishment of 'a reciprocally beneficial patron-client relationship' that 'steered [political] movements in a more specifically military direction'.[112] Further contributing to this trust and confidence was the

[106] Seegers, *The Making of the Military in Modern South Africa*, p. 169.
[107] Ibid., p. 167; Selfe, 'South Africa's National Management System', p. 155.
[108] Frankel, *Pretoria's Praetorians*, p. 53.
[109] Stiff, *Warfare by Other Means*, p. 82.
[110] Seegers, *The Making of the Military in Modern South Africa*, p. 168.
[111] Cornelius Bornman interview, Pretoria, South Africa, 9 June 2023.
[112] Frankel, *Pretoria's Praetorians*, p. 171.

SADF's combination of 'intimidation and legislative control' of the media, as well as 'genuine admiration for [its] effectiveness and professionalism'.[113]

SAP, for its part, saw itself as a guarantor of domestic stability. While an increasingly important part of NSMS, particularly after the Soweto riots, it never achieved the prestige or influence of SADF politically. The third leg of the NSMS system, the National Intelligence Service (NIS), never fully recovered from Bureau of State Security's (BOSS's) dismantling and was subject to additional competition from the SADF Directorate of Military Intelligence and SAP Special Branch. While their relatively small sizes led to cooperation, that rarely translated into effective national integration.

A higher level of integration was achieved in terms of special operations capabilities. As the leading edge of operations in Namibia and abroad, and reflecting the existing patron–client relationship, South African special operations fiercely guarded information about their activities. They did so not only for political reasons but also for practical ones. Divulging information not only threatened planned or ongoing operations, methods, and sources but it also ran contrary to Botha's priority of keeping South Africa's preventative security actions abroad as clandestine or covert as possible to preserve deniability. Integration was crucial only for the most *in extremis* actions of the state, or as Seeger concludes more generally about *VEIKOMS* activities, 'Co-ordination seemed to consist of supporting unpleasant elements of state business.'[114]

Covert or clandestine special operations abroad served several strategic purposes for Botha: providing an economy of force option, expanding the number of options he had beyond mere military intervention, and potentially limiting the escalation of conflict with neighbouring countries such as Tanzania and Botswana.[115] Compromise of a single element was often sufficient to justify closing down covert activities to minimize additional exposure and political risk, as occurred with operations in Tanzania.[116] Most importantly, special operations deployments served a crucial domestic goal: keeping SADF casualties to a minimum and limiting the risk and exposure of the Permanent and Citizen Forces to counterinsurgency operations in Namibia and intermittent large-scale cross-border offensives into Angola. As the pinnacle of the professional military, South African special operations stood at the opposite end of the competence spectrum from National Servicemen, who 'were often hard to motivate and slow to innovate', further endearing them to Botha.[117]

The patron–client relationship existing between Botha and the more trusted agencies within the SSC resulted in considerable freedom of action and allocation of additional resources. The movement of *SPESKOP* from the control of the chief of

[113] Kenneth Grundy, *The Militarization of South African Politics* (Bloomington, IN: Indiana University Press, 1987), p. 60.

[114] Seegers, *The Making of the Military in Modern South Africa*, p. 167.

[115] Three of the 'master claims' of the strategic utility of special operations made by Colin Gray in *Explorations in Strategy* (Westport, CT: Greenwood Press, 1996), pp. 168–174; 178–179.

[116] Frans Fourie interview, Pretoria, South Africa, 9 June 2023.

[117] Seegers, *The Making of the Military in Modern South Africa*, p. 264.

the Army to the chief of SADF on 1 January 1981 contributed to this freedom.[118] For episodic, high-risk missions outside of South Africa or Namibia, SADF Headquarters generated plans for special operations to conduct reconnaissance and intelligence gathering as well as raids in coordination with the intelligence agencies. The contract between commanders and staff to address risk was 'operators need to sort out risks first while senior [commanders] need to take them'.[119] Other plans were created within *SPESKOP* and subsequently staff through the senior staff officer for special operations (SSO-SO) for review and approval.[120] According to one senior operator's estimate, almost three-quarters of all special operations during the Border War period were generated from the bottom up, by special forces or other government departments.[121]

With the chief of SADF's approval, planned operations were subsequently discussed and approved or disapproved by the SSC or the state president. The one exception was cross-border activities in Angola, where the minister of defence could authorize missions 200 kilometres into the country and the chief of the Army or Air Force up to 150 kilometres, evidence of delegative control.[122] On a few occasions, Botha or the SSC ordered special operations missions scrubbed while operators were awaiting approval on the tarmac.[123] Such missions were cancelled because although the timing might be correct tactically, Botha determined it was the wrong time politically.[124] For other clandestine activities, Botha circumvented the SSC and went directly to *SPESKOP*, bypassing the service chiefs and the GOC.[125] For less risky sustained operations, particularly those in established sectors in Namibia, local sector commanders were the approval authorities for reconnaissance missions, patrols, and ambushes, particularly by 1 and 5 Recce, while larger missions or those requiring additional resources necessitated coordination and approval with either South West Africa Territorial Force (SWATF) or SADF HQs.

Based on the control models outlined in Chapter 3, the South African case under Botha appears to occupy a grey area between delegative, collaborative, and co-optive, which Figure 7.1 at the end of the chapter depicts. Botha supremely trusted the competence and advice of *SPESKOP* and gave its commanders and staff considerable bureaucratic and operational freedom of manoeuvre. The foundation of this trust was the credibility of special operators, combined with personal relationships they cultivated.[126] These, combined with Botha's role as state president, argue in favour of delegative control. Yet co-optive elements of control are apparent in the resources, rewards, and prestige he bestowed on special forces, including the provision of highly

[118] Els, *We Fear Naught but God*, p. 173.
[119] Irené Joubert interview, Langebaan, South Africa, 12 June 2023.
[120] Chris Serfontein interview, Pretoria, South Africa, 7 June 2023.
[121] Ibid.
[122] Hamann, *Days of the Generals*, p. 135.
[123] Les Rudman interview, Pretoria, South Africa, 8 June 2023; Frans Fourie interview, 9 June 2023.
[124] Les Rudman interview, Pretoria, South Africa, 8 June 2023.
[125] Cornelius Bornman interview, Pretoria, South Africa, 9 June 2023.
[126] Retired SADF special operations intelligence officer, Pretoria, South Africa, 7 June 2023; Irené Joubert interview, Langebaan, 12 June 2023.

specialized and expensive equipment as well as access to the Council on Scientific Research and bespoke equipment provided by EMLC.[127] The intelligence driving most special operations played a role as well. It was useful in 'getting the politicos to panic at the right time', suggesting a high degree of agency.[128] Entrepreneurial attitudes are also evident in another operator's observation: 'the law is there to provide parameters within which you can exercise your initiative'.[129]

As befits a Parliamentary system, there was a level of collaborative control as well. In some cases, Botha or the SSC's authorization was granted in a casual, unstated manner, suggesting collaborative control. This type of control appeared the closer special operations got to operating within South Africa against domestic opponents. As Kevin O'Brien outlines, 'general guidelines brought down from the top to deal with opponents of the state were interpreted as authorizing their elimination and other acts, some criminal (such as domestic bombings and the extra-legal killings of opponents), as a legitimate tool of the state'.[130] The subsequent actions by F.W. De Klerk in reining in and shedding light on special operations yet continuing their use within South Africa, the subject of considerable vitriol among some veterans, suggest delegative control with oversight mechanisms reimposed. What little collaborative control there was manifested for political purposes to shield principals from subsequent scrutiny and prosecution by maintaining sufficient deniability. In any event, collaborative control over special operations was the exception and occurred on a much smaller scale compared to the normative rules governing actions abroad. Even though Total Strategy in practice never lived up to the ideal, Botha's relationship with special forces suggests limitations in existing civil–military relations concepts.

## Special operations missions: size, type, and orientation

South African special operations experienced personnel shortages during the earliest phase of the Border War from 1978 until 1984. SWAPO incursions in the first three years increased in both scale and intensity, thanks to basing and sanctuary in MPLA-controlled Angola. A further shock occurred when South Africa lost its nearest regional ally when Rhodesia's Ian Smith succumbed to external pressure from the US and UK to accept the principle of 'one man, one vote' (1976), followed by an Internal Settlement (1978) with non-violent groups and later General Elections in 1980. One of the groups against whom the Rhodesians had been fighting, the Zimbabwe African National Union (ZANU), led by Robert Mugabe, received a majority of the votes, and changed the name of the country to Zimbabwe. This development alarmed Pretoria given the sympathies and support ZANU had provided to the ANC.

[127] Chris Serfontein interview, Pretoria, 7 June 2023; Johan Hechter interview, Cape Town, South Africa, 13 June 2023.
[128] Retired SADF special operations intelligence officer interview, Pretoria, South Africa, 7 June 2023.
[129] Pieter van Niekerk interview, Cape Town, South Africa, 13 June 2023.
[130] Kevin O'Brien, *The South African Intelligence Services: From Apartheid to Democracy, 1948–2005* (London: Routledge, 2010), pp. 90–91.

The loss of its ally Rhodesia removed South Africa's only remaining bulwark against African national liberation movements. While Tanzania continued to supply equipment and facilities for ANC fighters, Mozambique and Zimbabwe now offered more proximate options unsettlingly close to Pretoria and Johannesburg.

The Rhodesian elections offered an immediate, if only temporary, solution to South African special operations personnel shortages. Just as members of 2 Recce had been 'sheep dipped' by Rhodesia to become 'D Squadron SAS', the favour was returned for special operations refugees from the Rhodesian SAS Regiment. Closed negotiations between commanders coordinating joint SAS/Recce operations in the weeks and months before the Rhodesian election led to the offer of employment for former Rhodesian SAS personnel, numbering somewhere between eighty and one hundred, established as 6 Recce at Durban.[131] The experiment as a separate unit was short-lived, lasting less than a year. Language barriers and cultural difference proved sources of considerable tension between South Africans and Rhodesians. In addition to historical issues related to the Second Boer War (1897–1902), where Rhodesians had fought South Africans, the Rhodesians found the South Africans prudish and evangelical and the latter found the former unprofessional, sloppy, and 'not particularly well trained'.[132] Another potential solution, incorporate Selous Scouts that conducted pseudo operations as 7 Recce co-located with 5 Recce, was stillborn given a lack of sufficient numbers.[133] While some 6 Recce operators were absorbed into 1 and 5 Recce, other Rhodesians found work as staff officers in various SADF Headquarters.[134] Some of the former Selous Scouts joined the police special operations units.

The overall requirements facing the SADF, police, and South African special operations were daunting. They were driven by geographic proximity to SWAPO and FAPLA/PLAN bases in northern Angola, Angolan infrastructure targets, and ANC or dissident headquarters or individuals frequently located in neighbouring African states but also as far away as London. These daunting requirements drove a range of different actions: those by covert operatives working as singletons or small teams for the purposes of intelligence gathering, reconnaissance, subversion, hostage taking, or assassination; small teams conducting clandestine reconnaissance and/or indirect shock; and larger clandestine reconnaissance and indirect-shock missions. While the details of some long-range covert and clandestine 'externals' are publicly available, others remain shrouded in mystery or controversy.

Other 'externals', of larger scale and longer duration, occurred as part of offensive cross-border indirect-shock missions to deny SWAPO sanctuary, disrupt their support infrastructure, and destroy their personnel, supplies, and bases. Given

[131] Stiff, *The Silent War*, p. 296; Els, *We Fear Naught but God*, p. 231; Chris Serfontein interview, Pretoria, 7 June 2023.

[132] Stiff, *The Silent War*, pp. 310–311; Louis Bester interview, Langebaan, South Africa, 12 June 2023; Johan Hechter interview, Cape Town, South Africa, 13 June 2023.

[133] Stiff, *The Silent War*, pp. 311–312.

[134] Phyllis Johnson and David Martin, 'Mozambique: Victims of Apartheid', in Phyllis Johnson and David Martin, eds., *Frontline Southern Africa: Destructive Engagement* (New York: Four Walls Eight Windows, 1988), p. 12; Chris Serfontein interview, Pretoria, 7 June 2023.

the time of year when most occurred, when seasonal rains allowed vegetation to grow and made water more accessible, SADF operators sardonically referred to the strikes as 'the winter games.'[135] These actions, ranging far from South Africa's borders, consisted of tactical and operational offensive special operations in support of conventional forces. Specifically, 32 Battalion served in three distinct roles during these operations: reconnaissance in force for the main force, a flying 'guerrilla column' to outflank and strike the rear of advancing FAPLA or PLAN units, and 'stay behinds' once SADF main force withdrew.[136] From 1978 until 1983 there were seven cross-border named operations into Angola by conventional SADF forces: SCEPTIC (1980), PROTEA (1981), DAISY (1981), SUPER (1982), MEBOS (1982), PHOENIX (1983), and ASKARI (1983), the majority of which lasted less than a month.[137]

The increasing scale of cross-border missions, combined with their intensity, led them to shift from clandestine operations towards an overt and increasingly high-intensity conventional orientation. All served some purpose, but whether that consisted of 'destabilization with a purpose' or part of a cumulative strategy of accreted regional victories remains a subject of disagreement.[138]

The next set of operations occurred in the border area between Namibia and southern Angola. To apportion resources and facilitate coordination and control, SADF split this territory into four sectors (see Map 7.2). The first two running west to east were Sector 10 and 20. Sector 70 comprised the Caprivi Strip, a colonial artefact that extends the boundaries of Namibia almost 300 miles in a narrow

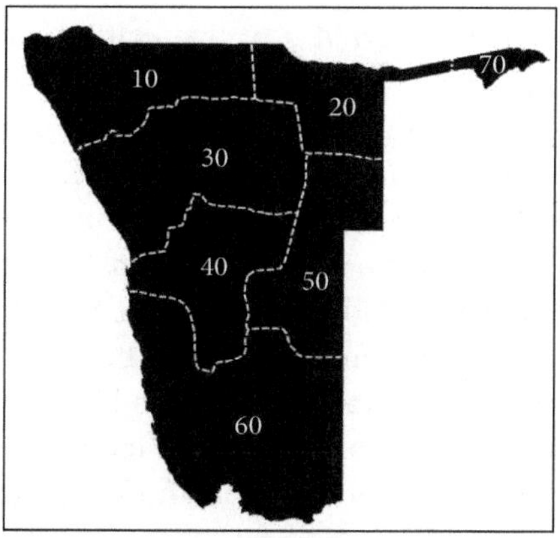


**Map 7.2** SADF military sectors, Southwest Africa.
*Source:* Author's work.

[135] Les Rudman interview, Pretoria, South Africa, 8 June 2023; Radies Rademeyer interview, Pretoria, South Africa, 10 June 2023.
[136] Les Rudman interview, Pretoria, 8 June 2023; Radies Rademeyer interview, Pretoria, 10 June 2023.
[137] Seegers, *The Making of the Military in Modern South Africa*, p. 226.
[138] Grundy, *The Militarization of South African Politics*, pp. 94–95.

corridor eastward to Zambia, with Angola bordering on the north and Botswana to the south. Sector 30, to the south of Sectors 10 and 20, consisted of much of the middle of Namibia including its major port and city, Walvis Bay and Windhoek. Special operations that occurred here were largely tactically and operationally defensive in nature. Sectors 10 and 20 were astride major SWAPO infiltration routes southward into Namibia and South Africa. SWAPO goals of infiltrations were the establishment of operating base areas, particularly in Sector 30, the centre of which was located the Etosha Pan and surrounding parks and nature reserves, natural sanctuary areas difficult to patrol. The frequency of ambushes and engagements here, incurring heavy costs on both sides, led to this Sector earning the nickname 'the triangle of death'.[139]

Other goals consisted of longer-range infiltrations into the largely administrative vice operational Sectors 40 and 50 by SWAPO SOF from the Typhoon and Volcano units to conduct sabotage, subversion, and indirect-shock missions.[140] Their sabotage actions consisted of mine laying missions to inflict casualties and force farmers to move. Subversion missions were undertaken to increase SWAPO's control over rural area populations. Finally, indirect shocks sought to impose costs, particularly among Boer farmers and National Servicemen, to provoke an overreaction by the SADF.[141] Operations in this area consisted of layered defence, decreasing in density and intensity the further south one went. For special operations, the focus in this area was asymmetric response: counter-guerrilla, counterterrorist, and counterinsurgency (COIN) missions. Given the seasonal and recurring nature of incursions, operators referred to their activities against SWAPO bases in this area as 'the summer games'.[142]

The final set of operations occurred near the South African border in Namibia or within the townships. These special operations were designed to react to any deep SWAPO incursions into Namibia that made it through the thick patrols of the border sectors and past Sectors 40 and 50 that threatened to move into South Africa. These occurred infrequently. Greater special operations energy and effort went into operations in the black townships. Missions here consisted of reconnaissance and intelligence gathering, kidnapping and assassination, and subversion in dense urban environments, the training for which began under Israeli auspices in 1981.[143] Their character and conduct reflected a combination of Afrikaner paranoia and cold, calculating realpolitik rationalizations. They remain the subject of considerable controversy and in the case of 'disappearances', a source of continuing moral outrage and tension.

[139] Steve Crump, *Koevoet, Volume 1: South West African Police Counter-Insurgency Operations During the South African Border War, 1978–1984* (Warwick, UK: Hellion, 2022), p. 37.

[140] Radies Rademeyer interview, Pretoria, South Africa, 10 June 2023; Scholtz, *The SADF in the Border War*, pp. 164, 199.

[141] Frans Fourie interview, Pretoria, South Africa, 9 June 2023.

[142] Jim Hooper, *Koevoet: Experiencing South Africa's Deadly Bush War*, rev. ed. (Solihull, UK: Helion Books, 2013), p. 101.

[143] Frans Fourie interview, Pretoria, 9 June 2023.

## Industrial counterinsurgency: Namibia

After 1975, the SAP in Namibia were increasingly outmanned, outgunned, and outfought by steadily improving SWAPO units infiltrated from bases in Angola and neutral Botswana. SWAPO lost Zambia as a sanctuary when the president of that country, Kenneth Kaunda, evicted the group leadership and denied it facilities, but the sudden Portuguese withdrawal encouraged SWAPO leaders to take make aggressive steps against farms, farmers, infrastructure, and security forces.

Despite the efforts of Hunter Group and staff colleges to institute counter-revolutionary warfare and COIN training throughout the force, most SADF units approached operations conventionally: large-scale sweeps and patrols. What these operations gained in their own security they lost in agility and initiative. SADF and SAP units travelling by road in vehicles were vulnerable to mining or ambushes, while in the bush they exercised poor sound and tracking discipline. SWAPO guerrillas, therefore, could engage with security forces at the time and place of their choosing. It would take years before the efforts of Jannie Geldenhuys, who was tasked with creating SWATF in 1978, would expand the successful recruitment and employment of bushmen trackers 31 Battalion and raise six more, which became 101, 102, 201, 202, 203, 701, and 911 Battalions.[144] Given this state of affairs, SWAPO's president, Samuel Nujoma, assessed that the time was ripe to move the conflict into its next phase of a Maoist insurgency, strategic stalemate, to achieve a positive correlation of forces and set the conditions for the final, decisive victory. Organized into the three separate military fronts, SWAPO would 'carry out military offensives on all ... at the same time, to confuse and over-stretch the enemy's military power which would result in heavy enemy casualties and expenditure'.[145] SWAPO were able to operate in platoon (thirty-plus) or company (one-hundred-plus) size elements, outmatching any SAP units and most SADF ones they encountered, as SAP Security Branch Lieutenant Eugene de Kock experienced during an ambush in 1978.[146]

Just as a group of SADF personnel were influenced by their experience in Rhodesia, so too were a cadre of SAP detailed to assist the British South Africa Police (BSAP) there. The SAP contingent to Rhodesia arrived earlier than SADF units, and with few exceptions stayed longer. The SAP presence, detailed to tribal trust lands in northern Rhodesia to help stem the tide of Zimbabwe African National Liberation Army (ZANLA) infiltrations from Mozambique, included nearly 3,000 officers and non-commissioned officers.[147] Relations between the Vorster and Smith governments worsened to the point that South Africa withdrew SAP forces in 1975. Experience in Rhodesia impacted the outlook of Commandant Johannes 'Hans' Dreyer and the

[144] Magnus Norval, *Death in the Desert: The Namibian Tragedy* (Washington, DC: Selous Foundation Press, 1989), pp. 155–156.

[145] Samuel Nujoma, *Where Others Wavered: The Autobiography of Sam Nujoma* (London: Panaf Books, 2001), pp. 315–316.

[146] Anemari Jansen, *Eugene de Kock: Assassin for the State* (Cape Town, South Africa: Tafelburg Publications, 2015), pp. 73–78.

[147] Paul Moorcraft and Peter McLaughlin, *The Rhodesian War: A Military History* (Mechanicsburg, PA: Stackpole Books, 2010), p. 32.

team he initially selected from Security Branch and Special Task Force to establish a new unit, Operation K, or *Koevoet.*

Dreyer's experiences in Rhodesia and Mozambique impressed upon him the need for the collection of better information and close cooperation with SADF SOF. In particular, Dreyer and his commanding officer at Security Branch, Lieutenant General Johann Coetzee, were struck by the close coordination between the Selous Scouts and BSAP Special Branch personnel who provided information. Coetzee's initial charter to Dreyer in mid-1978, driven by the deteriorating security situation in Ovamboland, was to develop supporting activities that allowed pseudo operations to achieve their impressive kill-rate against ZANLA. The basis for success was patient but specialized and methodical police information gathering and intelligence analysis to build up a picture of the insurgent organization and its members.[148] Special forces, specifically 5 Recce, would conduct pseudo operations to refine the intelligence picture and ambush guerrilla units or capture or kill specific SWAPO cadre.

To help facilitate cooperation, the Op-K team of four white and four black officers co-located their headquarters with 5 Recce. Other reasons drove the decision to move here, including an unexploited trove of information contained in the Security Branch archives. Through patient trawling through this information and analysis, Op-K developed a clear and refined intelligence picture of SWAPO's organization, operations, and personnel, although some significant gaps remained.[149]

Despite the best intentions of working together, cooperation broke down initially over an all-too-common cause related to special operations capabilities: who owns or controls it, in this case the pseudo force. Dreyer envisioned pseudo operations as a clandestine means of intelligence collection along the lines of the Selous Scouts during its early years. The operators of 5 Recce would perform the function the Rhodesian SAS had in that early relationship: killing but preferably capturing insurgents. Captured insurgents could not only provide timely and valuable information to update the intelligence picture but could also be turned and serve as pseudo operators against their former comrades.[150] Ops-K's own experiences detaining suspects and gaining perishable intelligence convinced Dreyer that he should develop a pseudo capability, separate from 5 Recce, able to work clandestinely and complete the intelligence and operations cycle in a single unit. He was doubtless unaware, even as he grew Ops-K by more than sevenfold to conduct pseudo operations, that 5 Recce's successes against SWAPO were being translated into a much broader and wider ranging clandestine capability first as D-40 and later, Barnacle, discussed below. *Koevoet* developed its own organizational structure and doctrine for pseudo operations-based COIN.

The basic structure of *Koevoet*, from 1980 until its disbanding in 1989, consisted of two units. The first was an Investigative Team, or IT, that patrolled in the bush for

[148] Peter Stiff, *The Covert War: Koevoet, Namibia, 1979–1989* (Alberton, South Africa: Galago Books, 2004), p. 53.
[149] Retired SADF special operations intelligence officer interview, Pretoria, South Africa, 7 June 2023.
[150] Frans Fourie interview, Pretoria, South Africa, 9 June 2023.

extended periods of time. The ITs, comprising a mix of police officers and turned insurgents, operated in denied or contested areas, interacted with local populations, addressed local crimes, and attempted to refine the intelligence picture. If they detected or suspected a SWAPO unit was in the vicinity, a localized area would be 'frozen' and the hunt would begin.[151] The second half of the SAP counterinsurgency equation in Namibia was the combat team. Combat teams provided both the rapid response capability missing in the abortive relationship with 5 Recce as well as mobility, mass, and firepower. A frozen area would act as a magnet, drawing in additional combat teams, to track and hunt the SWAPO unit.[152]

From these *Koevoet* operations an informal doctrine took form. This doctrine consisted of four phases: initial policing from the ITs, detection, tracking, and closing in, engaging and winning the firefight to 'neutralise the insurgent threat', and reorganization.[153] The latter phase consisted of two separate activities. One was returning to base of the ITs and/or combat team(s) to rest, refit, and replenish losses and conduct training to maintain readiness for future operations. The other was processing information on insurgents killed and any documents recovered, and interrogation of captured insurgents to update the intelligence picture and the understanding of SWAPOs evolving tactics, techniques, procedures, and organization.[154] Equally if not more importantly, interrogators sought to turn those captured to assist in manning the ITs, and later the combat teams as they expanded.[155]

In its initial operations *Koevoet* experienced considerable success in locating, tracking down, and killing or capturing SWAPO units. Success was based less on the need to conduct pseudo operations, whose principles 'Dreyer had already backed away from', than they were on three related elements.[156] The first was continuously updating the local intelligence picture. The intelligence cycle had two levels, based on duration: a long-term intelligence cycle that contained target dossiers on individuals and attempted to gain insight and understanding of SWAPO, and a short-term one 'based on 'hot', actionable intelligence'.[157] Both *Koevoet* and 51 Recce pseudo operators possessed 'detailed and specific intelligence' of a highly refined nature within the operating areas.[158] The second was a considerable freedom of action for those closest to the ground—junior officers and constables—to exercise their own judgement and initiative. Oversight, control, and direction from higher headquarters could only slow down the pace of operations against elusive, fleeting targets. Unlike other forms of special operations, which place a premium on repeated training and rehearsal to ensure a high standard of performance for a specific task in different circumstances, *Koevoet* operations were driven instead by responsiveness and results. The third,

[151] Retired SADF special operations intelligence officer interview, Pretoria, South Africa, 7 June 2023.
[152] Christo Roelfse interview, Pretoria, South Africa, 8 June 2023.
[153] Crump, *Koevoet, Volume 1*, p. 21.
[154] Retired SADF special operations intelligence officer interview, Pretoria, South Africa, 7 June 2023.
[155] Crump, *Koevoet, Volume 1*, p. 21.
[156] Hooper, *Koevoet!*, p. 111.
[157] Retired SADF special operations intelligence officer interview, Pretoria, South Africa, 7 June 2023.
[158] Ibid.

and perhaps most innovative, is what Steve Crump has termed 'matrix working'. In his words, the relationship between ITs and combat units, working together in a common cycle towards the same end, had the effect of 'enabling [them] to work both "vertically" (according to function) and "horizontally" (according to the specific COIN requirement)—that is evaluation of a possible spoor [tracks or other evidence], the mounting of a pursuit or searching the dead for intelligence, etc'.[159] What Crump draws attention to is an effective special operations capability based on a high degree of functional or tactical integration that increased tactical mission performance.

Mission performance and success had positive and negative consequences for *Koevoet*. In proving the concept *Koevoet* gained resources initially denied it, including armoured and mine-protected ground mobility vehicles—an initial batch of weary Hippos and later Casspirs—as well as communications equipment, direct combat and combat service support in the form of SAAF Alouette helicopters and gunships, and dedicated personnel from the South African Medical Service. Continuing results, in the form of kills, higher headquarters enthusiasm, and turned SWAPO, led to a dramatic expansion of *Koevoet*. The first two combat teams doubled in a year and then increased fivefold, to 20 teams, by 1983, and from 40 personnel to almost 1,000.[160] Overall numbers for *Koevoet* increased by 50 per cent in 1982 from 2,000 to 3,000 personnel.[161] *Koevoet* would peak at forty-two combat teams, or 'call signs', by 1989.[162] Such rapid expansion stressed the personnel pipeline over experience to the point where white SAP recruits, with little training and no experience, were put in command of combat teams veteran black trackers and former insurgents—one of the reasons Eugene de Kock gave for leaving the unit.[163] Increases in the number of units complicated the command and control problem on several levels, the most important of which was the expansion in the size and complexity of *Koevoet*'s doctrinal construct, doubling its number of steps and processes.[164] Most deleteriously, success led to a form of competition that removed the last vestige of possible integration at the operational level necessary for COIN to succeed; senior SADF ranks understood operations but not the requirements necessary for sustained, successful campaigning.[165]

Personal, professional, and institutional incentives determine the degree and nature of competition between units in countering asymmetric threats. Two incentives unique to Nambia corrupted operations there and ensured they served ends that

[159] Crump, *Koevoet*, p. 21.
[160] Ibid., p. 24.
[161] Denis Herbstein and John Evans, *The Devils Are Among Us: The War for Namibia* (London: Zed Books, 1989), p. 79.
[162] Crump, *Koevoet, Volume 1*, p. 26. Combat team call signs were derived from the radio phonetic alphabet, providing the titles for Arn Durand's popular series of personal accounts by *Koevoet* operators, including his own. See, for example, Arn Durand, *Zulu Zulu Golf: Life and Death with Koevoet* (Marina Beach, South Africa: JIM Publishers, 2011).
[163] Eugene de Kock, *A Long Night's Damage: Working for the Apartheid State* (Saxonwold, South Africa: Contra Press, 1998), pp. 78–79.
[164] Crump, *Koevoet, Volume 1*, pp. 44–46.
[165] Retired SADF special operations intelligence officer interview, Pretoria, South Africa, 7 June 2023.

overshadowed any comprehensive approach to countering revolutionary warfare. First, *Koevoet* paid its operators *kopgeld*, or 'head money'.[166] While notionally designed to reward the capture of SWAPO insurgents over their death, operators quickly concluded that they could receive the same payment for two dead insurgents as one live one.[167] Moreover, to confirm these kills, the practice evolved of strapping the bodies of dead SWAPO to vehicles and driving them through villages back to bases, which undermined local COIN efforts.[168] SAP members in *Koevoet* culturally were incentivized to put individual performance ahead of team or higher objectives, a difference between them and SADF personnel, according to Hilton Hamann.[169] Others arrive at a more blunt conclusion: 'the war was being privatized' in the entrepreneurial sense.[170]

One institutional incentive for this was the *Jakkalstrofee*, or Jackal Trophy awarded to units on the basis of the numbers of kills it had inflicted. There was fierce competition for the honour between SADF, SWATF, and SAP units.[171] The combination of personal and institutional incentives prioritized performance over effect in the same way that 'body counts' had done in Vietnam.[172] In the words of the members of the Truth and Reconciliation Commission, 'Koevoet in particular kept no proper or official records of the identities, numbers or whereabouts of people it killed. It seems that the unit was only really interested in keeping scorecards of those it killed for bounty.'[173] Whether noncombatants were killed to inflate numbers, as occurred in Vietnam, is unclear.

While the excesses of *Koevoet* will continue to spark debate, its effectiveness as a special operations capability is the primary concern here.[174] Its original personnel were highly selected and trained and individuals, particularly the remarkably capable Ovambos, as one *Koevoet* commander remembered: 'In their own environment they may be the best trackers in the world and their endurance and stamina are unbelievable.'[175] This qualitative edge was increasingly sacrificed as *Koevoet* grew by leaps and bounds. Growth, while detrimental to the development of a more cohesive COIN approach, did help solve one strategic problem for South Africa. By placing the onus of tracking and combat on black Namibians, and 'Namibianizing' the conflict, the issues surrounding limited white manpower and casualties were minimized, even if not entirely removed.[176] The earliest leadership and cadres within the organization survived and returned from Rhodesia and Mozambique with lessons learned and

[166] Christo Roelfse interview, Pretoria, South Africa, 8 June 2023.

[167] Crump, *Koevoet, Volume 1*, p. 60.

[168] Herbstein and Evans, *The Devils Are Among Us*, p. 76.

[169] Hamann, *Days of the Generals*, p. 65.

[170] Herbstein and Evans, *The Devils Are Among Us*, p. 76.

[171] Frans Fourie interview, Pretoria, 9 June 2023; Retired SADF special operations intelligence officer interview, Pretoria, South Africa, 7 June 2023.

[172] Leo McCann, '"Killing Is Our Business and Business Is Good": The Evolution of "War Managerialism" from Body Counts to Counterinsurgency', *Organization*, 24:4 (2007): 491–515.

[173] Truth and Reconciliation Commission, *The Truth and Reconciliation Commission of South Africa Report*, volume 2, p. 69.

[174] Scholtz, *The SADF in the Border War, 1966–1989*, pp. 211–214.

[175] Hooper, *Koevoet!*, p. 43.

[176] Norval, *Death in the Desert*, chapter 15 .

approaches to engage in counter-revolutionary warfare more effectively in Namibia. Despite some levels of friction between organizations and units, one veteran assessed that capabilities 'were not uneven but complementary at the tactical level'.[177]

Its law enforcement authorities contributed to its complementary capability yet also duplicated those found elsewhere, such as the seven SWATF battalions made of up indigenous trackers. These authorities, however, were not used to provide better governance or a more just society for Namibians. What was missing from the guidance in Pretoria down to the level of tactical implementation was a coherently directed COIN campaign. The same SADF veteran, in assessing one of the reasons for failure, concluded, 'we didn't understand the operational level of war'.[178] *Koevoet*'s incentive structure prioritized inflicting heavy casualties on SWAPO and engaging endlessly in the intelligence and operations cycle. In the final analysis, the casualties inflicted on SWAPO only struck at the tip of the insurgent iceberg. The much larger element below the surface, SWAPO's clandestine organization and its public relations and psychological warfare, remained relatively untouched.[179] This fact is borne out by the results of the 1989 election after independence. SWAPO, despite its considerable losses, captured 58 per cent of the popular vote.[180]

## Singleton and small-team externals: covert reconnaissance, indirect shock, and assassination

South Africa's covert special operations capability had its roots in the government reorganization that largely dismantled BOSS in favour of the SADF. Under Botha, BOSS was renamed first the Department of National Security and later the NIS country in 1980. Operation STARLIGHT in 1972 suggested a capability shortfall within the South African national security architecture: conducting covert actions abroad in quantity. The initial conduct of pseudo operations in Namibia and Angola, combined with successive covert deployments with the Rhodesian SAS on 'the Russian Front' in Mozambique from 1977 to 1979, led the commanding general of special forces, Major General Fritz Loots, to pitch a 'new covert organization as an adjunct to Special Forces' to Magnus Malan, Botha's minister of defence.[181] This new organization, initially named Section of Pseudo Operations, evolved into what Michael Schmidt has labelled a 'secret killing machine' or the 'third force' long suspected and only partially illuminated by the Truth and Reconciliation Commission.[182] From its

[177] Retired SADF special operations intelligence officer interview, Pretoria, South Africa, 7 June 2023.
[178] Ibid.
[179] Robert Thompson, 'Regular Armies and Insurgency', in Ronald Haycock, ed., *Regular Armies and Insurgency* (London: Croom Helm, 1979), p. 11.
[180] Christopher Wren, 'Namibia Rebel Group Wins Vote, Falls Short of Full Control', *The New York Times* (15 November 1989), p. A1.
[181] Stiff, *Warfare by Other Means*, p. 92.
[182] Michael Schmidt, *Death Flight: Apartheid's Secret Doctrine of Disappearance* (Cape Town, South Africa: Tafleberg, 2020), p. 61.

modest roots, the Section of Pseudo Operations would evolve into Project Barnacle and eventually the controversial Civil Cooperation Bureau (CCB), a covert hydra whose heads included intelligence gathering, targeted killing, 'disappearing' enemies of the state, and the use of chemical and biological weapons.[183]

The extent to which Rhodesia's considerable targeted killing and sabotage and subversion efforts in neighbouring countries influenced subsequent South Africa activities is unclear. Their successful conduct, drawing on retired special operators from the United Kingdom and elsewhere and compartmentalization to protect the 'hidden hand' of covert actions, limited South African awareness of or contact with them as a potential source of emulation.[184] Regardless of the initial inspiration for Barnacle, after its first year and a half of existence its mandate expanded into four different functions including targeted killing or 'eliminations' and 'ambushes against persons of strategic importance', special reconnaissance through either the 'gathering of combat intelligence' or 'gathering of information ... where other sources cannot be applied', counterintelligence or 'special security tasks for special forces', and an ill-defined catch-all function, 'other super-sensitive operations as ordered'.[185] The scope and scale of Barnacle's ambition was extraordinary given that these tasks were to be performed by between thirty and fifty personnel, more than half of whom commanded or oversaw operations.[186]

Covert eliminations stemmed from a vexing conundrum in irregular warfare and pseudo operations. To operate with any meaningful degree of success or effectiveness, they required the turning of former insurgents who possessed tactical and organizational knowledge. While the SADF in general and its special forces specifically were never short of former SWAPO insurgents to fill their ranks, the question became how to deal with those who refused to switch sides or who conducted acts of treachery once turned. Trying and failing, as occurred with the SWAPOs captured during and at Ongulumbashe, was out of the question for those refusing to cooperate after attempts to induce or coerce them had failed. During trials or after sentencing and jailing, their attempted recruitment would become a matter of public record and interest and this form of covert operation compromised as a result.[187] Those who had been turned would have remarkably discrete and useful information for SWAPO, including the locations of bases, names of South African special forces operators, and names of turned insurgents whose families could then be targeted. To address

[183] Jeffrey Bale, 'South Africa's Project Coast: "Death Squads," Covert State-Sponsored Poisonings, and the Dangers of CBW Proliferation', *Democracy and Security*, 2:1 (January–June 2006): 27–60.

[184] For unique insights into Rhodesian CIO recruitment and operational security measures for covert action operatives, see the reminiscences of one such operative in Peter Stiff, *I'll See You in November: Rhodesia's No-Holds-Barred Intelligence War* (Alberton, South Africa: Galago, 1985), pp. 96–101; 105–107; 336.

[185] Top secret instructions to Barnacle, dated 9 January 1981, cited in Schmidt, *Death Flight*, pp. 148–149.

[186] Numbers derived from organizational chart entitled 'D40 circa March 1980' in Ibid., p. 166.

[187] One SWAPO member captured in 1966 recounts his trial in considerable detail. See Hello Shityuwete, *Never Follow the Wolf: The Autobiography of a Namibian Freedom Fighter* (London: Kliptown Books, 1990), chapter 16.

this problem of '"redundant" SWAPO members', Colonel Johan Theron developed a ruthless yet not uncommon solution: killing them and dumping their bodies in the South Atlantic.[188]

The details of Operation DUAL, the codename given to these practices, do not make for easy reading. They involve bludgeoning to death, poisoning, anaesthetizing, and subsequent dumping at sea—violence of a highly personal or 'intimate' nature.[189] For most, such activities go well beyond the pale of special operations, which divides between moral and ethical 'good' operations and illegal, immoral, and unethical 'bad' ones. The systematic and institutionalized nature of DUAL separates it from incidents elsewhere.[190] Although precise figures for those eliminated are unavailable, estimates range from 'scores' suggesting between one to three or more dozen to hundreds by another source.[191] Higher estimates are difficult to reconcile given how few pilots and aircraft were available and their use in support of other special operations. Institutionalization is evident in DUAL eliminations through the euphemism they employed for those killed and dumped: 'packages'.[192] Another differentiator was the elimination of problematic special forces members, or deliberate 'own goals', as was the case for two members of 5 Recce, one unnamed and the other support staffer Corporal Fernando 'Mack' Anderson.[193] Anderson had 'fragged' five black pseudo operators and intimated he would kill white ones. He was drugged and his body dumped from an aircraft in a nature reserve in Mozambique in December 1987.[194] A possible cause for hope in an otherwise bleak corner of special operations results from how DUAL flights came to a halt. On becoming senior staff officer for Air Operations of Special Forces Command in late 1987, Colonel Hap Potgeier refused to support DUAL with SAAF aircraft from Barnacle. His refusal reached the General Officer Commanding, Special Forces, Abraham 'Joup' Joubert.[195] Numerous unanswered and likely unanswerable questions remain about DUAL elimination 'death flights' that occurred under other auspices, including those involving a clandestine covered aircraft company (i.e. civilian registered but government owned and operated) Wonder Air, about which few details are available.[196]

Barnacle and other small teams planned and conducted several assassination and abduction missions from 1979 until 1986. The latter were rarely conducted for strategic intelligence and source exploitation but limited to routine activities: the snatching of what Soviet naval special operations scouts referred to euphemistically as 'tongues'

[188] Stiff, *Warfare by Other Means*, p. 107.

[189] Ibid., pp. 110–115; Schmidt, *Death Flight*, pp. 154–160; 194–197; 206–207; Stathis Kalyvas, *The Logic of Violence in Civil War* (Cambridge: Cambridge University Press, 2006), chapter 10.

[190] De Kock alleges, for instance, that 5 Recce operators 'disappeared' captured SWAPO after interrogation, including throwing them out of helicopters at high altitude. De Kock, *A Long Night's Damage*, pp. 72–73.

[191] Schmidt, *Death Flight*, p. 239; O'Brien, *The South African Intelligence Services*, p. 130.

[192] Stiff, *Warfare by Other Means*, p. 110.

[193] Schmidt, *Death Flight*, pp. 235–237.

[194] Ibid., pp. 242–243.

[195] Ibid., pp. 238–239.

[196] Ibid., pp. 201–204.

for tactical intelligence gathering on nearby units.[197] According to one source, eighteen ANC members and two Swiss nationals were abducted from Botswana, Swaziland, or Lesotho; most were returned within a week.[198] One such mission was Operation JERICHO, a cross-border raid to snatch prisoners.[199] More frequent, and strategic in the mind of *SPESKOP* planners, were raids against what are now known as 'high-value targets', particularly ANC leaders, offices, and facilities outside of South Africa. The units conducting assassination missions most frequently consisted of one or two small teams of between three and five operators each, drawing on local agents and air delivered to their locations, although as many as fifteen teams did participate in a single mission. The small teams were selective with tight operational control. In one case, an ultimately abortive attempt to kill the ANC executive committee at a conference in Tanzania, the targeted killing team consisted of five personnel: the officer commanding 4 Recce and four operators, including a communications expert to retain links back to the tactical mission and *SPESKOP* headquarters.[200]

## Small-team externals: clandestine reconnaissance and indirect shock

Small-team reconnaissance operations became one of the hallmarks of South African Special Forces during the Border War period. The requirements driving such operations were straightforward. Shortfalls in conventional South African intelligence-gathering capabilities meant SWAPO bases, particularly larger training and staging facilities in southern Angola, went undetected by human intelligence sources or aerial photoreconnaissance. SWAPO military deception played a role as well. Thanks to Soviet training and the advice and assistance they received from Cuban and East German military advisors, as the Border War went on 'SWAPO became masters in camouflaging their bases', as small-team operator Jack Greeff relates.[201]

As the Border War intensified from 1985 until its conclusion in 1989, small-team reconnaissance units increasingly conducted base observation missions farther north in Angola, particularly the airfields at Menongue and Lubanga. In addition to the generally clandestine operational profile, the missions were conducted as long-range foot patrols to get close enough to the target to surveil it without being detected.

[197] See, for example, Viktor Leonov, *Blood on the Shores: Soviet Naval Commandos in World War II*, trans. James Gebhardt (Annapolis, MD: Naval Institute Press, 1993), pp. 19; 177n14; Yuriy Fedorovich Strekhin, *Commandos from the Sea: Soviet Naval Spetsnaz in World War II*, trans. James Gebhardt (Annapolis, MD: Naval Institute Press, 1996), p. 9.

[198] See table 3, 'Abductions', in Max Coleman, ed., *A Crime against Humanity: Analysing the Repression of the Apartheid State* (Braamfontein, South Africa: Human Rights Committee of South Africa, 1998), www.sahistory.org.za/archive/covert-operations.

[199] Frans Fourie interview, Pretoria, South Africa, 9 June 2023.

[200] Andre Diedricks, *Journey without Boundaries: The Operational Life and Experiences of a SA Special Forces Small Team Operator* (Durban, South Africa: Just Done Productions Publishing, 2007), pp. 99–100, Koos Stadler, *Recce: Small Team Missions behind Enemy Lines* (Oxford: Casemate, 2015), pp. 302–303.

[201] Jack Greeff, *A Greater Share of Honour: The Memories of a Recce Officer* (Durban, South Africa: Just Done Productions Publishing, 2008), p. 131.

Such patrols began after helicopter insertion by SAAF Pumas at landing zones up to 60 kilometres away. The teams, comprising between one and three operators, would move in marching bounds to various lay-up points until they got close enough to a target to visually survey it. They would take photographs, make sketches (as a backup in case photography failed), and note the layout and disposition of defences, movements of vehicles or aircraft, and other detailed information that only close persistent observation can reveal. Short reports were transmitted back via burst-transmitted high-frequency radios to prevent interception. Their mission accomplished, the operators would return to the landing zone for recovery by Pumas and return to base, where they filed detailed reports on what they had discovered.

The genesis of small-team clandestine externals was an operator experienced in Rhodesia, Andre 'Dieders' Diedricks. Diedricks passed selection and joined the Recces in 1974 under Jan Breytenbach. While a corporal during Operation SAVANNAH, he distinguished himself by calling fire on FAPLA artillery positions despite not having trained as a forward observer.[202] His concept of small teams was an outgrowth of South African experiences in Rhodesia and Mozambique. General Loots approved a proposal to send recruits to the Selous Scouts' reconnaissance course and courted a Selous Scouts major, Neil Kriel, to lead a small-team clandestine reconnaissance effort for the SADF.[203] Kriel came to the unit with formidable operational credentials. He commanded the Scouts' Reconnaissance Troop from January 1977 and led a demolitions team on a clandestine indirect-shock mission against a ZIPRA staging house for suitcase bombs, Operation IGNITION, in Francistown, Botswana.[204] The South African small-team reconnaissance effort was initially codenamed 'D40', a precursor to Project Barnacle.

The small-team reconnaissance concept was not universally embraced with the SADF. The greatest resistance, most surprisingly, came from within the Recces. It consisted, in Diedrick's words, of 'a fair dose of professional jealousy'.[205] For their detractors among the Recces, small-team operators 'just hid in the bush all the time' and avoided contacts.[206] From the operators' perspective, doing so was precisely the point: collecting, reporting, and returning with information was the focus of the mission, not killing the enemy. That could be done later with larger units and greater firepower.

Inclination and disposition played a crucial role in small-team operator selection and success. Instincts, some of which appeared uncanny or even supernatural such as heightened smell or hearing, could spell the difference between survival and 'a lonely death'.[207] In another paradoxical development, small-team operators fighting for a political system of racial exclusion were recruited from among the white and black

[202] Jan Breytenbach, *They Live by the Sword: 32 'Buffalo' Battalion—South Africa's Foreign Legion* (Alberton, South Africa: Lemur, 1990), pp. 61–62.

[203] Stiff, *The Silent War*, p. 312.

[204] Ron Reid Daly and Peter Stiff, *Selous Scouts: Top Secret War* (London: Galago, 1982), pp. 239–242; 250; 277.

[205] Diedricks, *Journey without Boundaries*, p. 52.

[206] Greeff, *A Greater Share of Honour*, p. 127.

[207] Ibid., p. 127; Frans Fourie interview, Pretoria, South Africa, 9 June 2023, p. 127.

operators based on ability, not their skin colour. The mental strain of operating on one's senses and wits required cunning and skill but the strain of doing so repeatedly could wear down the confidence of the most seasoned operator, as their luck could only last so long. In such cases, operators could crack emotionally, retire or request transfers, or develop a 'fear control plan' through discussions with trained special forces psychologists.[208] In addition, selection for small-team operational missions was done on a personal level, usually by the senior operator. The exclusion of peers or friends contributed to friction and tension individually and combined collectively to form negative perceptions among other Recce units.

These perceptions had organizational consequence as well. The desire of commanders to 'own' a niche unit led to 'fierce competition amongst the [Recce] unit commanders, who all wanted to house a centralized small team capability'.[209] D40 kept the small-team operations out of the standard Recce orbit for almost four years. During that time the reconnaissance missions for the unit and its operators were overtaken by more pressing clandestine special operations requirements. The unit began conducting reconnaissance missions, including Operation INTERRUPT (1980), surveying a SWAPO base near Luango in Angola in anticipation of SAAF airstrike.[210] Special Forces HQ tasked D40 subsequently with a range of different operations. They included reconnaissance and transport of supplies in advance of a 4 Recce sabotage mission in Mozambique in 1982 (Operation FORCE), unnamed operations with UNITA in ultimately unsuccessful attempts to down aircraft clandestinely in Angola in 1983, and independent sustained sabotage actions against the Angolan rail network under Operation BARRAGE the same year.[211] In 1984, the new special forces commander, Major General Kat Liebenberg, ordered D40 to settle formally under the administrative control of 5 Recce, where it became 54 Commando.

The unnamed and unsuccessful operations with UNITA developed into three separate, more successful ones between 1984 and 1985. The driver behind the operations was the increasing capability of MPLA air forces to contest what had up to that point been South African air superiority over most of Angola. The first, CATAMARAN I, targeted parked aircraft at Menongue using South Africa's newly developed long-range artillery, the G-5 howitzer. The reconnaissance operation in support of the mission failed but the bombardment occurred successfully.[212] The catalyst for the other two operations was the capture of a Soviet-supplied SA-9 mobile surface-to-air system. The system, mounted on a BRDM armoured car, consisted of four launchers and an optical targeting system. By using Soviet-supplied equipment, these special operations maintained a level plausible deniability for South Africa through the clandestine activities of 54 Commando operators. Two operations, CATAMARAN II (1984) and CERBERUS (1985), succeeded in downing several aircraft, including a

[208] Stadler, *Recce*, pp. 307–309; 313.
[209] Diedricks, *Journey without Boundaries*, p. 49.
[210] Cornelius Bornman interview, Pretoria, South Africa, 9 June 2023.
[211] Diedricks, *Journey without Boundaries*, pp. 55–89; Søderland and Steyn, *Iron Fist from the Sea* (2018), pp. 163–164. The earlier, rarer edition of Søderland and Steyn's work (2014) contains a wealth of operational detail excised from the more popular, mass market 2018 reprint.
[212] Diedricks, *Journey without Boundaries*, pp. 93–94.

Cuban-piloted Dornier, a Mi-8 helicopter, and a Russian An-12. The Dornier was transporting funds for MPLA forces while the An-12 carried almost a dozen-odd Russian military advisors along with seventeen other passengers and crew.[213]

The D40 concept of small, covert, team reconnaissance was revived in 4 Reconnaissance Regiment. An observation by Frans Fourie while on vacation led to the development of Project IVORY in 4 Reconnaissance Regiment. Having graduated from the military academy, Fourie noted the ease with which he could travel throughout Malawi while in civilian clothes.[214] His insight occurred at a precipitous time. Special Forces Headquarters was tasked by the chief of SADF with locating ANC headquarters in Tanzania after other attempts using more conventional intelligence means failed. Fourie conducted an initial proof of concept mission successfully in 1987 to establish credible cover.[215] From it, 4 Reconnaissance Regiment created a Covert Small Team, initially comprising four operators led by Fourie. The unit ultimately conducted numerous missions primarily in Tanzania, locating ANC training camps and headquarters, as well as Zambia and Uganda, providing reconnaissance information otherwise unavailable to Special Forces and SADF planners.[216]

In this vein, South Africa sometimes developed redundant and competing capabilities. Around the same time the Covert Small Teams were created, the minister of defence, Magnus Malan, authorized the CCB, the follow-up to Barnacle and its successor, 3VR. Developed in response to a change in the ANC's operational approach, its mission was clear: 'destroy the terrorists, their bases and their capabilities' through covert operations.[217] More specifically, its tasks were 'penetrating enemy organisations, gathering intelligence concerning them, and disrupting them'.[218] These tasks were carried out in eight different regions, including Region 6 (South Africa) discussed below, Region 1 (Botswana), Region 2 (Mozambique and Swaziland), Region 3 (Lesotho), Region 4 (Angola, Zambia, and Tanzania), Region 5 (International), Region 7 (Zimbabwe), and Region 8 (Southwest Africa).

Planning for CCB operations was divided along two lines: the development of self-funding front companies and through action-orientated 'red plans' to accomplish the missions. For external operations, CCB operatives, given their limited numbers, often required Recce support, such as mobility to and from targets located near coastlines by 4 Recce boat operators as planned in the stillborn Operation CURSOR (December 1986) and unsuccessful Operation CARNABY (May/June 1987) against MK facilities in Maputo, Mozambique.[219] The creation of another covert capability skirted what little control and oversight mechanisms existed. Depending on the nature and sensitivity of the target, the director of CCB could authorize a mission

[213] Ibid., pp. 94–98; Stadler, *Recce*, p. 220.

[214] Fourie interview, Pretoria, 9 June 2023.

[215] Ibid.

[216] Søderland and Steyn, *Iron Fist from the Sea* (2014), pp. 403–416.

[217] Magnus Malan, 'Statement to the Truth and Reconciliation Commission', section 15.2, 1997, www.justice.gov.za/trc/hrvtrans/submit/malan.htm.

[218] Magnus Malan, *My Life with the SA Defence Force* (Pretoria, South Africa: Protea Books, 2006), p. 328.

[219] Søderland and Steyn, *Iron Fist from the Sea* (2014), pp. 359–361.

without approval from higher authority. Notionally, control flowed from the chief, SADF down through the Operations Committee, Special Forces, to the director, CCB. Based on the chief, SADF's evaluation, particularly sensitive operations were referred to the SSC for president or minister of defence approval.[220] The opaque lines of authority and oversight led to charges of corruption and overreach, particularly when CCB's activities in Region 6 were unmasked.

## Integrated team externals: indirect shock

Most independent SOF operations from 1982 until 1989 consisted of large indirect-shock missions or raids against SWAPO or MPLA facilities in Angola as well as ANC ones abroad. Based on intelligence sources, primarily through Barnacle and 3VR missions or agents, the Directorate of Military Intelligence through senior staff officer, Intelligence would pass over collection files of potential targets. From these, the Operations Committee, Special Forces would develop plans passed to general officer commander, Special Forces for transmission to chief, SADF for approval by the SSC. Unless the target was limited to a single facility, more complex operations required considerable coordination, training, and rehearsal across different Recce units. A common phrase during several interviews with retired SOF personnel was augmented 'by elements of [specific number] Recce'. The mission focus during complex special operations meant personal differences and frictions, while present given the small size of the South African Special Forces community, were set aside in favour of accomplishing the mission.[221] Depending on its scale and complexity, whenever the officer commanding (OC) determined they 'need[ed] bodies' they would use less formal communications with the OCs of other Recces for additional personnel, often based on their capabilities and expertise or trust links based on skills and experience.[222]

A typical early complex special operation, and the largest South African Special Forces mission conducted up to that point, was the raid at Maseru, Lesotho in December 1982. Preparatory intelligence collection was conducted by Barnacle agents operating in the city and supplemented by aerial reconnaissance photographs. The number of targets dictated fifteen four-man teams from 1 and 5 Recce, who infiltrated on foot across the Caledon River and made their way to Maseru. There they struck more than a dozen targets simultaneously in the town over a three-hour period. More than three-quarters of the operators were black Africans led by white officers.[223] The raid included the elimination of more than two dozen ANC officials, including strategists, planners, and managers, as well as the capture of documentation for processing and intelligence analysis. During house clearing operations, operators

[220] O'Brien, *The South African Intelligence Services*, pp. 135–136.
[221] Chris Serfontein interview, Pretoria, South Africa, 7 June 2023.
[222] Johan Hechter interview, Cape Town, South Africa, 13 June 2023.
[223] Stiff, *The Silent War*, pp. 414–417.

killed nearly fifty people, including seven women and children.[224] The only casualties among the operators were a handful of wounded; none were killed. SAAF Pumas recovered and ferried the teams and their documents back to various bases in South Africa.

While successful operationally and tactically, the raid was problematic politically for South Africa. Not only was it widely condemned by the Organization of African Unity (OAU) and within the United Nations, but it also formed another, effective prong in the ANC's strategic messaging in the court of international opinion. Much like Operation REINDEER some five years prior, the 'massacre at Maseru' narrative featured several supporting arguments, labelling the raid illegal as a violation of Lesotho's sovereignty, the victims 'refugees', and the tactics and their timing (in the dead of night) as immoral and unethical.[225]

A less controversial and complex operation, but one requiring considerable planning, coordination, and rehearsal between various units, was Operation KERSLIG, conducted the year prior to Maseru. Seeking to build off the success of a similar amphibious indirect-shock attack against a different Angolan refinery in 1980, the target for this raid was the Petroangela refinery complex in the harbour of Luanda. 1 Recce was the mission lead for the operation ashore, supported by 4 Recce boat teams. The range of targets within the refinery complex required five demolitions teams and included storage tanks and distillation towers. Luanda's distance from Langebaan, almost 3,000 kilometres, required intensive planning and cooperation with SAN and SAAF. The plan was a seven-phase operation to get to the target, revise it based on updated reconnaissance information, and conduct the attack.[226] These resources included SAAF transportation of 1 Recce operators to a staging port in Southwest Africa, a satellite communications relay C-130 aircraft orbiting offshore to ensure adequate communications and control, a Wasp helicopter for search and recovery, three SAN vessels, including a support vessel for two strike craft to command the mission as well as ferry the 4 Recce boats nearer to the harbour, and medical personnel from 7 Medical Battalion in case of casualties.[227] The recently promoted General Officer Commanding, Special Forces, Brigadier General 'Kat' Liebenberg, was aboard as an observer to 'gain exposure to Special Forces['] ... seaborne operations'.[228]

Initial small-team reconnaissance by 4 Recce divers provided valuable last-minute updates, yet the mission was compromised due to equipment failure. Several charges either failed to explode or detonated prematurely, leading to the only

[224] Allister Sparks, 'S. African Commando Attack Kills 37 in Lesotho', *The Washington Post* (10 December 1982), www.washingtonpost.com/archive/politics/1982/12/10/s-african-commando-attack-kills-37-in-lesotho/ce5b3c91-3730-44cc-b7ec-8d08e43450ea.

[225] Pulane Mahula provides a useful survey on reporting that occurred after the raid in 'Memory, Trauma, Silences: Narratives of the 1982 Maseru Invasion', unpublished Master of Arts thesis, University of the Western Cape (November 2017), pp. 37–48, https://etd.uwc.ac.za/xmlui/handle/11394/6418.

[226] Alexander Strachan, *1 Recce, Volume 3: Through Stealth Our Strength* (Cape Town: Tafelburg, 2022), p. 25, Søderland and Steyn, *Iron Fist from the Sea* (2014), pp. 178–179.

[227] Søderland and Steyn, *Iron Fist from the Sea* (2014), pp. 177–178. Søderland was the naval group commander for KERSLIG while Steyn was the 1 Recce mission commander.

[228] Strachan, *1 Recce, Volume 3*, p. 24.

casualties of the otherwise well-planned and rehearsed operation: one dead and two wounded. The detonations served as a prearranged signal for other teams to cut short and abandon their demolition work. While the refinery was taken off-line for three months, the raid only temporarily disrupted the flow of truck-borne fuel to MPLA forces in Cunene or Cuando Cubango provinces. Attempts to attribute the operation to UNITA were a failure, particularly given its implausible cover story.

An indirect-shock mission against oil facilities in Cabinda in Angola in 1985, however, had a significant impact on the orientation and conduct of South African special operations that followed. This impact was unintentional but played directly into SSC concerns about the effects of casualties on popular support at home.[229] The Recces had established a reputation of professionalism and competence, in the words of one contemporary observer as 'the most highly trained troops in SADF' at home and abroad, thanks to their rigorous selection and training criteria.[230] This reputation would be challenged somewhat because of Operation ARGON in Angola in 1985. Given the presence of Cuban forces that could detect a surface fleet from a considerable distance, as well as the distance of Cabinda from SAN bases, a clandestine undersea approach via submarine was the only viable option, calling to mind the conditions of Operation STARLIGHT. Space within the submarine, limited the assault team to nine personnel and increased the risk to force and risk to mission, particularly as medical personnel also needed billeting aboard the vessel.

Somewhat mitigating these risks was the recent training conducted by the crew of the submarine with 4 Recce operators for a previously scrubbed joint operation. In part to minimize unnecessary transit time to the target, the submarine departed a week before political approval for the mission was granted on 12 May, loading up along the way, 'according to plan as this would avoid unnecessary changes to the plan once approval came through'.[231] Just over a week later, the assault forces made their approach but were delayed. Locals spotted them in the harbour and at a landing beach. After some discussion about modifying the plan or scrubbing it, the mission commander ordered it to proceed as conceived.[232]

A combination of unforeseeable factors contributed to the subsequent disastrous outcome. Recce operators, normally careful in removing their tracks, left a path of trampled grass that made their location identifiable. In addition, a lost cap discovered by a FAPLA patrol was identified by an alert junior officer 'as being of South African origin'.[233] Now alerted to their presence, FAPLA platoons descended on

[229] Chris Serfontein interview, Pretoria, South Africa, 7 June 2023; Frans Fourie interview, Pretoria, South Africa, 9 June 2023.

[230] Mark Philips, 'The Nuts and Bolts of Power: The Structure of SADF', in Jacklyn Cock and Laurie Nathan, eds., *War and Society: The Militarisation of South Africa* (Claremont, South Africa: David Phillip, Publisher (Pty] Ltd, 1989), p. 22.

[231] Søderland and Steyn, *Iron Fist from the Sea* (2014), p. 312.

[232] Ibid., p. 316.

[233] Ibid., p. 318.

their area. After firefights with the split elements of the small force, FAPLA managed to kill two 4 Recce troopers and wound three more. Two of the wounded managed to escape along with the help of four of their comrades and make it back to the sub.[234] One of the wounded, however, was the mission commander, Captain Wynand du Toit. Angolan doctors treated his wounds, Cubans subsequently interrogated him numerous times, and he remained imprisoned until September 1987.[235]

The timing of Du Toit's capture and his subsequent press confirmations could not have been worse for the SSC. Complex and delicate negotiations between South Africa, Cuba, UNITA, and MPLA, which were to implement a ceasefire and establish the withdrawal of Cuban and SADF forces from Angola and the removal of SWAPO bases from Namibia, collapsed partly due to revelations of the Recce role in the Cabinda attack.[236] The attack, which Du Toit confirmed was against Gulf Oil facilities, also had put American lives and goodwill at risk.[237] This special operation was followed three weeks later by news of another, in neighbouring Botswana, this time in the city of Gaborone against seventeen ANC and MK targets. The indirect-shock mission, Operation PLEXI, was a combined operation involving Barnacle reconnaissance forces and 5 Recce, augmented by 1 Recce's operators trained specifically in urban operations.[238] Tensions between CCB and Recce operators, as well as between operators and commanders, on how best to accomplish the mission plagued the planning process, in part due to personality clashes.[239] While militarily the raid featured a low number of casualties, strategically it had little effect. As Peter Stiff concludes: 'The infiltration of MK guerrillas into South Africa ... continued unabated.'[240] While the Americans subsequently repealed the amendment proscribing aid to UNITA, a positive development for South Africa, it was more than offset by the Cuban and Soviet doubling its equipment and personnel, which had increased by more than '$2 billion worth of military equipment and 7,000 additional Cuban forces'.[241] Externals continued against SWAPO bases but the character and intensity of the war in Angola required greater investment in proxy forces.

[234] Stiff, *The Silent War*, pp. 463–467.

[235] Du Toit and his wife, Louwna, provided their version of the capture, captivity, and release to three journalists, who dramatized the events. See Allan Soule, Gary Dixon, and René Richards, *The Wynand du Toit Story* (Johannesburg, South Africa: Hans Strydom Publishers, 1987). See also 'Ex-Special Forces Recce and POW, Wynand du Toit, Reflects on Operation Argon', Tales Less Told podcast, 1:24:30, www.youtube.com/watch? v=Ad6Xt-m9dhg.

[236] Noted by US assistant secretary of state for African affairs at the time, Chester Crocker, in *High Noon in Southern Africa: Making Peace in a Rough Neighborhood* (Johannesburg, South Africa: Jonathan Ball Publishers, 1993), p. 231.

[237] Gleijeses, *Visions of Freedom*, pp. 252–253.

[238] Stiff, *Warfare by Other Means*, pp. 472–504.

[239] Van Niekerk recounts a number of clashes, including with commanders Joe Vorster (CCB) and André Bestbier (1 Recce) in his characteristic caustic manner in Van Niekerk, *Born to Storm*, pp. 328–329. He attributes the raid to CCB, 1 Recce, 5 Recce, and 4 Recce, the latter a puzzling inclusion and likely a mistake, given the encyclopaedic history of the unit by Steyn and Søderland does not mention the operation at all.

[240] Stiff, *Warfare by Other Means*, p. 504.

[241] William Minter, 'The US and the War in Angola', *Review of African Political Economy*, 50 (1991): 135–144; Søderland and Steyn, *Iron Fist from the Sea* (2014), p. 329.

## By, with, and through: military advising and assistance

South Africa supported proxies in Angola and Mozambique. SADF had a long prior history of working with both. Its proxy in Mozambique, RENAMO, is best understood as an inheritance from Rhodesia after 1980. The other, UNITA in Angola, served the interest of South African national security objectives but also reflected the involvement of other powers in geopolitical competition. South Africa's support for both proxies relied on a high degree of secrecy, requiring clandestine or covert support. The inability of the NIS to provide capability or capacity to support either meant that liaison and assistance missions were conducted by operators from SADF's Directorate of Special Tasks (DST), many of whom had served in the Recces or 32 Battalion. Frictions between different groups, as well as the need to continue to supply and assist UNITA covertly after South Africa's withdrawal at the end of Operation SAVANNAH, led SADF created DST in November 1976. DST initially fell under the control of the Military Intelligence Directorate (MID) but almost a decade later was consolidated with other functions under the Directorate, Intelligence Operations.

Of the two proxies, South Africa had the longest ties with UNITA, headed by Jonas Savimbi. While Jan Breytenbach was working with the FNLA in September 1975, Lieutenant Colonel Willem van der Waals and another small cadre of SADF officers and enlisted were training and supplying UNITA.[242] Coordination between field elements of both forces left much to be desired, the result of hostility between the two proxies in their civil war. It led to some command headaches for Colonel Roelf van Heerden, the Task Force Zulu commander in charge of both. In one such incident, UNITA guides led one of Breytenbach's FNLA columns into an ambush just south of the town of Culevai. Breytenbach pledged 'never to trust UNITA again'.[243] South African support for UNITA continued after FNLA collapsed. UNITA and its bases served largely as a means to complicate SWAPO's attempts to infiltrate into Namibia, providing access and placement, in the form of forward operating bases, for Recce operations. Scholtz's concludes by 'limit[ing] the war as much as possible to Ovamboland ... the counterinsurgency war against SWAPO [was] so much easier'.[244]

To ensure deconfliction as well as a level of manageable control, DST divided regions of interest in two. DST 1 focused on Angola while DST 2 concentrated on Mozambique, Lesotho, and Zimbabwe. The field office in support of DST 1 was in Rundu, in the Kavango-East Region of Namibia on the border with Angola.[245] Both DSTs and the field office were under the command of Commandant, or Colonel.[246] DST 1 personnel split into three trained UNITA forces at primarily at three bases,

[242] Tyrone Groh, *Proxy War: The Least Bad Option* (Stanford, CA: Stanford University Press, 2019), p. 171.

[243] Breytenbach, *They Live by the Sword*, pp. 22–23.

[244] Scholtz, *The SADF in the Border War, 1966–1989*, p. 233.

[245] Les Rudman interview, Pretoria, South Africa, 8 June 2023.

[246] Truth and Reconciliation Commission, *The Truth and Reconciliation Commission of South Africa Report*, volume 2, p. 320.

including Rundu. As was the case with former FNLA and SWAPO cadres who operated with 32 Battalion, 1 and 5 Recce, and *Koevoet*, DST 1 personnel initially limited the scope of training to basic tasks and units and advanced guerrilla warfare skills. Many within UNITA's ranks were skilled in bushcraft, survival, and ambushes, the result of years of experience in hit-and-run guerrilla attacks.[247]

While the scope of DST 1's military advising and assistance was limited, its scale was anything but. DST 1 trained some 70,000 UNITA soldiers and on average flew one air supply mission a month after 1985 consisting of 2 C-130 loads of arms, ammunition, and supplies—between 22 and 26 tons of material.[248] DST 1 support increased after 1985 as a result of the combined Cuban and FAPLA offensive, codenamed Operation CONGRESSO II and involving five brigades with air support, designed to split and eradicate UNITA once and for all.[249] A timely intervention in the form of SAAF airlift and air support to UNITA forces, combined with the employment clandestinely of 32 Battalion and elements of 5 Recce, turned the tide during Operation MAGNETO.[250] From this point forward, DST personnel served as more than just advisors; they were 'military liaisons who were also involved in combat'.[251] Organized into mobile Liaison Teams, they deployed operationally with UNITA from January 1986 onwards and played a crucial role in detecting and disrupting FAPLA forces in subsequent operations. As occurred in several episodes during the Border War, elite forces, consisting of Pathfinders from the 1 Parachute Battalion, became SOF in large part due to their training and experience, including 'in clandestine operations'.[252]

UNITA compartmented its activities with the CIA and DST for sensible political reasons. The United States supported UN sanctions against South Africa and any American-provided equipment, including Stinger missiles, making its way into South African hands, had the potential of further escalation of Soviet support and involvement. The CIA 'did not want to get too far out in front' in its 'covert action program in Angola'.[253] It needed its support to remain covert and SADF's clandestine, preventing any coordination and indeed laying the foundation for competition. Savimbi had the upper hand politically: 'like suitors in a love triangle, both [the CIA and SADF] competed for UNITA's favours'.[254] Such competition led a senior UNITA commander to ask his DST-1 liaison: 'your special forces are good but why are you in such a hurry?'.[255]

[247] Groh, *Proxy War*, p. 174; Leopold Scholtz, *The SADF and Cuito Cuanavale: A Tactical and Strategic Analysis* (Johannesburg, South Africa: Delta Books, 2020), p. 202.
[248] Les Rudman interview, Pretoria, 8 June 2023.
[249] For details, see George, *The Cuban Intervention in Angola*, chapter 9.
[250] Piet Nortje, *32 Battalion: The Inside Story of South Africa's Elite Fighting Unit* (Cape Town: Zebra Press, 2004), pp. 215–218.
[251] Les Rudman interview, Pretoria, 8 June 2023.
[252] Helmoed-Römer Heitman, *War in Angola: The Final South African Phase* (Gibraltar: Ashanti Publishing, 1990), pp. 26–27.
[253] *Foreign Relations of the United States*, 1977–1980, volume XVI, Southern Africa, Myra Burton, ed. (Washington, DC: Government Printing Office, 2016), Document 37.
[254] Stiff, *The Silent War*, p. 532.
[255] Les Rudman interview, Pretoria, South Africa, 8 June 2023.

In contrast, DST-2's relationship with RENAMO in Mozambique remained limited to military support. Much of this situation was due to its adversary's (FRELIMO) very different relationship with its sponsors. FRELIMO's leaders courted socialist states prior to independence in the same manner as UNITA had done with the United States and South Africa. A pattern was established early between FRELIMO exiles and the Soviet Union involving cultural exchanges but little in the way of ideological penetration or control.[256] The group's leaders courted both the Soviet Union and PRC in hopes of obtaining aid and military support and benefitted from the former's desire to capitalize on US retrenchment following Vietnam. This desire translated into a windfall, with the Soviet Union providing independent African nations more than $4 billion in arms alone.[257] Soviet enthusiasm for more direct involvement, evident in a decades-long friendship treaty signed with Mozambique, was tempered by signs of internal strife and economic mismanagement within FRELIMO.[258] Soviet leaders received an additional shock when FRELIMO announced the Nkomati Accord with South Africa in March 1984, in which the former agreed to remove ANC bases from its territory in exchange for South Africa stopping its military support for RENAMO.[259]

South African support for RENAMO continued covertly and successfully, given its long involvement through special operations in Mozambique. RENAMO was a creation of the Rhodesian Central Intelligence Organisation (CIO) to deny ZANU and Zimbabwe African People's Union (ZAPU) bases and support in its immediate neighbour. To support RENAMO in some of its heaviest fighting, Rhodesia committed elements of its SAS. Given South Africa's support for Rhodesia, and the desire of its fledging special operators to gain experience in the field, a handful of 1 and 5 Recce operators were 'sheep dipped' upon their arrival to Rhodesia in 1977.[260] This process involved the operators being issued new identities and SAS equipment, and deployment under cover covertly as members of 'D Squadron, Rhodesian SAS'.[261] In total, over a two-year period, from 1977 to 1979, some 143 Recce operators from different units rotated through Rhodesia and saw action mostly in 'the Russian Front' of Mozambique in Gaza province, during which six were killed in action.[262]

With Rhodesia's transition of political power in 1980, FRELIMO support for the ANC, including allowing bases on its territory, guaranteed South African inheritance and sponsorship of RENAMO, which DST-2 began in March 1980. Military advice and assistance, along four different lines of operation—'military, political, financial,

[256] Andrew Ivaska, 'Leveraging Alternatives: Early FRELIMO, the Soviet Union, and the Infrastructure of African Political Exile', *Comparative Studies of South Asia, Africa, and the Middle East*, 41:1 (May 2021): 11–26.

[257] Christopher Andrew and Vasili Mitrokhin, *The World Was Going Our Way: The KGB and the Battle for the Third World* (New York: Basic Books, 2005), p. 455.

[258] Witney Schneidman, 'FRELIMO's Foreign Policy and the Process of Liberation', *Africa Today*, 25:1 (January–March 1978): 66; Andrew and Mitrokhin, *The World Was Going Our Way*, p. 467.

[259] 'What the Nkomati Accord Means for Africa', *The Black Scholar*, 15:6 (November/December 1984): 15–22.

[260] Els, *We Fear Naught but God*, pp. 311–312.

[261] Greeff, *A Greater Share of Honour*, pp. 83–85.

[262] Alexander Strachan, *1 Recce, Volume 2: Behind Enemy Lines* (Cape Town: Tafelburg, 2020), p. 67.

and sociological'—consisted of the training and equipping of RENAMO forces inside South Africa at four bases in the Northern Transvaal and Caprivi Strip.[263] A CIA analysis in 1986 assessed RENAMO as having 15,000 personnel who 'appear cohesive and highly motivated' and South African assistance 'a key factor in enabling the guerrillas to weather a variety of challenges, including pressure from both Mozambican and Zimbabwean forces, and the death of several key military leaders'.[264] In addition to military advice and training, South Africa provided numerous arms and personnel drops to equip RENAMO. Such drops could be substantial, including 'sixty palettes' dropped from 4 SAAF C-130s on 25 May 1983.[265] Over a four-year period, one author concludes RENAMO received 45 to 60 tons of supplies per month from South Africa by airlift alone.[266] To increase supply capacity and decrease risk to the transporting forces, SAN vessels conducted resupply and bulk RENAMO personnel extractions by sea, with 4 and 5 Recce support, during Operations AUTUMN (August 1981), BRISTOL (March 1983), MAHONIA (March 1984), APOSTEL (August 1984), and FISH (September 1984).[267] Operation BRISTOL, which saw 230 RENAMO guerrillas extracted for training and was approved by the Cabinet, was under the overall command of the GOC, Special Forces and the OC of 4 Recce.[268]

Details about DST personnel deployed to Mozambique remain vague. Coordination with RENAMO had to occur, but aside from the provision of advanced high-power radios and parachuting in of small numbers of Recces in advance of missions, no clear picture emerges.[269] Special operations along these lines formed one clandestine and covert prong of South Africa's strategic goals in Mozambique: military advice and assistance missions to RENAMO to destabilize the country, while also conducting indirect-shock missions to punish FRELIMO and disrupt and decapitate the ANC (Operations BEANBAG and VINE) in one of its presumed sanctuaries.

## Large-scale externals: indirect shock supporting conventional actions

Independent special operations, such as small-team reconnaissance missions or indirect-shock missions to locate and attack SWAPO bases in Angola, gave way from 1984 to 1988 to integration with conventional SADF columns during the 'winter

[263] Truth and Reconciliation Commission, *The Truth and Reconciliation Commission of South Africa Report*, volume 2, p. 93; Alex Vines, 'Renamo's Rise and Decline: The Politics of Reintegration in Mozambique', *International Peacekeeping*, 20:3 (April 2013): 376.

[264] Directorate of Intelligence, 'Mozambique: RENAMO Today—Profile and Prospects', Central Intelligence Agency (15 May 1986), p. 3, www.cia.gov/readingroom/docs/CIA-RDP86T01017R000707220001-2.pdf.

[265] Truth and Reconciliation Commission, *The Truth and Reconciliation Commission of South Africa Report*, volume 2, p. 93.

[266] Vines, 'Renamo's Rise and Decline', p. 376.

[267] Søderland and Steyn, *Iron Fist from the Sea* (2014), pp. 250–255; 269–273.

[268] Ibid., p. 252; Vines, 'Renamo's Rise and Decline', p. 376.

[269] The team that jumped in prior to BRISTOL was led by Captain Wynand du Toit, later captured in Angola and detailed elsewhere in the chapter. Søderland and Steyn, *Iron Fist from the Sea* (2014), p. 252.

games'. The Border War increasingly took on a higher pace and intensity thanks to the injection of additional Soviet equipment and Cuban, East German, and Russian advisors. These military advising and assistance missions, including tactical training and accompanying PLAN and occasionally SWAPO forces as well, had operational consequences for SADF. Thanks to them, MPLA and SWAPO became more competent in minor tactics, including patrolling and ambushes, as well as military deception and defensive preparations of their bases. During this phase, as noted above, South Africa also lost air superiority over Angola, which in turn compromised the SAAF's ability to support long-range air strikes against logistics and military support infrastructure. SAAF support to ground forces was increasingly restricted to mobility and resupply and tactical airstrikes against PLAN motorized or mechanized columns. SWAPO ambushes in Angola and Namibia became more complex and competent, particularly by their own SOF, increasingly inflicting losses on SADF units.

The capture and exposure of Captain du Toit also impacted SSC risk tolerance. Operations increasingly fell into one of two patterns: smaller and increasingly covert, or larger and more integrated with conventional forces. The smaller operations mitigated risk of detection and compromise of the mission but sacrificed capability, as the outcomes of Operations ABDUCT I and II suggest. At the same time, the formidable defensive base preparations and response of SWAPO forces required larger and larger special operations force packages, particularly when specific intelligence was lacking, as the outcome of Operations COLOSSEUM suggests. These force packages, and the increased size and complexity of conventional operations, had the effect of altering South African SOF. Those units, especially 32 Battalion, became increasingly conventional in organization yet asymmetric in operations.[270] In addition, larger operations relegated Recce units to pathfinder and combat support roles as occurred during Operations HOOPER, PACKER, and MODULAR.

Operations ABDUCT I (January–February 1987) and II (November–December 1987) targeted FAPLA MiG-21 and -23 aircraft provided by the Soviet Union, stationed at different airbases in southern Angola. Using SAAF aircraft to target the FAPLA forces was a risky proposition, given a limited stock of irreplaceable, advanced platforms flying against Soviet-provided early warning radars and air defence systems. Neither Botha nor the SSC were willing to risk Atlas Cheetahs (a French Mirage III derivative) or Impalas (Italian Aermacchi MB-326), much less pilots, as a result. The idea for the operation was conceived within 5 Recce, particularly 54 Recce Commando, its small-teams element, to reconnoitre and attack the FAPLA aircraft. No other 5 Recce elements were available for the mission as they were committed to preparations for Operation COLOSSEUM discussed below.[271] After insertion by Puma helicopter, the three operators for the mission, Diedricks, Stadler,

[270] Radies Rademeyer interview, Pretoria, South Africa, 10 June; Eeben Barlow interview, Pretoria, South Africa, 10 June 2023.

[271] Stadler, *Recce*, p. 280; Christo Roelfse interview, Pretoria, South Africa, 8 June 2023.

and Neves Mathias would proceed on foot 20 kilometres. They would conduct close reconnaissance into the base followed by a 'final stalk' onto the airfield to place explosives on the aircraft.[272]

Despite rehearsals and specialized equipment, such as first-generation night-vision googles and EMLC-designed charges resembling Second World War 'sticky bombs', ABDUCT I succeeded in only two regards: the operators were neither detected nor caught. Mission success was another matter. An uncompromising and unrealistic time schedule forced the operators to abandon their initial reconnaissance and attack. During a second attempt, only Diedricks and Stadler moved in, as Mathias manned the team's radio at their hide site. After the ordeal of getting onto Menongue airbase undetected, despite some close calls, Diedrick and Stadler discovered no worthwhile jet aircraft parked there that evening: only a transport helicopter and an inoperable transport.[273] Cutting their losses and supremely disappointed 'that all our efforts were in vain', the team returned via a similar route they had taken, and by helicopter back to Rundu.[274] The jet noise they heard the day prior was FAPLA evacuating the target aircraft to an airbase further north, a daily precaution against possible night attack, confirmed by intelligence reports Stadler acidly described as 'three weeks too late'.[275]

The failure of ABDUCT I did not invalidate the mission's rationale. In Diedrick's words, 'the strategic importance of delivering a decisive blow to the enemy's air capability was worth the risk'.[276] A follow-up effort was mounted, ABDUCT II, this time against Lubango airfield, also involving Diedricks and Stadler. The field was located roughly on the same latitude as Menongue but further west in Huila province, close to the border with Namibia province. Located on the eastern outskirts of the provincial capital, increasing the reconnaissance team's chances of being observed and detected, Lubango airfield was protected by two Cuban brigades, various FAPLA units, and a dedicated airfield security unit.[277] This time local intelligence reports confirmed the presence of military aircraft—more than fifty—although information required for tactical mission planning was either severely outdated or non-existent.

To mitigate the high level of risk to force for the small team, *SPESKOP* provided a range of available resources in support of the three men. Caches of supplies and equipment were stored along infiltration and exfiltration routes to minimize the weight the operators carried.[278] Other resources included a quick reaction force from 51 Recce to provide additional firepower and area security should the team require emergency extraction. In addition, an unspecified number and type of fighter aircraft, likely Impala Is, were assigned two tasks in support of the mission: escort duty for the

[272] Stadler, *Recce*, p. 281.
[273] Ibid., p. 293; Diedricks, *Journey without Boundaries*, p. 110.
[274] Stadler, *Recce*, p. 294.
[275] Ibid.; Diedricks, *Journey without Boundaries*, pp. 110–111.
[276] Diedricks, *Journey without Boundaries*, p. 111.
[277] Ibid.
[278] Chris Serfontein interview, Pretoria, 7 June 2023.

transport aircraft and radio relay or 'Telstar' duty.[279] To buy the operators more time by compressing the amount of travel time to and from their mission departure point, at least two Puma helicopters with extended range 'ferry tanks' would be used.[280] SAAF helicopter experience flying with night-vision googles had been perfected to the point of allowing nighttime insertion and extraction. Nevertheless, relying on tested and trusted small-team bushcraft and survival skills, the team for ABDUCT II planned for a 60-kilometre ingress on foot to reach their observation point, near a Roman Catholic church on a hill south of Lubango.[281]

ABDUCT II was no more successful than its predecessor. The preparations for it were intensive and included numerous rehearsals, some attended by senior *SPESKOP* personnel standing in as airfield guards, as well as liaison with their SAAF counterparts. Flying at night in blacked-out conditions at low level, to avoid radar, was a new and unsettling experience for the operators. Close contact and numerous previous deployments with the members of 19 Squadron, on whose pilots their lives depended, built trust.[282] FAPLA guarding the airfields had been put on alert thanks to previous attempts to destroy aircraft using special operations or long-range artillery, including Operation CATAMARAN I. As a result, Diedricks and Stadler managed to infiltrate Lubango despite the absence of vegetation but only made it close to one of the aircraft, a MiG-21, before guards spotted them. After some quick thinking, they managed to escape and decided not to chance fate anymore here.[283] Intelligence, too late or too dated, was the crucial factor between success and failure in both attempted indirect-shock missions against airfields in Angola.

Operation COLOSSEUM started with reconnaissance missions and rehearsals in late October and culminated in an attack on a SWAPO base in northeastern Angola, close to the town of Jamba in the province of Huila on 13 November 1986. According to South African military intelligence, the base contained the headquarters of SWAPO's Eastern Front command and could call upon nearby Cuban ground forces, estimated at more than 1,500 personnel, as well as advanced MiG-21 and -23 aircraft from a nearby air base at Menongue in support.[284] These support forces once again ruled out a more conventional South African airborne assault.

To confirm the presence of the command headquarters and gather tactically specific information, *SPESKOP* HQ dispatched a small reconnaissance unit consisting of Koos Stadler and José da Costa. After a harrowing 30-kilometre ingress from a nearby UNITA base, the small team confirmed the location of the SWAPO base. The OC, Commandant James Hills, cut the reconnaissance short. His reasons were sound tactically. To minimize the risk to force and risk to mission, the raiding force covered much of the distance in armoured and mine-protected Casspir fighting vehicles. The trade-off for protection and readiness, however, was in speed and detection. Fearing

[279] Stadler, *Recce*, p. 318; Helmoed-Römer Heitman, 'Equipment of the Border War', *Journal for Contemporary History*, Special Edition: Border War, 1966–1989, 31:3 (December 2006): 103.
[280] Stadler, *Recce*, p. 320.
[281] Ibid., p. 318.
[282] Diedricks, *Journey without Boundaries*, p. 114.
[283] Ibid., pp. 115–118.
[284] Stiff, *The Silent War*, p. 537.

he was losing the element of surprise, Hills curtailed the reconnoitre, including a more dangerous and time-consuming close-in reconnaissance of the base by Stadler and da Costa.[285] The base assault plan differed little from the tactical template used during PIGEON, except this time Special Forces through the regimental headquarters and 52 Recce would lead the assault with the 101 Battalion company and 2 Recce elements in support. Teams from 51 and 53 Recce provided the bulk of the blocking forces to prevent and SWAPO from escaping.[286]

The plan, sound enough tactically, did not survive contact with the enemy. Close-in reconnaissance might have revealed that most of the SWAPO cadres were in a training location 12 kilometres away. In addition, one of the two blocking forces was engaged by a heavy machine gun, sacrificing immediate surprise.[287] Some SWAPO cadre in the camp were killed; most managed to escape. An attempt to follow up the operation with an assault on the cadres away from the base turned into an ambush instead. Tactical discipline and training allowed the Reece-led forces to fight their way out and conduct additional follow-on operations.[288] While thirty-nine SWAPO were killed in the indirect-shock mission, the cost to the assault forces was not insubstantial: two Recce operators killed and an undisclosed number of wounded. Perhaps more substantial was the impact the mission had on the operators participating in it: it deflated their morale somewhat but also steeled their resolve to take the hard lessons learned during COLOSSEUM and improve training for future operations.[289]

Some lessons, particularly as they relate to the use of special operations in an increasingly firepower-intensive environment, appeared to go unlearned the following year during Operation FIREWOOD (September 1987). Prior to its authorization, SADF planners had ambitious goals: a large-scale sweep of Cunene province to rid it of SWAPO and PLAN bases. The impositions imposed by the threat of Cuban-assisted FAPLA to UNITA in the southeastern Angola limited the scope and scale of anti-SWAPO actions elsewhere. These concerns were warranted, as intelligence suggested to SADF a large-scale FAPLA offensive would occur in the coming months.[290] Small-team sabotage missions preceded the inevitable SADF response by cutting logistical arteries of FAPLA from the sea to the interior, specifically rail lines of communication, which occurred on 17 September although their effect on subsequent operations is unknown.[291]

The general mission and operating area for FIREWOOD was almost identical to COLOSSEUM. Its target, however, was different: SWAPO's Central Area headquarters, located north of the village of Indungo. The forces spearheading the action too were nearly duplicated: under Hills' command, a force of 5 Recce would be

[285] Stadler, *Recce*, pp. 270–271.
[286] Stiff, *The Silent War*, p. 537.
[287] Stadler, *Recce*, pp. 274–275.
[288] Stiff, *The Silent War*, pp. 540–542.
[289] Stadler, *Recce*, pp. 276–277.
[290] Scholtz, *The SADF in the Border War, 1966–1989*, p. 251.
[291] Marius Scheepers, 'The Battle at Indungo, 31 October 1987: An (*sic*) SADF Military Venture against SWAPO in the Angolan Border War', unpublished Master of Social Science dissertation, University of Pretoria (December 2019), p. 41.

augmented by operators from 2 Recce and troops and vehicles from 101 Battalion, who were experienced in *Koevoet* missions.[292] To increase the capabilities of the special operations-led task force, Ratel-81 armoured infantry fighting vehicles equipped with mortars, several Buffalo mine-protected vehicles, and two companies of 1 Parachute Battalion were added.

Increasing the size of the force to more than 1,000 personnel to decrease risks to personnel and mission greatly complicated and increased its logistical requirements and decreased its speed and manoeuvrability. Heavier armoured vehicles required spares, particularly when going off major roads or '*bundu* bashing' through brush, which damaged axles, stressed powertrains, and blew tyres. In addition, such vehicles consumed fuel at a considerable rate: an estimated 5.6 kilometres per gallon on a road for the Buffalos and likely half to one-quarter of that in sand.[293] To meet these transportation and supply requirements the overall number of vehicles in the task force ballooned to near 150.[294] To further decrease risk by increasing the amount of available firepower, the senior staff officer for Operations at South-West Africa Command and former 32 Battalion commander, Commandant Edward Viljoen, requested a unit of *Valkiri* rocket launchers. The request was allegedly dismissed disdainfully by the chief of the Army, Kat Liebenberg.[295] To help preserve surprise, a conventional task force, Daffodil, conducted manoeuvres and attacks and SAAF airstrikes on a range of targets 'to draw the attention of the enemy away from the advance of the main SADF Task Force'.[296]

The four phases of Operation FIREWOOD culminated in a plan for assaulting the base and conducting following operations. The movement of forces into position was delayed by 'frequent breakdowns' of vehicles, costing precious time.[297] Additional friction affected the rehearsed, coordinated plan: broken-down vehicles, severed radio cords, and the impact of vegetation on the cohesion of the line of advance led some assault units to close on the base quicker than others.[298] SWAPO within the base were prepared for the assault, tipped off by SAAF preparatory airstrikes, and personnel were waiting in trenches and in crew served weapons positions including heavy machine guns and recoilless rifles. SWAPO weapons fire destroyed three Casspirs. One Buffalo also subsequently overturned and had to be abandoned. The Recce assault, consisting of more than one hundred operators augmented by a company from 1 Para, succeeded in clearing the trenches at a cost of between one to almost two dozen SADF soldiers killed, including one Recce, and over four dozen more wounded.[299]

[292] Scholtz, *The SADF in the Border War, 1966–1989*, pp. 251–252.

[293] Jacques Gansler, William Lucyshyn, and William Varettoni, *Acquisition of Mine-Resistant, Ambush-Protected (MRAP) Vehicles: A Case Study*, UMD-AM-10-178 (College Park, MD: Center for Public Policy and Private Enterprise, University of Maryland, March 2010), pp. 27–28.

[294] Scheepers, 'The Battle at Indungo, 31 October 1987', p. 67.

[295] Scholtz, *The SADF in the Border War, 1966–1989*, p. 252.

[296] Scheepers, 'The Battle at Indungo, 31 October 1987', pp. 59–60.

[297] Ibid., p. 69.

[298] Ibid., pp. 79–81.

[299] Sources cite various numbers: twelve, fifteen, and nineteen dead and forty-seven, fifty-six, and sixty-four wounded. Scholtz, *The SADF in the Border War, 1966–1989*, pp. 252, 489n14; Marius Scheepers

Given the high number of casualties, the most 'the SADF suffered in a single day during the entire Angolan Border War', it is unsurprising that investigations and assignations of blame followed. As overall mission commander, much of the blame fell on Hills. Radies Rademeyer suggests its outcome 'cost him his career'.[300] Others, including Scheepers, rightly question the use of a regiment of highly trained special operators to conduct a conventional assault on a prepared defensive position. After an initial refusal, Hills was subsequently convinced to mount the assault when the chief of the Army, Kat Liebenberg, flew in personally and ordered him to do so.[301]

If correct, this suggests a remarkable command intervention and the conventionalizing of the special operations chain of command. The unit which suffered the bulk of the casualties, 101 Battalion, was the focus of subsequent attention. In Scheepers' estimation, its relatively poor performance was the consequence of poor integration, itself the result of a unit that had previously operated independently against mobile insurgent columns and not static defences. The full year that transpired between COLOSSEUM and FIREWOOD, using almost identical forces, is a perhaps the most damning indictment that can be drawn regarding the integration of the units into a coherent special operations capability. One operator from 52 Recce states succinctly: 'None of these [COLOSSEUM or FIREWOOD] should have been done by, or even involved Special Forces. There were typical mechanised Infantry [sic] attacks and not at all Special Forces style operations.'[302] The racial dimensions of apartheid governance limited its political fallout. By law, SADF only had to report casualty figures for white personnel, which consisted of twelve to thirteen casualties overall, not the five times or more losses actually incurred.[303]

FIREWOOD was destined to be overshadowed by events further east, in the conduct of three related operations: MODULAR, HOOPER, and PACKER. These three sequential phases of an overarching plan driven by the rotation of operational units into the theatre have since become known collectively as (the Battle of) Cuito Cuanavale, the final, decisive battle of the Border War.[304] While the Lusaka Accords imposed a notional ceasefire and suggested a phased withdrawal of SADF and Cuban and other Soviet Bloc forces, MPLA leaders remained committed to settling the matter militarily. In consequence, Soviet-provided aid and military advisors increased dramatically in 1985. Cuban advisors, the bulk of whom were organized into combat brigades to fight with FAPLA, also increased by 67 per cent, from an estimate 21,000 to 31,000. East Germany provided 3,500 advisors who mostly provided technical

and Frans Johan Pretorius, 'Loss of Military Equipment at the Battle of Indungo during the Border War, 31 October 1987', *Scientia Militaria, South African Journal of Military Studies*, 49:2 (2021): 56, 60.

[300] Radies Rademeyer interview, Pretoria, 10 June 2023.

[301] Ibid.

[302] Van Niekerk, *Born to Storm*, p. 366.

[303] Radies Rademeyer interview, Pretoria, 10 June 2023.

[304] Jannie Geldenhuys, *At the Front: A General's Account of South Africa's Border War* (Johannesburg, South Africa: Jonathan Ball, 2004), p. 228. It is worth pointing out that some participants and observers question the label in an exercise of postmodern deconstruction. The reality is that the moniker for a series of engagements is often the conurbation closest to them, in this case Cuito Cuanavale. For more on this point, see Scholtz, *The SADF and Cuito Cuanavale*, pp. 197–199.

advice and assistance.[305] South African intelligence reports in 1987 suggested almost a dozen flights were arriving in Monogue daily offloading equipment and observed lengthy truck convoys flowing from there to Cuito Cuanavale, the jumping-off point for a FAPLA offensive against UNITA.[306] Equipment and training turned FAPLA into a credible and substantial, if ponderous, conventional Soviet-style manoeuvre force. The seasonal weather patterns, however, constrained FAPLA to operations during the dry season to prevent tanks and other armoured vehicles from bogging down along the few roads available in southern and southeastern Angola. Soviet doctrine and training further restricted FAPLA operations to set march formations and response patterns, characterized in one intelligence analysis as 'slow moving, multibrigade, cautious, and well-supplied offensives' that 'take advantage of Luanda's traditional strengths in firepower, airpower, equipment, and numbers'.[307]

Given the buildup, the question was who would go on the offensive first—and when. Central Intelligence Agency estimates predicted a FAPLA conventional offensive in 1986 with UNITA responding with guerrilla warfare, erasing the former's gains.[308] That offensive failed to materialize for a number of reasons, including a UNITA pre-emptive raid against Cuito Cuanavale and the entirely unexpected outcome of Operation DROSDE by 4 Recce in Namibe harbour (May/June 1986), in which one Cuban and two Soviet cargo ships were sunk.[309] Further contributing factors were discord in Luanda between the Soviet commander and senior Cuban military advisors about how best to proceed.[310] The increased traffic and forward staging in March 1987 suggested to UNITA and SADF that an offensive was imminent. Within SADF Headquarters, several options were outlined and debated ranging from providing greater support clandestinely to UNITA all the way up to a pre-emptive strike against Cuito Cuanavale. As always, sensitivity to casualties and losses, and its political consequences, undercut further large-scale options.[311]

The options available for Operation MODULAR against the expected five FAPLA brigades ranged from clandestine and low level to an overt, pre-emptive strike. SADF commanders eventually submitted a compromise plan: special operations asymmetric attacks against advancing FAPLA columns to complement UNITA raids, combined with a composite battle group comprising mechanized manoeuvre forces (61 Mechanized Group), irregulars (32 Battalion), artillery, and other infantry operating 'in a clandestine covert way', to conceal South Africa's activities from domestic and

[305] Director of Central Intelligence, 'Angola: Short-Term Prospects for UNITA, Key Judgments', Special National Intelligence Estimate SNIE 71-86W, TOP SECRET (1986), p. 5; Hamann, *Days of the Generals*, p. 81.

[306] Scholtz, *The SADF and Cuito Cuanavale*, p. 20.

[307] Director of Central Intelligence, 'Prospects for the Angolan Civil War in 1987', Special National Intelligence Estimate, SECRET (1987), p. 8.

[308] Director of Central Intelligence, 'Angola: Short-Term Prospects for UNITA, Key Judgments', p. 3.

[309] Fred Bridgland, *The War for Africa: Twelve Months that Transformed a Continent* (Gibraltar: Ashanti Publishing, 1990), pp. 17–18; Søderland and Steyn, *Iron Fist from the Sea* (2014), p. 332–355; Johan Hechter interview, Cape Town, South Africa, 13 June 2023.

[310] Gleijeses, *Visions of Freedom*, pp. 393–394.

[311] Heitman, *War in Angola*, p. 31; Scholtz, *The SADF and Cuito Cuanavale*, pp. 22–25; Les Rudman interview, Pretoria, South Africa, 8 June 2023.

external audiences.[312] Helmoed-Römer Heitman concludes the plan for MODULAR was the worst of both worlds: 'This tendency to hedge bets and opt for compromise decisions was to plague this campaign to its end' because it 'was primarily a "political" campaign' serving the ends of the policy.[313] In essence, MODULAR would respond to a FAPLA operational offensive with tactical defensive actions leading to an operational counterstroke to blunt it.

Special operations initially played four supporting roles in MODULAR. The first predated the operation by a year and involved three Chief of Staff Intelligence (CSI) liaison or military advice and assistance teams to UNITA, staffed largely by 5 Recce personnel.[314] Pre-emptive guerrilla strikes by UNITA that same month, in Operation CHUVA, inflicted casualties on FAPLA but had a limited tactical impact. The second was the deployment of independent 5 Recce teams in July to report on advancing FAPLA columns, working with UNITA under Operation HUNTER. The Recce teams, distributed along the anticipated manoeuvre corridor between the Mianei and Lomba rivers, would gather information on advancing FAPLA units and conduct anti-tank attacks using handheld launchers and truck-mounted recoilless rifles.[315] 32 Battalion companies would advance and help delay advancing FAPLA units through blocking and flanking ambushes, but once the true scope of the offensive became clear, they were relegated to limited reconnaissance and ambush missions, infiltrating and protecting forward observation officers (FOOs) to spot for artillery, and once the fighting intensified, the rather diminutive mission: 'protect the artillery and supply convoys'.[316]

The third, also a delaying action, was Operation COOLIDGE. The target for COOLIDGE was a well-known bottleneck: a single-span bridge on the Cuanavale River, which was part of the main FAPLA supply route. An initial target survey (COOLIDGE) would potentially be followed up by an attack on the bridge itself (COOLIDGE B). A river survey dictated 4 Recce conduct the operation, but the fast-moving inland fresh water required the development and employment of 'a new capability'.[317] SAAF Transalls and Puma transported the operators from the bases to training bases and forward headquarters. Both teams, reconnaissance and raiding, received last-minute update briefings at UNITA headquarters by Commandant Les Rudman's 5 Recce staff and would be transported and escorted by UNITA forces to their dropping-off point.

COOLIDGE B, under the command of 4 Recce operator Major Irené 'Tuffy' Joubert, carried out the attack. It dropped almost half of the bridge spans that rested on wooden pilings, despite a company of FAPLA guarding the bridge.[318] Energetic FAPLA defence of the bridge wounded the tactical commander, Major Fred Wilke,

312 Gleijeses, *Visions of Freedom*, p. 395.
313 Heitman, *War in Angola*, p. 31.
314 Les Rudman interview, Pretoria, 8 June 2023.
315 Alexander Strachan, *1 Recce, Volume 3: Through Stealth Our Strength* (Cape Town: Tafelburg, 2022), p. 235; Heitman, *War in Angola*, p. 35.
316 Nortje, *32 Battalion*, pp. 237–249; 241.
317 Søderland and Steyn, *Iron Fist from the Sea* (2014), p. 394.
318 Ibid., p. 402; Geldenhuys, *At the Front* (2004), p. 229.

but crocodiles and other hazards, including fishing nets, caused the only other casualty of the mission.[319] Although repairs were made to the bridge, FAPLA forces had to conduct much more time-consuming ferrying operations across the river, starving their advance columns of fuel and heavy munitions, thanks to the bridge strike and UNITA's continued harassment of supply columns.[320] Two other factors contributed to their success: 'protection and support provided by 5RR and UNITA forces' and 'CSI Liaison Teams coordinating all planning with UNITA.'[321]

The fourth set of special operations support to MODULAR occurred after the key battle occurred on the Lomba River, which runs north of Mavinga. At Lomba, four FAPLA bridges were met by UNITA forces and the SADF's 20 Brigade, a battle group comprising three 61 Mechanized Group groups, three companies from 32 Battalion, and several artillery batteries, backed by air support. The operators from 5 Recce and 32 Battalion companies shadowed FAPLA brigade movements, reporting the size, position, and movement. After repeated attempts to cross the river, the three 61 Mech groups hit half of the FAPLA brigades despite their information regarding their disposition 'being all wrong' and they were 'up against a far stronger force.'[322]

Taking advantage of surprise and hitting 47 Brigade with a combination of anti-guided missiles, artillery, and later close-in fighting pitting Ratel-90 armoured cars against T-54/55 and T-34 tanks was devastating. In Roland de Vries' words, the 61 Mech combat groups 'destroyed the 47th Brigade of FAPLA and removed them form the Angolan troop list forever.'[323] Concentrated artillery fire and airstrikes proved particularly destructive against the other FAPLA brigades. This phase of MODULAR was marked by a control first as well: the visit to 20 Brigade Headquarters by several dignitaries including the most important half of the SSC, F.W. de Klerk, Pik Botha, and Malan.[324] The various actions that comprise the Battle at Lomba River from 9 September to 3 October ended the FAPLA offensive and set the stage for a series of SADF counteroffensive attacks against their isolated and strung-out brigade. The result of these second- and third-phase MODULAR operations resulted in FAPLA being driven back in a series of alternating blows to the eastern bank of the Cuito River. These operations achieved the primary aim for 20 Brigade of the second phase of MODULAR: 'using existing forces ... inflict the maximum casualties and losses on the Fapla [sic] forces east of the Cuito.'[325]

As a result of these operations, and in support of the final phase of MODULÊR designed to drive FAPLA across the Cuito, Recce teams would be integrated directly into 20 Brigade operations. This set of special operations would transfer the artillery observation mission from SOF to Special Forces, requiring the latter to learn new

[319] Irené Joubert interview, Langebaan, 12 June 2023; Pieter van Niekerk interview, Cape Town, 13 June 2023.
[320] Søderland and Steyn, *Iron Fist from the Sea* (2014), p. 403.
[321] Ibid., p. 402.
[322] Scholtz, *The SADF and Cuito Cuanavale*, p. 49.
[323] Roland de Vries, *Eye of the Firestorm: The Namibian-Angolan-South African Border War Memoirs of a Military Commander* (Solihull, UK: Helion & Company Limited, 2016), p. 404.
[324] Geldenhuys, *At the Front*, p. 228.
[325] Heitman, *War in Angola*, p. 31.

skills on the job during a scheduled force rotation.[326] The operators of 1 Recce relieved their 5 Recce brethren during this operation. Rather than conduct hunter-killer missions, for which 1 Recce was not trained, equipped, or experienced, they infiltrated to positions close to FAPLA brigade deployment areas, report on their dispositions and routines, and assist Army FOOs to conduct artillery strikes in Operation HUNTER B. In all, six five-man 1 Recce teams deployed to as high a ground as possible in concealed observation positions (OPs) for up to four weeks, conducting additional close-in 'stalks' of FAPLA brigade positions and reporting back using morse code to a Special Forces Tactical HQ collocated with 20 Brigade.[327] The longer the teams were in the field, the more tasks they received, including forward air control and ground site surveys.

One special operations task opened the successor to MODULAR, Operation HOOPER. Two teams, instructed to watch the Cuito bridge, conducted battlefield damage assessments. Tasked by Tac HQ 'that they had to keep watch on the bridge' without knowing why, Team 1 and Team 4 operators observed the first successful use of a precision guided bomb in the conflict, a stand-off gliding optically guided Kentron H-2 Raptor.[328] An antique SAAF Buccaneer, on its second attempt on 3 January 1988, launched a single bomb and inflicted half the damage on the Cuito bridge—a 20-metre span destroyed—with as much precision as 4 Recce in COOLIDGE B.[329] A final task included the construction of field positions to draw FAPLA airstrikes in the hope UNITA would use FIM-92 Stingers provided covertly by the Central Intelligence Agency, as well as attempt to down MiGs using their own SA-7s.[330] This mission set proved unpopular with 1 Recce team operators: the dust and smoke from launching the missiles immediately revealed their location to Angolan and Cuban artillery observers.[331]

The pause in operations allowed the seven remaining FAPLA brigades arranged in the 'Tumpo triangle' beside the Cuito to dig into defensive positions closer to their logistics, air, and artillery support and receive reinforcements. Their positions made them largely static targets for 20 Brigade artillery with 1 Recce observation support. Even so, 20 Brigade, augmented by UNITA, 32 Battalion, Special Forces, and artillery, stood little chance of tackling one brigade without incurring significant casualties. Casualties thus far had been inflicted largely by artillery and airstrikes.[332]

Any pretence of South Africa keeping its engagement in Angola hidden evaporated in late September, with drastic consequences. Botha and his Cabinet visited the forward deployed headquarters, to great fanfare and claims of forthcoming additional resources for SADF forces fighting there.[333] For HOOPER, the chief of SADF, with the approval of Botha, approved the dispatch of 4 South African Infantry, with

[326] Chris Serfontein interview, Pretoria, South Africa, 7 June 2023.
[327] Strachan, *1 Recce, Volume 3*, p. 244.
[328] Ibid., pp. 264–265.
[329] Ibid., p. 266; Heitman, *War in Angola*, p. 198.
[330] Strachan, *1 Recce, Volume 3*, p. 267.
[331] Ibid., p. 270.
[332] Heitman, *War in Angola*, pp. 157–163.
[333] Gleijeses, *Visions of Freedom*, p. 398.

a squadron of Oliphant (up-gunned and modified British Centurion) tanks.[334] Previously, Botha had balked at sending South African armour for fear of escalating the conflict. Such concerns prevented 61 Mech from deploying to Angola with their Oliphants for the first phase of MODULAR. The SADF counteroffensive, combined by the worrisome attrition of their FAPLA combat brigades, led MPLA leaders to request and receive even greater Cuban support, escalating the conflict.[335] Spearheading that support were 121 Cuban special forces officers and enlisted in a military advising and assistance mission.[336]

In doing so, Cuban reinforcements and advisors froze the conventional battlefield, leading to a duel within the Tumpo triangle that lasted another four months. The Cubans rallied and reinforced FAPLA, leading to three separate battles for a bridgehead at Tumpo south of the bridge at Cuito. Adding to the stalemate was Geldenhuys' orders to his forces on 28 January that comprised their own paradoxical trinity: relinquish no gains, run out the clock, and minimize casualties.[337] This continued to place a premium on the substitution of firepower for personnel.

Special operators continued to reconnoitre FAPLA positions and apply firepower in a variety of different ways during HOOPER. 1 Recce teams rotated in and out of the front lines where they conducted artillery and airstrike observation missions. Team members gradually learned fire control procedures and increasingly called in strikes without trained FOOs accompanying them.[338] Other sources suggest 1 Recce operators supported SAAF airstrikes on Cuito Cuanavale airfield by employing ground-released 'chaff' balloons to spoof Soviet-supplied radars, luring aircraft out of their hardened concrete shelters to be destroyed by long-range G-5 artillery fire.[339]

Another way special operations contributed to attrition during the battle was through a variation on road watch missions conducted by British and American SOF during the Second World War (in North Africa and occupied France) and Vietnam War (in Laos) respectively. The requirement came from 20 Brigade Headquarters to the 1 Special Forces Tac HQ and its commander at the time, Major Renier (Doiby) Coetzee. Team 5, under the leadership of Sgt. Johnny de Gouveia, comprising five members, were given two tasks. After infiltrating by Puma to a position off the Cuito Cuanevale and Menongue road, the teams were to split up after meeting up with UNITA operating in the area. From there, one half-team would scout and subsequently track and mark convoys moving out of Menongue. Should they make it past the first half-team, the second would do the same in relay fashion for convoys between the end of the first team's area to the Lomba road, the halfway point between Menongue and Cuito Cuanavale.[340]

334 Geldenhuys, *At the Front*, p. 235.
335 De Vries, *Eye of the Firestorm*, p. 405; Scholtz, *The SADF and Cuito Cuanavale*, pp. 60–61; 67–68.
336 Gleijeses, *Visions of Freedom*, p. 423.
337 Heitman, *War in Angola*, p. 220.
338 Strachan, *1 Recce, Volume 3*, p. 270.
339 Stiff, *The Silent War*, p. 549.
340 Strachan, *1 Recce, Volume 3*, p. 292.

Taking advantage of the FAPLA and Cubans' habit of laying up their vehicles for the evening from 1600 onwards, the operators marked the head and tail of the column with illumination flares.[341] If de Gouveia's account is accurate, the strikes conducted over the next several weeks destroyed almost 500 soft-skinned and armoured vehicles.[342] A second task for the team was to serve as a form of early warning for 20 Brigade of incoming FAPLA airstrikes against them. By monitoring radio traffic from Cuito Cuanavale airfield, they could determine not only the bearing and direction but also potentially the specific missions of the departing FAPLA aircraft.[343]

Resources determined the outcome and aftermath of the Battle of Cuito Cuanavale. The lack of more SADF manpower meant that any gains by the 3,000-odd forces under Geldenhuys' command, supplemented by an equal or greater number of UNITA, could not be consolidated, much less exploited.[344] South Africa and the SADF were, for all intents and purposes, locked in a battle of attrition whose best outcome was a stalemate. Years afterwards, Geldenhuys used sports metaphors to describe the outcome. He noted the South Africans were playing not for victory but rather winning either 'by technical knockout, or ... win[ning] on points'.[345] The bell ending the match rang on 23 March 1988. While extricated from long and costly regional involvements, and with Namibia on a path to freedom, the apartheid regime from 1987 onwards used special operations in an increasingly brutal manner internally against the ANC.

## Death squads: internal special operations

For many decades after the Sharpeville massacre in 1960, the internal threat posed to the apartheid regime by the ANC and other domestic revolutionary movements was minimal. In terms of internal security, police and intelligence service operational authorities, already formidable thanks to the passage of the Public Safety Act of 1953, increased dramatically first with the Terrorism Act of 1967 and later under Botha when he declared successive states of emergency from 1985 onwards.[346] So too did the number of police forces in the country. The various eleven police forces, including the SAP but the majority of which were 'homeland' and 'independent homeland' ones, were not much of a force to uphold domestic law and order. In Janine Rauch's assessment, they 'acquired a reputation for brutality, corruption and ineptitude' and organizationally 'were militarised, hierarchical, and ill-equipped to deal with "ordinary crime"'.[347]

[341] Ibid., pp. 293–294.
[342] Ibid.
[343] Ibid., p. 291.
[344] Geldenhuys, *At the Front*, p. 243.
[345] Ibid., pp. 251–252.
[346] Glenn Frankel, 'South Africa Imposes State of Emergency', *The Washington Post* (21 July 1985), www.washingtonpost.com/archive/politics/1985/07/21/south-africa-imposes-state-of-emergency/d125801e-fb8e-4069-9c7b-3626daf02cc5.
[347] Janine Rauch, 'Police Reform and South Africa's Transition', unpublished South African Institute for International Affairs conference paper (2000), p. 1, www.csvr.org.za/docs/policing/policereformandsouth.pdf.

The main revolutionary movements seeking to overthrow white rule were the ANC, the Pan Africanist Congress (PAC), and the Unity Movement (UMSA). These movements had their roots primarily among urban blacks working due to the shortage of white labour and initially formed as the result of the nexus between industrial workers and the urban intelligentsia.[348] All three movements, despite their ideological differences, suffered due to the crackdown after Sharpeville, including their proscription by the state. Anthea Jeffrey assesses its result: 'Internal resistance to [apartheid] measures had been muted, in part because the ANC and PAC—never strong in any event—had been banned and silenced since 1960'.[349]

In consequence, all three movements formed governments-in-exile and attempted to foment revolution from abroad through armed struggle and outside support. In addition, they sought to train the waves of exiles who fled the country to avoid arrest and prosecution. State control measures, including Pass Laws and networks of informers, rendered any significant effort at creating clandestine infrastructure necessary for internal revolution dangerous if not suicidal. The risks of training, organizing, and infiltrating revolutionaries back into South Africa from Tanzania, Botswana, and later Zimbabwe to agitate and conduct 'armed propaganda' paled in comparison to those within the country. Among the three groups, the ANC received the most support from external sponsors, thanks to the eloquence of its leaders. Sponsors ranged from individuals and non-governmental organizations in South Africa and the West to the leaders of governments sympathetic to the struggle for African liberation. This support was dwarfed by the level the ANC received from the Soviet Union from 1969 onwards in the form of arms, equipment, and money, including a 'subsidy' of almost $400,000 per year.[350]

South Africa's intelligence organizations, national, military, and police, remained well apprised of ANC and PAC efforts in other countries. Although often lacking the detailed and current information necessary to conduct externals, their web of informants and infiltrated agents provided considerable insights aiding security forces, particularly after the formation of the SSC and NSMS. Frustrated in its efforts to build a clandestine revolutionary structure within South Africa, the ANC committed a number of 'propaganda of the deed attacks' or 'spectaculars', alternating between infrastructure and military targets between 1980 and 1983, including a state-owned refinery (1980), SADF Headquarters (1981), nuclear power plant (1982), and SAAF headquarters (1983).[351] The attacks, conducted by MK's Special Operations division

[348] For a concise background and assessment by an African American journalist who worked for CBS News, Agence France-Presse, and Negro Press International, see Richard Gibson, *African Liberation Movements: Contemporary Struggles against White Minority Rule* (London: Oxford University Press, 1972), pp. 19–105. As an aside, Gibson was identified as CIA agent QRPHONE-1 in 2018. Jefferson Morley, 'CIA Reveals Name of Former Spy in JFK Files—And He's Still Alive', *Newsweek* (15 May 2018), www.newsweek.com/richard-gibson-cia-spies-james-baldwin-amiri-baraka-richard-wright-cuba-926428.

[349] Anthea Jeffrey, *People's War: New Light on the Struggle for South Africa* (Johannesburg, South Africa: Jonathan Cape, 2009), p. 13.

[350] Andrew and Mitrokhin, *The World Was Going Our Way*, pp. 443; 464–465.

[351] O'Brien, *The South African Intelligence Services*, pp. 96–97.

commanded by Joe Slovo, initiated a series of 'tit-for-tat' reprisals driven by emotion rather than meaningful purpose.[352]

Domestically, the events that began in the Vaal Triangle in 1984 gave rise to expanded South African special operations. The Triangle, south of Johannesburg, consists largely of an industrial area as well as black townships. Beginning on 3 September, 'Black Monday', violence spread to other towns including Sharpeville as protests over rent increases transformed from marches into riots that included the killing of the deputy mayor of Sharpeville. Riot control by SAP escalated into the use of deadly force. The riots and response prompted riots elsewhere, including the neighbouring township of Soweto. The mobilization of SADF in response appeared to quell the uprising and the Botha government made concessions to the various groups involved.

By early 1985, however, emboldened activist groups operating under the umbrella of the United Democratic Front (UDF) conducted protests elsewhere in South Africa. The subsequent Langa Massacre in March 1985, in which police fired on a crowd attending the funeral of a slain protestor on the twenty-fifth anniversary of the Sharpeville massacre, fuelling further protests and violence throughout the country. Over a two-month period, more than 108 protestors were killed, two-thirds by SAP.[353] The increasing violence led P.W. Botha to declare the first in a series of states of emergency on 20 July 1985. In keeping with the logic of violence in civil war, in the absence of any central directing authority controlling and orchestrating it, some local entrepreneurs instead used the protests and riots as a means to strike at the state but also as an opportunity to settle personal local scores by hacking or burning their enemies to death, including a trademark ANC technique of 'necklacing', or igniting a tire around the neck.[354]

The initial intensity of the violence surprised members of the SSC, most of whom believed internal violence was controllable. Immediately after Langa, the SSC concluded the ANC's revolutionary struggle was reaching its third, decisive phase: the long-anticipated 'general uprising ... sparked off by organised and well prepared guerrilla operations during the course of which the masses of the people will be drawn and armed'.[355] State monitoring and repression had largely denied MK its guerrilla forces but successive waves of cadres were given a 'rudimentary quick course' and then returned to South Africa for 'recruiting and training local black youth into

[352] Referred to in 'Umkhonto we Sizwe (MK) Operations Report', section 4.3, https://omalley.nelsonmandela.org/index.php/site/q/03lv02424/04lv02730/05lv02918/06lv02985.htm. See also Ian Liebenberg, 'The Integration of the Military in Post-Liberation South Africa: The Contribution of Revolutionary Armies', *Armed Forces & Society*, 24:1 (Fall 1997): 114; Hillel Frisch, 'Motivation or Capabilities? Israeli Counterterrorism against Palestinian Suicide Bombings and Violence', *Journal of Strategic Studies*, 29:5 (2006): 843–869.

[353] Truth and Reconciliation Commission, *The Truth and Reconciliation Commission of South Africa Report*, volume 2, p. 75.

[354] Ibid., p. 89; Sipho Mbuqe, 'Political Violence in South Africa: A Case Study of "Necklacing" in Colesberg', unpublished doctoral dissertation, Duquesne University, August 2010.

[355] From the text of MK's strategy document, 'Operation Mayibuye', captured during a 1963 raid. 'Operation Mayibuye' full text, http://law2.umkc.edu/faculty/projects/ftrials/mandela/mandelaoperationm.html.

becoming terrorists' using a variety of means.[356] The pace of attacks also escalated dramatically, from 44 ANC incidents in 1984 to 136 in 1985 and 118 in 1986.[357] In response, Botha directed tighter planning and connections, or close integration, between SAP, SADF, and NIS.[358]

To coordinate internal responses to the threat, one of the Committees of the SSC, Covert Collection, was modified to conduct covert action alongside its traditional mission of intelligence collection.[359] The new organization was known as the Counter-Revolutionary Intelligence Task Team, or the acronym *TREWITS* based on its Afrikaans title.[360] Following the second state of emergency in 1986, chief of staff, SADF tasked the GOC, Special Forces, Joup Joubert, with developing 'a plan by which Special Forces could support SAP operations inside the country'.[361] Joubert concluded some of that support could take the form of 'intensified action', covertly supplementing the urban information gathering and reconnaissance conducted by the specialists within 1.1 Recce, whose operations in this environment remain opaque.[362] Recce support included several liaison officers and staff to various regional headquarters and Special Branch commanders.[363] Pressures to expand missions drove Recces to accept conventional officers from other branches who had not passed selection or badging, diluting the overall quality of the force.[364] In addition, those who moved from Recces to covert units and back again were often distrusted by their peers as having been tainted by the former's ulterior motives and methods.[365]

The mechanism for clearer coordination to conduct internal covert special operations was established when SADF expanded the charter of the CCB in June 1988, which originally focused on covert action abroad, through the creation of a new operating area: Region 6. SAP expanded the charter of its existing capability, C1 or 'Vlakplaas', in reference to the location of its headquarters and main base.[366] Vlakplaas would subsequently come to public attention, and achieve a level of notoriety, thanks to its revealing and subsequent shocking confession of one of its senior leaders, Eugene de Kock. He stated C1's activities, as well as those by CCB in Region 6, comprised little more than a programme of selective capture, torture, and killing. De Koek is unrestrained in his apologia, paralleling in gruesome detail the means

[356] Morgan Norval, *Inside the ANC: The Evolution of a Terrorist Organization* (Washington, DC: Selous Foundation Press, 1990), p. 151.

[357] Seegers, *The Making of the Military in Modern South Africa*, p. 247.

[358] Les Rudman interview, Pretoria, South Africa, 8 June 2023.

[359] O'Brien, *The South African Intelligence Services*, pp. 148–151.

[360] *Teen Rewolusionere Inligting Taakspan.*

[361] O'Brien, *The South African Intelligence Services*, p. 147.

[362] Ibid., p. 148. Peter Stiff does not cover 1 Recce operations within South Africa in his three-volume trilogy of works, 'South Africa Secret Warfare', nor does 1 Recce historian Alexander Strachan. Els merely notes, 'They [1.1 Recce] could participate in the whole spectrum of Special Forces tasks, although they specialised in urban operations and it was for that reason that they remained in an urban environment.' Els, *We Fear Naught but God*, p. 205.

[363] O'Brien, *The South African Intelligence Services*, p. 148.

[364] Chris Serfontein interview, Pretoria, South Africa, 7 June 2023.

[365] Pieter van Niekerk interview, Cape Town, South Africa, 13 June 2023.

[366] For more details, see Robin Binckes, *Vlakplaas: Apartheid's Death Squads, 1979–1994* (Barnsley, UK: Pen and Sword Books Ltd, 2018).

and methods revealed by Paul Aussaresses in French Algeria a generation earlier.[367] C1 was initially constructed as a pseudo unit, trying to turn captured enemies of the regime into state assets. Members turned through entrapment, known as 'askaris', conducted most of the killings.[368] Like *Koevoet*, C1 expanded exponentially, particularly after 1985. Also like its predecessor, C1 became corrupt through the mixture of insufficient oversight combined with the filing of false claims, ostensibly to make use of available funds in a 'secret funds budget' but often to enrich individual operators.[369]

Similar corruption plagued CCB. Particularly prone were SAP officers, according to Eeben Barlow.[370] The control mechanisms, or command authority for CCB activities, were incredibly opaque. As Hamann ultimately concluded after a series of interviews with senior SADF commanders, 'even among the generals there seems to be some confusion about to whom the CCB was ultimately answerable'.[371] The lack of control is evident in other confusion related to CCB. Daniel Douek suggests the change in leadership in 1989 and overlapping webs of covert networks led to increasing incidents in which 'trigger-happy covert networks ... sometimes accidentally killed each other's operatives'.[372] F.W. de Klerk denied any knowledge of the existence of Region 6, much less CCB and C1's activities in his autobiography, while Magnus Malan acknowledges and defends them.[373]

Far more divisive and lethal were special operations that armed and mobilized township militias covertly, a form of treachery and subversion. This approach was driven, in large part, by the escalation of violence in the townships. More than 800 people were killed there, half by ANC's trademark 'necklacing', and many others were burned to death in their homes.[374] The internal state proxies would then engage in reciprocal local violence to counter rival ANC and UDF groups. Such proxies ranged from local independent armed groups, such as the 'Hippos' and 'The Black Cats', to the militias of established political parties who were bitter rivals of the ANC like the Inkatha Freedom Party (IFP).[375]

The IFP in KwaZulu received training and support by Directorate of Military Intelligence (DMI) between 1985 and 1988 under Operation MARION while in Ciskei and Transkei the Xhosa Resistance Movement had much the same as part of Operation KATZEN.[376] Such training included guerrilla warfare methods followed by extensive training in 'offensive, defensive, contra-mobilisation, and VIP

[367] Paul Aussaresses, *The Battle of the Casbah: Terrorism and Counter-Terrorism in Algeria, 1955–1957* (New York: Enigma Books, 2010). Unlike de Kock, Aussaresses was unrepentant about assassination and torture and defended its use until his death.

[368] Simon Taylor, 'Status Quo Terrorism: State-Terrorism in South Africa during Apartheid', *Terrorism and Political Violence*, 35:2 (2023): 304–320.

[369] De Koek, *A Long Night's Damage*, pp. 111–112.

[370] Eeben Barlow interview, Pretoria, South Africa, 10 June 2023.

[371] Hamann, *Days of the Generals*, p. 155.

[372] Douek, *Insurgency and Counterinsurgency in South Africa*, p. 93.

[373] F.W. de Klerk, *The Last Trek: A New Beginning* (New York: St Martin's Press, 1998), p. 204; Malan, *My Life with the SA Defence Force*, pp. 328–331.

[374] Malan, *My Life with the SA Defence Force*, p. 328.

[375] 'Goldstone Commission on Allegations of SADF Funding of Violence in Townships', Nelson Mandela Foundation, n.d., https://omalley.nelsonmandela.org/index.php/site/q/03lv02039/04lv02046/05lv02047/06lv02049/07lv02062.htm.

[376] Stiff, *Warfare by Other Means*, pp. 156–222.

protection'.[377] These groups and militias were identified by ANC leaders from the mid-1980s onwards as the hidden 'Third Force' of the security services.[378] The logic of 'Third Force' was simple: it distanced white South Africans from the violence, mobilized support from key black constituents by forcing them to choose a side, were an economical alternative to standing security forces, and also played into Afrikaner narratives that blamed blacks for internecine violence.[379] MK's shift in approach to 'people's army, people's war' in the townships led to limited tactical successes but suffered under the combination of covert special operations and police repression, enabled by the Emergency Powers Act, which convinced the ANC to abandon the armed struggle in favour of a negotiated settlement.[380]

A debilitating stroke led Botha to resign as state president in January 1989 and he was replaced by F.W. de Klerk. One of De Klerk's first actions was the dismantling of the NSMS. It was replaced by the National Coordinating Mechanism (NCM). In addition to changing the character of control fundamentally, with elected officials replacing those handpicked by Botha, the NCM also removed civilians from committees chaired by military officers, ensuring a division between those who set policy and those who conducted military strategy.[381] Officials from the SSC would also be elected and would have to submit their proposals to Cabinet, ensuring a degree of transparency and oversight. After reducing the power and influence of SADF, De Klerk also set his sights on the means of covert action.

The special operators of CCB and C1 understood which way the political winds were blowing. When it became obvious De Klerk was going to become state president, CCB and C1 operators were told to wrap up their activities by 1 September 1989.[382] De Klerk empowered the Harms Commission to investigate their activities, along with those of NIS and DMI. The Commission amounted to little and DMI continued fomenting violence in the townships. On 11 February 1990, De Klerk ordered Nelson Mandela, former leader of the ANC, released from prison.[383] The ANC, welcomed by De Klerk as a political actor in November 1990, threatened to walk away from negotiations based on the feeble outcome of the Commission's investigation and continued violence in the townships. De Klerk appointed another, the Goldstone Commission, in October 1991. The Goldstone Commission, reporting directly to the state president, uncovered evidence to varying degrees of SAP and SADF involvement in or support of state-sanctioned killings.[384]

[377] Ibid., p. 178.

[378] Stephen Ellis, 'The Historical Significance of South Africa's Third Force', *Journal of Southern Africa Studies*, 24:2 (June 1998): 261–299.

[379] Nicholas Haysom, 'Vigilantes and the Militarisation of South Africa', in Jacklyn Cock and Laurie Nathan, eds., *War and Society: The Militarisation of South Africa* (Claremont, South Africa: David Phillip, Publisher (Pty] Ltd, 1989), pp. 194–195.

[380] Van der Waag, *A Military History of Modern South Africa*, pp. 279–280.

[381] Seegers, *The Making of the Military in Modern South Africa*, pp. 266–267.

[382] Eeben Barlow interview, Pretoria, South Africa, 10 June 2023.

[383] Nelson Mandela, *Long Walk to Freedom* (Boston, MA: Little, Brown and Company, 1994), pp. 559–626.

[384] 'Goldstone Commission on Allegations of SADF Funding of Violence in Townships', Nelson Mandela Foundation, n.d., https://omalley.nelsonmandela.org/index.php/site/q/03lv02039/04lv02046/05lv02047/06lv02049/07lv02062.htm.

Having fostered domestic subversion through covert campaigns of targeted killings and domestic proxy sponsorship, South Africa faced how to put this genie back into the bottle from 1990 onward. With covert institutions either disbanded, under investigation, or deliberately lying low, and SAP implicated in much of the violence, De Klerk turned to the SADF. Most SADF commanders and soldiers, particularly from the Permanent Force, wanted nothing to do with putting down domestic unrest in the townships. They had little choice. UDF and trade union fighters clashed with neighbouring Inkatha in March 1990 in KwaZula Natal province, resulting in almost 100 dead and 20,000 persons displaced when homes were torched.[385] In addition to nine battalions of white SADF soldiers, troops from 32 Battalion, all but unemployed after the withdrawal from Namibia, were dispatched. The troops were sent to reconnoitre, patrol, or prevent violence from breaking out in five subsequent named operations, some independent and others in conjunction with SADF or SAP forces, including Operations EARDRUM, FIREFLY, PIKADEL, SABONA, and EXPOSE across Natal.[386]

The fortunes of the 32 Battalion turned irretrievably on 8 April 1992. Responding to the sound of gunfire in the Phola Park settlement in Alberton City, 32 Battalion elements conducted a cordon-and-sweep operation when they came under fire, where their operations had a 'mixed record' but generally were 'over aggressive'.[387] A firefight ensued with only a few casualties. The 'peacekeepers' of 32 Battalion, under scrutiny already and despised by many members of the ANC for fighting against other black Africans, were investigated during the Goldstone Commission enquiries.[388] The resulting findings, which remain bitterly contested by different sides and include allegations of murder and rape, played a role in the eventual inactivation of 32 Battalion on 26 March 1993.[389] The final missions of 32 Battalion saw the unit come full circle: from clandestine SOF outside of South Africa to stemming proxy clandestine special operations within its borders.

Internal special operations proved to be the last gasp of a dying apartheid regime. Their details continue to shock. Some will reject the notion that such activities deserve the title of 'special operations', which they equate with noble and honourable pursuits of arms. This attitude, however, rejects the realities of the context. The character of internal special operations was the logical culmination of South African special operations over two decades. Their logic had two unique qualities. First, numerous foreign operations increased the likelihood of a militarized response domestically. What had been used elsewhere to effect abroad could and would be

[385] South African Press Association, 'KwaZulu-Natal's Seven-Day War Claimed at Least 80 Lives', 18 November 1996, www.justice.gov.za/trc/media/1996/9611/s961118b.htm.

[386] Nortje, *32 Battalion*, pp. 260–267.

[387] Goldstone Commission of Enquiry in Phola Park, Exhibit 30, 'Verklaring ten Behoewe van 32 Batlajon', undated, 4 pp. University of the Witwatersrand, Goldstone Commission of Enquiry in Phola Park Records 1992–1993, Collection Number AK2702; Radies Rademeyer interview, Pretoria, South Africa, 10 June 2023.

[388] A spirited but partisan recounting of the details is contained in Jan Breytenbach, *Buffalo Soldiers: The Story of South Africa's 32 Battalion, 1975–1993* (Alberton, South Africa: Galago Publishing, Pty, 2002), pp. 332–337.

[389] Nortje, *32 Battalion*, p. 280.

used effectively at home against internal enemies. Those measures were born of the ruthless conduct of war over time. The second quality was the privatization of violence. The DMI and SAP, through the trifecta of *TREWITS*, CCB, and C1, in effect distanced themselves from the violence by 'subcontracting' it to local forces. In other words, the use of locally turned 'askaris' and others followed the logic underpinning the first use of special operations with 31 and 32 Battalions more than two decades prior: Afrikaners using black soldiers to fight a black insurgency, this time covertly instead of clandestinely.[390]

## Conclusion

South Africa's use of special operations from 1974 until 1994 provides an informative and unique case study. A recurring refrain among Western SOF over the past three decades has been for political leaders and lawmakers to fight with 'gloves off', abandoning Marquis of Queensbury restraints and control mechanisms to unleash capabilities. Such occurred in South Africa, particularly with P.W. Botha's ascent to first prime ministerial and later state presidential power. In addition, South Africa had a historical and cultural tradition of asymmetric war fighting but not of special operations in the formal sense of established units and organizations. Much of this reflected Afrikaner, and particularly Boer, political traditions and institutions.

Despite this lack of knowledge, within a relatively brief period, from 1975 until 1980, South Africa went from a 'cold start' to the establishment, fielding, and growth of a range of special operations capabilities with a reputation for operator skill and daring. Its operators were favourably compared to the top-tier SOF elsewhere in the world. Their fieldcraft, or bush skills, equalled those of Rhodesia and their ruthless effectiveness on a par with SOF in Israel and the Soviet Union. Operations such as ABDUCT I and II demonstrate the remarkable skill and endurance exhibited by individual South African operators. Their ability to move undetected over long distances, and penetrate a well-guarded airfield with little cover and concealment, speaks not only to their skill but also to operator self-confidence.

On a broader scale, South African special operations operated overtly, covertly, and clandestinely, domestically and abroad, performing the full range of special operations functions. Yet despite possessing the means to do so, South African special operations were never fully authorized along the lines of either 'the Argentinian route' or the Algerian solution: unconstrained counter-organization operations to eradicate their adversaries, particular the ANC and UDF.[391] This seems puzzling given the cosmology of Total Onslaught and the creation of a securocratic state optimized for crushing internal and external threats. One author argues that political disagreements within the SSC and Cabinet were the determining factor. Their

[390] Keith Gottschalk, 'The Rise and Fall of Apartheid's Death Squads, 1969–1993', in Bruce Campbell and Arthur Brenner, eds., *Death Squads in Global Perspective: Murder with Deniability* (New York: Palgrave Macmillan, 2000), pp. 253–254.

[391] O'Brien, *The South African Intelligence Services*, p. 141.

preferred route, selective assassination, was a function of National Party sensitivity to South Africa's reputation and desire to avoid even a whiff of disunity among political ranks.[392]

Political considerations played a role, but an incomplete one. Elements of the theory of special operations coherence offer additional insights. Limitations within South Africa's special operations capabilities restricted their effectiveness strategically and politically. The higher one moved away from the tactical level of execution, whose effectiveness was based on remarkably tight integration of tactical capabilities, the more distributed, diluted, uncontrolled, and ineffective South African special operations were. This state is represented in the bottom half of Figure 7.2. Also depicted is that despite the outward appearance of an optimal control arrangement, there was far too little in terms of oversight and control mechanisms to keep operators from exercising their agency or *exceeding* their authority.

While De Klerk is excoriated by veteran operators for being too weak and accommodating, he understood politics in both its domestic and international sense better than Vorster and Botha. After Vorster's ideological entrenchment and Botha's creation of an NSMS, which employed special operations extensively, the political pendulum swung back towards reform and concession when it became apparent that force could not quell South Africa's domestic opposition or international isolation. Against this background, and given considerable agency, South African special operations served the ends of the state as effectively and efficiently as possible but eventually exceeded the mandate entrusted to them by their polity in their domestic operations. The issue of 'victory', or winning, was problematic. Much like Rhodesia, the outcome was inevitable—the only questions were when and in what circumstances apartheid would end.

Their tactical successes resulted from high levels of horizontal integration. This integration was driven partially by their limited numbers. Limits in capacity meant that for larger-scale operations, Recce commanders were able to request and receive specific reinforcement from other regiments on an exigent basis given the mission requirements. While there was a strong sense of unit identity and coherence within and between the different Recce Regiments, these cultural and organizational lines were fungible. Their fungibility and the tight integration within task units was only possible given the close social ties and intimate familiarity that occurred within a small elite within an equally small professional military force. Rotational or contract employment sometimes limited the retention of a cornerstone of capability, the retention of corporate knowledge and best practices, but this was not debilitating.

Where special forces conducted missions outside South Africa, particularly in Angola, diverse special operations capabilities were well integrated on the battlefield. For intelligence-gathering and reconnaissance missions and indirect-shock missions in Angola, Mozambique, Botswana, and elsewhere, habitual working relations with supporting elements, particularly SAAF and SAN mobility platforms,

[392] Ibid., pp. 141–142.

greatly increased the operational reach and effectiveness of South African special operations. Even when South Africa's special operations were tested in the crucible of higher-intensity conventional war during the last stage of conflict in Angola, they initially exhibited high integration with conventional forces. This was especially true during the Battle of the Lomba River, arguably the greatest triumph of SADF arms during the Border War. Leading up to and after the main combat action, when the battlefield was at its most fluid, 5 Recce and 32 Battalion elements provided timely intelligence updates and harried and harassed FAPLA columns, preventing them from concentrating or massing in any meaningful manner.

When coordinated and the offensive, South African special operations succeeded thanks in part to the command relationships that existed between fielded *SPESKOP* headquarters and the conventional operational mission commander. When denied it and subordinated to conventional commanders in static and defensive circumstances in the Tumpo triangle, 1 Recce and 32 Battalion were relegated to conventional support roles any light infantry could perform. Tight integration, special operators' bias for action, immediate requirements and availability, and the desire of conventional commanders to impose unity of effort contributed to this misuse. The one subsequent bright spot, the use of two 1 Recce roadwatch teams integrating with SAAF airstrikes against FAPLA support columns, were both too little and too late to have any broader effect beyond the tactical.

Various tensions strained this integration. SADF cultural norms, shaped by decades of conformity with British structures and traditions, proved resistant to change. As late as 1986, senior SADF commanders continued to look up the Recces and other specialized units as a necessary if temporary reality. Other strains to integration occurred in Namibia as special forces countered the asymmetric threat posed by SWAPO. *Koevoet* and special forces, particularly 5 Recce, engaged in competition for 'kills' as a misguided and corrupting measure of success in countering insurgency.[393] Further stresses occurred in countering asymmetric threats later during the case study as special operations were used domestically against ANC and other enemies of state clandestinely and covertly. The reported fratricide of sources and agents in Region 6 between CCB and C1 is evidence of disintegration thanks to compartmentalized entrepreneurial security actions. This state is also depicted in Figure 7.2.

At the strategic level, special operations generated positive and negative effects. On the positive, 4 Recce indirect-shock missions interdicted supplies and damaged Angolan infrastructure that not only contributed to South Africa's policy of regional destabilizing of unfriendly regimes but also weakened the country militarily and economically prior to or during build-ups and offensives. These operations, driven as much by seasonal weather and opportunity, fell into a repetitive pattern that generated effects but were increasingly devoid of strategic purpose beyond inflicting damage.[394]

[393] Retired SADF special operations intelligence officer interview, Pretoria, South Africa, 7 June 2023.
[394] Retired SADF special operations intelligence officer interview, Pretoria, South Africa, 7 June 2023.

At other times these missions generated negative effects well outside of the region. The political consequences of such actions, although unintentional or unexpected, greatly outweighed their strategic and operational advantages. As South African special operations capabilities grew and matured, maintaining clandestinity was increasingly problematic. The presumed cover organizations behind which these operations hid, UNITA or RENAMO, could not pass scrutiny. Employing different uniforms and 'clean' equipment captured from SWAPO and the FNLA served the purpose of tactical military deception but not operational cover. Few were fooled by such efforts, and no one would reasonably believe UNITA could employ SA-9s or conduct complex maritime, littoral, or riverine sabotage missions. Indeed politically, after 1980 and Rhodesia's transition to Zimbabwe, any relatively reasonably complex, competent special operation could *only* have been conducted by South Africa. The unexpected sinking of Soviet and Cuban ships during Operation DROSDE escalated the involvement of both in Angola, to South Africa's strategic detriment. The capture of Wynand du Toit during Operation ARGON sent shock waves domestically in South Africa. Instead of showcasing competence, reassuring, and raising the morale of the South African polity, it lanced the idea of SA superiority of arms but also, even more importantly, of the invulnerability of the Recces.[395] Recce casualties had occurred before—as early as 1974—but the clandestine veil preventing political consequences grew increasingly transparent and was torn away after ARGON, destroying their aura.

The expansion of special operations into the covert realm stretched capabilities beyond reasonable limits, as Figure 7.2 suggests. In addition, they occurred in insufficient quantity or quality to have meaningful strategic effects. Such operations, first by D40 and Barnacle and later CCB and C1, were done for punishment, decapitation, or dissuasion purposes and evolved over the years in response to ANC and UDP actions. They were unable, however, to match the pace of ANC change. By 1986, when it was apparent that guerrilla warfare in South Africa had failed, the ANC changed the rules of the game. In backing mounting civil resistance and by mobilizing a new energized youth movement, the ANC turned the tide of the struggle. Expansion of covert operations appear, in hindsight, as the last gasp of a desperate regime, also evident in their uncoordinated uncontrolled conduct. The creation of Covert Small Team within 5 Recce further stretched the capability of the unit, which was already struggling to maintain requirements in Namibia and Angola.

South Africa's covert special operations capability is striking for its continuous evolution but also its expansion in size as well as autonomy. On the covert side, what made these special operations succeed also provided their downfall: insufficient oversight and control, whose lines are difficult to map, as Figure 7.2 suggests. Autonomy and agency allowed a high level of personal enrichment and corruption, particularly with CCB and C1. That corruption and agency came back to haunt political leaders and the regime once the scope and scale of special operations activities domestically was revealed to the public. The lack of formal authorization

[395] Gray, *Explorations in Strategy*, pp. 175–177.

mechanisms or controlling bodies allowed Botha, Malan, and others to distance themselves from such special operations during the Harms and Goldstone Commission hearings. When it was apparent to the covert operators that whatever top cover they expected had evaporated, individuals including Eugene de Kock disclosed the details of such special operations to a shocked South African public. Special operations increasingly *diverged* from strategic goals, with operators succumbing to the lure of personal or professional incentives.

As additional functions were included under special operations, limited numbers of operators could not keep up. Initially, South African special operations were *congruent* with desired political goals, control mechanisms, and risks, given the threats and stakes involved. Figure 7.1, referenced earlier but provided here for ease of comparison, maps these out schematically. Three forces eroded South African special operations capabilities in consequence: the increasing emphasis on covert treachery and subversion and intelligence-gathering operations to offset shortcomings, more overt, visible, and tactically irrelevant activities in conventional battles, and the inclusion but not integration of other elements such as the SAP into the covert architecture. These are represented in Figure 7.2 under calibration, capabilities, function, and effectiveness.

The eventual outcome in South Africa appears a foregone conclusion, according to theory. Gil Merom argues that the core constituency within democracies, the upwardly mobile middle class, play a decisive role in explaining why these polities

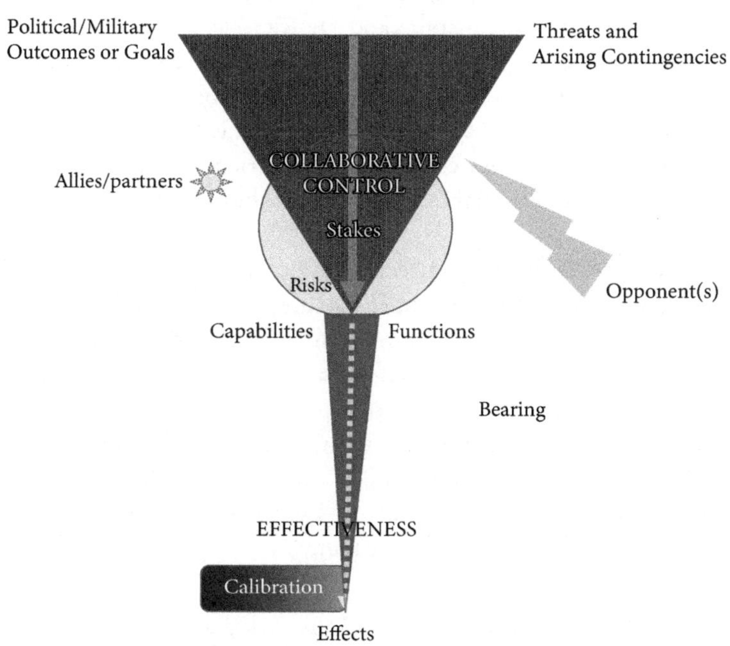


**Figure 7.1** South African special operations congruence, 1972.
*Source:* Author's work.

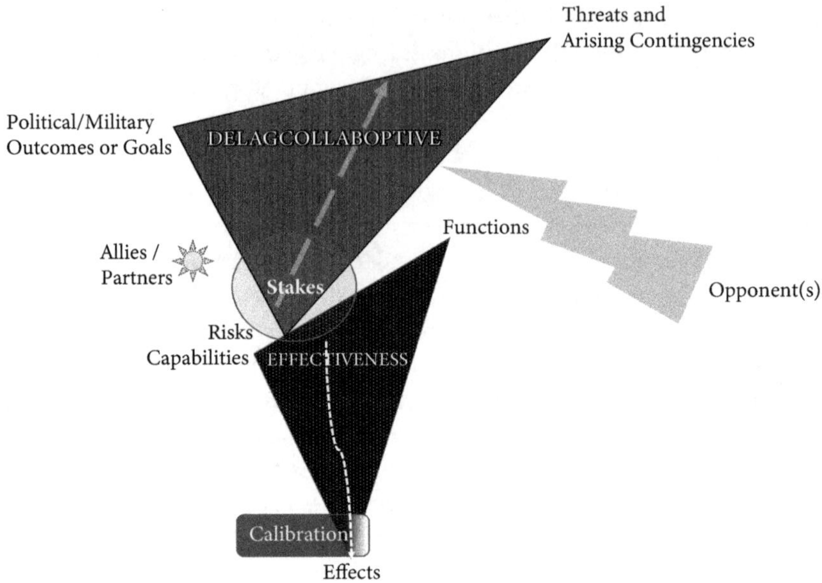


**Figure 7.2** South African special operations divergence, 1992.
*Source:* Author's work.

lose small wars.[396] While South Africa's security apparatus segregated the black African majority into the townships and Bantustans to control and prevent them from exercising political power, Botha and De Klerk were faced with increasingly vocal and coordinated opposition from a range of social and political organizations, including church and veterans groups. Increased needs for manpower for operations in Namibia and Angola to combat the Total Onslaught led to expansion of the terms of conscription of National Servicemen. Their experiences, reported back home to their families, were at odds with the official reports about the war's conduct from the National Party.

The increasing credibility gap between what Botha's regime was saying about the conduct of the war, and what influential members of the middle class were hearing from their sons, is consistent with Merom's overall finding. As J.H. Thompson observes, 'Many young men, straight out of school or university, were not staunchly patriotic and did not want to give two years of their lives to the military, mothers didn't want to lose sons, and South Africa's apartheid government was condemned internationally for fighting an unjust war.'[397] Within senior Permanent Force ranks, the primary enemy was increasingly seen as domestic, placing a serious strain on civil–military relations. Robin Renwick recounts:

[396] Gil Merom, *How Democracies Lose Small Wars: State, Society, and the Failures of France in Algeria, Israel in Lebanon, and the United States in Vietnam* (Cambridge: Cambridge University Press, 2003), pp. 229–250.

[397] J.H. Thompson, *An Unpopular War: From Afkak to Bosbefok* (Cape Town, South Africa: Zebra Press, 2006), p. viii.

> Visiting Namibia the end of the year (1987), I was briefed by a half-mad South African Colonel on the battle of the Lomba river, which was indeed a most impressive military exploit. On the struggle against SWAPO in Namibia, he took the view that victory was certain—but for the efforts of the enemy within. When I enquired who the enemy within were, he replied: "The churches, the trade unions and the teachers".[398]

Special operations played an outsized role in shaping those perceptions, not only ARGON but also when the character and scope of domestic covert actions became public knowledge.

South African special operations were their most effective in setting or changing the operational conditions within the different environments in which they were used. As with the SADF Permanent Force and SAP, however, they could never exist in sufficient quantity to meet all the security requirements of the state. Freedom of action permitted them to respond to local conditions in Namibia and branch out into covert operations. Strategically, however, special operations served as one of the only effective responses to buy time and space for other elements within the South African government to 'win' the counter-revolutionary fight at home and abroad. As former operator Van Niekerk observes, 'Killing people was so easy, especially when you had the authority, backing, and ways, and means to do so. Being used for political gain was the name of the game.'[399] The mission creep in the form of ever-increasing requirements, the political desire to limit risk and consequences, expanded agency of covert actions with little oversight, the intensification and escalation of the conflict, and adaptation by the ANC stretched South African special operations to their limit. These factors, combined with an exclusive ideology and ineffective strategic messaging campaigns, assured the currency of special operations was exchanged at the most limited rate politically possible.

[398] Robin Renwick, *Unconventional Diplomacy in Southern Africa* (London: Macmillan, 1997), p. 132.
[399] Van Niekerk, *Born to Storm*, p. 336.

# 8
# Redefining the System
## US Special Operations, 1996–2016

### Introduction

Explanations for the success of US special operations forces (SOF), epitomized by the bin Laden raid in 2011, are frequently overly simplified. Some attribute it directly to the Nunn-Cohen Amendment creating United States Special Operations Command (USSOCOM). Others credit success specifically to leadership, organizational culture, or process innovation, or generically identified attributes.[1] Others chalk up success to numbers or capabilities. USSOCOM is larger, in terms of personnel, than the active-duty militaries of the Netherlands and Denmark combined, or half the size of all British Armed Forces' regulars.[2] Moreover, the US military remains the only one funded, postured, and equipped to fight two major wars simultaneously.[3] Partners marvel at extravagant US SOF equipment supplies and combat support systems. The theory of special operations coherence permits analysis that evaluates the dynamic inherent in policy development and strategy and explains how success led to crisis exemplified by the Congressionally mandated comprehensive review of culture and ethics in 2020.

Events from 2017 until 2020 are detailed in Chapter 3. This chapter assesses US special operations the two decades prior, from 1996 until 2016, to show how they rapidly expanded, transformed, and, while congruent, ultimately stagnated. It starts with a brief historical background of the evolution of US special operations in the early Cold War, focusing on the impact of experiences in Vietnam and El Salvador. Capabilities and control mechanisms are not created in a vacuum. Rather, they develop in an evolutionary manner due to operational use and political intercession. Providing this background contextualizes the subsequent discussion of SOF capabilities and constraints during this period.

[1] See, for example, Linda Robinson, *Masters of Chaos: The Secret History of the Special Forces* (New York: PublicAffairs, 2004); Jon Lindsay, 'Reinventing the Revolution: Technological Visions, Counterinsurgent Criticism, and the Rise of Special Operations', *Journal of Strategic Studies*, 36:3 (June 2013): 440–445; Richard H. Shultz Jr, *Transforming US Intelligence for Irregular War: Task Force 714 in Iraq* (Washington, DC: Georgetown University Press, 2020); Mark Moyar, *Oppose Any Foe: The Rise of America's Special Operations Forces* (New York: Basic Books, 2017), pp. 327–345.

[2] International Institute for Strategic Studies, *The Military Balance 2023* (London: Routledge, 2023), pp. 72–149.

[3] Dakota Wood, ed., *2023 Index of U.S. Military Strength* (Washington, DC: The Heritage Foundation, 2023), p. 21.

*Special Operations Success*. James D. Kiras, Oxford University Press. © James D. Kiras (2024).
DOI: 10.1093/9780198902096.003.0008

The last sections look at capabilities, control mechanisms, mission evolution and conduct, and provide a concluding analysis. Much related to US control mechanisms structurally for its special operations is covered already in Chapter 4. Therefore, I elaborate on others as they have evolved, including permanent and temporary legal authorities, resources, and command-and-control mechanisms. Unlike capabilities, their details are publicly available but not always understood. The penultimate section and main body of the chapter is an analysis and review of US special operations conducted during this twenty-year period, in seven subsections, starting with activities conducting in special countries and eventually expanding to global ones. The final section of the chapter evaluates the success of US special operations from 1996 until 2016 using the theory of special operations coherence.

## Background

As Chapter 2 suggests, echoes of American special operations are identifiable in the earliest recorded history of political violence on the North American continent, including a variety of ranger and indigenous scouting and raiding units. The Founding Fathers of the United States employed covert actions, including the gathering of secret intelligence, subversion, and treachery, as well as sabotage missions.[4] From the country's foundational roots in irregular and asymmetric warfare, through the establishment of the Republic its territorial expansion and growth, special operations have been present in one form or another. Yet such special operations, while rich from the standpoint of insights they yield historically, did not comprise a capability as defined in Chapter 2. They were conducted by ad hoc units arbitrarily, largely in response to specific national security challenges and almost always as a means to respond to asymmetric threats internally, whether in the continental United States or its colonial territories such as Cuba and Philippines.[5] Russell Crandall concludes that while 'nothing in the American approach to fighting dirty wars is immutable', their very nature is nevertheless largely 'intractable', suggesting a persistent requirement for asymmetric responses including special operations.[6]

As outlined in Chapter 3, increasing demand for Green Berets in Vietnam led to dramatic expansion in their numbers, diluting their ability to conduct military advising and assistance missions. Initial growth was spurred by their successes in working with remote groups of minority Montagnard tribes as part of the Civilian Irregular Defense Group (CIDG) programme, particularly in the Buon Enao province in 1961 and 1962.[7] Racial and ethnic tensions between Montagnards and the Vietnamese,

[4] For details, see Stephen Knott, *Secret and Sanctioned: Covert Operations and the American Presidency* (New York: Oxford University Press, 1996).

[5] For an accessible overview, see John Tierney Jr, *Chasing Ghosts: Unconventional Warfare in American History* (Washington, DC: Potomac Books, Inc., 2006), Chapters 1–8.

[6] Russell Crandall, *America's Dirty Wars: Irregular Warfare from 1776 to the War on Terror* (Cambridge: Cambridge University Press, 2014), p. 469.

[7] Francis Kelly, *U.S. Army Special Forces, 1961–1971* (Washington, DC: Department of the Army, 1985), pp. 24–31.

insufficient numbers of capable South Vietnamese Armed Forces officers, American zeal for the anti-Communist mission, and political instability in Saigon changed the nature of the mission.[8] Special Forces, which bore the brunt of casualties in South Vietnam prior to the intervention of major ground forces in 1965, moved from Central Intelligence Agency (CIA) control to Military Assistance Command, Vietnam (MACV) as part of Operation SWITCHBACK in 1962.[9]

The shift in operational control authority had three consequences for US special operations capabilities. The first was the expansion of CIDG camps along the western border to detect and provide early warning of North Vietnamese supplies and units into South Vietnam from the Ho Chi Minh Trail, leading to parallel growth in Special Forces ODAs to man the camps, which peaked at 2,500 operators in-country.[10] Increasingly, this dilution led more experienced officers to note 'how poorly we were doing in the advisory business', a charge that fell on deaf ears institutionally.[11] The second was the development of a clandestine task force, MACV's Studies and Observation Group (MACV-SOG; 25 January 1964), to conduct a range of activities, including intelligence-gathering, indirect-shock, and subversion and treachery missions against North Vietnam.[12] The variety of missions, and their clandestine nature, drove the requirement for air and mobility platforms, in the form of C-130s, UH-1s, and Swedish 'Nasty' fast patrol boats.[13] In addition, insufficient SOF capability and capacity to deny the Viet Cong use of South Vietnam's littoral and riverine system for transport led to the creation of a maritime overt reconnaissance and indirect-shock capability, the Navy SEALs, on 1 January 1962.

MACV was chartered with preventing South Vietnam from falling and given considerable latitude in the conduct of its increasingly conventional war in that country. Part of it comprised an overt Special Forces role within the country. Covert and clandestine special operations, however, were another matter entirely. Initially, President Lyndon Johnson's administration sought to use such operations as one of several cost-effective, risk averse methods of coercing the North Vietnamese into halting their support and assistance for the insurgency in the south. Covert and clandestine actions were designed as a form of signalling out of the public's view that limited the likelihood of escalation. Inadvertently, they provided the pretext for

[8] Paul Elle, Richard Joyce, Robert Williams, and William Woodworth, *US Army Special Forces and Similar Internal Defense Advisory Operations in Mainland Southeast Asia, 1962–1967 (U)*, Technical Paper RAC-Tp-354 (McLean, VA: Research Analysis Corporation, June 1969), pp. 52–56.

[9] *Foreign Relations of the United States*, 1961–1963, Volume III, Part III, Vietnam, January–August 1963, eds. Edward Keefer and Louis Smith (Washington, DC: Government Printing Office, 1991), Document 136.

[10] Jeffrey Clarke, *United States Army in Vietnam: Advice and Support—The Final Years, 1965–1973* (Washington, DC: Center of Military History United States Army, 1988), p. 207.

[11] John Waghelstein, 'What's Wrong in Iraq? Or, Ruminations of a Pachyderm', *Military Review*, 86:1 (January–February 2006): 113–114.

[12] Kenneth Conboy and Dale Andradé, *Spies and Commandos: How America Lost the Secret War in North Vietnam* (Lawrence, KS: University Press of Kansas, 2000).

[13] Kenneth Sams and Lt Col Burt Aton, *USAF Support of Special Forces in SEA*, Project CHECO Report DOTECT-69-31 (San Francisco, CA: HQ PACAF, 10 March 1969); Kenneth Conboy and Dale Andradé, 'The Secret Side of the Tonkin Gulf Incident', *Naval History*, 13:4 (August 1999), www.usni.org/magazines/naval-history-magazine/1999/august/secret-side-tonkin-gulf-incident.

American escalation, when North Vietnamese patrol boats reacted to MACV-SOG raids in the Gulf of Tonkin and attacked the USS *Maddox*.[14] The attempt to use special operations coercively, however, suffered from the same restrictive control mechanisms that hampered the overt application of airpower: 'the structure built to oversee special operations in Southeast Asia was removed from the battlefield and placed in the hands of political decision makers'.[15] Less constrained controls, however, would not have made MACV-SOG operations any more effective, given their '[l]ack of imagination in planning, faulty execution of missions, and poor operational security'.[16]

US special operations within South Vietnam eventually were integrated together in an operational framework specifically designed to tackle the clandestine organizational infrastructure of the Viet Cong. This framework, known as the *Phung Hoang* or Phoenix programme, was an attempt to integrate US and South Vietnamese civil development and intelligence collection activities on the Viet Cong in South Vietnam with operations designed to eliminate cadres. Their specific target was the network comprising the Viet Cong's 'shadow government' that effectively 'out-administered' the Government of the Republic of Vietnam in the countryside.[17] This shadow government was not insubstantial, with some early estimates suggesting such cadres numbered from the tens of thousands up to as many as 100,000 personnel.[18] Operationally, CIA-backed paramilitary teams, operating with US special operations support including Special Forces and SEALs, conducted 'kill or capture' missions against identified cadres. The programme, which began in 1967, inflicted considerable losses on the Viet Cong cadre at a time when the war was shifting to its next, more conventional phase. Nevertheless, Phoenix operations, conducted by police strike teams termed Provincial Reconnaissance Units (PRU), appeared to external observers a programme of sanctioned assassination.[19]

The secretive nature of PRU activities, combined with the clandestine involvement of Green Beret teams supporting them, did little to change this appearance. Revelations from Vietnam confirmed a growing public perception of US special operations as a 'rogue elephant' out of control.[20] Thomas Adams' chapter subheading neatly

[14] Extracted, reviewed, and declassified chapter from in-house National Security Agency history, 'Chapter Six, The Gulf of Tonkin Incident, The DESOTO Patrols and OPLAN 34A', n.d., pp. 44–50, www.nsa.gov/portals/75/documents/news-features/declassified-documents/gulf-of-tonkin/articles/release-2/rel2_gulf_tonkin_incident_desoto.pdf.

[15] Conboy and Andradé, *Spies and Commandos*, p. 272.

[16] Ibid., p. 275.

[17] Orrin DeForest and David Chanoff, *Slow Burn: The Rise and Bitter Fall of American Intelligence in Vietnam* (New York: Simon & Schuster, 1990), p. 71; Bernard Fall, 'The Theory and Practice of Insurgency and Counterinsurgency', *Naval War College Review*, 18:3 (1965), https://digital-commons.usnwc.edu/nwc-review/vol18/iss3/4.

[18] William Rosenau and Austin Long, *The Phoenix Program and Contemporary Counterinsurgency*, Occasional Paper (Santa Monica, CA: RAND Corporation, 2009), p. 4.

[19] Andrew Finlayson, 'The Tay Ninh Provincial Reconnaissance Unit and Its Role in the Phoenix Program, 1969–1970', *Studies in Intelligence*, 51:2 (June 2007): 59–69; William Rosenau and Austin Long, *The Phoenix Program and Contemporary Counterinsurgency*, Occasional Paper (Santa Monica, CA: RAND Corporation, 2009), p. 13.

[20] Gerald Haines, 'The Pike Committee Investigations and the CIA: Looking for a Rogue Elephant', *Studies in Intelligence* (Winter 1998–1999), www.cia.gov/static/CIA-Pike-Committee-Investigations.pdf.

summarizes the conventional wisdom regarding covert special operations: 'immoral, criminal, and not a military function'.[21] Congressional inquiries prompted a series of post-Vietnam control measures, including an executive order (EO) placing severe restrictions on assassination (EO 11905, 1976; EO 12036, 1978; EO 12333, 1981) and outlining the scope of intelligence activities (EO 12333, 1981). Congress first passed a 1974 amendment (Hughes-Ryan) to the Foreign Assistance Act of 1961, imposing conditions and reporting requirements on the funding of covert action. Notably, the first programme eliminated in line with this amendment was for Angola in 1975 (see Chapter 7). In addition, Congress created specific committees to oversee intelligence, the first by the Senate (19 May 1976) and the second by the House of Representatives (14 July 1977).[22]

Legislators were not alone in viewing special operations capability suspiciously. A generation of general officers who fought at the junior and field grade in Vietnam continued to harbour 'deep-seated mistrust of Special Forces' based on their experience.[23] Several descriptive nouns and adverbs capture the view among these conventional leaders: 'cowboys' and 'snake-eaters' acting 'unilaterally' and 'unprofessionally'.[24] Control within the military took the form of resource removal and deprivation. Susan Marquis assesses this led to the 'near-eradication of an American special operations ability'.[25]

The post-Vietnam American desire to retrench and limit its foreign entanglements, including using rump special operations capabilities, was confronted by a new threat—international terrorism, detailed in Chapter 6. The story of Captain Charles Beckwith's crusade to fill a gap in the US Army's capabilities by modelling a new unit on the British Special Air Service, characterized by one of his first recruits as 'a ten-year slog through infighting, skepticism, power grabs, money grabs, and plain old bureaucratic foot-dragging', is well known.[26] The special mission unit he created and led was first tested in 1980 in the attempt to rescue the American hostages held in Tehran, Iran. The failure of that mission, and the reticence of political decision makers to take more forceful action to rescue hostages, led the United States to be seen globally as 'a paper tiger' and by its allies as 'inept'.[27]

A military advice and assistance effort to El Salvador by Special Forces from 7th Special Forces Group (SFG) between 1981–1988 succeeded due to control measures

[21] Thomas Adams, *US Special Operations Forces in Action: The Challenge of Unconventional Warfare* (London: Frank Cass, 1998), p. 116.

[22] US Senate, The Select Committee on Intelligence, *Legislative Oversight of Intelligence Activities: The U.S. Experience* (S. Prt. 103-88) (Washington, DC: Government Printing Office, 1994), pp. 5–6.

[23] General Norman Schwarzkopf's view during Operation DESERT SHIELD as noted in Peter de la Billiere, *Looking for Trouble: SAS to Gulf Command, the Autobiography* (London: HarperCollins, 1994), p. 410.

[24] Susan Marquis, *Unconventional Warfare: Rebuilding U.S. Special Operations Forces* (Washington, DC: Brookings Institution Press, 1997), pp. 34–35.

[25] Susan Marquis, *Unconventional Warfare: Rebuilding U.S. Special Operations Forces* (Washington, DC: Brookings Institution Press, 1997), p. 35.

[26] William Boykin, *Never Surrender: A Soldier's Journey to the Crossroads of Faith and Freedom* (New York: Faith Words, 2008), p. 88; Charlie Beckwith, *Delta Force* (London: Arms and Armour Press, 1984).

[27] Charlie Beckwith and Gary McCaleb, interview with Charlie Beckwith, 18 September 1990, video, Abilene, Texas (https://texashistory.unt.edu/ark:/67531/metapth779623); University of North Texas

despite these challenges. El Salvador was confronted by a formidable insurgency conducted by the *Frente Farabundo Marti de Liberacîon Nacional*, or FMLN, from 1980 until 1992. The FMLN efforts were 'without comparison the best-funded, best-organized, and best-supported guerrilla war ever fought on the American continent'.[28] Its forces, backed by Cuban sponsors with supply bases in neighbouring Honduras, were numerically equivalent to the conventionally trained and equipped *Forces Armada de El Salvador* (FAES).[29] The FMLN's January 1981 General Insurrection attempted to provoke a popular uprising in the country.[30] Two months later, the first 7th SFG Mobile Training Team arrived in-country in a military advising and assistance effort designed to assist FAES in responding to an asymmetric threat.

The 7th SFG had a legacy to build on but faced considerable obstacles impeding their military advice and assistance mission. Less than twenty years previously, 8th Special Forces Group teams operating under President Kennedy's hemispheric Alliance for Progress programme helped established a FAES airborne unit to counter FMLN mobility in the countryside.[31] Since then, however, the FAES had ossified thanks to officer intransigence, endemic corruption, and recruits abused by their superiors.[32] The FAES and El Salvadoran police routinely committed human rights abuses among the peasants, driving a portion of the population into the FMLN's arms.

Cuban and Vietnamese support for the FMLN turned it into a highly capable insurgent force, divided into various units including their own self-identified SOF, the special forces.[33] These specialized forces were divided into three categories: sappers, or forces designed to conduct assaults against heavily defended targets, frogmen, or combat swimmers, and urban commandos.[34] Some of their attacks were noteworthy. For example, eight commandos, with the assistance of two support personnel and employing special training received from Cuban advisors, destroyed fifteen helicopters and aircraft on Ilopango Air Base on 27 January 1982, some recently provided by the Reagan administration.[35] While host nation and guerrilla forces posed their own unique problems, perhaps the most considerable obstacle to the military advice and assistance was political and legal, in the form of resource constraints.

Libraries, The Portal to Texas History, https://texashistory.unt.edu, crediting Abilene Christian University Library; J. Paul de B. Taillon, *The Evolution of Special Forces in Counter-Terrorism: The British and American Experiences* (Westport, CT: Praeger, 2001), p. 115.

[28] José Angel Moroni Bracamonte and David Spencer, *Strategy and Tactics of the Salvadoran FMLN Guerrillas: Last Battle of the Cold War, Blueprint for Future Conflicts* (Westport, CT: Praeger, 1992), p. 7.

[29] Charles Briscoe, 'El Paraiso and the War in El Salvador, Part I (1981–1983)', *Veritas*, 3:1 (2007): 12–13; 15–16.

[30] Briscoe, 'El Paraiso and the War in El Salvador, Part I (1981–1983)', 16.

[31] Charles Briscoe, 'Building El Salvador's Airborne, Part I', *Veritas*, 4:1 (2008): 64–74.

[32] Joseph Stringham and Charles Carlton, interview with Joseph Stringham, US Army War College/Military History Institute Senior Officer Oral History Program, Session 1, Part 1, 29 May 1985, p. 28; John Waghelstein, 'El Salvador: Observations and Experiences in Counterinsurgency', unpublished programme paper, Unclassified/Limited Distribution, US Army War College, 1 January 1985, pp. 51–56.

[33] Bracamonte and Spencer, *Strategy and Tactics of the Salvadoran FMLN Guerrillas*, pp. 53–59; 67–71.

[34] David Spencer, *From Vietnam to El Salvador: The Saga of FMLN Sappers and Other Guerrilla Special Forces in Latin America* (Westport, CT: Praeger, 1996), pp. 4–9.

[35] Spencer, *From Vietnam to El Salvador*, pp. 80–82; Andrea Oñate, 'The Red Affair: FMLN–Cuban Relations during the Salvadoran Civil War, 1981–92', *Cold War History*, 11:2 (2011): 133–154.

Congress controlled the military advisory and assistance mission in El Salvador, including its special operations element, by limiting its capability and capacity. Only fifty-five advisors were authorized to operate in the Military Group (MILGRP) in the country. MILGRP personnel, however, found several methods to work around the personnel caps imposed by Congress, as these limits were only for those on one-year tours, not headquarters, support staff, or those on temporary duty.[36] 7th SFG rotated forces in-country on shorter tours, of between six and twelve weeks, for the purposes of conducting counterinsurgency (COIN) training. In comparison to their counterparts from other government agencies, special forces personnel 'were more reckless and more concerned with going where the action was'.[37]

Nevertheless, by 1984 special forces teams had made progress in training FAES units. The MILGRP could shift to the next of its training plan, sustainment, which required greater emphasis on advanced planning or larger-scale units as well as medical and other humanitarian support.[38] Although some issues regarding tactical execution still existed, the results of the training and equipping programme for the FAES was impressive: 'the army tripled in size, its soldiers were issued modern weapons and equipment, and the number of helicopters and counter-insurgency relevant fixed-wing aircraft in the Air Force increased sixfold'.[39]

Most participants and outside observers suggest the control measures put in place helped rather than harmed the special operations effort in El Salvador. In one former MILGRP commander's words, 'keeping the numbers small (US military involvement) has been one of the successes' given 'control is an important factor and I knew where every single person was all of time' by providing '[b]etter reports, more accurate information on what you need to do, and frequently better intelligence are additional spin-offs of small numbers'.[40] Additional constraints played a role in fostering the integration of US special operations with other efforts in the country, particularly 'the nature of the conflict' which 'demanded interagency mutual support', the unique strengths and attributes of the MILGRP and its primary interagency partner in El Salvador, the US Agency for International Development (USAID).[41] Other factors contributed to the eventual outcome as well, including the FMLN's appropriately named 'Final Offensive' in 1989 that broke its back militarily and the murder of civilians by a FAES battalion. The latter put considerable external political pressure on the Salvadoran government.[42] Constraints on military aid were necessary in prompting the Government of El Salvador as well as FAES to reform the military and

[36] Joseph Stringham and Charles Carlton, interview with Joseph Stringham, US Army War College/Military History Institute Senior Officer Oral History Program, Session 1, Part II, 29 May 1985, pp. 36–38.

[37] 1 Brigade OPATT Bob Watson, quoted in Scott Moore, 'Gold, Not Purple: Lessons from USAID–USMILGRP Cooperation in El Salvador, 1980–1992', unpublished MA thesis, Naval Postgraduate School, December 1997, p. 3 n5.

[38] Charles Briscoe, 'San Miguel: The Attack on El Bosque', *Veritas*, 3:3 (2007): 19.

[39] Walter Ladwig III, *The Forgotten Front: Patron–Client Relationships in Counterinsurgency* (Cambridge: Cambridge University Press, 2017), p. 244.

[40] Joseph Stringham and Charles Carlton, interview with Joseph Stringham, US Army War College/Military History Institute Senior Officer Oral History Program, Session 1, Part II, 29 May 1985, p. 38.

[41] Moore, 'Gold, Not Purple', p. 90.

[42] Ladwig III, *The Forgotten Front*, pp. 275–276.

hold it accountable.[43] Both sides grudgingly but willingly negotiated ceasefires and, in 1992, a UN-brokered peace agreement.

From 1986 until 1994, US special operations had a mixed record. One blunder was a failed covert action, conducted at the executive level of government, to work around Congressional restrictions of the Boland Amendment which prohibited funding to the Nicaraguan Contra rebels: 'Iran-Contra'.[44] It resulted in Congress further restricting Executive power, was less consequential for special operations than events in Somalia in 1993. The details of the events surrounding Operation GOTHIC SERPENT are now well known through the film *Black Hawk Down*. The mission, however, is synonymous with 'mission creep'. Responding to images of starving Somali civilians caught up in a civil war in a failed state, and with the violence escalating, President Clinton authorized the dispatch of US Army and Marine Corps to secure corridors for dispatch of humanitarian aid and help stabilize the country. Distribution of aid and its security was provided by United Nations peacekeeping forces.

On 5 June 1993, forces of the Somalia National Alliance (SNA) ambushed and killed two dozen Pakistani peacekeepers, prompting the dispatch of Task Force Ranger on 22 August 1993.[45] The mission of the Task Force was to capture and bring to trial senior members of the SNA, including its leader, Mohammed Farah Aideed. After several successful capture missions, and despite significant command-and-control problems, a raid by Task Force elements in the Bakara Market area of Mogadishu on 3 October 1993 was met by coordinated, local resistance on the return trip.[46] Task Force casualties, which were 'heavy' by special operations standards (sixteen killed and fifty-seven wounded), paled in comparison to those inflicted on the Somalis.[47] Photographs and video footage of cheering Somalis dragging the bodies of two soldiers and dancing on the burned-out hulk of a helicopter had an immediate effect on the American public and its elected leaders. The political and military fallout from GOTHIC SERPENT was significant: it 'spooked the Clinton administration as well as the brass [senior military leadership], and confirmed the Joint Chiefs in the view that SOF should never be entrusted with independent operations'.[48]

This brief background on modern US special operations highlighted special operations successes and failure up to 1996. The successes were driven largely by the type and nature of the specific threat confronting the United States, for which other existing capabilities, including conventional and later nuclear forces, were

[43] Ibid., pp. 284–287.

[44] For details, see John Tower, *The Tower Commission Report: The Full Text of the President's Special Review Board* (New York: Bantam/Times Books, 1987).

[45] Richard Stewart, *The United States Army in Somalia, 1992–1994*, CMH Pub 70-81-1 (Washington, DC: US Army Center of Military History, 2002), p. 18.

[46] Allard outlines the nature of the command-and-control problems, which included dual-hatting and parallel chains of command for US forces, which were further complicated by being outside UN control. Kenneth Allard, *Somalia Operations: Lessons Learned* (Washington, DC: National Defense University Press, 1995), pp. 57–59.

[47] Richard Stewart, *The United States Army in Somalia, 1992–1994*, CMH Pub 70-81-1 (Washington, DC: US Army Center of Military History, 2002), p. 23.

[48] Richard Shultz Jr, 'Showstoppers: Nine Reasons Why We Never Sent Our Special Operations Forces after al Qaeda before 9/11', *The Weekly Standard* (26 January 2004): 29.

too costly or unsuited. Special operations capabilities provided a means to address various threats—Soviet-sponsored treachery and subversion, in the form of state and non-state proxies, including insurgent and terrorist groups like the Viet Cong and FMLN—on an equal footing, either matching or countering their asymmetric approaches. This survey identifies how prior successes and failures shaped the capabilities and control measures in place by 1996, the subject of the next two sections.

## Capabilities

Special operations works are dominated by discussion of American units or roles and missions, as Chapter 2 made clear. Catalogues of these unit and roles and missions are available elsewhere.[49] Army doctrine published in 2012 identified two special operations 'capabilities': surgical strike and special warfare.[50] Their development reflected concerns regarding undue SOF focus on surgical strike. The existing lexicon 'was at best vague, and at times pejorative towards the practitioners of special warfare'.[51] Creating new terminology 'was a first step in remedying the long-standing and *unhealthy* [emphasis added] bifurcation of Army special operations'.[52] The split was further characterized by 'the nature of the funding, command-and-control structures, authorities, or operational approach'.[53] Changing terminology would allow Army SOF to evolve and 'address capability gaps and improve integration at the "seams" of capability between the interagency and conventional forces, to better enable the seamless application of combat power across the spectrum of responsibility'.[54] In addition, it would redress the atrophying of an Army SOF role central to their culture and identity: unconventional warfare.[55]

The terminology is useful for theory. Special warfare is defined 'that form of special operations in which the United States government defends its interests through training, material or even direct combat support to indigenous people and friendly governments whose interests coincide with those of the United States'.[56] Surgical

[49] See, for example, David Tucker and Christopher Lamb, *United States Special Operations Forces*, 2nd. ed. (New York: Columbia University Press, 2020), especially Chapters 2 and 6 and appendix 1; various USSOCOM posture statements, or semi-annual public reports to Congress, until their elimination and replacement by written testimony in 2008; the four editions of the Joint Special Operations University's *Special Operations Reference Manual*, whose last update was 2023; and various Congressional Research Service and Government Accounting Office reports. Previous noteworthy contributions include Thomas Adams, *US Special Operations Forces in Action: The Challenge of Unconventional Warfare* (London: Frank Cass, 1998), pp. 207–208; 210–214; John Collins, *Green Berets, SEALs, and Spetsnaz: U.S. and Soviet Special Military Operations* (London: Pergamon-Brassey's, 1987).

[50] The term 'surgical operations' was first used in 1982 of six cases in the use of military force in a limited capacity, four of which were special operations or led by them. See their definition and case study list in David Charters et al., 'Surgical Operations: Military Lessons from Selected Case Studies', unpublished report, The Centre for Conflict Studies, University of New Brunswick, Fall 1982, p. 2.

[51] Charles Cleveland and Daniel Egel, *The American Way of Irregular War: An Analytical Memoir*, RAND Publication PEA301-1 (Santa Monica, CA: RAND, 2020), p. 175.

[52] Cleveland and Egel, *The American Way of Irregular War*, p. 175.

[53] US Army Special Operations Command [USASOC], *ARSOF 2022: Part 2* (Fort Bragg, NC: US Army Special Operations Command, 2014), p. 4.

[54] USASOC, *ARSOF 2022: Part 2*, p. 17.

[55] Ibid., p. 6.

[56] Ibid., p. 12.

strike, in contrast, 'unilaterally or collaboratively ... extends operational reach and influence by engaging global targets discriminately and precisely. Strike operations are not always intended to be an isolated activity'.[57] These capabilities differ in terms of their intelligence, planning, training requirements, location of presence, and duration, although they share a common element: influence. Surgical strike sets the conditions for others to better comprehend and act against state and non-state threat networks, while special warfare seeks to influence populations to act in ways that support US national security priorities and objectives. At its simplest level, surgical strike special operations are episodic and reactive while special warfare ones are sustained and proactive. For reasons stated in Chapter 3, identifying surgical strike and special warfare as capabilities, as *ARSOF 2020* does, is misleading. What they better capture are overall categories of missions under which different, legally identified US roles and missions fall. In addition, these categories have an additional, yet unexploited value. When connected to the discussion of orientation and bearing in Chapter 5, they permit the inclusion of CIA special operations which occur covertly or clandestinely on a much smaller scale in both.

The historical mission sets from Chapter 2 map well, if incompletely, against the overarching categories of surgical strike and special warfare. The roles and missions most frequently ascribed to surgical strike, including direct action (raids), counterterrorism (CT), counterproliferation (CP), and recovery operations (hostages or materials), are indirect-shock missions. The forces useful in conducting such missions are 'uniquely designed to infiltrate austere, remote locations and rapidly mass combat power to seize, destroy, capture or recover designated targets and exfiltrate in a single period of darkness' in 'politically sensitive environments', including denied territory, 'ranging from sophisticated nation states with advanced anti-access, cyber, and defensive capabilities to (violent extremist organizations) operating inside failed states'.[58] The location of targets for surgical strike requires special reconnaissance (SR) missions in advance to provide sufficient information required to allow detailed mission planning, rehearsal, and execution, by personnel or uncrewed systems, in high-risk environments, elevating them to the operational and strategic level of importance. The combination of advanced information, equipment, training, and rehearsal is summarized in activities 'built to squeeze out uncertainty ... then execute—at the speed of war'.[59] Tactically, the missions are a form of asymmetric response by striking swiftly, precisely, and effectively in locations thought by an opponent to be secure or sufficiently defensible, given their location or forces assigned to protect them, giving them their unconventional or 'asymmetric' character.[60]

[57] Ibid., p. 14.

[58] Ibid., p. 15.

[59] Ibid., p. 16; US House of Representatives, House Armed Services Committee, Subcommittee on Emerging Threats and Capabilities, 'Statement of General Raymond A. Thomas, III, U.S. Army, Command, United States Special Operations Command', *Three Decades Later: A Review and Assessment of Our Special Operations Forces 30 Years after the Creation of U.S. Special Operations Command*, HASC No. 114-40 (Washington, DC: Government Printing Office, 2017), p. 43.

[60] James D. Kiras, *Special Operations and Strategy: From World War II to the War on Terrorism* (London: Routledge, 2006), p. 5.

Special warfare missions, in contrast, are military advising and assistance missions. The current roles and missions that map onto these are unconventional warfare (UW), foreign internal defence (FID), and counterinsurgency (COIN), with security assistance (SA) and humanitarian and disaster relief (HADR) forming tertiary roles. The three primary special warfare missions—UW, FID, and COIN—are long duration and differ in bearing and orientation from surgical strike missions. UW is offensive in nature, tactically to strategically, while FID and COIN primarily defensive strategically and operationally.[61] In addition, UW missions are conducted covertly and clandestinely, to keep the hand of the sponsor hidden for as long as possible, while FID and COIN missions are overt given the need to work with host nation security forces in an official capacity.[62] Part of this reflects their different operating environments. UW is continued in denied or contested territory or that over which state security forces often impose a high level of control; bearing, in the form of covert or clandestine presence, is done as much for SOF survival as it is political sensitivity.

FID and COIN, in contrast, are conducted where the host nation exerts the control over territory, or a much more permissive operating environment. They differ, however, in scale. FID missions tend to be conducted in circumspect manner, in terms of numbers of presence and often in the form of military training teams (MTTs), for one reason. The host nation may, given international and domestic political sensitivities, not want to broadcast the need for external assistance much less assistance from the United States. COIN, in contrast, involves military advising and assistance on a significant scale, specifically when host nation security forces need to be built from scratch, as occurred in Iraq and Afghanistan.[63] Here special operations play a subordinate role to conventional forces and often focus their efforts on training and advising host nation SOF. Special warfare is best understood as operations 'designed to wade into uncertainty ... and prevail, at the speed of partnered operations'.[64]

The historical special operations mission sets explain the similarities and differences between UW and FID and COIN. The combination of the non-state actor and their mission and purpose, subversion and sabotage through active insurgency, places UW squarely within the realm of asymmetric responses—in this case, doing so given the power of the state its forces are confronting. FID and COIN, in turn, conduct a variation of this response. Instead of an asymmetric one, FID and COIN missions are designed as a symmetrical response to the threat of an insurgency, countering irregulars by creating irregulars for the state. The historical mission sets

[61] A recent caveat to this statement is the return of 'resistance' into the special operations lexicon, which assumes an initial posture of defence strategically against a foreign invasion. For more, see Otto Fiala, Resistance Operating Concept (ROC) (MacDill Air Force Base, FL: JSOU Press, 2020), p. 1.

[62] The Joint Staff, *Joint Publication 3-05.1, Unconventional Warfare,* FOUO (Washington, DC: The Joint Staff, 2015), appendix A, Legal Considerations, p. A-1.

[63] For an insightful critique of COIN doctrine, specifically its underlying assumptions, see M. Chris Mason, 'COIN Doctrine Is Wrong', *Parameters*, 51:2 (Summer 2021): 19–34.

[64] USASOC, *ARSOF 2022: Part 1* (Fort Bragg, NC: US Army Special Operations Command, 2013), p. 16; US House of Representatives, House Armed Services Committee, Subcommittee on Emerging Threats and Capabilities, 'Testimony on Special Operations Forces', *The Future of Special Operations Forces*, HASC No. 112–139 (Washington, DC: Government Printing Office, 2014), p. 33.

offer a useful framework, in this case to separate and distinguish between roles and missions where confusion might exist. For example, categorizing UW, FID, and COIN primarily by the overall nature of the activity, military advising and assistance, acknowledges the aphorism that these roles and missions are 'two sides of the same coin', in terms of their overarching mission genus but different species in terms of their subordinate, functional missions. In addition, it places the emphasis for these special operations on where they should be: on the proxies, including sponsored groups and host nations, and the support and assistance provided to them, as opposed to the special operators themselves, an error evident in otherwise informed and insightful sources.[65] Underscoring mission sets in this way avoids a historical affliction with American special operations: impatience, primarily in pushing the proxy aside, subverting their agency or authority, or using them as a façade to conduct combat operations themselves. It does so by keeping in mind 'who pays the bill and who does most of the bleeding'.[66]

Separating and identifying American special operations according to historical mission sets clarifies relationships with other roles and missions, units, and capabilities as well. For example, intelligence and reconnaissance are both integral to surgical strike and special warfare but does not rise to the equivalent level of indirect shock or military assistance and advising. In addition, other SOF roles and missions nest as other supporting activities to primary mission sets. These include missions conducted by uniquely American special operations units such as Civil Affairs and Psychological Operations, whose activities primarily support special warfare military advising and assistance missions: civil affairs (CA) and HADR for the former and military support to information operations (MISO), primarily in the form of tactical or battlefield actions, in the case of the latter.

A final special operations mission category excluded thus far is covert action. Covert action is the purview of the CIA, according to legal definition and statute authority. These authorities are discussed subsequently. Covert action according to US Code is: 'an activity or activities of the United States Government to influence political, economic, or military conditions abroad, where it is intended that the role of the United States Government will not be apparent or acknowledged publicly'.[67] The scope of activities conducted as covert action most close resemble those under special warfare closely identified with UW, including training, equipping, and supporting proxy groups and militias. Those groups and militias can conduct a range of activities falling under the historical mission sets of treachery and subversion and indirect shock, ranging from propaganda and resistance through sabotage and targeted killing to regime overthrow.

Based on the preceding discussion, Figure 8.1 depicts the three overarching mission sets, their brief 'bumper sticker' descriptions, an overall description of what

[65] See, for example, Colin S. Gray, '"Doorkickers" and "Well-Diggers": Resolving the Tensions between SOF Roles', unpublished report, MDA 902-91-C-0030, National Institute for Public Policy, October 1992, pp. 4–5.

[66] Waghelstein, 'What's Wrong in Iraq?', p. 117.

[67] 50 USC §3093(e).

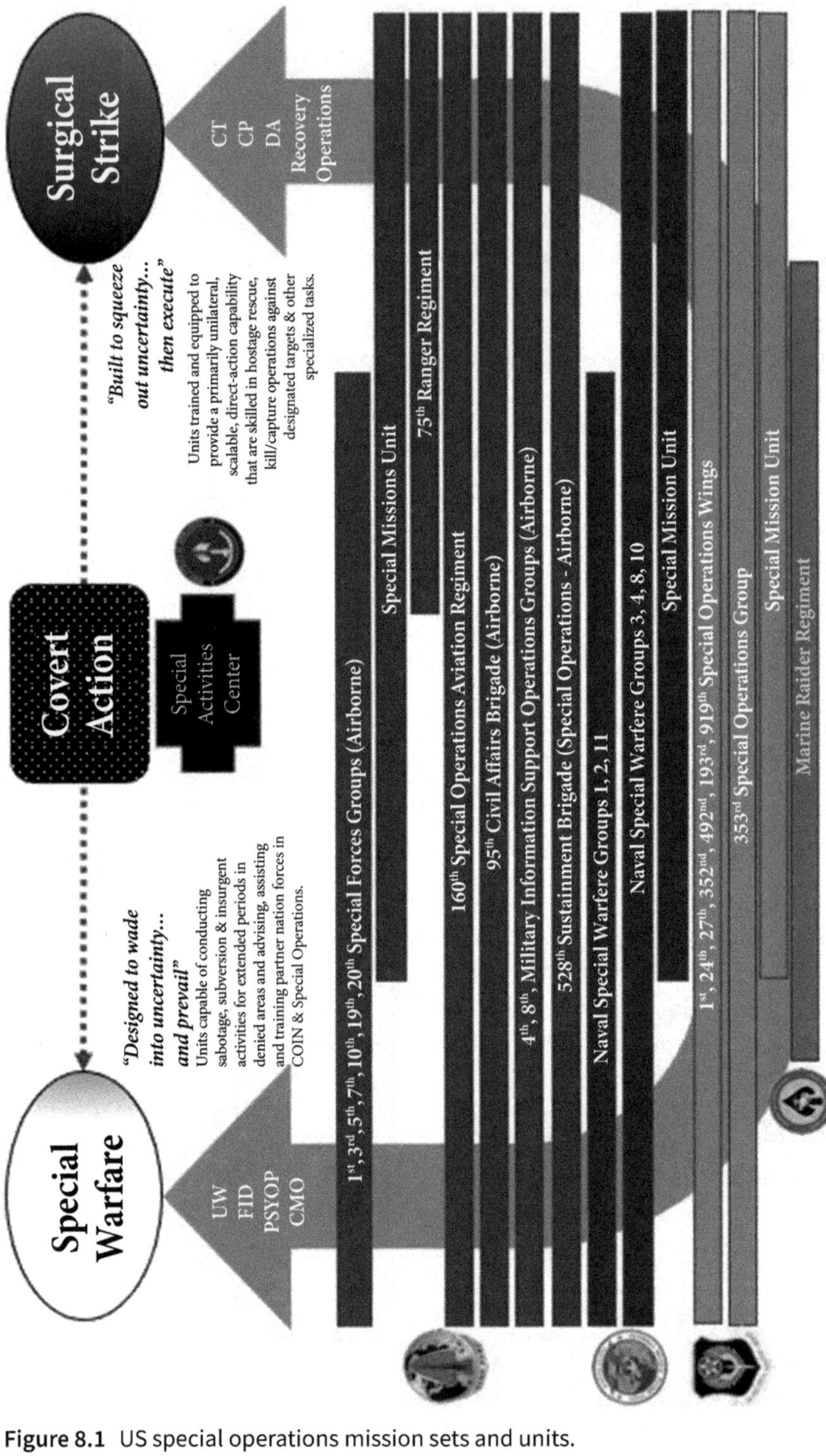


**Figure 8.1** US special operations mission sets and units.

*Source:* Author's work, adapted from 'mutually supporting special operations' graphic contained in *ARSOF 2022: Part 1* (Fort Bragg, NC: US Army Special Operations Command, 2013), p. 16.

units do within them, and specific missions as defined doctrinally underneath them. The double-ended arrow suggests that while each mission set has unique traits and characteristics, and can be conducted independent of each other, they can be mutually supporting rather than mutually exclusive. The same is true for covert action, as the subsequent case study analysis makes clear. Overlaid against these mission sets are various service component unit organizations, separated by colour and whose positioning indicates in which mission set their training, equipping, and exercising focus resides. Those unit organizations spanning both ends of the mission set spectrum provide support across it to other units, mostly but not exclusively on a service basis. Figure 8.6 illustrates more generally how US SOF capabilities and functions related to threats and opponents in 1996, as well as the subject of the next section, control.

## Control

Political leaders looking to control SOF can also legally constrain their permitted actions. In the American case, control is exerted through authorities ranging from fiscal and permanent to temporary and operational. Most but not all are fiscal authorities, which determines who receives control of what allocated funds, as well as how they can be spent. These authorities start by 'title' in US Code. The most widely cited title and section for US special operations is Title 10 § 167, created because of the 1987 Nunn-Cohen Amendment to the Goldwater-Nichols Act and codified in 1987 Defense Authorization Act. This section identifies special operations roles and missions, its executive agent (the unified command, USSOCOM), and its unique budget, Major Force Program or MFP-11, for 'SOF-peculiar' requirements, or purchases, as opposed to 'Service-common' ones funded instead by MFP-2.[68]

Fiscal authorities ensure that appropriated funds are spend only for the specified purpose, determined and monitored by DoD Comptroller-specified financial management regulation (FMR).[69] They are necessary to guide USSOCOM in one of its two designated roles: as a force provider and development authority to organize, train, and equip SOF. The web of force provider authorities is depicted in Figure 8.2.

[68] Susan Marquis, *Unconventional Warfare: Rebuilding U.S. Special Operations Forces* (Washington, DC: Brookings Institution Press, 1997), p. 146.

[69] Federal appropriations law guidelines and principles are covered in a three-volume set known informally as the 'Red Book' and are issued by the General Accounting Office. For an example of one such volume, temporary authorities, specified in annual or semi-annual budget authorizations, can become permanent. At times, Congress approves temporary authorities based on an identified emerging need not covered under another fiscal authority. Such authorities can be made permanent, either individually or bundled together, as occurred with Section 1208 (2006 NDAA), which became 10 USC § 127e, and 10 USC § 333, which gathered various authorities for building partner capacity for a variety of different missions, including CT, C-WMD, and counternarcotics (CN), under one. See Office of the General Council, *Principles of Federal Appropriations Law, Third Edition, Volume II* (Washington, DC: General Accounting Office, 2006), www.gao.gov/assets/2019-11/202819.pdf.

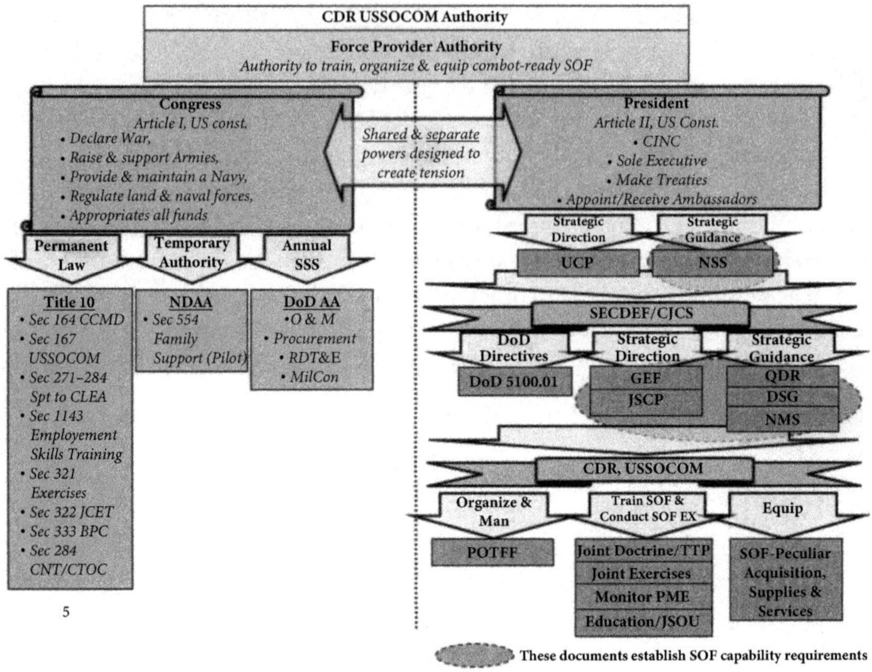


**Figure 8.2** USSOCOM force provider authorities, 2017.

*Source:* 'Force Provider Authorities', SOF Authorities Pocket Reference Guide (Tampa, FL: USSOCOM Headquarters, 2016). Author's collection.

Operational authorities exist alongside fiscal ones. In the simplest terms, they are specified for carrying out tasks and are often bounded by role, mission, region, or target. The authorization authority is specified in US Code and for special operations can range from the president, as is the case of a classified one for covert action or a lethal finding for the CIA under Title 50, or secretary of defense or chairman of the joint chiefs of staff (CJCS) for Title 10 ones. Title 10 authorities for SOF include the assignment of missions and their planning and conduct, trans-regional authorities for SOF activities, as well as support to others, including partner nations, proxy forces, as well as domestic civil authorities, including law enforcement agencies.

Various plans, orders, and programmes authorize the first two, including secretary of defense or CJCS tasking through overarching execute orders (EXORDs), plans, and contingency plans. Examples include a range of CT and preparation of the environment for C-WMD EXORDs. More specific, regionally focused EXORDs derived from these are developed by Combatant Commands (COCOM) for specific tasks in their area of responsibility. The contents of EXORDs are classified, so specific examples are problematic. Press accounts suggested a COCOM EXORD signed by General David Petraeus, first reported in the *New York Times* in 2010, was alleged to 'break [DoD's] dependence on the Central Intelligence Agency and other spy agencies' to send military forces to gather their own information and identify useful proxies.[70] COCOM EXORDs provide the basis for operations orders (OPORDs) and the creation of named operations, which range in scope, size, and number.[71]

Contingency plans (CONPLANs) are designed to provide rapid responses by delegating authority for planning and conduct of missions, on order, for specific tasks. Most related to special operations are classified but some notion of their scope can be gleaned from open sources.[72] Approval authority for certain tasks can be delegated down the chain of command, from secretary of defense and CJCS directly to USSOCOM, COCOM commanders, or agency directors through DoD Directives (DoDD) and Instructions (DoDI) or secretary of defense or their deputy's memoranda, as is the case of the conduct of CA missions within COCOMs (DoDD 2000.13) and approval authority to USSOCOM to provide support to civil authorities, including law enforcement (DoDD 3025.18, DoDI 3025.21, and various secretary of defense memoranda). The latter are particularly important because misperceptions exist, even within the military, about limitations to their activities domestically under the Posse Comitatus Act (1867). Military forces, including SOF, can be used but only under certain extreme conditions and circumstances.[73] Figure 8.3 outlines the hierarchy and structure of USSOCOM's operational authorities.

Authorities can be interpreted or extended in ways that benefit agents, operating beyond the intent of the principal. To guarantee compliance and ensure preference alignment, as opposed to divergence, principals must take two actions: create institutions and delegate resources for the expensive task of monitoring, and codify rules and regulations in the form of standard operating procedures. In the latter condition, expectations go up and down a chain of authority, or command, whereas in the former they serve to investigate possible divergence and notify principals. Monitoring for special operations, specifically for their fiscal and operational authority, exists in the form of various organizations and processes. For fiscal authority, the Office of Management and Budget (OMB) serves the purpose for the executive branch, the Congressional Budget Office (CBO) for the legislative, and within the DoD, the Capabilities and Program Evaluation (CAPE) office of the Office of the Secretary Defense. These offices review budget submissions and expenditures in general (OMB, CBO) and programmes and requirements (CAPE). To ensure compliance with regulations, USSOCOM has various mechanisms, including a Board of Directors (BOD) who provide 'comprehensive, concise, written guidance' to staffs, Assessment Directors (ADs) to provide 'a joint approach to capability assessment'

[70] Mark Mazetti, *The Way of the Knife: The CIA, a Secret Army, and a War at the Ends of the Earth* (New York: Penguin, 2013), pp. 206–208.

[71] 'Military Operations and Activities Necessary for the Safety of Life or the Protection of Property (National Security)', FOUO, 6 February 2018, https://irp.fas.org/agency/dod/milops-2018.pdf; Sharon Weinberger, Sean Naylor, and Jenna McLaughlin, 'Pentagon Launched New Classified Operation to Support Saudi Coalition in Yemen', Yahoo News (10 November 2018), www.yahoo.com/news/pentagon-launched-new-classified-operation-support-saudi-coalition-yemen-221305500.html.

[72] 'Distinguished Member of the Special Forces Regiment, Honorable Michael G. Vickers, Inducted August 2010', US John F. Kennedy Special Warfare Center and School, n.d., www.swcs.mil/Portals/111/sf_vickers.pdf.

[73] Joseph Nunn, 'The Posse Comitatus Act Explained', Brennan Center for Justice, 14 October 2021, www.brennancenter.org/our-work/research-reports/posse-comitatus-act-explained.

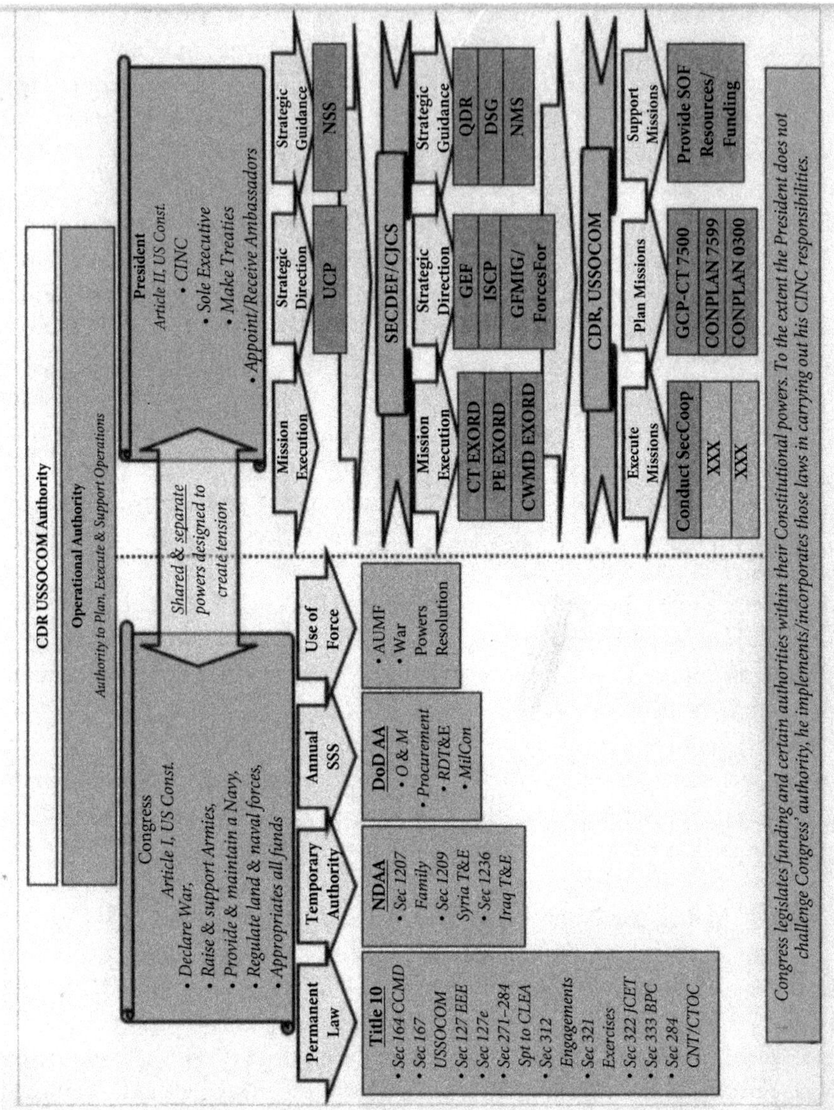


**Figure 8.3** USSOCOM operational authorities, 2017.

*Source:* 'Operational Authorities', SOF Authorities Pocket Reference Guide (Tampa, FL: USSOCOM Headquarters, 2016). Author's collection.

and 'priority programs and make recommendations to the BOD', as well as its own, unique Strategy Planning Process (SPP).[74] These mechanisms serve three goals: identify requirements as well as potential capability shortfalls, validate and priorities requirements by rank ordering them, and, more broadly, manage the MFP-11 'checkbook' responsibly.

[74] J. Michael Murphy Jr, 'Strategy-to-Task: USSOCOM's Strategic Planning Process, 1988–1996', unpublished paper, USSOCOM History and Research Office, 19 November 1997, pp. 19–20.

Other organizations and processes monitor the use of special operations operational authority. Suspected violations can be reported to any one of the following: component command inspector general offices (ARSOC, SPECWARCOM, Air Force Special Operations Command (AFSOC)), a unified command inspector general at USSOCOM headquarters (SOIG), COCOM inspector generals (per DoDD 5106.04), and up through the chain of command including the Office of the Secretary of Defense. The CIA also has its own inspector general. Federal guidelines dictate that offices conduct so-called 'peer reviews' periodically to ensure systemic and specific compliance of policies and processes. For example, the DoD Inspector General conducted a peer review audit of the monitoring office within SOIG in 2022, five years after a previous one, and found the latter to be compliant in its standards and reporting.[75] To prevent excesses from occurring, as well as to provide advice on recommendations of what is possible under existing authorities, special operations planners and staff can seek legal counsel ranging from the Office of General Council in the Pentagon down to the lowest level of control, the Joint Special Operations Task Force (JSOTF).[76] Although the reporting chains are clear, they are not always transparent. The sensitive nature of special operations ensures that most monitoring and reports remain classified, as a review of the reports listed on the DoD Inspector General's website suggests.[77]

In addition to monitoring elements, the legislative branch also imposes reporting requirements for special operations. Operational issues, as they pertain to new and ongoing activities, are reported on a quarterly basis primarily to select members of the HPSCI (House Permanent Select Committee on Intelligence) and SPSCI (Senate Permanent Select Committee on Intelligence). For one category of special operation, covert actions by the CIA, the president is responsible for disclosing the signing of a finding—its approval on his or her authority—in writing to Congress as soon as possible. In exceptional circumstances, to limit its distribution based on the sensitivity of the special operations, the president may invoke conditions requiring him or her to report only to select members of Congress, known as 'the Gang of Eight' and 'comprised of the top four leaders in Congress and the head Republican and Democrat from both the Senate and House Intelligence committees'.[78] This includes covert action using cyber means. Clandestine intelligence activities have no specified

[75] Inspector General, US Department of Defense, *External Peer of the United States Special Operations Command, Office of Inspector General, Audit Branch*, DODIG-2022-124 (Washington, DC: Department of Defense, September 2022), pp. 4–5.

[76] The Joint Staff, *Joint Publication 3-05.1: Joint Special Operations Task Force Operations* (Washington, DC: The Joint Staff, April 2007), appendix D: Joint Special Operations Task Force Headquarters Staff Checklists, pp. D-8; D-12; D-26; D-34.

[77] Review conducted 17 October 2023. Of the reports listed and available, only one was not redacted. The vast majority remain classified but can be requested for Freedom of Information Act (FOIA) review and release. The sensitive nature of special operations is evident in the titles of two IG report: 'Management Advisory: Evaluation of U.S. Special Operations Command's Joint Military Information Support Operations Web Operations Center (DODIG-2023-038)' and 'Investigation of a Questionable Intelligence Activity ASD/SOLIC/IC (Classified)'. www.dodig.mil.

[78] Michael DeVine, *Covert Action and Clandestine Activities of the Intelligence Community: Selected Congressional Notification Requirements*, Report R45191 (Washington, DC: Congressional Research Service, 2023), pp. 2–3; Rebecca Beitsch, '"Gang of Eight" Now Has Access to Classified Documents Found at Trump, Biden and Pence Homes', *The Hill*, 11 April 2023.

timeframe for reporting and are limited to half the members of the Gang of Eight: the chairs of the HPSCI and SPSCI, the ranking member of the former, and the vice chair of the latter. This group is known as either 'The Four Corners' or the 'Gang of Four'. An exception to the reporting period is for DoD sensitive (i.e., covert or clandestine) activities as part of 'operational preparation of the environment' (OPE), based on a fiscal threshold, which the Assistant Secretary of Defense for Special Operations and Low Intensity Conflict (ASD/SOLIC) reports on a quarterly basis to the Gang of Four.[79]

Other DoD operational activities must be reported to members of the HASC (House Armed Services Committee) and SASC (Senate Armed Services Committee) depending on their type. These include sensitive military activities targeting individuals, colloquially known as 'kill or capture' missions, those conducted in preventative self-defence against an imminent threat to members of the DoD or its partners, or for hostage rescue missions. In this case, the notifying authority to Congress is the secretary of defense, who must provide a written record within two days following his or her oral notification or once an activity is disclosed publicly. CT operations, in contrast, are reported on a monthly basis and include an impact assessment as well as summary of the 'authorities and legal issue as well as any related interagency activities'.[80] The nature of these activities ensures the briefings and reporting remain classified and therefore not open to public scrutiny, to protect information sources, methods of operation, and politically sensitivities.

Unlike its operational activities, which remain classified, USSOCOM's force provider/force development functions are a matter of public record. These include annual testimony or submissions of posture statements by the Commander, USSOCOM, to HASC and SASC subcommittee hearings on issues specific or related to special operations.[81] Members of the subcommittees are free to provide statements for the record in advance, expressing their concerns, as well as question the Commander, USSOCOM or ASD SO/LIC.[82] ASD SO/LIC also serves, metaphorically, as a canary in a coal mine for the military SOF community, raising to Congress concerns related

[79] Michael DeVine, *Covert Action and Clandestine Activities of the Intelligence Community: Selected Congressional Notification Requirements*, Report R45191 (Washington, DC: Congressional Research Service, 2023), pp. 4–5.

[80] DeVine, *Covert Action and Clandestine Activities of the Intelligence Community*, pp. 8–10.

[81] For a recent example, see US House of Representatives, House Armed Services Committee, Subcommittee on Intelligence and Emerging Threats and Capabilities, 'Statement for the Record, Mr. Mark E. Mitchell, Principal Deputy Assistant Security of Defense, Special Operations and Low-Intensity Conflict', 9 April 2019, https://congress.gov/116/meeting/house/109267/witnesses/HHRG-116-AS26-Wstate-MitchellM-20190409.pdf; also 'Fiscal Year 2024 Budget Estimates, United States Special Operations Command', March 2023, DoD Comptroller, https://comptroller.defense.gov/Portals/45/Documents/defbudget/fy2024/budget_justification/pdfs/01_Operation_and_Maintenance/O_M_VOL_1_PART_1/SOCOM_OP-5.pdf.

[82] For an example, see Jack Reed's statement expressing concerns about insufficient reporting or action on 'misconduct cases', as well as accountability and retention. US Senate, Senate Armed Services Committee, Defense Authorization Request for Fiscal Year 2024 and the Future Years Defense Program, 'Opening Statement of U.S. Senator Jack Reed, Chairman, Senate Armed Services Committee', 7 March 2023, www.armed-services.senate.gov/imo/media/doc/SOCOM_CYBERCOM%20Posture%20Hearing%203.7.23_final.pdf.

to Services not fulfilling their obligations or taking detrimental actions. While there were sound historical reasons and evidence behind such concerns, there has been little recently, save for recent Army signalling of shedding Army SOF personnel.[83] Given its legislative and budgetary authority, Congress can impose control on special operations in one of several ways: through commissioning critical studies; imposing new reporting requirements; suspending, revoking, or rewriting operational or fiscal authorities; or constraining fiscal resources, including cancelling programs of record or reducing the budget.

Control mechanisms for American special operations have evolved over time based on legal changes and experience, often resulting from SOF overreaching or exceeding their given authorities as occurred in Vietnam and elsewhere. Others, such as the OASD SO/LIC and its policy representation and reporting authorities, were put in place by Congress to guarantee special operations positions were represented in strategic discussions, as well as protect military ones from abuse and neglect. Congress also ensured paramilitary special operations were not only reported to some of its members but for covert action specifically, the president had to provide a signed authority for it, preventing repeats of Iran-Contra. Roles and missions between military and paramilitary special operations were clearly delineated in US Code. To monitor compliance and report transgressions, a host of Inspectors General exist internally within special operations organizations or their parent services. Given these control measures, at the start of the case study control measures followed a directive model, *acceding* to decision makers' preferences, with SOF capabilities in *congruence* with their expanded functions (depicted in Figure 8.6).

The central element in any discussion of control of American special operations is authorities. Previous analyses frame authorities in broad legal terms: international law, the Constitution of the United States, executive guidance, and legislative statutes.[84] While necessary for cultural understanding, this level of analysis obscures how special operations are actually enabled or controlled. According to one definition, an authority, once conferred by the executive or legislative branch, 'the execution of policy ... an expression of executive or legislative policy, and a request to modify or gain and authority is a request to modify policy'.[85] Authorities, permanent or temporary, provide the domestic legal basis for action by government agents. Permanent authorities are codified by both chambers of Congress in the form of sections of the US Code, the legal body which derives its basis from the founding legal document of the country, the Constitution.

Gaps and seams exist nonetheless and to fill them, SOF actively promote or benefit from the granting of authorities, permanent and temporary. Authorities are rarely discussed in the context of American special operations beyond generalities, which is

[83] Davis Winkie and Leo Shane III, 'Pentagon Briefs Hill on Army Special Ops Cuts, Vacant Slots Targeted', *Army Times* (12 October 2023), www.armytimes.com/news/your-army/2023/10/12/pentagon-briefs-hill-on-army-special-ops-cuts-vacant-slots-targeted.

[84] Colin S. Gray, 'Special Operations and American Culture', unpublished report, MDA 903-91-C-0030, National Institute for Public Policy, December 1992, pp. 13–28.

[85] USSOCOM, 'SOF Authorities Reference Guide', Version 2, January 2017.

unfortunate given their importance. As one operator observed, 'SOF live and breathe based on their given authorities. Master authorities and you determine what's possible with special operations.'[86] SOF engagement in developing and making authorities permanent is part of policy development and fits within one of three process streams identified by John Kingdon: the development of policy proposals.[87] As subsequent sections suggest, US SOF expanded from virtual obscurity to a global juggernaut from 1996 through 2016.

## US special operations: 1996–2016

To make sense of special operations activities that started discretely and expanded globally, this section is divided into seven parts. Rather than recounting specific tactical information or attempt a detailed history of US special operations, the building blocks of this case study exploration are the historical missions and the examples of them, to assess the level of integration present. The seven parts are the period from 1996 to 2001 in a serious of regional activities, Afghanistan from 2001 until 2010, the Philippines from 2001 until 2011, Iraq from 2003 until 2007, global special operations and more local ones in Afghanistan from 2007 until 2016, and Libya, Iraq, and Syria from 2011 to 2016.

In the mid-1990s, the US Armed Services were still recovering from the budget cuts and restructuring designed to reap a peace dividend with the recent end of the Cold War with the Soviet Union. These cuts, roughly $15 billion, occurred despite the overwhelming and, to many, unexpected triumph of American arms in Iraq during Operation DESERT STORM (1991).[88] Future uncertainty and a desire to demonstrate their utility led USSOCOM leaders to take on a variety of additional roles and missions. They ballooned from ten in 1990 to seventeen by 1998, through the addition of ancillary or collateral activities.[89] Such activities included information/command-and-control warfare, countermine, and counter drugs/narcotics (CN). CN blurred lines between military (Title 10) and law enforcement (Title 18), potentially militarizing the latter. In words as sibylline today as then, '[d]iffusion of energy is a continuing threat to the small SOF community. Accordingly, its leaders may want to exercise considerable discipline regarding the scope and intensity of peacetime operations.'[90] By 1996, USSOCOM had managed to restore almost one-third of its budgetary funding lost due to Clinton-era retrenchment, at a time when the overall DoD was still

[86] Mike Curry, conversation with author, 18 February 2019.

[87] John Kingdon, *Agendas, Alternatives, and Public Policies*, 2nd ed. (New York: Longman, 2003), p. 87.

[88] 'Memorandum for the Record, from: Mick Miller, Ray Hall, and Bill Myers, Subject: The Administration's Defense Budget for 1994', Congressional Budget Office, 21 April 1993, www.cbo.gov/sites/default/files/103rd-congress-1993-1994/reports/93doc05b.pdf.

[89] US Government Accounting Office, *Special Operations Command: Progress in Implementing Legislative Activities*, GAO/NSAID-90-166 (Washington, DC: Government Accounting Office, September 1990), pp. 37–38; USSOCOM, *United States Special Operations Forces, 1996 Posture Statement* (MacDill Air Force Base, FL: USSOCOM, 1996), pp. 31–32.

[90] Charles Dunlap Jr, 'Special Operations Forces after Kosovo', *Joint Forces Quarterly, Number 28* (Spring/Summer 2001): 11.

declining, largely by acceding to control.[91] In support of its budget request, US special operations could point to 986 'SOF employments' in 1991 to 2,765 by 1996, primarily in Latin America and the Pacific.[92] Such training missions carried little risk, hence their approval.

The threat from terrorism evolved during this time into its globalized form, al-Qaeda. The group increasing asserted itself through declarations of war (1996), religious justifications (fatwas, 1996 and 1998), and attacks against American targets in 1998 and 2000. The Clinton administration and some within the military chain of command, including the secretary of defense and CJCS, cancelled planned special operations reprisals for al-Qaeda attacks, in part due to risk aversion but for other reasons as well. Two prevailing attitudes among senior conventional officers about their special operations counterparts, reflected the tension between regular and specialized or elite forces. If not uncontrollable 'cowboys', SOF were 'aloof and insubordinate, and ... thought they were better than everyone else'.[93] SOF were frustrated at not being used for the primary purpose. A former special mission unit commander observed, 'It was like having a brand-new Ferrari in the garage, and nobody wants to race it because you might dent the fender.'[94]

To respond to terrorism, the Clinton administration considered a range of paramilitary and military special operations against al-Qaeda. Media reports suggest the signing by Clinton of a 'special activities' finding for the capture Osama bin Laden, involving different proxy organizations funded, equipped, and trained by the CIA.[95] Steve Coll, however, suggests a joint CIA and FBI team arresting Mir Amal Kasi in Karachi spurred plans to employ a proxy force of Afghan nationals. They were known unofficially as 'the tribals' but dubbed TRODPINT by the Agency, to capture or kill bin Laden. Another series of activities, codenamed JAWBREAKER, would make use of the Taliban's nemesis, the Northern Alliance, in more aggressive action against al-Qaeda bases and camps in Afghanistan. Among the options considered were roadside ambushes and raids against the camps.[96] The options, however, appealed only to those planning them in the bin Laden cell, whose possibilities overshadowed the inherent risks and consequences, facts considered and weighed by senior CIA and Clinton administration officials, as well as Afghan tribal leaders such as Ahmed Shah Massoud.

[91] US Government Accounting Office, *Special Operations Command*, pp. 25–26; USSOCOM, *United States Special Operations Forces, 1996 Posture Statement*, p. 29.

[92] USSOCOM, *United States Special Operations Forces, 1996 Posture Statement*, p. 9.

[93] William Boykin, *Never Surrender: A Soldier's Journey to the Crossroads of Faith and Freedom* (New York: Faith Words, 2008), p. 295.

[94] Shultz Jr, 'Showstoppers', p. 26.

[95] Will Irwin cites a Fletcher School MA thesis which in turn cites a *Washington Post* article from 2001. Will Irwin, *Support to Resistance: Strategic Purpose and Effectiveness*, JSOU Report 19-2 (MacDill Air Force Base, FL, JSOU Press, 2019), p. 122; Abigail Linnington, 'Unconventional Warfare as a Strategic Foreign Policy Tool: The Clinton Administration in Iraq and Afghanistan', unpublished Master's thesis, the Fletcher School of International Law and Diplomacy, Tufts University, 18 February 2004, p. 55.

[96] Steve Coll, *Ghost Wars: The Secret History of the CIA, Afghanistan, and Bin Laden, from the Soviet Invasion to September 10, 2001* (New York: Penguin, 2004), pp. 371–396.

Other options included indirect-shock missions relying on a partnership between CIA and SOF to develop the intelligence necessary to carry out either a raid or an airstrike using an AC-130 gunship.[97] Several, however, were unappealing for practical reasons, including time, support assets and basing requirements, and collateral damage concerns. The primary reason these options fizzled out, however, was the translation of risk from the operational to the strategic and political. The members of the 9/11 Commission found:

> the CIA considered the possibility of putting US personnel on the ground in Afghanistan. The CIA had been discussing this option with Special Operations Command and found enthusiasm on the working level but reluctance at higher levels. CIA saw a 95 percent chance of Special Operations Command forces capturing Bin Laden if deployed—but less than a 5 percent chance of such a deployment. Sending CIA officers into Afghanistan was to be considered "if the gain clearly outweighs the risk"—but at this time no such gains presented themselves to warrant the risk.[98]

## Afghanistan: 2001–2010

The 9/11 attacks on the World Trade Center and the Pentagon, which killed more than 3,000 people, opened a new period for special operations. Will Irwin and Ike Wilson refer to the decade and a half after 9/11 as the 'third age of SOF', one characterized by 'regeneration and validation'.[99] Others describe it as a 'golden era of special operations', in which SOF proved its enduring strategic utility.[100] If this 'third age' indeed exists, it began by accident rather than design.

In hindsight, the facts about the special operations element of the United States' campaign against al-Qaeda after 9/11, termed Operation ENDURING FREEDOM, are well established. The primary problem confronted the Bush administration was forcing the Taliban to surrender the senior leaders of al-Qaeda to the United States. Bin Laden was convinced that the United States was unlikely to respond to 9/11 given Afghanistan's location and geography. The country not only features numerous mountain ranges and valleys to its east and north but its infrastructure stagnated during decades of civil war, limiting options for conventional force invasion and logistical sustainment. A traditional US advantage, seapower and the mobility and power projection it confers were negated by Afghanistan's landlocked geography.

[97] National Commission on Terrorist Attacks, *The 9/11 Commission Report* (New York: W.W. Norton, 2004), p. 135.

[98] National Commission on Terrorist Attacks, *The 9/11 Commission Report*, p. 143.

[99] Will Irwin and Isiah 'Ike' Wilson III, *The Fourth Age of SOF: The Use and Utility of Special Operations Forces in a New Age*, JSOU Report 22-1 (MacDill Air Force Base, FL: JSOU Press, 2022), pp. 37–47.

[100] The phrase is attributed to William McRaven. Dan Lamothe, 'Retiring Top Navy SEAL: "We Are in the Golden Age of Special Operations"', *The Washington Post*, 29 August 2014, www.washingtonpost.com/news/checkpoint/wp/2014/08/29/retiring-top-navy-seal-we-are-in-the-golden-age-ofspecial-operations-2. Bernd Horn concludes for the reasons mentioned that 'the golden age of SOF is unlikely to end anytime soon'. Horn, 'The End of "The Golden Age of SOF"? Is There a Role for Special Operations Forces in the Renewed "Great Power Competition"?', *The Journal of Future Conflict*, 2 (Fall 2020): 21.

In geopolitical terms Afghanistan presented further obstacles given the attitudes of leaders in the country surrounding it: Iran, Pakistan, Uzbekistan, Turkmenistan, Tajikistan, and China. Bin Laden and others reasoned, as did informed contemporary policy analysts, that the United States had two options open to it: 'commando strikes' or 'weapons and technical assistance to the Afghan resistance'.[101]

When it was evident that no more attacks on the United States were imminent, President George W. Bush and senior members of his cabinet sought to retaliate against al-Qaeda and bring those responsible to justice. Of the first three military options presented to President Bush, only one appeared to be anything more than desultory response strikes using missiles or airpower. This SOF-centred option, presented by Chairman of the Joint Chiefs of Staff Hugh Shelton, had several advantages: it was more responsive and timely; had a 'small footprint'; and, in Secretary of Defense Donald Rumsfeld's eyes, it was both complementary to but also competed with the aggressive options for covert action put forward by the director of central intelligence, George Tenet.[102] Bush approved the option, in slightly modified form, which involved combined air and missile strikes with American military and paramilitary special operations supporting local Afghan Northern Alliance forces. In his words, responding in this manner would serve two purposes: counter the image that America was a 'paper tiger' and 'put boots on the ground, and keep them there until the Taliban and al-Qaeda were driven out and a free society could emerge'.[103]

Much of what occurred over the next three months in scattering Taliban out and al-Qaeda is well known. Ground and carrier-based aircraft began Operation ENDURING FREEDOM on 7 October and a company of Army Rangers seized an airfield near Kandahar on 19 October, an event recorded and publicized at a press conference. The chairman of the joint chiefs of staff suggested the seizure sent a message to the Taliban that the United States was 'capable of, at a time of our choosing, conducting the kind of operations we want to conduct'.[104] In other words, the United States was using its SOF overtly to showcase its competence and signal that the initiative had passed from the Taliban and al-Qaeda to the US and its coalition partners.

That same day, the first two American Army Special Forces Operational Detachment Alpha teams (ODAs) with attached Air Force Special Tactics (ST) personnel were airlifted into Afghanistan. They formed Task Force Dagger, the designation for JSOTF-North, under the command of Colonel John Mulholland Jr. They linked up with an estimated one hundred CIA paramilitary operatives in northern Afghanistan from the Special Activities Division and the Directorate of Support, who were in

[101] Michael O'Hanlon, 'Symbolic Missile Strikes Won't Work; We Need Sterner Measures', Brookings Institution, 14 September 2001, www.brookings.edu/articles/symbolic-missile-strikes-wont-work-we-need-sterner-measures.

[102] Bob Woodward, *Bush at War* (New York: Simon & Schuster, 2002), pp. 79–80.

[103] George W. Bush, *Decision Points* (New York: Crown, 2010), p. 191.

[104] Charles Briscoe, Richard Kiper, James Schroeder, and Kalev Sepp, *Weapon of Choice: U.S. Special Operations Forces in Afghanistan* (Fort Leavenworth, KS: Combat Studies Institute Press, 2010), pp. 109–117.

contact with various Northern Alliance warlords.[105] A total of 316 ODA SOF and ST operators proceeded to gain their trust and confidence. The American special operators brought four valuable assets with them: money, the willingness and ability to fight and other critical skills, such as battlefield medical care, the supply of weapons and other badly needed equipment, and access to a seemingly endless supply of airstrikes. The latter not only evened the combat odds by killing the more numerous Taliban and al-Qaeda forces but also destroyed or disrupted fixed positions, allowing Northern Alliance forces to manoeuvre and close with the enemy. In one engagement, a single ODA team (ODA 555) 'called in 25 air attacks, which resulted in the destruction of 29 enemy tanks and six command posts and 2,200 enemy casualties'.[106]

The intensity and accuracy of the airstrikes, combined with the speed and different axes of attack by Northern Alliance forces, demoralized and confused the Taliban and al-Qaeda defenders. A southern JSOTF, designated Task Force K-Bar under the command of Captain (N) Robert Harward and comprising Navy SEALs, ST, a battalion from 3rd Group, Special Forces, and coalition SOF from seven other nations, was dispatched to Kandahar in November to conduct SR missions.[107] In keeping with long-standing Afghan survival strategy, Taliban numbering into the thousands defected and switched sides in the face of the combined SOF, proxy, and airpower offensive. After forty-five days the Taliban stronghold Mazar-i-Sharif and the capital, Kabul, fell to Northern Alliance forces.

How American SOF forces came to be in Afghanistan and how they remained there provides insights into US SOF capability and integration. Three dimensions of special operations capability offer an explanation: authorization and control, basing and mobility, and relationships. While authorization and control lines were clear, the special operations command structure at the tactical level was anything but. Legally, Congress almost unanimously passed the short Authorization to Use Military Force (AUMF) against al-Qaeda and its affiliates on 14 September 2001. From these flowed the execute (EXORD), deployment (DEPORD), and OPORD. There was a clear line of authorization, in other words, from the executive and legislative through the secretary of defense and CJCS down through the COCOM and Theater Special Operations Command (TSOC) to the JSOTFs and tactical units.

The command structure and processes, however, needed to be rethought and improvised in combat. As the ST commander for Task Force Dagger suggests:

> We weren't trying to fit SOF under a conventional/doctrinally correct C2 structure; we were trying to scale and match a conventional C2 structure and all the associated components to a SOF fight. The JSOTF/JSOAC spent significant resources

[105] 'About CIA in Afghanistan', Exhibits, On the Frontlines: CIA in Afghanistan, Central Intelligence Agency, n.d., www.cia.gov/legacy/museum/exhibit/on-the-front-lines-cia-in-afghanistan.

[106] Ben Lambeth, *Air Power against Terror*, MG-166 (Washington, DC: RAND, 2005), p. 129; Linda Robinson, *Masters of Chaos: The Secret History of the Special Forces* (New York: PublicAffairs, 2004), pp. 156–158.

[107] Micheal Krivdo, 'Combined Joint Special Operations Task Force-Afghanistan (CJSOTF-A): A Short History, 2002–2014', *Veritas*, 12:2 (2016): 11.

> trying to reverse engineer much of what we knew about traditional relationships and processes while the bullets were flying.[108]

Some of this 'reverse engineering' took the form of reconsidering command relationships between platform 'owners' and managers and ground SOF. At issue was whose should be the primary, supported effort and which the secondary or supporting one. The entire basis of what was subsequently termed 'the Afghan model' of SOF, airpower, and proxies rested on resolving this historically contentious issue.[109] In essence, a forward-deployed headquarters liaison element, as well as flexible interpretation or redesignation of air support ensured that sufficient airpower was always available for ground SOF and their Northern Alliance partners throughout.

Basing was essential for success given Afghanistan's geography. Without the Karshi Khanabad (K-2) Air Base, 160th SOAR, 9th Special Operations Squadron (SOS), and CIA Directorate of Support aircraft could not have initially infiltrated ground special operations units, much less supported and sustained them. The next nearest accessible airfields were Diego Garcia, in the Indian Ocean, almost 5,000 kilometres distant, or Incirlik Air Base in Turkey, more than 2,500 kilometres away. K-2 is less than 150 kilometres from the northern Afghanistan border and provided both a location for forward-deployed headquarters units, including JSOTF, and served as Special Forces Forward Operating Base for Alpha Forward Support Company, 528th Support Battalion's efforts to provide 'initial supply, food service, ammunition, water, and fuel points and contracted for additional supplies from local sources'.[110] Access to K-2, while diplomatically negotiated, was facilitated as part of prior and ongoing special operations training missions. In particular, a detachment from the 6th SOS, fortuitously on a joint combined exercise training (JCET) in Uzbekistan, assisted SOF ODAs and ST teams in 'rolling in' to this military airbase, due to their working relationships with their Uzbek counterparts, as opposed to a less secure civilian airfield.[111]

K-2 provided a forward basing area 'lily pad' to receive supplies and forces flowing into the theatre early on in Operation ENDURING FREEDOM.[112] Shahbaz Air Base near Jacobabad, Pakistan, served a similar function for 20th SOS and 8th SOS rotary and fixed-wing SOF aircraft, although few details are available about its negotiation and initial use.[113] The 7th, 15th, and 550th SOS delivered 900 tons of other supplies in night air-drops to Task Force Dagger while in communication with ground forces, a

[108] Frank Kisner, 'Evolution of a JSOAC in a Non-Linear War', *Air Commando Journal*, 8:3 (February 2020): 22.

[109] For details on how this was resolved, see James D. Kiras, 'T. Michael Moseley: Air Power Warrior', in John Andreas Olsen, ed., *Air Commanders* (Dulles, VA: Potomac, 2013), pp. 395–428.

[110] Robert Jones Jr, 'A Legacy of Support: The 528th Sustainment Brigade', *Veritas*, 5:3 (2009): 25.

[111] Calvin Markham, 'ENDURING FREEDOM: An Air Force Combat Controller's Experience', unclassified briefing, 2002; Kisner, 'Evolution of a JSOAC in a Non-Linear War', p. 25.

[112] Stephen Schwalbe, 'The World's Fireman and Its Lily Pads: The Case of K2', *Air and Space Power Journal Chronicles*, online journal (2007): 4.

[113] Kisner, 'Evolution of a JSOAC in a Non-Linear War', p. 22; Forrest Marion, 'Building USAF "Expeditionary Bases" for Operation ENDURING FREEDOM—AFGHANISTAN, 2001– 2002', *Air and Space Power Journal Chronicles*, online journal (2005): 3–5.

legal requirement for 'lethal aid drops', in round trips from Incirlik lasting 14 hours.[114] SOF air also flew recovery and casualty evacuation missions. An MH-53 from the 20th SOS, Knife 04, recovered the personnel from one of their crashed brethren and the 'next day, with much better weather conditions, a civilian contract helicopter was able to finally retrieve the ailing Soldier from ODA 555' suffering from altitude sickness and meningitis.[115] Neither recovery would have been possible without close coordination from airborne warning and control (AWACs) air battle management aircraft or US Navy F-18s flying combat air patrols above, as well as special operations MC-130E refuelers.[116] Control mechanisms in place at the TSOC level ensured field requirements were met as quickly as possible, both smoothing out and working through bureaucratic process restrictions while simultaneously translating and filtering out unrealistic requests from fielded ODAs, such as delivery of 'all-terrain vehicles to airdropped in Afghanistan in 18 hours' despite their unavailability, commercially or otherwise.[117]

Relationships, the third dimension of capability, resulted from habitually working together. Training, while no substitute for experience, contributed to American special operations capability early in Afghanistan in several ways. Calvin Markham, a 23rd Special Tactics Squadron (STS) combat controller, associated two values with joint combined exchange training, or JCETs. The first, mentioned above, was access and placement in countries that was essential with K-2. The second was month-long training developed familiarity, through habitual working relationships, across different special operations service boundaries. Such relationships, in Markham's words, created 'familiarity with their cross-service brethren' that proves invaluable; they 'smooth transitions', or rather quicken them, between different operating environments and support.[118] The team members of ODA 555 selected Markham because they knew him, meaning they trained together previously.[119]

Other relationships occurred due to training via proximity. In its after-action briefing, Task Force Dagger pointed out the value of its collocation with the 160th SOAR at Fort Campbell, Kentucky, which 'allowed for an easy transition during OEF'.[120] Representation is another form of relationship building that serves another function: translation. A Special Forces senior NCO, Tom Rosenberg, was stationed at the Combined Air Operations Center for the Combined Forces Air Component Commander (CFACC), serving as a liaison officer, 'to help translate the army and air force's different terminology, equipment, units of measure, and any other communications problems that came up'.[121] Preparation and planning for operations served

[114] Briscoe et al., *Weapon of Choice*, p. 131.

[115] Darrel Whitcomb, *On a Steel Horse I Ride: A History of the MH-53 Pave Low Helicopters in War and Peace* (Maxwell Air Force Base, AL: Air University Press, 2012), pp. 514; 519; Robinson, *Masters of Chaos*, p. 156.

[116] Whitcomb, *On a Steel Horse I Ride*, pp. 515–517.

[117] Briscoe et al., *Weapon of Choice*, p. 131, Kisner, 'Evolution of a JSOAC in a Non-Linear War', p. 29.

[118] Markham, 'ENDURING FREEDOM'.

[119] Ibid.

[120] 5th Special Forces Group (Airborne), 'Task Force Dagger, Operation Enduring Freedom: Key Lessons Learned', briefing slides, n.d., slide 16.

[121] Robinson, *Masters of Chaos*, p. 157.

as an additional, invaluable contributor to capability. While JCETs were helpful in creating joint familiarity and being at the right place and the right time outside of Afghanistan, inside the country it was the relationships developed by the CIA the Northern Alliance that were an essential precondition for operations there. Although there was little stomach at the political and strategic level for CIA-backed proxy operations in Afghanistan between 1998 and 2000, the activities of paramilitary special operators in contacting Northern Alliance and other tribal leaders proved invaluable in Operation ENDURING FREEDOM.[122] Such contact not only eased the job of follow-on special forces working particularly delicate negotiations and providing support to warlords suspicious or hostile to one another, but Agency personnel also established a network of safe houses in key locations, including north of Bagram airfield that served as a hub for three local observation posts from which a range of airstrikes were called in.[123]

The collapse of Taliban rule at the end of November marked a new stage in Operation ENDURING FREEDOM. SOF were split into two roles: stabilizing portions of Afghanistan formerly under Taliban rule, and conducting SR missions, alone or in conjunction with Afghan proxies, to locate, capture, or kill the al-Qaeda senior leadership or so-called '2 plus 7': bin Laden, Ayman al-Zawahiri, plus seven others. For the stabilizing mission, ODAs such as 595 at Mazar-e-Sharif worked together with Civil Affairs (CAT-A33) and Psyop teams (TPT 922) to conduct the sort of ancillary activities they had been doing in Bosnia-Herzegovina and elsewhere over the previous decade. Tactically, the missions succeeded thanks to 'integration' between these SOF units that 'had a synergistic effect' in caring for the survival needs and interests of the local population.[124] On the broader operational or campaign level, however, efforts to integrate information operations, specifically public information, psyops, and civil-military information, foundered in Afghanistan for the same reasons they did in Bosnia: a lack of integration at the operational planning level with ongoing indirect-shock missions, or raids, which answered to different chains of command.[125]

If the first eight weeks of Operation ENDURING FREEDOM displayed high levels of integration tactically and horizontally within and between paramilitary and military special operations, specifically their capability to conduct offensive military advising and assistance missions to the Northern Alliance and other warlords, the subsequent Battle of Tora Bora (6–17 December 2001) as well as Operation ANACONDA (1–18 March 2002) suggested that conventional and SOF vertical integration was tenuous at best.

The collapse of the Taliban and al-Qaeda occurred so rapidly that once they were driven from the major cities, SOF needed to operate in a more distributed and decentralized manner. By late November, sustained, distributed activities exhausted

[122] Mark Folse, *The U.S. Army in Afghanistan: Operation Enduring Freedom, September 2001–March 2002*, Pub 70-83-1 (Washington, DC: Center for Military History, 2022), p. 38.

[123] Gary C. Schroen, *First In: An Insider's Account of How the CIA Spearheaded the War on Terror in Afghanistan* (New York: Ballantine Books, 2005), pp. 182–183, Markham, 'ENDURING FREEDOM'.

[124] Briscoe et al., *Weapon of Choice*, p. 186.

[125] Pascale Siegel, *Target Bosnia: Integrating Information Activities in Peace Operations* (Washington, DC: National Defense University Press, 1998), p. 146.

the supply of 5th Group operators: 'there was only a Special Forces company from 5th SFG at Fort Campbell' to help their activated brethren from a National Guard unit, the 19th SFG, prepare in advance to heading to Afghanistan.[126] In short order another National Guard unit, the 20th SFG, was similarly activated. As Taliban and al-Qaeda forces became increasingly dispersed and mobile, accurate intelligence was a scarce commodity. In addition, the small number of special operators, which had responded so admirably to initial conditions by forging non-doctrinal command, control, and fire support procedures, shifted from being a tactical asset to a liability. During actions in and around the cave complex at Tora Bora, a mission designed to capture or kill Osama bin Laden, there were too few special operators to engage al-Qaeda forces quickly and effectively.[127] 5th SFG teams, specifically ODA 572, conducted SR missions from two observation posts on the cave complex while proxy forces under local warlord General Hazaret Ali surrounded it and provided quick reaction and response forces, rather than US SOF.[128] Despite integrated military and intelligence planning efforts, as well as operations involving 63 SOF from ODA 572 and Task Force 11 backed by air support that killed an estimated 250 al-Qaeda fighters while suffering no casualties themselves, special operators did not manage to find bin Laden.[129] While sources continue to dispute his presence there before or during the battle, the Senate Foreign Relations Committee unequivocally spelled out in 2009 the strategic consequences of this early, single special operations battle:

> Removing the Al Qaeda leader from the battlefield eight years ago would not have eliminated the worldwide extremist threat. But the decisions that opened the door for his escape to Pakistan allowed bin Laden to emerge as a potent symbolic figure who continues to attract a steady flow of money and inspire fanatics worldwide. The failure to finish the job represents a lost opportunity that forever altered the course of the conflict in Afghanistan and the future of international terrorism, leaving the American people more vulnerable to terrorism, laying the foundation for today's protracted Afghan insurgency and inflaming the internal strife now endangering Pakistan.[130]

After the Battle of Tora Bora, the Taliban and al-Qaeda consolidated and regrouped in the Shah-i-Kot Valley in Paktia Province, the location for Operation ANACONDA. This operation was the first in Afghanistan in which SOF would support conventional forces, instead of being supported by them. Information gathered

[126] Briscoe et al., *Weapon of Choice*, pp. 216–217.

[127] According to Woodward, President Bush signed the 'capture or kill order' for bin Laden, presumably a finding, on 17 September 2001. Woodward, *Bush at War*, p. 100.

[128] Dalton Fury (pseudo.), *Kill Bin Laden: A Delta Force Commander's Account of the Hunt for the World's Most Wanted Man* (New York: St Martin's Press, 2008), pp. 99, 121; Briscoe et al., *Weapon of Choice*, pp. 213–216.

[129] United States Special Operations Command History and Research Office, *United States Special Operations Command History, 20th Anniversary, 1987–2007* (MacDill Air Force Base, FL: USSOCOM, 2007), p. 98.

[130] US Senate, Committee on Foreign Relations, *Tora Bora Revisited: How We Failed to Get Bin Laden and Why It Matters Today* (S. Prt. 111-1) (Washington, DC: Government Printing Office, 2009), p. 1.

by SOF ODAs conducting SR identified the concentration of enemy forces in the valley, estimated at upwards of 1,000 al-Qaeda.[131] Combined Joint Task Force Mountain (CJTF-Mountain) led the planning efforts, in which elements of the 10th Mountain, 101st Airborne Division, and Canadian 3rd Princess Patricia Canadian Light Infantry of Task Force Rakassan would provide airlifted blocking forces, or 'anvil', onto the 'Whale' mountain. The 'hammer' used to drive Taliban and al-Qaeda forces onto it would comprise SOF working in two task forces, Task Force Hammer to the northeast and Task Force Anvil to the south and east, working in conjunction with 1,600 Afghan militia forces. Other SOF, specifically the Australian Special Air Service Regiment in Task Force 64, provided overwatch from the mountain heights in a range of SR hide sites. A final SOF element were three unidentified teams sent in days before the operation to 'validate ... [landing zones], verify presence of High Value Targets [HVTs], and provide terminal guidance for [close air support]'.[132]

From the outset, coordination, timing, sequencing, command and control, and supporting issues bedevilled the operation. The small geographic area in which ANACONDA took place and its proximity to the Pakistan border complicated combat support for special operations teams and the infantry, particularly air support.[133] Some special operations teams were compromised, several AH-64s and an MH-47 helicopter were shot down or badly damaged, turning a clandestine SR infiltration into a rescue operation. SOF's numerous Afghan militia partners proved well suited to combat in open valleys and plains, with plenty of room to manoeuvre, but less so in sustained close-quarter battle against a well-entrenched enemy.[134] The higher headquarter desire to 'give combat experience to as many forces as possible', while sound in theory, also led to dangerous practices: 'the on-site special operations reconnaissance task force (headquartered in Bagram) was cut out of its surveillance mission so the TF K-Bar personnel (headquartered 1,000 miles away) could get combat experience' which resulted in 'the untimely insertion of (SEAL unit) Mako 30 on Takur Ghar and the deaths of seven dedicated American servicemen'.[135]

In addition, the battle revealed serious control, coordination, and information-sharing problems between overt TF K-Bar personnel and clandestine TF-11 special operations. While an on-scene SOF commander, Pete Blaber 'had long believed in the power of combinations in order to maximize the potential of each', such integration worked only at the tactical level and only so long as it worked to the benefit of specific operations, capturing or killing HVTs.[136] He felt a liaison officer offer at

[131] United States Special Operations Command History and Research Office, *United States Special Operations Command History, 20th Anniversary*, p. 98.

[132] Ibid.

[133] For more on the air perspective of this engagement, see Kiras, 'T. Michael Moseley', pp. 395–428.

[134] Stephen Biddle, *Afghanistan and the Future of Warfare: Implications for Army and Defense Policy* (Carlisle Barracks, PA: Strategic Studies Institute, 2002), p. 14.

[135] Lester Grau and Dodge Billingsley, *Operation Anaconda: American First Major Battle in Afghanistan* (Lawrence, KS: University Press of Kansas, 2011), p. 352. For specific details on the battle, see Malcolm MacPherson, *Roberts Ridge: A Story of Courage and Sacrifice on Takur Ghar Mountain, Afghanistan* (New York: Delacorte Press, 2005).

[136] Sean Naylor, *Not a Good Day to Die: The Untold Story of Operation Anaconda* (New York: Berkeley, 2005), pp. 38–39.

Bagram headquarters did little to coordinate with TF K-Bar, while providing consistent updates to CJTF-Mountain.[137] While Army historians consider this battle a military success, Operation ANACONDA is not a strategic one. An unknown, but significant number of Taliban and al-Qaeda fighters eluded blocking forces and, in an all-too-frequent occurrence over the next seventeen years, escaped into Pakistan.

The two geographically distinct JSOTFs, Task Force Dagger (JSOTF-N) and Task Force K-Bar (JSOTF-S), were combined and amalgamated in March 2002 into a single command, Combined Joint Special Operations Task Force-Afghanistan, or CJSOTF-A. The timing roughly coincided with the rotation of 5th SFG out of the Afghanistan and the rotation of 3rd SFG, whose area of specialization and culture and language focus had been, over the previous decade, the Caribbean and West Africa, suggesting an uphill cultural learning curve for its deploying operators.[138] The standup of CJSOTF-A and rotation in of 3rd also coincided with a change of command for the former, which occurred in late May under the leadership of Joseph Celeski. For the next three years, CJSOTF-A would be under the tactical control of a conventional army headquarters, Combined Joint Task Force 180, and no longer independent.[139]

More troublesome was the shift that occurred operationally. Forces along the Pakistan border moved from the tactical and operational offence to the operational defensive, as Gary Schroen notes:

> Over the next eighteen months, operational bases in the border area grew in size, taking on the aspect of permanency, and operations tended to focus less on sending out small teams and more on attempting larger sweeping operations throughout the areas to locate terrorists, weapons, and ammunition. More often than not, these sweeps resulted in the alienation of the local tribal populations in the area, who deeply resented the destructive searches of their homes and the frightening of their women and children.[140]

Worse still, airstrikes killed an estimated 400 Afghan civilians by the summer of 2002, including a SOF-directed AC-130 strike that reportedly hit a wedding party.[141] Real or not, such incidents and casualties handed the Taliban a propaganda theme they could exploit that played to Afghan fears of foreign occupation.[142]

[137] Blaber blames the planning and outcome of the battle on poor information sharing, inadequate command experience, conventional force inertia, and higher headquarters meddling—the standard tropes of SOF misuse—in his self-serving account of Anaconda. Pete Blaber, *The Mission, the Men, and Me* (New York: Berkeley, 2008), pp. 223–297.

[138] Briscoe et al., *Weapon of Choice*, p. 256.

[139] Micheal Krivdo, 'Combined Joint Special Operations Task Force-Afghanistan (CJSOTF-A): A Short History, 2002–2014', *Veritas*, 12:2 (2016): 9.

[140] Gary C. Schroen, *First In: An Insider's Account of How the CIA Spearheaded the War on Terror in Afghanistan* (New York: Ballantine Books, 2005), p. 360.

[141] Brian Neumann, Lisa Mundey, and Jon Mikolashek, *The U.S. Army in Afghanistan: Operation Enduring Freedom, March 2002–April 2005*, Pub 70-122-1 (Washington, DC: Center of Military History, 2013), p. 15.

[142] Folse, *The U.S. Army in Afghanistan*, p. 78.

With the Taliban and al-Qaeda largely removed from Afghanistan, the next strategic priority for the United States and its coalition partners was the creation of stable governance for the country. The operational emphasis for SOF, in consequence, shifted to CA and security force assistance. Special operations missions focused on helping Afghans establish the mechanisms of governance, build infrastructure, and create security forces capable of addressing foreign and domestic threats. A SOF sustainment plan, developed by CJSOTF-A with USSOCOM assistance, saw the nearly continuous rotation of 3rd SFG and 7th SFG battalions into and out of the country. SOF continued to conduct military advising and assistance missions with the fledgling Afghan National Army (ANA), as well as accompany its units on missions to build their confidence.

To create an ANA quickly and at scale, the CJTF-180 commander, Lieutenant General Lloyd Austin, made it a conventional forces mission first for a brigade of the 10th Mountain Division and later, National Guard brigades from Oklahoma, Indiana, Florida, and other states under Task Force Phoenix.[143] Although energetic, such forces lacked the working relationships with the various warlords, cultural intelligence, and language skills of the special operators. This missed opportunity for special operations to assume responsibility for Afghan stability and FID had strategic consequences later. In one assessment, 'Although the international community was committed to training a "national" Afghan army and the central government promulgated the need for ethnic balance, the achievement of that goal was problematic ... Primary loyalty, as it had been for centuries, was to the tribe, not the nation.'[144]

The geographic separation of the country into four regional commands, combined with the shift of the ANA training mission to conventional forces, relegated SOF to four activities in Afghanistan, part of a pattern that continued over the next eleven years. Overt Special Forces dispersed along the 750-mile border with Pakistan in forward operating bases to detect Taliban and al-Qaeda infiltration via 'rat-lines', or traditional tribal trails and smuggling routes across the border through the mountains.[145] The twenty-five distributed ODAs tasked with this mission recalled, in form and function, the outposts used to detect Viet Cong and North Vietnamese Army (NVA) infiltration into South Vietnam, with equally unsatisfying results.[146] The second activity was the incorporation of Civil Affairs and Psyop units with provincial reconstruction teams (PRTs) to provide 'non-lethal fires' in support of 'three integrated lines of effort' but primarily the Herculean task of addressing sources of instability.[147] A third was the continued overt and clandestine hunt for al-Qaeda and Taliban leaders and facilitators through a mixture of reconnaissance

[143] Neumann et al., *The U.S. Army in Afghanistan*, p. 28.
[144] Briscoe et al., *Weapon of Choice*, p. 390.
[145] Neumann et al., *The U.S. Army in Afghanistan*, pp. 14–15.
[146] Krivdo, 'Combined Joint Special Operations Task Force-Afghanistan (CJSOTF-A)', p. 12.
[147] Robert Herndon et al., 'Effects-Based Operations in Afghanistan: The CJTF-180 Method of Orchestrating Effects to Achieve Objectives', *Field Artillery* (January–February 2004): 26; 28–29.

and indirect-shock capture or kill missions. These 'night raids' became increasingly controversial within the United States and internationally, as well as a source of political tension between the Government of the Islamic Republic of Afghanistan (GIRoA) and International Stabilization Force (ISAF) as media reports surfaced.[148] The controversy only grew when reports surfaced about CIA or SOF-trained Afghan militias were conducting the missions, calling to mind Phoenix operations in Vietnam.[149] The fourth and final mission set, connected to the last, was the tasking of CJSOTF-A in 2006 by Special Operation Command Central (SOCCENT) was the creation of a Commando battalion, or *Kandak*, as the elite striking force of the ANA. The success of the initial effort led to an expansion of the programme to nine *Kandak* battalions into a Commando Corps.[150] According to media reporting, the burden of combating the Taliban within the ANA increasingly fell on the *Kandaks* who, despite their small numbers, were estimated to be conducting 70–80 per cent of combat against them.[151] Operation ENDURING FREEDOM officially ended on 28 December 2014, although special operations would continue to play a role until the collapse of the GIRoA on 30 August 2021.

## Iraq: 2003–2004

The shift of the ANA training mission from SOF to conventional forces in Afghanistan occurred as the Bush administration focused increasingly on Iraq. Four days after the 11 September attacks, in the same meeting in which the principal members of the National Security Council discussed the various military options for Afghanistan, Deputy Secretary of Defense Paul Wolfowitz argued to strike Iraq instead. His rationale was that Saddam may have played a role in the terrorist attacks and that 'Iraq was a brittle, oppressive regime that might break easily'.[152] In addition, Saddam flaunted United Nations weapons inspections and despite Operation DESERT FOX, he was thought to be developing even more powerful weapons of mass destruction (WMD).

President Bush identified Iraq as one of three members of 'an axis of evil' in January 2002, regimes that possessed both WMD and terrorists who might use them. CENTCOM Commander General Tommy Franks was tasked in November 2001 to review the existing operational plan, and provide Rumsfeld with his 'Commander's Concept'. The subsequent plan, OPLAN 1003V, came together in

[148] Theo Farrell, *Unwinnable: Britain's War in Afghanistan, 2011–2014* (London: The Bodley Head, 2014), pp. 329–331.

[149] For example, see Andrew Quilty, 'The CIA's Afghan Death Squads: The U.S.-Backed Militia That Kills Children May Be America's Exit Strategy from Its Longest War', The Intercept, 18 December 2020, https://theintercept.com/2020/12/18/afghanistan-cia-militia-01-strike-force/; Matthew Hoh, 'CIA in Afghanistan: Operation Phoenix Redux?" CounterPunch, 24 October 2017, www.counterpunch.org/2017/10/24/cia-in-afghanistan-operation-phoenix-redux.

[150] Krivdo, 'Combined Joint Special Operations Task Force-Afghanistan (CJSOTF-A)', p. 15.

[151] Helene Cooper, 'Afghan Forces and Praised, Despite Still Relying Heavily on U.S. Help', *New York Times* (20 August 2017), www.nytimes.com/2017/08/20/world/asia/afghanistan-military-strategy.html.

[152] Woodward, *Bush at War*, p. 83.

earnest in January 2002, just a few weeks after the Battle at Tora Bora ended.[153] Planning at CENTCOM went through several iterations as Secretary Rumsfeld and his staff challenged the size of the force required to overthrow Saddam. Success in Operation ENDURING FREEDOM with small numbers led Rumsfeld and other DoD officials to question CENTCOM's submitted plans and force consideration of lower troop estimates. Senior Bush administration officials, in turn, believed 'that most of Saddam's army and police would switch sides once the regime was gone' and 'we would draw on the rest of the Saddam-era forces to form the foundation of the new Iraqi military and police' to guarantee internal security.[154] The OPLAN was codenamed 'COBRA II', with the overall mission labelled Operation IRAQI FREEDOM.

Special operations played a sizeable role during the twenty-one-day campaign. They were conducted along four major lines of effort. The first was an attempt by paramilitary and military SOF to conduct both treachery and subversion as well as intelligence-gathering and reconnaissance missions. These efforts commenced well in advance of the campaign, some eight months prior.[155] The second served two operational purposes: deny Saddam use of western Iraq to avoid missile launches there against Israel, and confuse the Iraqi leadership as to the main invasion effort. The third, building off efforts of the first, was designed to present a threat from the north by mobilizing Kurdish forces protected by a no-fly zone. The fourth and final line of effort was the seizure of infrastructure vital to the restoring of the Iraqi economy once Saddam was deposed, to avoid the destruction of oil wells and refineries as had occurred during Operation DESERT STORM.

Combined with airstrikes against key communications facilities, special operations in the west and north were designed to confuse Saddam and his generals and prevent the movement of reinforcements. Their net effect was to induce a level of operational paralysis into Iraqi decision making and responses. Special operations alone could not do so, given the number of targets, their location, and redundancy within the system.[156] While special operations have long been used for deception, their scale in Operation IRAQI FREEDOM sets them apart. In western Iraq, a combined American, British, and Australian task force, Combined Joint Special Operations Task Force (West) (CJSOTF-W), was commanded by Colonel John Mulholland. Informally known as Task Force Dagger, it contained elements from 5th and 19th SFG, 23rd STS, and 160th SOAR operated in conjunction with three British squadrons (A & D, 22nd SAS and C, Special Boat Service) as well as a composite

[153] For a first-hand account of the deliberations, see Tommy Franks, *American Soldier* (New York: Regan Books, 2004), pp. 329–363.

[154] George W. Bush, *Decision Points* (New York: Crown, 2010), p. 249.

[155] Dale Herspring, *The Pentagon and the Presidency: Civil–Military Relations from FDR to George Bush* (Lawrence, KS: University Press of Kansas, 2005), pp. 398–400.

[156] Alastair Finlan treads old ground in his 'anaphylaxis model', which is nothing more than strategic paralysis by another name, with all of its logical flaws. See Alastair Finlan, 'A Dangerous Pathway? Toward a Theory of Special Forces', *Comparative Strategy*, 38:4 (2019): 255–275 [264–266]. For more on strategic paralysis and its flaws, specifically as they related to special operations, see James D. Kiras, *Special Operations and Strategy: From World War II to the War on Terrorism* (London: Routledge, 2007), Chapter 2.

Australian force (one squadron, SASR and a company from Commando battalion of the Royal Australian Regiment).

Inserting from bases in neighbouring Jordan, Task Force Dagger performed three tactical missions: destroyed Iraqi early warning radar systems and outposts; denied Saddam access to the western desert's major highway, supply, potential escape route to Jordan, and launch areas for his Scud missiles; and, seized airfields to forward-employ attack aircraft and provide armed overwatch for the mobile SOF, including the H-2 and H-3 airfields.[157] To complete the deception effort, another special operations task force based around a special mission unit, using ground mobility vehicles, had attached to a handful of conventional forces, including five tanks. Any Iraqi field units contacting the force would report seeing these vehicles, suggesting they were the advance reconnaissance units of a more sizeable conventional armoured thrust coming from the west. In the words of the force commander, who modelled the operation after the Trojan Horse, 'Our ruse had worked way beyond even our most optimistic expectations.'[158] Beyond the ruse, this task force also managed to secure a crucial piece of infrastructure, the Haditha Dam, despite it being guarded by four companies of Iraqi armour, backed by infantry and artillery.[159] In a short but fierce battle from 1 to 3 April 2003, a company from the 75th Ranger Regiment assaulted, secured, and defended the dam from Iraqi counterattacks, with the support of 160th SOAR helicopters and airstrikes.[160] The dam, or rather its destruction and flooding, concerned Secretary Rumsfeld sufficiently that it appeared in one of his infamous 'snowflake' memos. The deception succeeded in diverting an Iraqi division tasked with guarding an important manoeuvre corridor, the Karbala Gap, for conventional units of the Army's V Corps advancing from the south.[161]

In northern Iraq, the SOF deception operation was driven by several factors, including sanctuary and diplomatic considerations. Per the COBRA II plan, the 4th Infantry Division was to invade from Turkey in the North. The Turkish Congress, however, denied the US request to use their country as a staging area for an invasion of Iraq. This decision affected more than conventional forces; Army SOF were to be based at Diyarbakir, where advanced SOF site surveyors already scouting the location but had to abandon their effort in favour of sites in Romania.[162] From that location, the Combined Joint Special Operations Task Force (North) (CJSOTF-N), under the command of Colonel Charles Cleveland, commenced planning for its operations. Cleveland's command, comprising two battalions from 10th SFG augmented another battalion from 3rd SFG.[163]

[157] Briscoe et al., *All Roads Lead to Baghdad: Army Special Operations Forces in Iraq* (Fort Bragg, NC: USASOC History Office, 2006), pp. 32–36.

[158] Blaber, *The Mission, the Men, and Me*, pp. 2–9.

[159] Sean Naylor, *Relentless Strike: The Secret History of Joint Special Operations Command* (New York: St Martin's Press, 2015), pp. 213–218.

[160] For details, see James Schroder, 'The Rangers Take Haditha Dam', *Veritas*, 1:1 (Winter 2005): 55–60.

[161] Mark Moyar, *Oppose Any Foe: The Rise of America's Special Operations Forces* (New York: Basic Books, 2017), p. 260.

[162] Briscoe et al., *All Roads Lead to Baghdad*, p. 82. For a different perspective on the events and their significance, see the recollections of the deployment special operations air wing commander in O.G. Mannon, 'The Power of Political National Interest', *Air Commando Journal*, 10:1 (June 2021): 41–44.

[163] Kenneth Finlayson, 'Operation Viking Hammer: 3/10 SFG against the Ansar al-Islam', *Veritas*, 1:1 (2005): 15.

CJSOTF-N, labelled Task Force Viking, had two primary missions. The first was linking up and providing military advice and assistance to some 50,000 local Kurdish *peshmerga* militia members.[164] TF Viking and their Kurdish allies were tasked initially with destroying known al-Qaeda affiliated terrorist camps (Operation VIKING HAMMER). In addition, they would also threaten Mosul and seize the city of Kirkuk, thereby focusing the attention of three Iraqi corps, comprising thirteen divisions, with their estimated 100,000 personnel.[165] To assist in the deception, elements from the 173rd Airborne Brigade parachuted in and were attached to CJSOTF-N, with a secondary mission: guarding the oil fields between Irbil and Kirkuk. The attachment of conventional ground forces in a supporting role to SOF was a first for American special operations. CJSOTF-N's actions were designed to prevent three divisions of Saddam's infantry from moving either northward or southward by keeping them engaged as part of Operation NORTHERN DELAY.[166] By fixing the thirteen Iraqi divisions in place along 'the Green Line' separating Iraqi from Kurdistan, distributing SOF teams with *peshmerga* companies along it and calling in precision airstrikes, special operations destroyed numerous Iraqi battalions 'between deaths and desertions' while suffering limited casualties in localized offensives against prepared defenses.[167]

As in Operation ENDURING FREEDOM, military special operations were preceded by paramilitary ones. These efforts occurred along two primary lines of effort: one with Kurdish *peshmerga* and the other with a group of recruited and trained Sunni Iraqis from the centre of the country. CIA contact with *peshmerga* began in February 2002. The CIA liaison effort, which Bob Woodward identifies as the Northern Iraq Liaison Element (NILE) team, was 'to establish contact with the two main political parties in Iraqi Kurdistan, the Kurdish Democratic Party (KDP) and the Patriotic Union of Kurdistan (PUK)' and to 'begin military advising and assistance missions, in the form of training and equipping in two primary bases'.[168] The team comprised fourteen members, not unlike the composition of a Special Forces ODA. Almost half of the NILE team members were case officers with specific language and cultural skills and nearly a quarter of the team were senior 10th SFG NCOs to provide specific training instruction.[169] The small size and covert nature of the operation convinced Turkish leaders to allow the team to deploy into northern Iraq from Turkey, supporting the notion that special operations can go where conventional forces cannot.[170]

164 Moyar, *Oppose Any Foe*, p. 258.

165 Finlayson, 'Operation Viking Hammer', pp. 15–19; Michael Gordon and Bernard Trainor, *Cobra II: The Inside Story of the Invasion and Occupation of Iraq* (New York: Pantheon, 2006), p. 450; Robinson, *Masters of Chaos*, p. 299.

166 Briscoe et al., *All Roads Lead to Baghdad*, pp. 83–4; 460.

167 Richard Andres, 'The Afghan Model in Northern Iraq', *Journal of Strategic Studies*, 29:3 (Summer 2006): 412.

168 Woodward, *Plan of Attack*, p. 109; Andrew 'L', Mick Mulroy, and Ken Tovo, *Irregular Warfare: A Case Study in the CIA and US Army Special Forces Operations in Northern Iraq, 2002–03* (Washington, DC: Middle East Institute, August 2021), p. 6.

169 Woodward, *Plan of Attack*, pp. 208–209. For a first-hand account, see Sam Faddis, *The CIA War in Kurdistan: The Untold Story of the Northern Front of the Iraq War* (Havertown, PA: Casemate, 2020).

170 Woodward, *Plan of Attack*, p. 208; Colin S. Gray, *Explorations in Strategy* (Westport, CT: Greenwood Press, 1996), pp. 146; 149.

While NILE was a covert advising and assistance mission, the other line of effort, allegedly codenamed DBANABASIS, was designed from the outset to conduct treachery and subversion missions. The team, nicknamed the Scorpions and according to one source consisting of one-hundred-odd members, was trained and equipped to operate in Iraq prior to the start of the invasion and give the impression of widespread popular resistance through nonviolent means as well as well as 'light sabotage'.[171] As with some prior covert actions, enthusiasm and support for such programs was unmatched by capability on the ground, with field officers suggesting their charges were motivated by money and staffed by criminals and other unsavoury elements. In addition, a late start, overly ambitious goals, and logistical gremlins consigned the Scorpions to insignificance during the war.[172] More problematic was the lack of control exerted over the Scorpions after the fall of Baghdad. US journalists subsequently reported that 'CIA control over the unit became weaker as chaos grew in Iraq'.[173]

The substantial tactical and operational accomplishments of Task Forces Dagger and Viking overshadowed special operations in southern Iraq. In the Al-Faw Peninsula, Navy SEALs from Teams 3, 8, and 10, working in conjunction with Polish operators from *Grupa Reagowania Operacyjno-Manewrowego* (GROM), seized a number of oil platforms and terminals to prevent their destruction.[174] Their success can be attributed to the close working relationship the SEALs forged with GROM, including training and exchanges that resulted in what one Polish operator described as 'a true partnership'.[175] These operations were a precursor to a larger maritime forcible entry operation onto the peninsula between a combined US Marine Corp and Royal Navy Marines and Commando task force.[176]

Although joint operations in IRAQI FREEDOM may not have achieved the ideal of an organic whole, special operations contribution to their effectiveness would not have been possible without considerable horizontal integration of capabilities. A key contributor to that effectiveness was air power, both conventional and special. The complete air dominance achieved by the United States and its coalition partners was crucial to the outcome of Operation IRAQI FREEDOM. The fact that so few airstrikes killed special operators or their proxies during mobile operations, save for one tragic incident during the Battle of Debecka Pass that claimed fifteen lives due

[171] Mike Tucker and Charles Faddis, *Operation Hotel California: The Clandestine War Inside Iraq* (Guilford, CT: The Lyons Press, 2009), p. 42; Dana Priest and Josh White, 'Before the War, CIA Reportedly Trained a Team of Iraqis to Aid U.S.', *Washington Post* (3 August 2005), www.washingtonpost.com/wp-dyn/content/article/2005/08/02/AR2005080201579_pf.html.

[172] Tucker and Faddis, *Operation Hotel California*, pp. 43–44; 124–126; James Risen, *State of War: The Secret History of the CIA and the Bush Administration* (New York: Simon and Schuster, 2006), pp. 138–140; 169.

[173] Priest and White, 'Before the War, CIA Reportedly Trained a Team of Iraqis to Aid U.S.'.

[174] Mike Ryan, *Special Operations in Iraq* (London: Pen & Sword, 2004), pp. 37–42; Dariusz Dachowicz and Hubert Królikowski, *Polish Special Operations Forces and Special Operations: Historical Outline and Present Days* (Warsaw: Taktyka i Strategia, 2016), pp. 62–65.

[175] Naval Polska (pseud.), 'GROM and SEAL Joint Operation: Umm Qasr Offshore Oil Terminal', SOFREP, 26 March 2014, https://sofrep.com/news/grom-seal-joint-operation-umm-qasr-offshore-oil-terminal.

[176] Gordon and Trainor, *Cobra II*, pp. 194–195.

to the incompatibility of service radios, can be attributed to remarkable efforts to integrate via technology, processes, training, exercises, and lessons learned.[177] Technology included 'blue force trackers' which provided unprecedented abilities for higher headquarters to monitor the precise location of specific teams, as well as a 'chat' function connecting them together for secure, instantaneous coordination.[178] In terms of improved processes, communications were simplified between distributed teams, which could coordinate with JSOTF headquarters and their organic Joint Fires Element (JFE) cell and its related Joint Air Coordination Element (JACE), which in turn were linked at the theatre level with Joint Air Coordination Element within the CFACC headquarters.[179]

While some friction nevertheless existed, the end result was a system that 'could distribute the air to whoever needed it first of whoever needed it most' and the number of available air platforms ensured a 'nonstop' supply of airstrikes.[180] Close coordination, in terms of smoothing the interface between the operational and tactical level of special operations upwards and downwards, occurred in JSOTF-W's Joint Operations Center (JOC). The JOC positioned the JFE centrally and linked it in close proximity to field liaison officers in direct contact with distributed teams.[181] Higher up the chain of command, within the CFACC, a range of platform expert weapons officers numbering close to sixty personnel in a Special Operations Liaison Element (SOLE) cell coordinated SOF air assets for mobility and resupply of fielded teams and to quickly and efficiently address unanticipated, potential sources of friendly fire casualties. Such sources included cruise missile and long-range artillery fires with conventional close air support aircraft and SOF air.[182] In addition to the credibility members of the SOLE team established with the Combined Air Operations Center (CAOC) counterparts, a defining aspect that improved coordination between SOF and conventional air were the 'relationships' its personnel developed vital to creating mutual respect and trust.[183]

Relationships built on this foundation were also developed training and exercising together—military SOF conduct pre-deployment as well as isolated training on arrival lasting several weeks, in common with their conventional counterparts. Where SOF differ from them, however, is in the integration into this training of other embedded elements. For IRAQI FREEDOM this included Army Tactical Air Control Party personnel, with whom they do not always exercise. The small size of specialized

[177] The battle is detailed in the official history in Briscoe et al., *All Roads Lead to Baghdad*, pp. 251–259. The friendly fire incident, caused by several factors, is detailed specifically on p. 257.

[178] Bryan Eovito, 'An Assessment of Joint Chat Requirements from Current Usage Patterns', unpublished Master's thesis, Naval Postgraduate School, 1 June 2006.

[179] Jody Jacobs et al., *Enhancing the Integration of Special Operations and Conventional Air Operations: Focus on the Air-Surface Interface*, U//FOUO, MG-535-AF (Santa Monica, CA: RAND, July 2007), p. 52.

[180] Quoted in Steve Call, *Danger Close: Tactical Air Controllers in Afghanistan and Iraq* (College Station, TX: Texas A&M Press, 2007), p. 226.

[181] Jacobs et al., *Enhancing the Integration of Special Operations and Conventional Air Operations*, pp. 53–58.

[182] Paul Harmon, 'Special Operations Liaison Element in OIF: SOLE, Part II', *Air Commando Journal*, 9:2 (October 2020): 23, 25; Randy O'Boyle, '16th Special Operations Group: Early Action in Operation IRAQI FREEDOM', *Air Commando Journal*, 9:2 (October 2020): 32.

[183] Harmon, 'Special Operations Liaison Element in OIF', p. 23.

communities meant distributed units with assigned specialists could form immediate trust bonds based on prior personal relationships.[184] Exercises conducted at the operational and tactical level prior to IRAQI FREEDOM, including Millennium Challenge 02, Joint Force Exercise 02, Internal Look 02, the Early Victor series, and efforts of the Counter-SCUD Working Group experimented and validated concepts of operations (CONOPS) later used to great effect, including dealing with distributed command and control and prosecuting mobile or fleeting time sensitive targets.[185] At the level of execution, particularly in empowering more junior officers to develop solutions, 'it fostered an uncommon level of trust ... to quickly develop a shared awareness of the mission' that spread across the various combined and coalition partners, including SOF and air operators.[186]

Technology, training, and exercises ensured that paramilitary and military special operations were prepared for contingencies once they arrived. Getting there, staying there, and moving around, however, could not have been performed without the efforts of supporting elements working together to create a cohesive capability. CIA and SOF teams initially infiltrated into northern and western Iraq through unspecified various covert and clandestine air and ground means. Closer to the start of IRAQI FREEDOM, SOF aviators prepared to support the three different JSOTFs in different ways, given their unique requirements. One battalion of the 160th SOAR, fourteen helicopters in all, was lifted by Air Force C-17s first to Kuwait and then to its forward base at the H-5 airfield to support Task Force Dagger.[187] Elements of the 16th Special Operations Group, including MH-53s and MC-130s, staged out of Kuwait and 'a partner nation out west' first to infiltrate Task Force Dagger and unspecified CIA proxy teams into Iraq while simultaneously supporting JSOTF-S on the Al-Faw Peninsula, creating a command-and-control headache for the group commander.[188]

Infiltrating Task Force Viking was shaped by the Turkish refusal to allow basing. The subsequent plan, from Romania to Jordan and onward to Iraq the following evening, lent it its eventual name: 'Ugly Baby'.[189] Iraqi air defences precluded high-level flight; instead, an infiltration by MC-130Hs of the 7th SOS and 15th SOS relied on speed and surprise while flying 100 feet above ground at night on a 5-hour flight, doing their best to avoid Iraqi units.[190] Although a 'high-risk option', the plan was approved for two reasons: 'detailed and intensive planning' reduced risk to acceptable levels but, more importantly, 'it was the only option available'.[191] Despite two aircraft being significantly damaged, Ugly Baby accomplished its mission, inserting more than nineteen ODAs and a handful of company headquarters into northern

[184] See, for example, the anecdote related in Call, *Danger Close*, p. 229.

[185] Briscoe et al., *All Roads Lead to Baghdad*, p. 459; Kiras, 'T. Michael Moseley', pp. 417–418.

[186] Brook Leonard, 'How the West Was Won: The Essence of Network Centric Operations (NCO)', unpublished Master's thesis, School of Advanced Air and Space Studies, June 2006, p. 33.

[187] Briscoe et al., *All Roads Lead to Baghdad*, pp. 96–98.

[188] Randy O'Boyle, '16th Special Operations Group: Early Action in Operation IRAQI FREEDOM', *Air Commando Journal*, 9:2 (October 2020): 28–29.

[189] Robinson, *Masters of Chaos*, p. 298.

[190] Briscoe et al., *All Roads Lead to Baghdad*, pp. 116–121.

[191] Cory Peterson, 'Task Force Viking and the Ugly Baby Mission', *Air Commando Journal*, 9:2 (October 2020): 36–37.

Iraq, which 'allowed the United States to keep its promise to the Kurds' to support them 'in force'.[192] More importantly, the risks accepted by the force—almost losing a SOF aircraft and its onboard operators—likely convinced Turkish military leaders to allow overflights and limited use of basing in the country.[193]

After IRAQI FREEDOM began in earnest, SOF air and increasingly conventional C-17s resupplied special operations ground teams with fuel, munitions, and spares, which were trucked forward in composite convoys thanks to the advanced work by forward support companies of the 528th Sustainment Brigade.[194] Secure field communication networks, involving satellite communication up- and downlinks, were created virtually overnight, linking deployed headquarters back to higher echelons due to the efforts of the 112th Signals Battalion. Their efforts, along the with Joint Communications Signal Element, allowed SOF to connect and link in ways previous commands could only dream of.[195]

Other special operations during IRAQI FREEDOM resulted from crisis as opposed to deliberate planning. On 23 March 2003, three days into the operation, Private First Class Jessica Lynch and eight others were captured after being ambushed.[196] Various information sources suggested Lynch was being held in a hospital in Nasiriya, close to the site of the ambush, due to her injuries. This information led to planning for a hostage rescue operation. On 1 April 2003, 2nd Battalion, 75th Rangers and the 160th SOAR, along with additional Marine Corps helicopters, effected her rescue.[197] Although not the first US female prisoner of war, she was the first to be rescued while hostilities were still underway since the Second World War.[198] Less well reported, during the same mission the Rangers also recovered the remains of other Americans discovered on the scene, giving credibility to the creed 'no American left behind', despite the risks of remaining on-scene for several hours to do so.[199] Operation IRAQI FREEDOM ended with Saddam Hussein's overthrow twenty-one days after it began and special operations made several important contributions to the campaign. On the other side of the world, a very different special operations effort was underway.

[192] Peterson, 'Task Force Viking and the Ugly Baby Mission', p. 41; Briscoe et al., *All Roads Lead to Baghdad*, pp. 120–121.

[193] Peterson, 'Task Force Viking and the Ugly Baby Mission', p. 41.

[194] Briscoe et al., *All Roads Lead to Baghdad*, pp. 163–165.

[195] Ibid., p. 45.

[196] For details, see Rick Bragg, *I Am a Soldier, Too: The Jessica Lynch Story* (New York: Vintage Books, 2004). The incident has become an inspiration for future learning. For details on the ambush and the learning objectives for a 'virtual' staff ride, see '507 Maintenance Company Staff Ride', July 2022, US Army Press, www.armyupress.army.mil/Portals/7/educational-services/staff-rides/VSR/507th-Maintenance-Company/Staff-Ride-Readahead-Guidance-507-Maintenance-July-2022.pdf.

[197] Briscoe et al., *All Roads Lead to Baghdad*, pp. 246; 315–316; Naylor, *Relentless Strike*, pp. 211–213.

[198] Nicholas Reynolds, *U.S. Marines in Iraq, 2003: Basrah, Baghdad, and Beyond* (Washington, DC: History Division, US Marine Corps, 2007), p. 109; Specialist Shoshana Johnson, also taken prisoner, was rescued along with six others on 13 April as Marines entered the city of Tikrit, near the end of offensive operations in IRAQI FREEDOM. She was the first African American prisoner of war rescued. Peter Baker and Mary Beth Sheridan, 'Marines Rescue Seven Prisoners', *Washington Post* (14 April 2003), www.washingtonpost.com/archive/politics/2003/04/14/marines-rescue-seven-us-prisoners/5e64984d-6a60-4835-b8c2-b9b1eccf09c3.

[199] Briscoe et al., *All Roads Lead to Baghdad*, p. 318.

## The Philippines, 2001–2014

US special operations in the Philippine archipelago were conducted under the banner of Operation ENDURING FREEDOM but as an adjunct: OEF-Philippines, or OEF-P. These operations share much in common with the military advising and assistance mission to El Salvador. Both comprised a supporting effort in a secondary theatre of interest in a global struggle, conducted to assist an important but flawed partner nation. Operationally, both missions were authorized and acted within significant restrictions and constraints against adaptive and elusive irregular opponents. Unlike El Salvador, OEF-P took place at the height of US global dominance. The restrictions imposed on force levels and conditions placed on aid in El Salvador were a function of US legislative controls put in place to avoid a repeat of Vietnam. In the Philippines, they reflected instead host nation domestic political sensitivities about a perceived return of American forces to the archipelago less than a decade after their withdrawal following nearly a century of occupation there.[200]

In the Western Pacific region, the Philippines government was under siege by three different terrorist groups in the southeastern most group of islands in the archipelago, including the Abu Sayyaf Group (ASG), the Moro National Liberation Front (MNLF), and the Moro Islamic Liberation Front (MILF). The groups were united by a range of overlapping motivations, including hostility against outside rule and rule by a Christian-dominated government that favoured those closest to the governing centre (the island of Luzon), at the expense economically of islands and minorities on the periphery, among others.

Increasingly aggressive and politically embarrassing raids by ASG groups sought foreign revenue by targeting tourist resorts. Commencing its first raid on a Malaysian resort in April 2000, which saw the abduction of twenty-one people, ASG followed this up in August with the capture of American Jeffrey Schilling and the demand of a \$10 million ransom for his release.[201] Eleven months later ASG struck another resort on Palawan Island in the Philippines, capturing three Americans and a number of Filipinos. Military ties between the United States and the Philippines were formally reestablished in 1999 through a Visiting Forces Agreement.[202] Although acts of terrorism, including bombings, were conducted by MILF and armed groups throughout this period, the al-Qaeda plot to assassinate Pope John II during a 1995 visit focused American and Filipino attention on the links ASG, MILF, and MNLF had with global terrorism.[203] Libyan leader Muamar Gaddafi's involvement in negotiating the terms of release of ASG's first set of hostages, for an alleged \$15 million ransom payment,

[200] William Branagin, 'U.S. Military Ends Role in Philippines', *Washington Post* (24 November 1992), www.washingtonpost.com/archive/politics/1992/11/24/us-military-ends-role-in-philippines/a1be8c14-0681-44ab-b869-a6ee439727b7.

[201] Sender redacted to DIA Washington, 'Update on Abu Sayyaf Group (U)', Document ASG01-7, 13 September 2000, p. 3, www.dia.mil/FOIA/FOIA-Electronic-Reading-Room/FileId/162167.

[202] Larry Niksch, 'Philippine–U.S. Security Relations', Report RS 20,697 (Washington, DC: Congressional Research Service, 2000), p. 1.

[203] 'Plan Bojinka Revisited: The Plot and Its Legacy', *Transportation Security International* (16 October 2019), www.tsi-mag.com/oplan-bojinka-revisited-the-plot-and-its-legacy.

raised US suspicions about links to state sponsors given his history of supporting different terrorist groups.[204]

American military advising and assistance support to the Armed Forces of the Philippines (AFP) focused initially on hostage rescue efforts but evolved rapidly after the Palawan raid. The advising and assistance efforts were centred first around the creation of an AFP capability to conduct indirect-shock missions, including hostage rescues. The proposal to create Light Reaction Companies (LRC) started with an enterprising Special Forces officer, Joseph Felter, who took advantage of his previous experience with the AFP and contacted SOF in a bid to conduct more substantial assistance and advising efforts. His plans were submitted and approved up the chain of command from the embassy level to the COCOM and also took advantage of an opportunity to brief a senior visiting State Department official.[205] These efforts yielded negotiations with the government of the Philippines that led to their approval of a mobile training team, which started its activities with a series of JCET efforts, involving personnel from the 6th SOS, as well as different companies from 1st Battalion, 1st SFG. Exercises between AFP and the US, including Balance Piston 01 'to improve the interoperability between the [AFP] and U.S. special forces'.[206]

The JCETs expanded into a broader exercise marking the formal start of OEF-P, Exercise Balikatan, in 2002, focused on the small island of Basilan west of Mindanao. This exercise, which saw 1st SFG practise nighttime insertions onto an island as part of a script, was marred by the crash of a MH-47E special operations helicopter from the 160th SOAR, killing the ten personnel onboard.[207] In a change for most American operators used to working unobtrusively if not clandestinely, the terms of agreement between the Philippines government and the United States required considerable media coverage of special operations activities to allay domestic fears among Filipinos of resurrected colonialism. This coverage helped satisfy an important line of operation for OEF-P: information operations designed to 'enhance government legitimacy' by showcasing progress in the two other lines, building AFP capacity and focused civil-military operations.[208] David Maxwell, who commanded 1/1 SFG and later JSOTF-P, adds intelligence to the list, arguing that the Philippines had sufficient 'intelligence support to enhance the effectiveness of host-nation operations', some of where were provided by 'enabling elements from all of the services' as well as 'the US intelligence community'.[209]

[204] Steven Lambakis, James Kiras, and Kristin Kolet, *Understanding 'Asymmetric' Threats to the United States* (Fairfax, VA: National Institute for Public Policy, September 2002), p. 46n71.

[205] Linda Robinson, Patrick Johnston, and Gillian Oak, *U.S. Special Operations Forces in the Philippines, 2001–2014*, Report RR 1236 (Santa Monica, CA: RAND Corporation, 2016), pp. 13–14.

[206] Kenneth Finlayson, 'Night Stalkers in the Philippines: Tragedy and Triumph in Balikatan 02-1', *Veritas*, 2:1 (2006): 55; David Allen, 'Balance Piston Exercise Begins in Philippines', *Stars and Stripes* (2 February 2003), www.stripes.com/news/balance-piston-exercise-begins-in-philippines-1.1530.

[207] Finlayson, 'Night Stalkers in the Philippines', pp. 56–57.

[208] Gregory Wilson, 'Anatomy of a Successful COIN Operation: OEF-Philippines and The Indirect Approach', *Military Review*, 86:6 (November–December 2006): 6.

[209] David Maxwell, 'Lessons from the Philippines: Irregular Warfare in Action', Modern Warfare Institute at West Point, 10 July 2021, https://mwi.westpoint.edu/lessons-from-the-philippines-irregular-warfare-in-action.

The initial military mission was labelled Joint Task Force-510 but subsequently evolved into a JSOTF. Unlike several other American special operations, the shift in headquarters was accompanied by a decrease in the number of fielded forces. It is partially explained by a shift in focus as well as context. JTF-510 provided military advice and assistance from 2002 to 2004, focusing initially on developing the capabilities of the LRC, the release of the American hostages, and denying ASG the use of Basilan Island as a sanctuary and base of operations.[210] A raid by Philippine Scout Rangers, prompted by ASG beheading several prisoners including one American, Guillermo Sobero, subsequently managed to free only one of the remaining two Americans, Gracia Burnham, in June 2002.[211] Jeffrey Schilling was rescued in another raid the year prior, in April 2001.[212] In addition, JTF-510 ensured training activities, which included helicopter and waterborne insertions, were done as overtly as possible. The rationale for such a move was projecting strength to threaten ASG and reassure the local population.[213] At its height, JTF-510 comprised almost 1,300 personnel in a mix of SOF and conventional forces.

After these initial successful efforts, JSOTF-P expanded its area of operations but shrank the number of deployed Americans by two-thirds to approximately 500–600 personnel.[214] JSOTF-P contained other conventional force elements, including engineers from different armed services and a Marine Corps counterintelligence detachment. Credible media reports suggest CIA support, in the form of intelligence and surveillance, but this has not been publicly acknowledged.[215] SOF were split into three different geographic regions to support AFP in continuing what was termed 'the Basilan COIN model'. These regions were centred on the two main southern islands, Mindanao and Sulu, as well as the maritime environment between them with its range of small islands. Each task force contained detachments to pursue all three lines of operations simultaneously: ODAs working as liaison coordination elements at the AFP battalion, brigade, and division level (as well other joint SOF performing the same function with the Philippine Air Force [PAF] and Navy), military information support teams (MIST), and CA advising their AFP counterparts.[216] In addition to badly needed upgrades to existing AFP platforms, such as the UH-1 Huey helicopters and OV-10 Bronco armed observation aircraft, liaison combat

[210] Greggory Hayes, 'NCO Leadership and OEF-P', unpublished Army Sergeants Major Academy class paper, 5 October 2006, pp. 5–6. Sergeant Major Gregg served a four-month rotation to JTF-510 with ODA-172.

[211] Burnham wrote an account of her rescue, as did one of her rescuers. See Gracia Burnham and Dean Merrill, *In the Presence of My Enemies* (Carol Stream, IL: Tyndale House, 2003) and Oliver Almonares and Phil Fortuno, *The Rescue: God's Amazing Grace* (Manila: MindSpark, 2019).

[212] Associated Press, 'American Hostage Freed in Philippines', *The Washington Post* (13 April 2001), p. 6.

[213] Roger Thompson, 'Special Forces ODA-192 and OEF-P', unpublished Army Sergeants Major Academy class paper, 20 September 2006, p. 4. Sergeant Major Thompson recounted his five-month experience in ODA-192 with JTF-510, including the rescue of Gracia Burnham.

[214] Linda Robinson, 'The SOF Experience in the Philippines and the Implications for Future Defense Strategy', *Prism*, 6:3 (December 2016): 152.

[215] Mark Bowden, 'Jihadists in Paradise', *The Atlantic* (March 2007): 54–75.

[216] Barry Stentiford, *Success in the Shadows: Operation Enduring Freedom-Philippines and the Global War on Terror, 2002–2015* (Fort Leavenworth, KS: Combat Studies Institute, 2018), p. 63.

elements assisted PAF in the planning and execution of joint operations, which were accompanied or followed by MIST and CA support to meet objectives set in an internal defence and development (IDAD) plan.

This plan was developed by Colonel David Fridovich, who led the initial assessment team in 2001 and was selected as the first JTF-510 commander. It was subsequently modified as the AFP fielded an operationally capable LRC, night-vision capable helicopter pilots to deny ASG and later MILF sanctuary after nightfall, small boat handling and maritime night operations, medical evacuation, and precision close air support. Information sharing between different partner organizations, always a problem but more so within the Philippines government, was overcome thanks to the efforts of Colonel Fran Beaudette, JSOTF-P commander from 2011 to 2012, through determined effort over a period and with a mind to creating processes and institutions the host nation could sustain on their own.[217]

In assessing the success of OEF-P, RAND researchers credited several factors. Unsurprisingly they include established COIN principles, including allowing local forces to carry the burden of actions and supporting and enabling them to do so, conducting frequent assessments to ensure plans reflect realities on the ground, and ensuring that all aspects, not just military or security ones, were used in a coherent and coordinated manner. Perhaps more central to the outcome of OEF-P was its function as 'a synergistic interagency effort', which was only possible due to the formal and informal horizontal linkages that already existed between various SOF components due to training and exercises and were extended and sustained through considerable liaising and partnering with conventional support units, different US government departments and agencies, and the AFP.[218]

JSOTF-P also forged connections vertically, again through liaison efforts, to support both the government of the Philippines and the approving and overseeing authority for the Americans in the islands, the Ambassador and their country team. The control measures put in place for OEF-P, particularly tight country team control and restrictive rules of engagement regarding the use of force, likely mitigated the tendency of American advisors to want to solve problems for their host nation partners, as occurred in Vietnam and elsewhere. The agreement allowing US forces into the island to conduct military advising and assistance ensured clear but separate national chains of command between the Philippines and the United States, a proscription on independent US operations, and limited use of force exclusively to self-defence.[219] Few tactical and operational command-and-control problems existed during OEF-P, as from its very beginning it followed doctrinal constructs for how a joint task force and later JSOTF should be constructed. Furthermore, the JTF commander offered direct liaison authorized (DIRLAUTH) among subordinate units,

[217] Konrad Trautman, *Special Operations Intelligence: Guiding the Tip of the Spear*, JSOU Special Report, U//FOUO//REL to USA, FVEY (MacDill Air Force Base, FL: JSOU Press, November 2014), p. 37.

[218] Robinson et al., *U.S. Special Operations Forces in the Philippines*, p. xxv.

[219] Joseph Tomczak, *Parallel Lives in the Indo-Pacific: Edward Lansdale, Donald Wurster, and the Irregular Warfare Mind-Set*, Kenney Paper No. 5 (Maxwell Air Force Base, AL: Air University Press, 2023), p. 57.

granting them the ability to conduct and liaise as the situation required, trusting in their judgement for coordination and achieving a level of 'centralized planning and decentralized execution' in OEF-P.[220]

As a military advising and assistance mission, OEF-P was as successful as the AFP and governance institutions it helped develop. Politically, OEF-P had limited success, as observed by renewed ASG activity including kidnappings, bombings, and coordinated attacks from 2010 onwards. Brian Burton offers a lack of success has its roots in deep tension in Philippines civil-military relations, particularly the fear that military leaders will create power bases they can use to depose civilian ones, as well as persistent corruption within the AFP.[221] Filipino officers echo these points in their writing.[222] This partially explains why political leaders curtailed AFP operations in 2010, only to resume them in 2016 after the ASG beheaded a Filipino hostage.[223] Nonetheless, ASG lost its most senior operational commanders and had the number of its armed combatants curtailed by almost two-thirds, from an estimated 1,200 to 300–400 fighters, posing 'a diminished terrorism threat'.[224] MILF leaders were convinced by the IDAD efforts to negotiate with the government of the Philippines in 2014.[225] OEF-P is the quintessential special warfare success, only working as effectively as the partner with whom one is working 'by, with, and through' and whose results are only as enduring as their sustainment and commitment to addressing the structural and economic inequities once American focus, advice, and resources dry up.

One measure of the operational and strategic effectiveness of US special operations capabilities in Afghanistan, Iraq, and the Philippines is the few casualties they suffered. From October 2001 until November 2004, a twenty-six-month period of nearly continuous operations, American military special operators suffered a mere eighty-two battlefield fatalities. Of these, only fifty-eight were combat fatalities, the others occurring because of accidents such as the crash of the MH-47E off Basilan mentioned above.[226] In contrast, Task Force Ranger suffered eighteen fatalities during a 24-hour period in Mogadishu in 1993. Some of this effectiveness can be ascribed to using proxies, particularly the Northern Alliance or the AFP, to bear the brunt of the fighting. Successes in Afghanistan and the Philippines led Rumsfeld, the Bush administration, and Congress to expand funding and authorities for SOF.

[220] Denis Doty, 'Command and Control of Special Operations Forces for 21st Contingency Operations', unpublished course paper, 3 February 2003, pp. 16–17.

[221] Brian Burton, 'The Promise and Peril of the Indirect Approach', *Prism*, 3:1 (January 2012): 53.

[222] Leonardo Peña, 'Finding the Missing Link to a Successful Philippine Counterinsurgency Strategy', unpublished Master's thesis, Naval Postgraduate School, June 2007, p. 63; Jimmy Larida, 'Isolating the Terrorists: The Abu Sayyaf Case Study', unpublished Master's thesis, Marine Corps Command and Staff College, 16 February 2011, pp. 5–6.

[223] Tino Maslan Jr, 'Combined Kinetic and Non-Kinetic Approaches in Defeating the Sword of God', unpublished Master's thesis, Marine Corps Command and Staff College, 28 April 2017, p. 1. Maslan Jr was a major in the Philippine Marine Corps.

[224] Stentiford, *Success in the Shadows*, p. 83; Zachary Abuza, 'The Demise of the Abu Sayyaf Group in the Southern Philippines', *CTC Sentinel*, 1:7 (June 2008): 10.

[225] Malin Åkebo, '"Coexistence Ceasefire' in Mindanao", *Peace & Change: A Journal of Peace Research*, 44:4 (October 2019): 468–496.

[226] John Holcomb et al., 'Causes of Death of U.S. Special Operations Forces in the Global War on Terrorism, 2001–2004', *Annals of Surgery*, 245:6 (June 2007): 986–987.

Afghanistan seemingly validated special warfare activities and its combined paramilitary and military special operations approach. Surgical strike, however, transformed in response to an insurgent threat in Iraq.

## Iraq: 2004–2007

The initial elation of Iraqis once Saddam was deposed gave way to a degree of local confusion, frustration, and, increasingly, violence. Toppling Saddam consumed so much energy and effort that both senior political and military leaders gave relatively little thought or preparation to how to stabilize Iraq afterwards, as several authors have subsequently made clear.[227] Although President Bush authorized the Office of Reconstruction and Humanitarian Affairs (ORHA) prior to the invasion, neither it nor its successor, the Coalition Provisional Authority (CPA), possessed the expertise or resources to meet Iraqi needs and expectations.

The lack of expertise within CPA is evident in its first two decisions, which ostracized all members of Saddam's Ba'athist party and disbanded the Iraqi army. These decisions had a dramatic destabilizing effect on Iraqi internal security. They removed the primary means of upholding order within Iraq and in the process, alienated those with the skills, contacts, and resources to organize into groups and conduct an insurgency against the drastically outnumbered coalition forces. The small numbers of forces that made Operation IRAQI FREEDOM (OIF) so successful now were a liability as their mission shifted from an offensive to a defensive one that prioritized military advice and humanitarian assistance, especially CA and governance. Further complicating matters was a rotation of commanders and forces out of Iraq, to allow those committed to OIF time to recover and refit. For one SOF task force, Task Force Dagger, 80 per cent of its assigned forces to OIF rotated out during this period.[228] In consequence, larger numbers of conventional forces were committed in a series of stratified and hierarchical commands, growing from a combined joint task force (CJTF-7) after the invasion into a national multinational headquarters, Multi-National Force-Iraq (MNF-I), in ten months.

To prevent 'former regime elements' from coordinating effectively, US special operations units were committed to the hunt for Saddam and his officials. SOF first located Saddam's sons, Uday and Qusay; both were killed when attached forces used anti-tank missiles fired into the building in July 2003.[229] Saddam was

[227] See, for example, Gordon and Trainor, *Cobra II*; Thomas Ricks, *Fiasco: The American Military Adventure in Iraq* (New York: Penguin, 2006); and, more recently, Robert Draper, *To Start a War: How the Bush Administration Took America into Iraq* (New York: Penguin, 2021).

[228] Briscoe et al., *All Roads Lead to Baghdad*, p. 426.

[229] Kevin Sullivan and Rajiv Chandrasekaran, 'Hussein's Two Sons Killed in a Firefight with U.S. Troops', *Washington Post* (23 July 2003), www.washingtonpost.com/archive/politics/2003/07/23/husseins-two-sons-killed-in-firefight-with-us-troops/214c57b8-9c7c-4123-8d33-60e2385d1bb5; 'Mission Saddam: Inside TF 20 operations', *Al Jazeera*, 29 July 2003, www.aljazeera.com/news/2003/7/29/mission-saddam-inside-tf-20-operations.

captured in December, also during a SOF-led operation. In one sense, special operations missions against HVTs in Iraq had measurable success. Crawford states that, 'as of November 2005 ... 80 percent of the Most Wanted Ba'ath Party leaders' had been either captured or killed.[230] Such operations helped clear the way for two Iraqi elections in 2005: the first in January to ratify the country's constitution, and the second in December as a general election to elect members of parliament.

The security vacuum created by CPA was filled by various Iraqi insurgent groups.[231] According to their communiqués and other propaganda, the groups collectively pursued four main strategies: conducting a war of attrition against coalition forces to drive them out; thwarting any political normalization; fighting Iraqi government and security forces; and, conducting spectacular operations and shows of force.[232] Special operations missions increasingly shifted focus from hunting former members of Saddam's regime to capturing or killing insurgent HVTs as Iraqi groups conducted increasingly numerous and technically sophisticated attacks against patrolling coalition forces. Thanks to the efforts of Abu Musab al-Zarqawi, the leader of al-Qaeda's franchise organization in Iraq, the country descended into a civil war that overwhelmed the fledging Iraqi government, its security forces, and the more than 130,000 American and coalition forces there.[233]

While any one of the four main Islamist groups could have assumed leadership over the Iraqi insurgency in 2004, it was Zarqawi's repeated willingness to up the ante, in terms of shock and naked violence or 'brute force against the Americans', that catapulted him to the top.[234] This rise started with the capture and videotaped beheading of contractor Nicholas Berg on 11 May. Zarqawi's networks previously targeted the UN headquarters in August 2003, killing the Brazilian high commissioner for human rights and twenty-one others. Zarqawi increasingly divided al-Qaeda in Iraq (AQI) attacks against four different types of targets: foreign nationals, including Egyptians, Japanese, Russians, and Americans; symbolic targets, specifically Shi'a mosques in Najaf, Karbala, and Ramadi; Iraqi and American security forces; and more indiscriminate bombings of other Shi'a gathering places including markets. The symbolic targets were part of Zarqawi's self-declared 'all-out war' on Shi'a Muslims that culminated in the seizure and systematic destruction of one of the holiest sites of the faith, the Al-Aqsa Mosque in Samarra, in February 2006.[235] The destruction of the Al-Aqsa Mosque was one of a number of special operations conducted by AQI, which also

[230] George Crawford, *Manhunting: Counter-Network Organization for Irregular Warfare*, JSOU Report 09-7 (Hurlburt Field, FL: JSOU Press, 2009), p. 57.

[231] International Crisis Group, *The Next Iraqi War? Sectarianism and Civil Conflict*, Report 52 (New York: International Crisis Group, February 2006), pp. 1–3.

[232] International Crisis Group, *The Next Iraqi War?*, pp. 23–25.

[233] General Accounting Office, *Rebuilding Iraq: Resource, Security, Governance, Essential Services, and Oversight Issues*, Report GAO-04-902R (Washington, DC: General Accounting Office, One 2004), www.gao.gov/assets/a92675.html.

[234] Jean-Charles Brisard, *Zarqawi: The New Face of Al-Qaeda* (New York: Other Press, 2005), p. 133.

[235] 'Al-Zarqawi Declares War on Iraqi Shia', *Al Jazeera*, 14 September 2005, https://web.archive.org/web/20110303163626/http://english.aljazeera.net/archive/2005/09/200849143727698709.html.

included an attempted assault on Abu Ghraib prison the previous April.[236] The Abu Ghraib attack, while unsuccessful militarily, struck at a source of Arab and Muslim humiliation.

The Samarra mosque destruction, considered a special operation by AQI, succeeded in accomplishing the group's main goals: destabilize the Iraqi government by unleashing sectarian violence. The rationale for the violence was clear: mobilizing other Sunnis and forcing them into fighting. As Zarqawi suggested in 2004: 'If we succeed in dragging them into the arena of sectarian war, it will become possible to awaken the inattentive Sunnis as they feel imminent danger and annihilating death at the hands of these Sabeans.'[237] Social norms of tolerance and avoidance were replaced by reprisals and revenge killings, which increased monthly from between triple to sevenfold over the next eleven months, as Figure 8.4 suggests. Within the United States and elsewhere, the increase in violence contributed to mounting criticism of the war in Iraq and its conduct, as well as questions about its outcome. Such criticism translated into political pressure on the Bush administration for a decisive response.

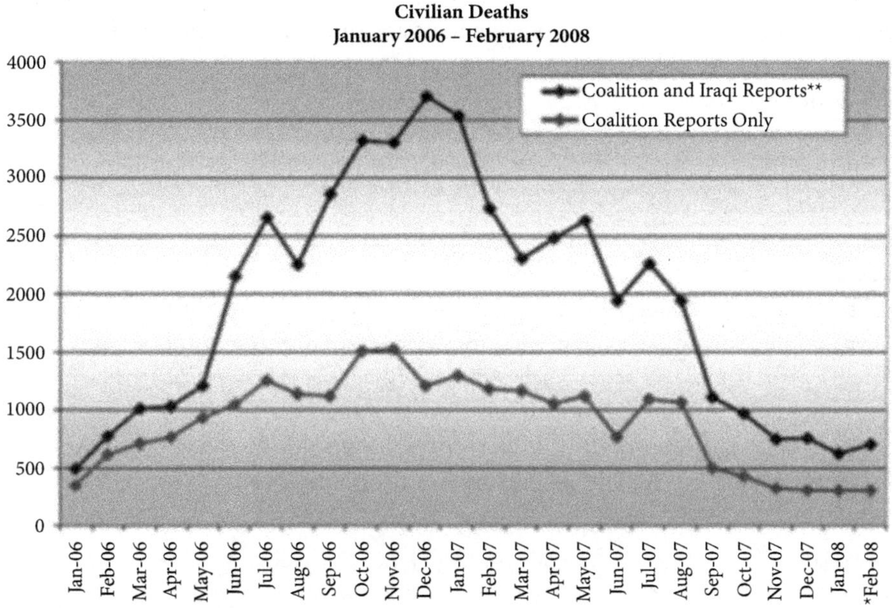


**Figure 8.4** Iraqi civilian deaths, 2006–2008.

*Source:* Hannah Fisher, Congressional Research Service, based on *Measuring Security and Stability in Iraq*, Multi-National Corps—Iraq Strategic Plans Assessments Iraq Significant Activities (SIGACTS) III database (Coalition Reports Only) and Coalition and Iraqi Reports, as of 23 February 2008. https://en.m.wikipedia.org/wiki/File:Iraqcivcas.png.

[236] These attacks fit the criteria for non-state special operations developed by Ian Rice and Craig Whiteside, 'The Strategic Utility of SOF from a Non-State Perspective', in James D. Kiras and Martijn Kitzen, eds., *Into the Void: Special Forces after the War on Terror* (London: Hurst, 2024), pp. 107-125.

[237] 'Zarqawi Letter', February 2004 Coalition Provisional Authority English translation of terrorist Musab al-Zarqawi letter obtained by United States Government in Iraq, n.d., US State Department archive, https://2001-2009.state.gov/p/nea/rls/31694.htm.

With the benefit of hindsight, it is possible to discern the confluence of several social and military streams that collectively marked the end of the AQI. The first, and perhaps most important, was discontent among the Sunni tribes in Anbar province in western Iraq. An insurgency of a different sort formed the basis for the second stream. Networks of military officers and civilians within the Department of Defense were increasingly frustrated with the seeming inability to adapt to the character of the conflict in Iraq. This stream formed the basis of 'the surge' of 30,000 personnel into Iraq to create the conditions for local security and stability, announced by President Bush in January 2007.[238]

The important stream for special operations was a transformation in surgical strikes based on both horizontal and vertical integration. Its roots lay in a dysfunctional portion of the relationship between American paramilitary and military special operations. Tension between the CIA and special mission units was based in perceived encroachment of the latter into mission areas historically the preserve of the former. This included intelligence collection and analysis conducted clandestinely or covertly, leading to 'turf struggles' that 'exacerbated SOF-CIA animosities'.[239] These struggles took on new life during the tenure of Secretary of Defense Donald Rumsfeld. He had little patience for bureaucratic equivocation, particularly if led to obstruction and delay. The president's stated priority was 'killing and capturing the enemy'; despite his expressed reservations, Rumsfeld proceeded with his characteristic drive to fulfil it.[240] His frustration over delays imposed by CIA led Rumsfeld to create a coordinating official under his direct control in March 2003, the under-secretary of defense for intelligence.[241] Another purpose was to provide the authority to expand intelligence gathering and collection under the auspices of traditional military activities to pursue terrorists through more expanded clandestine and covert means.

The heart of transformation lay in closer coordination of various authorities, resources, techniques, and expertise that resided across organizations not traditionally associated with special operations. While special operations coordination had existed for specific missions and cross-training between the Federal Bureau of Investigation, military SOF, and the CIA, it had not for other potential partners such as the Department of Treasury, State, or individual state law enforcement

[238] Nicholas Schlosser, *The Surge, 2007–2008*, CMH Pub 78-1 (Washington, DC: Center of Military History, 2017), p. 92; William Knowlton Jr, *The Surge: General Petraeus and the Turnaround in Iraq*, Industrial College of the Armed Forces Case Study (Washington, DC: National Defense University Press, December 2010); Thomas Ricks, *The Gamble: General David Petraeus and the American Military Adventure in Iraq, 2006–2008* (New York: Penguin, 2009).

[239] Thomas Henriksen, *Eyes, Ears and Daggers: Special Operations Forces and the Central Intelligence Agency in America's Evolving Struggle against Terrorism* (Stanford, CA: Stanford University Press, 2016), p. 103.

[240] Eric Schmitt and Thom Shankar, *Counterstrike: The Untold Story of America's Secret Campaign against Al Qaeda* (New York: Times Books, 2011), p. 42; Rowan Scarborough, *Rumsfeld's War: The Untold Story of America's Anti-Terrorist Commander* (Washington, DC: Regnery, 2004), pp. 22; 28.

[241] Douglas Feith, *War and Decision: Inside the Pentagon at the Dawn of the War on Terrorism* (New York: Harper, 2008), pp. 60; 104; Naylor, *Relentless Strike*, pp. 98–99; Michael DeVine, 'Defense Primer: Under Secretary of Defense for Intelligence & Security', IF10523, Congressional Research Service, November 2022.

agencies such as New York's Joint Terrorism Task Force.[242] The globally distributed and networked nature of contemporary terrorism suggested the need for greater coordination, a point reinforced by the 9/11 Commission's call for greater 'unity of effort'.[243] Experiments early in OEF at the operational or COCOM level proved successful enough to become a standing interagency task force, Task Force Bowie. Additional funding and personnel led to the growth in scope of effectiveness and the creation of the first 'all-source fusion cells' that formed the basis for special operations 'intelligence-operations fusion'.[244]

This fusion, which provided the type of information necessary to conduct surgical strikes more frequently and accurately than before, in line with historical practice elsewhere, remained episodic in nature.[245] For example, in a thirty-day period in Iraq in 2004, special missions units were only able to conduct eighteen indirect-shock missions.[246] More problematically, their actions were insufficiently coordinated tactically and operationally, leading to additional tension between units and organizations ostensibly working towards the same operational and strategic goals. The three most visible manifestations of this tension were the clash in objectives between those conducting surgical strikes intermittently in areas where others were working towards providing security and stability, residual suspicion based on different organizational cultures and operating methods, and an inability to coordinate or share specific information based on standard operating procedures and unique security classification and protection methods.[247]

The commander of Task Force (TF) 714, Stanley McChrystal, discerned that success in conducting surgical strikes against individuals was impotent against AQI's burgeoning terrorist network. So too did his chief of intelligence, Michael Flynn. In McChrystal's words:

> We realized we had to have the rapid ability to detect nuanced changes, whether the emergence of new personalities and alliances or sudden changes in tactics. And we had to process that new information in real time—so we could act on it. A stream of hot cinders was falling everywhere around us, and we had to see them, catch

[242] Christopher Lamb and Evan Munsing, *Secret Weapon: High-Value Target Teams as an Organizational Innovation*, Institute for National Strategic Studies Strategic Perspectives 4 (Washington, DC: National Defense University Press, March 2011), p. 11.

[243] National Commission on Terrorist Attacks, *The 9/11 Commission Report* (New York: W.W. Norton, 2004), pp. 399–419.

[244] Lamb and Munsing, *Secret Weapon*, p. 12.

[245] Graham Turbiville Jr, *Hunting Leadership Targets in Counterinsurgency and Counterterrorist Operations: Selected Perspectives and Experience*, JSOU Report 0706 (Hurlburt Field, FL: JSOU Press, 2007), pp. 77–78.

[246] Richard Shultz Jr, *Military Innovation in War: It Takes a Learning Organization, a Case Study of Task Force 714 in Iraq*, JSOU Report 16-6 (MacDill Air Force Base, FL: JSOU Press, 2016), p. 7.

[247] US House of Representatives, The Subcommittee on Intelligence, Information Sharing, and Terrorism Risk Assessment, Committee on Homeland Security, *The Over-Classification and Pseudo-Classification, Part I, II, and III* (S. 110-20) (Washington, DC: Government Printing Office, 2009); General Accounting Office, *Managing Sensitive Information: DOD Can More Effectively Reduce the Risk of Classification Errors*, Report GAO-06-706 (Washington, DC: General Accounting Office, June 2006); Jon Lindsay, 'Target Practice: Counterterrorism and the Amplification of Data Friction', *Science, Technology, and Human Values*, 42:6 (November 2017): 1072.

> those we could, and react instantly to those we had missed that were starting to set the ground on fire.[248]

Changes in four areas were necessary to realize McChrystal's insight that 'it takes a network to fight a network': the nature of leadership in command, the command relationship with fielded units, enhanced intelligence sharing (which meant changing organizational norms and breaking down structural and cultural barriers), and the relationship between intelligence and special operations.

The nature of command, the first such change, involved embracing an adaptive mindset and approach. For TF 714 to combat a distributed network, its leaders needed to embrace the key features of a learning organization: not being paralyzed by surprise, problem solving as a core competency, no 'sacred cows' in terms of organizational practices, learning lessons and transforming information into knowledge, and stimulating learning throughout the organization.[249] Other key features included clear mandates and purpose, access to decision makers, training, and rewards, split across the organizational, team, and individual level.[250] Above all else, McChrystal sought to break an approach to management that sought efficiency through repetitive processes to hedge against chaos and uncertainty. This leadership culture not only trusted its subordinates but also 'rewarded individual initiative and critical thinking, as opposed to simple execution of commands' through 'patient, yet relentless nurturing of competence and adaptability', or 'the Nelson touch'.[251]

TF 714 was trying to accomplish what Horatio Nelson did over decades, while simultaneously fighting AQI. McChrystal already had existing integrated teams within his SOF units: ODAs, SEAL teams, and AFSOC crews and Special Tactics teams. What he wanted instead were 'superteams' whose success was based on relationships rather than hierarchy: 'The first step of this is constructing a strong lattice of trusting relationships ... In a command, the connections that matter are vertical ties; team building, on the other hand, is all about horizontal connectivity.'[252] Such superteams had to react to circumstances without waiting for orders to do so.

The second of these changes in command relationships sought to put into practice an ideal of empowerment. This ideal is expressed militarily in the term 'mission command', pursued for the presumed advantage it confers: the quickening of decision making and being more responsive to changes in the environment, as well as the exploitation of unanticipated opportunities. McChrystal conferred agency to his subordinates, while assuming responsibility for their actions, to increase the

[248] Stanley McChrystal, 'It Takes a Network: The New Front Line of Modern Warfare', *Foreign Policy*, 21 February 2011, https://foreignpolicy.com/2011/02/21/it-takes-a-network.

[249] Richard Shultz Jr, *Transforming U.S. Intelligence for Irregular War: Task Force 714 in Iraq* (Washington, DC: Georgetown University Press, 2020), pp. 179–181.

[250] Lamb and Munsing, *Secret Weapon*, p. 37.

[251] In reference to the Admiral Horatio Nelson, commander of several decisive British naval victories during the Napoleonic Wars. Stanley McChrystal, *Team of Teams: New Rules of Engagement for a Complex World* (New York: Portfolio, 2015), pp. 30–34.

[252] McChrystal, *Team of Teams*, pp. 96–97.

tempo of surgical strikes.[253] Subordinates at lower levels could leapfrog their chain of command in pursuit of an objective, if sufficiently relevant. If not, the episode provided a teachable moment. In other words, commanders conferred on their subordinates the power to act as well as fail. McChrystal could trust in the judgement and decisions of his subordinates based on their demonstrated skills and abilities.[254] McChrystal termed his idea for coordinated operations 'decentralized control, empowered execution', ones based on 'emergent, adaptive intelligence'.[255] Other scholars have labelled the desired organizational effect of empowerment as 'swarming' or 'chaoplexic war'.[256]

Changes in intelligence, the third requirement, were ongoing but insufficient. Intelligence may have been shared and 'fused' tactically from a range of different sources but operations were often delayed waiting for outside intelligence agencies to locate specific individuals and determine their importance in a network based on seven analytic criteria.[257] A similar organization was necessary to facilitate information fusion at the nexus between the operational and the strategic level. In pursuit of this, McChrystal built a JOC that literally broke down walls between different operational sections with collectors, analysts, and others '[w]orking coffee-breath close with operators' to create access and a level of transparency.[258] Another was ensuring that other organizations were sufficiently represented and treated as partners in the surgical strike mission.

To this end, McChrystal developed an initial joint interagency task force (JIATF) for countering terrorism, based on McRaven's initial idea, that served as a bridge between the strategic and operational by facilitating coordination through inclusiveness. Partner organizations contributing expertise or information would see their value, as they had a 'stake in policy and its outcome' as it related to special operations surgical strikes.[259] More importantly, the JIATF concept was necessary to 'break down information silos' that traditionally inhibited information sharing between different intelligence agencies.[260] Once implemented, JIATFs were scalable from the strategic (JIATF-NCR) down through the operational level of combating terrorism. By creating valued partners vested in the success of the mission, gaps in the intelligence process could be minimized if not eliminated altogether.[261]

[253] Shultz Jr, *Transforming U.S. Intelligence for Irregular War*, pp. 184–185.
[254] Stanley McChrystal, *My Share of the Task: A Memoir* (New York: Portfolio, 2013), pp. 99; 109.
[255] McChrystal, *Team of Teams*, p. 198.
[256] John Arquilla and David Ronfeldt, *Swarming and the Future of Conflict*, Report DB311 (Santa Monica, CA: RAND Corporation, 2005); Antoine Bousquet, *The Scientific War of Warfare: Order and Chaos on the Battlefields of Modernity*, 2nd ed. (Oxford: Oxford University Press, 2022).
[257] Trautman, *Special Operations Intelligence*, p. 29; Matthew Ford, 'Precision Targeting and the Industrialisation of the Intelligence Cycles on the Battlefields of Iraq and Afghanistan', *Inteligencia y seguridad*, 13 (May–June 2013): 226.
[258] Brian Keller, JSOTF J-2, quoted in Trautman, *Special Operations Intelligence*, p. 28.
[259] Dennis Skocz, 'A Front-Line View of "The" Interagency: The Practice of Policy Coordination Inside the Government', in Gabriel Marcella, ed., *Affairs of State: The Interagency and National Security* (Carlisle Barracks, PA: Army War College Press, 2008), p. 379.
[260] Patrick Bury, 'U.S. Special Forces Transformation: Post-Fordism and the Limits of Networked Warfare', *International Affairs*, 98:2 (2022): 589.
[261] McChrystal, *Team of Teams*, pp. 208–210.

Flynn summarized the effect the changes in intelligence holistically: 'By assembling small groups of bright, capable individuals under the same roof, Fusion Centers were able to coordinate classified [signals intelligence] and [human intelligence], and real-time surveillance video, allowing commanders to "action" the information with airstrikes and special operations that led to the death or capture of notorious terrorists.'[262]

Sufficient intelligence is crucial to the success of surgical strike operations. Changing the relationship between intelligence and special operations was the fourth and final requirement. The beating heart of the JIATF was an established targeting cycle frequently revised and revisited: first 'find, fix, finish, and analyze' later modified to add 'exploit' and 'disseminate', or the F3EAD cycle.[263]

Four aspects of this cycle and its relationship to special operations are noteworthy. First, five-sixths of the cycle is intelligence and operations related, requiring rapid identification and processing. This process, however, necessitated a mindset change among operators. Instead of the focus in time and attention being placed on operations with support from intelligence, the ratio for TF 714 was inverted, according to Flynn: 20 per cent on operations and 80 per cent on intelligence.[264] Second, as Aki Peritz and Eric Rosenbach point out, the 'finish' phase is effect- and not unit- or platform-specific. Stated differently, while special operations were often the preferred method of finishing, the effect—when a threat is neutralized—might be better performed by another relevant and accessible unit or system, using force or through other available means.[265] Third, the nature of a networked opponent meant time took on a greater meaning when it came to countering asymmetric threat success. Fourth, the focus of special operations shifted from killing AQI members to capturing them for immediate interrogation to exploit the initial shock of the raid.[266] Those collecting and analysing the information needed to influence and change the mindset of operators. 'Sensitive site exploitation' (SSE) consisted of recovering material of strategic importance but in Iraq it shifted away from sensitivity to focus on what was important and exploitable.[267] Wayne Barfoot, a TF 16 intelligence chief, brought best practices and lessons learned from OEF-P to integrate together operators and intelligence analysts, forcing the former to perform the latter's tasks while off-duty and forging a level of mutual respect along the way.[268] The intensity of the last two phases of the cycle

[262] Michael Flynn, Matt Pottinger, and Paul Batchelor, *Fixing Intel: A Blueprint for Making Intelligence Relevant in Afghanistan* (Washington, DC: Center for a New American Security, January 2010), p. 21.

[263] Ford, 'Precision Targeting and the Industrialization of the Intelligence Cycle on the Battlefields of Iraq and Afghanistan', pp. 224–225.

[264] Shultz Jr, *Transforming U.S. Intelligence for Irregular War*, p. 187.

[265] Aki Peritz and Eric Rosenbach, *Find, Fix, Finish: Inside the Counterterrorism Campaigns that Killed Bin Laden and Devastated Al-Qaeda* (New York: PublicAffairs, 2012), p. 8.

[266] Matthew Alexander, *Kill or Capture: How a Special Operations Task Force Took Down a Notorious Al Qaeda Terrorist* (New York: St Martin's Press, 2011), p. 271.

[267] An example of a task force conducting SSE is 75th Exploitation Task Force, deployed to Iraq in 2003 to search for WMD. For details, see Major Kevin Brown, interview by Operational Leadership Experiences Project team with Combat Studies Institute, digital recording, 17 October 2005. Fort Leavenworth, Kansas. www.armyupress.army.mil/Portals/7/online-publications/documents/interview-with-maj-kevin-brown-75th-exploitation-task-force-operation-iraqi-freedom.pdf.

[268] McChrystal, *My Share of the Task*, pp. 138–139.

could be relentless, not just on AQI but on the analysts as well. Sean Naylor suggests the goal was to get these phases down to 'within an hour of coming off target'.[269]

Rather than separate processes which imposed delays or blinks, the modified F3EAD process utilized by TF 714 became an organic whole. The only break was the need for decision making enabled by a communications architecture linking together main and subordinate headquarters, together with partners and later allies, and routinely synchronized despite the distances involved through battle and commander update briefings conducted through video teleconferences (VTCs).[270] The net effect of shared intelligence feeding common processes in a collaborative environment was a 'shared consciousness' across TF 714 and the entire capability 'enterprise'.[271]

The resulting changes in TF 714's approach to special operations took time and experimentation to yield their results in the 'living laboratory' of Iraq.[272] The initial efforts were modest. The original JIATF experiment, conducted by McRaven at the tactical level out of a tent in Bagram, Afghanistan, comprised between twenty-five and thirty-five people. Its success prompted the creation of a larger counterpart, at scale, at Balad in 2005.[273] As McChrystal stated in a 2013 interview, special operations indirect-shock missions occurred at a rate of just over one every other day in 2004, increasing to ten per night in a single month in 2006.[274] More importantly, the changes in TF 714 yielded an important operational results and strategic effects. Under siege politically given the dire state of security in Iraq, TF 714's methodology led to the 'finish' of the symbol of al-Qaeda's violence in Iraq, Zarqawi. Through meticulous contact tracing, 'pattern of life' development, and despite some 'blinks' in coverage, special operators were able to fix and verify his location in a building on the outskirts of Baghdad and call in an airstrike by a pair of National Guard F-16s on 7 June.[275] Two years later, TF 714's activities evolved into 'a high-volume, awe-inspiring machine' in Iraq.[276] More pointedly, what had been a capability labelled as 'manhunting' or 'hunter-killer teams' had evolved into a different capability entirely: the 'hunter gatherer', blurring the line between operations and intelligence.[277]

To maintain the tempo of operations in a sustainable manner, the task force extended its collaborative efforts to include trusted partner nations. Although operators worked, cross-trained, or exercised together to develop skills or share expertise,

[269] Naylor, *Relentless Strike*, p. 289.

[270] Shultz Jr, *Military Innovation in War*, pp. 49–50. Lloyd Edwards, Matt Gibson, and David McCarthy, 'From Fix to Finish: The Impact of New Technologies on the Special Operations Approval Process', Harvard Kennedy School, M-RCBG Associate Working Paper No. 56 (Cambridge, MA: Harvard Business School, May 2016), p. 72.

[271] McChrystal, *Team of Teams*, p. 197.

[272] Peritz and Rosenbach, *Find, Fix, Finish*, p. 215.

[273] McChrystal, *My Share of the Task*, p. 119.

[274] Gideon Rose, 'Generation Kill: A Conversation with Stanley McChrystal', *Foreign Affairs*, 92:2 (March/April 2013): 2–3.

[275] Peritz and Rosenbach, *Find, Fix, Finish*, pp. 130–132.

[276] Lamb and Munsing, *Secret Weapon*, p. 50.

[277] Joseph Celeski, *Hunter-Killer Teams: Attacking Enemy Safe Havens*, JSOU Report 10-1 (Hurlburt Field, FL: JSOU Press, 2010); Crawford, *Manhunting*; John Hardy, 'Hunters and Gatherers: The Evolution of Strike and Intelligence Functions in Special Operations Forces', *International Journal of Intelligence and CounterIntelligence*, 36:4 (2023): 1143–1163.

national sensitivities as well as the 'stovepipes' mentioned above limited closer coordination during operations. The standard operating practice, drawing upon doctrine and used since OEF, was to allocate other nationalities to their own joint operations area.[278] Informal mechanisms outside of operator preparation were useful in bridging the divide and forging collaborative relationships. For example, in 1999 OASD SO/LIC created the Counter Terrorism Technical Support Office (CTTSO) comprising several functional area subgroups to share resources and expertise between the United States, Australia, Canada, Israel, Singapore, and the United Kingdom.[279]

These preexisting informal mechanisms, combined with recent experience, eased the integration of the SAS into the TF 714 enterprise architecture by mid-January 2006. Closer working relationships had been stymied thanks to national sensitivities, particularly over detention practices and facilities, as well as personality clashes with senior British officers.[280] Once integrated into the architecture, British media reports suggest that over a two-year period the SAS killed or captured '[m]ore than 3,500 insurgents' in Iraq, having tapped specifically into American intelligence and mobility capabilities, although these figures cannot be verified.[281] Other relationships resulted from the dispatch of observation teams, or TF 714 liaison officers to partner organizations as well as countries including Israel, Saudi Arabia, and Jordan.[282] Some of the best personnel were selected for such duty, which could involve reciprocal practice and information sharing, and commanders stressed the need to understand issues from partner's perspectives as well as forge and build relationships with them.[283]

Assessing the success of TF 714 against AQI is difficult; its supporters and critics both make their cases based on incomplete information. Some use the term 'industrial counterterrorism', positively or negatively, to characterize its effectiveness.[284] Supporters point to evidence of declining AQI influence throughout the country, including civilian deaths, numbers of attacks, and other measurements of performance.[285] Others are more guarded, suggesting that while TF 714 'took out a staggering number of insurgents', its effects were no substitute for a more comprehensive

[278] Joint Staff, Joint Publication 3-05, *Special Operations* (Washington, DC: The Joint Staff, 16 July 2014), p. III-7; Development, Concepts, and Doctrine Centre, *Special Forces Operations*, JDP 3-05, Restricted (Swindon: DCDC, 2008), p. I-2.

[279] For details on IWTSD, its partners, and its technical support working groups, see 'About IWTSD', Irregular Warfare Technical Support Directorate, n.d., www.cttso.gov/About.html.

[280] Mark Urban, *Task Force Black* (London: Little, Brown, 2010), pp. 86–87; 90–91; 117.

[281] Sean Rayment, 'SAS Kills Hundreds of Terrorists in "Secret War" against al-Qaeda in Iraq', *The Sunday Telegraph*, 30 August 2008, www.telegraph.co.uk/news/newstopics/onthefrontline/2652496/SAS-kill-hundreds-of-terrorists-in-secret-war-against-al-Qaeda-in-Iraq.html.

[282] Shultz Jr, *Transforming U.S. Intelligence for Irregular*, pp. 194–200.

[283] McChrystal, *Team of Teams*, p. 177.

[284] Stanley McChrystal, 'Becoming the Enemy: To Win in Afghanistan, We Need to Fight More Like the Taliban', *Foreign Policy*, 185 (2011): 66–70; Matthew Ford, 'Precision Targeting and the Industrialization of the Intelligence Cycle on the Battlefields of Iraq and Afghanistan', *Inteligencia y seguridad*, 13 (January–June 2013): 221–233; Rebecca Jensen, 'The Diffusion of Innovation: SOF in Iraq', in Jessica Turnley, Kobi Michael, and Eyal Ben-Ari, eds., *Special Operations Forces in the 21st Century: Perspectives from the Social Sciences* (London: Routledge, 2018), pp. 227–228.

[285] Lamb and Munsing, *Secret Weapon*, pp. 50–51.

approach 'integrating the surgical strikes of [SOF] with the population-security and area-security operations of conventional forces'.[286] At a critical juncture in the campaign, Shultz Jr argues TF 714 came to grips with the another crucial element of an insurgency: its 'clandestine infrastructure or secret underground apparatus' that comprises a 'largely invisible ecosystem'.[287] Some entirely overlook the task force's contribution and attribute success to parallel covert efforts by the CIA and overt ones by Marines in Anbar Province and AQI infighting.[288] Others observe that success was hollow and short-lived since the terrorists captured, interrogated, and detained by TF 714 were subsequently released between 2011 and 2012 and gave rise to the Islamic State.[289] Despite the critiques, TF 714 reduced AQI to a husk by 2009.

TF 714's transformation influenced US and global special operations for the better part of the next decade. The F3EAD process became the model for success to emulate elsewhere. That record has been mixed. In terms of successes, the formation of a coordinating headquarters for NATO in 2006, the NATO SOF Coordination Centre (NSCC; now Allied Special Operations Forces Command), continues to work towards information sharing and 'synchronis[ing] capability development across NATO'.[290] Prompted by McRaven and initially staffed with officers chartered to improve coordination along the lines of JIATF model, the NSCC published its inaugural study in 2008. Although the study authors acknowledge 'one universally applicable organisational model does not exist', they nevertheless identify common characteristics of 'a world class special operations organisation':

- Comprising 'expeditionary SOF tactical units capable of performing special operations in harsh, uncertain, hostile, denied, and politically sensitive environments in concert with other SOF', a deployable headquarters to direct them, and SOF-specific combat support forces and capabilities in a lean organisational architecture to facilitate agility;
- Capable of:
  - 'Providing centralized stewardship, authority, and direction';
  - 'Accessing senior defense leaders directly and advising them on SOF';
  - 'Direct access to the senior defence leadership'.[291]

In sum, NATO partner members should have a TF 714-like structure. Other successes include the creation of a regional JIATF-like 'information sharing

[286] Moyar, *Oppose Any Foe*, pp. 279–280.

[287] Richard Shultz Jr, '"The Irreducible Minimum": An Evolution of Counterterrorism Operations in Iraq', *Prism*, 7:3 (2018): 107.

[288] Seth Jones, *Hunting in the Shadows: The Pursuit of Al Qa'Ida since 9/11* (New York: Norton, 2012), p. 260.

[289] Paul Rogers, 'Not Just the CIA: Special Forces in Iraq Fought a Shadow War That Led to the Rise of IS', *The Conversation*, 15 December 2014, https://theconversation.com/not-just-the-cia-special-forces-in-iraq-fought-a-shadow-war-that-led-to-rise-of-is-35507.

[290] 'Special Operations Forces', NATO, 23 July 2023, www.nato.int/cps/en/natohq/topics_105950.htm#:~:text=operations%20and%20exercises.-,Allied%20Special%20Operations%20Forces%20Command%20(SOFCOM),Allied%20Commander%20Europe%20(SACEUR).

[291] NATO Special Operations Coordination Centre (NSCC), *The North Atlantic Treaty Organization Special Operations Forces Study* (Mons, Belgium: NSCC, December 2008), pp. 28–29.

effort', or what the New Zealand government terms an 'intelligence fusion centre' in Jordan named Operation GALLANT PHOENIX.[292] First created to track Islamic State foreign fighters, the effort 'provide[s] a platform for coalition partners to collect, monitor and process material regarding potential and existing terrorist threats and trends at home and globally'.[293] In addition, it 'has grown from fewer than 20 people representing two nations, to more than 250 people representing 25 nations', according to the chairman of the joint chiefs of staff.[294]

Strategic successes in exporting TF 714 practices have not been matched tactically. Their spread elsewhere, to include other SOF operating in Iraq and Afghanistan, did not necessarily take into account the unique factors that made them work for TF 714 in Iraq. The author of an official study noted:

> success [against AQI] was predicated on an unprecedented level of direct action operations that were enabled by a host of critical factors: a very high level of special operations, intelligence, and monetary resources; almost complete freedom of movement and action across the entirety of the country (most of which also featured easily navigable terrain); authorities for action pushed down to essentially the tactical level; a high degree of tolerance for the (at least temporary) detention of suspected members of Al-Qaeda and collateral damage; and a number of significant innovations in technical means of gathering and analyzing intelligence.[295]

Jon Lindsay provides a different perspective, drawing on observations from a six-month deployment supporting a SEAL Special Operation Task Force (SOTF) in Anbar Province starting in 2007. This period saw 'over 200 raids resulting in the capture of over 300 detainees'.[296] Lindsay outlines the basis for 'data friction' in the F3EAD system:

> More machinery was in place for building up a target than for inspecting the quality of construction. Intelligence techs at the lowest echelons faced a great deal of pressure from operators to produce targets. The suppression of skepticism about target quality and disinterest in alternatives to direct action was rarely the result of explicit coercion by operators and more often a matter of self-censorship. By

[292] Jim Garamone, 'Dunford Asks Defense Chiefs to Guard against Complacency', Joint Chiefs of Staff, n.d., www.jcs.mil/Media/News/News-Display/Article/1664622/dunford-asks-defense-chiefs-to-guard-against-complacency; 'Jordan: Operation Gallant Phoenix', New Zealand Ministry of Defense, n.d., www.defence.govt.nz/what-we-do/diplomacy-and-deployments/deployment-map/jordan.

[293] From Office of the Minister of Defence to Chair, Cabinet National Security Committee, 'New Zealand Participation in Operation Gallant Phoenix', March 2017, redacted documents released under the Official Information Act 1982, www.defence.govt.nz/assets/publication/file/1dbb25a7e7/NZ-Participation-in-OGP.pdf.

[294] Garamone, 'Dunford Asks Defense Chiefs to Guard against Complacency'.

[295] Zack Gold, *Al-Qaeda in Iraq (AQI): An Al-Qaeda Affiliate Case Study*, Report DIM-2017-U-016118-2Rev (Alexandria, VA: Center for Naval Analyses, October 2017), p. 22.

[296] Jon Lindsay, 'Target Practice: Counterterrorism and the Amplification of Data Friction', *Science, Technology, and Human Values*, 42:6 (November 2017): 1063.

> producing targets, moreover, techs could participate in the 'hunt' along with operators, which improved solidarity throughout the military organization. Questioning assumptions or methodologies behind the initial target designation became harder as the hunting process gained momentum, as more collection effort was invested, and as the representational residua of tracking accumulated. The time and effort spent going after a target then reinforced a sense of the target's value.[297]

The remainder of his article suggests features key to the success of HVT teams identified previously existed only in varying degrees or were entirely absent from the SOTF. Most notably among them was a failure to break down and equalize status within the task force, which in turn ensured that bias clouded analysis and impeded more critical inquiry of activities and practices.[298] Worse, 'the SOTF reflexively pursued direct action when it had other options'.[299] Success in Iraq lead to expansion globally and export to Afghanistan, with disappointing outcomes.

## 2008–2014: Expansion, Global SOF, Convergence, Afghanistan

This subsection differs from previous ones in focus and content. It shifts away from analysis of US special operations in named operations and looks more broadly at force developments and control measures. The patterns set in Afghanistan and Iraq, and to a lesser extent the Philippines, of increasing horizontal integration of capabilities continued and evolved as US special operations expanded. More intriguing, the theory of special operations coherence suggests what occurred in control measures and risk calculations was vertical integration. This subsection examines four matters related to vertical integration: the resourcing and force management issues, the expansion of the TF 714 model as the dominant American special operations paradigm into a global SOF network, increasing professional and political concerns regarding the convergence of SOF into intelligence functions, and the split between surgical strike and special warfare activities that occurred in Afghanistan from 2010 onwards.

SOF were rewarded for their successes with greatly expanded resources and authorities. The most visible was the expansion in the number of special operations personnel. Overall personnel strength increased from 46,134 in 1999, which includes National Guard, service Reserve, and civilians, to 58,000 by 2008. Some occurred in operators, such as a fourth battalion to each of the SFGs in a plan approved by the secretary of defense in May 2008.[300] Another battalion added four additional companies and eighteen more ODAs to each group, relieving some of the stress of

[297] Lindsay, 'Target Practice', pp. 1078–1079.
[298] Ibid., pp. 1070; 1076; 1085.
[299] Ibid., p. 1088.
[300] Jared Tracy, 'Victoria Ex Umbra: Activating 4th Battalion, 1st Special Forces Group (Airborne)', *Veritas*, 14:2 (2008): 5.

continuous deployments. The additional battalions were programmed for activation in a rolling annual sequence, beginning with 5th Group in 2008 and ending with 7th Group in 2012.[301] This sequencing was done in a bid to avoid violating the 2nd and 3rd SOF Truths regarding maintaining quality and not expanding, or mass producing, operators. The challenges in doing so were considerable, thanks in part to cuts in the potential recruiting pool for SF and despite the implementation of novel programmes, such as 18X, which allowed candidates to attempt selection via the Q-course without prior military service or skills development.[302] Growth occurred in the other service components as well: NAVSPECWARCOM added two new SEAL teams in 2002, the 160th SOAR added a battalion in 2005 (one of whose MH-47Es crashed in OEF-P), and a Marine Special Operations Command added to USSOCOM's roster in 2005, adding another 3,500 personnel to the SOF roster.

Most growth, however, consisted of staff and operator enablers, including special air assets. AFSOC, for example, grew from three wings (one active duty, one National Guard, and one Reserve) and three groups (two forward-deployed to the United Kingdom and Japan and the other consisting of Special Tactics personnel) in 1998 to eight wings and a group: five active duty, two National Guard, and one Reserve.[303] The growth in wings accompanied a growth in numbers of squadrons, from twenty-five to forty over the same period, mostly to fulfil demands for intelligence, surveillance, and reconnaissance (ISR), armed overwatch, and mobility. Other growth areas included combat support, with the expansion of the 528th Sustainment from a Battalion to a Brigade, intelligence with the creation of a new Joint Special Operations Intelligence Brigade in 2008, as well as additional headquarters staff. Civilian numbers increased from almost 2,800 in 1998 to just under 6,500 by 2018, augmented by almost another 6,000 contractors.[304] While the number of SEAL teams remained constant after its initial expansion, its parent headquarters more than doubled from three Naval Special Warfare Groups in 1998 to seven by 2014, suggesting an increase in the number of commanders and staff, vice operators.[305]

The unrelenting demands on SOF as a force of choice against terrorism hid some of the costs, both direct and indirect, of their extended use. Critics charged, for example, that selection criteria had been diluted over the years and focused on the

[301] Tracy provides the sequence of activation orders in a footnote to his article. Tracy, 'Victoria Ex Umbra', p. 8, n.27.

[302] Jim Thomas and Chris Dougherty, *Beyond the Ramparts: The Future of U.S. Special Operations Forces* (Washington, DC: Center for Strategic and Budgetary Assessment, 2013), p. 31.

[303] USSOCOM, *United States Special Operations Forces Posture Statement 1998* (MacDill Air Force Base, FL: USSOCOM, 1998), pp. 52–53; Andrew Feickert, *U.S. Special Operations Forces (SOF): Background and Considerations for Congress*, Report RS 21,048 (Washington, DC: Congressional Research Service, 2023), p. 5.

[304] USSOCOM, *United States Special Operations Forces Posture Statement 1998*, p. 17; Mark Cancian, *U.S. Military Forces in FY2019: The Buildup and Its Limits* (Washington, DC: Center for Strategic and International Studies, 2018), p. 59.

[305] USSOCOM, *United States Special Operations Forces Posture Statement 1998*, pp. 50–51; Andrew Feickert, *U.S. Special Operations Forces (SOF)*, p. 6.

wrong skill sets.[306] Others observed that selection had been abandoned in favour of direct accession, threatening not only operational effectiveness but also collective identity.[307] Other direct costs were starker. Fatalities among operators jumped dramatically, doubling in 2005 and peaking again in 2009, with AFSOC and U.S. Army Special Operations Command (USASOC) suffering their deadliest years in 2005, NAVSPECWARCOM in 2011, and MARSOC in 2012, for a combined total of 425 between 2005 and 2014.[308] The indirect costs, in terms of physically wounded and psychologically damaged through moral injury, will likely remain unknown for some time, although one study suggests operators suffer post-traumatic stress disorder (PTSD) at double the rate of their conventional counterparts.[309]

The sacrifices borne by SOF, as well as their responsiveness and seeming suitability to combat the asymmetric threat of terrorism, prompted successive Congresses and Executives to approve almost all requests for funding and authorities during this period. Congress approved funding for specific and temporary funding authorities as well as expanded USSOCOM's base budget. Other sources of funding included the Global Security Contingency Fund (GCSF, formerly Section 1207 and later made permanent), the Cooperative Threat Reduction programme, Emergency Extraordinary Expense Funds, Confidential Military Purpose Funds, and the Commander's Emergency Response Program (CERP).[310] In its overall funding, USSOCOM's allocation increased by a 'substantial' margin: 400 per cent growth over thirteen years, from $2.3 billion to $10.4 billion.[311] USSOCOM's allocation was supplemented by Overseas Contingency Operations (OCO) funding, approved by Congress as an emergency allocation after 9/11 but subsequently normalized. The rationale behind Congressional approval and continuing support of OCO, until recently, was ensuring that ongoing operations and emerging crises could be paid for without military organizations such as USSOCOM having to do so through their operations, equipment, and maintenance budgets. To ensure oversight of funds, 'Congress has

[306] While some of the accusations reflect personality or leadership style clashes, such as those directed against Kurt Sonntag, or to serve as evidence in straw man arguments, the number of reports from different quarters over the period suggest fire might accompany this smoke. See Todd South, '2-Star Responds to Anonymous Email Blasting Watered-Down Special Forces Training Standards', *Army Times*, 30 November 2017, www.armytimes.com/news/your-army/2017/11/30/2-star-responds-to-anonymous-email-blasting-watered-down-special-forces-training-standards and R.D. Hooker Jr, 'America's Special Operations Problem', *Joint Forces Quarterly*, 108 (1st Quarter, January 2023): 52.

[307] Patrick Wnetrzak, 'Finding Apollo's Warriors: The Future of the Air Commando Heritage', unpublished Master's thesis, School of Advanced Air and Space Studies, June 2013 and Kelly Hobgood, 'Training Blueprint for an Air Commando', unpublished Master's thesis, School of Advanced Air and Space Studies, June 2016, pp. 44–45.

[308] Ross Kotwal et al., 'A Descriptive Study of US Special Operations Command Fatalities, 2001–2018', *Journal of Acute Trauma Care Surgery*, 87:3 (2019): 655.

[309] Matthew Hing, Jorge Cabrera, Craig Barstow, and Robert Forsten, 'Special Operations Forces and Incidence of Post-Traumatic Stress Disorder Symptoms', *Journal of Special Operations Medicine*, 12:3 (Fall 2012): 23–35; Deane-Peter Baker, Roger Herbert, and David Whetham, *The Ethics of Special Ops: Raids, Recoveries, Reconnaissance, and Rebels* (Cambridge: Cambridge University Press, 2024), pp. 202–204.

[310] For a summary of these funding authorities, their limitations, and reporting requirements, see Elvira Loredo et al., *Authorities and Options for Funding USSOCOM Operations*, Report RR-360-SOCOM (Santa Monica, CA: RAND Corporation, 2014), pp. 47–52.

[311] Jim Thomas and Chris Dougherty, *Beyond the Ramparts: The Future of U.S. Special Operations Forces* (Washington, DC: Center for Strategic and Budgetary Assessment, 2013), p. 31.

required DOD to provide separate annual budget justification documents detailing the costs of US Armed Forces' participation in all named contingency operations where the total cost of the operation exceeds $100 million or is staffed by more than 1,000 U.S. military personnel.'[312] USSOCOM 'dependence on OCO funding' steadily increased over the years, reaching a high in 2019 of one-third 'of its total funding, three times the [DoD]'s overall rate'.[313]

Congress, civilian decision makers, and leaders did have some limits in special operation approvals. While comfortable with USSOCOM's 2005 request to plan and coordinate activities against terrorists globally, Rumsfeld balked at its argument for the next logical step: the authority to approve and execute those missions. Representatives of the chairman of the joint staff argued that doing so would subvert his approval authority and regardless, 'SOCOM was not set up to handle [the] mission ... for worldwide planning and operating'.[314] An exception written in the coordination authorization, and the subject of considerable debate within the Pentagon for a year and a half, was a conditional: 'as directed'. Translated, this meant only the secretary or the president had the direct authority to approve USSOCOM requests to execute missions.

Nevertheless, from 2008 until 2014, the relationship between US special operations, the executive, and the legislative branches can be described as comfortable, in terms of resources and authorities. Members of Congress received regular reporting updates on funding allocations and requests for temporary authorities to be extended or codified, and new ones added. USSOCOM submitted its annual budget requests, including requests for increases, through the joint staff and the secretary of defense. These overt processes largely functioned as designed as part of the system of checks and balances. Continued SOF successes seemingly justified the relatively small investment, in comparison to armed service MFP-2 programs that occupied Congress' attention, including ones seemingly irrelevant to CT and COIN fights such as the US Air Force's F-22 and F-35 programs.[315] The coziness of the relationship between SOF and Congress is evident: requests to Congress' research arm, the Congressional Research Service, were largely limited to annual 'backgrounder' reports on SOF, General Accounting Office reports and audits indicated few concerns beyond basic force management and sustainment issues, and there were almost no hearings specifically on special operations, save for the annual testimony of the USSOCOM commander and the ASD SO/LIC as part of the normal budgetary process.

[312] Brendan McGarry and Emily Morgenstern, *Overseas Contingency Operations Funding: Background and Status*, Report R44519 (Washington, DC: General Accounting Office, 2019), p. 44.

[313] Mark Cancian, *U.S. Military Forces in FY2019: The Buildup and Its Limits* (Washington, DC: Center for Strategic and International Studies, 2018), p. 58.

[314] Feith, *War and Decision*, p. 112. The case for retained authority with the CJCS was made in Michael Bell, *The Exigencies of Global, Integrated Warfare: The Evolving Role of the CJCS and his Dedicated Staff* (Carlisle Barracks, PA: Army War College Press, 2011).

[315] Kristina Sherry, 'Senate Votes to Stop Making More F-22 Raptor Fighter Jets', *Los Angeles Times* (22 July 2009), www.latimes.com/archives/la-xpm-2009-jul-22-na-f-22-plane22-story.html; Sharon Weinburger, 'Gates: F-22 Has No Role in War on Terror', *Wired* (7 February 2008), www.wired.com/2008/02/gates.

The expansion of SOF and successes of TF 714 in Iraq paved the way for more grandiose special operations plans. USSOCOM shifted its focus from being a global synchronizer of planning and plans to building a 'global SOF network'. The establishment of the NSCC within NATO provided a proof of this concept through the creation of a regional JIATF. Despite successes in Iraq, the phenomenon of global terrorism seemed to be growing, not just in the United States but elsewhere—a point driven home by high-profile attacks in India (Mumbai), Sri Lanka, Indonesia (2009), Russia (2009 and 2011), and Nigeria (2011), as well as Iraq and Afghanistan. The expansion of al-Qaeda franchises in western and northern Africa, the Arabian Peninsula, and elsewhere led some authors to cast global terrorism as a phenomenon rendering quaint concepts like states and borders irrelevant. Global terrorism, two influential authors argued, is best conceptualized as either a mental or cognitive archipelago comprised of various islands, or a viral contagion, thanks to information technologies.[316] Studies conducted by think tanks, including RAND, focused less on al-Qaeda and more on 'the broader jihadist movement', which in turn informed subsequent revisions to USSOCOM's CONPLAN in 2008.[317] If it took a network to fight a network in Iraq, then a global one was necessary to fight global networked terrorism.

This fight took two primary forms. One was unilateral US efforts in countries unable or unwilling to do so themselves. TF 714 continued its operations under several distributed task forces. To do so, however, stretched authorities and oversight given the fleeting nature of global terrorist cells and opportunities to strike them. The other, however, resulted from two realizations. The first was a maths problem. Even with its increased resources, USSOCOM could not be everywhere at all times. While the value of partners in the struggle was apparent during first phase of OEF, only uncoordinated and haphazard efforts were made to engage them. USSOCOM's embrace of its role as the global combatant commander for countering terrorism, along with the resources it possessed, meant it led efforts to build partner capacity, including SOF capable of coordinating and integrating with its forces. The vision for the global SOF network, as well as its operators, was clearly stated in 2012:

> The Global SOF Enterprise will become a globally networked force of Special Operations Forces, Services, Interagency, Allies and Partners able to rapidly and persistently address regional contingencies and threats to stability.
>
> The SOF Operator will remain the strategic cornerstone of the Global SOF Enterprise ... [one who] understands and practices the principles of Mission Command and decentralized operations. They cultivate a bias for action in their subordinates, develop mutual trust and understanding, and exercise moral nerve and restraint.[318]

[316] John Mackinley, *The Insurgent Archipelago: From Mao to Bin Laden* (London: Hurst, 2009); David Kilcullen, *The Accidental Guerrilla: Fighting Small Wars in the Midst of a Big One* (Oxford: Oxford University Press, 2009), pp. 169–179.

[317] Angela Rebasa, Peter Chalk et al., *Beyond al-Qaeda, Parts 1 and 2: The Global Jihadist Movement* and *The Outer Rings of the Terrorist Universe* (Santa Monica, CA: RAND Corporation, 2005, 2006).

[318] United States Special Operations Command, *SOCOM 2020: Forging the Tip of the Spear* (MacDill Air Force Base, FL: USSOCOM, 2012), p. 8.

The vision, however, faced several political and economic realities, succinctly captured in the subtitle of a chapter from conference proceedings on the subject of the global SOF network: little money, unclear ends, and big ideas.[319]

Creating a globally synchronized network of SOF would take time, if it was ever achievable, but civilian decision makers continually relied on three unilateral activities to fight terrorism: remote strikes, or targeted killing, cyber operations, and special operations conducted covertly, clandestinely, or overtly. All three formed the basis of newly elected President Barack Obama's approach to the use of force overseas, a variation of the 'Afghan model' comprising a '"light footprint" alternative to the large and expensive deployments of the Bush era'.[320] Overtly, indirect-shock missions diminished in Iraq while they ramped up in Afghanistan, conducted by TF 714 as well as other SOTFs. Their use expanded to a range of other countries, in all 'more than 80', 'supporting 10 named operations' by 2015, according to USSOCOM Commander Joseph Votel.[321] Although SOF were not deployed in the same volume as in the Bush administration, more than 10,500 were 'deployed in support of a variety of [Geographic Combatant Command; COCOMs] requirements on any given day'.[322] In addition to minimized strategic and political risk compared to other options, SOF appeared effective and responsive to threats posed by al-Qaeda's affiliates and the Taliban.

Measuring SOF effectiveness has always been problematic. Surgical strike and special warfare missions share a common affliction in this regard: both tend to rely too heavily on measures of performance, as opposed to measures of effectiveness. Measures of performance range from 'body counts' at their most basic and unhelpful to raids conducted, civilians not killed, and items seized or recovered. Measures of effectiveness, more difficult to grasp, include the impact of an incident or event on a system and a determination of its significance. Doing so requires considerable information on how that system works and changes over time, which is difficult to obtain against an opponent operating primarily in a clandestine manner to avoid being found, fixed, and finished. The difficulties in assessing the overall effectiveness of TF 714 in Iraq are a case in point. Shultz Jr, relied on interviews with senior commanders and their judgements to answer the question of success in his study. Even then, the interviewees' assessments were vague and based on intuitive conclusions.[323] Barnett Koven and Katy Lundquist suggest combined longitudinal and latitudinal assessment, as necessary to evaluate counterterrorism effectiveness properly.[324]

[319] Chuck Ricks, ed., *The Role of the Global SOF Network in a Resource Constrained Environment* (MacDill Air Force Base, FL: JSOU Press, November 2013), Chapter 1.

[320] Jeff Goldsmith and Matthew Waxman, 'The Legal Legacy of Light-Footprint Warfare', *The Washington Quarterly*, 39:2 (2016): 7.

[321] Statement of General Joseph L. Votel, Commander, US Special Operations Command, before the House Armed Services Committee Subcommittee on Emerging Threats and Capabilities, 18 March 2015, https://irp.fas.org/congress/2015_hr/031815votel.pdf.

[322] Statement of General Joseph L. Votel, 18 March 2015.

[323] Shultz Jr, *Transforming US Intelligence for Irregular War*, pp. 167–172.

[324] Barnett Koven and Katy Lundquist, *Barriers to Special Operations Forces-Led Counterterrorism Effectiveness*, JSOU Report 21-7 (MacDill Air Force Base, FL: JSOU Press, 2021), pp. 192–195.

Measures of performance brief well, however, and provide evidence that SOF are taking meaningful action to address a political and strategic problem.[325]

If effectiveness is difficult to measure, responsiveness is not. It is relative to other available options and can be combined with measures of performance to give the illusion of effectiveness. Chapter 4 developed the idea of intended acceleration to capture this phenomenon. Shades of SOF *exceeding* are observable during this period of US special operations activity. In several cases, SOF leaders lead-turned their conventional military or other competitors in providing proposed courses of action and potential solutions to problems. Some of this was facilitated by taking advantage of the information gleaned from their constellation of liaison officers, begun under McChrystal but subsequently expanded by USSOCOM commanders including Eric Olson, William McRaven, and Joseph Votel. In addition to the USSOCOM satellite office in Washington, DC, liaison consisted of the following:

> Efforts such as our Special Operations Liaison Officers, or SOLOs, are helping us build this network of international partners. Now present in embassies in 15 nations and operating in every geographic combatant command area of responsibility, SOLOs help us facilitate coordination across [COCOM] boundaries to address challenges that span the globe. We also have liaison officers from 13 partner nations that work with us at our Headquarters in Tampa. Similarly, our Special Operations Support Teams (SOSTs) help us interface more effectively with the interagency.[326]

This network within a network facilitated horizontal and vertical integration, or 'unprecedented horizontal cooperation', but it also forged relationships capable of accessing privileged information to provide advance warning of impending issues.[327]

A problem could and indeed did occur during this period where policy goals were slow, nebulous, contradictory, or non-existent. In such cases, operators became not only 'masters of chaos' but 'masters of process', in terms of seeking and gaining approval for their actions. All bureaucracies attempt to manipulate processes to their advantage, even if the ultimate purpose remains debated.[328] What differentiated SOF efforts, however, was their redefinition of them as a function of their collective cultural bias for action.

In the case of TF 714, mission focus and the nature of the crisis in Iraq, combined with the results of its operators' exhaustive training, served as 'ethical armouring'.[329]

[325] See, for example, the range of conclusions and opinion in a special issue of *Foreign Affairs* published for the twentieth anniversary of the 9/11 attacks. 'Special Issue: Who Won the War on Terror?', *Foreign Affairs*, 100:5 (September/October 2021).

[326] Statement of General Joseph L. Votel, 18 March 2015, p. 14.

[327] Steve Niva, 'Disappearing Violence: JSOC and the Pentagon's New Cartography of Networked Warfare', *Security Dialogue*, 44:3 (2013): 187

[328] See, for example, Graham Allison and Philip Zelikow, *Essence of Decision: Explaining the Cuban Missile Crisis*, 2nd ed. (New York: Longman, 1999), pp. 294–313; Todd Kunioka and Lawrence Rothenberg, 'The Politics of Bureaucratic Competition: The Case of Natural Resource Policy', *Journal of Policy Analysis and Management*, 12:4 (Autumn 1993): 700–725; Cheryl Eavey, 'Bureaucratic Competition and Agenda Control', *The Journal of Conflict Resolution*, 31:3 (September 1987): 503–524.

[329] Baker, Herbert, and Whetham, *The Ethics of Special Ops*, pp. 207–208.

More specifically, their professionalism stemmed from serving in a 'calling', the creation of a strong institutional identity (an 'institution within an institution'), one with a unique jurisdiction.[330] The process they perfected, F3EAD, became a vehicle that worked to their advantage. As Chapter 4 made clear, SOF prefer to work rather than shirk—monitoring is necessary to ensure they do not work too much. After Iraq TF 714 sought to fill the policy vacuum through continuous employment and action. The goal was not so much recognition as it was relevance, along with influence and the resources that accompanied it.

To that end, the product they created and perfected—the storyboard—became a means to accomplish it. A storyboard is just as it sounds: a distillation of information, separated by panels, that conveys a narrative, in this case a target intelligence package, or TIP. TF 714 storyboards, later reduced to so-called 'baseball cards', became a method of distilling complex information from a variety of different sources into a target deck, including visual images, that conveyed a target, its significance, and the risks to force that existed.[331] This collection of storyboards served as a decision aid for a concept of operations (CONOP) that outlined the bigger picture: force package, risk to mission, means of mitigation risks, and so on. Sent up the chain of command, such storyboards became an efficient and effective method of approval, given their apparent clarity and simplicity.[332]

The problem existed not with the product but with the F3EAD process it served. Focusing on terrorist and insurgent leaders is problematic for a variety of reasons, including the security measures they take, the time it takes to 'find' and 'fix' them, not to mention the uncertain effectiveness or consequences of killing or capturing them.[333] For these reasons, McChrystal targeted the 'middle management' of AQI: those 'who made the networks run' and who occupied the intervening space between leaders and foot soldiers.[334] In shifting to Afghanistan, Somalia, and elsewhere where policy goals were less well-defined, every target TF 714 killed or captured refined understanding of the terrorist network and, when combined with new information, perpetually identified new targets of opportunity. The result, as it was applied to other SOTFs, was a 'commando organization ... primed to look for targets and its collection was biased accordingly. Data potentially relevant to a deeper understanding of the social milieu dropped out as people stabilized the pathways that led to actionable targets'.[335] Worse still, the accumulation of various approved actions in a policy vacuum became *de facto* policy. A phrase from within the SOF community captures this phenomenon: 'policy by CONOP'. This section, and the discussion in Chapter

[330] C. Anthony Pfaff, 'Professionalizing Special Operations Forces', *Parameters*, 52:3 (2022): 113–114.

[331] Josh Begley, 'A Visual Glossary: Decoding the Language of Covert Warfare', *The Intercept* (15 October 2015), https://theintercept.com/drone-papers/a-visual-glossary.

[332] For a short primer on storyboards, see Peter Molin, 'Military PowerPoint Storyboards', Acolytesofwar.com, 9 January 2023, https://acolytesofwar.com/2023/01/09/military-powerpoint-storyboards. Molin reproduces two leaked classified storyboards that have not been redacted or cleared for public release.

[333] Hardy, 'Hunters and Gatherers', p. 1152.

[334] Shultz Jr, *Transforming US Intelligence for Irregular War*, pp. 168–169.

[335] Lindsay, 'Target Practice', p. 1076.

4, argues that far from being in the 'the unenviable position of developing "policy by CONOP"' where '[c]ommanders and their staffs, often with minimal guidance about US strategy, have to tease out how to advance American policy as they gauge it from their own professional experience and judgment', SOF leaders and institutions took advantage of emerging opportunities, and developed products to serve the processes they created with an end of serving institutional and national goals as they understood them.[336]

The net effect was that from 2008 to 2014, US special operations constructed their own strategy against terrorism. Their approach, based on indirect action, is what Robert Chia and Robin Holt have termed 'strategy without design', focusing on emergent behaviour in a system as opposed to specified outcomes.[337] This was not done unethically, immorally, or maliciously. Instead, the strategy was the outgrowth of a culture biased toward action, used to solving problems in an unorthodox manner, and filling a void in direction. This situation was only remedied by President Obama's 2013 revised criteria for targeting killing, which accepted greater risk by seeking to capture terrorists using SOF.[338]

Fewer concerns were expressed about the ends special operations were designed to achieve strategically, the means used, or even their ways than with an emerging trend. What troubled observers was the closeness resulting from SOF's horizontal integration with other departments and agencies. They termed it 'convergence', or an inability to determine where special operations and SOF end and where the activities of other government departments and agencies begin. First raised in response to the SOF use of non-standard uniforms in Afghanistan that blurred the distinction between military and paramilitary as well as combatant and non-combatant, the initial focus was on the legality of such practices as potentially voiding law of armed conflict protections for operators.[339] Subsequent discussions included convergence of law enforcement on land and at sea, as well as the seeming indistinguishability between special operations and intelligence activities.

Convergence between special operations and law enforcement activities, first noted in 2008, was driven by three specific factors, two related to capability and one to control issues. The first reflected the evolution and adaptation of terrorist and criminal groups. While members of terrorist groups had historically engaged in criminal activities as a source of funding and resourcing, their abilities had increased exponentially given access to information technologies, weak or non-existent governance and

[336] Water Haynes, 'The Hidden Costs of Strategy of Special Operations', War on the Rocks, 17 April 2019, https://warontherocks.com/2019/04/the-hidden-costs-of-strategy-by-special-operations.

[337] Robert Chia and Robin Holt, *Strategy without Design: The Silent Efficacy of Indirect Action* (Cambridge: Cambridge University Press, 2009).

[338] President Barack Obama, 'Remarks by the President at the National Defense University', speech transcript, 23 May 2013, https://obamawhitehouse.archives.gov/the-press-office/2013/05/23/remarks-president-national-defense-university. An unclassified version of this policy directive was published as 'U.S. Policy Standards and Procedures for the Use of Force in Counterterrorism Operations Outside the United States and Areas of Active Hostilities' on the same day. https://obamawhitehouse.archives.gov/the-press-office/2013/05/23/fact-sheet-us-policy-standards-and-procedures-use-force-counterterrorism.

[339] W. Hays Parks, 'Special Forces' Wear of Non-Standard Uniforms', *Chicago Journal of International Law*, 4:2 (Fall 2003): 493–496.

border controls, and a host of other factors.[340] While specific means varied nationally and regionally, their funding sources included state sponsorship, narcotics trafficking and protection rackets, the levying of taxes, front companies and funds siphoning, and witting or unwitting donations.[341] Although criminal groups had long used terror in pursuit of control and profit, what concerned law enforcement officials was the increasingly lethal equipment and tactics employed by them, as well as an increasing willingness of perpetrators to kill as many innocents as possible in the commission of their acts.[342] Domestic critics, who observed little difference in the outward appearance of SOF and law enforcement tactical teams, decried convergence as the militarization of law enforcement.[343]

The second factor was the capacity of existing law enforcement capabilities to counter these evolving threats. Mirroring American SOF, law enforcement units at the federal or national level were separated by skill and specific functions into designated 'tiers' of capability: FBI Hostage Rescue Teams, Enhanced Special Weapons and Tactics (SWAT) Teams, and Field Division SWAT teams. In 1997, these teams comprised a total of just over 1,100 agents, divided into 57 teams. To them, the US Coast Guard added the creation of 14 Maritime Safety and Security Teams that perform similar indirect-shock missions at sea, in rivers and littorals, as well as in port facilities.[344] Such capabilities were sufficient to deal with individual domestic incidents, but problems existed in deploying them overseas, including both legal barriers and insufficient numbers to investigate or respond to counterintelligence, counternarcotics, counterterrorism, and organized crime. Such capacity issues became particularly acute after OIF, when SOF were increasingly used to train Iraqi police forces, a task that some operators did not see as special operations missions and for which they were insufficiently prepared to conduct.[345]

The third factor reflected specific control mechanisms. These mechanisms, put in place by Congress in the form of specified legal authorities, were designed to separate powers and clearly distinguish between primary agents and secondary supporting ones. Authorities for different departments and agencies were separated by Title in US Code: Title 10 for DoD, Title 18 for Department of Justice, and Titles 14, 19, and later 46 for the Coast Guard.[346] The designation of USSOCOM as executive agent to counter terrorism outside of the United States was a tacit recognition of the twin insufficiencies of law enforcement capacity and approach to terrorism, which is

[340] Lambakis et al., *Understanding 'Asymmetric' Threats to the United States*, pp. 36–37.

[341] See the typology developed by Michael Freeman in Chapter 2 and used in the various case study chapters contributed by others in Michael Freeman, *Financing Terrorism: Case Studies* (Farnham, Surrey: Ashgate, 2012), pp. 12–22.

[342] John Alexander, *Convergence: Special Operations Forces and Law Enforcement*, JSOU Report 10-6 (MacDill Air Force Base, FL: July 2010), pp. 7–15.

[343] For contrasting perspectives on this issue, see Radley Balko, *Rise of the Warrior Cop: The Militarization of America's Police Forces* (New York: PublicAffairs, 2013) and Scott Phillips, *Police Militarization: Understanding the Perspectives of Police Chiefs, Administrators, and Tactical Officers* (London: Routledge, 2018).

[344] Gary Bowen, *Coast Guard SOF*, JSOU Report 05-7 (Hurlburt Field, FL: JSOU Press, 2006), pp. 48–49.

[345] Alexander, *Convergence*, p. 2.

[346] For Coast Guard authorities, see Bowen, *Coast Guard SOF*, appendix B.

responsive and not preventative or proactive. Stated simply, there were too few law enforcement officers focused too passively on the problem of global terrorism.

Given newfound agency, USSOCOM leaders sought and obtained authority to become executive anti-terrorist agent within the national security apparatus and DoD, codified through DoD Directive 5205.14 in 2010, with ASD SOLIC providing reports to civilian leadership.[347] This provides an example of SOF acting to expand the 'box' of its authorities to encompass their preferred gray zone of operations 'where the bureaucratic structures are weaker and the rules and regulations are less clear', or the realm of 'unstructured problems, in situations of high uncertainty, with only vague guidance', in line with Searle's assertion.[348] USSOCOM tried to develop a level of certainty and structure via its strategic assessment process, but this ultimately only confirmed how complex and interconnected global terrorism is with a host of political, social, and economic issues, as Figure 8.5 depicts.

This complexity, far beyond the scope, mandate, and resources of USSOCOM to address, led some to conclude that convergence was a symptomatic of a deeper-rooted malady: the structural barriers and inefficiency inherent in American government. A completely new organization to address terrorism was needed or entire interagency process required restructuring, centred around. The popular phrase for the latter was 'a second Goldwater-Nichols Act'.[349]

Concerns over special operations convergence were voiced more forcefully in the realm of intelligence activities. Jennifer Kibbe aptly summarized these concerns in 2004: 'Thanks to the vagueness of US law governing covert action, using the military for such an operation is—at least under one interpretation of the law—much easier than using the CIA. And this facility seems to appeal to Rumsfeld.'[350] Her unease was seemingly based on SOF expanding Searle's box more quickly than Congress could effectively monitor and constrain their activities. While only time and future inquiry, aided by access to documentary evidence, will determine whether such disquiet was warranted, the absence of scandals comparable to those in previous eras, including in Vietnam and elsewhere, suggest two possible inferences: SOF pushed the boundaries of rules and regulations and just have not been caught doing so, or alternatively, they rewrote them to operate more effectively.

Informed insiders suggest that rather than seeking to acquire separate authorities to remove restrictions, SOF instead partnered more effectively. Tom Henriksen, for example, describes the SOF/CIA partnership as one of jointness, a metaphorical

[347] USD(P), Department of Defense Directive 5205.14, 19 August 2010, pp. 4–5, www.esd.whs.mil/Portals/54/Documents/DD/issuances/dodd/520514p.pdf? ver=2019-02-05-084012-033.

[348] Tom Searle, *Outside the Box: A New General Theory of Special Operations*, JSOU Report 17-4 (MacDill Air Force Base, FL: JSOU Press, 2017), p. 38.

[349] See, for example, David Kilcullen, *The Accidental Guerrilla: Fighting Small Wars in the Midst of a Big One* (Oxford: Oxford University Press, 2009), pp. 169–179; Catherine Dale, Nina Serafino, and Pat Towell, *Organizing the U.S. Government for National Security: Overview of the Interagency Reform Debates*, Report RL34455 (Washington, DC: Congressional Research Service, 2008), pp. 4, n.6; 8; Clark Murdock and Richard Weitz, 'Beyond Goldwater-Nichols: New Proposals for Defense Report', *Joint Forces Quarterly*, 38 (2005): 34–41; Joint Staff J-7, *Decade of War, Volume 1: Enduring Lessons from the Past Decade of Operations* (Suffolk, VA: J-7 Joint and Coalition Operational Analysis, 2012), p. 27.

[350] Jennifer Kibbe, 'The Rise of the Shadow Warriors', *Foreign Affairs*, 83:2 (March/April 2004): 103.

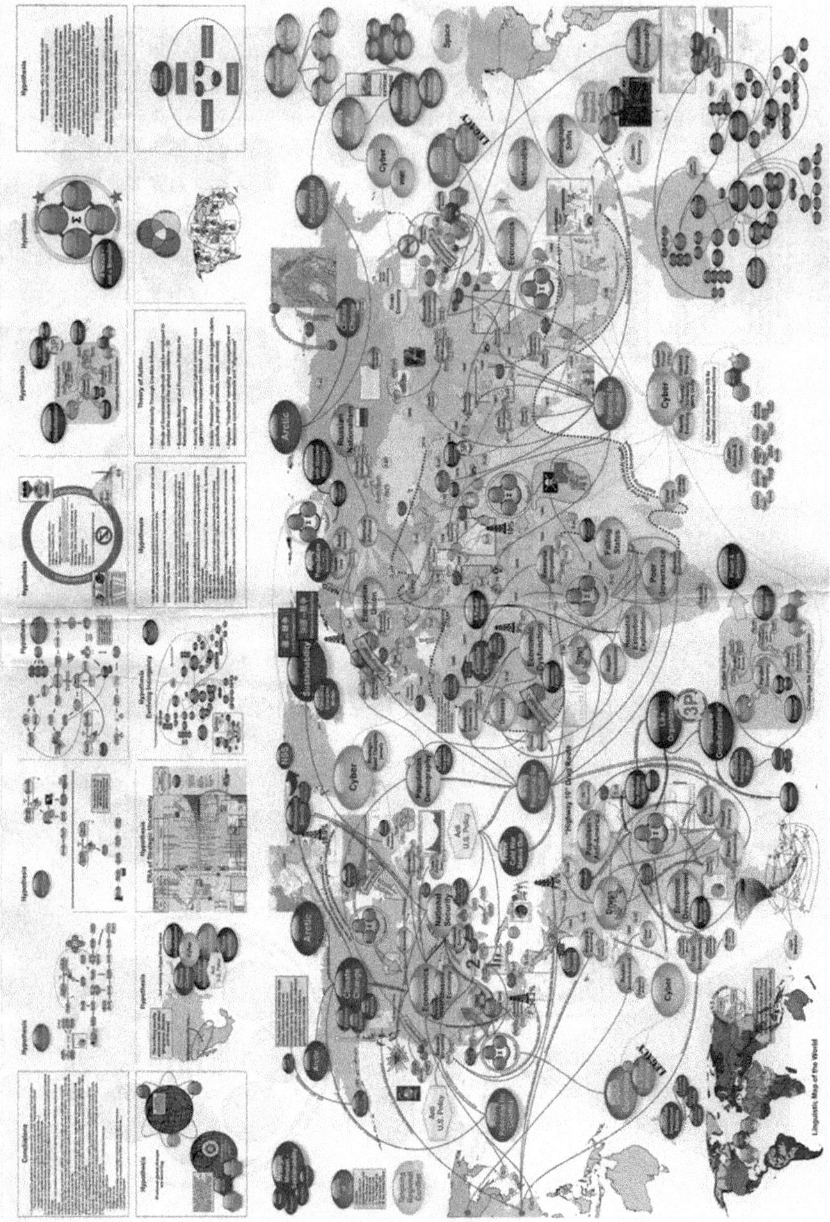

**Figure 8.5** USSOCOM strategic assessment chart, 2014.
*Source:* USSOCOM strategic assessment chart. Author's collection.

'wrestling tag team', in which 'the CIA replaced SOF ... when circumstances dictated a substitution'.[351] One such substitution occurred in 2011, with the withdrawal of US

[351] Henriksen, *Eyes, Ears and Daggers*, p. 140.

forces from Iraq save a small contingent of Counter Terrorism Service (CTS) advisors and trainers, which SOF grudgingly accepted. Others argued convergence of intelligence and special operations was part of a natural evolutionary process, occurring over decades, resulting from an imperfect patchwork of authorities and directives cobbled together to meet emerging circumstances, including those related to threats such as terrorism and new operating environments such as cyberspace.[352] Too often, expressed reservations reflected policy concerns, such as communal reputation or organizational control.[353]

Members of Congress expressed reservations about special operations expansion for different reasons. From 2008 onwards, members of the HPSCI and SPSCI, for example, publicly stated that SOF conduct on activities labelled as 'operational preparation of the environment', or OPE, which were not under their oversight and scrutiny should be, as they were nothing less than intelligence gathering.[354] Rumsfeld's creation of USD(I) was little less than open competition with the CIA to conduct similar activities outside Congressional scrutiny, raising the potential for a repeat of the embarrassments of the scandal-laden, Reagan-era Army covert collection unit, the Intelligence Support Activity.[355] As Michael DeVine points out, for members of Congress the distinction between OPE and covert action is similar in action as well as consequence if compromised. What separates the two is authority, authorization, and oversight.[356] The benefits of operating covertly are not worth the risks to the operators or the nation should the mission be compromised.[357] Those working on behalf of special operations point out the crux of the issue is not SOF overreach but instead Congress' inability to rewrite laws to reflect contemporary realities, which 'sows confusion over statutory authorities and causes Executive Branch attorneys to waste countless hours distinguishing distinct lines of authority and funding'.[358] Convergence therefore is less a matter of agency and exceeding authority than it is special operations navigating existing frameworks in pursuit of the goals specified by principals.

No case better illustrates this point than Operation NEPTUNE SPEAR, the raid that resulted in Osama bin Laden's death, on 1 May 2011. Several accounts exist of this indirect-shock mission, by its participants as well as informed journalists such

[352] Robert Chesney, 'Military-Intelligence Convergence and the Law of the Title 10/Title 50 Debate', *Journal of National Security Law & Policy*, 5:2 (January 2012): 544–554.

[353] Andru Wall, 'Demystifying the Title 10-Title 50 Debate: Distinguishing Military Operations, Intelligence Activities, and Covert Action', *Harvard National Security Journal*, 3 (2011): 88–91.

[354] Chesney, 'Military-Intelligence Convergence and the Law of the Title 10/Title 50 Debate', pp. 604–605; Michael DeVine, *Covert Action and Clandestine Activities of the Intelligence Community: Selected Definitions in Brief*, Report R45175 (Washington, DC: Congressional Research Service, 2019), pp. 7–9.

[355] On the creation of ISA, see declassified and redacted 'Brief History of Unit', n.d., https://nsarchive2.gwu.edu/NSAEBB/NSAEBB46/document11.pdf. On its scandals, see Michael Smith, *Killer Elite: The Inside Story of America's Most Secret Special Operations Team*, rev. ed. (New York: St Martin's Griffin, 2008), pp. 54–66.

[356] DeVine, *Covert Action and Clandestine Activities of the Intelligence Community*, p. 1.

[357] Michael McAndrew, 'Wrangling in the Shadows: The Use of United States Special Forces in Covert Military Operations in the War on Terror', *Boston College International and Comparative Law Review*, 29:1 (2006): 164.

[358] Wall, 'Demystifying the Title 10-Title 50 Debate', p. 141.

as Chris Wallace and Mark Bowden.[359] In addition, senior individuals within the Obama administration gave director Kathryn Bigelow access to various CIA and DoD officials involved in the planning and conduct of the raid for her film *Zero Dark Thirty*.[360] Its details do not require recounting. In providing a media interview about the mission after its completion, CIA director of central intelligence, Leon Panetta, contributed to ongoing discourse about convergence when talking about the authorization of the mission. Two days after the raid, in response to a question from PBS Newshour's Jim Lehrer about his location, Panetta stated:

> Since this was what's called a 'title 50' operation, which is a covert operation, and it comes directly from the president of the United States who made the decision to conduct this operation in a covert way, that direction goes to me. And then, I am, you know, the person who then commands the mission. But having said that, I have to tell you that the real commander was Adm. McRaven because he was on site, and he was actually in charge of the military operation that went in and got bin Laden.[361]

Observers were confused by the apparent contradiction inherent in his statement: was it a covert or a military operation? US SOF, operating overtly under the direct command and control of a local military commander, William McRaven, were nevertheless under the legal authority of the Director of Central Intelligence (DCI) as a covert action. This convoluted arrangement, done to expedite the mission as quickly as possible, raised more questions than it answered. They included voiding the protections of the operators involved in the mission, should they have been captured, as well as a fig leaf of deniability should a repeat of Desert One—a mission abort by the on-scene commander—occur. Yet one of the reasons for undertaking the mission as a Title 50 covert action, plausible deniability, was almost immediately demolished with the announcement made by President Obama in the East Wing about the raid. As Joseph Berger assesses, the authorization and disclosure in this manner solved a short-term problem while creating a longer-term one for future covert actions.[362]

Convergence could become a reality for several reasons in this case. Legal divisions yet subtle overlaps between authorities constitute a 'grey area' exploitable

[359] Mark Owen, pseudo. Matt Bissonnette, *No Easy Day: The Autobiography of a Navy SEAL* (New York: Dutton, 2012); Robert O'Neill, *The Operator: Firing the Shots that Killed Osama bin Laden and My Years as a SEAL Team Warrior* (New York: Scribner, 2018), pp. 271–314; Mark Bowden, *The Finish: The Killing of Osama bin Laden* (New York: Atlantic Monthly Press, 2012); Chris Wallace, *Countdown bin Laden* (New York: Avid Reader Press, 2021).

[360] Hillary Busis, 'Kathryn Bigelow Had White House Access on bin Laden Film, Says Watchdog Group', *Entertainment Weekly*, 23 May 2012, https://ew.com/article/2012/05/23/white-house-kathryn-bigelow; Julie Miller, 'The Pentagon Gave Kathryn Bigelow High-Level Information about the Osama bin Laden Raid, Documents Reveal', *Vanity Fair*, 23 May 2012, www.vanityfair.com/hollywood/2012/05/white-house-president-obama-kathryn-bigelow-osama-bin-laden-classified-information.

[361] 'CIA Chief Panetta: Obama Made "Gutsy" Decision on Bin Laden Raid', PBS Newshour, 3 May 2011, www.pbs.org/newshour/show/cia-chief-panetta-obama-made-gutsy-decision-on-bin-laden-raid.

[362] Joseph Berger III, 'Covert Action: Title 10, Title 50, and the Chain of Command', *Joint Forces Quarterly*, 67 (4th Quarter, 2012): 38.

by operators when possible and by the Executive when necessary.[363] Loose and contradictory definitions of key terms, such as clandestine, covert, covert action, and traditional military authorities further contribute to confusion.[364] Operations also suggest expansion of Executive power and agency, stretching terms such as 'imminent threat' to take consistent action while meeting the minimal oversight and reporting requirements to the Gang of Four and Gang of Eight.[365] Concern over convergence subsided in the years following the bin Laden raid for several reasons, mostly other operational or political distractions.

Despite a decade having elapsed, Congress, the DoD, or the CIA have done little to clarify or codify the distinctions and boundaries between covert and clandestine activities between military and paramilitary special operations, suggesting satisfaction or ambivalence at the current state of convergence—until a sequence of disclosures and events forces the matter. Reasserting tighter control through legislative oversight occurs in American special operations because of negative publicity and failure. In the face of success, particularly in bringing bin Laden 'to justice', Congress showcased capabilities and their integration instead, quoting the Director of National Intelligence, James Clapper: 'Never have I seen a more remarkable example of focused integration, seamless collaboration, and sheer professional magnificence as was demonstrated by the Intelligence Community in the ultimate demise of Osama bin Laden.'[366] Members of the SPSCI, rather than conducting a hearing into convergence, instead requested the minimum reporting requirement for the mission, an official record of it, which will remain classified for the foreseeable future.[367]

In Afghanistan divergence in special operations missions was more problematic than convergence of roles and capabilities. There were two primary schools of thought regarding policy for Afghanistan. One is termed 'the COIN school', which reflected successes in Iraq in reducing terrorist and insurgent violence and bringing a level of security and stability there. Advocates of this approach included Generals Stanley McChrystal and David Petraeus, successive ISAF commanders. This approach aligned with the views of several senior administrative officials, including President Obama, who wanted to withdraw US forces from Iraq and had campaigned for election on prioritizing Afghanistan and winning the war there. Of the various options produced by ISAF staff, a surge of forces, numbering 30,000 troops and trebling the number of civilian experts, could accomplish in Afghanistan what occurred

[363] Michael Curry, 'A Grey Area: Congressional Oversight of the Middle Ground between Title 10 and Title 50', unpublished Master's thesis, School of Advanced Air and Space Studies (Maxwell Air Force Base, AL, June 2012), pp. 71–74.

[364] Berger III, 'Covert Action: Title 10, Title 50, and the Chain of Command', pp. 33–35.

[365] Chris Wallace, drawing heavily upon interviews with senior Obama administration officials, suggests the latter tried to 'notify congressional leaders' of the impending raid 72 hours before its execution despite the potential risks associated with a leak. Wallace, *Countdown bin Laden*, p. 208.

[366] Public Law 112-87, 'Intelligence Authorization Act for Fiscal Year 2012', 3 January 2012, 125 Stat 1892, www.intelligence.senate.gov/sites/default/files/laws/11,287.pdf.

[367] As a 2011 USSOCOM Freedom of Information Act log of requests suggests, most related to NEPTUNE SPEAR were rejected the same day they were received. USSOCOM, '2011 FOIA Log', redacted spreadsheet, n.d., pp. 5–8, www.socom.mil/FOIA/FOIA%20Logs/FOIA%20Log%20FY%202,011.pdf. While some media requests could reasonably be considered 'in the public interest', others are nothing less than sensational and ghoulish.

in Iraq in just eighteen months. Notionally, these forces would assist in providing security and stability but while training a credible and capable ANA, at increased risk to US forces. This option was eventually selected by the Obama administration.

The second school, espoused by then-Vice President Joseph Biden, was labelled the 'CT-plus' option. In this option, a variation of the 'Afghan model' would be employed, with a goal of imposing costs on the Taliban and al-Qaeda while limiting risk to US forces. SOF indirect-shock missions, known colloquially as 'night raids', would inflict losses on Taliban networks within Afghanistan as TF 714 had done to AQI in Iraq. In addition, CIA strikes by remotely piloted aircraft (RPA), would kill Taliban and al-Qaeda leadership in areas they felt safe and had sanctuary, specifically Pakistan's largely ungoverned Federally Administered Tribal Areas (FATA).[368] The division of labour between military and paramilitary special operations reflected political sensitivities in Afghanistan and Pakistan. Although not particularly popular in Pakistan, RPA strikes generated fewer political consequences than cross-border raids by SOF, as occurred in the wake of a September 2008 in south Waziristan, killing between eighteen and twenty-four al-Qaeda members.[369] While tactically successful and showcasing the integration of raiding SEALs with helicopter lift as well as SOF air support, the political fallout from the violation of Pakistan's sovereignty by overt SOF led to threats of closing the border to logistics convoys into Afghanistan supplying ISAF forces—what one unnamed official termed a 'strategic miscalculation'.[370] According to some media reporting, with the overt option closed special mission units opted instead to run proxy militia forces into Pakistan, in parallel with similar CIA efforts, in a variant of the Vietnam-era Phoenix programme, providing additional evidence of convergence.[371] Documents on Wikileaks suggested 'night raids' caused higher-than-reported Afghan civilian casualties, leading the government of Hamid Karzai to pressure ISAF to limit their use.[372] Covert actions in Pakistan ramped up in consequence, with a senior Pakistani general suggesting 'that between 2007 and 2011 about 164 predator [sic] strikes had been carried out and over 964 terrorists had been killed. Of those killed, 793 were locals and 171 foreigners, including Arabs, Uzbeks, Tajiks, Chechens, Filipinos and Moroccans'.[373]

While surgical strike missions engaged in repetitive 'lawn mowing' raids against Taliban and al-Qaeda targets, leading operators to question their value, a special warfare effort began in Afghanistan in support of the COIN surge.[374] CJSOTF-A

[368] Bob Woodward, *Obama's Wars* (New York: Simon & Schuster, 2010), pp. 159–201.

[369] Sean Murphy, 'The International Legality of US Military Cross-Border Operations from Afghanistan into Pakistan', in Michael Schmitt, ed., *The War in Afghanistan: A Legal Analysis*, International Law Studies, vol. 85 (Newport, RI: Naval War College Press, 2009), p. 117.

[370] Sean Naylor, 'Spec Ops Raids into Pakistan Halted', *Navy Times*, 29 September 2008.

[371] Mark Mazzetti et al., 'SEAL Team 6: A Secret History of Quiet Killings and Blurred Lines', *New York Times* (6 June 2015), https://archive.nytimes.com/www.nytimes.com/2015/06/07/world/asia/the-secret-history-of-seal-team-6.html.

[372] ISAF Public Affairs, 'ISAF Issues Guidance on Nights Raids in Afghanistan', Defense Visual Information Distribution Service (DVIDS), 5 March 2010, www.dvidshub.net/news/46,182/isaf-issues-guidance-night-raids-afghanistan.

[373] Brian Glyn Williams, 'Accuracy of U.S. Drone Campaign: The Views of a Pakistani General', *CTC Sentinel*, 4:3 (March 2011): 10.

[374] Conversations with special mission unit operators, 12 September 2012.

efforts to create capable Afghan security forces had mixed results, including success in developing Afghan SOF but failure to develop local security forces under the Afghan National Auxiliary Police (ANAP) programme.[375] Like similar programs in Iraq, the ANAP programme suffered from an industrial-scale effort at capacity building through military advice and assistance efforts in a short period of time.[376] The creation of a new SOF command headquarters in Afghanistan, Combined Force Special Operations Component Command, Afghanistan (CFSOCC-A), was in line with McChrystal's efforts to create unity of command and purpose in what to then had been fractured and largely uncoordinated regional and national efforts.[377] In addition, CFSOCC-A would put special operations considerations on an equal footing with other task force efforts underway, including counternarcotics, anti-corruption, and detainee handling, among others.[378] To ensure more effective command and control, tour durations would be extended from 6–9 to 18–24 months to provide continuity of effort, and the operational control of CFSOCC-A was delegated 'down' from COCOM level to the more localized theatre one, falling under US Forces in Afghanistan or USFOR-A.[379] The special warfare programme, known as Village Stability Operations and Afghan Local Police (VSO/ALP), was established by CFSOCC-A in 2010 after initial efforts under other labels failed. In perhaps a display of the flexibility and adaptability of special operators, or historical irony, two of the most senior special operations leaders championing the effort, McChrystal and Austin Miller, had worked together building the surgical strike machine that crippled AQI in Iraq.[380] In the words of one operator, 'it took a black ops guy from [the special mission units, Miller,] to understand the limits of kill and capture'.[381]

Political support for VSO/ALP resulted from convergence of ideas within the SOF community. A Special Forces major, Jim Gant, wrote and distributed a short paper, 'One Tribe at a Time', that suggested a different approach in Afghanistan could be successful. Gant arrived deductively and intuitively at several insights related to irregular warfare reached by T.E. Lawrence and David Galula. Gant concluded tribal

[375] Mark Moyar, *Village Stability Operations and the Afghan Local Police*, JSOU Report 14-7 (MacDill Air Force Base, FL: JSOU Press, 2014), pp. 7–8.

[376] Richard Pregent, 'Rule of Law Capacity Building in Iraq', in Raul Pedrozo, ed., *The War in Iraq: A Legal Analysis*, International Law Studies, vol. 86 (Newport, RI: Naval War College Press, 2010), pp. 329–330; Jesse Wozniak, 'Iraq and the Material Basis of Post-Conflict Police Reconstruction', *Journal of Peace Research*, 54:6 (November 2017): 806–818.

[377] Charles Cleveland offers that the idea for CFSOCC-A was his, one resulting from what he termed a 'special operations *shura*' in Afghanistan in 2008. Cleveland and Egel, *The American Way of Irregular War*, pp. 158–160.

[378] Frederick Kagan et al., *Defining Success in Afghanistan* (Washington, DC: American Enterprise Institute, 2011), p. 16; US House of Representatives, House Committee on Oversight and Government Reform, Subcommittee on National Security, Homeland Defense, and Foreign Operations, 'Statement of Brigadier General Stephen J. Townsend, Director, Pakistan-Afghanistan Coordination Cell, J-5, the Joint Staff', *Defense Department Contracting in Afghanistan: Are We Doing Enough to Combat Corruption?*, Serial No. 112-80 (Washington, DC: Government Printing Office, 2012), p. 9.

[379] William Knarr and Mark Nutsch, *Village Stability Operations and the Evolution of SOF Command and Control in Afghanistan: Implications for the Future of Irregular Warfare*, JSOU Report 20-2 (MacDill Air Force Base, FL: JSOU Press, 2020), p. 7.

[380] Moyar, *Village Stability Operations and the Afghan Local Police*, pp. 9–10.

[381] Linda Robinson, *One Hundred Victories: Special Ops and the Future of American Warfare* (New York: PublicAffairs, 2013), p. 25.

engagement was crucial to success, but this in turn required military advising and assistance missions working from the local level, not imposed from Kabul. Gant's 'Tribal Engagement Strategy' required extended presence, trust- and confidence-building, cultural understanding, and exposing special operators to much higher levels of risk by embedding them in local communities.[382] Gant's efforts were aided locally by Miller but also in the United States by journalist Ann Scott Tyson and historical fiction author Steven Pressfield, who posted the paper on his blog.[383] Subsequently, both Generals Petraeus and McChrystal, as well as USSOCOM commander Eric Olson, endorsed Gant's ideas as the foundation for VSO/ALP.[384]

VSO/ALP, combined with leadership changes and a surge of political interest and resources, revitalized efforts to bring security and governance to portions of rural Afghanistan. In 2009 McChrystal managed a seemingly impossible task, integrating different SOF elements together horizontally, including surgical strike and special warfare as well as coalition special operations.[385] Miller, as CFSOCC-A, succeeded in a much more challenging task in 2010: vertical integration of local VSO/ALP efforts with national-level ones emanating from Kabul, the 'unique contribution' of the former to stability operations in the country by providing a heretofore missing link in the chain of governance.[386] These connected security efforts with stability and development ones, including those of PRTs. Doing so bolstered the legitimacy, in part, of the Karzai government in providing for the needs and requirements of the Afghan population. SOF ODAs, split in half, were paired with conventional forces from the 101st and 82nd Airborne to provide the fifty-eight teams necessary for VSO/ALP.[387] The ambitious schedule for employment, approved by General Petraeus and Hamid Kharzai and driven in part by political time limits placed on the surge, 'demanded more Americans than SOF could provide'.[388]

Several factors ultimately contributed to the demise of the VSO/ALP programme. Within the programme, perhaps the most important was relationships, influenced by individual and collective personalities. The dynamic tension inherent in sustaining vertical and horizontal integration was frustrated by the constant rotation of ISAF personnel in and out of theatre. In the assessment of one District Augmentation Team (DAT) leader, between various military and civilian units 'the synergy was difficult to maintain' with the rapid and frequent changeover of partnering and rotational

[382] Jim Gant, *One Tribe at a Time: A Strategy for Success in Afghanistan* (Los Angeles, CA: Nine Sisters Imports, 2009), pp. 6–8.

[383] Steven Pressfield, 'The Full Document at Last!', 26 October 2009, https://stevenpressfield.com/2009/10/one-tribe-at-a-time-4-the-full-document-at-last.

[384] Ann Scott Tyson, 'Jim Gant, The Green Beret Who Could Win the War in Afghanistan', *The Washington Post* (17 January 2010); Montgomery McFate, *Military Anthropology: Soldiers, Scholars and Subjects at the Margins of Empire* (New York: Oxford University Press, 2018), p. 178.

[385] In technical terms, 'By April 2010 General McChrystal had achieved unity of command of the three SOF tribes with NATO [operational control; OPCON] of ISAF SOF, [tactical control; TACON] of the [Direct Action Task Force] and OPCON of CFSOCC-A'. Knarr and Nutsch, *Village Stability Operations and the Evolution of SOF Command and Control in Afghanistan*, p. 100.

[386] Knarr and Nutsch, *Village Stability Operations and the Evolution of SOF Command and Control in Afghanistan*, p. 95.

[387] Ibid., p. 117.

[388] Moyar, *Village Stability Operations and the Afghan Local Police*, p. 14.

forces. 'Relationships were severed much more quickly than established and each incoming team approached key Afghan personalities differently; some were effective and others less so.'[389] Others note the differences in interest levels and approaches to local tribal leaders among various conventional forces rotated in, based on training, specialty, and leader personalities.[390]

Special operations leadership qualities, as well as their connection to sufficient combat support for VSO/ALP, also played a role. Miller's successor, Tony Thomas, was known for his impatience and repeatedly pushed his subordinates to value the creation of Afghan capacity *en masse* over capability.[391] Another contributing factor to failure was the inability to develop the crucial link between VSO/ALP and national-level Afghan supporting resources such as communications, quick reaction forces, and airpower in sufficient quantity. Local VSO/ALP detachments were dependent on US and coalition support when they encountered Taliban forces. With it, VSO/ALP units had a commendable record. Deprived of it, however, they were frequently overrun or melted away.

The military advising and assistance effort to develop ANA and Afghan National Police (ANP) capabilities started far too late to overcome significant hurdles, including competition for limited local talent, institutionalized normative practices among a generation of military officers, human capital issues such as widespread illiteracy, and contracting difficulties. Even when such efforts started bearing fruit, Taliban actions and political reactions could and did often erase progress made. Forrest Morgan, for example, outlines the severe consequences that occurred after a handful of insider attacks on military advising and assistance efforts to develop the ANA Air Force.[392] VSO/ALP, designed as an interim measure until sufficient ANA and ANP were fielded, was corrupted for political reasons in much the way as local security efforts were in Vietnam a generation prior. This corruption, based on good intentions, sought industrial outputs within a short period of time in line with American political constraints, allowing predatory local Afghan leaders an avenue to enrich themselves.[393]

Special operations nonetheless achieved some noteworthy successes in Afghanistan. Most were limited to the tactical and operational level; all were transitory. Politically, special operations were necessary to buy time and space for legitimate Afghan governance to develop. Surgical strikes, embodied in 'night raids', became an increasing source of tension politically between the Karzai government and ISAF until their proscription in 2011. The raids, while occurring at a similar

[389] Lieutenant Commander Ty Bathurst, quoted in Knarr and Nutsch, *Village Stability Operations and the Evolution of SOF Command and Control in Afghanistan*, p. 91.

[390] Moyar, *Village Stability Operations and the Afghan Local Police*, pp. 14–15.

[391] Robinson, *One Hundred Victories*, p. 256.

[392] For details, see Forrest Morgan, *Flight Risk: The Coalition's Air Advisory Mission in Afghanistan, 2005–2015* (Annapolis, MD: Naval Institute Press, 2018), pp. 96–117.

[393] Sam Wilkins, 'The Rise and Fall of Village Stability Operations in Afghanistan: Lessons for Future Irregular Warfare Campaigns', Modern War Institute blog, 9 August 2022, https://mwi.westpoint.edu/the-rise-and-fall-of-village-stability-operations-in-afghanistan-lessons-for-future-irregular-warfare-campaigns.

pace to those in Iraq against AQI, did not impact either the Taliban or al-Qaeda in the same way. Sanctuary in Pakistan, combined with proselytizing in madrassas, ensured a steady source of replacements. The F3EAD process, perceived as so successful in Iraq, lost some of its focus in Afghanistan as its target sets expanded to include lower-level Taliban and al-Qaeda members.[394] The same evening as the bin Laden raid, '12 other missions' were conducted in Afghanistan that 'captured or killed between 15 to 20 targets' and reached totals estimated at 600 per month.[395] In Helmand province, 'striking ever-lower individuals in the "Taliban" hierarchy ... had very little strategic effect on the Taliban movement as a whole', according to field research.[396] Covert actions in Pakistan, specifically drone strikes, solved one political problem at the expense of another. While 'creating more insurgents than they killed' is an oft-used expression related to drone strikes, backed in some cases by questionable data or subjective assessments, their unpopularity in Pakistan was a grievance Taliban recruiters exploited.[397]

Special operations in Afghanistan from 2007 until 2014 oscillated between two historical approaches to COIN. They are 'enemy-centric' targeting of an insurgent organization to dismantle its overt and clandestine infrastructure, and 'population-centric', which prioritizes engaging with communities and mobilizing them to defend themselves and the government. Indirect-shock surgical strike missions, both overt in Afghanistan and clandestine in Pakistan, filled the role of the former and VSO/ALP the latter. The debate that occurred in the White House in 2008 over which path to take was instead a false choice—both are necessary at different times and in different locations. The approach was further undercut by continued American subsidizing of Afghan governance and security—almost $12 billion annually, or three-quarters of the country's gross domestic product—with little of the necessary conditionality to hold Afghan leaders responsible.[398] Special operations of any stripe, overt, covert, or clandestine, were incapable of addressing the fundamental political problems in Afghanistan: providing sufficiently responsive governance while at the same time coopting or subverting the traditional source of power in the country, local tribal patronage networks.

Tribal patronage networks served as an important enabler of success in OEF in 2001/2 and barrier to it from 2003 onward. Special operations raids succeeded in reinforcing the grip tribal patronage networks had locally and delegitimized the government, while at the same time enriching them thanks to a rewards scheme

[394] Mazzetti et al., 'SEAL Team 6'.

[395] Niva, 'Disappearing Violence', pp. 186; 196.

[396] Mike Martin, *An Intimate War: An Oral History of the Helmand Conflict, 1978–2012* (Oxford: Oxford University Press, 2014), p. 201.

[397] International Human Rights and Conflict Resolution Clinic and Global Justice Clinic, *Living Under Drones: Death, Injury, and Trauma to Civilians from U.S. Drone Practices in Pakistan* (Stanford, CA: Stanford Law School, 2012), pp. 131–138; Akbar Ahmed, *The Thistle and the Drone: How America's War on Terror Became a Global War on Tribal Islam* (Washington, DC: Brookings Institution Press, 2013). A more recent, more objective, and less conclusive assessment is Patrick Johnston and Anoop Sarbahi, 'The Impact of Drone Strikes on Terrorism in Pakistan', *International Studies Quarterly*, 60:2 (June 2016): 203–219.

[398] Stephen Biddle and Peter Feaver, 'Assessing Strategic Choices in the War on Terror', in James Burk, ed., *How 9/11 Changed Our Ways of War* (Stanford, CA: Stanford University Press, 2013), p. 46.

put in place for denunciations.[399] Stathis Kalyvas explores this 'locus of agency' phenomenon, suggesting the tribal leaders practiced strategy more effectively and achieved their desired ends at the expense of a weaker central authority unable to control them, in this case the government in Kabul and ISAF.[400] American special operators perfected processes but never developed sufficient understanding of the local tribal politics to navigate their nuances, a trait shared by those working elsewhere in the military and civil government.[401] Carter Malkesian, who worked for two years as a PRT advisor for the Department of State, identified two crucial areas of misunderstanding in Garmsar province: 'deep social cleavages' and land issues tied to the 'after-effects of a great modernization scheme' undertaken decades prior as part of an American-sponsored development effort, about which most were entirely unaware.[402] Implicit in his observations was a source of tension and opportunity: navigating political cleavages adroitly, through sufficient understanding, and using conditionality in discussions with district, province, and national government to ensure the most effective Afghan leaders were in place and the local patronage networks broken up in consequence. Issues in understanding local politics also affected the Taliban. They faced similar problems in trying to consolidate resources with Helmandi tribal leaders. Their efforts at trying to develop *nezami,* or patronage from outside, fell victim to local *mahaz*, or seeking patrons 'to help them fight local enemies'.[403]

Better understanding these realities by policy makers and operators alike might have saved considerable SOF effort in the Sisyphean task of dismantling the Taliban infrastructure. This infrastructure was not an organization in the classical sense but instead 'a plurality of opinions' and a shifting series of core interests articulated by the Quetta Shura, the Tehrik-e Taliban Pakistan (TTP), and the Haqqani network, as well as actions by regional warlords and local opportunists engaged at their convenience.[404] Flynn's problem statement regarding intelligence collection in Afghanistan in 2010 suggests coalition planners understood the scale of the problem:

> the vast intelligence apparatus is unable to answer fundamental questions about the environment in which U.S. and allied forces operate and the people they seek to persuade. Ignorant of local economics and landowners, hazy about who the powerbrokers are and how they might be influenced, incurious about the correlations between various development projects and the levels of cooperation

[399] Martin, *An Intimate War*, p. 135.

[400] Stathis Kalyvas, *The Logic of Violence in Civil War* (Cambridge: Cambridge University Press, 2006), pp. 376–381.

[401] The 'battle rhythm' of special operations in Afghanistan, specifically during the early years of CJSOTF-A, is outlined clearly and concisely for the non-specialist in Adrian Bogart III, *One Valley at a Time*, JSOU Report 06-6 (Hurlburt Field, FL: JSOU Press, 2006), pp. 44–51.

[402] Carter Malkesian, *War Comes to Garmsar: Thirty Years of Conflict on the Afghan Frontier* (Oxford: Oxford University Press, 2013), pp. 268–269.

[403] Martin, *An Intimate War*, pp. 204–205.

[404] Thomas Ruttig, 'Have the Taliban Changed?', *CTC Sentinel*, 14:3 (March 2021): 12; Vahid Brown and Don Rassler, *Fountainhead of Jihad: The Haqqani Nexus, 1973–2012* (New York: Columbia University Press, 2013), pp. 105; 126.

> among villagers, and disengaged from people in the best position to find answers—whether aid workers or Afghan soldiers—U.S. intelligence officers and analysts can do little but shrug in response to high level decision-makers seeking the knowledge, analysis, and information they need to wage a successful counterinsurgency.[405]

The only reasonable conclusion reachable based on this quote and given its date, one ripe for further study and assessment, is that comprehending the barrier to understanding was one thing but fixing it bureaucratically quite another—beyond even the power of reformers such as David Petraeus over the next dozen years.

## Libya (2011) and Iraq redux (2014–2016)

Relatively little can be said about the conduct of American special operations in Libya, against the Islamic State in Iraq and Syria (ISIS), or globally. Reliable information about them is unavailable. In lieu of SOF providing targeting information and coordinating airstrikes, NATO planners received an overwhelming volume of the former from networks of indigenous spotters relaying information using commercial applications and methods. According to Libyan sources, only the arrival of an unspecified and unidentified number of 'advisors', presumably paramilitary officers from one or more national intelligence agency, succeeded in reducing the time from report to airstrikes from days down to minutes.[406] Political risk calculations and unclear political and strategic objectives, along capacity commitments to Afghanistan, argued against the commitment of American special operators to Libya for ODYSSEY DAWN (OOD).

Although Libya only partially followed the pattern of operationally offensive special operations established in OEF and OIF, Operation INHERENT RESOLVE (OIR), the operation against ISIS, mirrored them. ISIS, formed out of the remaining rump of AQI, grew initially out of the conditions of the Syrian Civil War. This war and the First Libyan Civil War grew out of the 2011 Arab Spring movement that saw autocratic leaders removed by spontaneous, popular uprisings in Egypt, Tunisia, and Yemen and political or economic changes or concessions granted in eleven other North African or Middle Eastern countries. In Syria, ISIS, under the leadership of Abu Bakr al-Baghdadi, initially was part of Jabhat al Nusra, the umbrella organization for Sunni jihadist groups fighting against the Shi'a regime of Bashir al Assad. The group and its leader later rose to prominence in 2013 for its split with al Nusra as well as al-Qaeda in February 2014. Few at the time assessed ISIS as a serious threat, as President Obama's infamous 2014 analogy of the group as a '[junior varsity] team',

[405] Flynn et al., *Fixing Intel*, p. 5.

[406] Frederic Wehrey, 'The Libyan Experience', in Karl Mueller, ed., *Precision and Purpose: Airpower in the Libyan Civil War*, Report RR676 (Santa Monica, CA: RAND, 2015), pp. 61–64.

with local objectives and capabilities, rather than professional and global one like al-Qaeda suggested.[407]

The kindling for ISIS to expand rapidly existed in neighbouring Iraq, where the government of Nouri al-Maliki had increasingly persecuted and isolated significant portions of the Sunni population, including the tribal leaders in western provinces bordering Syria, including Anbar.[408] In June 2014, ISIS launched an offensive which rapidly spread into northern and eastern portions of Iraq, capturing Mosul and Tikrit and threatening Baghdad, and scattering four US-trained and -equipped Iraqi divisions, some of whom defected or fled.[409] By August, ISIS forces threatened the Kurdish capital of Erbil and were poised to wipe out minority Yazidis who fled its capture of Sinjar, prompted an initial US response in the form of air strikes.

ISIS presented the Obama administration with problems on several fronts. Legally, the administration constructed a fine if tenuous web of justifications for its use of force, including Article 51 of the UN Charter in support of Iraq and self-defence, collectively, nationally, and later individually, at the unit level, when SOF were introduced, based on the notion of imminent threat.[410] In addition, the Obama administration drew upon connections between ISIS and al-Qaeda to fold action under the 2001 AUMF, thereby bypassing the need for Congressional approval.[411] ISIS inspired or supported attacks in Paris and San Bernadino in 2015 provided reasonable evidence in support of the argument for collective and national self-defence. Politically, the administration's reticence to reengage in another war in the Middle East reflected the mood of much of the American people. Their support for airstrikes and the use of SOF against ISIS was high but did not extend to the introduction of ground troops.[412] Diplomatically, the US was looking to draw upon support from regional partners such as Jordan while at the same time avoiding escalating the conflict with Iraq and Russia. One grievance among the Sunni population of Iraq was al-Maliki's increasingly close ties to the Shi'a regime in Iran, most evident in the presence of units of the Iranian Revolutionary Guard Corps Quds Force providing military advice and assistance to several proxy militias such as Kata'ib Al-Imam Ali.[413] The intervention of Russia into the Syrian Civil War in September 2015 raised the real possibility of inadvertent escalation given the close proximity of American and Russian advisors, proxies, and aircraft to one another.

[407] David Remnick, 'Going the Distance', *The New Yorker* (19 January 2014), www.newyorker.com/magazine/2014/01/27/going-the-distance-david-remnick.

[408] Fawaz Gerges, *ISIS: A History* (Princeton, NJ: Princeton University Press, 2016), pp. 113–116.

[409] Joby Warrick, *Black Flags: The Rise of ISIS* (New York: Doubleday, 2015), pp. 302–306.

[410] Douglas Cantwell, 'The ETF and the Legality of U.S. Intervention in Syria under International Law', Lawfare, 28 March 2016, www.lawfaremedia.org/article/etf-and-legality-us-intervention-syria-under-international-law.

[411] Jessica Michek, 'What the AUMF Is and Why You Should Care', Bipartisan Policy Center, 18 April 2018, https://bipartisanpolicy.org/blog/what-the-aumf-is-and-why-you-should-care-2018.

[412] Adam Taylor, 'Poll: Most Americans Support U.S. Military Action in Syria—But Only Want a Limited Amount', *The Washington Post* (15 August 2016), www.washingtonpost.com/news/worldviews/wp/2016/08/15/poll-most-americans-support-u-s-military-action-in-syria-but-only-want-a-limited-amount.

[413] Michael Knights, 'Iran's Expanding Militia Army in Iraq: The New Special Groups', *CTC Sentinel*, 12:7 (August 2019): 1–12.

To help address these problems and concerns, the Obama administration drew on ongoing paramilitary special operations and authorized the dispatch of military SOF to both Iraq and Syria under OIR, according to media reports. Various reporters have ascribed a name to the paramilitary effort, Timber Sycamore, which they allege began as 'modest' covert military advice and assistance programme in 2012, one distilled down from a much more ambitious plan.[414] Members of Congress, spurred at first by seeming Executive inaction and later concerns over escalation with Russia, put forward and later approved an overt military advice and assistance programme, the Syria Train and Equip Program (STEP), in 2014. Funded using existing Section 1206 authorities, STEP was designed 'to overtly train and lethally equip vetted members of the Syrian opposition for select purposes'.[415] These forces were designed to reach upwards of 10,000 proxy fighters over a more than three-year period.

STEP was later cancelled by the Donald Trump administration after producing meagre results, largely due to onerous legal constraints associated with State Department vetting. Its history offers valuable insights relating to control.[416] The programme was originally proposed in Congressional legislation in 2013, its funding mechanisms were discussed and revised in 2014, and in a series of back-and-forth negotiations between members of the Executive and Legislative branches, approved by both legislative branches and signed into law by the president on 16 December 2014.[417] This process, as well as the subsequent development of a modified AUMF specifically for ISIS, suggest control mechanisms and processes vertically, at the highest levels, remain effective if inefficient.

Specific details regarding the employment of other military special operations in Syria from 2014 through 2016 are fragmentary due to classification. Despite these barriers, some aspects of military special operations can be highlighted even if only in general terms, through specific episodes or inferentially. There are four such aspects, two related to historical mission sets, one for control, and a final for integration of capabilities. The first historical mission set, indirect-shock missions, were conducted using the F3EAD methodology increasingly perfected, conducted, and refined by TF 714. Surgical strike missions were conducted for multiple purposes simultaneously. They were designed to capture or kill ISIS HVTs and senior leaders. The best-known mission in this period was the raid against Abu Sayyaf, an ISIS financial emir, in May 2015. It was conducted 'deep in ISIS-controlled territory' in the

[414] Austin Carson and Michael Poznansky, 'The Logic for (Shoddy) U.S. Covert Action in Syria', War on the Rocks, 21 July 2016, https://warontherocks.com/2016/07/the-logic-for-shoddy-u-s-covert-action-in-syria. For details, based in part on a publicly released and CIA-vetted personal account by one of its planners, see Ken Dilanian and Kevin Monahan, 'Obama Nixed CIA Plan That Could Have Stopped ISIS: Officials', NBC News, 2 April 2016, www.nbcnews.com/news/us-news/obama-nixed-cia-plan-could-have-stopped-isis-officials-n549111.

[415] Christopher Blachard and Amy Belasco, *Train and Equip Program for Syria: Authorities, Funding, and Issues for Congress*, Report R43727 (Washington, DC: Congressional Research Service, June 2015), pp. 1–2.

[416] Joseph Osborne, 'Syria Train and Equip: Who Left the Interns in Charge?', *Special Operations Journal*, 7:2 (2021): 166–189; Øystein H. Rolandsen and Kjetil Selvik 'Disposable Rebels: US Military Assistance to Insurgents in the Syrian War', *Mediterranean Politics*, 29:1 (2024): 1–22.

[417] Details and a concise chronology are available in Blachard and Belasco, *Train and Equip Program for Syria*, p. 4.

town of Deir Ezzor, resulted in Sayyaf's death and the capture of his wife, along with 'a rich trove of documents'.[418] These documents proved essential in mapping out the chokepoints in ISIS's web of its captured oil and gas resources for a subsequent air campaign, Operation TIDAL WAVE II. This air campaign, 'the most extensive deep-fight effort of OIR' or use of force for purposes other than direct support of proxy and SOF forces, was designed to aid in the targeting process to prevent unnecessary civilian casualties, while at the same time denying ISIS access to one of its considerable sources of revenue: black market oil and gas.[419] In addition, the raid was designed to send several political and strategic messages: first, to ISIS leaders, that nowhere are you safe, even in territory you control, and second, by raising the morale of those fighting against ISIS, not to mention domestic and international support for the operation, by striking 'another significant blow'.[420] The surgical strike was also notable for its size. According to media reporting, more than one hundred special operators, or roughly 20 per cent of those acknowledged to be in Syria, were used for the operation, backed by a 'gorilla' force package including 'Blackhawk helicopters and V-22 aircraft', suggesting force overmatch to buy down tactical mission risk.[421]

Likely for reasons related to political risk, the Obama administration limited the number of indirect-shock missions to three from 2014 through 2016. Possible reasons including avoiding large single event losses and escalation resulting from accidental contact with Russian or Syrian forces. In July 2014 SOF attempted a mission to rescue journalist James Foley and other hostages but were unsuccessful in their efforts.[422] Another to capture Abu Ali al-Anbari, an ISIS deputy commander, ended with his death and the recovery of documents and equipment subsequently exploited in the campaign. Although his capture and interrogation was the preferred outcome, the SOF raid managed to put to rest confusion surrounding his alleged death twice before in previous airstrikes.[423]

In military advice and assistance missions, SOF worked once again with the Kurdish *peshmerga* in northern Iraq. A new proxy was the People's Protection or Defence

[418] Barbara Starr, Laura Smith-Spark, and Ray Sanchez, 'Abu Sayyaf, a Key ISIS Figure in Syria, Killed in U.S. Raid', CNN.com, 17 May 2015, www.cnn.com/2015/05/16/middleeast/syria-isis-us-raid/index.html; Becca Wasser et al., *The Air War against the Islamic State: The Role of Airpower in Operation Inherent Resolve*, Report RA388-1 (Santa Monica, CA: RAND Corporation, 2021), p. 69.

[419] Wasser et al., *The Air War against the Islamic State*, p. 217. TIDAL WAVE was named as a modern successor to the bold initial airstrikes against German-controlled resources in the oilfields in and around Ploesti, Romania, during the Second World War.

[420] 'Statement by Secretary of Defense Ash Carter on Counter-ISIL Operation in Syria', Defense.gov, 16 May 2015, www.defense.gov/News/Releases/Release/Article/605,506/statement-by-secretary-of-defense-ash-carter-on-counter-isil-operation-in-syria.

[421] Starr et al., 'Abus Sayyaf, a Key ISIS Figure in Syria, Killed in U.S. Raid'.

[422] Spencer Ackerman, 'James Foley: US Reveals Failed Special Forces Rescue Mission within Syria', *The Guardian* (21 August 2014), www.theguardian.com/world/2014/aug/21/elite-forces-us-raid-syria-james-foley-failed.

[423] Damian Paletta, Adam Entous, and Ben Kesling, 'Senior Islamic State Leader Killed', *The Wall Street Journal* (25 March 2016), www.wsj.com/articles/senior-islamic-state-leader-killed-1,458,921,361. Much of the confusion had to do with al-Anbari's use of various *noms de guerre*. Michael Weiss and Hassan Hassan, 'Everything We Knew about This ISIS Mastermind Was Wrong', *Daily Beast* (15 April 2016), www.thedailybeast.com/everything-we-knew-about-this-isis-mastermind-was-wrong.

Units (*Yekîneyên Parastina Gel*, or YPG). The YPG, based primarily in northern Syria, provided a proxy nucleus around which smaller anti-Assad militias could orbit, collectively forming the Syrian Democratic Forces or SDF. Paramilitary and military special operations supported different elements of the SDF based on geographic area, political sensitivity, and the different bearing the latter required: covert in the case of CIA-backed units west of the Euphrates River and overt and clandestine for military SOF east of it. Media reports in 2016 mistakenly attributed internecine fighting between different SDF factions to interagency friction between paramilitary and military SOF. Frictions instead related to control, specifically maintaining unity of effort over a geographically spread out and federated collection of groups with different interests.[424]

In Syria, paramilitary and military special operations conducted in OIR followed a modified Afghan model previously used for successful operational and tactical SOF-led offensives. Unlike OEF and OIF, paramilitary and military SOF were split into separate areas of authority to ease their control. In addition, coalition special operations partners, including those from the Gulf Arab states (Qatar and United Arab Emirates, for example), acted as an additional social bridge between American special operators where preexisting relationships did not exist. The rough estimate of SOF in OIR numbers around 2,000, although official numbers of US SOF in CJOSTF-Syria (CJSOTF-S) are 500.[425]

In Iraq, similar numbers of SOF worked with two forces: Kurdish peshmerga, noted above, as well as formerly US-trained Iraqi SOF in three brigades of the CTS. In 2015, the Obama administration approved a specific funding authority, Section 1236, for them to provide equipment 'to fulfill immediate train-and-equip needs in the current fight against ISIS', which was subsequently extended and expanded.[426] Military advisory and assistance SOF were collectively termed by the administration an 'expeditionary targeting force', one of a number of 'accelerants' designed to improve the capabilities of proxy forces while at the same time limiting risk to US ground forces.[427]

Such special operations proved useful as a control measure as command of airstrikes was delegated down from the four-star general CENTCOM commander initially in August 2014 gradually down to one-star general strike cell commanders. The early restraints on airstrikes, particularly the presidential directive for a non-combatant casualty cutoff threshold initially of zero, were a source of considerable criticism of the Obama administration at the time and in subsequent

[424] Sam Heller, 'Are CIA-Backed Syrian Rebels Really Fighting Pentagon-Backed Syrian Rebels?' War on the Rocks, 28 March 2016, https://warontherocks.com/2016/03/are-cia-backed-syrian-rebels-really-fighting-pentagon-backed-syrian-rebels.

[425] Wasser et al., *The Air War against the Islamic State*, p. 129; Shannon Collins, 'Carter: U.S. Will Deploy More Troops to Syria to Combat ISIL', DoD News, 10 December 2016, www.defense.gov/News/News-Stories/Article/Article/1,026,625/carter-us-will-deploy-more-troops-to-syria-to-combat-isil.

[426] Inspector General of the US DoD, *Assessment of U.S. and Coalition Plans and Efforts to Train, Advise, Assist, and Equip the Iraqi Counterterrorism Service and the Iraqi Special Operations Forces*, Report DODIG-2017-074 (Washington, DC: DoD IG, April 2017), pp. 5; 7.

[427] Wasser et al., *The Air War against the Islamic State*, p. 72.

campaign assessments.[428] A range of political and operational considerations, including coalition support, limited numbers of RPAs providing full-motion video, and the virtual absence of human intelligence reporting drove this risk consideration.[429] Operational control measures for OIR, and their gradual relaxation, reflected changing political risk calculations as well as increased tactical-level risk being assumed by SOF.

At the heart of operational control measures for special operations in OIR was the development of SOF strike cells. These strike cells had their roots in the Joint Air Ground Integration Center (JAGIC), resulting from lessons learned during OEF and the friction that occurred in Operation ANACONDA.[430] Static cells were embedded in the two Combined Joint Operations Centers (CJOCs) in Iraq: one in Baghdad and the other in Erbil. SOF units acted as a form of mobile ground sensor, identifying targets and passing this information along to the CJOCs. Host nation constraints, rather than US risk calculations, determined where, under what conditions, and when airstrikes could occur. As RAND study authors acknowledge, these constraints affected the use of all munitions or target engagement authority (TEA) in Iraq, with Iraqi or Kurdish officers giving their approval.[431] As one Iraqi SOF commander noted, 'Sometimes we locate the targets, and sometimes it's the Americans, but they don't hit it until they get our permission.'[432]

SOF strike cells in Syria, in contrast, could receive strike authority from the CJSOTF-S commander. Relaxed controls, offering SOF greater flexibility, reflected the different operating environment. Syrian host nation approval was unnecessary and unlikely given the SDF's ultimate objectives. SOF units moved with their SDF counterparts in distributed teams within the fluid and dynamic battle space in Syria. Initially SOF strike cells were comprised of just three personnel but gradually grew in scale as the campaign gathered momentum, providing an integrating capability between the SDF and air power comparable to the CJOCs in Iraq, in demanding circumstances.[433] The fluid and complex nature of the battle space in Syria and Iraq imposed significant pressures on those authorizing strikes. Paul Birch reflected on them:

> You've got a lot of ISR assets ... You have at least three sectors of artillery that are able to provide fires into that fight. You have at least three ... ground commanders. You'd have the [coalition partners] and maybe an Iraqi division is on scene. And then ... some sort of third party, Kurdish or someone else, involved in the fight. And you're making these calls for fires that the ground commander wants deconflicted

[428] See, for example, David Deptula, 'How to Defeat the Islamic State', *Washington Post* (5 June 2015) and Ben Lambeth, *Airpower in the War against ISIS* (Annapolis, MD: Naval Institute Press, 2021), pp. 178–185.

[429] Wasser et al., *The Air War against the Islamic State*, p. 55.

[430] Center for Army Lessons Learned, *Joint Air Ground Integration Center: Lessons and Best Practices*, CALL Handbook No. 17-04, U//FOUO (Fort Leavenworth, KS: CALL, January 2017); Joint Chiefs of Staff, *Joint Publication 3-09: Joint Fire Support* (Washington, DC: The Joint Staff, April 2019), p. II-18.

[431] Wasser et al., *The Air War against the Islamic State*, p. 25.

[432] Ibid., p. 165.

[433] Ibid., p. 133.

> in a matter of minutes or seconds ... It made for a very interesting physics problem, which became a command and control problem.[434]

The campaign momentum and tempo of operations, as well as incidents during the battle for Mosul, led to changes in the rules of engagement, including TEA delegated from one-star commanders and colonels down to lieutenant colonels in Iraq and SOF strike cell majors in Syria.[435]

SOF strike cells and advisors were necessary based on changing technology. In OOD, militias and proxies overwhelmed JOCs with information collected and/or transmitted using personal electronic devices such as mobile phones and laptops. To ensure sufficient control of information, SDF forces could report an ISIS target's GPS coordinates via SOF-provided tablets and pass this information back to field headquarters and, eventually, SOF strike cells.[436] The tablets, loaded with specific applications including Android Team Awareness Kit and a chat function, were paired with encrypted cell phones and required some training, as well as translation, in order to function in an integrated manner.[437] Pushing intelligence reporting and collection forward allowed SOF to minimize risk of casualties by operating further behind advancing SDF. SOF provided the SDF not only with access to coalition airstrikes but also valuable ISR information, including FMV.

In Iraq similar means were employed. This was especially true during fierce fighting to retake cities, where CTS forces took almost 3,500 casualties, half of their overall force.[438] In addition, the creation and fielding of non-standard, smaller composite units comprised of various SOF and conventional forces specialties, hinted at in media reports, suggests integration occurred horizontally and vertically, both in Syria and Iraq, from the tactical level upwards.[439]

Integration of capabilities at the tactical level was not always matched vertically at the operational. Persistent sources of friction existed, including for special operations. Tactical-level issues became sources of operational friction as battle space ownership and control converged, particularly when SDF moved to the edge of ISIS-controlled and defended cities, specifically its capital, Raqqa. RAND's analysts relate the following:

[434] Quoted in Ibid., p. 178.

[435] Ibid., p. 26; Stephen Renner, 'Air Power in the Battle of Mosul', in Phil Haun, Colin Jackson, and Timothy Schultz, eds., *Air Power in the Age of Primacy: Air Warfare since the Cold War* (Cambridge: Cambridge University Press, 2022), p. 269. Renner served in CJOC-Erbil during the Battle of Mosul.

[436] Sofia Barbarani, 'Calling the Shots: The Syrians behind Air Attacks on Raqqa', *Al Jazeera* (1 July 2019), www.aljazeera.com/features/2019/7/1/calling-the-shots-the-syrians-behind-air-attacks-on-raqqa.

[437] Kyle Rempfer, 'Air Force JTACs Training Syrians to Help Call in Her Airstrikes Raises Questions', *Army Times* (5 April 2021), www.armytimes.com/news/your-air-force/2021/04/05/air-force-jtacs-training-syrians-to-help-call-in-helo-airstrikes-raises-questions.

[438] Jeffrey Martini et al., *Operation Inherent Resolve: U.S. Ground Force Contributions*, Report RRA719-1 (Santa Monica, CA: RAND Corporation, 2022), p. 172.

[439] 'Soldier Killed in Iraq Raid Belonged to Delta Force', ABC News, 23 October 2015, https://abcnews.go.com/Politics/soldier-killed-iraq-raid-belonged-delta-force/story? id=34,676,330; Barbara Starr and Veronica Stracqualursi, '4 Americans Killed in Syria Bombing Identified', CNN, 18 January 2019, www.cnn.com/2019/01/18/politics/syria-attack-americans-killed/index.html.

> As the SDF closed in on Raqqa from three sides, control over the airspace became a source of tension between the SOF-run strike cells and the CFACC. Until that point, the CJTF-OIR Commander had delegated control of the airspace over Raqqa to the CFACC, which launched strikes in the city to soften up ISIS targets and keep pressure on the organization. CJSOTF was controlling the airspace in the territory outside Raqqa, because it was supporting the SDF's ground maneuver and thus had the best situational awareness. As the SDF approached Raqqa city, the SOF strike cell commanders had discussions with the CFACC about assuming control over that airspace so that they could better support the SDF's offensive. From the perspective of the strike cell, the CFACC was often using scarce air assets and weapons to hit valid ISIS targets, but ones that did not directly support the ground scheme of maneuver.[440]

This suggests that operations tensions will flare up, despite high degrees of integration, if effective transitions of command-and-control authorities do not occur rapidly enough to reflect realities on the ground. These tensions, dating back to the first employment of air power and ground manoeuvre forces, were particularly acute in this case as campaign level requirements were compressed in space and time with tactical-level ones.

Other operational tensions required the direct intervention of civilian control measures at the strategic level. For example, while the military was trying to distribute command authorities downward, Secretary of Defense Ash Carter was aware of measures' disjointed nature. To address what he saw as a 'fractured' relationship between CENTCOM and CJTF-OIR, Carter pushed command authority of all efforts against ISIS down to the latter in 2015. The sole exception to this authority remained special mission units, which remained under his direct authority after approval from President Obama.[441] Such units, which included cyber efforts against ISIS under Operation GLOWING SYMPHONY, were effective thanks to remarkable degrees of collaboration of special mission units and national-level assets and agencies as well as coalition partners, according to media reporting.[442] Official Cyber Command (CYBERCOM) documents reviewed and released suggest that GLOWING SYMPHONY achieved considerable effects in disrupting media operations vital for ISIS' recruiting efforts.[443] Joint Task Force-Ares, the instrument for Operation GLOWING

[440] Wasser et al., *The Air War against the Islamic State*, p. 104.

[441] Ibid., p. 69; Ash Carter, 'Remarks by Secretary Carter at the 2016 IISS Manama Dialogue, Manama, Bahrain', transcript, 10 December 2016, www.defense.gov/News/Transcripts/Transcript/Article/1,026,655/remarks-by-secretary-carter-at-the-2016-iiss-manama-dialogue-manama-bahrain.

[442] Dina Temple-Raston, 'How the U.S. Hacked ISIS', National Public Radio, 26 September 2019, www.npr.org/2019/09/26/763,545,811/how-the-u-s-hacked-isis; Garrett Graff, 'The Man Who Speaks Softly—and Commands a Big Cyber Army', *Wired* (13 October 2020), www.wired.com/story/general-paul-nakasone-cyber-command-nsa.

[443] USCYBERCOM, 'Operation Glowing Symphony Assessment Framework Briefing Version 1.0, November 10 2016'. Declassified and redacted. https://nsarchive.gwu.edu/sites/default/files/documents/6,655,592/National-Security-Archive-1-USCYBERCOM-Operation.pdf; USCYBERCOM, 'USCYBERCOM 120-Day Assessment of Operation Glowing Symphony, April 12 2017'. Declassified and redacted. https://nsarchive.gwu.edu/sites/default/files/documents/6,655,597/National-Security-Archive-6-USCYBERCOM.pdf.

SYMPHONY, demonstrated an integration of special operations, cyber, and other capabilities unprecedented in the history of conflict, heights likely unsurpassable for some time.

While control mechanisms at the operational level proved initially vexing in OIR, they operated smoothly and as designed. Congressional unease over expansion of presidential powers, driven in part by lobbying efforts, led President Obama to submit a draft AUMF in 2015 specifically for ISIS. The draft, however, was rejected by the House of Representatives.[444]

Even though political consensus was not unified, monitoring and control mechanisms were. Congress could not agree on amending the overall legal basis for OIR but its members received quarterly reports from an appointed Lead Inspector General for Overseas Contingency Operations, which oversaw efforts of the DoD, the Department of State, and the Agency for International Development. These updates, including one from 2016 that provided a summary of the eight submitted that quarter about the mission, also referred to an additional twelve investigations opened to ensure the $11.7 million spent per day on OIR was done responsibly.[445] The rapid collapse of the Iraqi military in the face of ISIS in 2014 spurred Congress to demand additional reporting on previous as well as ongoing train and equip missions, including the CTS.[446] Later reports tracked and published data on previous DoD Inspector General (IG) recommendations left unimplemented, demonstrating greater scrutiny and closer monitoring of the mission.[447] While many reports and audits were open for public release, others were restricted or classified, including some from the General Accounting Office, and they generated recommendations to close oversights, gaps, or overreaches. For example, related to ongoing special operations training, equipping, and advisory mission in Iraq, in 2017 the DoD IG offered six findings, two of which were classified, and four subsequent recommendations.[448] In sum, there was a wealth of monitoring and reporting occurring on OIR to guarantee tight control vertically down to the point of mission execution.

Integration of special operations capabilities in OIR, and the delegation of control down to such low levels would not have been possible without several contributing factors. First and foremost was the general political consensus that ISIS represented more than just a security threat but also comprised a menace to civilization. While

[444] Joint Resolution, 11 February 2015, https://obamawhitehouse.archives.gov/sites/default/files/docs/aumf_02112015.pdf; Patricia Zengerle, 'House Rejects Measure to Force Vote on Islamic State Fight', Reuters, 11 June 2015, www.reuters.com/article/us-mideast-crisis-usa-congress/house-rejects-measure-to-force-vote-on-islamic-state-fight-idUSKBN0OR2LC20150611.

[445] Lead Inspector General for Overseas Contingency Operations, *Operation Inherent Resolve: Report to the United States Congress* (Washington, DC: DoD Inspector General, 2016), p. 3.

[446] Inspector General of the US DoD, *Assessment of U.S. and Coalition Plans and Efforts to Train, Advise, Assist, and Equip the Iraqi Counterterrorism Service and the Iraqi Special Operations Forces*, Report DODIG-2017-074 (Washington, DC: DoD IG, April 2017).

[447] Lead Inspector General for Overseas Contingency Operations, *Operation Inherent Resolve: Report to the United States Congress, July 1, 2017–September 30, 2017* (Washington, DC: DoD Inspector General, 2017), pp. 67–68.

[448] Inspector General of the US DoD, *Assessment of U.S. and Coalition Plans and Efforts to Train, Advise, Assist, and Equip the Iraqi Counterterrorism Service and the Iraqi Special Operations Forces*, Report DODIG-2017-074 (Washington, DC: DoD IG, April 2017), pp. 7–29.

some termed ISIS's goals as millenarian, based on its statements, others looked no further than its recruiting and propaganda videos, which featured mass beheadings, immolations, and other 'spectacles of sovereignty in digital time'.[449] In addition, the group's external actions, sponsoring or inspiring terrorist attacks in 29 countries killing more than 2,000 people, as well as regional groups switching their affiliation from al-Qaeda to ISIS, elevated perceptions globally of its power and influence.[450] This clear and present danger was also acutely felt by regional states as well as proxy militias. In theoretical terms, all aspects for successful proxy war were present: coherence and control existed in equal measure for proxy as well as sponsor, an insight that can be extended to include the coalition and partners as well.[451] In terms of vertical control, successive presidents and their appointed officials retained approval authorities, Congress had monitoring mechanisms, but authorities as well as risk were pushed to the lowest levels to prosecute the campaign, with sufficient but minimally intrusive oversight. In turn integration of capabilities at the tactical level—SOF, conventional forces, proxies, and coalition partners—occurred that acknowledged their strengths as well as limits. The foundation of this remarkable integration, with special operations at its locus, was trust.

Given its intangible nature, trust is difficult or impossible to quantify. Some inferences regarding it, however, can be drawn. The Obama administration was increasingly confident in employing SOF, in part due to the options its leaders provided for Afghanistan in 2008. In addition, the convergence of paramilitary and military SOF led to Osama bin Laden's death, the most consequential special operation in modern history, despite significant political and operational risks. Although there were limits to the executive and legislative branches' largess that prevented USSOCOM from obtaining all its leaders submitted for approval, they nevertheless trusted SOF to deliver when it came to operations against terrorist and irregular threats overseas. This dimension of civil-military relations remains open for future scholars to examine.

Two anecdotes provide insight into how this trust facilitated integration and enhanced capability. In the fight for Raqqa in Syria, ISIS's means most feared by SDF forces were their ability to inflict massive casualties suddenly through difficult to counter vehicle-borne improvised explosive devices, or VBIEDs. Termed 'Mad Max vehicles', VBIEDs used by ISIS utilized captured vehicles as a form of mobile heavy artillery protected by bolted or welded on, and later hidden, armour plating and guided by a driver.[452] Given SDF anxieties, 'SOF Joint Terminal Air Controllers

[449] Simone Molin Friis, '"Beyond Anything We Have Ever Seen": Beheading Videos and the Visibility of Violence in the War against ISIS', *International Affairs*, 91:4 (July 2015): 725–746; Roxanne Euben, 'Spectacles of Sovereignty in Digital Time: ISIS Executions, Visual Rhetoric and Sovereign Power', *Perspectives on Politics*, 15:4 (December 2017): 1007–1033.

[450] Tim Lister et al., 'ISIS Goes Global: 143 Attacks in 29 Countries Have Killed 2,043', CNN, 25 July 2016, www.cnn.com/2015/12/17/world/mapping-isis-attacks-around-the-world/index.html.

[451] Tyrone Groh, *Proxy War: The Least Bad Option* (Stanford, CA: Stanford University Press, 2019), p. 123.

[452] Reuters, 'Islamic State's "Mad Max" Weapons of War', CNBC.com, 19 July 2017, www.cnbc.com/2017/07/18/isis-weapons-of-war-suicide-cars-.html; Hugo Kaaman, *Car Bombs as Weapons of War: ISIS's*

(JTACs) needed to be able to call in fires in a matter of minutes to interdict VBIEDs to protect the light SDF forces and enable them to maneuver. If the strike cells failed to do so, the rapport between the SDF fighters and US SOF advisers suffered.'[453] The time element was particularly crucial as earlier in OIR, against dynamic targets in the same city of Raqqa, the approval process for some air strikes 'routinely took more than 30 minutes and sometimes even lasted hours.'[454]

In Iraq, trust was built due to historical training and advising as well as demonstrated commitment and capabilities. During the difficult fighting retaking Ramadi and Mosul from ISIS, SOF and conventional JTAC-guided US airstrikes turned the tide of the battle and prevented some Iraqi units from being wiped out.[455] In consequence,

> coalition forces that had formed strong relationships with particular ISF units (e.g., US SOF with the Iraqi CTS) to remain the trusted conduit for coordinating air support, albeit with the strike cell at CJOC-B operating as the ultimate coordination body sitting above these individual cells. Ramadi was also a further demonstration of the reassurance and confidence Iraqi forces derived from the promise of airpower. Looking back at the operation, CJFLCC Commander MG Gary Volesky noted, "the [Iraqi Security Forces] needed some confidence. They had no confidence. Ramadi started it".[456]

Trust, combined with an ongoing evolution of capabilities centred around special operations that were the product of a long period of convergence between paramilitary and military special operations and conventional and proxy forces, provided the foundation of successful horizontal integration. The net result was a realization of a stated policy vision and objective identified by Secretary of Defense Ash Carter in 2016: 'enabling capable, motivated, local forces to take the fight to ISIS, and also bringing down to bear the full weight of US forces around the theater of operations like the funnel of a giant tornado.'[457] Later, he used the example of a type of special operations mission to illustrate his point: 'By combining U.S. capabilities with local partners, they have been squeezing ISIS with simultaneous pressure from all sides and across domains through a series of deliberate actions to continue to build momentum', the secretary said. 'For example, when U.S. and coalition special operators conduct raids, free hostages, gather intelligence and capture ISIS leaders, it creates a virtuous cycle of better intelligence. This, in turn, generates more targets, more raids, more airstrikes and more opportunities that can be seized to generate

*Development of SVBIEDs, 2014–2019*, Policy Paper 2019-7 (Washington, DC: The Middle East Institute, 2019), p. 9.

453 Wasser et al., *The Air War against the Islamic State*, pp. 102–103.

454 Wasser et al., *The Air War against the Islamic State*, p. 55.

455 Renner, 'Air Power in the Battle of Mosul', p. 265.

456 Wasser et al., *The Air War against the Islamic State*, p. 166.

457 Shannon Collins, 'Carter: U.S. Will Deploy More Troops to Syria to Combat ISIL', DoD News, 10 December 2016, www.defense.gov/News/News-Stories/Article/Article/1,026,625/carter-us-will-deploy-more-troops-to-syria-to-combat-isil.

even more momentum.'[458] Only by integrating capabilities horizontally, with control sufficiently relaxed at the right level of command, could ISIS be constricted and reduced in this manner. Special operations were the key to that success.

## Conclusion

Chapter 4 suggested that SOF have two choices in terms of control: *accede*, or conform to the constraints placed on their use, or knowingly *exceed* them. While much information remains inaccessible, the analysis here suggests four overall conclusions that collectively point to SOF *acceding* to constraints. The conclusions pertain to capabilities and integration, convergence, mission focus, process and control mechanisms.

The story of capability and integration for the United States during the period from 1996 until 2016 is one of incremental evolution based on experimentation, punctuated by spikes in innovation and adaptation. In terms of capabilities, US special operations were divided along three lines at the beginning of the case studies. The initial time period was characterized by risk aversion. Covert special operations, the purview of the CIA, were generally focused on intelligence gathering for political reasons, which emphasized transparency and cooperation. Overt military special operations, ostensibly focused on advising and assistance missions or offensive and defensive special warfare devoted much of their time and effort to justifying their existence militarily and politically. Clandestine surgical strike forces were the 'Ferrari in the garage', optimized for manhunting or hostage rescue missions but employed rarely pre-2001 given the perceived risk of failure, reputation for exceeding their mandate, and military bureaucratic hostility. Yet SOF continued to exercise and train to traditional mission sets, such as surgical strike and special warfare, individually, jointly, or with alliance or individual partner nations in JCETs. For these reasons and others, special operations capabilities, functions, and effectiveness were in a state of *divergence* (Figure 8.6).

The 2001 terrorist attacks sparked a transformation in US special operations capabilities, driven by integration. The access and placement of covert paramilitary SOF with capable proxy militias, forged several years prior in the hunt for bin Laden, allowed overt military SOF to integrate or plug into them. Overt SOF, in turn, connected and integrated tactically across the services, gaining access to mobility, lift, and precision munitions delivered from the air. The CIA provided the entry and the infrastructure, while augmented ODAs, backed by special operations aviation with whom they had exercised intermittently, provided massed accurate firepower that allowed the Northern Alliance and other tribal militias to manoeuvre, close in, and finish Taliban forces in only a few weeks. Subsequent efforts to capture bin Laden and other leaders exposed the limitations of this approach in defensive terrain manned by experienced, motivated troops. However, this model of employment

[458] Collins, 'Carter: U.S. Will Deploy More Troops to Syria to Combat ISIL'.

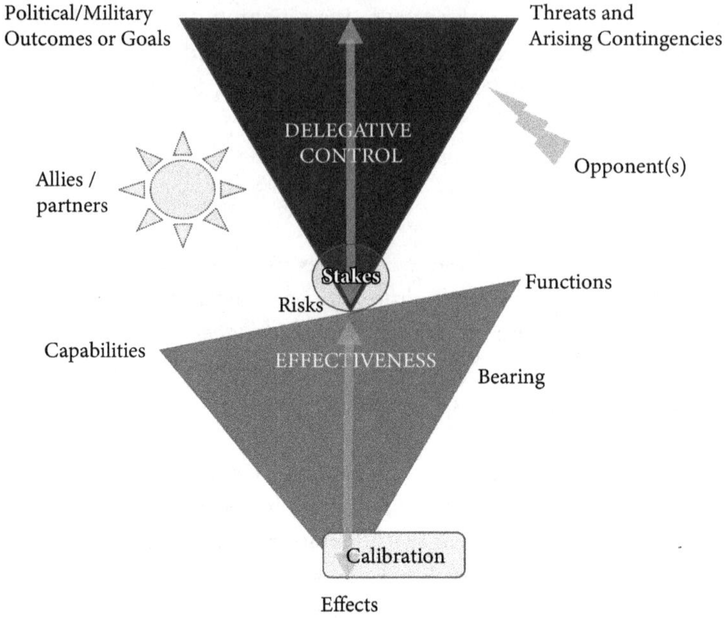


**Figure 8.6** US special operations divergence, 1996.
*Source:* Author's work.

was subsequently refined in training and exercises and expanded for use in western and northern Iraq in a more conventional offensive manoeuvre campaign as supporting efforts to present Saddam Hussein with an intractable problem: identifying which of three possible axes of approach was the main one. This integration of capabilities fusing indirect shock with military advice and assistance missions was used once again in OIR to considerable effect against ISIS through proxies in Iraq and Syria. The result was a very different form of offensive special warfare capability, whose basis was integration, a fact not lost on Chinese and Russian military observers.

Integration did not rest alone on the foundation of training, equipping, and exercise and experimentation. Instead, its bedrock was an intangible: trust. Chapter 3 identified SOF's habitual working relationships as the basis for trust, growing out of informal ties from training and exercises across units. The innovations that occurred in OEF, OIF, within TF 714 against AQI and elsewhere that contributed to high levels of integration was trust, through the building of natural or force habitual working relationships. McChrystal doubled down on them, breaking down traditional barriers that existed between operators and intelligence personnel internally but also externally with those who could provide value to the mission through their own resources and capabilities. These began, in a ripple-like fashion, with the CIA and overt SOF, expanded out to the conventional force, beyond it to other government departments and agencies through the SOLO network, and eventually to partner nation SOF. For McChrystal, the basis of trust was not merely a function of sharing and openness. For him, credibility was key in its establishment. Subsequently he codified a formula for developing it: 'Credibility = Proven Competence + Integrity + Relationships.'[459] Integration can be outlined in similar terms: Capability = Integration + Trust + Unique Skills and Resources. McChrystal used this approach to innovate in the face of a new and pressing threat, in the form of AQI. What resulted was more than just an innovation, but a revolution in surgical strikes operations, pace, intensity, and effectiveness.

Capabilities, no matter how highly integrated, can meet their Waterloo given their scope of activities, scale and nature of threat, and size and composition of operating area. US special operations capabilities for surgical strike and special warfare converged in the period from 2008 onwards in favour of the former, as the focus shifted to global operations against surging al-Qaeda affiliates. It functions, relative to its capabilities, declined (Figure 8.7) SOF capability and capacity, although expanded substantially, was nevertheless insufficient. Successive USSOCOM and service component commanders deny that they violated the second SOF Truth, that quality is better than quantity. Mid-career operators at the time, opined standards declined or changed too much to meet demand. Quality can be subjective but evidence from 2016 onwards, detailed in Chapter 4, suggests its gradual erosion.

The capabilities for surgical strike and special warfare that succeeded so well in Iraq and Afghanistan on the operational offensive failed in Afghanistan defensively from 2008 onwards for a variety of reasons, including a greatly expanded scale and complexity of interconnected problem sets, lack of credible governance outside of Kabul, and a relentless, self-fulfilling commitment to cyclical special operations. Special operations capabilities are part of a comprehensive solution to complex problems—they cannot be *the* solution. VSO/ALP suggests significant limits to the effectiveness of special operations for two other reasons: they cannot overcome national structural deficiencies, nor can they engage in local politics without elegant culture-linguistic training. Stated simply: there were never enough 'special warfare' personnel sufficiently versed in Pashto or other dialectics, or possessing enough expertise in local politics, to make a difference. Special warfare capabilities had become additional surgical strike ones to meet capacity demands, and optimized their training and preparation as such; they could not flip a switch and become instant cultural experts. A third, related concern is that geography and terrain still matter in special operations and its nature in Afghanistan absorbed SOF at rates far higher than Iraq.

Convergence, unwelcome to some, developed organically to fill gaps and seams in control mechanisms, as operations (and operators) abhor a void. The intersection of paramilitary and military SOF explores their legal and functional boundaries in terms of mission areas and authorities. In certain periods of resource scarcity and

[459] McChrystal, *My Share of the Task*, p. 142.

bureaucratic competition, avenues of convergence can be extremely useful in accomplishing given missions by using informal connections to surmount individual and institutional barriers based on relationships, trust, and credibility. Congruence is more likely to occur in the absence of specifically stated policy goals or outcomes, or when vague ones are issued, yet operators are given a mission to accomplish (depicted in Figure 8.7). They will seek to fill the gap where outcomes and terms of reference are unavailable or elliptical.

Convergence reflects two specific aspects of special operations. First, what is special is relative as Searle points out: military special operations exist outside of the normative military 'box' inhabited by conventional forces. In different terms, paramilitary SOF are more agile, less visible, and have greater flexibility than military SOF, thanks to their ability to operate covertly. The cost, however, is increased risks associated with covert action, due to loss of control of efforts and proxies, and elegant covert capability at the expense of capacity. Military SOF, both overt and clandestine, are less 'special' than their paramilitary counterparts but have considerably greater capacity and resources to draw upon. Yet they are 'special' in their own right as they are more responsive, less visible, and less accountable than conventional forces. The authors of the 9/11 Commission Report alluded to the need for convergence to take advantage of the unique attributes of the paramilitary and military special operations

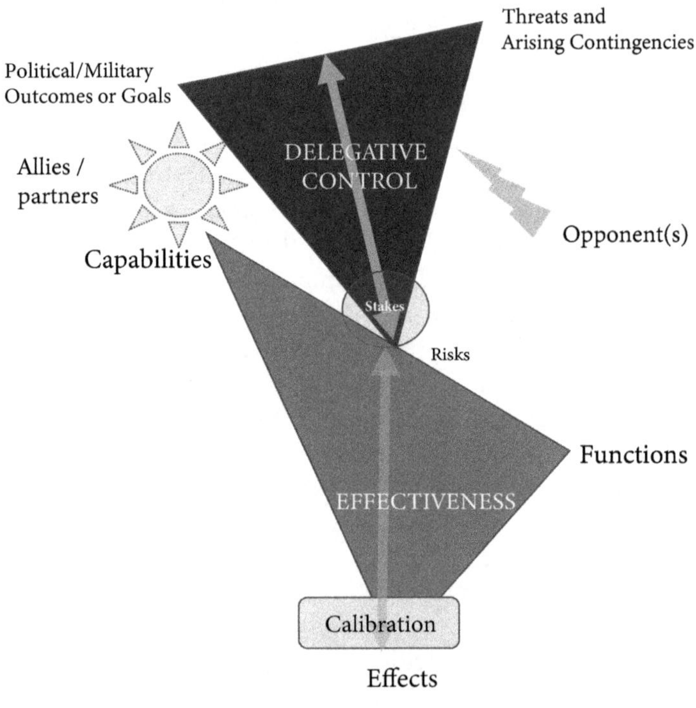


**Figure 8.7** US special operations congruence, 2016.

*Source:* Author's work.

and intelligence communities.[460] Doing so requires framing some aspects of convergence differently, in relation to Title 10 and 50, as 'mutually supporting, not mutually exclusive authorities'.[461] Convergence can be thought of as the outcome of integration of capabilities and control measures taken to their natural conclusion. Yet as the case of the bin Laden raid in 2011 suggests, such integration can be taken too far when control transcends its premise.

The sustained use of special operations relies on a combination of capabilities, resources, political will, clear guidance, and effective processes. OEF succeeded despite the absence of process because of SOF's inherent flexibility and adaptability. The limits of this approach were fully exposed at Tora Bora and in Operation ANACONDA. Process refinement and improvement, in the creation of new TTPs and organizations such as the JAGIC, addressed these limits, contributing in no small part to the outcome of OIF. The surgical strike process modified by TF 714 to address the adaptive threat posed by AQI placed the terrorist group on the defensive, buying space for the Surge to bolster security, stability, and confidence in the Iraqi government. The accession of TF 714 leaders to higher command in Afghanistan translated into an attempt to import and expand these processes there. More broadly, the task force-based F3EAD process and its trappings became the template for activity wherever clandestine SOF were employed. Much the same was true of the 'softly, softly' special warfare approach so successful in OEF-P and its attempted recreation in Afghanistan in the form of VSO/ALP.[462] The refinement of processes from OEF, OIF, through to OIR stretched special operations capabilities to the breaking point globally and in Afghanistan.

Processes for special operations can take on a life of their own as the author has argued elsewhere. This case study builds on this point. Sustained effort employing processes perfected elsewhere is not always productive, as the US experience in Afghanistan from 2008 onwards suggests. In the absence of modified policy guidance, however, such processes will continue not so much to avoid control but rather so long as their perceived success pays off in the form of continuing institutional influence and resources. 'Policy by CONOP' developed as the speed of process greatly exceeded that of policy. In addition, such a *de facto* policy, with its low cost and apparent success, assured its continuation until interest groups raised concerns. These concerns, about the policy's transparency, legality, morality, and control, led to a publicly announced policy change in 2013.

In Afghanistan, fewer accessible targets of significance led to a shift in the focus of action to lower ranks within the Taliban or served the political agendas of local Afghan leaders. Special operators need to reassess their models and processes periodically and evaluate their effectiveness lest they be used reflexively and out of habit.

[460] National Commission on Terrorist Attacks, *The 9/11 Commission Report*, pp. 402; 416.

[461] Trautman, *Special Operations Intelligence*, p. 75.

[462] The 'softly, softly' term is associated with the less coercive approach to irregular warfare ascribed to the British historically, in contrast to other approaches, including the American, that are 'heavy-handed' in contrast. See, for example, Sean Rayment and Philip Sherwell, '"Heavy-Handed" US to Adopt British Softly-Softly Line', *The Telegraph* (26 March 2006), www.telegraph.co.uk/news/worldnews/northamerica/usa/1514030/Heavy-handed-US-to-adopt-British-softly-softly-line.html.

Such assessments should occur during ongoing campaigns but is essential whenever the focus shifts elsewhere or expands in scope. If not, it is guaranteed to occur as control is exerted by outside authorities. SOF can take the action themselves and reevaluate their own processes before they go awry or wait for someone else to do it. The historical record suggests the latter most frequently results in diminished resources, drastically curtailed agency, and peremptory control measures.

Control measures and mechanisms are necessary but imperfect. As this case study suggests, their imperfections result from structural inefficiencies within the American machinery of government subject to manipulation for bureaucratic advantage. Such manipulation is a natural result of bureaucratic politics. SOF seek to gain as much flexibility and adaptability in their activities as possible. Among the 'principals' for whom SOF work, there is only as much control as there is relative interest in monitoring and desire to know the details. As was the case with SOF processes, with special operations appearing to solve national security problems in a cost-effective and responsive manner, there is little likelihood of senior policy makers wanting to inquire about the details. In addition, special operations conducted within the area of convergence could do so comfortably in the knowledge that they were fulfilling necessary monitoring requirements, in reporting their clandestine activities quarterly, within a blizzard of other reports to Congress.

The final point in this case study highlights the precariousness of congruence. US SOF achieved a high degree of it for reasons outlined above. What the theory of special operations coherence suggests, when its elements are used heuristically, is the tenuous nature of congruence. Figure 8.7 depicts the relationships between these elements by 2016. When viewed as intended, as a snapshot of a dynamic process, the combined forces of over optimized capabilities, decreased functions, orientation favouring clandestine activities, and ill-defined strategic objectives could not continue given changing circumstances and external influences. Serious scrutiny of the control of special operations occurred in 2014 with Congressional concerns about Executive overreach in extending the original AUMF to include ISIS, not about the use of special operations. Questions regarding control only occurred in 2017 following the Tonga Tonga ambush and apparent Congressional bewilderment over the presence of operators there. The line from this event and other incidents to the Comprehensive Review was short but its impact on SOF decisive. Congruence can turn into divergence quickly if preventative actions are not taken. The theory of special operations coherence provides the means of diagnosing when such actions are necessary.

# 9 Conclusion

This work developed a theory, termed 'special operations coherence', that connects several elements together. While I make no claims about its originality, it is novel. It utilizes concepts identified by others but synthesizes and connects them together to form a theory. Its purpose is not prescription, telling operators how to sharpen their tactical activities. Rather, this descriptive theory provides analytic tools in its explanatory yet dynamic model to fulfil two functions. One allows for deeper, more objective evaluations of historical special operations experience. The other, more important and difficult one is for operators to anticipate circumstances where apparent successes could lead to failure based on the information they have available. Converting the currency of special operations success, its tactical actions, into desired strategic and policy outcomes is their primary challenge identified and discussed in Chapter 4. The theory of special operations coherence not only explains how this happened but also provides insights as to when it might in ongoing campaigns.

A foundational premise of a general theory of special operations is its universality, or ability to explain events far in the past. In looking back at the pre-modern period, I avoid, in Mao's words, 'cutting the feet to fit the shoes' by imposing theory on ancient warfare. I am not a classics expert; those who are face daunting evidentiary challenges. Given the written sources and scholastic work to which I had access, I look to the past to identify precedents not in example but rather by type of activity. What emerges in Chapter 2 from these precedents across different civilizations is identifiable with contemporary special operations: intelligence gathering and reconnaissance, subversion and treachery, direct and indirect shock, military assistance and advising, and asymmetric response. At times these activities were conducted by specialists, while at others they were done in response or reaction, in an ad hoc manner, by those best qualified and suited to perform them. Looking at the past in this manner serves two purposes for special operations scholarship. The first assists contemporary and future scholars in drawing upon ancient history in a more responsible manner, providing them with different categories of activities and not favoured examples. The second avoids the cognitive traps associated with the terminology of existing special operations roles and missions, such as the Trojan Horse. The risk in employing new and different terms is rejection but these are preferable to the pedantic discussions, circular reasoning, and tautologies that accompany self-identification with a task. Stated simply, far too much energy and effort has been expended defending or attacking ideas associated with specific special operations

*Special Operations Success*. James D. Kiras, Oxford University Press. © James D. Kiras (2024).
DOI: 10.1093/9780198902096.003.0009

forces (SOF) communities and their identities, or these have been held prisoner by lexicon and experience of the dominant paradigm: the United States.

Yet this paradigm necessarily serves as the basis for Chapters 3 and 4, capability and control. Capability is a term used loosely and excessively within the military and in external analyses, from individual weapons at the tactical level to collective ones nationally or strategically. Theories or concepts which have explored special operations capabilities focus on SOF and a particular phenomenon associated with them: their misuse or disbandment. Roger Beaumont examines the selection–destruction cycle of military elites while George Monroe looks specifically at military advice and assistance missions, and their similar cyclical pattern, within the US Air Force. Both perpetuate a myth convenient for special operations: destruction or misuse stems from the inability or unwillingness of hidebound conventional minds to understand the unique qualities and value of SOF. Left unanswered are questions surrounding the sustained use of SOF that does not lead to destruction or disbandment.

The 'myth of SOF misuse', which served a useful purpose in forging a collective US SOF identity in the 1980s, robs them of agency. The myth paints SOF as hapless victim rather than active, willing participant. To escape it, as well as the prison of roles and missions and gross assessments based on numbers, Chapter 3 provides a different frame of reference for special operations capabilities. It looks across various SOF units and their functional specializations, links them to the historical precedent missions, and adds conditions such as distance and complexity. Assessments of capabilities are possible after one examines evidence of formal and informal institutional and social connections between different units. Such connections, including doctrine, training, exercises, equipment, personnel exchanges, and operations, and their closeness are integration horizontally tactically. To succeed in complex circumstances, SOF often require the support of other resources. These resources, from conventional militaries and other government departments, can also be evaluated according to such connections to determine horizontal integration at the operational or campaign level. That assessment, which can change over time, can be ascribed a value, from low to high. To test the soundness of this evaluative tool, Chapter 3 concluded with an assessment of Indian SOF and suggested that while comprising effective units and operators, its level of horizontal integration was low. Assessing SOF capabilities horizontally, useful in its own right, forms the first of two analytic approaches to special operations.

The second approach looks at how SOF and special operations are controlled, the subject of Chapter 4. The body of scholarship on civil–military relations, or the type and extent of civilian relationship to and oversight of the military, is the departure point here. Here again, the American experience dominates the discourse, this time scholastically. To broaden it out, and consider other forms of government, I identify different control arrangements in addition to the prevailing ones and come up with five possibilities: assertive, delegative, collaborative, competitive, and co-optive control.

While interesting, these different types of control arrangement between civil authorities and the military provide few insights on special operations particularly.

What sets special operations apart from other activities are their inherent risks. These risks vary across mission type. Other scholars have explored risk and special operations. The unique contribution here is the development of a risk model that explores risk at the tactical level and connects it to calculations at the various levels upwards, including the operational, strategic, and political. The insight here is risk calculations assume different forms at different levels. How they are interpreted and clearly communicated upwards and downwards can complicate efforts at control. Control itself can take different forms, involving orders, monitoring and oversight mechanisms, resources, and legal authorities. The soundness and clarity of these control measures vertically, or their integration, comprises a second analytic approach to special operations. In addition to mapping out in detail various control mechanisms for American special operations, Chapter 4 provides illustrations of control arrangements in other systems, including competitive and co-optive. This chapter ends with an examination of the 2017 Tongo Tongo ambush in Niger from the perspective of vertical integration, concluding that although shortfalls in monitoring and oversight existed, control measures worked as designed. While acknowledging gaps do exist, the examination suggests a high degree of vertical integration for the United States.

Horizontal integration of capabilities in Chapter 3 and vertical integration in terms of control in Chapter 4 each provide insights into successful special operations use. Neither, however, comprise a general theory. Linking them together, or superimposing one over the other, is the first step towards such a theory, which is presented in Chapter 5. They also solve the challenge of currency conversion posed by Gray, much less account for the full range of special operations activities. The inclusion of four factors reflecting special operations in use gets us closer. These factors are the use of capabilities over time, vice-specific missions, the political stakes surrounding the use of special operations, the effects generated by interacting against an opponent or enemy as well as partners and allies, and how special operations are employed operationally and functionally. The first three are inherent in competition and conflict and contribute to the difficulties and dynamic nature of strategy. The fourth, however, reflects specifically how special operations are used in a general sense. I develop two categories of use: bearing, or offensive and defensive, and orientation, as well as calibration, which includes the full spectrum of special operations employment, including overt, covert, and clandestine.

Two trinities are at the heart of the general theory of special operations coherence. The first synthesizes special operations functions, or the historical precedents, with the employment of horizontally integrated capabilities, including their bearing and calibration, to assess special operations effectiveness. The second, which determines control, incorporates desired political and military outcomes and goals, calculations of risks and level of vertical integration, and threats and contingencies, influenced by opponents, allies and partners, and the stakes involved. These two trinities, once connected together, with control extended into capability effectiveness and looked at over time, comprise the theory of special operations coherence.

All theories should have a central proposition: their core idea. The central proposition of the theory of special operations coherence is that successful special operations use lies in balancing the tension between the political and military appetite for their use against SOF's own ambitions against their need to be effective and adaptive in response to circumstances over time. This theory is consistent with Art Lykke's formulation of strategy and Harold Winton's functions of military theory. The theory of special operations coherence accounts for how the means of special operations are used in ways to accomplish specific ends, bearing in mind the risks involved, to fulfil the former. For the latter, this theory defines the central phenomenon, categorizes its component elements (capabilities, functions, control mechanisms, etc.), explains their relationship and interaction together, and connects various fields of study together (political science and history, as well as their subfields, including civil–military relations, etc.), but also anticipates future developments without being invalidated by them. The theory of special operations coherence is comprehensive in scope and can be used as such for analytic and predictive purposes. In addition, specific aspects of the theory can form the basis for an assessment of special operations use, depending on the quality of information available. To test its value fully, a specific national special operations experience exploring as many facets of the theory as possible is required.

For the period of the UK case study in Chapter 6, 1971 to 1991, much is known about specific episodes but explanations why, or how they link with other political developments, are unsatisfying. The theory of special operations coherence validates the assessments of some former operators, challenges others, and offers new insights. For example, the failures that occurred during the Falklands and First Gulf Wars were not merely the result of British parsimony in defence spending, weather, or bad luck. Rather, they reflected several systemic factors: informal social ties relaxing military control mechanisms, allowing the hubris of commanders and leaders to go unchecked; senior Special Air Service (SAS) leaders desiring additional influence, through political and military lobbying at the strategic and operational level, ahead of military risk considerations; insufficient attention to capability development or integration with conventional forces; and at the tactical level, insufficient horizontal integration between SAS and Special Boat Squadron (SBS) that led to friendly fire casualties. Weather conditions and bad luck played a role in the South Atlantic and Iraqi western desert; the creation of a coordinating headquarters, Directorate of Special Forces (DSF), in between the two should have fixed the obvious problems identified in the Falklands War. The theory helps answer the question why and is summarized graphically in Figures 6.1 and 6.2.

Three particular insights from the British experience stem from the theory. Few case studies highlight the benefits and downsides of control and integration based on informal social ties as well. Cohesion within the SAS and SBS is remarkable and the skill of its operators unquestionable. Rotation of officers in and out of the SAS, with expertise and corporate knowledge residing in its experienced non-commissioned officer (NCOs) and troopers, created unnecessary tensions within the unit when it was submitted to pressure. Senior SAS officers, in seeking to expand the influence and resources of their unit, became increasingly out of touch with the operational realities within it. External pressures related to control put additional strains on capability; the gutting of the UK's covert action capability in SIS led the SAS to assume most of its functions as well, along with private military companies. The case study also illustrates the frangibility of informal social ties. SAS leaders, particularly Peter de la Billière, fully exploited these ties once they were able to do so, including the colonel commandant, given loose control mechanisms. These allowed de la Billière to deflect Parliamentary oversight and scrutiny when there was evidence that loose control mechanisms were resulting in criminal enterprise within the SAS during successive rotations in Oman. British SOF had the confidence of Harold Wilson but also the trust of Margaret Thatcher. Both subsequently evaporated with the election of John Major as prime minister, as well as the retirement of a generation of politically ambitious SAS officers. From the perspective of theory, political ambition combined with increased mission demand (functions) without growth or tighter horizontal integration caused British special forces to steadily devolve from coherence to divergence.

Second, British special operations succeeded most often when the strengths of its cohesion based on informal ties were exploited in circumstances of congruence, when capabilities were aligned with functions. The SAS and SBS excelled when playing to their strengths: the skill of individuals or small teams. These ties provided the SAS in particular with access to communications and cutting-edge technology used to good effect in the Falklands—to overcome gaps in their own capabilities. During the First Gulf War, those ties were meaningless as American special mission units and conventional commanders focused on their own requirements. The successes of British special operations occurred when employed clandestinely and covertly in small teams: to Chile during the Falklands to install radars, in manning observation posts, or in close surveillance operations in Northern Ireland. Inadequate ties and connections, personnel exchanges, and the like ensured insurmountable problems compounded in attempts to scale up special operations capabilities for more conventional conflict. The exception to this was the SBS, which remained directly connected to and serviced the requirements of the Royal Marines and Royal Navy.

For small-scale asymmetric responses, indirect-shock, and intelligence-gathering and reconnaissance missions conducted covertly and clandestinely, British capabilities were high. The creation of other SOF to fill capability shortfalls in Northern Ireland contributed to these effective small-scale capabilities, at a cost. At the operational or campaign level, however, their capabilities were low relative to their functions, as evidenced by the failures that arose, steadily leading to divergence with political outcomes. Failure never occurred sufficiently at this level to have the political or strategic consequences necessary to compel change, as it did with the Americans in Iran in 1980. Secrecy policies, informal ties, and the opacity of the British political system largely ensured tactical failures rarely saw the light of day. These failures were shielded institutionally from public scrutiny, or when publicized, spun as a triumph of British pluck, perseverance, and grit, stymieing necessary control and oversight.

Third, the political consequences of SOF overreach close to home in Northern Ireland imposed necessary and sufficient control on special operations and their scale. SAS operators were placed initially under sector control and later, the Tasking and Co-ordination Group (TCG). The integration of various SOF capabilities, together with conventional forces, was ensured through the establishment of Int & Sy Group in 1980 ensuring congruence. Other evidence of tight control is apparent in the trials of operators for the death of John Boyle, the Gibraltar Four, and others. The results of the trials were unsatisfying to Provisional Irish Republican Army (PIRA) supporters and those critical of Thatcher, the British government, and SOF. Other trials of PIRA members caught, imprisoned, and subsequently acquitted demoralized SOF operators but drove home an additional point: their agency was limited and justice impartial. Future scholars with access to documents will resolve the question surrounding 'shoot to kill' in Northern Ireland but the theory of special operations coherence suggests restrictive controls were in place serving political ends, loosened occasionally to respond to PIRA attacks in a 'tit-for-tat' manner but also to cripple individual active service units (ASUs) and send a clear and unequivocal message to PIRA leaders.

The South Africa case study in Chapter 7 overlaps with the preceding British one in several ways. They occur in roughly the same decades and share a common military culture and heritage. Both feature strong informal social ties among individuals and between units. In addition, they saw an expansion of special operations capabilities into the covert and clandestine realm after overt beginnings. Yet they also share consider differences, in capabilities, control, and conditions, making South African special operations worthy of examination, including the system they were designed to serve. The National Security Management System (NSMS) and long tenure of P.W. Botha created a control system for all aspects of national security, including special operations, that was only dismantled in 1990. Control appears constant and assured given this system, allowing more detailed exploration of the range of South Africa's special operations activities.

Four themes emerge from this case study. The first relates to differences in primary terms of reference. South African Defence Forces (SADF) leaders distinguish between the actions being taken—special operations—and different actors conducting them. South African Recces alone conduct special operations as their primary mission. Other units, such as 32 Battalion or *Koevoet*, are SOF—those not specifically selected, trained, and equipped to conduct special operations but who may be called upon to do so. The South African definition of special operations is expansive and extends to conventional forces as well—in short, it looks at capabilities as well as units. Some may get caught up in semantics or etymological hair-splitting. The theory of special operations coherence, specifically the historically identified functions and expansion of capabilities, avoids this problem. Functions change the frame of reference from action to specified role and look to evaluate the degree of integration existing within it. At the beginning of the case study, a high degree of congruence was evident between capabilities and functions, political outcomes and risk, relative to the threats South Africa faced (Figure 7.1). Actions preceding and following the

Battle of Lomba River suggest a high degree of integration between conventional forces, the Recces, and 32 Battalion while on the defensive for asymmetric response and intelligence and reconnaissance missions. Operations PIGEON, FIREWOOD, and others, in contrast, point to significant integration shortcomings between Recces and other SOF when on the offensive against prepared positions.

Within the Recces, and between them and supporting elements, informal social ties along with their small size overall made tight integration tactically a necessity, the second theme of Chapter 7. This constant held true regardless of the special operations function performed, its calibration, or its scale. Recce units borrowed from each other for more complex operations based on the working relationship between commanders. Often, specific operators were requested based on their skills, reputation, and prior working history. Although not universally true, many operators interviewed or who have subsequently written about their experiences identified cultivating relationships with supporting or neighbouring units formally in training exercises or informally in operational areas. Resource limitations, combined with an understanding of mutual dependence to ensure survival, created trust as well as some contradictions with official state policy. Pragmatism and self-survival trumped state policy.

The third theme, in common with the British case study, is the opportunities and limits of informal social ties and control mechanisms. Despite NSMS, SADF and South African Police (SAP) forces had considerable agency in carrying out their activities. In borrowing command-and-control constructs from Rhodesian special operations, South African special operations appeared coordinated and integrated operationally. The reality was very different. In Namibia, sector and unit commanders, as well as higher headquarters, encouraged competition. The problem existed not so much with it as its incentives, payments, and awards, which ensured a disintegrated response to the asymmetric threat posed by People's Liberation Army of Namibia (PLAN) incursions and bases. Informal control mechanisms also led to some of the worst excesses of the apartheid era, committed by Civil Cooperation Bureau (CCB) and others. Political risks drove Botha to change the calibration of special operations increasingly from clandestine to covert. In the subsequent free-for-all, with little oversight, unscrupulous individuals engaged in uncoordinated criminal enterprise domestically, including embezzlement, racketeering, and murder. The assertive control of NSMS was an illusion; what emerged, due to the agency of organizations and individuals and the desire of principals to distance themselves from state actions, was a hybrid form of control with blurred lines between delegation, collaboration, and co-option, as Figure 7.2 suggests.

Political risks are the basis for the fourth theme. Despite their vigorous support for apartheid, seemingly at all costs in their political rhetoric, Botha and the State Security Council (SSC) were sensitive to crossing the line between pariah nation and rogue one. They may have had little choice, particularly as increased Cuban, Soviet, American, and other involvement increased the stakes and consequences of their actions. These considerations impacted special operations in several ways. Increasingly, Botha and the SSC tried to conceal South Africa's conduct of special operations, forcing them into the covert and clandestine realm, with consequences for control identified previously. In addition, countless external special operations were called off given potential political consequences. Others planned and executed, such as DROSDE and ARGON, had unintended political consequences that likely influenced Botha and the SSC's risk calculations for future operations. Complacency regarding the African National Congress (ANC) led the SSC, based in part on the attention paid to the endgame in Namibia, to miss signs that the threat was evolving into something new and dangerous. This threat, youth-led violence in the townships, eventually led to the use of special operations capabilities to counter them.

Much as in the British case in Northern Ireland, when stripped of their clandestine or covert protection and applying approaches used to effect elsewhere, South African special operations came under increasing attention, scrutiny, and disapproval. Also in common with the British experience, South Africa's special operations capabilities struggled in responding to conditions at opposite ends of the spectrum in a relatively short period of time. Optimized for small-scale reconnaissance, indirect-shock, and asymmetric response missions in episodic, short-duration externals where integration remained high, they were faced with two sets of entirely different conditions stretching these capabilities to their breaking point: high-intensity conventional combat in Angola, in a static battle of attrition at one end, to covert reconnaissance and overt response to asymmetric threats internally in so-called 'peacekeeping' missions. For domestic asymmetric response missions, including Namibia, as well as defensive high-intensity conflict where they were robbed of mobility, South Africa's special operations capabilities can be assessed as low and divergent, as Figure 7.2 depicts.

The final case study, the United States from 1996 until 2016 in Chapter 8, is just as complex, for different reasons. Its special operations training activities were global prior to 9/11 but its operations relatively limited. Relative to its functions, threats, and capabilities, American SOF were divergent given desired outcomes and goals: the risks of their use were rarely worth the rewards, as Figure 8.6 suggests. From 9/11 onwards, US special operations were increasingly employed overtly, covertly, and clandestinely in pursuit of al-Qaeda and its affiliates. A pattern of US special operations solidified by 2008, one divided by bearing and calibration. Operationally and strategically, covert and clandestine special operations engaged in a seemingly endless process of identifying, monitoring, and striking terrorist groups and cells. These operations, which suited political and strategic risk calculations no matter which political party was in charge, complemented US SOF's bias for action. An overabundance of targets, driven in part by the increasing regional distribution of active al-Qaeda affiliates and inclusion of others such as the Taliban, strained the capacity of special mission units. Increasingly, overt SOF were drawn into a surgical strike enterprise of global proportions. Capacity limits drove overt SOF on a separate path, special warfare, or military advising and assistance missions for Iraqi and later Afghan security forces. The Village Stability Operations (VSO)/Afghan Local Police (ALP) missions commencing in 2011 in Afghanistan were a decade too late.

Four themes emerge from this global, more contemporary case study. The first relates to orientation. US SOF succeeded best when on the offensive tactically and operationally, when they seized the initiative and dictated the tempo and pace of operations against al-Qaeda, Saddam's regime, and elsewhere. This theme is consistent with the findings of the theories of William McRaven and Robert Spulak, specifically the value of the initiative in creating relative superiority and reacting to and increasing friction in an opponent's system. The theory of special operations coherence, however, expands the aperture to the operational, strategic, and political across time and suggests the same is true in campaigns and achieving outcomes and goals. When special operations cede the initiative to an opponent and are forced to act defensively, they have difficulty doing so where host nation security forces are weak. The experience in Operation Enduring Freedom-Philippines (OEF-P) suggests that where stakes are limited and security forces reasonably competent, controls and constraints on SOF politically and militarily are beneficial. They serve two purposes: maintain SOF focus on political objectives and decelerate momentum resulting from their bias for action.

Integration, vertically and horizontally, is the scope of the second theme. Horizontal integration takes considerable energy, effort, and time to build but relatively less to maintain. Bureaucracy, counterintuitively, complements tactical and operational formal and informal initiatives at integration. United States Special Operations Command (USSOCOM) headquarters ensured resources, including personnel, were assigned or used to foster and strengthen these initiatives as well as sustain them through programmes such as SOLO. In some instances, SOF ambitions outran capabilities—the failure to create a functioning global SOF network is a case in point. Where goals were more modest, integration efforts have been successful, as in the task-focused regional information sharing of GALLANT PHOENIX or where a solid foundation already existed, as in NATO and the develop of an SOF headquarters there. At another level, US SOF experienced unprecedented, high levels of integration with conventional forces and other government agencies for two reasons. The first was a political focus on terrorism as the primary threat to national security, leading to the hyper-optimization of American special operations capabilities at the expense of other functions. The second, specifically for the conventional Department of Defense (DoD), was a surplus of capability and questions surrounding their utility. Conventional military and other department and agency leaders had a compelling reason to approve requests to support SOF and integrate with them, in the absence of pressing crises or imminent threats from the People's Republic of China (PRC), Russia, Iran, North Korea, or elsewhere: continuing value and relevance.

A third theme relates to a cornerstone of success in this period: process transformation. In line with theories of military innovation, special operations require a mixture of internal mavericks, external pressure, and culture shifts in order to do so. The experience of McChrystal and TF 714 is one such example. Its creation merits study and emulation. So too does its expansion and ostensible corruption, serving its own ends and in the absence of strategic guidance, becoming a strategy in its own right. Beneath the TF 714 example, a more mundane one is evident. The convergence of military and paramilitary special operations did not occur because of 9/11 or the conduct of global special operations. It was the result of a steady evolutionary process, where each recognized their own value and shortcomings: paramilitary in their covert authorities, different risk calculations, and authorities but limits in capacity and equipment and military the opposite. Personality-depended competition organizationally for political influence and attention rarely, if ever, affected effective coordination between paramilitary and military operators operationally or tactically in Iraq, Afghanistan, and elsewhere. A historical stumbling block related to integration, the transition or handover of responsibilities, appeared seamless in most cases, if existing accounts are accurate. In looking at a broader swathe of special operations activities over time, through the different lenses of the theory of special operations coherence, other patterns of process transformation emerge.

The fourth and last relates to control. Chapter 4 suggested there are two options open to special operators given the mechanisms put in place to control them. They can either *accede* to them, submit to monitoring and punishment, or *exceed*, and find ways to work around them. The findings of Chapter 8 suggest SOF evaded, explored, or modified the boundaries of control but rarely exceeded them. Concerns about elites running amok, while based on historical precedents, appear overinflated in the US case based on the range of existing control mechanisms and legal safeguards. A number of them were identified in Chapters 4 and 8, including standing congressional committees and investigative and auditing functions. Overreach or abuses can occur but seldom for long and without outside scrutiny or public debate. Much like special operations, control mechanisms, or at least their sufficiency, are relative and a product of their political environment. In a 2023 study comparing UK, Australian, and US SOF, a team of British human rights researchers reached the following conclusion about American control measures:

> even in the most sudden and tragic failures, the congressional oversight mechanism in the United States enables clear identification of problems and rapid policy changes to be made regarding Special Forces during times of trouble that may arise unexpectedly.
>
> As an additional external oversight mechanism, the Government Accountability Office, an independent and non-partisan agency, can evaluate US government programmes and activities. With the authority to conduct audits and investigations into USSOF activities, the Government Accountability Office provides an avenue for making recommendations for improvements which may inform the Congressional Committees in strengthening their oversight.[1]

This case study has identified other control mechanism in place to maintain different forms of control over special operations, including legal, financial, and ethical ones.

[1] Hamza Bozkurt et al., *Accountability and Oversight of UK Special Forces* (Colchester, UK: Human Rights Centre Clinic, June 2023), p. 27.

The theory of special operations suggests control is not static but dynamic, expanding and contracting for various reasons.

A more interesting finding related to all three case studies, for which the theory of special operations coherence can provide insights, relates to their overextension. In all three cases capability problems arose due to insufficient control or focus between the political, strategic, and operational levels. Writing on the efficacy of covert action over several decades, B. Hugh Tovar concluded in 1981:

> Policy, leadership, and continuity—collectively, these are the *sine qua non* of effective covert action … If the people and policies change frequently or precipitately, the impetus is lost and confusion reigns. Skillful operators cannot prevail against it.
>
> Operational failure may also coincide with and quite possibly derive from a deteriorating domestic political environment, national uncertainty, a leadership vacuum, and bureaucratic confusion … If covert action is to have any reasonable chance of succeeding, it must bear a coherent relationship to the main thrust of US foreign policy. Once it transcends the latter's premises or begins to probe the limits of commitment its outlook is dubious.
>
> The most impressive achievements in American covert action have developed in situations of strength … Conversely, the worst episodes in covert history have coincided with periods of national uncertainty, days of transition and confusion.[2]

Tovar emphasizes clarity and consistency of political leadership and objectives, implying sufficient control. The theory of special operations coherence explicitly recognizes control but offers a different warning from all three case studies. Skilful operators will fill voids where they exist, a reflection of their entrepreneurial qualities and attributes. When employed across too many functions over extended periods of times, the primary risks are not rebellion against civil authorities but outcomes much lower down the chain of command: leaders who become more political in nature, causing them to lose sight of what matters for the future health, welfare, and relevance of the force; to meet demand, those leaders justify compromising selection criteria, training standards, and equipment, putting quantity over quality; persistent activities across a spectrum, straining operators to the breaking point and risking the collapse of capabilities; and, in the process, removing the boundaries between overt, covert, and clandestine activities and blurring or obscuring legal and ethical lines in place. By the end of the case study in Chapter 8, American SOF paradoxically (in contrast to the other two cases) were congruent with outcomes, threats, and contingencies. Yet this conclusion is no cause for celebration. As Figure 8.7 points out, congruence was only possible as SOF capabilities rose relative to ill-defined policy goals and rested on a knife's edge, balanced in part with the influence of allies and partners. The events in Tongo Tongo in 2017, along with publicized moral and ethical lapses leading to the

[2] B. Hugh Tovar, 'Strengths and Weaknesses in Past US Covert Action', in Roy Godson, ed., *Intelligence Requirements for the 1980s: Covert Action* (Washington, DC: National Strategy Information Center Inc., 1981), p. 193.

self-grading Comprehensive Review that dissatisfied Congressional leaders, suggest SOF have returned to divergence. Evidence of this are the recent mandated cuts of 5,000 personnel from USSOCOM.

Heeding these warnings requires special operators to educate those employing them as well as themselves. Self-education in the absence of formal education is one method. Formal education through professional military means or civilian universities is another. The theory of special operations coherence provides one tool in aiding the education of those interested in, supporting, conducting, or controlling special operations. Too many myths persist about special operations; too little serious, systematic inquiry is undertaken conceptually. When such inquiries are undertaken, consequential scholarship replaces pedantic internal debates, and special operators are able to articulate their limitations as clearly as their capabilities and courses of action, more effective use of special operations will occur and the study of them taken more seriously. The theory of special operations coherence provides the means for operators to challenge their assumptions, evaluate activities and campaigns along a much broader spectrum, spur alternative explanations, and frame current and future challenges. Doing so requires investments in several ways, including time for sufficient education and critical reflection to occur, by individuals and institutions. Those who better understand themselves, including how a bias for action and the desire for influence affect operations and leadership, become more informed products that can better inform their consumers. If the theory of special operations coherence performs this function, it will have realized its purpose.

# Bibliography

## Official histories


Briscoe, Charles et al. *Weapon of Choice: U.S. Special Operations Forces in Afghanistan.* Fort Leavenworth, KS: Combat Studies Institute Press, 2010.

Briscoe, Charles et al. *All Roads Lead to Baghdad: Army Special Operations Forces in Iraq.* Fort Bragg, NC: USASOC History Office, 2006.

Chief of the General Staff. *Operation Banner: An Analysis of Military Operations in Northern Ireland.* London: Chief of the General Staff, July 2006.

Clarke, Jeffrey. *United States Army in Vietnam: Advice and Support: The Final Years, 1965–1973.* Washington, DC: Center of Military History United States Army, 1988.

Folse, Mark. *The U.S. Army in Afghanistan: Operation Enduring Freedom, September 2001–March 2002.* Pub 70-83-1. Washington, DC: Center for Military History, 2022.

*Foreign Relations of the United States, 1977–1980. Volume XII: Afghanistan.* Washington, DC: Government Printing Office, 2018.

*Foreign Relations of the United States, 1977–1980. Volume XVI: Southern Africa.* Washington, DC: Government Printing Office, 2016.

*Foreign Relations of the United States, 1981–1988. Volume XIII: Conflict in the South Atlantic, 1981 -1984.* Washington, DC: Government Printing Office, 2015.

*Foreign Relations of the United States, 1961–1963. Volume V: Soviet Union.* Washington, DC: Department of State, 1998.

*Foreign Relations of the United States, 1961–1963. Volume III, Part III: Vietnam.* Washington, DC: Government Printing Office, 1991.

Freedman, Lawrence. *The Official History of the Falklands Campaign.* 2 vols. London: Routledge, 2005.

Kelly, Francis. *U.S. Army Special Forces, 1961–1971.* Washington, DC: Department of the Army, 1985.

Kretchik, Walter, Robert Baumann, and John Fishel. *Invasion, Intervention, 'Intervasion': A Concise History of the U.S. Army in Operation Uphold Democracy.* Fort Leavenworth, KS: US Army Command and General Staff College Press, 1998.

Neumann, Brian, Lisa Mundey, and Jon Mikolashek. *The U.S. Army in Afghanistan: Operation Enduring Freedom, March 2002–April 2005.* Pub 70-122-1. Washington, DC: Center of Military History, 2013.

Postgate, Malcolm. *Operation Firedog: Air Support in the Malayan Emergency, 1948–1960.* London: Her Majesty's Stationery Office, 1992.

Reynolds, Nicholas. *U.S. Marines in Iraq, 2003: Basrah, Baghdad, and Beyond,* Washington, DC: History Division, US Marine Corps, 2007.

Richard, Brooks. *Secret Flotillas.* 2 vols. London: Routledge, 2004.

Romjue, John. *From Active Defense to Air Land Battle: The Development of Army Doctrine, 1973–1982.* Fort Monroe, VA: Historical Office, TRADOC, June 1984.

Sams, Kenneth and Burt Aton. *'USAF Support of Special Forces in SEA'.* Project CHECO Report DOTECT-69-31. San Francisco, CA: HQ PACAF, 10 March 1969.

Schlosser, Nicholas. *'The Surge, 2007–2008'.* CMH Pub 78-1. Washington, DC: Center of Military History, 2017.

Sihotang, K. Joy et al. *Pengabdian Korps Baret Merah Abad XX.* Jakarta: Kopassus, 2000.

Stewart, Richard. *The United States Army in Somalia, 1992–1994.* CMH Pub 70-81-1. Washington, DC: US Army Center of Military History, 2002.

Stone, James, ed. *Crisis Fleeting: Original Reports on Military Medicine in India and Burma in the Second World War.* Washington, DC: Office of the Surgeon General, Department of the Army, 1969.

## Government documents and briefings

*Information used in this manuscript from source documents that are marked U//FOUO or U//FOUO/REL to USA, FVEY is not of a critical or sensitive nature, and was not extracted from any portions of the source documents with restricted classification markings.*

5th Special Forces Group Airborne. *'Task Force Dagger, Operation Enduring Freedom: Key Lessons Learned'.* Briefing slides. Undated.

Assessing Revolutionary and Insurgent Strategies (ARIS) Project. *'Little Green Men': A Primer on Modern Russian Unconventional Warfare, Ukraine 2013–2014*, Fort Bragg, NC: US Army Special Operations Command, 2014.

Barondes, Arthur. *The Accomplishments of Airpower in the Malayan Emergency (1948–1960).* Maxwell Air Force Base, FL: Aerospace Studies Institute, 1963.

Bartholomees Jr, J. Boone, ed. *U.S. Army War College Guide to National Security Issues, Volume I: Theory of War and Strategy.* Carlisle Barracks, PA: Strategic Studies Institute, 2012.

Center for Army Lessons Learned. *Joint Air Ground Integration Center: Lessons and Best Practices.* CALL Handbook No. 17-04. U//FOUO. Fort Leavenworth, KS: CALL, January 2017.

Chairman of the Joint Chiefs of Staff. *Universal Joint Task List (UJTL). CJCSM 3500.04c*, Washington, DC: The Joint Staff, 1 July 2002.

Chilcot, John. *The Report of the Iraq Inquiry.* 12 vols. London: Her Majesty's Stationery Office, 2016.

Defense Intelligence Agency. *North Korea Military Power: A Growing Regional and Global Threat.* Washington, DC: Government Printing Office, 2021.

Department of the Navy. *Lessons of the Falklands: Summary Report.* Washington, DC: Department of the Navy, February 1983.

Development, Concepts, and Doctrine Centre. *Special Forces Operations, JDP 3-05* (Restricted). Swindon, Wiltshire: DCDC.

Grdovic, Mark. *A Leader's Handbook to Unconventional Warfare.* SWCS Pub 09-1. Fort Bragg, NC: John F. Kennedy Special Warfare Center and School, November 2009.

Headquarters, Department of the Army. *FM 3-24/MCWP 3-33.5, Insurgencies and Countering Insurgencies.* Washington, DC: HQ, Department of the Army, 2014.

Headquarters, Department of the Army. *Special Forces Detachment Mission Planning Guide*, GTA 31-01-003. Washington, DC: HQ, Department of the Army, January 2020.

Inspector-General of the Australian Defence Force (ADF). Report of Inquiry Under Division 4A of Part 4 of the Inspector-General of the Australian Defence Force Regulation 2016 into Questions of Unlawful Conduct Concerning the Special Operations Task Group in Afghanistan, Part 1—The Inquiry, Part 3—Operational, organisation and cultural issues, OFFICIAL (Redacted). Canberra: ADF HQ, 2020.

Inspector General of the US DoD. Assessment of U.S. and Coalition Plans and Efforts to Train, Advise, Assist, and Equip the Iraqi Counterterrorism Service and the Iraqi Special Operations Forces, Report DODIG-2017-074, Washington, DC: DoD IG, April 2017.

Fraser, C.A. *'Revolutionary Warfare: Basic Principles of Counterinsurgency (Being a Review of "Lessons Learnt from Past Revolutionary Wars")'*, n.d., RESTRICTED.

Joint Chiefs of Staff. *Joint Publication 3-09: Joint Fire Support.* Washington, DC: The Joint Staff, April 2019.

Joint Chiefs of Staff. *Charter of the Joint Requirements Oversight Council (JROC) and Implementation of the Joint Capabilities Integration and Development System (JCIDS).* CJCSI 5123.01H, Washington, DC: The Joint Staff, 31 August 2018.

Joint Chiefs of Staff. *Joint Publication 3-0: Operations.* Washington, DC: The Joint Staff, 2018.

Joint Chiefs of Staff. Joint Publication 3-05.1. *Unconventional Warfare.* FOUO. Washington, DC: The Joint Staff, 2015.

Joint Chiefs of Staff. *Joint Publication 3-05: Special Operations.* Washington, DC: The Joint Staff, 2014.

Joint Chiefs of Staff. *Joint Publication 3-26: Counterterrorism.* Washington, DC: The Joint Staff, October 2014.

Joint Special Operations University. *Special Operations Forces Reference Manual,* 4th ed. MacDill Air Force Base, FL: JSOU Press, June 2015.

Joint Staff J-7. *Decade of War, Volume 1: Enduring Lessons from the Past Decade of Operations.* Suffolk, VA: J-7 Joint and Coalition Operational Analysis, 2012.

Lead Inspector General for Overseas Contingency Operations. *Operation Inherent Resolve: Report to the United States Congress, July 1, 2017–September 30, 2017.* Washington, DC: DoD Inspector General, 2017.

Lead Inspector General for Overseas Contingency Operations. *Operation Inherent Resolve: Report to the United States Congress.* Washington, DC: DoD Inspector General, 2016.

Liang, Qiao, and Wang Xiangsui. *Unrestricted Warfare.* Foreign Broadcast Information Service Translation. Beijing: People's Liberation Army Literature and Arts Publishing House, 1999.

Markham, Calvin. 'ENDURING FREEDOM: An Air Force Combat Controller's Experience'. Unclassified briefing, 2002.

Ming, Guo. *Lectures on the Science of Special Operations.* Academy of Military Sciences Master's Postgraduate Series Teaching Materials, 2nd ed. U//FOUO translation, 2016. SAIC Selected Book Translations CVI, Beijing: Military Science Press, 2013.

Ministry of National Defence. *Report of the Somalia Commission of Inquiry, Executive Summary.* Ottawa, ON: Ministry of National Defence, 1997.

National Commission on Terrorist Attacks, *'The 9/11 Commission Report',* New York: W.W. Norton, 2004.

OSS Assessment Staff, *Assessment of Men: Selection of Personnel for the Office of Strategic Services* (New York: Rinehart & Co, 1948; reprint, New York: Johnson Reprint Corp., 1978), pp. 264–281.

Special Operations Review Group. *Rescue Mission Report.* Washington, DC: Joint Chiefs of Staff, 1980.

Steury, Donald, ed. *Intentions and Capabilities: Estimates on Soviet Strategic Forces, 1950–1983,* Washington, DC: Center for the Study of Intelligence, 1996.

Thomas, Jim and Chris Dougherty. *Beyond the Ramparts: The Future of U.S. Special Operations Forces.* Washington, DC: Center for Strategic and Budgetary Assessment, 2013.

Tianliang, Xiao, ed. *The Science of Military Strategy*(2020). China Aerospace Studies Institute, Trans. Maxwell Air Force Base, AL: CASI, January 2022.

Tianliang, Xiao, ed. 战略学 (*The Science of Strategy*). Translated by SAIC. U//FOUO. Beijing, China: National Defense University Press, 2015.

Tower, John. *The Tower Commission Report: The Full Text of the President's Special Review Board.* New York: Bantam/Times Books, 1987.

Truth and Reconciliation Commission. *The Truth and Reconciliation Commission of South Africa Report.* 5 vols. Pretoria: Truth and Reconciliation Commission, 1998.

United States Africa Command. 'Oct 2017 Niger Ambush, Summary of Investigation'. Stuttgart: Africa Command, 10 May 2018.

United States Army Special Operations Command. *ARSOF 2022.* 2 parts. Fort Bragg, NC: US Army Special Operations Command, 2013.

United States Cyber Command. 'USCYBERCOM 120-Day Assessment of Operation Glowing Symphony, April 12, 2017'. Declassified and redacted.

United States Cyber Command. *'Operation Glowing Symphony Assessment Framework Briefing Version 1.0*, November 10, 2016'.

United States Special Operations Command. 'Comprehensive Review', unclassified, 23 January 2020.

United States Special Operations Command. *SOCOM 2020: Forging the Tip of the Spear.* MacDill Air Force Base, FL: USSOCOM, 2012.

United States Special Operations Command, United States Special Operations Command History and Research Office. *United States Special Operations Command History*, 6th ed. MacDill Air Force Base, FL: USSOCOM, 2008.

United States Special Operations Command. *United States Special Operations Command History, 20th Anniversary, 1987–2007.* MacDill Air Force Base, FL: USSOCOM, 2007.

United States Special Operations Command. *United States Special Operations Forces Posture Statement, 2003–2004: Transforming the Force at the Forefront of the War on Terrorism.* Washington, DC: USSOCOM, 2004.

United States Special Operations Command. *United States Special Operations Command History, 15th Anniversary*. MacDill Air Force Base, FL: USSOCOM, 2002.

United States Special Operations Command, *United States Special Operations Forces Posture Statement 1998.* MacDill Air Force Base, FL: USSOCOM, 1998.

United States Special Operations Command. *United States Special Operations Forces, 1996 Posture Statement.* MacDill Air Force Base, FL: USSOCOM, 1996.

## Government reports

Blachard, Christopher and Amy Belasco. *'Train and Equip Program for Syria: Authorities, Funding, and Issues for Congress'*. Report R43727. Washington, DC: Congressional Research Service, June 2015.

Congressional Research Service. *'Niger: Frequently Asked Questions about the October 2017 Attack on U.S. Soldiers'*. Report R44995. Washington, DC: Congressional Research Service, 27 October 2017.

Dale, Catherine, Nina Serafino, and Pat Towell. *'Organizing the U.S. Government for National Security: Overview of the Interagency Reform Debates'*. Report RL34455. Washington, DC: Congressional Research Service, 2008.

DeVine, Michael. *'Covert Action and Clandestine Activities of the Intelligence Community: Selected Definitions'*. Report R45195. Washington, DC: Congressional Research Service, 29 November 2022.

Elle, Paul, Richard Joyce, Robert Williams, and William Woodworth. *'US Army Special Forces and Similar Internal Defense Advisory Operations in Mainland Southeast Asia, 1962–1967 (U)'*. Technical Paper RAC-Tp-354. McLean, VA: Research Analysis Corporation, June 1969.

Fasick, J. Kenneth. *Profiles of Military Assistance Advisory Groups in 15 Countries.* ID-78-51, Washington, DC: General Accounting Office, 1 September 1978.

Feickert, Andrew. *U.S. Special Operations Forces* (SOF)*: Background and Issues for Congress.* Congressional Research Service RS21048. 28 June 2007, updated 2023.

Flynn, Michael, Matt Pottinger, and Paul Batchelor. *Fixing Intel: A Blueprint for Making Intelligence Relevant in Afghanistan.* Washington, DC: Center for a New American Security, January 2010.

Lamb, Christopher and Evan Munsing. *'Secret Weapon: High-Value Target Teams as an Organizational Innovation'*. Institute for National Strategic Studies Strategic Perspectives 4. Washington, DC: National Defense University Press, March 2011.

Niksch, Larry. '*Philippine–U.S. Security Relations*'. Report RS20697. Washington, DC: Congressional Research Service, 2000.

North Atlantic Treaty Organization (NATO) Special Operations Coordination Centre (NSCC). *The North Atlantic Treaty Organization Special Operations Forces Study*. Mons, Belgium: NSCC, December 2008.

Plater-Zyberk, Henry. *Russia's Special Forces*. Watchfield, Swindon: Conflict Studies Research Centre, 2005.

Stewart, Nora Kinzer. '*South Atlantic Conflict of 1982: A Case Study in Military Cohesion*'. Research Report 1469. Alexandria, VA: US Army Research Institute for the Behavioral and Social Sciences, 1988.

US Government Accountability Office. *Special Operations Forces: Better Data Necessary to Improve Oversight and Address Command and Control Challenges*, GAO–23–105163. Washington, DC: Government Accountability Office, October 2022.

US Government Accountability Office. *Special Operations Forces: Additional Actions Are Needed to Effectively Expand Management Oversight*. GAO-19-386. Washington, DC: Government Accountability Office, May 2019.

US Government Accountability Office. *Overseas Contingency Operations Funding: Background and Status*. Report R44519. Washington, DC: Government Accountability Office, 2019.

US Government Accountability Office. *Managing Sensitive Information: DOD Can More Effectively Reduce the Risk of Classification Errors*. GAO-06-706. Washington, DC: Government Accountability Office, June 2006.

US Government Accountability Office. *Rebuilding Iraq: Resource, Security, Governance, Essential Services, and Oversight Issues*. GAO-04-902R. Washington, DC: Government Accountability Office, 2004.

US Government Accountability Office. *Special Operations Command: Progress in Implementing Legislative Activities*. GAO/NSAID–90–166. Washington, DC: Government Accountability Office, September 1990.

US House of Representatives, House Armed Services Committee, The Subcommittee on Intelligence, Information Sharing, and Terrorism Risk Assessment, Committee on Homeland Security. *The Over-Classification and Pseudo-Classification, Part I, II, and III*, 110–120. Washington, DC: Government Printing Office, 2009.

US Senate, The Select Committee on Intelligence. *Legislative Oversight of Intelligence Activities: The U.S. Experience*, 103–188. Washington, DC: Government Printing Office, 1994.

Waghelstein, John. 'El Salvador: Observations and Experiences in Counterinsurgency', unpublished program paper, Unclassified/Limited Distribution, US Army War College, 1 January 1985.

Zabrodskyi, Mykhaylo, Jack Watling, Oleksandr V. Danylyuk, and Nick Reynolds. *Preliminary Lessons in Conventional Warfighting from Russia's Invasion of Ukraine: February–July 2022*. London: Royal United Services Institute for Defence and Security Studies, 2022.

## Official testimony

Brown, Bryan. 'Statement of Lietenant [sic] General Bryan D. Brown, U.S. Army, Deputy Commander, United States Special Operations Command, Before the Senate Committee on Armed Services, Subcommittee on Emerging Threats and Capabilities on the State of Special Operations Forces', Senate Committee on Armed Services, Subcommittee on Emerging Threats and Capabilities on the State of Special Operations Forces, 9 April 2003.

Clarke, Richard. 'Statement of General Richard Clarke, USA, Commander, United Special Operations Command before the 117th Congress Senate Armed Services Committee'. 5 April 2022.

McRaven, William. 'Posture Statement of Admiral William McRaven, USN Commander, United Special Operations Command before the 113th Congress House Armed Services Committee'.

Szymanski, Timothy. 'Statement of Vice Admiral Timothy Szymanski, Deputy Commander, United Special Operations Command before the 117th Congress House Armed Services Committee Subcommittee on Intelligence and Special Operations'.

Thomas, Raymond. 'Statement of General Raymond A. Thomas, III, U.S. Army, Command, United States Special Operations Command', *Three Decades Later: A Review and Assessment of our Special Operations Forces 30 Years after the Creation of U.S. Special Operations Command*, HASC No. 114-40. Washington, DC: Government Printing Office, 2017.

Votel, Joseph. 'Advance Policy Questions for Lieutenant General Joseph L. Votel, USA Nominee for Commander, United States Special Operations Command', Headquarters, USSOCOM (10 July 2014).

## Books (primary)

Akehurst, John. *We Won a War: The Campaign in Oman, 1965–1975*. Wilton, UK: Michael Russell Publishing Ltd, 1982.

Allard, Kenneth. *Somalia Operations: Lessons Learned*. Washington, DC: National Defense University Press, 1995.

Almonares, Oliver and Phil Fortuno. *The Rescue: God's Amazing Grace*. Manila: MindSpark, 2019.

Aristotle. *Aristotle in 23 Volumes*. H. Rackham, trans. Cambridge, MA: Harvard University Press, 1944.

Armstrong, John, ed. *Soviet Partisans in World War II*. Madison, WI: University of Wisconsin Press, 1964.

Aussaresses, Paul. *The Battle of the Casbah: Terrorism and Counter-Terrorism in Algeria, 1955–1957*. New York: Enigma Books, 2010.

Bagnold, Ralph. *Libyan Sands: Travel in a Dead World*. London: Eland, 2010, repr. 1935.

Beckwith, Charlie. *Delta Force*. London: Arms and Armour Press, 1984.

Blaber, Peter. *The Mission, the Men, and Me*. New York: Berkeley, 2008.

Botten, Stanley. *We Dare: History of the 2nd Reconnaissance Regiment, South African Special Forces*. Scotts Valley, CA: CreateSpace, 2017.

Boykin, William. *Never Surrender: A Soldier's Journey to the Crossroads of Faith and Freedom*. New York: Faith Words, 2008.

Breytenbach, Jan. *Forged in Battle*. Pretoria, South Africa: Protea, 2014.

Breytenbach, Jan. *Buffalo Soldiers: The Story of South Africa's 32 Battalion, 1975–1993*. Alberton, South Africa: Galago Publishing, Pty, 2002.

Breytenbach, Jan. *They Live by the Sword: 32 'Buffalo' Battalion—South Africa's Foreign Legion*. Alberton, South Africa: Lemur, 1990.

Brown, Robert. *Special Forces Operator: Serving with the SAS and MRF*. Barnsley: Pen & Sword, 2021.

Burnham, Gracia and Dean Merrill. *In the Presence of My Enemies*. Carol Stream, IL: Tyndale House, 2003.

Bush, George W. *Decision Points*. New York: Crown, 2010.

Caesar. 1804. *Caesar Delphini*. Philadelphia, PA: Classics Press.

Cheshire, Leonard. *The Face of Victory*. London: Hutchinson, 1961.

Christensen, Gitte Hjostrup, ed. *Interdisciplinary Perspectives on Special Operations Forces, Conference Proceedings* 4(a). Copenhagen: Royal Danish Defence College, 2017.

Clarke, A.F.N. *Contact*. London: Secker & Warburg, 1983.

Cleveland, Charles and Daniel Egel. *The American Way of Irregular War: An Analytical Memoir*. RAND Publication PEA301-1. Santa Monica, CA: RAND, 2020.
Cocks, Chris. *Fireforce: One Man's War in the Rhodesian Light Infantry*. 2nd ed. Roodepoort, South Africa: Covos Books, 1997.
Cooper, Johnny. *One of the Originals: The Story of a Founder Member of the SAS*. London: Pan Books, 1991.
Crocker, Chester, *High Noon in Southern Africa: Making Peace in a Rough Neighborhood*. Johannesburg, South Africa: Jonathan Ball Publishers, 1993.
Daly, Ron Reid and Peter Stiff. *Selous Scouts: Top Secret War*. London: Galago, 1982.
Davies, Barry. *Shadow of the Dove*. London: Bloomsbury, 1996.
Davies, Barry. *Fire Magic: Hijack at Mogadishu*. London: Bloomsbury, 1994.
DeForest, Orrin and David Chanoff. *Slow Burn: The Rise and Bitter Fall of American Intelligence in Vietnam*. New York: Simon & Schuster, 1990.
de Klerk, F.W. *The Last Trek: A New Beginning*. New York: St Martin's Press, 1998.
de Kock, Eugene. *A Long Night's Damage: Working for the Apartheid State*. Saxonwold, South Africa: Contra Press, 1998.
de la Billère, Peter. *Looking for Trouble: SAS to Gulf Command, the Autobiography*. New York: HarperCollins, 1994.
de la Billère, Peter. *Storm Command: A Personal Account of the Gulf War*. London: HarperCollins, 1992.
de Pizan, Christine. *The Book of Deeds of Arms and of Chivalry*, Sumner Willard, trans. and Charity Cannon Willard, ed. University Park, PA: Penn State Press, 1999.
de Vries, Roland. *Eye of the Firestorm: The Namibian-Angolan-South African Border War Memoirs of a Military Commander*. Solihull, UK: Helion & Company Limited, 2016.
Delves, Cedric. *Across an Angry Sea: The SAS in the Falklands War*. London: Hurst and Company, 2020.
Diedricks, Andre. *Journey without Boundaries: The Operational Life and Experiences of a SA Special Forces Small Team Operator*. Durban, South Africa: Just Done Productions Publishing, 2007.
Edwards, Sidney. *My Secret Falklands War*. Lewes, Sussex: Book Guild Publishing, 2014.
Falconer, Duncan. *First into Action: A Dramatic Personal Account of Life in the SBS*. London: Little, Brown and Company, 1998.
Farran, Roy. *Winged Dagger*. London: Fontana Books, 1956.
Feith, Douglas, *War and Decision: Inside the Pentagon at the Dawn of the War on Terrorism*. New York: Harper, 2008.
Fiennes, Ranulph. *Where Soldiers Fear to Tread*. London: Hodder and Stoughton, 1975.
Firmin, Rusty. *The Regiment: 15 Years in the SAS*. Oxford: Osprey, 2015.
Flower, Ken. *Serving Secretly: Rhodesia's CIO Chief on Record*. Alberton, South Africa: Galago, 1987.
Ford, Sarah. *One Up: A Woman in Action with the S.A.S*. London: HarperCollins, 1997.
Franks, Tommy. *American Soldier*. New York: Regan Books, 2004.
Fury, Dalton (Pseudonym). *Kill Bin Laden: A Delta Force Commander's Account of the Hunt for the World's Most Wanted Man*. New York: St Martin's Press, 2008.
Gant, Jim, *One Tribe at a Time: A Strategy for Success in Afghanistan*. Los Angeles, CA: Nine Sisters Imports, 2009.
Geldenhuys, Jannie. *At the Front: A General's Account of South Africa's Border War*. Johannesburg, South Africa: Jonathan Ball, 2004.
Greeff, Jack. *A Greater Share of Honour: The Memories of a Recce Officer*. Durban, South Africa: Just Done Productions Publishing, 2008.
Guderian, Heinz. *Panzer Leader*. New York: E. Dutton, 1957.
Guthrie, Charles. *Peace, War and Whitehall: A Memoir*. Oxford: Osprey, 2021.

Haney, Eric, *Inside Delta Force: The Story of America's Elite Counterterrorist Unit*. New York: Delacorte Press, 2002.

Hoare, Tony. *Born for War*. London: Welbeck, 2022.

Hooper, Jim. *Koevoet! Experiencing South Africa's Deadly Bush War*, revised ed. Solihull, UK: Helion Books, 2013.

Hutchings, Richard. *Special Forces Pilot: A Flying Memoir of the Falklands War*. Barnsley: Pen & Sword, 2008.

Jeapes, Tony. *SAS: Secret War*. London: HarperCollins, 1996.

Josephus, Flavius. *The Works of Flavius Josephus*, vol. II, William Whiston, trans. Philadelphia, PA: J.B. Lippencott & Co., 1878.

Kautilya, L. *The Arthashastra*, L. N. Rangarajan, ed. and trans. New Delhi, India: Penguin Books, 1992.

Kennedy, Paul. *Soldier 'I' SAS*. London: Bloomsbury, 1989.

Kitson, Frank. *Bunch of Five*. London: Faber and Faber, 1977.

Large, Lofty. *Soldier against the Odds: From Infantry to SAS*. Edinburgh: Mainstream Publishing, 1999.

Leonov, Viktor. *Blood on the Shores: Soviet Naval Commandos in World War II*. James Gebhardt, trans. Annapolis, MD: Naval Institute Press, 1993.

Lewis, Rob. *Fishers of Men*. London: Hodder and Stoughton, 1999.

Linford, Delville. *As the Crow Flies: My Bushman Experience with 31 Battalion*. Pretoria, South Africa: Protea, 2014.

Lloyd Owen, David. *The Desert: My Dwelling Place*. London: Cassell and Co., 1957.

Locher III, James R. *Victory on the Potomac: The Goldwater-Nichols Act Unifies the Pentagon*. College Station, TX: Texas A&M Press, 2002.

Luttrell, Marcus and Patrick Robinson. *Lone Survivor: The Eyewitness Account of Operation Redwing and the Lost Heroes of SEAL Team 10*. New York: Back Bay Books, 2007.

Machiavelli, Niccolo. *The Art of War*, Henry Neville, trans. 1675. https://oll.libertyfund.org/title/machiavelli-the-art-of-war-neville-trans.

Machiavelli, Niccolo. *The Prince*, W.K. Marriott, trans. Project Gutenberg eBook, 1998. www.gutenberg.org/files/1232/1232-h/1232-h.htm#pref03.

Machiavelli, Niccolo. *Discourses on Livy*, Harvey Mansfield and Nathan Tarcov, trans. Chicago, IL: University of Chicago Press, 1996.

MacKenzie, Alastair. *Pilgrim Days: A Lifetime of Soldiering from Vietnam to the SAS*. London: Osprey, 2019.

Malan, Magnus. *My Life with the SA Defence Force*. Pretoria, South Africa: Protea Books, 2006.

Mandela, Nelson. *Long Walk to Freedom*. Boston, MA: Little, Brown and Company, 1994.

Maurice. *Maurice's Strategikon: Handbook of Byzantine Military Strategy*. George Dennis, trans. Philadelphia, PA: University of Pennsylvania Press, 1984.

McAleese, Peter. *No Mean Soldier: The Autobiography of a Professional Fighting Man*. London: Orion Books, 1993.

McChrystal, Stanley. *My Share of the Task: A Memoir*. New York: Portfolio Penguin, 2013.

Mercer, Peter. *Not by Strength, by Guile*. London: Blake, 2000.

Morgan, William. *The O.S.S. and I*. New York: W.W. Norton & Co., 1957.

Nujoma, Samuel. *Where Others Wavered: The Autobiography of Sam Nujoma*. London: Panaf Books, 2001.

Obama, Barack. *A Promised Land*. New York: Crown, 2020.

O'Neill, Robert. *The Operator: Firing the Shots That Killed Osama bin Laden and My Years as a SEAL Team Warrior*. New York: Scribner, 2018.

Owen, Mark (Pseudonym Matt Bissonnette). *No Easy Day: The Autobiography of a Navy SEAL*. New York: Dutton, 2012.

Panetta, Leon. *Worthy Fights: A Memoir of Leadership in War and Peace.* New York: Penguin Press, 2014.

Parker, John. *SBS: The Inside Story of the Special Boat Service.* London: Headline Book Publishing, 1997.

Pin, Sun. *The Art of Warfare*, D.C. Lay and Roger Ames, trans. New York: Ballantine, 1996.

Plato. *Plato in Twelve Volumes*, vols. 5 and 6 Paul Shorey, trans. Cambridge, MA: Harvard University Press, 1969 (*Republic*, Book 7), www.perseus.tufts.edu/hopper/text?doc=Perseus%3Atext%3A1999.01.0168%3Abook%3D7%3Asection%3D539b.

Plutarch. *The Age of Alexander: Nine Greek Lives by Plutarch*, Ian Scott-Kilvert, trans. London: Penguin Books, 1973.

Plutarch. *The Parallel Lives of Plutarch*, vol. 10, Bernadotte Perrin, trans. Cambridge, MA: Loeb, 1923.

Polyaenusk. *Strategems of War*, R. Shepherd, trans. Chicago, IL: Ares Publishers Inc., 1974.

Po-wen, Liu. *Unorthodox Strategies: 100 Lessons in the Art of War.* Ralph Sawyer, trans. Boulder, CO: Westview Press, 1996.

Powell, Des and Damien Lewis. *SAS: Bravo Three Zero.* London: Quercus Editions Ltd, 2021.

Prado Salmón, Gary. *The Defeat of Che Guevara: Military Response to Guerrilla Challenge in Bolivia.* Westport, CT: Praeger, 1987.

Radcliffe, Peter. *Eye of the Storm: Twenty-Five Years in Action with the SAS.* Miami, FL: Lewis International, Inc., 2000.

Rennie, James. *The Operators: Inside 14 Intelligence Company—The Army's Top Secret Elite.* London: Century, 1996.

Renwick, Robin. *Unconventional Diplomacy in Southern Africa.* London: Macmillan, 1997.

Rodriguez, Felix and John Weisman. *Shadow Warrior: The CIA Hero of a Hundred Unknown Battles.* New York: Simon & Schuster, 1989.

Schroen, Gary. *First In: An Insider's Account of How the CIA Spearheaded the War on Terror in Afghanistan.* New York: Ballantine Books, 2005.

Shaw, Tony. *SAS: South Georgia Boating Club.* Barnsley: Pen & Sword, 2022.

Shityuwete, Helao. *Never Follow the Wolf: The Autobiography of a Namibian Freedom Fighter.* London: Kliptown Books, 1990.

Slim, William. *Defeat into Victory: The Magnificent Account of a Great Campaign in the Second World War.* London: Cassell and Co., 1956.

Smuts, J.C. *Jan Christian Smuts.* London: Cassell & Company, 1952.

Stadler, Koos. *Recce: Small Team Missions Behind Enemy Lines.* Oxford: Casemate, 2015.

Stockwell, John. *In Search of Enemies: A CIA Story.* New York: W.W. Norton & Co., 1984.

Strassler, Robert, ed. *The Landmark Herodotus: The Histories.* Andrea Purvis, trans. New York: Pantheon, 2007.

Strassler, Robert, trans. and ed. *The Landmark Thucydides: A Comprehensive Guide to the Peloponnesian War.* New York: Free Press, 1996.

Strekhin, Yuriy Fedorovich. *Commandos from the Sea: Soviet Naval Spetsnaz in World War II*, James Gebhardt, trans. Annapolis, MD: Naval Institute Press, 1996.

Sutherland, David. *He Who Dares: Recollections of Service in the SAS, SBS, and MI5.* Annapolis, MD: Naval Institute Press, 1999.

Tacitus, Aeneas. *Siege Warfare*, Illinois Greek Club, trans. Cambridge, MA: Harvard University Press, 1928.

Thatcher, Margaret. *The Downing Street Years.* New York: HarperCollins, 1993.

Thompson, Julian. *No Picnic: 3 Commando Brigade in the South Atlantic.* London: Guild, 1985.

Thucydides. *The Peloponnesian Wars*, vol. 3, William Smith, trans. London: A.J. Valpy, 1831.

Tse-Tung, Mao. *Selected Military Writings of Mao Tse-Tung.* Beijing: Foreign Languages Press, 1963.

Tzu, Sun. *The Illustrated Art of War*, Samuel B. Griffith, trans. Oxford: Oxford University Press, 2005.

van Niekerk, Hermann. *Born to Storm: A Courageous Boy Became a Fearless Soldier.* Pretoria, South Africa: Group 7 Drukkers en Uitgewers [Pty] Ltd., 2019.

Whitcomb, Darrel. *On a Steel Horse I Ride: A History of the MH-53 Pave Low Helicopters in War and Peace.* Maxwell Air Force Base, AL: Air University Press, 2012.

Xenophon. *The Persian Expedition.* Rex Warner, trans. Baltimore, MD: Penguin Books, 1969.

## Books (secondary)

Adam, Barbara, Ulrich Beck, and Joost van Loon. *The Risk Society and Beyond: Critical Issues for Social Theory.* London: Sage, 2000.

Adams, James, Robin Morgan, and Anthony Bambridge. *Ambush: The War between the SAS and the IRA.* London: Pan, 1988.

Adams, Thomas. *U.S Special Operations Forces in Action: The Challenge of Unconventional Warfare* London: Frank Cass, 1998.

Adonis, Andrew. *Making the Aristocracy Work: The Peerage and the Political System in Britain, 1884–1914.* Oxford: Oxford University Press, 1993.

Ahmed, Akbar. *The Thistle and the Drone: How America's War on Terror Became a Global War on Tribal Islam.* Washington, DC: Brookings Institution Press, 2013.

Aid, Matthew. *Intel Wars: The Secret History of the Fight against Terror.* New York: Bloomsbury Press, 2012.

Alexander, Matthew. *Kill or Capture: How a Special Operations Task Force Took Down a Notorious Al Qaeda Terrorist.* New York: St Martin's Press, 2011.

Allen, Joel. *Hostages and Hostage-Taking in the Roman Empire.* Cambridge: Cambridge University Press, 2006.

Allison, Graham and Phillip Zelikow. *The Essence of Decision: Explaining the Cuban Missile Crisis*, 2nd ed. New York: Longman, 1999.

Andrew, Christopher. *Her Majesty's Secret Service: The Making of the British Intelligence Community.* New York: Viking, 1986.

Andrew, Christopher and Vasili Mitrokhin. *The World Was Going Our Way: The KGB and the Battle for the Third World.* New York: Basic Books, 2005.

Andrew, Christopher and Vasili Mitrokhin. *The Sword and the Shield: The Mitrokhin Archive and the Secret History of the KGB.* New York: Basic Books, 1999.

Anglim, Simon. *Orde Wingate, Unconventional Warrior: From the 1920s to the Twenty-First Century.* London: Pen and Sword, 2015.

Arkin, William. *Code Names: Deciphering U.S. Military Plans, Programs, and Operations in the 9/11 World.* Hanover, NH: Steerforth Press, 2005.

Arkless, David. *The Secret War: Dhofar, 1971/1972.* London: William Kimber, 1988.

Arquilla, John and David Ronfeldt. *Swarming and the Future of Conflict.* Report DB311. Santa Monica, CA: RAND Corporation, 2005.

Arquilla, John and David Ronfeldt, eds. *From Troy to Entebbe: Special Operations in Ancient and Modern Times.* Lanham, MD: University Press of America, 1996.

Arreguin-Toft, Ivan. *How the Weak Win Wars: A Theory of Asymmetric Conflict.* Cambridge: Cambridge University Press, 2005.

Ashdown, Paddy. *A Brilliant Little Operation: The Cockleshell Heroes and the Most Courageous Raid of WW2.* London: Aurum, 2012.

Asher, Michael. *The Regiment: The Definitive Story of the SAS.* London: Penguin, 2008.

Aston, Mark. *SAS: Sea King Down.* London: Penguin, 2022.

Austin, N.J.E. and N.B. Rankov. *Exploratio: Military and Political Intelligence in the Roman World from the Second Punic War to the Battle of Adrianople.* London: Routledge, 1995.

Bailey, Jonathan, Richard Iron, and Hew Strachan. *British General in Blair's Wars.* London: Ashgate, 2013.

Bailey, Roderick. *The Wildest Province: SOE in the Land of the Eagle.* London: Jonathan Cape, 2008.

Baker, Deane-Peter, Roger Herbert, and David Whetham. *The Ethics of Special Ops: Raids, Recoveries, Reconnaissance, and Rebels.* Cambridge: Cambridge University Press, 2024.

Balasevicius, Tony, ed. *Casting Shadows on the Light: Canadian Perspectives of Special Operations Forces.* Kingston, ON: Canadian Defence Academy Press, 2007.

Balko, Radley. *Rise of the Warrior Cop: The Militarization of America's Police Forces*, New York: PublicAffairs, 2013.

Barlow, Eeben. *Composite Warfare: The Conduct of Successful Ground Force Operations in Africa.* Pinetown, South Africa: 30° South Publishers [Pty] Ltd, 2016.

Barnett, Corelli. *Britain and Her Army, 1509–1970: A Military, Political, and Social Survey.* London: Allen Lane, 1970.

Barnett, Frank, B. Hugh Tovar, and Richard Shultz Jr, eds. *Special Operations in U.S. Strategy.* Washington: National Defense University Press, 1984.

Barno, David and Nora Bensahel. *Adaptation under Fire: How Militaries Change in Wartime.* New York: Oxford University Press, 2020.

Basinger, Franz. *Mehmed the Conqueror and His Time.* William Hickman, ed. and Ralph Manheim, trans. Princeton, NJ: Princeton University Press, 1978.

Baylis, John et al., eds. *Strategy in the Contemporary World: An Introduction to Strategic Studies.* Oxford: Oxford University Press, 2002.

Beaufre, André. *An Introduction to Strategy, with Particular Reference to Problems of Defense, Politics, Economics, and Diplomacy in the Nuclear Age.* R.H. Barry, trans. New York: Frederick Praeger, 1965.

Beaumont, Roger. *Military Elites: Special Fighting Units in the Modern World.* New York: Bobbs-Merrill, 1974.

Beck, Ulrich. *Risk Society: Towards a New Modernity.* New York: Sage Publications, 1992.

Bell, J. Bowyer and Barton Whaley. *The Dynamics of the Armed Struggle.* London: Frank Cass, 1998.

Bell, J. Bowyer and Barton Whaley. *Cheating and Deception.* London: Routledge, 1991.

Bell, Michael. *The Exigencies of Global, Integrated Warfare: The Evolving Role of the CJCS and his Dedicated Staff.* Carlisle Barracks, PA: Army War College Press, 2011.

Belote, James and William Belote. *Typhoon of Steel: The Battle for Okinawa.* New York: Harper & Row, 1970.

Bermudez Jr, Joseph. *North Korean Special Forces*, 2nd ed. Annapolis, MD: Naval Institute Press, 1998.

Best, Jan. *Thracian Peltasts and Their Influence on Greek Warfare.* Leiden: Brill, 1969.

Bhakari, Surinker Kumar. *Indian Warfare.* New Delhi: Munshiram Manoharial Publishers, 1981.

Biddle, Tami Davis. *Rhetoric and Reality in Air Warfare: The Evolution of British and American Ideas about Strategic Bombing, 1914–1945.* Princeton, NJ: Princeton University Press, 2002.

Binckes, Robin. *Vlakplaas: Apartheid's Death Squads, 1979–1994.* Barnsley: Pen & Sword Books Ltd, 2018.

Black, Jeremy. *War: Past, Present and Future.* New York: St Martin's Press, 2000.

Blackstock, Paul. *The Strategy of Subversion: Manipulating the Politics of Other Nations.* Chicago, IL: Quadrangle Books, 1964.

Bloom, Mia. *Dying to Kill: The Allure of Suicide Terror.* New York: Columbia University Press, 2005.

Bond, Brian. *Liddell Hart: A Study of His Military Thought*. London: Cassell, 1977.
Boot, Max. *Invisible Armies: An Epic History of Guerrilla Warfare from Ancient Times to the Present*. New York: Liveright, 2013.
Borden, William. *There Will Be No Time: The Revolution in Strategy*. New York: Macmillan, 1946.
Bousquet, Antione. *The Scientific War of Warfare: Order and Chaos on the Battlefields of Modernity*. New York: Columbia University Press, 2009.
Bowden, Mark. *The Finish: The Killing of Osama bin Laden*. New York: Atlantic Monthly Press, 2012.
Bowden, Mark. *Killing Pablo: The Hunt for the World's Greatest Outlaw*. New York: Atlantic Monthly Press, 2001.
Bowden, Mark. *Blackhawk Down: A Story of Modern War*. New York: Atlantic Monthly Press, 1999.
Bracamonte, José Angel Moroni and David Spencer. *Strategy and Tactics of the Salvadoran FMLN Guerrillas: Last Battle of the Cold War, Blueprint for Future Conflicts*. Westport, CT: Praeger, 1992.
Bracewell, Catherine Wendy. *The Uskoks of Senj: Piracy, Banditry, and Holy War in the Sixteenth-Century Adriatic*. Ithaca, NY: Cornell University Press, 1992.
Bradford, Ernle. *The Great Siege, Malta 1565*. London: Hodder & Stoughton 1961.
Bragg, Rick. *I Am a Soldier, Too: The Jessica Lynch Story*. New York: Vintage Books, 2004.
Bridgland, Fred. *The War for Africa: Twelve Months That Transformed a Continent*. Gibraltar: Ashanti Publishing, 1990.
Brisard, Jean-Charles. *Zarqawi: The New Face of Al-Qaeda*. New York: Other Press, 2005.
Brock, George et al. *Siege: Six Days at the Iranian Embassy*. London: Macmillan, 1980.
Brodie, Bernard. *War and Politics*. London: Cassell, 1973.
Brown, David. *The Royal Navy and the Falklands War*. London: Arrow Books, 1989.
Brown, Don. *Call Sign Extortion 17: The Shoot-Down of SEAL Team Six*. Guilford, CT: Lyons Press, 2015.
Brown, Vahid and Don Rassler. *Fountainhead of Jihad: The Haqqani Nexus, 1973–2012*. New York: Columbia University Press, 2013.
Burk, James, ed. *How 9/11 Changed Our Ways of War*. Stanford, CA: Stanford University Press, 2013.
Burke, David. *Frank Kitson's Irish War: Mastermind of the Dirty War in Ireland*. Cork, Ireland: Mercier Press, 2021.
Burkhart, Muller-Hillebrand. *Germany and Its Allies in World War II: A Record of Axis Collaboration Problems*. Lanham, MD: University Publications of America, 1980.
Byman, Daniel. *A High Price: The Triumphs and Failure of Israeli Counterterrorism*. New York: Oxford University Press, 2011.
Byman, Daniel and Matthew Waxman. *The Dynamics of Coercion: American Foreign Policy and the Limits of Military Might*. Cambridge: Cambridge University Press, 2002.
Cable, James. *Political Institutions and Issues in Britain*. London: Macmillan, 1987.
Caddick-Adams, Peter. *Snow and Steel: The Battle of the Bulge, 1944–1945*. Oxford: Oxford University Press, 2015.
Call, Steve. *Danger Close: Tactical Air Controllers in Afghanistan and Iraq*. College Station, TX: Texas A&M Press, 2007.
Callwell, Charles. *Small Wars: Their Principles and Practice*, 3rd ed. London: Parnell Book Services Ltd, 1976.
Campbell, Bruce and Arthur Brenner, eds. *Death Squads in Global Perspective: Murder with Deniability*. New York: Palgrave Macmillan, 2000.
Cann, John. *Counterinsurgency in Africa: The Portuguese Way of War, 1961–1974*. Westport, CT: Greenwood, 1997.

Capua, James and Carl Kruger. *Jåger, Europe's First Special Operations Forces: History, Organization, Arms and Equipment of the Austro-Hungarian Empire's Elite Light Infantry to 1866*. New York: Signal Horn Publishers, 2020.

Carroll, Rory. *There Will Be Fire: Margaret Thatcher, the IRA, and Two Minutes That Changed History*. New York: Putnam, 2023.

Castex, Raoul. *Strategic Theories*. Eugenia Kiesling, trans. and ed. Annapolis, MD: Naval Institute Press, 1994.

Cate, Arthur ten and Martijn van der Vorm. *Callsign Nassau: Dutch Army Special Forces in Action in the 'New World Disorder'*. Leiden: Leiden University Press, 2016.

Cesarani, David. *Major Farran's Hat: Murder, Scandal, and Britain's War against Jewish Terrorism, 1945–1948*. London: William Heinemann, 2008.

Chambers, James. *The Devil's Horsemen: The Mongol Invasion of Europe*. New York: Atheneum, 1985.

Charters, David. *Whose Mission, Whose Orders? British Civil-Military Command and Control in Northern Ireland, 1968–1974*. Montreal: McGill-Queen's University Press, 2017.

Chia, Robert and Robin Holt. *Strategy without Design: The Silent Efficacy of Indirect Action*. Cambridge: Cambridge University Press, 2009.

Chilcote, Ronald. *The Portuguese Revolution: State and Class in the Transition to Democracy*. Lanham, MD: Rowman & Littlefield, 2010.

Clausewitz, Carl von. *On War*. Peter Paret and Michael Howard, trans. and eds. Princeton, NJ: Princeton University Press, 1984.

Cline, Lawrence. *Pseudo Operations and Counterinsurgency: Lessons from Other Countries*. Carlisle Barracks, PA: Strategic Studies Institute, June 2005.

Clutterbuck, Richard. *Protest and the Urban Guerrilla*. London: Abelard-Schuman Ltd, 1973.

Cobb, Paul, ed. *The Lineaments of Islam: Essays in Honor of Fred McGraw Donner*. Leiden: Brill, 2012.

Cock, Jacquelyn and Laurie Nathan, eds. *War and Society: The Militarisation of South Africa*. Claremont, South Africa: David Phillip Publisher [Pty] Ltd, 1989.

Cockburn, Andrew. *Kill Chain: The Risk of High-Tech Assassins*. New York: Henry Holt, 2015.

Cohen, Eliot. *Supreme Command: Soldiers, Statesmen, and Leadership in Wartime*. New York: Simon & Schuster, 2002.

Cohen, Eliot. *'Commandos and Politicians: Elite Military Units in Modern Democracies'*. Harvard Studies in International Affairs No. 40. Boston, MA: Center for International Affairs, 1978.

Coker, Christopher. *War in an Age of Risk*. New York: Polity, 2009.

Coll, Steve. *Ghost Wars: The Secret History of the CIA, Afghanistan, and Bin Laden, from the Soviet Invasion to September 10, 2001*. New York: Penguin, 2004.

Collins, John. *Green Berets, SEALs, and Spetsnaz: U.S. and Soviet Special Military Operations*. London: Pergamon-Brassey's, 1978.

Conboy, Kenneth and Dale Andradé. *Spies and Commandos: How America Lost the Secret War in North Vietnam*. Lawrence, KS: University Press of Kansas, 2000.

Connor, Ken. *Ghost Force: The Secret History of the SAS*. London: Weidenfeld & Nicolson, 1998.

Cookridge, E.H. *Inside S.O.E.: The First Full Story of Special Operations Executive in Western Europe, 1940–1945*. London: Arther Barker, 1966.

Corbett, Julian. *Some Principles of Maritime Strategy*. Annapolis, MD: Naval Institute Press, 1988.

Cordesman, Anthony and Abraham Wagner. *The Lessons of Modern War, Volume III: The Afghan and Falklands Conflicts*. Boulder, CO: Westview Press, 1990.

Cormac, Rory. *Disrupt and Deny: Spies, Special Forces, and the Secret Pursuit of British Foreign Policy*. Oxford: Oxford University Press, 2018.

Crandall, Russell. *America's Dirty Wars: Irregular Warfare from 1776 to the War on Terror*. Cambridge: Cambridge University Press, 2014.

Cronin, Audrey Kurth. *Power to the People: How Open Technological Innovation is Arming Tomorrow's Terrorists*. New York: Oxford University Press, 2020.

Crump, Steve. *Koevoet, Volume 1: South West African Police Counter-Insurgency Operations During the South African Border War, 1978–1984*. Warwick, UK: Helion, 2022.

Cutler, Thomas. *Brown Water, Black Berets: Coastal and Riverine Warfare in Vietnam*. Annapolis, MD: Naval Institute Press, 1988.

Dachowicz, Dariusz and Hubert Królikowski. *Polish Special Operations Forces and Special Operations: Historical Outline and Present Days*. Warsaw: Taktyka i Strategia, 2016.

Daftary, Farhad. *The Assassin Legends: Myths of the Isma'ilis*. London: I.B. Tauris, 1994.

Daniels, Robert. *Year of the Heroic Guerrilla: World Revolution and Counterrevolution in 1968*. Cambridge, MA: Harvard University Press, 1987.

Danielson, Tone. *Making Warriors in a Global Era: An Ethnographic Study of the Norwegian Naval Special Operations Commando*. Lanham, MD: Lexington Books, 2018.

Danilovic, Vesna. *When the Stakes Are High: Deterrence and Conflict among Major Powers*. Ann Arbor, MI: University of Michigan Press, 2002.

Danner, Mark. *Spiral: Trapped in the Forever War*. New York: Simon & Schuster, 2016.

Darack, Ed.. *Victory Point: Operations Red Wings and Whalers—The Marine Corps' Battle for Freedom in Afghanistan*. New York: Berkley Books, 2009.

David, Saul. *SBS: Silent Warriors*. London: William Collins, 2021.

Davies, Barry and Nigel McCrery. *The Complete History of the SAS*. London: Andre Deutsch, 2011.

de B. Taillon, Paul. *Hijacking and Hostages: Government Responses to Terrorism*. Westport, CT: Greenwood Press, 2002.

de B. Taillon, Paul. *The Evolution of Special Forces in Counter-Terrorism: The British and American Experiences*. Westport, CT: Greenwood Press, 2000.

Debray, Régis. *Revolution in the Revolution*. Bobbye Ortiz, trans. London: Penguin, 1967.

DeLay, Brian. *War of the Thousand Deserts: Indian Raids and the U.S.–Mexican War*. New Haven, CT: Yale University Press, 2008.

Desch, Michael. *Cult of the Irrelevant: The Waning Influence of Social Science on National Security*. Princeton, NJ: Princeton University Press, 2019.

Dewar, Michael. *The British Army in Northern Ireland*. London: Arms and Armour Press, 1985.

Dodge, Theodore Ayrault. *Hannibal: The History of The Art of War Among the Carthaginians and the Romans, Down to the Battle of Pydna, 168 BC, with a Detailed Account of the Second Punic War*. London: Greenhill Books, 1994.

Dolman, Everett. *Pure Strategy: Power and Principle in the Space and Information Age*. London: Frank Cass, 2005.

Donnithorne, Jeffrey. *Four Guardians: A Principled Agent View of American Civil–Military Relations*. Baltimore, MD: Johns Hopkins University Press, 2018.

Douek, Daniel. *Insurgency and Counterinsurgency in South Africa*. London: Hurst, 2020.

Draper, Robert. *To Start a War: How the Bush Administration Took America into Iraq*. New York: Penguin, 2021.

Dunnigan, James. *The Perfect Soldier: Special Operations, Commandos, and the Future of War*. New York: Citadel Press, 2003.

Eckert, Nicholas. *Fatal Encounter: The Story of the Gibraltar Killings*. Dublin: Poolbeg, 1999.

Edwards, Aaron. *Defending the Realm: The Politics of Britain's Small Wars since 1945*. Manchester: Manchester University Press, 2012.

Egnell, Robert. *Complex Peace Operations and Civil–Military Relations: Winning the Peace*. London: Routledge, 2011.

Els, Paul. *We Fear Naught but God: The Pictorial Edition of the South African Special Forces*, 4th rev. ed. Cape Town, South Africa: PelsA Books, 2014.

Els, Paul. *Ongulumbashe: Where the Bushwar Began*. Wandsbeck, South Africa: Reach Publishers, 2007.

Elton, Hugh. *Warfare in Roman Europe, AD 350–425*. Oxford: Clarendon Press, 1997.

Emerson, Steven. *Secret Warriors: Inside the Covert Military Operations of the Reagan Era*. New York: G.P. Putnam's, 1988.

English, Richard. *Armed Struggle: The History of the IRA*. Oxford: Oxford University Press, 2003.

Ensminger, Thomas. *Spies, Supplies, and Moonlit Skies*, 2 volumes. Fairborn, OH: Hillborn Wildlife Refuge, 2001.

Erfurth, Waldermar. *Surprise*, Sean Possony and Daniel Vilfroy, trans. Harrisburg, PA: Military Service Publishing Company, 1943.

Eriksson, Gunilla and Ulrica Petterson, eds. *Special Operations from a Small State Perspective: Future Security Challenges*. New York: Palgrave Macmillan, 2017.

Ethell, Jeffrey and Alfred Price. *Air War South Atlantic*. London: Sidgwick & Jackson, 1983.

Evans, Richard. *The Coming of the Third Reich*. New York: The Penguin Press, 2003.

Faddis, Sam. *The CIA War in Kurdistan: The Untold Story of the Northern Front of the Iraq War*. Havertown, PA: Casemate, 2020.

Farrell, Theo. *Unwinnable: Britain's War in Afghanistan, 2011–2014*. London: The Bodley Head, 2014.

Feaver, Peter. *Armed Servants: Agency, Oversight, and Civil–Military Relations*. Cambridge, MA: Harvard University Press, 2003.

Feaver, Peter. *Guarding the Guardians: Civilian Control of Nuclear Weapons in the United States*. Ithaca, NY: Cornell University Press, 1992.

Finer, Samuel. *The Man on Horseback: The Role of the Military in Politics*. London: Routledge, 2002.

Finlan, Alastair. *Special Forces, Strategy, and the War on Terror: Warfare by Other Means*. London: Routledge, 2009.

Frankel, Philip. *Pretoria's Praetorians: Civil–Military Relations in South Africa*. Cambridge: Cambridge University Press, 1984.

Freedman, Lawrence. *Strategy: A History*. Oxford: Oxford University Press, 2013.

Freeman, Michael. *Financing Terrorism: Case Studies*. Farnham: Ashgate, 2012.

Friedman, Norman. *Terrorism, Afghanistan, and America's New Way of War*. Annapolis, MD: Naval Institute Press, 2003.

Fuller, J.F.C. *The Generalship of Alexander the Great*. London: Eyre & Spottiswoode, 1958.

Fuller, J.F.C. *Generalship: Its Diseases and Their Cure*. Harrisburg, PA: Military Service Publishing Co., 1936.

Ganor, Boaz and Liram Koblentz-Stenzler, *Israel's Targeted Killing Policy: Moral, Ethical, and Operational Dilemmas*. New York: Palgrave Macmillan, 2022.

Gat, Azar. *War in Human Civilization*. Oxford: Oxford University Press, 2006.

George, Edward. *The Cuban Intervention in Angola: 1965–1991*. London: Frank Cass, 2005.

Geraghty, Tony. *Beyond the Front Line: The Untold Exploits of Britain's Most Daring Cold War Spy Mission*. London: HarperCollins, 1996.

Geraghty, Tony. *Who Dares Wins: The Story of the Special Air Service, 1950–1980*. London: Arms & Armour Press, 1980.

Gerges, Fawaz. *ISIS: A History*. Princeton, NJ: Princeton University Press, 2016.

Glantz, David. *The Military Strategy of the Soviet Union: A History*. London: Frank Cass, 1992.
Glantz, David. *Soviet Military Operational Art: In Pursuit of Deep Battle*. London: Frank Cass, 1991.
Gleijens, Piero. *Conflicting Missions: Havana, Washington, and Africa, 1959–1976*. Chapel Hill, NC: University of North Carolina Press, 2002.
Godkin, Edwin Lawrence. *The History of the Hungary and the Magyars: From the Earliest Period to the Close of the Late War*. New York: Alexander Montgomery, 1853.
Godson, Roy, ed. *Intelligence Requirements for the 1980s: Covert Action*. Washington, DC: National Strategy Information Center Inc., 1981.
Gordon, Michael and Bernard Trainor. *Cobra II: The Inside Story of the Invasion and Occupation of Iraq*. New York: Pantheon, 2006.
Grau, Leslie and Dodge Billingsley. *Operation Anaconda: American First Major Battle in Afghanistan*. Lawrence, KS: University Press of Kansas, 2011.
Gray, Colin. *Theory of Strategy*. Oxford: Oxford University Press, 2018.
Gray, Colin. *Airpower for Strategic Effect*. Maxwell Air Force Base, AL: Air University Press, 2012.
Gray, Colin. *The Strategy Bridge: Theory for Practice*. Oxford: Oxford University Press, 2010.
Gray, Colin. *Explorations in Strategy*. Boulder, CO: Westview Press, 1996.
Grenier, John. *The Far Reaches of Empire: War in Nova Scotia, 1710–1760*. Norman, OK: University of Oklahoma Press, 2008.
Grenkevich, Leonid. *The Soviet Partisan Movement, 1941–1944: A Critical Historiographical Analysis*. London: Frank Cass, 1999.
Groh, Tyrone. *Proxy War: The Least Bad Option*. Stanford, CA: Stanford University Press, 2019.
Grundy, Kenneth. *The Militarization of South African Politics*. Bloomington, IN: Indiana University Press, 1987.
Guertner, Gary, ed. *The Search for Strategy: Politics and Strategic Vision*. Westport, CT: Greenwood Press, 1993.
Gwynn, Charles. *Imperial Policing*. London: Macmillan and Co. Ltd, 1934.
Hamann, Hilton. *Days of the Generals*. Cape Town, South Africa: Zebra Press, 2001.
Hammes, Thomas. *The Sling and the Stone: On War in the 21st Century*. New York: Zenith Press, 2006.
Harclerode, Peter. *Fighting Dirty: The Inside Story of Covert Operations from Ho Chi Minh to Osama bin Laden*. London: Cassell, 2002.
Harari, Yuval Noah. *Special Operations in the Age of Chivalry, 1100–1550*. Woodbridge, UK: The Boydell Press, 2007.
Hargreaves, Andrew. *Special Operations in World War II: British and American Irregular Warfare*. Norman, OK: University of Oklahoma Press, 2013.
Hart-Davis, Duff. *The War That Never Was: The True Story of the Men Who Fought Britain's Most Secret Battle*. London: Arrow, 2012.
Haun, Phil, Colin Jackson, and Timothy Schultz, eds. *Air Power in the Age of Primacy: Air Warfare since the Cold War*. Cambridge: Cambridge University Press, 2022.
Haycock, Ronald, ed. *Regular Armies and Insurgency*. London: Croon Helm, 1979.
Heitman, Helmoed-Römer. *War in Angola: The Final South African Phase*. Gibraltar: Ashanti Publishing, 1990.
Heng, Yee-Kuang. *War as Risk Management: Strategy and Conflict in an Age of Globalised Risks*. London: Routledge, 2006.
Hennessy, Peter and James Jinks. *The Silent Deep: The Royal Navy Submarine Service since 1945*. London: Allen Lane, 2015.
Henriksen, Thomas. *Eyes, Ears, and Daggers: Special Operations Forces and the Central Intelligence Agency in America's Evolving Struggle against Terrorism*. Stanford, CA: Hoover Institution Press, 2016.

Herbstein, Denis and John Evans. *The Devils Are among Us: The War for Namibia*. London: Zed Books, 1989.

Herspring, Dale. *The Pentagon and the Presidency: Civil–Military Relations from FDR to George Bush*. Lawrence, KS: University Press of Kansas, 2008.

Hiranandani, G.M. *Transition to Guardianship: The Indian Navy, 1991–2000*. New Delhi: Lancer, 2013.

Hoe, Alan and Eric Morris. *Re-Enter the SAS: The Special Air Service and the Malayan Emergency*. London: Leo Cooper, 1994.

Hoffman, Bruce. *Inside Terrorism*, 3rd ed. New York: Columbia University, 2017.

Hogan Jr, David. *Raiders or Elite Infantry? The Changing Role of the U.S. Army Rangers from Dieppe to Grenada*. Westport, CT: Greenwood Press, 1992.

Horn, Bernd, John Paul de B. Taillon, and David Last, eds. *Force of Choice: Perspectives on Special Operations*. Kingston, ON: McGill-Queen's University Press, 2004.

Howard, Michael. *Soldiers and Governments: Nine Studies in Civil–Military Relations*. Bloomington, IN: Indiana University Press, 1959.

Huntington, Samuel. *The Soldier and the State: The Theory and Politics of Civil–Military Relations*. Cambridge, MA: Belknap, 1959.

Hutchinson, Robert. *Elizabeth's Spy Master: Francis Walsingham and the Secret War That Saved England*. London: Weidenfeld & Nicholson, 2006.

Ikenberry, John. *Liberal Leviathan: The Origins, Crisis, and Transformation of the American World Order*. Princeton, NJ: Princeton University Press, 2011.

Ion, A. Hamish and Keith Nelson. *Elite Military Formations in War and Peace*. Westport, CT: Praeger, 1996.

Ingham, Sarah. *The Military Covenant: Its Impact on Civil–Military Relations in Britain*. London: Routledge, 2016.

Issac, Benjamin. *The Limits of Empire: The Roman Army in the East*. Oxford: Clarendon Press, 1990.

Janis, Irving. *Groupthink: Psychological Studies of Policy Decisions and Fiascoes*, 2nd ed. Boston, MA: Wadsworth, 1982.

Janowitz, Morris. *The Professional Soldier: A Social and Political Portrait*. New York: Free Press, 1960.

Jansen, Anemari. *Eugene de Kock: Assassin for the State*. Cape Town, South Africa: Tafelburg Publications, 2015.

Jeffrey, Anthea. *People's War: New Light on the Struggle for South Africa*. Johannesburg, South Africa: Jonathan Cape, 2009.

Jeffrey, Keith. *The Secret History of MI6: 1909–1949*. New York: Penguin, 2010.

Jervis, Robert. *Perception and Misperception in International Politics*. New ed. Princeton, NJ: Princeton University Press, 1976.

Johnson, Phyllis and David Martin, eds. *Frontline Southern Africa: Destructive Engagement*. New York: Four Walls Eight Windows, 1988.

Johnston, Alastair Iain. *Cultural Realism: Strategic Culture and Grand Strategy in Chinese History*. Princeton, NJ: Princeton University Press, 1998.

Jones, Archer. *The Art of War in the Western World*. New York: Barnes and Noble Press, 1997.

Jones, Clive. *Britain and the Yemen Civil War, 1962–1965: Ministers, Mercenaries, and Mandarins: Foreign Policy and the Limits of Covert Action*. Brighton, UK: Sussex Academic Press, 2004.

Jones, Seth. *Three Dangerous Men: Russia, China, Iran, and the Rise of Irregular Warfare*. New York: W.W. Norton & Co., 2021.

Jones, Seth. *Hunting in the Shadows: The Pursuit of Al Qa'ida since 9/11*. New York: Norton, 2012.

Jones, Tim. *SAS: The First Secret Wars*. London: I.B. Tauris, 2005.

Jones, Tim. *Postwar Counterinsurgency and the SAS, 1945–1952: A Special Type of Warfare.* London: Frank Cass, 2001.

Jumper, Roy. *Death Waits in the 'Dark': The Senoi Praaq, Malaysia's Killer Elite.* Westport, CT: Greenwood Press, 2001.

Kahn, Herman. *On Escalation: Metaphors and Scenarios.* New York: Hudson Institute, 1965.

Kahneman, Daniel. *Thinking, Fast and Slow.* New York: Farrar, Straus and Giroux, 2013.

Kalyvas, Stathis. *The Logic of Violence in Civil War.* Cambridge: Cambridge University Press, 2006.

Kamen, Henry. *The Spanish Inquisition: A Historical Revision.* London: Weidenfeld & Nicolson, 1997.

Katoch, P.C. and Saikat Datta. *India's Special Forces: History and Future of Indian Special Forces.* New Delhi: Vij Books, 2013.

Keeley, Lawrence. *Constant Battles: The Myth of the Peaceful, Noble Savage.* New York: St Martin's Press, 2003.

Keeley, Lawrence. *War before Civilization: The Myth of the Peaceful Savage.* Oxford: Oxford University Press, 1996.

Keen, Maurice, ed. *Medieval Warfare: A History.* Oxford: Oxford University Press, 1999.

Kemp, Anthony. *The SAS: Savage Wars of Peace, 1947 to the Present.* London: John Murray, 1994.

Kern, Paul. *Ancient Siege Warfare.* Bloomington, IN: Indiana University Press, 1999.

Kier, Elizabeth. *Imagining War: French and British Military Doctrine between the Wars.* Princeton, NJ: Princeton University Press, 1997.

Kilcullen, David, *Out of the Mountains: The Coming Age of the Urban Guerrilla.* New York: Oxford University Press, 2013.

Kilcullen, David. *The Accidental Guerrilla: Fighting Small Wars in the Midst of a Big One.* Oxford: Oxford University Press, 2009.

King, Anthony. *Command: The Twenty-First-Century General.* Cambridge: Cambridge University Press, 2019.

Kingdon, John. *Agendas, Alternatives, and Public Policies*, 2nd ed. New York: Longman, 2003.

Kinzer, Stephen. *Overthrow: America's Century of Regime Change from Hawaii to Iraq.* New York: Times Books, 2006.

Kiras, James D. and Martijn Kitzen, eds. *Into the Void: Special Forces after the War on Terror.* London: Hurst, 2024.

Kiras, James D. and Martijn Kitzen, eds. *Special Operations and Strategy: From World War II to the War on Terrorism.* London: Routledge, 2006.

Kittrie, Orde. *Lawfare: Law as a Weapon of War.* Oxford: Oxford University Press, 2016.

Knott, Stephen. *Secret and Sanctioned: Covert Operations and the American Presidency.* New York: Oxford University Press, 1996.

Kometer, Michael. *Command in Air War: Centralized Versus Decentralized Control of Combat Airpower.* Maxwell Air Force Base, AL: Air University Press, 2007.

Kosto, Adam. *Hostages in the Middle Ages.* Oxford: Oxford University Press, 2012.

Kurowski, Franz. *The Brandenburgers: Global Mission.* David Johnston, trans. Winnipeg, MB: J.J. Fedorowicz, 1997.

Kynaston, David. *Austerity Britain, 1945–1951.* London: Bloomsbury, 2008.

Ladd, James. *SAS Operations.* London: Robert Hale, 1986.

Ladd, James. *SBS: The Invisible Raiders.* London: Arms and Armour Press, 1983.

Ladwig III, Walter. *The Forgotten Front: Patron–Client Relationships in Counterinsurgency.* Cambridge: Cambridge University Press, 2017.

Lambeth, Ben. *Airpower in the War against ISIS.* Annapolis, MD: Naval Institute Press, 2021.

Lambeth, Ben. *Air Power against Terror, MG-166.* Washington, DC: RAND, 2005.

Lartéguy, Jean. *The Centurions.* Xan Fielding, trans. Paris: Presses de la Cité, 1960.

Lazenby, J.F. The *First Punic War: A Military History*. London: University College of London Press, 1996.

Lazenby, J.F. *Hannibal's War: A Military History of the Second Punic War*. Norman, OK: University of Oklahoma Press, 1978.

Le Bohec, Yann. *Histoire militaire des guerres puniques, 264–146 av. J.C.* Paris: Editions Tallandier, 2014.

Leary, William. *Perilous Missions: Civil Air Transport and CIA Covert Operations in Asia.* Tuscaloosa, AL: University of Alabama Press, 1984.

Leebaert, Derek. *To Dare and to Conquer: Special Operations and the Destiny of Nation, from Achilles to Al Qaeda*. New York: Little, Brown & Company, 2006.

Leonhard, Robert. *Fighting by Minutes: Time and the Art of War*. Westport, CT: Praeger, 1994.

Lewis, Bernard. *The Assassins: A Radical Sect in Islam*. New York: Basic Books, 2003.

Liddell Hart, B.H. *Scipio Africanus: Greater Than Napoleon*. London: Greenhill Books, 1992.

Liddell Hart, B.H. *The Revolution in Warfare*. New Haven, CT: Yale University Press, 1947.

Li Feng. *Early China: A Social and Cultural History*. Cambridge: Cambridge University Press, 2013.

Logan, F. Donald. *The Vikings in History*, 3rd ed. London: Routledge, 2005.

Loo, Bernard, ed. *Military Transformation and Strategy: Revolutions in Military Affairs and Small States*. Abingdon: Routledge, 2009.

Lucas Phillips, C.E. *Cockleshell Heroes: Epic Exploit of the War*. London: Pan, 1956.

Luttwak, Edward. *The Grand Strategy of the Byzantine Empire*. Cambridge, MA: Belknap Press, 2009.

Luttwak, Edward. *Strategy: The Logic of War and Peace*, rev. ed. Cambridge, MA: Belknap Press, 2001.

Luttwak, Edward. *The Grand Strategy of the Roman Empire: From the First Century A.D. to the Third*. Baltimore, MD: The Johns Hopkins University Press, 1976.

MacDonald, Michael. *Children of Wrath: Political Violence in Northern Ireland*. Cambridge: Polity Press, 1986.

MacGregor, Knox and Williamson Murray, eds. *The Dynamics of Military Revolution, 1300–2050*. Cambridge, MA: Cambridge University Press, 2001.

MacKenzie, Alastair. *Special Force: The Untold Story of 22nd Special Air Service Regiment (SAS)*. London: I.B. Tauris, 2011.

MacKinley, John. *The Insurgent Archipelago: From Mao to Bin Laden*. London: Hurst, 2009.

Mahadevan, Prem. *An Eye for an Eye: Decoding Global Special Operations and Irregular Warfare—A Vision for India*. New Delhi: Wisdom Tree, 2014.

Mahan, Alfred Thayer. *The Influence of Sea Power Upon History, 1660–1783*. Boston, MA: Little, Brown and Co., 1890.

Makarov, Stepan. *Discussion of Questions in Naval Tactics*. John Bernadou, trans. Annapolis, MD: Naval Institute Press, 1990.

Malkesian, Carter. *War Comes to Garmsar: Thirty Years of Conflict on the Afghan Frontier*. Oxford: Oxford University Press, 2013.

Malone, Patrick. *Skulking Way of War: Technology and Tactics Among the New England Indians*. Lanham, MD: Madison, 1991.

Manela, Erez and Heather Streets-Salter, eds. *The Anticolonial Transnational: Imaginaries, Mobilities, and Networks in the Struggle against Empire*. Cambridge: Cambridge University Press, 2023.

Marcella, Gabriella, ed. *Affairs of State: The Interagency and National Security*. Carlisle Barracks, PA: Army War College Press, 2008.

Marquis, Susan. *Unconventional Warfare: Rebuilding U.S. Special Operations Forces*. Washington, DC: Brookings Institution Press, 1997.

Marsh, Christopher, James Kiras, and Patricia Blocksome, eds. *Special Operations: Out of the Shadows*. Boulder, CO: Lynne Rienner, 2020.

Martin, Mike. *An Intimate War: An Oral History of the Helmand Conflict, 1978–2012*. Oxford: Oxford University Press, 2014.

Mazarr, Michael. *Rethinking Risk in National Security: Lessons of the Financial Crisis for Risk Management*. New York: Palgrave, 2016.

Mazzetti, Mark. *The Way of the Knife: The CIA, a Secret Army, and a War at the Ends of the Earth*. New York: Penguin, 2013.

McCallion, Harry. *Undercover War: Britain's Special Forces and their Secret Battle against the IRA*. London: John Blake, 2020.

McChrystal, Stanley and Anna Burtico. *Risk: A User's Guide*. New York: Portfolio Penguin, 2021.

McChrystal, Stanley and Anna Burtico. *Team of Teams: New Rules of Engagement for a Complex World*. New York: Portfolio, 2015.

McFate, Montgomery. *Military Anthropology: Soldiers, Scholars and Subjects at the Margins of Empire*. New York: Oxford University Press, 2018.

McFate, Sean. *The New Rules of War: Victory in the Age of Durable Disorder*. New York: HarperCollins, 2019.

McGladdery, Gary, *The Provisional IRA in England: The Bombing Campaign, 1973–1997*. Dublin: Irish Academic Press, 2006.

McRaven, William. *Spec Ops, Case Studies in Special Operations Warfare: Theory and Practice*. Novato, CA: Presidio Press, 1995.

Megargee, Geoffrey. *Inside Hitler's High Command*. Lawrence, KS: University Press of Kansas, 2000.

Merom, Gil. *How Democracies Lose Small Wars: State, Society, and the Failures of France in Algeria, Israel in Lebanon, and the United States in Vietnam*. Cambridge: Cambridge University Press, 2003.

Merrick, K.A. *Flights of the Forgotten: Special Duties Operations in World War Two*. London: Arms and Armour Press, 1989.

Middlebrook, Martin. *The Fight for the Malvinas: The Argentine Forces in the Falklands War*. London: Penguin, 1989.

Mockler, Anthony. *The New Mercenaries: The History of the Mercenary from the Congo to the Seychelles*. London: Guild Publishing, 1985.

Moloney, Ed. *A Secret History of the IRA*. New York: W.W. Norton & Company, 2002.

Moorcraft, Paul and Peter McLaughlin. *The Rhodesian War: A Military History*. Mechanicsburg, PA: Stackpole Books, 2010.

Morgan, Forrest. *Flight Risk: The Coalition's Air Advisory Mission in Afghanistan, 2005–2015*. Annapolis, MD: Naval Institute Press, 2018.

Mortimer, Gavin. *David Stirling: The Phoney Major*. London: Constable, 2022.

Moyar, Mark. *To Oppose Any Foe: The Rise of America's Special Operations Forces*. New York: Basic Books, 2017.

Mueller, Karl, ed. *Precision and Purpose: Airpower in the Libyan Civil War*. Report RR676. Santa Monica, CA: RAND, 2015.

Muenger, Elizabeth. *The British Military Dilemma in Ireland: Occupation Politics, 1886–1914*. Lawrence, KS: University Press of Kansas, 1991.

Murray, Raymond. *The SAS in Ireland*. Dublin: Mercier Press, 1990.

Nagl, John. *Learning to Eat Soup with a Knife: Counterinsurgency Lessons from Malaya and Vietnam*. Chicago, IL: University of Chicago Press, 2005.

Naylor, Sean. *Relentless Strike: The Secret History of Joint Special Operations Command*. New York: St Martin's Press, 2015.

Naylor, Sean. *Not a Good Day to Die: The Untold Story of Operation Anaconda.* New York: Berkeley, 2005.

Neilsen, Suzanne and Don Snider, eds. *American Civil–Military Relations: The Soldier and the State in a New Era.* Baltimore, MD: Johns Hopkins University Press, 2009.

Neumann, Peter. *Britain's Long War: British Strategy in the Northern Ireland Conflict, 1969–1998.* New York: Palgrave Macmillan, 2003.

Newsinger, John. *Dangerous Men: The SAS in Popular Culture.* London: Pluto, 1997.

Nicolle, David. *Crusader Warfare, Volume II: Muslims, Mongols, and the Struggle against the Crusades.* London: Hambleton Continuum, 2007.

Nolan, Cathal. *The Allure of Battle: A History of How Wars Have Been Won and Lost.* Oxford: Oxford University Press, 2017.

Nortje, Piet. *32 Battalion: The Inside Story of South Africa's Elite Fighting Unit.* Cape Town: Zebra Press, 2004.

Norval, Magnus. *Inside the ANC: The Evolution of a Terrorist Organization.* Washington, DC: Selous Foundation Press, 1990.

Norval, Magnus. *Death in the Desert: The Namibian Tragedy.* Washington, DC: Selous Foundation Press, 1989.

O'Brien, Kevin. *The South African Intelligence Services: From Apartheid to Democracy, 1948–2005.* London: Routledge, 2010.

O'Brien, Phillips Payson. *How the War Was Won: Air-Sea Power and Allied Victory in World War II.* Cambridge: Cambridge University Press, 2015.

Olsen, John Andreas, ed. *Air Commanders.* Dulles, VA: Potomac, 2013.

Oman, Charles. *The History of the Art of War in the Sixteenth Century.* London: Greenhill Books, 1991.

Oppenheimer, A.R. *IRA Bullets and Bombs: A History of Deadly Ingenuity.* Dublin: Irish Academic Press, 2009.

Osgood, Robert E. *Limited War: The Challenge to American Strategy.* Chicago, IL: University of Chicago Press, 1957.

Osinga, Frans. *Science, Strategy and War: The Strategic Theory of John Boyd.* New York: Routledge, 2007.

Ostovar, Afshon. *Vanguard of the Imam: Religion, Politics, and Iran's Revolutionary Guards.* Oxford: Oxford University Press, 2016.

Paddock Jr, Alfred. *U.S. Army Special Warfare: Its Origins*, rev. ed. Lawrence, KS: University Press of Kansas, 2002.

Paret, Peter, Gordon Craig, and Felix Gilbert, eds. *Makers of Modern Strategy: From Machiavelli to the Nuclear Age.* Princeton, NJ: Princeton University Press, 1986.

Parker, John. *Death of a Hero: Captain Robert Nairac, GC, and the Undercover War in Northern Ireland.* London: Metro Books, 1999.

Parkinson, Alan. *1972 and the Ulster Troubles.* Dublin: Four Courts Press, 2010.

Parnell, Ben. *Air Commandos: The Saga of the Carpetbaggers of World War II.* New York: iBooks, 2004.

Patel, Ketan. *The Master Strategist: Power, Purpose, and Principle.* New York: Hutchinson, 2005.

Payne, Keith. *The Fallacies of Cold War Deterrence and a New Direction.* Lexington, KY: University Press of Kentucky, 2001.

Pedrozo, Raul, ed. *The War in Iraq: A Legal Analysis.* International Law Studies, vol. 86. Newport, RI: Naval War College Press, 2010.

Peritz, Aki and Eric Rosenbach. *Find, Fix, Finish: Inside the Counterterrorism Campaigns That Killed Bin Laden and Devastated Al-Qaeda.* New York: PublicAffairs, 2012.

Perkins, Roger. *Operation Paraquat: The Battle for South Georgia.* Chippenham: Picton, 1986.

Perlmutter, Amos. *The Military and Politics in Modern Times: On Professionals, Praetorians, and Revolutionary Soldiers.* New Haven, CT: Yale University Press, 1977.

Peterson, J.F. *Oman's Insurgencies: The Sultanate's Struggle for Supremacy.* London: Saqi, 2007.

Petit, Brian. *Going Big by Getting Small: The Application of Operational Art by Special Operations in Phase Zero.* Parker, CO: Outskirts Press, 2013.

Phillips, Scott. *Police Militarization: Understanding the Perspectives of Police Chiefs, Administrators, and Tactical Officers.* London: Routledge, 2018.

Phillips, Thomas, ed. *Roots of Strategy: A Collection of Military Classics.* London: John Lane, the Bodley Head, 1943.

Pocock, Tom. *Fighting General: The Public and Private Campaigns of General Sir Walter Walker.* London: Collins, 1973.

Pokrant, Marvin. *New Principles of War: Enduring Truths and Timeless Examples.* Lincoln, NE: Potomac Books, 2021.

Porter, Patrick. *Military Orientalism: Eastern War through Western Eyes.* New York: Columbia University Press, 2009.

Posen, Barry. *The Sources of Military Doctrine.* Ithaca, NY: Cornell University Press, 1984.

Pottinger, Brian. *The Imperial Presidency: P.W. Botha in the First 10 Years.* Johannesburg, South Africa: Southern Book Publishers [Pty] Ltd., 1988.

Price, Neil. *Children of Ash and Elm: A History of the Vikings.* New York: Basic Books, 2020.

Priest, Dana. *The Mission: Waging War and Keeping Peace with America's Military.* New York: W.W. Norton & Co., 2003.

Ramsay, Jack. *SAS: The Soldier's Story.* London: Macmillan, 1996.

Rasmussen, Mikkel. *The Risk Society at War: Terror, Technology and Strategy in the Twenty-First Century.* Cambridge: Cambridge University Press, 2006.

Ricks, Thomas. *The Gamble: General David Petraeus and the American Military Adventure in Iraq, 2006–2008.* New York: Penguin, 2009.

Ricks, Thomas. *Fiasco: The American Military Adventure in Iraq.* New York: Penguin, 2006.

Rid, Thomas. *Active Measures: The Secret History of Disinformation and Political Warfare.* London: Farrar, Straus, & Giroux, 2020.

Robinson, Linda. *One Hundred Victories: Special Ops and the Future of American Warfare.* New York: PublicAffairs, 2013.

Robinson, Linda. *Masters of Chaos: The Secret History of the Special Forces.* New York: PublicAffairs, 2004.

Robinson, Stephen. *The Blind Strategist: John Boyd and the American Art of War.* Dunedin, New Zealand: Exisle Publishing Pty Ltd, 2021.

Rothstein, Hy. *Afghanistan and the Troubled Future of Unconventional Warfare.* Annapolis, MD: Naval Institute Press, 2006.

Runciman, Steven. *A History of the Crusades, Volume II: The Kingdom of Jerusalem and the Frankish East, 1100–1187.* London: Folio Society, 1994.

Russell, Frank. *Information Gathering in Classical Greece.* Ann Arbor, MI: University of Michigan Press, 2000.

Ryan, Mike. *Special Operations in Iraq.* London: Pen & Sword, 2004.

Rynning, Sten. *NATO in Afghanistan: The Liberal Disconnect.* Stanford, CA: Stanford University Press, 2012.

Sacquety, Troy. *The OSS in Burma: Jungle War against the Japanese.* Lawrence, KS: Kansas University Press, 2013.

Sanger, David. *Confront and Conceal: Obama's Secret Wars and Surprising Use of American Power.* New York: Crown, 2012.

Sawyer, Ralph. *The Seven Military Classics of Ancient China, Including the Art of War.* Boulder, CO: Westview Press, 1993.

Scahill, Jeremy. *Dirty Wars: The World Is a Battlefield.* New York: Nation Books, 2013.

Scarborough, Rowan. *Rumsfeld's War: The Untold Story of America's Anti-Terrorist Commander.* Washington, DC: Regnery, 2004.

Schelling, Thomas. *Arms and Influence.* New Haven, CT: Yale University Press, 1966.

Schmid, Alex, ed. *The Routledge Handbook of Terrorism Research.* London: Routledge, 2011.

Schmidt, Michael. *Death Flight: Apartheid's Secret Doctrine of Disappearance.* Cape Town, South Africa: Tafleberg, 2020.

Schmitt, Erik and Thomas Shankar. *Counterstrike: The Untold Story of America's Secret Campaign against Al Qaeda.* New York: Times Books, 2011.

Schmitt, Michael, ed. *The War in Afghanistan: A Legal Analysis.* International Law Studies, vol. 85. Newport, RI: Naval War College Press, 2009.

Scholtz, Leopold. *The SADF in the Border War, 1966–1989.* Solihull, UK: Helion, 2015.

Serena, Chad. *It Takes More Than a Network: The Iraqi Insurgency and Organizational Adaptation.* Stanford, CA: Stanford University Press, 2014.

Seymour, William. *British Special Forces.* London: Sidgwick & Jackson, 1985.

Shapiro, Jacob. *The Terrorist's Dilemma: Managing Violent Covert Organizations.* Princeton, NJ: Princeton University Press, 2013.

Sheldon, Rose Mary. *Intelligence Activities in Ancient Rome: Trust the Gods, but Verify.* London: Routledge, 2005.

Shultz Jr, Richard. *Transforming U.S. Intelligence for Irregular War: Task Force 714 in Iraq.* Washington, DC: Georgetown University Press, 2020.

Shultz Jr, Richard. *The Secret War Against Hanoi: Kennedy's and Johnson's Use of Spies, Saboteurs, and Covert Warriors in North Vietnam.* New York: HarperCollins, 1999.

Shultz Jr, Richard et al., eds. *Special Operations Forces: Roles and Missions in the Aftermath of the Cold War.* Washington, DC: Department of Defense, 1995.

Siegel, Pascale. *Target Bosnia: Integrating Information Activities in Peace Operations.* Washington, DC: National Defense University Press, 1998.

Simons, Anna. *The Company They Keep: Life Inside the U.S. Army Special Forces.* New York: Free Press, 1997.

Simpson, Emile. *War from the Ground Up: Twenty-First Century Combat as Politics.* Oxford: Oxford University Press, 2013.

Simpson III, Charles. *Inside the Green Berets, the First Thirty Years: A History of U.S. Army Special Forces.* Novato, CA: Presidio Press, 1983.

Sircar, Kaushik. *Guns and a Butterfly: The Black Cats' Strike—Operation Ashwamedh.* Mumbai: Himalaya Books Pty Ltd, 2009.

Slepyan, Kenneth. *Stalin's Guerrillas: Soviet Partisans in World War II.* Lawrence, KS: University Press of Kansas, 2006.

Smith, Michael. *Killer Elite: The Inside Story of America's Most Secret Special Operations Team,* rev. ed. New York: St Martin's Griffin, 2008.

Smith, M.L.R. *Fighting for Ireland? The Military Strategy of the Irish Republican Movement.* London: Routledge, 1995.

Søderland, Arne and Douw Steyn. *Iron Fist from the Sea: Top Secret Seaborne Recce Operations (1978–1988).* Johannesburg, South Africa: Delta Books, 2018.

Soule, Allan, Gary Dixon, and René Richards. *The Wynand du Toit Story.* Johannesburg, South Africa: Hans Strydom Publishers, 1987.

Southby-Tailyour, Ewen. *Exocet Falklands: The Untold Story of Special Forces Operations.* Barnsley: Pen & Sword, 2014.

Southby-Tailyour, Ewen, ed. *Jane's Amphibious and Special Forces,* issue 17. Coulsdon, Surrey: Jane's Information Group, January 2007.

Southby-Tailyour, Ewen. *Blondie: A Biography of Lieutenant Colonel H.G. Hasler, DSO, OBE, Croix de Guerre, Royal Marines.* London: Leo Cooper, 1998.

Spence, Cameron. *Sabre Squadron*. London: Michael Joseph, 1997.
Spencer, David. *From Vietnam to El Salvador: The Saga of FMLN Sappers and Other Guerrilla Special Forces in Latin America*. Westport, CT: Praeger, 1996.
Spencer, Emily, ed. *Special Operations Forces: A National Capability*. Kingston, ON: Canadian Defence Academy Press, 2011.
Spulak Jr, Robert. *A Theory of Special Operations: The Origins, Qualities, and Use of SOF*. Hurlburt Field, FL: Joint Special Operations University Press, 2007.
Stadtmüller, Georg, ed. *Ungarn-Jahrbuch: Zeitschrift für die Kunde Ungarns und verwandte Gebiete*. Band 21. Jahrgagn 1993–1994. Munich: Verlag Ungarisches Institut, 1995.
Staniland, Paul. *Networks of Rebellion: Explaining Insurgent Cohesion and Collapse*. Ithaca, NY: Cornell University Press, 2014.
Stanton, Shelby. *Green Berets at War: U.S. Army Special Forces in Southeast Asia, 1956–1975*. Novato, CA: Presidio Press, 1985.
Stein, Jeff. *A Murder in Wartime: The Untold Spy Story That Changed the Course of the Vietnam War*. New York: St Martin's Press, 1992.
Stentiford, Barry. *Success in the Shadows: Operation Enduring Freedom—Philippines and the Global War on Terror, 2002–2015*. Fort Leavenworth, KS: Combat Studies Institute, 2018.
Stevenson, Jonathan. *Losing Mogadishu: Testing U.S. Policy in Somalia*. Annapolis, MD: Naval Institute Press, 1995.
Stiff, Peter. *The Covert War: Koevoet, Namibia, 1979–1989*. Alberton, South Africa: Galago, 2004.
Stiff, Peter. *Warfare by Other Means: South Africa in the 1980s and 1990s*. Alberton, South Africa: Galago, 2001.
Stiff, Peter. *The Silent War: South African Recce Operations, 1969–1994*. Alberton, South Africa: Galago, 1999.
Stiff, Peter. *I'll See You in November: Rhodesia's No-Holds-Barred Intelligence War*. Alberton, South Africa: Galago, 1985.
Stoker, Donald. *Why America Loses Wars: Limited War and US Strategy from the Korean War to the Present*. Cambridge: Cambridge University Press, 2019.
Strachan, Alexsander. *1 Recce*, 3 volumes. Cape Town: Tafelburg, 2018.
Strachan, Hew. *The Politics of the British Army*. Oxford: Clarendon Press, 1997.
Strawson, John. *A History of the S.A.S. Regiment*. London: Guild Publishing, 1985.
Strenkhnen, Iurii. *Commandos from the Sea: Soviet Naval Spetsnaz in World War II*, James Gebhardt, trans. Annapolis, MD: Naval Institute Press, 1996.
Subramaniam, Arjun. *A Military History of India since 1972: Full Spectrum Operations and the Changing Contours of Modern Conflict*. Lawrence, KS: University Press of Kansas, 2021.
Summers Jr, Harry. *On Strategy: A Critical Analysis of the War*. Novato, CA: Presidio Press, 1982.
Taylor, F.L. *The Art of War in Italy, 1494–1529*. Cambridge: Cambridge University Press, 1921.
Taylor, Peter. *Brits: The War against the IRA*. London: Bloomsbury, 2001.
Thomas, Martin. *Empires of Intelligence: Security Services and Colonial Disorder after 1914*. Berkeley, CA: University of California Press, 2008.
Thompson, J.H. *An Unpopular War: From Afkak to Bosbefok*. Cape Town, South Africa: Zebra Press, 2006.
Tierney, Jr., John. *Chasing Ghosts: Unconventional Warfare in American History*. Washington, DC: Potomac Books Inc., 2006.
Todman, Daniel. *Britain's War: A New World, 1942–1947*. Oxford: Oxford University Press, 2020.
Tomczak, Joseph. *Parallel Lives in the Indo-Pacific: Edward Lansdale, Donald Wurster, and the Irregular Warfare Mind-set*. Kenney Paper No. 5. Maxwell Air Force Base, AL: Air University Press, 2023.

Tophoven, Rolf. *GSG 9: German Response to Terrorism.* Koblenz: Bernard & Graefe Verlag, 1985.
Tourison, Sedgwick. *Secret Army, Secret War: Washington's Tragic Special Operation in North Vietnam.* Annapolis, MD: Naval Institute Press, 1995.
Towle, Philip. *Pilots and Rebels: The Use of Aircraft in Unconventional Warfare, 1918–1988.* London: Brassey's, 1989.
Tucker, David and Christopher Lamb. *United States Special Operations Forces*, 2nd rev. ed. New York: Columbia University Press, 2020.
Tucker, David and Christopher Lamb. *United States Special Operations Forces.* New York: Columbia University Press, 2007.
Tucker, Mike and Charles Faddis. *Operation Hotel California: The Clandestine War Inside Iraq.* Guilford, CT: The Lyons Press, 2009.
Turner, John. *Continent Ablaze: The Insurgency War in Africa, 1960 to the Present.* London: Arms and Armour Press, 1998.
Turnley, Jessica, Kobi Michael, and Eyal Ben-Ari, eds. *Special Operations Forces in the 21st Century: Perspectives from the Social Sciences.* London: Routledge, 2018.
Turse, Nick. *The Changing Face of Empire: Special Ops, Drones, Spies.* Chicago, IL: Dispatch Books, 2012.
Tyerman, Christopher. *God's War: A New History of the Crusades.* Cambridge, MA: Belknap Press, 2006.
Ullman, Harlan and James Wade Jr. *Shock and Awe: Achieving Rapid Dominance.* Washington, DC: National Defense University Press, November 1996.
Unnithan, Sandeep. *Black Tornado: The Three Sieges of Mumbai, 26/11.* Noida: HarperCollins India, 2020.
Urban, Mark. *Task Force Black: The Explosive True Story of the Secret Special Forces War in Iraq.* London: St Martin's Griffin, 2012.
Urban, Mark. *UK Eyes Alpha: Inside British Intelligence.* London: Faber & Faber, 1997.
Urban, Mark. *Big Boys' Rules: The Secret Struggle against the IRA.* London: BCA, 1992.
van Creveld, Martin. *Command in War.* Cambridge, MA: Harvard University Press, 1985.
Vandenbroucke, Lucien. *Perilous Options: Special Operations as an Instrument of U.S. Foreign Policy.* New York: Oxford University Press, 1993.
van der Waag, Ian. *A Military History of Modern South Africa.* Johannesburg: Jonathan Cape, 2015.
Van Wagner, R.D. *Any Place, Any Time, Any Where: The 1st Air Commandos in World War II.* Atglen, PA: Schiffer Military, 1998.
van Wees, Hans. *Greek Warfare: Myths and Realities.* London: Duckworth, 2004.
Venter, Al. *Vorster's Africa: Friendship and Frustration.* Johannesburg, South Africa: Ernest Stanton Publishers, 1977.
Vertzberger, Yaakob. *Risk Taking and Decisionmaking: Foreign Military Intervention Decisions.* Stanford, CA: Stanford University Press, 1998.
Waldman, Thomas. *War, Clausewitz, and the Trinity.* Burlington, VT: Ashgate, 2013.
Wallace, Chris with Mitch Weiss. *Countdown Bin Laden: The Untold Story of the 247-Day Hunt to Bring the Mastermind of 9/11 to Justice.* New York: Avid Reader Press, 2021.
Walt, Stephen. *The Hell of Good Intentions: America's Foreign Policy Elite and the Decline of U.S. Primacy.* New York: Farrar, Straus and Giroux, 2018.
Walton, Calder. *Empire of Secrets: British Intelligence, the Cold War, and the Twilight of Empire.* New York: Overlook Press, 2013.
Waltz, Michael. *Warrior Diplomat: A Green Beret's Battles from Washington to Afghanistan.* Lincoln, NE: Potomac Press, 2014.
Warner, Philip. *The Special Air Service.* London: William Kimber, 1972.
Warrick, Joby. *Black Flags: The Rise of ISIS.* New York: Doubleday, 2015.

Wasserman, James. *The Templars and Assassins: The Militia of Heaven*. New York: Simon & Schuster, 2001.

Weale, Adrian. *Secret Warfare: Special Operations Forces from the Great Game to the SAS*. London: Coronet Books, 1998.

Weiss, Mitch and Kevin Maurer. *Hunting Che: How a U.S. Special Forces Team Helped Capture the World's Most Famous Revolutionary*. New York: Berkeley, 2013.

West, Nigel. *The Secret War for the Falklands: The SAS, MI6, and the War Whitehall Nearly Lost*. London: Little, Brown & Co., 1997.

Westad, Odd Arne. *The Global Cold War: Third World Interventions and the Making of Our Times*. Cambridge: Cambridge University Press, 2007.

Wheeler, Everett. *Stratagem and the Vocabulary of Military Trickery*. Leiden: E.J. Brill, 1988.

White, Rowland. *Storm Front*. London: Corgi Books, 2012.

Winton, Harold. *Corps Commanders of the Bulge: Six American Generals and Victory in the Ardennes*. Lawrence, KS: University Press of Kansas, 2007.

Woodward, Bob. *Obama's Wars*. New York: Simon & Schuster, 2010.

Woodward, Bob. *Bush at War*. New York: Simon & Schuster, 2002.

Wylie, J.C. *Military Strategy: A General Theory of Power Control*. Annapolis, MD: Naval Institute Press, 1989.

Yuen, Derek. *Deciphering Sun Tzu: How to Read 'The Art of War'*. Oxford: Oxford University Press, 2014.

Zeiger, Henry. *The Seizing of the Santa Maria*. New York: Popular Library, 1961.

Zweibelson, Ben. *Understanding the Military Design Movement: War, Change, and Innovation*. London: Routledge, 2023.

## Monographs and thinktank reports

Alexander, John. *Convergence: Special Operations Forces and Law Enforcement*. JSOU Report 10-6. MacDill Air Force Base, FL: July 2010.

Bazin, Aaron, ed. *On Competition: Adapting to the Contemporary Strategic Environment*. JSOU Report 21-5. MacDill Air Force Base, FL: JSOU Press, 2021.

Bogart III, Adrian. *One Valley at a Time*. JSOU Report 06-6. Hurlburt Field, FL: JSOU Press, 2006.

Bos, Nathan, ed. *Human Factors Considerations of Undergrounds in Insurgencies*, 2nd ed. Fort Bragg, NC: United States Army Special Operations Command, 2013.

Bozkurt, Hamza et al. *Accountability and Oversight of UK Special Forces*. Colchester, Essex, UK: Human Rights Centre Clinic, June 2023.

Cancian, Mark. *U.S. Military Forces in FY2019: The Buildup and Its Limits*. Washington, DC: Center for Strategic and International Studies, 2018.

Celeski, Joseph, Timothy Slemp, and John Jogerst. 'The Evolution of Special Operations Power: Purpose, Application, and Theory'. Unpublished draft report. Joint Special Operations University, December 2012.

Celeski, Joseph. *Hunter-Killer Teams: Attacking Enemy Safe Havens*. JSOU Report 10-1. Hurlburt Field, FL: JSOU Press, 2010.

Charters, David et al. 'Surgical Operations: Military Lessons from Selected Case Studies'. Unpublished report. The Centre for Conflict Studies, University of New Brunswick, Fall 1982.

Crawford, George. *Manhunting: Counter-Network Organization for Irregular Warfare*. JSOU Report 09-7. Hurlburt Field, FL: JSOU Press, 2009.

Dalton, Melissa et al. *Shifting the Burden Responsibly: Oversight and Accountability in U.S. Security Sector Assistance*. Washington, DC: Center for Strategic and International Studies, 2019.

Ebright, Katherine Yon. *Secret War: How the U.S. Uses Partnerships and Proxy Forces to Wage War Under the Radar*. New York: Brennan Center for Justice, 2022.

Engstron, Jeffrey. *Systems Confrontation and System Destruction Warfare: How the Chinese People's Liberation Army Seeks to Wage Modern Warfare*, RAND Report RR1708. Santa Monica, CA: RAND Corporation, 2018.

Fiala, Otto. *Resistance Operating Concept (ROC)*. MacDill Air Force Base, FL: JSOU Press, 2020.

Gold, Zack. *Al-Qaeda in Iraq AQI: An Al-Qaeda Affiliate Case Study*. Report DIM-2017-U-016118-2Rev. Alexandria, VA: Center for Naval Analyses, October 2017.

Gray, Colin S. *Tactical Operations for Strategic Effect: The Challenge of Currency Conversion*. JSOU Special Report. MacDill Air Force Base, FL: JSOU Press, 2015.

Gray, Colin S. 'Special Operations and American Culture'. Unpublished report. MDA 903-91-C-0030. National Institute for Public Policy, December 1992.

Gray, Colin S. '"Doorkickers" and "Well-Diggers": Resolving the Tensions Between SOF Roles'. Unpublished report. MDA 902-91-C-0030. National Institute for Public Policy, October 1992.

Gray, Colin S. 'Special Operations: What Succeeds and Why? Lessons of Experience, Phase I'. Unpublished report. National Institute for Public Policy, 1992.

Haddick, Robert. *How Do SOF Contribute to Comprehensive Deterrence?* JSOU Report 17-11, MacDill Air Force Base, FL: JSOU Press, 2017.

Horn, Bernd, Ed. *Risk: SOF Case Studies*. Ottawa, ON: Canadian Special Operations Forces Command, 2020.

Horn, Bernd, Ed. *Risk and Decision-Making*, Ottawa, ON: Canadian Special Operations Forces Command, 2019.

Horn, Bernd. *'We Will Find a Way': Understanding the Legacy of Canadian Special Operations Forces*. JSOU Report 12-2, MacDill Air Force Base, FL: JSOU Press, 2012.

Horn, Bernd, James D. Kiras, and Emily Spencer, eds. *The (In)Visible Hand: Strategic Sabotage*. Ottawa, ON: Canadian Special Operations Forces Command, 2021.

International Human Rights and Conflict Resolution Clinic and Global Justice Clinic. *Living under Drones: Death, Injury, and Trauma to Civilians from U.S. Drone Practices in Pakistan*. Stanford, CA: Stanford Law School, 2012.

International Institute for Strategic Studies. *The Military Balance 2023*. London: Routledge, 2023.

Irwin, Will and Isiah 'Ike' Wilson III. The Fourth Age of SOF: The Use and Utility of Special Operations Forces in a New Age. JSOU Report 22-1. MacDill Air Force Base, FL: JSOU Press, 2022.

Irwin, Will. *Support to Resistance: Strategic Purpose and Effectiveness*. JSOU Report 19-2. MacDill Air Force Base, FL: JSOU Press, 2019.

Jacobs, Jody et al. *Enhancing the Integration of Special Operations and Conventional Air Operations: Focus on the Air-Surface Interface*. U//FOUO. MG-535-AF. Santa Monica, CA: RAND, July 2007.

Kaaman, Hugo. *Car Bombs as Weapons of War: ISIS's Development of SVBIEDs, 2014–2019*. Policy Paper 2019-7. Washington, DC: The Middle East Institute, 2019.

Kagan, Frederick et al. *Defining Success in Afghanistan*. Washington, DC: American Enterprise Institute, 2011.

Knarr, William and Mark Nutsch. *Village Stability Operations and the Evolution of SOF Command and Control in Afghanistan: Implications for the Future of Irregular Warfare*. JSOU Report 20-2. MacDill Air Force Base, FL: JSOU Press, 2020.

Komer, Robert. *Bureaucracy Does Its Thing: Institutional Constraints on U.S.-GVN Performance in Vietnam*. Report R-967-ARPA. Santa Monica, CA: RAND Corporation, 1927.

Koven, Barnett and Katy Lundquist. *Barriers to Special Operations Forces-Led Counterterrorism Effectiveness.* JSOU Report 21-7. MacDill Air Force Base, FL: JSOU Press, 2021.

Lambakis, Steven, James Kiras, and Kris Kolet. *Understanding 'Asymmetric' Threats to the United States.* NIPP Report. Fairfax, VA: National Institute for Public Policy, October 2002.

Loredo, Elvira et al. *Authorities and Options for Funding USSOCOM Operations.* Report RR-360-SOCOM. Santa Monica, CA: RAND Corporation, 2014.

Luttwak, Edward, Steven Canby, and David Thomas. 'A Systematic Review of "Commando", Special, Operations, 1939–1980, Part I'. Unpublished report. C & L Associates, 24 May 1982.

Mahnken, Thomas. *Secrecy and Stratagem: Understanding Chinese Strategic Culture.* Sydney: Lowy Institute for International Policy, 2011.

Malvesti, Michele. *Policy Considerations in Combating Terrorism: Decision-Making under Conditions of Risk and Uncertainty.* West Point, NY: Combating Terrorism Center, January 2012.

Martini, Jeffrey et al. *Operation Inherent Resolve: U.S. Ground Force Contributions.* Report RRA719-1. Santa Monica, CA: RAND Corporation, 2022.

McCabe, Peter and Paul Lieber, eds. *Special Operations Theory*, JSOU Report 17-6. MacDill Air Force Base, FL: JSOU Press, 2017.

Mulroy, Mick, Andrew 'L', and Ken Tovo. *Irregular Warfare: A Case Study in the CIA and US Army Special Forces Operations in Northern Iraq, 2002–03.* Washington, DC: Middle East Institute, August 2021.

Paul, Christopher, Isaac Porche, and Elliot Axelband. *The Other Quiet Professionals: Lessons for Future Cyber Forces from the Evolution of Special Forces.* Report RR-780-A. Santa Monica, CA: RAND Corporation, 2014.

Rabasa, Angela et al. *The Lessons of Mumbai.* Occasional Paper OP249. Santa Monica, CA: The RAND Corporation, 2009.

Rabasa, Angela et al. *Beyond al-Qaeda, Parts 1 and 2: The Global Jihadist Movement and The Outer Rings of the Terrorist Universe.* Santa Monica, CA: RAND Corporation, 2005, 2006.

Ricks, Chuck, ed. *The Role of the Global SOF Network in a Resource Constrained Environment.* MacDill Air Force Base, FL: JSOU Press, November 2013.

Robinson, Linda, Daniel Egel, and Ryan Andrew Brown. *Measuring the Effectiveness of Special Operations.* Report RR2504. Santa Monica, CA: RAND Corporation, 2019.

Robinson, Linda, Daniel Egel, and Ryan Andrew Brown. *The Future of U.S. Special Operations Forces.* Council Special Report No. 66. New York: Council on Foreign Relations, April 2013.

Robinson, Linda, Daniel Egel, Ryan Andrew Brown, Patrick Johnston, and Gillian Oak. *U.S. Special Operations Forces in the Philippines, 2001–2014.* Report RR 1236. Santa Monica, CA: RAND Corporation, 2016.

Rosenau, William and Austin Long. *The Phoenix Program and Contemporary Counterinsurgency.* Occasional Paper. Santa Monica, CA: RAND Corporation, 2009.

Rubright, Richard. *A Unified Theory for Special Operations.* JSOU Report 17-1. MacDill Air Force Base, FL: JSOU Press, 2017.

Searle, Tom. *Outside the Box: A New General Theory of Special Operations.* JSOU Report 17-4. MacDill Air Force Base, FL: JSOU Press, 2017.

Shashank, Joshi. *Indian Power Projection: Ambitions, Arms, and Influence.* Whitehall Papers, 85:1, 2015.

Shultz Jr, Richard. *Military Innovation in War: It Takes a Learning Organization—A Case Study of Task Force 714 in Iraq.* JSOU Report 16-6. MacDill Air Force Base, FL: JSOU Press, 2016.

Trautman, Konrad. *Special Operations Intelligence: Guiding the Tip of the Spear.* JSOU Special Report. U//FOUO//REL to USA, FVEY. MacDill Air Force Base FL: JSOU Press, November 2014.

Turbiville Jr, Graham. *Hunting Leadership Targets in Counterinsurgency and Counterterrorist Operations: Selected Perspectives and Experience.* JSOU Report 07-06. Hurlburt Field, FL: JSOU Press, 2007.

Vick, Alan et al. *Air Power in the New Counterinsurgency Era: The Strategic Importance of USAF Advisory and Assistance Missions.* RAND Report MG-509. Santa Monica, CA: RAND Corporation, 2006.

Wehrey, Frederic et al. *The Rise of the Pasdaran: Assessing the Domestic Roles of Iran's Islamic Revolutionary Guard Corps.* MG–821–OSD. Santa Monica, CA: RAND, 2009.

Wood, Dakota, ed. *2023 Index of U.S. Military Strength.* Washington, DC: The Heritage Foundation, 2023.

Yarger, Harry. *21st Century SOF: Toward an American Theory of Special Operations.* JSOU Report 13-1. MacDill Air Force Base, FL: JSOU Press, April 2013.

## Articles

Abrahamian, Ervand. 'The 1953 Coup in Iran'. *Science & Society*. 65:2. Summer 2001: 182–215.

Acciarini, Chiara, Federica Brunet, and Paolo Boccardeill. 'Cognitive Biases and Decision-Making Strategies in Times of Change: A Systematic Literature Review'. *Management Decision*. 59:3. 2021: 638–652.

Åkebo, Malin. '"Coexistence Ceasefire" in Mindanao'. *Peace & Change: A Journal of Peace Research*. 44:4. October 2019: 468–496.

Andres, Richard. 'The Afghan Model in Northern Iraq'. *Journal of Strategic Studies*. 29:3. 2006: 395–422.

Anglim, Simon. 'Orde Wingate and the Theory behind the Chindit Operations: Some Recent Findings'. *The RUSI Journal*. 147:2. 2002: 92–98.

Ansbacher, Yair and Ron Schleifer. 'The Three Ages of Modern Western Special Operations Forces'. *Comparative Strategy*. 41:1. 2022: 32–45.

Arkin, William. 'Nuclear Backpacks'. *Bulletin of the Atomic Scientists*. 41:4. April 1985: 4–5.

Asher, Jeffrey. 'An Unworthy Foe: Heroic Ἔθη, Trickery, and an Insult in Ephesians 6:11'. *Journal of Biblical Literature*. 130:4. Winter 2011: 729–748.

Asklund Johnsen, Anton and Gitte Højstrup Christensen. 'Clarifying the Antisystemic Elements of Special Operations: A Conceptual Inquiry'. *Special Operations Journal*. 2:2. Fall 2016: 106–123.

Bale, Jeffrey. 'South Africa's Project Coast: "Death Squads", Covert State-Sponsored Poisonings, and the Dangers of CBW Proliferation'. *Democracy and Security*. 2:1. January–June 2006: 27–60.

Bamford, Bradley. 'The Role and Effectiveness of Intelligence in Northern Ireland'. *Intelligence and National Security*. 20:4. 2005: 581–607.

Barclay, Glen St J. '"Butcher and Bolt": Admiral Roger Keyes and British Combined Operations, 1940–1941'. *Naval War College Review*. 35:4. July–August 1982: 99–101.

Baudez, Claude-François and Nicolas Latsanopoulos, 'Political Structure, Military Training and Ideology at Chichen Itza'. *Ancient Mesoamerica*. 21:1. Spring 2010: 1–20.

Berger III, Joseph. 'Covert Action: Title 10, Title 50, and the Chain of Command'. *Joint Forces Quarterly*. 67. 4th quarter 2012: 32–39.

Betts, Richard. 'Is Strategy an Illusion?' *International Security*. 25:2. Fall 2000: 5–50.

Biddle, Stephen and Robert Zirkle. 'Technology, Civil–Military Relations, and Warfare in the Developing World'. *Journal of Strategic Studies*. 19:2. June 1996: 171–212.

Bland, Douglas. 'A Unified Theory of Civil–Military Relations'. *Armed Forces and Society*. 26:1. Fall 1999: 7–26.

Bowden, Mark. 'Jihadists in Paradise'. *The Atlantic*. March 2007: 54–75.

Brands, Hal. 'Rethinking America's Grand Strategy: Insights from the Cold War'. *Parameters*. 45:4. Winter 2015: 7–16.

Brennan, Brian. 'Limited vs. Total War'. *Armor*. 11:5 September–October 2002: 8–11.

Bridge, Cyprian. 'Sub-Aqueous Warfare, Ancient and Modern'. *Fraser's Magazine*. 18. October 1878: 458–470.

Brinkerhoff, Thomas. 'Reexamining the Lore of the "Archetypal Conquistador": Hernán Cortés and the Spanish Conquest of the Aztec Empire, 1519–1521'. *The History Teacher*. 49:2. February 2016: 169–187.

Briscoe, Charles. 'Operation ZUMMARI 241: Braving Fire Ambushes in Afghanistan'. *Veritas*. 12:2. October 2016: 51–63.

Briscoe, Charles. 'Welcome to Bolivia, MTT-BL 404-67X'. *Veritas*. 4:4. 2008: 62–67.

Briscoe, Charles. 'El Paraiso and the War in El Salvador, Part I 1981–1983'. *Veritas*. 3:1. 2007.

Brulé, David. 'Explaining and Forecasting Leaders' Decisions: A Poliheuristic Analysis of the Iran Hostage Rescue Decision'. *International Studies Perspectives*. 6:1. February 2005: 99–113.

Burch, Martin and Ian Holliday. 'The Prime Minister's and Cabinet Offices: An Executive Office in All but Name'. *Parliamentary Affairs*. 52:1. January 1999: 32.

Burchall Henningsen, Troels. 'Small State Strategy in Practice: The Formulation and Execution of Danish Theories of Success in the United Nations Security Council, 2005–2006'. *Scandinavian Journal of Military Studies*. 5:1. 2022: 129–142.

Burgos, Russell A. 'Pushing the Easy Button: Special Operations Forces, International Security and the Use of Force'. *Special Operations Journal*. 4:2. 2018: 109–128.

Burton, Brian. 'The Promise and Peril of the Indirect Approach'. *Prism*. 3:1. January 2012: 47–62.

Bury, Patrick. 'U.S. Special Forces Transformation: Post-Fordism and the Limits of Networked Warfare'. *International Affairs*. 98:2. 2022: 587–607.

Caferro, William. 'Warfare and Economy in Renaissance Italy, 1350–1450'. *The Journal of Interdisciplinary History*. 39:2. August 2008: 167–209.

Campbell, Kenneth. 'Otto Skorzeny: The Most Dangerous Man in Europe'. *American Intelligence Journal*. 30:1. 2012: 142–150.

Chace, Jesse. 'Defining Asymmetric Warfare: A Losing Proposition'. *Joint Forces Quarterly*. 61. 2nd Quarter 2011: 115–120.

Charters, David. 'Professionalizing Clandestine Military Intelligence in Northern Ireland: Creating the Special Reconnaissance Unit'. *Intelligence and National Security*. 33:1. 2017: 130–138.

Cherry, Robert. 'The Initiative in War'. *Royal United Services Institution Journal*. 66:461. 1921: 87–100.

Chesney, Robert. 'Military-Intelligence Convergence and the Law of the Title 10/Title 50 Debate'. *Journal of National Security Law and Policy*. 5. 2012: 539–629.

Connelly, M.L. 'The Army, the Press, and the "Curragh Incident", March 1914'. *Historical Research*. 84:225. August 2011: 535–557.

Constable, Trevor. 'Bagnold's Bluff'. *The Journal of Historical Review*. 18:2. March/April 1999.

Cormac, Rory. 'Coordinating Covert Action: The Case of the Yemen Civil War and the South Arabian Insurgency'. *Journal of Strategic Studies*. 36:5. 2013: 692–717.

Crane, Conrad. 'The Lure of Strike'. *Parameters*. 43:2. 2013: 5–12.

Daftary, Farhad. 'The "Order of the Assassins": J. von Hammer and the Orientalist Misrepresentations of the Nizari Ismaelis'. *Iranian Studies*. 39:1. March 2006: 71–81.

Dandecker, Christopher. 'National Security and Democracy: The United Kingdom Experience'. *Armed Forces and Society*. 20:3. Spring 1994: 353–374.

Davies, Huw. 'The Influence of Intelligence on Wellington's Art of Command'. *Intelligence and National Security*. 22:5. October 2007: 619–643.

Davies, Philip. 'From Special Operations to Special Political Action: The "Rump SOE" and SIS Post-War Covert Action Capability, 1945–1977'. *Intelligence and National Security*. 15:3. 2000: 55–76.

Diehl, Glen and Solomon Major. 'MOE vs. M&E: Considering the Difference between Measuring Strategic Effectiveness and Monitoring Tactical Evaluation'. *Military Medicine*. 180:1. January 2015: 77–82.

Doig, Alan et al. 'Marching in Time: Alliance Politics, Synchrony, and the Case for War in Iraq, 2002–2003'. *Australian Journal of International Affairs*. 61:1. 2007: 23–40.

Dunlap Jr, Charles. 'Special Operations Forces after Kosovo'. *Joint Forces Quarterly*. 28. Spring/Summer 2001: 7–11.

Dyson, Stephen L. 'Native Revolts in the Roman Empire'. *Historia: Zeitschrift für Alte Geschichte*. 20:2/3. 2nd Quarter 1971: 239–274.

Eavey, Cheryl. 'Bureaucratic Competition and Agenda Control'. *The Journal of Conflict Resolution*. 31:3. September 1987: 503–524.

Edmunds, Timothy. 'British Civil–Military Relations and the Problem of Risk'. *International Affairs*. 22:2. Summer 2012: 265–282.

Edwards, Aaron. 'Misapplying Lessons Learned? Analysis of British Counterinsurgency Strategy in Northern Ireland, 1971–1976'. *Small Wars & Insurgencies*. 21:2. 2010: 303–330.

Eisenstadt, Michael. 'Tribal Engagement Lessons Learned'. *Military Review*. 88:5. September–October 2007: 16–31.

Ellis, Stephen. 'The Historical Significance of South Africa's Third Force'. *Journal of Southern Africa Studies*. 24:2. June 1998: 261–299.

Euben, Roxanne. 'Spectacles of Sovereignty in Digital Time: ISIS Executions, Visual Rhetoric and Sovereign Power'. *Perspectives on Politics*. 15:4. December 2017: 1007–1033.

Ezov, Amiram. 'The "Missing Dimension" of C. Julius Caesar'. *Zeitschrift für Alte Geschichte*. 45:1. 1996: 66–67.

Fearon, James. 'Rationalist Explanations for War'. *International Organization*. 49:3. Summer 1995: 379–414.

Feaver, Peter. 'The Civil-Military Problematique: Huntington, Janowitz, and the Question of Civilian Control'. *Armed Forces & Society*. 23:2. Winter 1996: 149–178.

Field, Alexander. 'Schelling, von Neumann, and the Event That Didn't Occur'. *Games*. 5:1. 2014: 53–89.

Figueira, Thomas. 'Herodotus on the Early Hostilities between Aegina and Athens'. *The American Journal of Philology*. 106:1. Spring 1985: 72–73.

Finegan, Rory. 'Shadowboxing in the Dark: Intelligence and Counter-Terrorism in Northern Ireland'. *Terrorism and Political Violence*. 28:3. 2016: 479–519.

Finlan, Alastair. 'A Dangerous Pathway? Toward a Theory of Special Forces'. *Comparative Strategy*. 38:4. Summer 2019: 255–275.

Finlan, Alastair. 'The (Arrested) Development of UK Special Forces and the Global War on Terror'. *Review of International Studies*. 35. 2009: 971–982.

Finlayson, Andrew. 'The Tay Ninh Provincial Reconnaissance Unit and Its Role in the Phoenix Program, 1969–1970'. *Studies in Intelligence*. 51:2. June 2007: 59–69.

Finlayson, Kenneth. 'The 2nd Ranger Battalion and the Capture of Che Guevara'. *Veritas*. 4:4. 2008: 94–97.

Finlayson, Kenneth. 'Night Stalkers in the Philippines: Tragedy and Triumph in Balikatan 02-1'. *Veritas*. 2:1. 2006: 54–59.

Finlayson, Kenneth. 'Operation Viking Hammer: 3/10 SFG against the Ansar al-Islam'. *Veritas*. 1:1. 2005: 15–19.

Fogarty, Stephen and Jamie Nasi. 'Special Operations Forces Truth: Cyber Truths'. *The Cyber Defense Review*. 1:2. Fall 2016: 19–28.

Foot, M.R.D. 'Was SOE Any Good?' *Journal of Contemporary History*. 16:1. January 1981: 167–181.

Ford, Matthew. 'Precision Targeting and the Industrialization of the Intelligence Cycle on the Battlefields of Iraq and Afghanistan'. *Inteligencia y seguridad*. 13. January–June 2013: 221–243.

Friis, Simone Molin. '"Beyond Anything We Have Ever Seen": Beheading Videos and the Visibility of Violence in the War against ISIS'. *International Affairs*. 91:4. July 2015: 725–746.

Frost, Frank. 'Scyllias: Diving in Antiquity'. *Greece & Rome*. 15:2. October 1968: 180–185.

Fuller, J.F.C. 'Tanks Won't Do Much'. *The War Weekly*. 3. 10 November 1939.

Gerasimov, Valery. 'The Value of Science Is in the Foresight: New Challenges Demand Rethinking the Forms and Methods of Carrying Out Combat Operations'. *Military Review*. January–February 2016: 23–29.

Giffone, Benjamin D. '"Special Forces": A Stereotype of Benjaminite Soldiers in the Deuteronomistic History and Chronicles'. *Scandinavian Journal of the Old Testament*. 30:1. 2016: 16–29.

Gill, Paul and Horgan, John. 'Who Were the Volunteers? The Shifting Sociological and Operational Profile of 1240 Provisional Irish Republican Army Members'. *Terrorism and Political Violence*. 25:3. 2013: 435–456.

Goldsmith, Jeff and Matthew Waxman. 'The Legal Legacy of Light-Footprint Warfare'. The *Washington Quarterly*. 39:2. 2016: 7–21.

Gray, Colin. 'Handfuls of Heroes on Desperate Ventures: When Do Special Operations Succeed?' *Parameters*. 29:1. Spring 1999: 2–24.

Gray, Colin. 'Strategic Culture as Context: The First Generation of Theory Strikes Back'. *Review of International Studies*. 25:1. January 1999: 45–69.

Gregorian, Raffi. 'CLARET Operations and Confrontation, 1964–1966'. *Conflict Quarterly*. Winter 1991: 48–49.

Griffiths, Robert. 'South African Civil–Military Relations in Transition: Issues and Influences'. *Armed Forces & Society*. 21:3. Spring 1995: 395–410.

Grundy, Kenneth. 'A Black Foreign Legion in South Africa?' *African Affairs*. 80:318. January 1981: 101–114.

Gutteridge, William. 'South Africa's Defence Posture'. *The World Today*. 36:1. January 1980: 26–31.

Hack, Karl. 'The Malayan Emergency as a Counter-Insurgency Paradigm'. *The Journal of Strategic Studies*. 32:3. June 2009: 384–414.

Harmon, Paul. 'Special Operations Liaison Element in OIF: SOLE, Part II'. *Air Commando Journal*. 9:2. October 2020: 21–27.

Heitman, Helmoed-Römer. 'Equipment of the Border War'. *Journal for Contemporary History, Special Edition: Border War, 1966–1989*. 31:3. December 2006: 91–108.

Hermann, Irene and Daniel Palmieri. 'A Haunting Figure: The Hostage through the Ages'. *International Review of the Red Cross*. 87:857. March 2005: 135–145.

Herndon, Robert et al. 'Effects-Based Operations in Afghanistan: The CJTF-180 Method of Orchestrating Effects to Achieve Objectives'. *Field Artillery*. January–February 2004: 26–30.

Hing, Matthew et al. 'Special Operations Forces and Incidence of Post-Traumatic Stress Disorder Symptoms'. *Journal of Special Operations Medicine*. 12:3. Fall 2012: 23–35.

Holcomb, John et al. 'Causes of Death of U.S. Special Operations Forces in the Global War on Terrorism, 2001–2004'. *Annals of Surgery*. 245:6. June 2007: 968–991.

Hooker Jr, R.D. 'America's Special Operations Problem'. *Joint Forces Quarterly*. 108. 1st Quarter 2023: 50–55.

Horn, Bernd. 'The End of "The Golden Age of SOF": Is There a Role for Special Operations Forces in the Renewed "Great Power Competition"?'. *The Journal of Future Conflict*. 2. Fall 2020: 1–33.

Horsely, Richard. 'The Sicarii: Ancient Jewish "Terrorists"'. *The Journal of Religion.* 59:4. October 1979: 435–458.

Howard, Michael. 'Civil–Military Relations in Great Britain and the United States, 1945–1958'. *Political Science Quarterly.* 71:1. March 1960: 35–46.

Hughes, Geraint. 'A Proxy War in Arabia: The Dhofar Insurgency and Cross-Border Raids into South Yemen'. *Middle East Journal.* 69:1. Winter 2015: 91–104.

Hughes, Geraint. 'The Use of Undercover Military Units in Counter-Terrorist Operations: A Historical Analysis with Reference to Contemporary Anti-Terrorism'. *Small Wars & Insurgencies.* 21:4. 2010: 561–590.

Ioannidou, Eleftheria. 'Mythologies of Genesis and Neo-Nazi Palingenesis: Commemorating the Battle of Thermopylae in the Political Rites of the Golden Dawn'. *Humanities.* 11:88. 2022: 1–21.

Ivaska, Andrew. 'Leveraging Alternatives: Early FRELIMO, the Soviet Union, and the Infrastructure of African Political Exile'. *Comparative Studies of South Asia, Africa, and the Middle East.* 41:1. May 2021: 11–26.

Izadi, Roya. 'State Security or Exploitation: A Theory of Military Involvement in the Economy'. *Journal of Conflict Resolution.* 66:4–5. 2022: 729–754.

Jacobs, C.J. 'The Forward Defence Strategy of the South African Defence Force (SADF), 1978–1989'. *The Southern Journal for Contemporary History.* 31:3. 2006: 23–41.

Johnson, Jesse. 'The Cost of Security: Foreign Policy Concessions and Military Alliances'. *Journal of Peace Research.* 52:5. September 2015: 655–679.

Johnston, Patrick and Anoop Sarbahi. 'The Impact of Drone Strikes on Terrorism in Pakistan'. *International Studies Quarterly.* 60:2. June 2016: 203–219.

Jones Jr, Robert. 'A Legacy of Support: The 528th Sustainment Brigade'. *Veritas.* 5:3. 2009: 21–28.

Jordan, Jenna. 'When Heads Roll: Assessing the Effectiveness of Leadership Decapitation'. *Security Studies.* 18. 2009: 719–755.

Khoon Liat, Colflesh. 'Whither Special Forces? The Strategic Relevance of Special Operations'. *Pointer.* 39:2. 2013: 31–40.

Kibbe, Jennifer. 'The Rise of the Shadow Warriors'. *Foreign Affairs.* 83:2. March/April 2004: 102–115.

Kimenyi, Mwangi and John Mukum Mbaku 'Rents, Military Elites, and Political Democracy'. *European Journal of Political Economy.* 11:4. April 1996: 699–708.

Kimura, Jun et al. 'Naval Battlefield Archaeology of the Lost Kublai Khan Fleets'. *The International Journal of Nautical Archaeology.* 41:1. 2013: 1–11.

King, Anthony. 'The Special Air Service and the Concentration of Military Power'. *Armed Forces and Society.* 35:4. July 2009: 646–666.

Kiras, James. 'Future Tasks: Threats and Missions for SOF'. *Special Operations Journal.* 5:1. 2019: 15–16.

Kiras, James. 'A Theory of Special Operations: "These Ideas Are Dangerous"'. *Special Operations Journal.* 1:2. 2015: 75–88.

Kiras, James. 'Special Operations and Strategies of Attrition'. *Infinity Journal.* 2:4. Fall 2012: 18–21.

Kiras, James. 'The Role of Special Operations Forces: Past, Present, and Future'. *Pointer.* 37:2. Winter 2011: 90–95.

Kisner, Frank. 'Evolution of a JSOAC in a Non-Linear War'. *Air Commando Journal.* 8:3. February 2020: 22–30.

Korski, Daniel. 'British Civil-Military Integration: The History and Next Steps'. *The RUSI Journal.* 154:6. 2009: 14–24.

Kotwal, Ross et al. 'A Descriptive Study of US Special Operations Command Fatalities, 2001–2018'. *Journal of Acute Trauma Care Surgery.* 87:3. 2019: 645–657.

Krivdo, Michael. 'CJSOTF-A (Combined Joint Special Operations Task Force-Afghanistan): A Short History, 2002–2014'. *Veritas*. 12:2. October 2016: 1–28.

Kumar, Amit. 'India's Special Forces: An Appraisal'. *Journal of Defence Studies*. 9:2. April–June 2015: 157–178.

Kunioka, Todd and Lawrence Rothenberg. 'The Politics of Bureaucratic Competition: The Case of Natural Resource Policy'. *Journal of Policy Analysis and Management*. 12:4. Autumn 1993: 700–725.

Kuperman, Alan. 'The Stinger Missile and U.S. Intervention in Afghanistan'. *Political Science Quarterly*. 114:2. Summer 1999: 219–263.

Kuyers, Joshua. '"Operational Preparation of the Environment": "Intelligence Activity" or "Covert Action" by Any Other Name?'. *National Security Law Brief*. 4:1. 2013: 21–40.

Laborie, Géraud. 'The Afghan Model More Than 10 Years Later'. *ASPJ Africa & Francophonie*. 3rd Quarter 2013: 49–60.

Lambakis, Steven. '"Forty Selected Men Can Shake the World": The Contributions of Special Operations to Victory'. *Comparative Strategy*. 13:2. Summer 1994: 211–221.

Lautenschläger, Karl. 'Technology and the Evolution of Naval Warfare'. *International Security*. 8:2. Fall 1983: 3–51.

Leary, John. 'Searching for a Role: The Special Air Service (SAS) Regiment in the Malayan Emergency'. *Journal of the Society for Army Historical Research*. 73:296. Winter 1995: 251–269.

Lebow, Richard Ned. 'Thomas Schelling and Strategic Bargaining'. *International Journal*. 51:3. Summer 1996: 555–576.

Legro, Jeffrey. 'Military Culture and Inadvertent Escalation in World War II'. *International Security*. 18:4. Spring 1994: 108–142.

Lindsay, Jon. 'Target Practice: Counterterrorism and the Amplification of Data Friction'. *Science, Technology, and Human Values*. 42:6. November 2017: 1061–1099.

Lindsay, Jon. 'Reinventing the Revolution: Technological Visions, Counterinsurgent Criticism, and the Rise of Special Operations'. *Journal of Strategic Studies*. 36:3. June 2013: 422–453.

Marsh, Christopher, Mike Kenny, and Nathanael Joslyn. 'SO What? The Value of Scientific Inquiry and Theory Building in Special Operations Research'. *Special Operations Journal*. 1:2. 2015: 89–104.

Mason, M. Chris. 'COIN Doctrine Is Wrong'. *Parameters*. 51:2. Summer 2021: 19–34.

McAndrew, Michael. 'Wrangling in the Shadows: The Use of United States Special Forces in Covert Military Operations in the War on Terror'. *Boston College International and Comparative Law Review*. 29:1. 2006: 153–164.

McCann, Leo. '"Killing Is Our Business and Business Is Good": The Evolution of "War Managerialism" from Body Counts to Counterinsurgency'. *Organization*. 24:4. 2007: 491–515.

McCann, Stuart. 'What's in a Name? The Evolving Role of the *Frvmentarii*'. *The Classical Quarterly*. 69:1. May 2019: 340–354.

McChrystal, Stanley. 'It Takes a Network: The New Front Line of Modern Warfare'. *Foreign Policy*, 21. February 2011.

Mehmood, Nasir and Usman Haider. 'Understanding India's "Surgical Strike" Special Operations and Pakistan's Response'. *Journal of Security and Strategic Analyses*. 8:1. Summer 2022: 49–70.

Milevski, Lukas. 'The Mirage of Post-Clausewitzianism: Understanding War and Politics on the Frontier of Clausewitzian Thought'. *Military Strategy Magazine*. Special Edition. 2021: 16–21.

Milevski, Lukas. 'Asymmetry Is Strategy, Strategy Is Asymmetry'. *Joint Forces Quarterly*. 75. 4th Quarter 2014: 77–83.

Miller, Matthew. 'Special Operations in the 21st Century: Revisiting the Falklands War'. *Inter Populum: The Journal of Irregular Warfare and Special Operations.* 1:1. Spring 2023: 57–68.

Monroe, George. 'The Rebirth of the Outback Air Force'. *Armed Forces Journal.* 1. February 2008.

Murdock, Clark and Richard Weitz. 'Beyond Goldwater-Nichols: New Proposals for Defense Report'. *Joint Forces Quarterly.* 38. 2005: 34–41.

Niva, Steve. 'Disappearing Violence: JSOC and the Pentagon's New Cartography of Networked Warfare'. *Security Dialogue.* 44:3. 2013: 185–202.

O'Boyle, Randy. '16th Special Operations Group: Early Action in Operation IRAQI FREEDOM'. *Air Commando Journal.* 9:2. October 2020: 28–34.

Oñate, Andrea. 'The Red Affair: FMLN–Cuban Relations during the Salvadoran Civil War, 1981–92'. *Cold War History.* 11:2. 2011: 133–154.

Osborne, Joseph. 'Syria Train and Equip: Who Left the Interns in Charge?'. *Special Operations Journal.* 7:2. 2021: 166–189.

Parks, W. Hays. 'Special Forces' Wear of Non-Standard Uniforms'. *Chicago Journal of International Law.* 4:2. Fall 2003: 493–560.

Peterson, Cory. 'Task Force Viking and the Ugly Baby Mission'. *Air Commando Journal.* 9:2. October 2020: 34–41.

Pfaff, C. Anthony. 'Professionalizing Special Operations Forces'. *Parameters.* 52:3. 2022: 111–126.

Plümper, Thomas and Eric Neumayer. 'Free-Riding in Alliances: Testing an Old Theory with a New Method'. *Conflict Management and Peace Science.* 32:3. 2015: 247–268.

Pritchett, W. Kendrick. 'Xerxes' Fleet at the "Ovens"'. *American Journal of Archaeology.* 67:1. January 1963: 1–6.

Ranft, D.D. 'Parachuting in Malaya'. *Army Quarterly.* 66. April and July 1953: 205–207.

Rawnsley, Adam and David Brown. 'The Littlest Boy'. *Foreign Policy.* 30 January 2014.

Reiter, Dan and Paul Poast. 'The Truth about Tripwires: Why Small Force Deployments Do Not Deter Aggression'. *Texas National Security Review.* 4:3. Summer 2021: 33–53.

Richmond, J.A. 'Spies in Ancient Greece'. *Greece & Rome.* 45:1. April 1998.

Robinson, Linda. 'The SOF Experience in the Philippines and the Implications for Future Defense Strategy'. *Prism.* 6:3. December 2016: 151–167.

Rogers, Clifford. 'Clausewitz, Genius and the Rules'. *Journal of Military History.* 66:4. October 2002: 1167–1176.

Rogin, Josh. 'McRaven to OPSEC: Zip It'. *Foreign Policy.* 24 August 2012.

Rolandsen, Øystein and Kjetil Selvikm. 'Disposable Rebels: US Military Assistance to Insurgents in the Syrian War'. *Mediterranean Politics.* 29:1. 2024: 1–22.

Rose, Gideon. 'Generation Kill: A Conversation with Stanley McChrystal'. *Foreign Affairs.* 92:2 March/April 2013.

Sacquety, Troy. 'The Evolution of the Special Forces (SF) Operational Detachment-Alpha (ODA)'. *Veritas.* 19:1. 2023: 23–27.

Sanders, Andrew. 'Operation Motorman (1972) and the Search for a Coherent British Counterinsurgency Strategy in Northern Ireland'. *Small Wars & Insurgencies.* 24:3. 2013: 465–492.

Scheepers, Marius and Frans Johan Pretorius. 'Loss of Military Equipment at the Battle of Indungo during the Border War, 31 October 1987'. *Scientia Militaria, South African Journal of Military Studies.* 49:2. 2021: 43–66.

Schneidman, Witney. 'FRELIMO's Foreign Policy and the Process of Liberation'. *Africa Today.* 25:1. January–March 1978: 57–67.

Shaw, Alexander. 'British Counterinsurgency in Brunei and Sarawak, 1962–1963: Developing Best Practices in the Shadow of Malaya'. *Small Wars & Insurgencies.* 27:4. 2016: 702–725.

Shaughnessy, Edward. '"New" Evidence on the Zhou Conquest'. *Early China*. 6. 1980–1981: 57–79.
Shkolnik, Michael. '"Mowing the Grass" and Operation Protective Edge: Israel's Strategy for Protracted Asymmetric Conflict with Hamas'. *Canadian Foreign Policy Journal*. 23:2. Summer 2017: 185–189.
Shoham, Dany and Michael Liebig. 'The Intelligence Dimension of Kautilyan Statecraft and Its Implications for the Present'. *Journal of Intelligence History*. 15:2. 2016: 119–138.
Shultz Jr, Richard. '"The Irreducible Minimum": An Evolution of Counterterrorism Operations in Iraq'. *Prism*. 7:3. 2018: 103–117.
Sinai, Tamir. 'Eyes on Target: "Stay-Behind" Forces during the Cold War'. *War in History*. 23:3. 2021: 681–700.
Smith, Hugh. 'Clausewitz's Definition of War and Its Limits'. *Military Strategy Magazine*. Special Edition. 2021: 9–15.
Smith, Hugh. 'The Womb of War: Clausewitz and International Politics'. *Review of International Studies*. 16:1. January 1990: 39–58.
Smith, M.L.R. and Peter Neumann. 'Motorman's Long Journey: Changing the Strategic Setting in Northern Ireland'. *Contemporary British History*. 19:4. December 2005: 413–435.
Smith, Morton. 'Zealots and Sicarii, Their Origins and Relations'. *Harvard Theological Review*. 64:1. January 1971: 1–19.
Smith, Steve. 'Groupthink and the Hostage Rescue Mission'. *British Journal of Political Science*. 15:1. January 1985: 117–123.
Spark, Alasdair. 'The Soldier at the Heart of the War: The Myth of the Green Beret in the Popular Culture of the Vietnam Era'. *Journal of American Studies*. 18:1. April 1984: 29–48.
'Special Issue: Who Won the War on Terror?'. *Foreign Affairs*. 100:5. September/October 2021.
Stapleton, Tim. 'Bush Tracking and Warfare in Late Twentieth-Century East and Southern Africa'. *Historia*. 59:2. November 2014: 229–251.
Strachan, Hew. 'Strategy in Theory; Strategy in Practice'. *Journal of Strategic Studies*. 42:2. 2019: 171–190.
Stringer, Kevin. 'Jomini and Naval Special Operations Forces: An Applied-Competition Approach to Russia'. *Naval War College Review*. 74:4. Autumn 2021: 79–93.
Sumaryani, Ni Made. 'Kautilya's Views on Secret Activities of Intelligence in *Arthasastra* and Its Current Relevance'. *International Journal of Hindu Science and Religious Studies*. 5:1. June 2021: 104–110.
Szabados, Gyorgy. 'The Hungarian National Defense during the German Wars 1030–1052'. *Chronica*. 6. January 2006: 72–81.
Taylor, Simon. 'Status Quo Terrorism: State-Terrorism in South Africa during Apartheid'. *Terrorism and Political Violence*. 35:2. 2023: 304–320.
Thijs, Simon. 'Hostages of Rome'. *Athens Journal of History*. 2:3. July 2016: 199–212.
Thomas, Timothy. 'Russia's Reflexive Control Theory and the Military'. *Journal of Slavic Military Studies*. 17. 2004: 237–256.
Tobin, P.A. 'The Bertrand Stewart Prize Essay, 1952: British Special Forces'. *Army Quarterly*. 65. October 1952/January 1953: 161–184.
Toros, Harmonie. '"We Don't Negotiate with Terrorists!": Legitimacy and Complexity in Terrorist Conflicts'. *Security Dialogue*. 39:4. August 2008: 407–408.
Tracy, Jared. '"Building the Airplane in Flight": PSYOP in Operation DESERT SHIELD, Part 2'. *Veritas*. 19:1. 2022: 49–71.
Tracy, Jared. 'Victoria Ex Umbra: Activating 4th Battalion, 1st Special Forces Group Airborne'. *Veritas*. 14:2. 2008: 4–8.
Tuck, Christopher. 'Shaping Hearts and Minds: Claret Operations in Borneo, 1965–1966'. *Small Wars & Insurgencies*. 34:4. 2023: 803–827.

Tuck, Christopher. 'The Limits of Covert Action: SAS Operations during "Confrontation", 1964–66'. *Small Wars & Insurgencies*. 27:6. 2016: 996–1018.

Tweedie, June and Tony Ward. 'The Gibraltar Shootings and the Politics of Inquests'. *Journal of Law and Society*. 16:4. Winter 1989: 464–476.

ul-majeed, Ikram. 'The Farce of the "Surgical Strike"'. *Defence Journal*. 20:3. October 2016: 76–77.

Vigne, Randolph. 'SWAPO of Namibia: A Movement in Exile'. *Third World Quarterly*. 9:1. January 1987: 85–107.

Vines, Alex. 'Renamo's Rise and Decline: The Politics of Reintegration in Mozambique'. *International Peacekeeping*. 20:3. April 2013: 375–393.

Waghelstein, John. 'What's Wrong in Iraq? Or, Ruminations of a Pachyderm'. *Military Review*. 86:1. January–February 2006: 112–117.

Wagner, R. Harrison. 'Bargaining and War'. *American Journal of Political Science*. 44:3. July 2000: 469–484.

Walcot, Peter. 'Cattle Raiding, Heroic Tradition, and Ritual: The Greek Evidence'. *History of Religions*. 18:4. May 1979: 326–351.

Wall, Andru. 'Demystifying the Title 10-Title 50 Debate: Distinguishing Military Operations, Intelligence Activities, and Covert Action'. *Harvard National Security Journal*. 3. 2011: 85–142.

Wesbrouck, Jason, Glenn Harned, and Preston Plous. 'Special Operations Forces and Conventional Forces Integration, Interoperability, and Interdependence'. *Prism*. 6:3. December 2016: 85–95.

'What the Nkomati Accord Means for Africa'. *The Black Scholar*. 15:6. November/December 1984: 15–22. http://www.jstor.org/stable/41067114.

White, Kenton. 'Mearsheimer's Folly: NATO's Cold War Capability and Credibility'. *Infinity Journal*. 6:4. Summer 2017: 22–31.

White, Taylor. 'Security Cooperation: How It All Fits'. *Joint Forces Quarterly*. 72. 1st Quarter 2014: 106–108.

Wilkenfield, Johnathan et al. 'Escalation Management in Gray Zone Crises: The Proxy Factor'. *International Studies Quarterly*. 66:3. September 2022.

Williams, Mary Francis. 'Philopoemen's Special Forces: Peltasts and a New Kind of Greek Light-Armed Warfare (Livy 35:27)'. *Historia: Zeitscrifte für Alte Geschichte*. 53:3. 2004: 257–277.

Wills, Craig, Thomas Griffith Jr, and Richard Andres. 'Winning with Allies: The Strategic Value of the Afghan Model'. *International Security*. 30:3. December 2005: 124–160.

Wilson, Gregory. 'Anatomy of a Successful COIN Operation: OEF-Philippines and The Indirect Approach'. *Military Review*. 86:6. November–December 2006: 2–12.

Winton, Harold. 'An Imperfect Jewel: Military Theory and the Military Profession'. *The Journal of Strategic Studies*. 34:6. December 2011: 853–877.

Woodhouse, J.M. 'Some Personal Observations on the Employment of Special Forces in Malaya'. *Army Quarterly*. 66. April and July 1953: 69–74.

Wozniak, Jesse. 'Iraq and the Material Basis of Post-Conflict Police Reconstruction'. *Journal of Peace Research*. 54:6. November 2017: 806–818.

Yang, Lien-Shang. 'Hostages in Chinese History'. *Harvard Journal of Asiatic Studies*. 15:3/4. December 1952: 507–521.

Yates, Robin D.S. 'New Light on Ancient Chinese Military Texts: Notes on Their Nature and Evolution, and the Development of Military Specialization in Warring States China'. *T'oung Pao*. 74:4/5. 1988: 211–248.

# Index

*For the benefit of digital users, indexed terms that span two pages (e.g., 52–53) may, on occasion, appear on only one of those pages.*

Tables and figures are indicated by an italic *t* or *f* following the paragraph number.